MICROBIAL ENZYMES
AND
BIOCONVERSIONS

ECONOMIC MICROBIOLOGY

Series Editor

A. H. ROSE

ECONOMIC MICROBIOLOGY
Volume 5

MICROBIAL ENZYMES AND BIOCONVERSIONS

edited by

A. H. ROSE

School of Biological Sciences
University of Bath,
Bath, England

1980

ACADEMIC PRESS

A Subsidiary of Harcourt Brace Jovanovich, Publishers

LONDON NEW YORK TORONTO SYDNEY SAN FRANCISCO

ACADEMIC PRESS INC. (LONDON) LTD.
24/28 Oval Road
London NW1

United States edition published by
ACADEMIC PRESS INC.
111 Fifth Avenue
New York, New York 10003

British Library Cataloguing in Publication Data

Microbial enzymes and bioconversions. – (Economic
  microbiology; vol. 5).
  1. Microbial enzymes
  I. Rose, Anthony Harry   II. Series
  576'.11'925   QR90   77-77361

  ISBN 0-12-596555-9

Printed in Great Britain by Spottiswoode Ballantyne Ltd.
Colchester and London

CONTRIBUTORS

KNUD AUNSTRUP, Novo Industri A/S, Novo Alle, DK-2880 Bagsvaerd, Denmark.

S. A. BARKER, Department of Chemistry, University of Birmingham, Birmingham, England.

J. ERIKSEN, Institute of Microbiology, University of Bergen, 5015 Bergen, Norway.

WILLIAM F. FOGERTY, Department of Industrial Microbiology, University College, Stillorgan Road, Dublin 4, Eire.

JOSTEIN GOKSØYR, Institute of Microbiology, University of Bergen, 5015 Bergen, Norway.

K. KIESLICH, Gesellschaft fur Biotechnologische Forschung mbH, D-3300 Braunschweig-Stockheim, West Germany.

W. PILNIK, Department of Food Science, Agricultural University, De Dreijen 12, 6703 BC Wageningen, Netherlands.

F. M. ROMBOUTS, Department of Food Science, Agricultural University, De Dreijen 12, 6703 BC Wageningen, Netherlands.

A. H. ROSE, School of Biological Sciences, University of Bath, Claverton Down, Bath, England.

OLDRICH K. SEBEK, The Upjohn Company, Kalamazoo, Michigan 4900, U.S.A.

J. SHIRLEY, Department of Biochemistry, University of Birmingham, Birmingham, England.

E. J. VANDAMME, Laboratory of General and Industrial Microbiology, University of Gent, Coupure 533, 9000-Gent, Belgium.

LEO C. VINING, Department of Biology, Dalhousie University, Halifax, Nova Scotia, Canada.

DR. DAVID PERLMAN

This volume of *Economic Microbiology* is dedicated to the memory of Dr. David Perlman (1920–1980).

Dr. Perlman received his doctorate from the University of Wisconsin where he studied under Marven J. Johnson and William H. Peterson. After brief spells in the research laboratories of Hoffman–LaRoche Inc. and Merck and Co., he joined the Squibb Institute for Medical Research in New Brunswick, New Jersey, U.S.A. During his 20 years at Squibb, he developed several fermentation processes and his name will always be associated with streptomycin production, a fermentation in which he made a number of outstanding discoveries. Finally, Dr. Perlman moved back to the University of Wisconsin, as Professor of Pharmaceutical Biochemistry, where he also served as Dean of the School of Pharmacy.

Throughout his career, Dr. Perlman contributed regularly and richly to the literature on industrial microbiology. He has his name on 28 patents and more than 350 papers. His contribution to the review literature in fermentation science was especially powerful, most notably in his editorship of *Advances in Applied Microbiology* and, more recently, of *Annual Reports on Fermentation Processes*. Dr. Perlman had promised to contribute a short chapter to the present volume on conversions by *Gluconobacter* spp. Sadly, his last illness prevented this. He will be remembered as one of the great contributors to fermentation science.

A.H.R.

PREFACE

Production by microbiological processes of alcoholic beverages, and of a vast range of primary and secondary products of microbial metabolism, was covered in the first three volumes in this series. The fourth volume dealt with manufacture of microbial biomass, which in physiological terms can be viewed as production of cellular protein for use as food and fodder. All of the processes covered in these first four volumes exploit the metabolic idiosyncracies of micro-organisms for commercial use. The topics covered in the present volume differ from those described in the first four volumes of the series in that they are concerned with industrial exploitation not of entire metabolic pathways in microbes, but of individual enzymes or short sequences of enzymes.

Microbial enzymes have been used industrially for centuries, and until comparatively recently without any knowledge of the nature of enzymes. Commercial use of microbial enzymes received a tremendous fillip in the 1960s when proteases were incorporated into washing powders, and this did much to encourage further industrial use of microbial enzymes. The majority of microbial enzymes currently used in industry are extracellular depolymerizing or hydrolytic enzymes, but other classes of enzyme are beginning to find commercial uses, particularly in diagnostic kits. Additional uses for microbial enzymes are being actively researched in many countries of the World, and this constitutes one of the major growth areas in industrial microbiology.

In exploiting intact micro-organisms to carry out particular enzyme-catalysed reactions, or a chemical conversion that involves just a few enzymes, the industrial microbiologist is essentially using microbes as chemical reagents. The number of these processes operated commercially is now quite large. This particular group of processes became widely known when microbes were first used to bring about structural alterations in steroid molecules, a breakthrough that greatly lowered the cost of cortisone production and, later, was extensively used in the manufacture of oral contraceptives. This too is a major growth area in economic microbiology, one closely allied with the chemical and pharmaceutical industries.

I am extremely grateful for the advice given to me during the preparation of this volume by my colleague Alan Rayner.

August, 1980   ANTHONY H. ROSE

CONTENTS

1. History and Scientific Basis of Commercial Exploitation of Microbial Enzymes and Bioconversions

ANTHONY H. ROSE

2. Proteinases

KNUD AUNSTRUP

3. Amylases, Amyloglucosidases and Related Glucanases
WILLIAM M. FOGARTY and CATHERINE T. KELLY

4. Glucose Oxidase, Glucose Dehydrogenase, Glucose Isomerase, β-Galactosidase and Invertase
S. A. BARKER and J. A. SHIRLEY

5. Pectic Enzymes

FRANK M. ROMBOUTS and WALTER PILNIK

6. Cellulases

JOSTEIN GOKSØYR and JONNY ERIKSEN

7. Immobilized Enzymes

S. A. BARKER

8. Steroid Conversions

KLAUS KIESLICH

9. Penicillin Acylases and *Beta*-Lactamases

ERICK J. VANDAMME

10. Conversions of Alkaloids and Nitrogenous Xenobiotics

L. C. VINING

11. Microbial Transformations of Antibiotics

O. K. SEBEK

1. History and Scientific Basis of Commercial Exploitation of Microbial Enzymes and Bioconversions

ANTHONY H. ROSE

School of Biological Sciences, University of Bath, Bath, Avon, England

I. INTRODUCTION

Commercial production of microbial primary and secondary products of metabolism, as well as of alcoholic beverages and biomass, involves a highly co-ordinated series of enzyme-catalysed reactions which are associated with growth of the producing micro-organism. In addition, there are other industrially important microbiological processes which involve just one enzyme or a small number of enzymes. Where the process involves just one microbial enzyme, this can be extracted from cultures of the producing micro-organism, and used under conditions where the extent to which the substrate of the reaction is converted into product is under the more or less total control of the operator. Alternatively, activity of the enzyme—or there may be two or more—can be exploited using intact microbes, again under conditions that allow the operator to control the extent to which the substrate is converted into the desired product. In both types of process, the isolated enzyme or intact micro-organism is being used essentially as a chemical reagent to effect economically important molecular trans-formations. Individual chapters in this volume describe the more important of these enzymes and bioconversions.

II. HISTORY OF INDUSTRIALLY IMPORTANT MICROBIAL ENZYMES AND BIOCONVERSIONS

A. Enzymes

The concept of catalysis, the process by which the rate at which chemical reactions proceed is speeded up by the presence of another chemical compound that classically does not itself undergo chemical change, was developed rapidly in the early part of the Nineteenth Century. The first catalytic reaction described in the literature was the observation made by Vogel (1812) that, at low temperatures, oxygen and hydrogen react chemically when mixed in the presence of charcoal. Berzelius, a few years later in 1837, defined catalysis as a force apart 'such that material may, by their mere presence, and not on account of their chemical affinities, awaken in a substance such affinities as are latent at the temperature in question'; this translation comes from Keilin (1966). Presciently, Berzelius (1837) went on to assume that 'in living plants and animals, thousands of catalytic processes take place between tissues and fluids, thus giving rise to multitudes of chemical compounds'. Again, I am indebted to Keilin (1966) for this translation.

Perhaps, after all, the prescience was not as great as appeared, for Berzelius made these comments a few years after Payen and Persoz (1833) had reported what we now recognize as amylase activity from animal materials, and after Schwann (1836) had discovered pepsin. Be that as it may, these reports quickly led to the discovery of a number of other hydrolytic enzymes, including invertase (β-fructofuranosidase) from *Saccharomyces cerevisiae* (Berthelot, 1860).

The existence of not only extracellular but also intracellular enzymes, particularly non-hydrolytic ones that require cofactors or coenzymes, was suggested early on by Berthelot (1856) and Hoppe-Seyler (1876), but their role in the activity and metabolism of living tissues was the subject of a great controversy, at the centre of which was alcoholic fermentation by yeast. Even after Louis Pasteur had provided convincing evidence in favour of the 'vitalistic theory' of fermentation, originally proposed by Cagniard-Latour, Schwann and Kützing, rather than the 'chemical theory' of fermentation advocated by Liebig, Wöhler and Berzelius, one important question remained unanswered, namely the precise cause of decomposition of the sugar molecule during alcoholic fermentation. Supporters of the vitalistic theory of fermentation proposed that the sugar molecule was decomposed inside the yeast cell by enzymes, while their opponents favoured the view that there occurred a transference of decomposing activities from the cell plasm to the sugar molecule. The answer came in 1897 from experiments carried out by the Buchner brothers and Hahn (Buchner *et al.*, 1897). They discovered that, by grinding yeast cells with a mixture of kieselguhr and sand, it was possible to obtain a juice that brought about an alcoholic fermentation of sugar in the absence of living cells. In so doing, they essentially proved the vitalistic theory of fermentation, and provided the first demonstration of cell-free action of intracellular enzymes.

Rapid strides were made in enzyme chemistry during the first half of the present century. Notable landmarks were the first crystallization of an enzyme—urease from the jack bean—by Sumner (1926), and further evidence for the protein nature of a variety of other enzymes, which came largely from Northrop's laboratory (Northrop *et al.*, 1948). In more recent years, research has been concentrated on discovering the primary, secondary and tertiary structures of enzyme polypeptides, as well as the mechanism of enzyme action. The Enzyme Commission, set up by the International Union of Biochemistry (1965), has published a system of enzyme classification, which is universally adhered to throughout biochemistry.

Commercial exploitation of microbial enzymes began well before an understanding of their nature and properties had been arrived at. Extracts of plants had for centuries been used to bring about, more often than not, hydrolysis of polymeric material; the use of extracts of malted barley in beer brewing is the best known example. Another ancient use of enzymes is the application of animal dung in the bating of hides in leather manufacture. As an understanding of the properties of enzyme-catalysed reactions began to emerge, so new uses were developed. Examples are the use of crude preparations from certain animal tissues, such as pancreas and stomach mucosa, as sources of amylolytic enzymes, and of extracts of papaya fruit to obtain proteolytic activity. However, these sources of enzyme were unreliable and expensive, and searches began to be made for alternative sources. These were found very largely in microbial cultures.

Pride of place in the development of commercially important microbial enzymes must go to Jokichi Takamine. Born in Japan, but emigrating to work in the U.S.A., Takamine was primarily interested in moulds as sources of industrial enzymes and, as early as 1894 (Takamine, 1894), he patented a method for making a preparation of diastatic enzymes from moulds which was marketed under the name Takadiastase. His method, which was to continue in use for many years, involved growing the mould on the surface of a solid substrate, usually wheat bran although other fibrous materials can be used. Additional nutrients were added, as well as a salts solution to buffer the pH value. After inoculation with mould spores, the solid substrate was spread out on trays, and kept in a cabinet with controlled temperature and humidity. After growth of the mould on the surface of the solid substrate had become extensive, the bran was soaked in water or salts solution to extract the extracellular enzymes. The early processes, based on the Takamine patent (Takamine, 1914), mainly used strains of *Aspergillus oryzae*. Plants were still operating in the United States in the mid 1940s making mould hydrolytic enzyme preparations by the Takamine process. According to Underkofler *et al.* (1946), the Mold Bran Company, at Eagle Grove in Iowa, U.S.A. was, in 1946, producing 10 tonnes of mouldy bran each day, using incubation of the inoculated bran in shallow trays. However, towards the end of the

1950s, submerged methods for cultivating the moulds were introduced, methods that had been developed largely at the Northern Regional Research Laboratory of the United States Department of Agriculture at Peoria, Illinois, and which had been used successfully in production of penicillin.

The use of bacteria as sources of commercially important enzymes was introduced some years after Takamine's pioneer work. Patents awarded in 1917 to the Frenchman Auguste Boidin and the Belgian Jean Effront (Boidin and Effront, 1917a, b) described the use of strains of *Bacillus subtilis* and *B. mesentericus* to produce amylases and diastases. These workers recommended media rich in nitrogenous nutrients, and prepared from soybean cake and peanut cake. Modified versions of the method were published by Schultz and Atkin (1931) and by Waller-stein (1939). All involved growing the bacteria as a surface pellicle in shallow trays of semisolid medium, and extracting the enzyme after a suitable period of incubation. Details of these and other modifications can be found in the text by Tauber (1949). As with the mould processes, these surface bacterial cultures were ultimately replaced, at least in part, by deep-culture methods.

The 1950s saw a steady, but hardly spectacular, increase in the industrial production and use of microbial enzymes. A report by Underkofler and his colleagues in 1958 described the then current state of the microbial enzyme industry in some detail. These authors listed ten microbial enzymes that were then being produced commercially, eight of which were hydrolytic enzymes. Amylases had the most extensive application, with uses in baking and brewing, production of fruit juices and starch syrups, and in preparation of digestive aids. Next came various proteases, with uses that varied from chillproofing of beer to bating of hides. The remaining six hydrolytic enzymes listed by Underkofler *et al.* (1958)—pectinase, lactase (β-galactosidase), invertase, lipase and cellulase—had far fewer applications. The two non-hydrolytic microbial enzymes listed in the report were catalase and glucose oxidase. Glucose oxidase, which catalyses conversion of glucose and molecular oxygen into gluconic acid and hydrogen peroxide, was listed as being used, among other applications, to remove glucose from preparations of dried egg, thereby increasing the shelf life of the preparations.

While listing the many advantages that microbial enzymes have over other reagents, Underkofler *et al.* (1958) failed to anticipate two applications of microbial enzymes which, in the decades that followed, were to assume major proportions. These are the use of microbial proteases in detergents and as rennet substitutes. They did, however, correctly foresee the more extensive application of microbial enzymes in clinical and pharmaceutical test kits. Nevertheless, the review by Underkofler *et al.* (1958) is a most valuable document in the history of the industrial use of microbial enzymes.

Commercial production of microbial enzymes received its greatest fillip around 1960, when proteolytic enzymes produced by strains of *Bacillus* began to be extensively incorporated into domestic detergents. Microbial protease preparations had for many years been used to remove stains and spots from clothing, an application that had been recognized since 1913 when it was introduced by Dr. Röhm in Germany. But widespread incorporation of these enzymes into detergents gave this application a totally new dimension. However, this use of microbial proteases was not without its problems, most of which arose from allergic reactions developed by workers who regularly handled finely powdered enzyme preparations. These problems were ultimately overcome, but only after some recession in the extent to which microbial proteases were used in detergent preparations. This application of microbial enzymes was seen at the time as the first of many novel applications, a view that stimulated increased research and development in the area. Viewed some 20 years later, it has to be admitted that this optimism was unjustified, for over this period no really large-scale novel application, comparable in size to the use of proteases in detergents, has materialized. Many major manufacturers of microbial enzymes nevertheless continue research in the hope of just such another windfall.

B. Bioconversions

Although micro-organisms, or more correctly materials containing micro-organisms, have been used for centuries in processes such as retting and unhairing of hides, the use of pure cultures of microbes to effect specific and strictly defined conversions, or series of conversions, of chemical compounds only became possible after the discoveries of the late Nineteenth Century in microbiology, and the birth of biochemistry at the end of that Century. The technique is now widely employed in the chemical and pharmaceutical industries and, essentially, it uses micro-organisms to carry out single-step or multi-step transformations of organic compounds that are not easily accomplished by conventional chemical methods.

1. Manufacture of Ephedrine

The first important use of a micro-organism as a 'living' chemical reagent was discovered in the early 1930s and was concerned with the manufacture of ephedrine. Ephedrine is a plant alkaloid, which occurs in species of *Ephedra*, particularly *E. sinica* and *E. equisetina*. Extracts of these plants were used medicinally many thousands of years ago in China and, in the mid 1880s, the active principle was isolated and named ephedrine. The drug is used as a vasoconstrictor in relieving bronchial congestion, allergic disorders and hypertensive states. The ephedrine molecule contains two asymmetric carbon atoms, so that there are four optically active forms and two racemic mixtures. Of the six forms the D(−) isomer has by far the greatest pressor activity. This is the isomer that occurs in species of *Ephedra*. Ephedrine can be synthesized chemically, but the product of the synthesis is a racemic mixture, which can, however, be resolved into individual isomers by mandelic acid.

A partial chemical synthesis of the active isomer, which included a microbiological step, was described in a patent awarded to

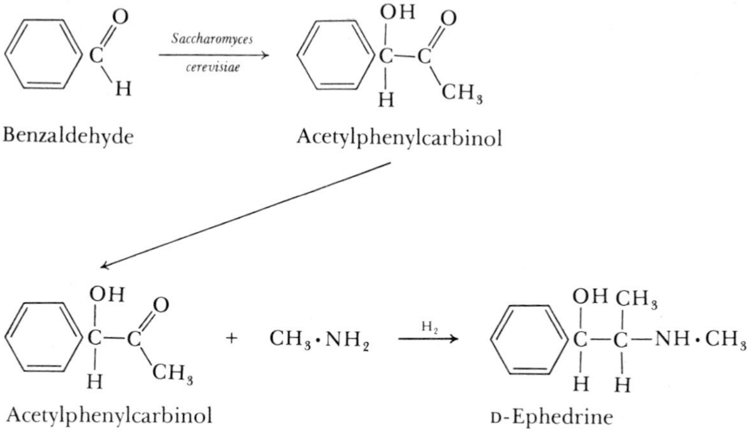

Fig. 1. Reactions involved in synthesis of D-ephedrine from benzaldehyde. See text for details.

Hildebrandt and Klavehn in 1934. The method exploited the discovery, made by Neuberg (1921, 1922), that strains of *Sacch. cerevisiae*, in a glucose-containing suspension, can convert benzaldehyde into acetylphenylcarbinol (Fig. 1). The process described by Hildebrandt and Klavehn (1934) starts with benzaldehyde, and the product of the yeast-mediated transformation is condensed with monomethylamine to yield D(−)-ephedrine (Fig. 1). The microbiological step involves condensation of benzaldehyde with acetyl-CoA (Smith and Hendlin, 1953). In the yeast suspension, some benzaldehyde is converted, instead, into benzyl alcohol, in a reaction involving NADH and catalysed by an ethanol dehydrogenase. In a second paper, Smith and Hendlin (1954) reported experiments carried out to minimize the extent of this side reaction. Addition to the yeast suspension of compounds that might preferentially oxidize NADH did not meet with success, although it was found that certain structural analogues of nicotinamide lowered yields of the alcohol.

Although ephedrine is now made very largely by chemical synthesis, and resolution of the resulting racemic mixture, the synthesis involving a microbiological step was an important discovery, not least because it focused attention on the potential for microbes as chemical reagents.

2. Synthesis of Ascorbic Acid

The 1930s also saw the development of a second industrially important synthetic process which included a step mediated by a micro-organism; this involved synthesis of ascorbic acid or vitamin C. This vitamin, which was first isolated crystalline in 1928, can be synthesized by a purely chemical route. However, on a commercial scale, it is still manufactured by the Reichstein process (Fig. 2), the second reaction of which is effected by a bacterium. The starting material is D-glucose, which is converted into D-sorbitol by electrolytic reduction. The next step, conversion of D-sorbitol into L-sorbose, is carried out using strains of *Gluconobacter suboxydans* (erstwhile referred to as *Acetobacter suboxydans*). Subsequent steps in the Reichstein process are entirely chemical, and involve preparation of the diacetone derivative of L-sorbose, which is then oxidized to the corresponding carboxylic acid before the acetone residues are removed by acid hydrolysis. The methyl ester of 2-oxo-L-gulonic acid is then converted into L-ascorbic acid by enolization and ring closure using sodium hydroxide.

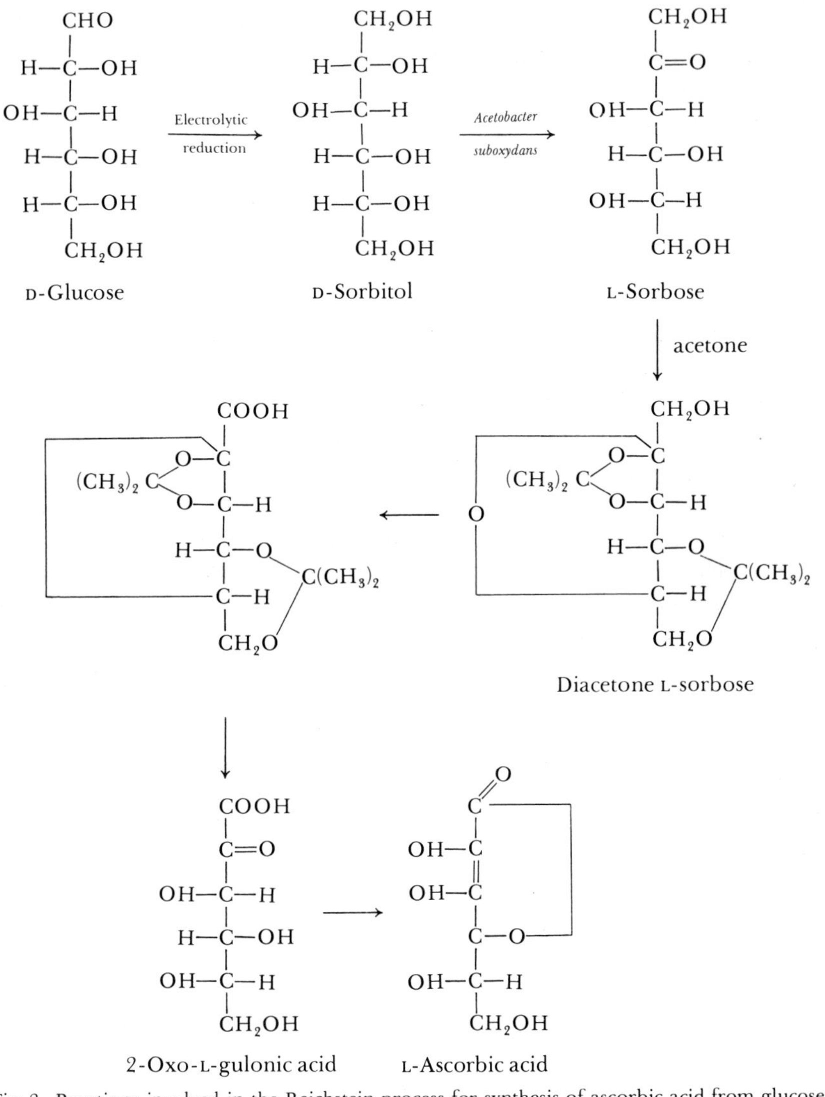

Fig. 2. Reactions involved in the Reichstein process for synthesis of ascorbic acid from glucose. See text for details.

In the original Reichstein process, strains of *G. xylinum* rather than *G. suboxydans* were employed in the conversion of D-sorbitol. Because *G. xylinum* forms tough cellulosic pellicles when grown in static culture, the early conversion processes were rather inefficient. However, following studies at the Northern Regional Research Laboratory of the

United States Department of Agriculture in Peoria, Illinois, on submerged culture of acetic-acid bacteria, strains of *G. suboxydans* began to be used (Wells *et al.*, 1937, 1939). Although several modifications have been examined, the production methods described by Wells and his colleagues (1937, 1939) are basically those still used in industrial practice.

In an attempt to shorten incubation times and increase the overall efficiency of the conversion, several groups of investigators have examined the possibility of using continuous cultures of *G. suboxydans*. Details of this research can be found in the publications of Elsworth *et al.* (1959), Muller (1966) and Bull and Young (1977). On the whole, use of continuous culture, in the laboratory, does increase efficiency but, because of economic considerations, the technique has not been introduced on a large scale.

The efficiency of the conversion mediated by *G. suboxydans* in the Reichstein process for synthesizing ascorbic acid has prompted research on possible alternative uses of bacteria in the overall synthesis. An alternative route to 2-oxo-L-gulonic acid was suggested by Hori and Nakatani (1953). This involved oxidation of D-glucose to 5-oxo-D-gluconic acid by strains of *G. suboxydans*, chemical conversion of this keto acid to L-idonic acid, and transformation of L-idonic acid into 2-oxo-L-gulonic acid by strains of *Aerobacter* spp., *Gluconobacter* spp. or *Pseudomonas* spp. The efficiency of the chemical conversions in the Reichstein process has prevented commercial exploitation of this alternative route. This applies also to microbiological conversion of L-sorbose into 2-oxo-L-gulonic acid which has been the subject of several research reports. First examined by Tengendy (1961), and patented by Huang (1962), it was later claimed that the conversion involved idose and L-idonic acid as intermediates (Okazaki *et al.*, 1968, 1969; Kanzaki and Okazaki, 1970). The suggestion that L-sorbosone might be an intermediate in the conversion has come from Makover *et al.* (1975) who used *Klebsiella pneumoniae*, *Pseudomonas putida* and *Serratia marcescens*.

3. Production of Dihydroxyacetone

Strains of *G. suboxydans* are also used for converting glycerol into dihydroxyacetone, a compound that for some years has been recognized as a useful intermediate in chemical synthesis of pharma-

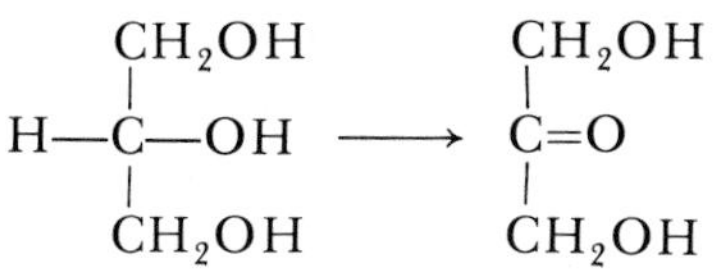

$$
\begin{array}{ccc}
\text{CH}_2\text{OH} & & \text{CH}_2\text{OH} \\
| & & | \\
\text{H}-\text{C}-\text{OH} & \longrightarrow & \text{C}=\text{O} \\
| & & | \\
\text{CH}_2\text{OH} & & \text{CH}_2\text{OH} \\
\text{Glycerol} & & \text{Dihydroxyacetone}
\end{array}
$$

ceutical compounds, but which has more recently been shown to be a commercially important sun-tanning agent (Andreadis and Miklean, 1960). Details of the conversion process can be found in the publications by Low and Krema (1928), Underkofler and Fulmer (1937) and Flickinger and Perlman (1977). The commercial importance of the conversion is very small compared with that involved in synthesis of ascorbic acid.

4. Steroid Conversions

Although the use of microbes in the manufacture of ephedrine and ascorbic acid illustrated the potential of micro-organisms as chemical reagents, realization of the extent of this potential came only after micro-organisms had been used with such success, and in so many diverse ways, in the synthesis of industrially important steroids.

As the role of steroid hormones as 'chemical messengers' in regulating animal metabolism became clear, it was natural that attempts should be made to use steroids as therapeutic agents in treating various diseases, particularly those, such as Addison's disease, that are caused by a deficiency of one or more steroid hormones. The whole field of steroid chemotherapy was revolutionized following the announcement by Hench and his coworkers at the Mayo Clinic in Rochester, Minnesota, U.S.A., in April 1949, of the dramatic effects of cortisone in treating rheumatoid arthritis. Other steroids, related to cortisone, were subsequently shown by clinical trials to possess similar therapeutic properties. Towards the end of 1949, cortisone was made available to physicians in the U.S.A. at around $200 per gram. The starting material in the manufacture was deoxycholic acid obtained from saponified bile and, according to the published data, the synthesis required some 37 separate chemical operations, 10 of which were needed to shift the oxygen from position 12 to position 11; the overall yield was only a few tenths of one per cent. Since the daily dose being prescribed at the time was as high as 100 mg, and as the drug has

an alleviative and not a curative effect, it was obvious that this synthesis
starting with bile acids could hardly be expected to meet the demand.
A search was therefore made for other steroids that could be used as
starting materials in the synthesis. For a long time, ergosterol from
yeast was used, but then the cheaper plant steroids, such as hecogenin
and diosgenin, were discovered.

Nevertheless, even starting with these plant steroids, the synthesis
was still protracted, the principal difficulty being the introduction of a
hydroxyl group on carbon 11, for an oxygen function in this position
had been shown by clinical trials to be essential for antiarthritic
activity. In an attempt to overcome this difficulty, a group of workers at
the Upjohn Company, Kalamazoo, Michigan, U.S.A., led by Durey
Peterson, set out to find a micro-organism capable of hydroxylating a
steroid in position 11, and in 1952 they reported that they had been
successful and had isolated a strain of *Rhizopus arrhizus* which converted
progesterone into 11-hydroxyprogesterone in appreciable yield
(Peterson *et al.*, 1952). As progesterone can be obtained readily from
diosgenin, while 11-hydroxyprogesterone can be converted into
cortisone in approximately six steps, this was clearly a considerable
improvement over the purely chemical synthesis.

These discoveries by the Upjohn group spurred numerous other
drug companies throughout the world to examine the role of
micro-organisms as chemical reagents in steroid transformations. In
the last two decades, this has led to synthesis of a vast range of new
steroid drugs, which have found applications in many areas of
chemotherapy, one of the most important being the development of
oral contraceptives.

5. Antibiotic Conversions

The dramatic achievements that came from research on steroid
conversions inevitably encouraged drug companies and other research
groups concerned with antibiotic chemotherapy to examine the
possibility of using micro-organisms to bring about structural
alterations to known antibiotics in the hope that these conversions
would produce structurally modified compounds with superior properties (higher antimicrobial activity, broader spectrum, lower toxicity,
and better absorption) compared with the parent antibiotic. In general,
five methods are available for manufacturing analogues of antibiotics

(Sebek, 1979). The first is purely chemical alteration of molecular structure; the remaining four are biological. Firstly, there is directed biosynthesis, in which a precursor of an antibiotic is introduced into the culture of the producing organism in order to maximize synthesis of a particular analogue containing that precursor (Perlman, 1973; Rose, 1979). Secondly, there is mutational biosynthesis, in which mutants of the producing organism are isolated which are unable to complete synthesis of the antibiotic unless supplied with the precursor moiety, which they cannot synthesize. By supplying the culture with utilizable analogues of the moiety, structural analogues of the antibiotic can be produced. This approach, which was pioneered in Kenneth Rinehart's laboratory in the University of Illinois at Urbana (Shier et al., 1969), has been most successfully applied to the aminoglycoside antibiotics (Claridge, 1979). Thirdly, there are purely genetical methods, in which mutants of the producing organisms are isolated by a variety of means. Finally, there is the possibility that analogues of antibiotics can be obtained by chemical modification by a micro-organism other than the producing organism.

There are numerous reports of active antibiotics being produced as a result of structural modifications by micro-organisms other than that which synthesized the parent antibiotic. However, with one important exception, none of these conversions, often because of purely economic reasons, has yet proved to be of commercial values. An example is the ability of Curvularia lunata, an organism known to be capable of hydroxylating several steroids in the 11α position, to hydroxylate 12α-deoxytetracycline to tetracycline (Holmlund et al., 1959). Rather more is known, however, of the capacity of micro-organisms to bring about structural modifications to antibiotics as a result of which activity of the antibiotic is lost. These conversions form the basis of the resistance of many bacteria to antibiotics. A summary of the reactions involved is given by Sebek (1977).

The only commercially successful enzyme-mediated conversion in antibiotic synthesis is hydrolysis of penicillins to give 6-amino-

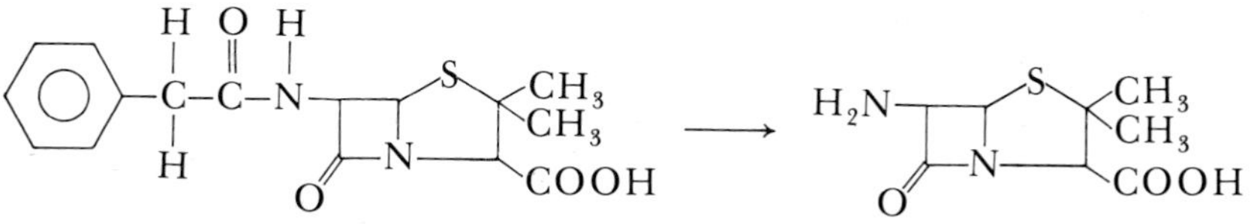

Penicillin G   →   6-Aminopenicillanic acid

penicillanic acid, a reaction catalysed by penicillin acylase, and the first step in synthesis of semisynthetic penicillins.

Some of the very early research on penicillin production established the importance of the side-chain structure on the antibiotic molecule for activity of the drug. Workers at the Northern Regional Research Laboratory of the United States Department of Agriculture at Peoria, Illinois, found that, by incorporating various side-chain precursors in the production medium, it was possible to steer the mould to produce different varieties of penicillin. This discovery led naturally to a search for a much wider range of new penicillins, with possibly more attractive properties than those produced directly by fermentation, by chemical rather than microbiological attachment of new side chains onto 6-aminopenicillanic acid. The lead in this approach to production of new penicillins was taken by researchers at the Beecham Research Laboratories in Surrey, England, in collaboration with workers at the Bristol Laboratories in Syracuse, New York, U.S.A. and another group at the International Research Centre for Chemical Microbiology in Rome, Italy, led by Ernst Boris Chain. They developed methods for attaching new side chains onto 6-aminopenicillanic acid. At first, 6-aminopenicillanic acid was produced in cultures of *Penicillium chrysogenum* that were not supplemented with a side-chain precursor. Later, it was found more convenient to produce the precursor from penicillin G or penicillin V, using the enzyme penicillin acylase which is synthesized by a variety of micro-organisms. 6-Aminopenicillanic acid can also be produced from penicillins by chemical means (Carrington, 1971). Some commercial drug houses are known to have favoured the enzymic method, others the chemical approach (Queener and Swartz, 1979).

There are a variety of sources of penicillin acylases, and these have been reviewed by Huber *et al.* (1972) and Moss (1977). Two general classes of the enzyme are recognized, namely those that cleave penicillin G more rapidly than penicillin V, and those that have the opposite action (Lemke and Brannon, 1972). In general, bacteria usually synthesize the former type, and fungi the latter.

6. Other Conversions

As a result of nearly half a century of research into microbial transformations of compounds that have actual or potential commercial value, there exists a vast library of information on the types of reactions that can be carried out by different classes and species of micro-organism, and this is now being applied in many other branches of the pharmaceutical and chemical industry. Useful compilations of type reactions that can be carried out by micro-organisms have been assembled by Wallen *et al.* (1959), Iizuka and Naito (1968) and Fonken and Johnson (1972). To date, there has not been a development in any of the fields entered comparable with that witnessed in steroid conversions. Greatest interest has probably been shown in microbial conversions of alkaloids, and it could well be in this area that the next major breakthrough will come.

III. BIOCHEMISTRY AND PHYSIOLOGY OF ENZYME PRODUCTION BY MICRO-ORGANISMS

A. Mechanism of Enzyme Synthesis

Proteins occupy a unique position in cell metabolism since each of the metabolic reactions that takes place in the cell is catalysed by a specific protein, i.e. an enzyme. When considering the biosynthesis of proteins, it is necessary therefore to explain not only how amino acids are joined together to form polypeptides but how the order in which they are joined is specified since the sequence of amino acids is different in each protein and determines the enzymic activity of the protein.

It has been appreciated from the very early days of biochemistry that the characteristics of a living organism are determined by its ability to carry out certain metabolic reactions and so, therefore, by its capacity to synthesize specific enzymes. However, geneticists have shown that these characteristics are also controlled by the DNA in the cell. The essential identity of these two points of view became clear when George Beadle formulated his one gene–one enzyme hypothesis, which states that the synthesis of any enzyme (or more correctly any polypeptide) is controlled by one particular gene. Formulation of the one gene–one enzyme hypothesis immediately established a fundamental relationship between nucleic acids and proteins, and it led inevitably to the question: is the gene directly involved in synthesis of a specific protein? Experiments with *Amoeba proteus* and the huge unicellular alga *Acetabularia mediterranea*, in which it was shown that protein synthesis continued even after nuclei had been removed experimentally, strongly suggested that DNA is not directly concerned in protein synthesis.

Information about the nature of the intermediary between DNA and protein came from experiments that showed a direct correlation between the RNA content of a cell and its growth rate, while the DNA content remained approximately constant.

The DNA in a cell carries a set of coded information for the 'running' of the cell. This information is passed on to a species of RNA, namely messenger RNA (mRNA), in a process referred to as *transcription*. In both types of nucleic acid, the information is coded in the form of a language of nucleotide bases, each sequence of three bases coding for an amino acid or a full stop. Transference of information from a mRNA molecule to a polypeptide takes place on intracellular organelles known as ribosomes, in a process termed *translation*. Since the late 1940s, research on the reactions that go to make up the transcription and translation processes has proceeded at a feverish pace. Nevertheless, firm information is available for the processes in only a limited number of micro-organisms, principally *Escherichia coli* among prokaryotes, and *Sacch. cerevisiae*, *Neurospora crassa* and certain slime moulds among eukaryotes. More recently, the main research thrust in this central area of molecular biology has been the manner in which transcription and translation are regulated.

1. Transcription

Pathways for synthesis of the four nucleoside triphosphates required for synthesis of RNA, namely ATP, CTP, GTP and UTP, are well charted and can be obtained from any basic text in biochemistry (Lehninger, 1970) or microbial physiology (Rose, 1976; Mandelstam and McQuillen, 1973). Briefly, the purine residue in purine nucleotides is built up on a residue of phosphoribosyl pyrophosphate starting with glutamine. The end product of the pathway is inosinic acid (IMP), the ribonucleotide corresponding to the purine hypoxanthine. This, in turn, is converted into AMP or GMP, and thence to ATP and GTP. The pathway leading to synthesis of pyrimidine ribonucleotides differs from that leading to purine ribonucleotides not only in being shorter but also in that phosphoribosyl pyrophosphate enters the biosynthetic sequence at a later stage when it combines with the pyrimidine orotic acid to give orotidine 5-phosphate. Uridylic acid (UMP) is formed by decarboxylation of orotidine 5-phosphate. Cytidine triphosphate is produced in a reaction that involves UTP and ammonium ions.

Deoxyribonucleotides, required in synthesis of DNA, are formed from ribonucleotides in reactions catalysed by ribonucleotide reductases:

$$\text{NADPH} + \text{H}^+ + \text{ribonucleotide} \longrightarrow$$
$$\text{NAPD}^+ + \text{Deoxyribonucleotide} + \text{H}_2\text{O}$$

Two classes of reductase have been described (Thelander and Reichard, 1979). The first, represented by the enzyme from *Lactobacillus leichmannii*, is monomeric and uses adenosylcobalamin as a cofactor. The other, which is synthesized by *E. coli* among other micro-organisms, does not require this cofactor, and consists of two non-identical subunits.

Replication of DNA, which is a prerequisite for transcription, is a unique event in the life of a cell since it is the process by which the genetic information in the cell is copied before the transfer of a genome to a daughter cell. Since this genetic information is encoded on the DNA molecule, DNA synthesis must involve an exact replication of the parent DNA. According to the Watson–Crick hypothesis, the DNA molecule consists of two unbranched polynucleotide chains wound round each other along one axis to form a double-stranded helix. Since the model postulates a pair of templates each complementary to the other, there is no need for the production of a 'negative' copy of the DNA during replication.

Synthesis of DNA *de novo* using an enzyme from *E. coli* was reported in 1956 by Arthur Kornberg and his colleagues at Stanford University in California, U.S.A. The enzyme, known as DNA nucleotidyltransferase or DNA polymerase, catalyses a pyrophosphate-releasing reaction involving all four deoxyribonucleoside triphosphates and requires the presence of template DNA in addition to Mg^{2+}:

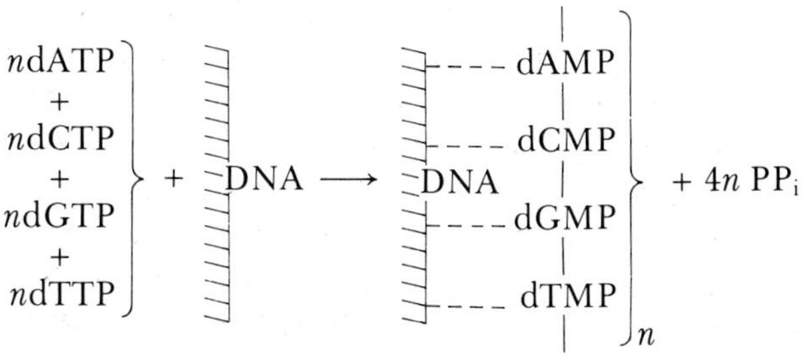

$$
\left.
\begin{array}{c}
n\text{dATP} \\
+ \\
n\text{dCTP} \\
+ \\
n\text{dGTP} \\
+ \\
n\text{dTTP}
\end{array}
\right\}
+ \text{DNA} \longrightarrow
\left[
\begin{array}{l}
\text{---- dAMP} \\
\text{---- dCMP} \\
\text{DNA} \\
\text{----dGMP} \\
\text{---- dTMP}
\end{array}
\right]_n
+ 4n\,\text{PP}_i
$$

The enzyme discovered by Kornberg and his colleagues, and now referred to as DNA polymerase I, is capable of synthesizing only short strands of DNA, and soon after its discovery doubts were cast on its role in DNA replication. Two other DNA polymerases, dubbed II and III, have since been discovered, DNA polymerase II interestingly by the son of Arthur Kornberg, which explains why it has been called the Kornberg Jr enzyme. It has also been shown that DNA replication is preceded by, and dependent on, a short burst of RNA synthesis. Protein synthesis is also known to be essential for the onset of replication. Currently, DNA replication is thought to involve synthesis of a number of stretches of DNA which are then linked together. These stretches are known as 'Okazaki pieces' after their discovery in Japan by the late Reiji Okazaki. The process involves a plethora of enzymes: RNA polymerase to synthesize RNA primers, a DNA polymerase (thought to be polymerase III) to elongate the DNA primers, a ribonuclease to excise the RNA fragments, another DNA polymerase (possibly the I and II polymerases) to fill in the gaps with DNA, and finally a joining enzyme or ligase to stitch the whole chain together. For an up-to-date account of DNA replication in prokaryotes, stressing the initiation phase, the reader should consult Tomizawa and Selzer (1979).

Rather less is known of the reactions involved in DNA replication in eukaryotic micro-organisms, although preliminary evidence suggests that a series of reactions, similar to those that operate in prokaryotes, also participate in eukaryotic DNA synthesis (Sheinin et al., 1978). Information is most extensive for the DNA polymerases of eukaryotic micro-organisms. Some organisms, including Sacch. cerevisiae, Euglena gracilis and Ustilago maydis, have two distinct polymerases. In these organisms, the polymerases are usually designated I and II, since they have different properties from the DNA polymerases of higher organisms, which are conventionally designated α, β and γ. Other eukaryotic microbes, however, seem to synthesize just one DNA polymerase. This group includes Dictyostelium discoideum.

Transcription involves synthesis of rRNA, mRNA and tRNA, but it is in relation to mRNA synthesis that the term is most commonly used, simply because a large part of the genome codes for mRNA. This fact was revealed by experiments on E. coli in which RNA molecules and homologous DNA were annealed or hybridized. Hybridization revealed that only 0.2 and 0.02%, respectively, of the DNA in E. coli codes for rRNA and tRNA.

Although the various types of RNA have different functions and are coded for by different regions on the genome, their synthesis in bacteria at least is catalysed by the same enzyme, namely RNA nucleotidyltransferase or RNA polymerase. The enzyme catalyses a pyrophosphate-releasing reaction in which all four ribonucleoside triphosphates take part, and which requires template DNA.

Information on the properties of RNA polymerase is most extensive for the enzyme from *E. coli*, although it is thought that the polymerases in other bacteria are similarly constituted. The enzyme in *E. coli* is made up of five subunits. There are two α chains, each with a molecular weight of 41,000, one β chain of 155,000, one β' chain of 165,000 and a smaller (86,000) unit known as the σ component. The complex constitutes the holo-enzyme. However, the σ component readily dissociates from the complex to give the core polymerase which retains polymerizing ability. The functions of the individual subunits are far from being understood, but it is believed that the β' subunit contains the DNA-binding site and possibly the active site of the enzyme. The σ component, which is a heat-labile acidic protein, locates the polymerase at the correct starting point on the genome.

The activity of RNA polymerase is associated with several other components known as *transcriptional factors*. Termination of transcription is effected by the action of the ρ factor, a protein with a molecular weight of 200,000. Another factor, the ψ factor, regulates the amounts of the different types of RNA that are synthesized. As much as 40% of the RNA made by *E. coli* is rRNA, although we have seen only 0.2% of the genome codes for this RNA. Clearly, there must be a very efficient initiation of rRNA synthesis in the bacterium, and it is believed that this is effected by the ψ factor. The existence of several other transcriptional factors has been postulated, but their functions are less clearly understood.

Synthesis of RNA in eukaryotic micro-organisms is a rather more complicated process. All eukaryotes so far examined contain several forms of DNA-dependent RNA polymerase (Roeder, 1974). Moreover, they are located in specific organelles of the cell. One enzyme occurs in the nucleolus, and would seem to function in rRNA synthesis. The enzyme located in the nucleoplasm is believed to catalyse synthesis of the bulk of the cell's mRNA. Other RNA polymerases have been shown to be present. One in the nucleus has a function that has, as yet, not been identified. An RNA polymerase is found in mitochondria, and this has a lower molecular weight than those located in the

nucleus. These various RNA polymerases differ in their sensitivity to
α-amanitin.

2. Translation

Truly spectacular progress has been made during the past couple of
decades in elucidating the mechanism by which amino acids are
polymerized into proteins. Once methods had been devised for
preparing cell-free extracts capable of polypeptide synthesis, the way
was paved for an analysis of the crude extracts to discover which of the
components were involved in the polymerization process. In recent
years, this phase has given way to one in which details of the reaction
mechanisms and the function of various factors are being studied.
Animal and plant tissue, as well as micro-organisms, have been used in
these studies, but there is every reason to believe that the mechanisms
of protein synthesis are basically the same in all types of organism.

The most important fact concerning the translation process is that it
is accomplished not by direct interaction of amino acids with mRNA,
but through intermediate adapter molecules, namely tRNA species.
Polymerization of amino-acid residues attached to tRNA molecules
takes place on polysomes, which consist of a group of 70S or 80S
ribosomes, lying closely together and connected by a molecule of
mRNA. Ribosomes do not have genetic specificity and they function
quite non-specifically in the translation process. The mRNA molecule
binds to the smaller 30S or 40S ribosomal subunits. Ribosomes begin
translating mRNA while it is still being transcribed from DNA (Fig. 3).

The necessary preamble to polymerization of amino-acid residues
on polysomes is the formation of aminoacyl-tRNA molecules. In every
microbe that has been examined, there exist at least 20 different
aminoacyl-tRNA synthetases which catalyse formation of an ester
linkage between the carboxyl group of an amino acid and the
3′-hydroxyl group of the terminal adenosine residue of a specific tRNA
molecule. Formation of an aminoacyl-tRNA molecule is a two-stage
process, involving first activation of the amino acid and then transfer to
a tRNA molecule. The first reaction involves the amino acid and ATP,
resulting in release of inorganic pyrophosphate:

Amino acid + Synthetase → Synthetase–(aminoacyl-AMP) + PP$_i$

In the aminoacyl-AMP complex, which remains attached to the

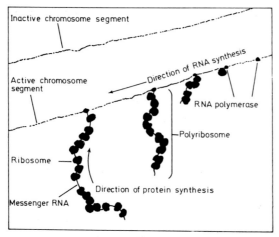

Fig. 3. Diagrammatic representation of the relationship between the chromosome and related organelles during transcription in bacteria, drawn from an electron micrograph (Rose, 1976). The diagram shows two strands of the bacterial chromosome with only the lower one involved in transcription. Also depicted are molecules of RNA polymerase, the one on the far right of the diagram being at the approximate initiation site. Successively transcribed mRNA molecules are seen to peel off towards the left of the diagram.

synthetase, there is a mixed anhydride linkage between the carboxyl group of the amino acid and the phosphate of AMP. In a second reaction, the enzyme-bound aminoacyl-AMP reacts with the specific tRNA molecule to form the aminoacyl-tRNA:

Synthetase–(aminoacyl-AMP) + tRNA →

Aminoacyl-tRNA + AMP + Synthetase

The specificity of aminoacyl-tRNA synthetases, a prerequisite for faithful translation of the genetic message, must be very high. It is, in fact, exercised in both the activation and transfer reactions, the specificity in the latter reaction being higher than that in the former.

Polypeptide-chain synthesis involves three distinct processes, namely initiation, elongation and termination. A valuable review of these processes as they operate in bacteria has recently come from Grunberg-Manago *et al.* (1978).

The first or *N*-terminal amino-acid residue in all polypeptides synthesized on ribosomes is *N*-formylmethionine (fMet). The first step in initiation of polypeptide synthesis is formation of Met-tRNA^{fMet}. There exist two species of tRNA that accept methionine residues, one an initiator tRNA (tRNA^{fMet}) and the other an elongation tRNA

(tRNA^{fMet}). Formylation of the methionine residue is favoured when the residue is attached to tRNA^{fMet}. The reaction is catalysed by a transformylase which transfers a formyl residue from a N-formyl-tetrahydrofolate donor to Met-tRNA^{fMet}. *In vivo*, the formyl group initially present on the methionine residue at the beginning of a polypeptide chain is rapidly removed by a deformylase.

Thereafter, an initiation complex, made up of 70S ribosomes, fMet-tRNA^{fMet} and mRNA, is formed in a reaction that requires hydrolysis of GTP. It also involves the operation of at least three proteins or initiation factors. Two of these factors (IF I and III) regulate association of 30S and 50S ribosomes, whereas IF II directs binding of the fMet-tRNA^{fMet} part of the complex to the P (or peptide-formation) site on the 50S ribosome (Fig. 4).

Elongation of the polypeptide chain also involves specific proteins known as elongation factors (EF). The first step involves formation of a complex between one of these factors, EF-Tu, GTP and the amino-acyl-tRNA which is to supply the next amino-acid residue. As a result of a reaction between this complex and the peptidyl-tRNA at the P site on the 50S ribosome, GTP is hydrolysed allowing EF-Tu-GDP to dissociate from the ribosome, and leaving peptidyl-tRNA at the P site and aminoacyl-tRNA at the A site. The peptidyl moiety is then transferred to the aminoacyl-tRNA at the A site as a result of the activity of a peptidyl transferase located on the 50S ribosome. Another elongation factor (EF-G) and GTP, probably as a complex, now bind to the ribosomal complex. The uncharged tRNA at the P site is then

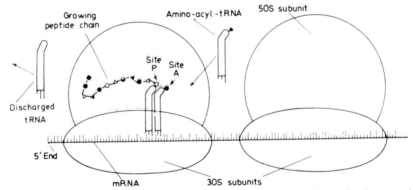

Fig. 4. Diagram showing an early stage in the translation process on a bacterial polysome, with site A occupied by an aminoacyl-tRNA molecule and site P by a polypeptide-tRNA molecule. See text for an explanation of the process.

displaced by the newly formed peptidyl-tRNA, and mRNA is translated by one codon. Factor EF-G and GDP dissociate from the ribosome, which is then ready for a further step in the elongation process. In a shuttle process catalysed by a third factor, EF-Ts, the complex aminoacyl-tRNA·EF-Tu·GTP is reformed from EF-Tu·GDP.

Three codons cause release of the polypeptide, or chain termination, by hydrolysis of the ester bond by which it is linked to tRNA. Translocation is required before this process can take place at the P site. Termination codons are recognized by yet other proteins (termination factors) which bind to the ribosome and modify the specificity of the peptidyl transferase so that the polypeptide is transferred to water rather than an aminoacyl-tRNA at the A site.

Details of the translation process in eukaryotic micro-organisms have still to be fully clarified, but it is known that, in several respects, it is similar to that which operates in prokaryotes, especially with regard to the involvement of specific protein factors.

B. Regulation of Enzyme Synthesis

1. Genetic Control

From the account given in the previous section, it is clear that the microbial genome determines qualitatively the enzymic composition of that microbe. If the gene for a particular enzyme is not present in the genome, then that enzyme cannot be synthesized by the micro-organism, no matter what the nature of the medium in which it is grown. In any industrial microbiological process, including the use of microbes as a source of enzymes and to bring about chemical transformations, an immensely powerful tool is strain selection. Traditionally, this has been carried out very largely in an empirical manner, using techniques such as hybridization, isolation of variants and chemical mutagenesis. These techniques have proved the test of time, and most of the strains currently used as sources of industrially important enzymes have arisen in this way (Vaněk et al., 1973; MacDonald, 1976; Elander et al., 1977).

The last few years have seen several important advances in the use of techniques which could, eventually, revolutionize strain improvement in industrial microbiology. Protoplast fusion, which can be followed by integration of the genetic material of the two protoplasts, has been explored for some years (Ferenczy *et al.*, 1976). Potentially of greater importance, however, is the discovery that, through the agency of phage or plasmid vectors, genetic information from conceivably any organism can be introduced into any other micro-organism. This is the practice of genetic engineering, and theoretically it provides a means whereby the appropriate genetic information for producing a particular enzyme can be introduced into a micro-organism that is the ideal vehicle because, for example, of its ease of handling (Glover, 1978).

2. Metabolic Control

It is self-evident that the metabolism of micro-organisms, and indeed of all living organisms, is under strict control *in vivo*. Micro-organisms show a general reproducibility in structure and composition, one that on the whole is maintained even when they are grown in media of widely different composition. Moreover, rapidly growing micro-organisms usually accumulate intracellularly, or excrete into the environment, only very small quantities of compounds that could be used in the synthesis of new cell material. One of the major achievements of molecular biology has been the discovery of the mechanisms whereby microbes effect this efficient control over their metabolism. Basically, there are two types of metabolic control. Firstly, microbes are able to control the activity of enzymes that have been synthesized by devices that switch enzymes on and off as the need for the end product of the pathway changes. This is a short-term device, since it can be viewed as an inefficient use of enzyme protein. A long-term and more efficient mechanism is that by which synthesis of enzymes is switched on or induced when they are required, and switched off or repressed when there is not a metabolic requirement for the enzymes.

a. *Induction of enzyme synthesis.* Some enzymes are synthesized by micro-organisms irrespective of the chemical composition of the environment. These are known as *constitutive enzymes* to distinguish them from *induced enzymes*, which are synthesized by a micro-organism only in response to the presence in the environment of an inducer, which is usually the substrate for the enzyme or some structurally related compound. When an inducer is present in the environment, not one but a number of inducible enzymes may be formed by a micro-organism. This occurs when the pathway for metabolism of the compound is mediated by a sequence of inducible enzymes. The micro-organism responds to the inducer by synthesizing the appropriate enzyme, and the intermediate formed as a product of the reaction induces synthesis of a second enzyme; the intermediate formed in the reaction catalysed by the second enzyme then induces synthesis of a third enzyme, and so on. This phenomenon has been termed *sequential induction.*

Induced synthesis of enzymes by micro-organisms was reported during the 1930s but it was not until a decade later that serious thought was given to the mechanism of the process. Most of the work was concerned with induced synthesis of β-galactosidase in *E. coli.* When this bacterium is grown in a lactose-free medium it is unable to break down lactose, but when it is grown in a lactose-containing medium it synthesizes an enzyme, β-galactosidase, which catalyses hydrolysis of lactose into glucose and galactose. Two important facts were established regarding induction of β-galactosidase synthesis in *E. coli.* First, growth of the bacterium in a medium containing lactose as the sole carbon source was shown not to be the result of the selection of mutant strains that all along had the capacity to synthesize the enzyme. Rather, it was shown that each of the bacteria present synthesized β-galactosidase. Secondly, the enzyme was shown to be synthesized by the bacteria *de novo* from amino acids and not by some modification of preformed protein.

b. *Repression of enzyme synthesis.* One of the first examples of repression of enzyme synthesis was discovered during studies on the pathway of methionine synthesis in *E. coli*, which proceeds from homoserine via cystathionine and homocysteine. When the bacterium was grown in a medium containing only glucose and inorganic salts, extracts of the bacteria contained an enzyme that catalyses conversion of homocysteine into methionine. But when methionine was included in the growth medium, extracts of the bacteria were devoid of this enzyme. Thus, synthesis of the enzyme had been prevented by growing the bacterium in the presence of methionine, the bacterium preferring to use the methionine supplied rather than synthesize its own. Numerous other examples of end-product repression of enzyme synthesis have since been reported.

End-product repression of enzyme synthesis on a pathway differs from feedback inhibition of enzyme activity in that all of the enzymes involved on the pathway are usually affected. On some pathways, such as that leading to synthesis of histidine in *Salmonella typhimurium*, synthesis of each of the enzymes is repressed by the end product to the same extent; this has been termed *co-ordinate repression*. Other examples have been reported in which synthesis of each of the enzymes on a pathway is repressed to a different extent. One such pathway is that leading to arginine synthesis in *E. coli*. With several of the pathways that show co-ordinate repression, the genes that direct synthesis of the enzymes on the pathway are thought to lie clustered together on the chromosome. End-product repression of enzyme synthesis also differs from feedback inhibition of enzyme action in that, with pathways leading to synthesis of certain amino acids including histidine and valine, the effector molecules that activate the repression mechanism are tRNA derivatives of the amino acids rather than the free amino acids.

An interesting example of enzyme repression has been found to operate on pathways leading to synthesis of the branched-chain amino acids leucine, valine and isoleucine and the B-vitamin pantothenate. These two separate pathways involve a homologous series of reactions, and enzymes catalyse reactions on both pathways. Studies on enzyme repression on these pathways in *E. coli* and *S. typhimurium* have shown that there is no repression of synthesis of the common enzymes until each of the end products, valine, leucine, pantothenate and isoleucine, is simultaneously present in excessive concentration. This type of concerted action has been termed *multivalent repression*.

Repression of enzyme synthesis is not confined to processes in which the effector is the end-product of a pathway. Synthesis of many enzymes is known to be repressed by low molecular-weight compounds that are not directly associated with those pathways on which reactions are catalysed by the enzymes concerned. The best studied of these processes is *catabolite repression*. It has been known for well over 30 years that synthesis of certain enzymes by micro-organisms can be prevented by including glucose in the growth medium, which explains why the effect was originally referred to as the *glucose effect*. For example, if *E. coli* is grown in a mineral salts medium containing glycerol as the carbon source for growth, addition of the compound isopropylthiogalactoside, which like lactose induces synthesis of the enzyme β-galactosidase, induces a high rate of synthesis of this enzyme. If glucose is subsequently added to the culture, there is a rapid and severe repression of synthesis of β-galactosidase. After several generations of growth, there may be a partial recovery to a somewhat higher rate of enzyme synthesis, but this is always much lower than the originally observed induced rate. These two types of repression are referred to as, respectively, *transient catabolite repression* and *permanent catabolite repression*.

Catabolite repression of enzyme synthesis operates on many metabolic pathways. It operates simultaneously with induction of enzyme synthesis and with repression caused by end-products of pathways. Operation of these various mechanisms for regulating synthesis of enzymes in micro-organisms is very well illustrated by the pathways that pseudomonads (*Pseudomonas aeruginosa* and *Ps. putida* have been the favoured species in these studies) use for breaking down aromatic compounds. Synthesis of the enzymes on these pathways is regulated by a process of sequential induction of a group of enzymes by intermediates on the pathways as well as by catabolite repression. Starting with the aromatic compound mandelic acid, synthesis of each of the five enzymes that catalyse steps in the conversion of mandelate into benzoate is induced by mandelate and, at the same time, synthesis is subject to catabolite repression by benzoate, catechol and succinate. Benzoate itself can be used as a growth substrate, and it induces synthesis of an enzyme that catalyses its conversion into catechol. Synthesis of this enzyme is subject to catabolite repression by catechol or succcinate. Further steps in the breakdown of catechol involve conversion into *cis*, *cis*-muconate and β-oxoadipate, before this last intermediate is cleaved to give acetate and succinate. Compounds that induce or repress synthesis of enzymes that catalyse these reactions are indicated on Figure 5.

The advantage of this simultaneously operating system of sequential induction and repression of enzyme synthesis is that intermediates on the pathway for degradation of mandelate can induce synthesis of sufficient amounts of the enzymes required for their own metabolism, without necessitating synthesis of enzymes that catalyse earlier steps in the breakdown. Also different starting substrates can induce synthesis of their own specific enzymes, again with the operation of catabolite repression, in amounts that are needed.

Regulation of enzyme synthesis can also take place in microbes by mechanisms which, although related to catabolite repression, differ in

INDUCER   CATABOLITE   REPRESSOR

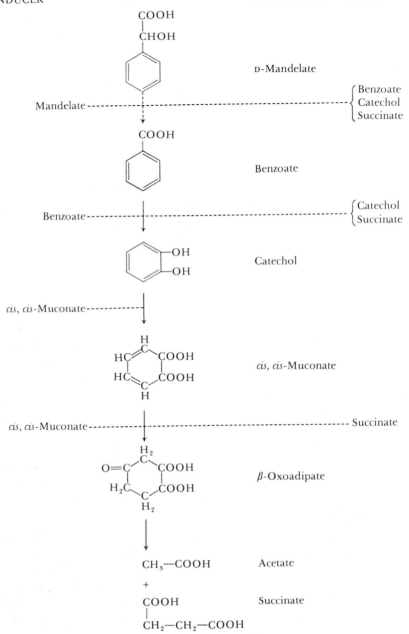

Fig. 5. Diagram showing sites for regulation of enzyme synthesis on the pathway used by *Pseudomonas putida* for degradation of mandelate.

that the effector is not an intermediate on a catabolic pathway. In species of *Hydrogenomonas*, using molecular hydrogen as the energy source, synthesis of several enzymes is subject to catabolite repression with molecular hydrogen being the effector. Moreover, in many fungi, the ammonium ion regulates synthesis of some enzymes and transport proteins, a phenomenon that has been termed *ammonia metabolite repression*.

3. Molecular mechanisms in regulation of enzyme synthesis

Induction and repression of enzyme synthesis have in common that each involves regulation of synthesis of high molecular-weight proteins by relatively low molecular-weight effector molecules. It was obvious from the start that the effectors must act at the transcriptional or translational level of protein synthesis alongside the regulatory mechanisms (mainly involving protein factors) that have already been described. Some brilliant genetic experimentation by Francois Jacob and Jacques Monod, at the Institut Pasteur in Paris, led these workers to formulate a hypothesis in 1961 for the molecular mechanisms involved in induction and repression of enzyme synthesis in bacteria. The Jacob–Monod hypothesis has, in the intervening years, been subjected to the most rigorous experimental examination, and is now firmly accepted as the only established explanation for induction and repression of enzyme synthesis in prokaryotes. Its discovery must surely represent the most important biological achievement of the Twentieth Century.

Jacob and Monod (1961) proposed that synthesis of enzymic proteins (or more accurately polypeptides) is controlled by *structural genes*, related groups of which lie closely together on the genome. They suggested, further, that there exist separate *regulator genes* responsible for producing compounds (which we now know are proteins) which act by controlling the expression of a group of structural genes. These are the *repressor proteins*. The well established fact, that synthesis of enzymes with quite different catalytic activities is quite frequently co-induced or co-repressed to the same extent by a low molecular-weight compound, suggested that repressor proteins act not on individual structural genes but on a separate region of the genome where groups of genes can be switched on or switched off. This site at which repressor proteins act on the genome is known as the *operator gene*, and each of these genes controls transcription of a group of structural genes. Another genetic element, the existence of which was not proposed by Jacob and Monod but which was later discovered from the results of genetic experiments, is the *promotor region* which lies alongside the operator gene and is the site on the genome at which RNA polymerase becomes attached. A group of structural genes, together with the associated regulator and operator genes and promotor region, is known as an *operon*. Operons on bacterial genomes differ tremendously in size and the degree to which the different genes in the operon are juxtaposed. Figure 6 shows an indealized structure for a small operon.

An operon very similar in structure to the idealized arrangement shown in Figure 6 is the lactose (or *lac*) operon in *E. coli* which formed the subject of the pioneer research by Jacob and Monod. Since their early studies, the *lac* operon in *E. coli* has been researched very intensively, and is now by far the best understood part of any cellular genome. The elements in the *lac* operon in *E. coli*, reading from left to right in Figure 6, are the regulator (or *i* gene), the promotor region (*p*) and the operator gene (*o*), followed by three structural genes, namely β-galactosidase (the *z* gene), the lactose-transport protein gene (*y*) and a

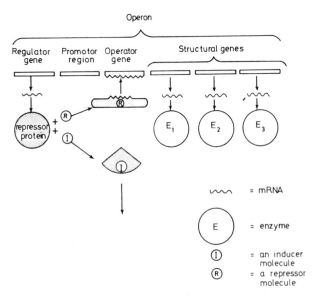

Fig. 6. Diagrammatic representation of the Jacob and Monod (1961) scheme for the molecular basis of induction and repression of enzyme synthesis in bacteria. See the text for a detailed explanation.

gene that specifies thiogalactoside transacetylase (a) an enzyme which, although it is regulated in concert with the first two structural genes, has no known metabolic role in *E. coli*.

The fine structure of the *lac* operon in *E. coli* is rapidly being revealed, largely as a result of research by Walter Gilbert and his colleagues at Harvard University in Boston, U.S.A. The operator gene in this operon is only 27 base pairs long. Interestingly, the arrangement of these base pairs in the gene shows a remarkable symmetry. Gilbert and his colleague Benno Müller-Hill have also isolated the repressor protein for this operon. This is a tetramer protein with a molecular weight of about 150,000 and 347 amino-acid residues.

Other bacterial operons are very different in structure from the *lac* operon in *E. coli*. In the same species of bacterium and in *S. typhimurium*, the histidine (*his*) operon is truly a giant with 10 structural genes, all close together, and one operator gene; this operon is sometimes referred to as the *histidon*. At the opposite extreme, the arginine (*arg*) operon in *E. coli* has structural genes scattered all over the genome, but with only one operator gene. In the operon that specifies enzymes for galactose utilization in *E. coli* (the *gal* operon) there are two operator genes.

The next question, of course, is exactly how are genes switched on or off? Work on this has proceeded at a feverish pace over the past 15 years. As a result it has been possible to recognize two separate types of control mechanism, namely *negative* and *positive control of protein synthesis*.

a. *Negative control of protein synthesis.* In their original scheme, Jacob and Monod (1961) proposed that induction of protein synthesis is possible only when the repressor protein of the operon is not in contact with the operator gene. When an inducer is not available, either because it is not supplied in the medium or because it cannot be synthesized by the microbe, the repressor protein remains in contact with the operator gene, thereby preventing expression of, that is mRNA synthesis on, associated structural genes. The repressor protein is thought to exert this effect by occluding part of the promotor region next to the operator gene, so preventing RNA polymerase from making contact with the promotor region. When an inducer is available to a micro-organism, either by being provided in the medium or because it has been synthesized in the cell, it combines with the repressor protein which, since it is an allosteric protein, changes conformation and loses its affinity for the operator gene.

Repression of enzyme synthesis was explained by Jacob and Monod (1961) by assuming that, in the absence of the effector molecule, the repressor protein has no affinity for the operator gene, but that it acquires an affinity when it undergoes a conformational change after interaction with the effector. The common feature to these suggested mechanisms is that gene expression is possible only when the operator gene is free. For this reason, these control mechanisms are said to be negative in nature. Negative control of protein synthesis is well established at the molecular level in many bacterial operons, notably induction of synthesis of proteins of the *lac* operon in *E. coli* grown in the presence of lactose and in the absence of glucose. Induction of enzyme synthesis can therefore be viewed as a derepression process (Rose, 1976).

b. *Positive control of protein synthesis.* This is said to occur when an association between a protein and a part of the regulatory region of an operon, which may not necessarily be the operator gene, is essential for expression of related structural genes in the operon. Examples of positive control are not as well documented as those for negative control. One of the better understood relates to expression of the genes concerned with utilization of arabinose by *E. coli* (the *ara* operon), mainly because of work on this operon by Ellis Englesberg and his associates of the University of California at Santa Barbara, U.S.A. The first three genes on this operon specify synthesis of, respectively, an epimerase, an isomerase and a kinase; the respective genes are *ara D*, *ara A* and *ara B*. Another linked gene, *ara C*, does not determine the structure of an enzyme but behaves as a regulator gene. It is suggested that the product of the *ara C* gene is a protein that combines with the inducer arabinose to form an activator molecule which acts at the initiator site on the operon. This activator molecule is therefore responsible for positive control of expression of the *ara* structural genes. When the protein is unable to combine with arabinose, it acts as a repressor protein, like the product of the *i* gene in the *lac* operon, and combines with the operator gene in the *ara* operon to prevent expression of structural genes.

Positive control of protein synthesis also operates during catabolite repression. For a long time, the molecular basis of the effect of glucose on synthesis of enzymes in microbes puzzled microbial physiologists. It was suspected that the effector, which presumably acts on the genome during catabolite repression, might be a compound, or a group of compounds, whose synthesis is regulated by the concentration of glucose in the growth medium. It is now known that this assumption was correct, but identification of the elusive effector (or group of effectors) was greatly helped by the results of research on the effect of mammalian hormones on cells. These researches pointed to a compound, cyclic adenosine monophosphate (cAMP), as a possible candidate for the role of effector in microbial catabolite repression because of the similar role that this compound plays in the control of enzyme synthesis in mammalian cells. The first evidence that cAMP was important in regulating enzyme synthesis in bacteria was the finding that addition of this compound to growth media can overcome transient catabolite repression of β-galactosidase synthesis in *E. coli* (Robison *et al.*, 1971).

It has since been established that cAMP combines with a protein, called the catabolite gene activator (CGA) protein, to give a complex that facilitates binding of RNA polymerase to the promotor region on the operon. The CGA protein from the *lac* operon in *E. coli* has been isolated and shown to be a protein dimer, with a strong affinity for DNA, which is enhanced considerably when the protein is combined with cAMP. The situation is further complicated in that cells synthesize a second catabolite repression effector, cyclic guanosine monophosphate (cGMP). This compound acts in exactly the opposite way to cAMP, and a combination of cGMP with CGA protein lowers the affinity of the protein for the promotor region on the operon.

Cyclic AMP and cGMP are made from the corresponding nucleoside triphosphates (ATP, GTP) in reactions catalysed, respectively, by adenyl cyclase and guanyl cyclase. Moreover, both cAMP and cGMP can be broken down in reactions catalysed by esterases. Clearly, therefore, the activities of these enzymes concerned with synthesis and breakdown of cAMP and cGMP are extremely important in the regulation of enzyme synthesis.

Regulation of enzyme synthesis in eukaryotic micro-organisms is less completely understood. It is known, however, that certain basic types of control mechanism, such as induction and repression of enzyme synthesis, operate. Less is known, however, of the molecular mechanisms by which gene expression is regulated at the chromosome.

Moreover, it is far from clear whether a Jacob–Monod type of regulation operates.

C. Location and Secretion of Enzymes

1. Location of Enzymes

One of the major achievement of biochemistry over the first half of this Century was to ascertain the location of different types of enzyme in the prokaryotic and eukaryotic cell. This information is now largely complete, and it can be found in any basic text of biochemistry (Lehninger, 1970) or microbial physiology (Rose, 1976; Mandelstam and McQuillen, 1973). Roodyn (1967) has produced a useful text on enzyme cytology in living cells.

Not surprisingly, the situation differs in eukaryotes and prokaryotes. In the former organisms, with their distinctive intracellular compartmentation, many enzymes are characteristically associated with particular organelles. For example, enzymes involved in DNA replication are located in the nucleus, and many of those associated with generation of ATP are in the mitochondria. Many lytic and depolymerizing enzymes, including a large number of industrial importance, are located in intracellular low-density vesicles including vacuoles (Matile and Wiemken, 1967). The metabolic role of these lytic enzymes is unclear, but it is thought that they are involved in growth of the envelope layer in eukaryotic micro-organisms (Cartledge et al., 1977). They also convey packets of enzymes to the cell envelope as a prelude to secretion.

In prokaryotes, enzymes are either bound to the plasma membrane (or in some instances the wall) or occur in a soluble form in the cytosol. However, in both prokaryotes and eukaryotes, the so-called cytosol enzymes should not be envisaged as being homogeneously dissolved in cell water. It is far more likely that they lie loosely associated with membranes and organelles in a fairly ordered manner.

2. Secretion of Enzymes

Many industrially important microbial enzymes are exo-enzymes, that is they are secreted from the the producing organisms into the external medium. This association is explained by the fact that enzymes are often required in industrial processes to catalyse depolymerization of natural polymers (polysaccharides, proteins), and micro-organisms frequently secrete these enzymes to catalyse the initial stage in utilization of naturally occurring polymers as a source of nutrients. Because they are easily isolated from microbial cultures, extracellular enzymes have been studied for many years. A very useful account of these enzymes, with a comprehensive compilation, was provided by Davies (1963). More recently, a useful account of bacterial extracellular enzymes has come from Priest (1977).

In physiological terms, it is important to define precisely what is meant by a microbial extracellular enzyme. This has been defined by Pollock (1962) as one 'which exists in the medium around the cell, having originated from the cell without any alteration to cell structure greater than the maximum compatible with the cell's normal processes of growth and reproduction'. The significance of this definition lies in its distinction between exo-enzymes and periplasmic enzymes, which have also passed through the plasma membrane.

Extracellular enzymes from both prokaryotic and eukaryotic micro-organisms have several properties in common (Davies, 1963; Glenn, 1976). In general, they are more often than not hydrolytic enzymes, and their synthesis is sensitive to end-product and catabolite repression. For example, synthesis of exoproteases by several bacterial species, particularly *Bacillus* spp., is repressed by amino acids, especially isoleucine and proline (Millet *et al.*, 1969). The majority of microbial exo-enzymes lack, or contain a very low proportion of, cysteine residues, a property that was first noted by Pollock and Richmond (1962). Some microbial extracellular enzymes are glyco-proteins, but this is not a universal property (Glenn, 1976). When carbohydrate residues are attached to these enzymes, they are most frequently N-acetylglucosamine and mannose.

Research into the mechanism of enzyme secretion has been in progress for many years. Not surprisingly, it has concentrated largely on secretion by mammalian cells, although useful progress has recently been made in explaining the process in bacteria. The general view of the process in mammalian cells (Palade, 1975) is that the polypeptide chain of the enzyme that is to be secreted is formed on ribosomes attached to the rough endoplasmic reticulum. During peptide-chain formation, the protein is gradually passed into the lumen of the rough endoplasmic reticulum by a process involving formation of an initial

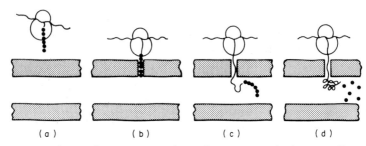

Fig. 7. Diagram showing four stages in synthesis of secretory proteins in mammalian cells. (a) depicts polypeptide-chain synthesis on a free ribosome. The chain begins with a signal sequence of residues indicated by closed circles. (b) shows the signal sequence attaching itself to an internal membrane through which it begins to pass. (c) before synthesis of the polypeptide chain is complete, the signal sequence is removed. (d) shows extension of polypeptide-chain synthesis after removal of the signal sequence, and continued passage of the chain through the membrane.

signal sequence of amino-acid residues which, after attaching itself to the membrane and passing through it, is released and broken down. Figure 7 gives a diagrammatic representation of the process. Once inside the rough endoplasmic reticulum, sugar residues may be attached to the polypeptide chain. Thereafter, the polypeptide passes to the smooth endoplasmic reticulum, often through a Golgi apparatus, and then to storage granules from which they are released by fusion with the plasma membrane.

Unfortunately, there is very little experimental evidence to show that this sequence of events operates during protein secretion in eukaryotic micro-organisms. The only firm data are on the enzymic activities of intracellular low-density vesicles isolated from strains of *Sacch. cerevisiae*, which have been shown to contain a variety of lytic enzymes including possibly the periplasmically located invertase (Matile and Wiemken, 1967; Cartledge *et al.*, 1977).

Secretion of enzymes by bacteria has been studied in rather greater detail (Glenn, 1976), with penicillinase secretion by *Bacillus licheniformis* and amylase secretion by *Bacillus amyloliquefaciens* being the favourite systems for study. Work in Lampen's laboratory (Lampen, 1976, 1978) has begun to provide a picture of the sequence of molecular events during penicillinase secretion by *B. licheniformis*. Two major types of this enzyme have been isolated from early- to mid-exponential phase cultures, namely a hydrophobic membrane-bound form, called the membrane enzyme, and an exo-enzyme. The membrane enzyme is a single polypeptide containing 292 amino-acid residues, and the exo-enzyme can be considered as a cleavage product of this, containing all but 25 amino-acid residues at the amino terminus. The hydrophobic nature of the membrane enzyme is attributable to the presence of a phosphatidylserine residue at the amino terminus. A protease has been isolated from B. *licheniformis* which cleaves the membrane penicillinase to give the exo-enzyme and an oligopeptide. Lampen and his colleagues suggest that secretion of penicillinase by B. *licheniformis* is caused by proteolytic cleavage of the membrane-bound enzyme, a process formally analogous to the signal mechanism believed to be involved in passage of proteins across mammalian membranes (Fig. 7). Clearly, there is a need to learn more about the process of enzyme secretion in both eukaryotic and prokaryotic micro-organisms, for this might suggest ways in which the process could be maximized in industrial fermenters.

D. Specificity of Enzymes

Enzymes have long been known not to act discriminately but to possess a degree of selectivity or specificity towards their substrates (Dixon and Webb, 1964). Indeed, it is clear that this specificity is at the basis of all regulatory processes that operate in living cells. However, the degree of specificity does vary among different enzymes. In general, it can be stated that the size and complexity of the molecular groupings required in a compound to make it a substrate are a measure of the specificity of the enzyme. Thus, an enzyme that required in its substrate only a relatively simple grouping, such as an ester linkage, will be relatively unspecific compared with an enzyme that requires a complex grouping, such as NAD^+. At one end of the spectrum, therefore, we have dehydrogenases, methyl transferases, kinases and synthetases, which in general have a high degree of specificity; at the other end there are enzymes, such as esterases and phosphatases, which have a low specificity towards their substrates. Not surprisingly, there are exceptions. Alcohol dehydrogenase, for example, has a lower degree of specificity than many other dehydrogenases, as does aldehyde dehydrogenase. Moreover, phospholipases show much greater substrate specificity than other types of esterase. These generalizations apply to both chemical specificity and stereospecificity.

Enzyme specificity has a particular relevance in a discussion of microbial biotransformations, since many of the substrates used in these transformations are not compounds which a micro-organism would metabolize during growth. For example, steroids that are modified chemically by a variety of micro-organisms (see Chapter 8) are foreign to the normal metabolism of these organisms. The only microbial cell components that structurally resemble steroids are sterols, which occur in the membranes of eukaryotic micro-organisms. It might be thought that the microbial enzymes that act on steroids would be those that are involved in synthesis of membrane sterols, and which show a reasonably broad specificity. This may hold for eukaryotic micro-organisms, but it cannot for prokaryotes that lack sterols, although they frequently have the ability to carry out steroid conversions. The only explanation for this ability must be that these prokaryotic microbes possess certain enzymes with a sufficiently broad specificity to be able to use certain steroids as substrates.

IV. CURRENT STATUS AND FUTURE PROSPECTS FOR COMMERCIAL EXPLOITATION OF MICROBIAL ENZYMES AND BIOCONVERSIONS

A. Current Status

1. Enzymes

Reference has already been made to the discovery, around 1969, that workers involved in large-scale manufacture of finely powdered enzyme preparations tended to develop allergies, and to the recessive effect that this discovery had on development of the microbial-enzyme industry World-wide. Fortunately, the problem has now been solved, largely as a result of introduction of measures that lower dust levels in production establishments and improve ventilation, coupled with more extensive use of protective clothing. Juniper et al. (1977) have reported on the medical control programme that was introduced.

In his valuable check list on World-wide current products of the fermentation industries, Perlman (1977) lists 25 different enzymes as being retailed in bulk quantities. Clearly, the enzyme-producing industries are in a buoyant state, a fact reflected in the two recent annual reviews that have come from Aunstrup (1977, 1978). The literature on microbial enzymes is vast and burgeoning. Aunstrup (1978) calculated, however, that only a small percentage of the approximately 10,000 papers which appear annually on the subject are of potential interest for the industrial development of enzymes.

a. *Proteases.* Production of proteolytic enzymes accounts for the greatest bulk of enzyme with several hundred tonnes being produced annually. The predominant application, with production at about 500 tonnes each year, is use of the alkaline serine protease from *Bacillus licheniformis* in detergents, followed by use of Mucor proteases in cheese manufacture; the latter application is currently the third largest microbial enzyme industry (Aunstrup, 1978). Perlman (1977) lists 21 industrial organizations that retail proteolytic enzymes; in Chapter 2 of this volume, K. Aunstrup describes production of these enzymes in detail.

b. *Carbohydrases.* A wide variety of microbial enzymes that catalyse reactions involving carbohydrates and related compounds are manufactured on a commercial scale. Over 12 such enzymes were listed by Perlman (1977), ranging from amylases (18 producers) to 'gumase' (one producer). The microbiology and biochemistry of production of the more important of these carbohydrases are dealt with in Chapters 3 and 4 in the present volume. In Chapter 3, W. F. Fogarty describes production of amylases, amyloglucosidases (eight producers) and glucanases (three), while in Chapter 4, S. A. Barker deals with glucose oxidase (five producers), glucose dehydrogenase (one), glucose isomerase (nine), β-glucosidase (three) and invertase (seven). Pectinases find several applications, mainly in industries that process fruit material. Perlman (1977) lists nine manufacturing firms, and the microbiology and biochemistry that underpin these processes are described by W. Pilnik and F. M. Rombouts in Chapter 5. Hydrolysis of cellulose, the main permanently renewable organic energy source on the planet, has long interested microbiologists, and prompted considerable research on cellulase production. Although ten producers of cellulases are listed by Perlman (1977), present-day commercial exploitation of this enzyme is limited largely to inclusion in digestive aids (Aunstrup, 1977). J. Goksøyr and J. Eriksen, in Chapter 6, give an account of microbial cellulases.

Perlman's (1977) check list includes two other carbohydrases which are not the subjects of chapters in this volume, namely hemicellulases and pentosanases, two closely related classes of enzyme. Dekker and Richards (1976) reviewed these enzymes up to 1973. The list includes four manufacturers of hemicellulases, two of them also retailing pentosanases. Although commercial exploitation of these enzymes is, as yet, very limited, considerable interest has been shown in their potential, particularly because they catalyse hydrolysis of polymers which are abundant in plant material to give sugars which can be fermented or used in alternative ways. An interesting application for hemicellulases was described by Oi *et al.* (1977). They were able to accelerate a methane fermentation process in which soybean seed coat was the raw material by adding a hemicellulase preparation from *Aspergillus niger*. Xylan hydrolysis has been examined particularly in relation to manufacture of the sweetener xylitol. Recent research on xylanases is described by Aunstrup (1978).

c. *Other Enzymes.* Six other enzymes are listed by Perlman (1977) none of which is described in detail in this volume because of their limited commercial application. Over the years, considerable interest has been shown in exploitation of microbial lipases of which there are currently five producers World-wide. Commercial interest in these enzymes goes back as far as 1914 when they were used in detergent preparations. This interest continues, and a recent patent (Stewart *et al.*, 1976) advocates coupling these enzymes in detergent preparations with synthetic activators such as naphthalene sulphonates. Other suggested applications are to promote lipolysis in cheeses, and to improve the flavour of pet foods (Haas and Lugay, 1976).

Catalase, of which Perlman (1977) lists four manufacturers, is another enzyme that has interested industrial enzymologists for many years but which has yet to be used on a large scale to break down hydrogen peroxide. Significantly, perhaps, the enzyme is not referred to in either of Aunstrup's recent (1977, 1978) reviews. Publicity has been given in recent years to the use of asparaginase in treating some types of leukemia cells that have a nutritional requirement for asparagine. However, success has been limited. Perlman (1977) lists just two manufacturers of this enzyme, which has been the subject of a review by Wriston and Yellin (1973). Anticyanase, glutamate decarboxylase and uricase are, according to Perlman (1977), each manufactured on a commercial scale by one different manufacturer. Industrial interest in these enzymes does not seem to be increasing.

2. Bioconversions

Great interest is currently being shown in the use of micro-organisms in bioconversions. The topic can be divided roughly into two areas of interest, namely the use of microbes to bring about specific chemical modifications in relatively high molecular-weight compounds, and their use in the synthesis of low molecular-weight compounds.

Developments in the first area of activity were influenced very profoundly by the successful use of microbes in effecting steroid conversions. These processes, which are discussed by K. Kieslich in Chapter 8, are still extensively used and being developed. The vast body of knowledge that accumulated as a result of research on steroid conversions, particularly information concerning group reactions that can be carried out by individual micro-organisms or groups of micro-organism, has more recently been applied to many other classes of high molecular-weight compounds. These developments are discussed in detail by O. K. Sebek in Chapter 9 and by L. C. Vining in Chapter 11. At present, many of the bioconversions described for antibiotics and alkaloids are of potential rather than actual importance (Shibata and Uyeda, 1978). The potential for using microbes as chemical reagents in transformations of a variety of alicyclic compounds has also attracted attention (Sebek and Kieslich, 1977). In this area, bioconversions of prostaglandins are most likely in the future to have industrial significance. Kieslich (1976) has compiled a very useful list of microbial transformations of non-steroid cyclic compounds, and Abbott (1979) has recently reviewed the application of microbial bioconversions.

Several processes are in commercial operation for synthesis of low molecular-weight compounds by micro-organisms, processes that may involve just one or a series of enzyme-catalysed reactions. Reference has already been made to the use of micro-organisms in synthesis of ephedrine, ascorbic acid and dihydroxyacetone, and these remain commercially viable. They have been joined by the use of penicillin acylase activity in microbial cells to convert penicillin into 6-amino-penicillanic acid, a process described by A. J. Vandamme in Chapter 10, and by use of glucose isomerase activity in cells to convert glucose into fructose, of which there is an account by S. A. Barker and J. A. Shirley in Chapter 4. The latter bioconversion is used in the manufacture of high-fructose corn syrups, which have been produced in increasingly large volumes in recent years (Abbott, 1977).

Other processes in this category which have attracted attention recently, but which have a restricted economic importance that is if they are commercially viable, include organic-acid and sugar bioconversions. A group of the Tanabe Seiyaku Company in Japan have developed a process using *E. coli* to convert fumaric acid into L-aspartic acid (Chibata *et al.*, 1974). Reports on other bioconversions that lead to production of amino acids have been reviewed by Abbott (1977).

Another bioconversion that uses an amino acid as a substrate has also been developed by the Tanabe Seiyaku Company. This uses L-histidine to produce urocanic acid, exploiting the histidine ammonia lyase activity of *Achromobacter liquidum* (Yamamoto *et al.*, 1974). Urocanic acid is used in cosmetic and pharmaceutical preparations because of its sun-screening properties.

Yet another bioconversion that would appear to be just about commercially viable is hydrolysis of raffinose in beet sugar to a mixture of sucrose and galactose using *Mortierella vinacea*. In the Steffen process for removal of sucrose from beet molasses, raffinose gradually accumulates in the molasses stream and ultimately prevents crystallization of sucrose thereby halting the process and obliging the operator to discard large quantities of raffinose-rich molasses (McGinnis, 1975). Suzuki and his colleagues (1963, 1964) removed raffinose from molasses by exploiting the melibiase activity of *M. vinacea*. Other bioconversions involving carbohydrate as substrate have been reviewed by Abbott (1977).

3. Production Technology

No review of the current status of industries using microbial enzymes and bioconversions would be complete without reference to developments in production technology. On the whole, developments in processing and plant in most branches of industrial microbiology have been minimal, revealing, some maintain, a woeful lack of imagination. But there are exceptions, and as regards plant and equipment, reference must be made to new designs in fermenters used to grow microbes as single-cell protein, particularly the pressure-cycle fermenter used by Imperial Chemical Industries in Great Britain (Hamer, 1979). In the use of microbial enzymes and bioconversions, considerable progress has been made with the new technology of cell and enzyme immobilization. Basically, the technique consists of incorporating microbial cells or enzymes into solid support materials, and using these materials in a suitably designed reactor, often some form of column. Compared with traditional batch-fermentation processes, immobilized systems offer very considerable gains in operating costs. The techniques are described fully by S. A. Barker in Chapter 7. They have also been recently reviewed elsewhere (Abbott, 1978; Bernath *et al.*, 1977).

B. Future Prospects

In the current World economic climate, making forecasts is a decidely hazardous business. This is particularly true for industries concerned with microbial enzymes and bioconversions, firstly because of the very wide range of different applications, and secondly because of the events of around 1970 when allergies were first reported on a large scale in workers handling powdered enzymes, despite the fact that these problems have been solved. Fortunately, a limited amount of data on prospects for the future of microbial enzymes are available in the form of reports released by the Battelle Geneva Research Centre (Battelle, 1974) and by Wolnak (1972, 1974). Both reports forecast increased activity, but only in certain industries and with a limited number of enzymes. The Battelle report believes that the medical, pharmaceutical and food technology industries are likely to gain most from use of enzymes in future years, with many fewer applications likely to arise in the treatment of waste water and of solid wastes and in synthesis of fine chemicals. The Wolnak report forecasts increased production of glucose isomerase (six-fold), cellulase (four-fold), detergent protease (three- to five-fold) and microbial rennet (two-fold).

REFERENCES

Abbott, B. J. (1977). *In* 'Annual Reports on Fermentation Processes' (D. Perlman, ed.), vol. 1, p. 205. Academic Press, New York.
Abbott, B. J. (1978). *In* 'Annual Reports on Fermentation Processes' (D. Perlman, ed.), vol. 2, p. 91. Academic Press, New York.
Abbott, B. J. (1979). *Progress in Industrial Microbiology* **20**, 345.
Andreadis, J. T. and Miklean, S. (1960). United States Patent 2,949,403.
Aunstrup, K. (1977). *In* 'Annual Reports on Fermentation Processes' (D. Perlman, ed.), vol. 1, p. 181. Academic Press, New York.
Aunstrup, K. (1978). *In* 'Annual Reports on Fermentation Processes' (D. Perlman, ed.), vol. 2, p. 125. Academic Press, New York.

Battelle. (1974). Geneva Research Centre. A Forecast of Industrial Application Possibilities of Enzymes (1st Round).

Bernath, F. R., Venkatasubramanian, K. and Vieth, W. R. (1977). In 'Annual Reports on Fermentation Processes' (D. Perlman, ed.), vol. 1, p. 235. Academic Press, New York.

Berthelot, K. (1856). Compte Rendus Hebdomadaire des Seances de l'Academie des Sciences 50, 980.

Berthelot, M. (1860). Compte Rendus Hebdomadaire des Seances de l'Academie des Sciences 51, 980.

Berzelius, J. J. (1837). Lehrbuch de Chemie, vol. VI, Leipzig.

Boidin, A. and Effront, J. (1917a). United States Patent 1,227,374.

Boidin, A. and Effront, J. (1917b). United States Patent 1,227,525.

Buchner, E., Buchner, H. and Hahn, M. (1897). 'Die Zymasegärung'. R. Oldenberg, Munchen.

Bull, D. N. and Young, M. D. (1977). Abstracts of Papers at the 174th Meeting of the American Chemical Society, p. 01.

Carrington, T. R. (1971). Proceedings of the Royal Society B 179, 321.

Cartledge, T. G., Rose, A. H., Belk, D. M. and Goodall, A. A. (1977). Journal of Bacteriology 132, 426.

Chibata, I., Tosa, T. and Sato, T. (1974). Applied Microbiology 27, 878.

Claridge, C. A. (1979). In 'Economic Microbiology' (A. H. Rose, ed.), vol. 3, p. 151. Academic Press, London.

Davies, R. (1963). In 'Biochemistry of Industrial Micro-organisms' (C. Rainbow and A. H. Rose, eds.), p. 68. Academic Press, London.

Dekker, R. F. H. and Richards, G. N. (1976). Advances in Carbohydrate Chemistry and Biochemistry 32, 277.

Dixon, M. and Webb, E. C. (1964). 'Enzymes'. Longmans, London.

Elander, R. P., Chang, L. T. and Vaughn, R. W. (1977). In 'Annual Reports on Fermentation Processes' (D. Perlman, ed.), vol. 1, p. 1. Academic Press, New York.

Elsworth, R., Telling, R. C. and East, D. N. (1959). Journal of Applied Bacteriology 22, 138.

Ferenczy, L., Kevei, F. and Szegedi, M. (1976). In 'Microbial and Plant Protoplasts' (J. F. Peberdy, A. H. Rose, H. J. Rogers and E. C. Cocking, eds.), p. 177. Academic Press, London.

Flickinger, M. C. and Perlman, D. (1977). Applied and Environmental Microbiology 33, 706.

Fonken, G. and Johnson, R. S. (1972). 'Chemical Oxidations with Micro-organisms'. Marcel Dekker Inc., New York.

Glenn, A. R. (1976). Annual Review of Microbiology 30, 41.

Glover, S. W. (1978). Advances in Microbial Physiology 18, 235.

Grunberg-Manago, M., Buckingham, R. H., Cooperman, B. S. and Hershey, J. W. B. (1978). Symposium of the Society for General Microbiology 28, 27.

Haas, G. J. and Lugay, J. C. (1976). United States Patent 3,968,255.

Hamer, G. (1979). In 'Economic Microbiology' (A. H. Rose, ed.), vol. 4, p. 315. Academic Press, London.

Hildebrandt, G. and Klavehn, W. (1934). United States Patent 1,956,950.

Holmlund, C. E., Andres, W. W. and Shay, A. J. (1959). Journal of the American Chemical Society 81, 4750.

Hoppe-Seyler, F. (1876). Archiv für die Gesamte Physiologie 12, 1.

Hori, I. and Nakatani, T. (1953). Journal of Fermentation Technology 31, 72.

Huang, H. T. (1962). United States Patent 3,043,749.

Huber, F. M., Chauvette, R. R. and Jackson, B. G. (1972). In 'Cephalosporins and

Penicillins: Chemistry and Biology', (E. H. Flynn, ed.), p. 27. Academic Press, New York.

Iizuka, H. and Naito, A. (1968). 'Microbial Transformations of Steroids and Alkaloids'. University Park Press, Baltimore, Maryland.

International Union of Biochemistry (1965). Commission of Editors. Enzyme Nomenclature. Elsevier Publishing Co., Amsterdam.

Jacob, F. and Monod, J. (1961). *Journal of Molecular Biology* 3, 318.

Juniper, C. P., How, M. J., Goodwin, B. F. J. and Kinshot, A. K. (1977). *Journal of Social and Occupational Medicine* 27, 3.

Kanzaki, T. and Okazaki, H. (1970). *Agricultural and Biological Chemistry* 34, 432.

Keilin, D. (1966). 'The History of Cell Respiration and Cytochrome'. Cambridge University Press, Cambridge.

Kieslich, K. (1976). 'Microbial Transformations of Non-Steroid Cyclic Compounds'. Georg Thieme Publishers, Stuttgart.

Lampen, J. O. (1976). *In* 'Microbiology – 1976' (D. Schlessinger, ed.), p. 540. American Society for Microbiology, Washington, D.C.

Lampen, J. O. (1978). *Symposium of the Society for General Microbiology* 28, 231.

Lehninger, A. L. (1970). 'Biochemistry. The Molecular Basis of Cell Structure and Function'. Warth Publishers Inc., New York.

Lemke, P. A. and Brannon, D. R. (1972). *In* 'Cephalosporins and Penicillins: Chemistry and Biology' (E. H. Flynn, ed.), p. 370. Academic Press, New York.

Low, A. and Krema, A. (1928). *Klinische Wochenschrift* 7, 2432.

MacDonald, K. D. ed. (1976). 'Second International Symposium on the Genetics of Industrial Microorganisms'. Academic Press, London.

Makover, S., Ramsey, G. B., Vane, F. M., Witt, C. G. and Wright, R. B. (1975). *Biotechnology and Bioengineering* 17, 1485.

Mandelstam, J. and McQuillen, K. (1973). 'Biochemistry of Bacterial Growth', 2nd edition. Blackwells Scientific Publications, Oxford.

Matile, P. and Wiemken, A. (1967). *Archiv für Mikrobiologie* 56, 148.

McGinnis, R. (1975). *Sugar Journal* 38, 8.

Millet, J., Archer, R. and Aubert, J. P. (1969). *Biotechnology and Bioengineering* 11, 1233.

Moss, M. O. (1977). *Topics in Enzyme and Fermentation Biotechnology* 1, 111.

Muller, J. (1966). *Zentralblatt für Bakteriologie, Parasitenkunde, Infektionskrankheiten und Hygiene, Abteilung II* 120, 349.

Neuberg, C. (1921). *Biochemische Zeitschrift* 115, 282.

Neuberg, C. (1922). *Biochemische Zeitschrift* 128, 610.

Northrop, J. H., Kunitz, M. and Herriott, R. M. (1948). 'Crystalline Enzymes'. John Wiley, New York.

Oi, S., Matsui, Y., Iizuka, M. and Yamamoto, T. (1977). *Journal of Fermentation Technology* 55, 114.

Okazaki, H., Kanzaki, T., Doi, M., Nara, K. and Motizuki, M. (1968). *Agricultural and Biological Chemistry* 32, 1250.

Okazaki, H., Kanzaki, T., Susajima, K. and Terada, K. (1969). *Agricultural and Biological Chemistry* 33, 207.

Palade, G. (1975). *Science, New York* 189, 347.

Payen, A. and Persoz, J. F. (1833). *Annales de Chimie et Physique* 53, 73.

Perlman, D. (1973). *Process Biochemistry* 8(7), 18.

Perlman, D. (1977). *American Society for Microbiology News* 43, 82.

Peterson, D. H., Murray, H. C., Epstein, S. H., Reinecke, L. M., Wintraub, A., Meister, P. D. and Leigh, H. M. (1952). *Journal of the American Chemical Society* 74, 5933.

Pollock, M. R. (1962). *In* 'The Bacteria' (I. C. Gunsalus and R. Y. Stanier, eds.), vol. 4, p. 121. Academic Press, New York.

Pollock, M. R. and Richmond, M. H. (1962). *Nature, London* **194**, 446.

Priest, F. G. (1977). *Bacteriological Reviews* **41**, 711.

Queener, S. and Swartz, R. (1979). *In* 'Economic Microbiology' (A. H. Rose, ed.), vol. 3, p. 35. Academic Press, London.

Robison, G. A., Butcher, R. W. and Sutherland, E. W. (1971). 'Cyclic AMP'. Academic Press, New York.

Roeder, R. G. (1974). *Journal of Biological Chemistry* **249**, 241.

Roodyn, D. B. (1967). 'Enzyme Cytology'. Academic Press, London.

Rose, A. H. (1976). 'Chemical Microbiology. An Introduction to Microbial Physiology', 3rd edition. Butterworths, London.

Rose, A. H. (1979). *In* 'Economic Microbiology' (A. H. Rose, ed.), vol. 3, p. 1. Academic Press, London.

Schwann, T. (1836). *Archiv für Anatomie und Physiologie* **90**, 128.

Schultz, A. S. and Atkin, L. (1931). United States Patent 2,159,678.

Sebek, O. K. (1977). *In* 'Biotechnological Applications of Proteins and Enzymes' (Z. Bohak and N. Sharon, eds.), p. 203. Academic Press, New York.

Sebek, O. K. (1979). *In* 'Microbiology-1979' (D. Schlessinger, ed.), p. 318. American Society for Microbiology, Washington, D.C.

Sebek, O. K. and Kieslich, K. (1977). *In* 'Recent Reports on Fermentation Processes (D. Perlman, ed.), vol. 1, p. 267. Academic Press, New York.

Sheinin, R., Humbert, J. and Pearlman, R. E. (1978). *Annual Review of Biochemistry* **47**, 277.

Shibata, M. and Uyeda, M. (1978). *In* 'Annual Reports on Fermentation Processes' (D. Perlman, ed.), vol. 1, p. 267. Academic Press, New York.

Shier, W. T., Rinehart, K. L. and Gottlieb, D. (1969). *Proceedings of the National Academy of Sciences of the United States of America* **63**, 198.

Smith, P. F. and Hendlin, D. (1953). *Journal of Bacteriology* **65**, 440.

Smith, P. F. and Hendlin, D. (1954). *Applied Microbiology* **2**, 294.

Stewart, R. L., McCune, H. W. and Diehl, F. L. (1976). United States Patent 3,950,277.

Sumner, J. B. (1926). *Journal of Biological Chemistry* **69**, 435.

Suzuki, H., Ozawa, Y. and Tanabe, O. (1963). *Nippon Nogeikagaku Kaishi* **37**, 623.

Suzuki, H., Ozawa, Y. and Tanabe, O. (1964). *Nippon Nogeikagaku Kaishi* **38**, 334.

Takamine, J. (1894). United States Patents 525,820 and 525,823.

Takamine, J. (1914). *Industrial and Engineering Chemistry* **6**, 824.

Tauber, H. (1949). 'The Chemistry and Technology of Enzymes'. John Wiley and Sons, New York.

Tengendy, R. P. (1961). *Journal of Biochemical and Microbiological Technology and Engineering* **3**, 241.

Thelander, L. and Reichard, P. (1979). *Annual Review of Biochemistry* **48**, 133.

Tomizawa, J. and Selzer, G. (1979). *Annual Review of Biochemistry* **48**, 999.

Underkofler, L. A. and Fulmer, E. I. (1937). *Journal of the American Chemical Society* **59**, 301.

Underkofler, L. A., Severson, G. M. and Goering, K. J. (1946). *Industrial and Engineering Chemistry* **38**, 980.

Underkofler, L. A., Barton, R. R. and Rennert, S. S. (1958). *Applied Microbiology* **6**, 212.

Vaněk, Z., Hošťálek, K. and Cudlín, J. eds. (1973). 'Genetics of Industrial Micro-organisms'. Elsevier Publishers Inc., Amsterdam.

Vogel, F. C. (1812). *Journal of Chemistry and Physics* **4**, 142.

Wallen, L. L., Stodola, F. H. and Jackson, R. W. (1959). 'Type Reactions in Fermentation Chemistry'. Agricultural Research Service. United States Department of Agriculture.

Wallerstein, L. (1939). *Industrial and Engineering Chemistry* **31**, 1218.

Wells, P. A., Stubbs, J. J., Lockwood, L. B. and Roe, E. T. (1937). *Industrial and Engineering Chemistry* **29**, 1385.

Wells, P. A., Lockwood, L. B., Stubbs, J. J., Roe, E. T., Porges, N. and Gastrock, E. A. (1939). *Industrial and Engineering Chemistry* **31**, 1518.

Wolnak, B. (1972). Report PB-219636.N.T.I.S. U.S. Department of Commerce, Springfield, Virgina.

Wolnak, B. (1974). *In* 'Enzyme Engineering' (E. Pye and L. B. Wingard, eds.), vol. 2, p. 363. Plenum Press, New York.

Wriston, J. C. and Yellin, T. O. (1973). *Advances in Enzymology* **39**, 185.

Yamamoto, K., Sato, T., Tosa, T. and Chibata, I. (1974). *Biotechnology and Bioengineering* **16**, 1601.

2. Proteinases

KNUD AUNSTRUP

Novo Research Institute, Bagsvaerd, Denmark

I. INTRODUCTION

Production and processing of proteins have always been an important part of man's industrious activities. The primary objective is an adequate supply of the essential amino acids in food, but preparation of items suitable for clothing and various other utensils also plays an important role.

Proteolytic processes are part of many of these operations, but, in most cases, the enzymes used are traditionally of plant or animal origin (Table 1) in the West. In the Orient, there is an ancient tradition for the

Table 1

Traditional applications for plant and
animal proteases

Application	Protease
Cheese manufacture	Rennin
Bating	Trypsin
Chill haze prevention	Papain
Meat tenderization	Papain
Brewing	Malt proteases
Protein hydrolysis	Pancreatin
	Pepsin

Table 2

Oriental fermented food, the raw materials and the proteolytic
fungi used in their preparation

Food	Raw materials	Fungi
Hamanatto	Soybean, wheat, barley	*Aspergillus oryzae*
Katsoubashi	Fish	*Aspergillus glaucus*
Lao-chao	Rice	*Rhizopus chinensis*
Meitanza	Soybean	*Actinomucor elegans*
Miso	Soybean, rice, barley	*Aspergillus oryzae*
Ontjom	Peanut meal	*Neurospora sitophila*
Soya sauce	Soybean, wheat	*Aspergillus oryzae*
		Aspergillus sojae
Sufu	Soybean	*Actinomucor elegans*
		Mucor sp.
Tempeh	Soybean	*Rhizopus oligosporus*
		Rhizopus sp.

use of microbial proteases in the preparation of various food substances, primarily from soybeans, rice and fish (Table 2).

During the last 50 years, the fermentation industry has developed methods for low-cost production of proteolytic enzymes of high purity from microbial sources. These enzymes have to some extent replaced the traditional proteases from animal and plant sources, and they have found new applications because of their special properties.

This chapter will be limited to the commercially available microbial proteases, and furthermore to those enzymes that now find industrial application. The extensive literature on proteases from various micro-organisms that have not been commercially utilized will not be dealt with here.

II. DEVELOPMENT OF THE INDUSTRIAL USE OF MICROBIAL PROTEASES

Microbial enzyme production on an industrial scale was initiated by the Japanese Takamine who, in 1890, settled in the U.S.A. and started production of the enzyme preparation Takadiastase. This product was made by semisolid fermentation of *Aspergillus oryzae* and was based on traditional Japanese technology improved by an important new development which Takamine introduced, namely the use of wheat bran instead of rice as the main medium component (Takamine, 1894). In this chapter, the nomenclature of Raper and Fennell (1965) is followed as closely as possible. Because of the small differences between many of the species within the *Aspergillus niger* and *Aspergillus flavus-oryzae* groups, the author finds it advisable not to use terms like *A. awamori* and *A. phoenicis* but rather *A. niger* for the black *Aspergillus* strains used for enzyme production, and *A. oryzae* for the yellow-green *Aspergillus* strains used for enzyme production (*A. oryzae* and *A. sojae*).

Takadiastase was mainly an α-amylase preparation and was marketed as such, but it contained a substantial amount of protease and was soon recommended for various proteolytic applications, primarily as a digestive aid. Preparations similar to Takadiastase are still produced. The methods used have in principle not been changed much, although new strains and new medium compositions have been introduced. In this way it has become possible to develop products that predominantly contain protease.

In 1913, a patent was granted to Boidin and Effront in France on a process for production of an enzyme preparation from *Bacillus subtilis* (Boidin and Effront, 1917). The bacterium was grown in still culture as a surface film on a broth prepared from a grain mash. The enzymes were secreted into the broth and could easily be recovered. This enzyme preparation contained mainly α-amylase, but like Takadiastase it held a substantial proteolytic activity as well. The enzyme preparation was primarily used as a substitute for malt or pancreatic amylases in desizing of textiles, i.e. removal of starch applied to the warp to protect it during weaving. Little use was made of the proteolytic component in the composition, although many ideas were tested in the following years. The most important applications were within the tanning industry, where the protease was used in bating and to a small extent in a dewooling of sheep skin.

The development after World War II of submerged fermentation technology for production of antibiotics was also adopted by the enzyme industry and, during the 1950s, a switch from the old still culture to submerged fermentation took place, together with a modest development in strains and media so that it became possible to make preparations with a predominant content of protease. Sales of microbial proteases were, however, limited until their use in detergents was introduced about 1960. Enzyme-containing detergents have been used since 1913 when Dr. Röhm in Germany invented a presoaking agent containing sodium carbonate and pancreatic enzymes. The preparation owed its value as a presoaking agent mainly to the sodium carbonate content which gave a pH value in the suds of about 10, but unfortunately the pancreatic enzyme did not work very well at that high pH value, so the enzymic effect was less than expected.

During World War II, when resources were limited, there was a renewed interest in enzymic presoaking agents, and in Switzerland, Dr. Jaag of the company Gebrüder Schnyder developed a new product, Bio 38, which was tested in 1945. This preparation was based on pancreatic enzymes, too, and although the composition was adjusted so that the pH value in the washing suds was sufficiently low for the pancreatic proteases to work, the preparation still did not have the correct properties to make a real breakthrough. In 1959, a new preparation, Bio 40, based on a protease from *B. subtilis* was marketed. This was a considerable improvement, but a year later a much better enzyme was introduced, namely Subtilisin Carlsberg, produced by *B. licheniformis*.

This enzyme had been studied from a scientific point of view by Ottesen and his coworkers at the Carlsberg laboratory in Denmark, and it was introduced as a commercial enzyme by the Danish company Novo Industri A/S in 1960.

At Novo, work had been going on for several years to develop an enzymic washing process, and several microbial enzymes were tested for this purpose. The protease of *B. licheniformis* proved to be superior to, and was consequently used as a substitute for, the enzyme from *B. subtilis* in Bio 40. The preparation enjoyed commercial success and, in 1963, the Dutch company Kortman and Schulte marketed the preparation Biotex, containing protease from *B. licheniformis*. This preparation also became a great success and, during the following years, enzyme was added to more and more detergents so that, in 1969, almost 50% of all detergents manufactured in Europe and in the U.S.A. contained proteolytic enzymes. Thus, protease from *B. licheniformis* had become by far the most important microbial enzyme from an economic point of view (Dambmann et al., 1971).

The enzyme preparations produced at that time were made in the traditional way by precipitation from the fermentation broth with salts or solvents, and consisted of a finely ground powder which gave rise to a substantial amount of dust during handling. Like other proteins, enzymes are allergenic especially when inhaled as a fine dust, and it was known that precautions should be taken when handling enzyme concentrates. Some incidents of pulmonary disease caused by allergic reactions among workers handling enzymes in detergent factories were reported in 1967–70, and these caused much public concern about the safety of these new products. Therefore, in 1971, an investigation was initiated in the U.S.A. under the direction of the National Academy of Sciences. The conclusion of this investigation was that the use of enzyme detergents imposed no risk on the consumer. The dust problems in the detergent factories were soon alleviated by the development of dust-free enzyme granulates. Public concern about the possible risk of detergent enzymes was dramatic, and enzymes were taken out of many detergents until the investigation had shown that their use was safe.

With the commercial success of the detergent proteases, a search began for new, improved enzymes with better stain-removing properties and better stability in the washing suds. This resulted in the development of a new series of proteases from alkalophilic *Bacillus* species. Three such enzymes have been marketed but, despite their technical advantages, they have so far only been used to a limited extent. The reason for this is primarily that new enzymes for detergents must undergo a thorough and lengthy testing so that all consumer risks are eliminated. Another application for these new enzymes, which can be used up to pH 12, is dehairing in the tanning industry. An enzyme product for this purpose was developed in 1972 by Novo. However, the use was limited for, although the technical performance of the process was satisfactory, it was uneconomic compared with the traditional lime sulphide process. In the last few years, an improved process has been developed by Röhm in Germany; this process appears to be more promising.

The most important animal protease is calf rennet which is used in the manufacture of cheese. Because of the high price of this enzyme, numerous attempts have been made to make a microbial substitute. Milk can be coagulated by almost all proteases, but most enzymes will hydrolyse the casein further thus making it impossible to prepare a palatable cheese. The problem has therefore been to produce a microbial protease with a specificity similar to that of calf rennet. This was first achieved by K. Arima and his coworkers who, in 1960, screened a large number of micro-organisms for milk-coagulating activity and found that the protease of the thermophilic fungus *Mucor pusillus* had satisfactory properties. A commercial preparation was subsequently marketed by the Meito Sangyo Co. of Japan around 1970.

Mucor pusillus produces the protease in high yields only in semisolid culture which is a drawback since submerged cultivation from an economical and hygienic point of view is advantageous. In 1965, I found that *Mucor miehei* produces a similar milk-coagulating enzyme in good yields in submerged culture. This enzyme was marketed by Novo Industry in 1972. Both enzymes are widely used in the dairy industry and will produce cheese of the same quality as that produced with calf rennet when used under proper circumstances.

A different approach was taken by the Pfizer Company, where Sardinas in 1963 found that *Endothia parasitica* produces a protease that can be used for cheese manufacture. The enzyme has a somewhat higher proteolytic activity than the proteases from *Mucor* spp. but this is compensated for by a very low stability so that proteolytic degradation of the casein is prevented by inactivation of the enzyme.

Proteases are also used in a number of applications of minor economic importance. These will be described in Section VI (p. 104) of this chapter.

III. PROTEOLYTIC ENZYMES PREPARATIONS AND PRODUCERS

A list of the commercially important microbial protease preparations is given in Table 3. The list contains only enzymes for industrial use. Enzymes for analytical or medical purposes are not included. The major producers of microbial proteases are listed in Table 4. The list includes only companies that produce enzymes; trading companies are not included, although some of these companies have large sales of enzymes and thus are of importance to the industry. It is known that there is a substantial production of microbial enzymes in the Soviet Union and several other east European countries, but the information available to the author about their enzyme industries is fragmentary, so these countries are not mentioned here. Of the companies listed in Table 4, Gist Brocades and Novo are by far the largest producers of

Table 3

Commercially-important proteases, trade names and microbial sources. The companies are identified by the numbers indicated in Table 4

Aspergillus niger	Proctase (7); Pamprosin (12).
Aspergillus oryzae	Veron P (20); Panazyme (3); Prozyme, Biozyme A (5); Sanzyme (10); Sumyzyme AP (11); Fungal Protease (15).
Rhizopus sp.	Newlase (5).
Bacillus amyloliquefaciens (*Bacillus subtilis*)	Neutrase (1); Rapidermase (2); Proteinase 18 (3); Protin (6); Bioprase, Nagase (9); Rhozyme (18).
Bacillus licheniformis	Alcalase (1); Optimase (19); Maxatase P (4).
Bacillus thermoproteolyticus	Termoase, Thermolysin (6).
Bacillus sp.; alkalophilic	Esperase, Savinase (1); Highly alkaline protease (4).
Endothia parasitica	Surecurd, Suparen (17).
Mucor miehei	Rennilase (1); Fromase (4); Marzyme (15); Morcurd (17).
Mucor pusillus	Emporase, Meito rennet, Noury lab (8).

Table 4

Companies producing microbial extracellular proteases for industrial use

Denmark	1.	Novo Industri A/S, Bagsvaerd
France	2.	Soc. Rapidase, Seclin (subsidiary of 4)
Great Britain	3.	Associated British Maltsters, Stockport, Cheshire
Holland	4.	Gist Brocades N.V., Delft
Japan	5.	Amano, Nagoya
	6.	Daiwa Kasei, Osaka
	7.	Meiji Seika, Tokyo
	8.	Meito Sangyo, Nagoya
	9.	Nagase, Osaka
	10.	Sankyo, Tokyo
	11.	Shin Nihon, Tokyo
	12.	Yakult Biochemicals, Nishinomiya
Switzerland	13.	Swiss Ferment Ltd. (a subsidiary of 1)
U.S.A.	14.	GB Fermentation Industries, Kingstree, South Carolina (a subsidiary of 4).
	15.	Miles, Elkhart Indiana
	16.	Novo Biochemical Industries Inc., Franklinton, North Carolina (subsidiary of 1)
	17.	Pfizer, New York, New York
	18.	Rohm and Haas, Philadelphia, Pennsylvania
West Germany	19.	Miles Kali-Chemie GmbH, Nienburg a.d. Weser (a subsidiary of 15)
	20.	Röhm GmbH, Darmstadt

microbial proteases. Together they account for the major part of the detergent protease sales. Total sales of microbial proteases in 1977 were estimated to be about U.S. $100 million. Approximately 75% is used in the detergent industry and 10% in the dairy industry. The residual 15% is used in a number of minor applications. Uses of proteases in digestive-aid preparations and for other medical purposes are not included. Moreover, it should be borne in mind that estimates of this sort are subject to large uncertainty.

IV. PROPERTIES OF PROTEASES AND THEIR MICROBIAL SOURCES

Proteolytic enzymes can be found in any micro-organism. The few species that are used in industrial production have been selected because they can be easily made to produce protease in good yields or because the enzyme formed has specific properties. Since they are readily available in large quantities, industrial proteases have been extensively used as model systems for scientific investigations. Therefore the molecular properties of some of the enzymes are known in detail. When using basic physicochemical or enzyme kinetic constants of proteases, one should be aware that the data often are uncertain due to autodigestion of the protease and contamination with peptides adsorbed to the enzyme molecule.

Microbial enzymes are usually grouped according to their active centre as serine-, metallo- and acid proteases. Enzymes similar to the thiol proteases of plants have been found in micro-organisms, but they are not used industrially.

A. Serine Proteases

1. Subtilisin Carlsberg

In 1952, Güntelberg and Ottesen at the Carlsberg laboratories described the preparation and crystallization of a protease that had the ability to convert ovalbumin into plakalbumin (Güntelberg and Ottesen, 1952). This enzyme was later known as Subtilisin Carlsberg, and it has become the most important microbial protease economically because it is the predominant enzyme used in detergents. It has been in commercial production since 1960, when the first industrial production method was developed at Novo. Annual production of the enzyme is at present equivalent to approximately 500 tonnes of pure enzyme protein.

a. *The microbial source.* Because of the unclear systematics of the genus *Bacillus* at the time of their discovery, Güntelberg and Ottesen (1952) classified the organism as *Bacillus subtilis*. They might have added aff. Ford strain since it was later found that the proper classification of the strain is *Bacillus licheniformis* (unpublished results by Helle Outtrup, Novo Industry). Production of proteases by several other strains of *B. licheniformis* has been described (Table 5).

No thorough comparative investigations have been published on the properties of proteases from the various strains of *B. licheniformis*. Unpublished results obtained at Novo (Keay *et al.*, 1970; Zuidweg *et al.*, 1972) indicate that the known proteases from different *B. licheniformis*

Table 5

Strains of *Bacillus licheniformis* that have been studied
for protease production

Bernlohr and Novelli (1963)	N.C.T.C. 8720
Damodaran *et al.* (1955)	A.T.C.C. 11560
Viccaro (1973)	A.T.C.C. 21424
Wouters and Buysman (1977)	A.T.C.C. 25972
Zuidweg *et al.* (1972)	A.T.C.C. 9445A

strains are practically identical, so that it may be expected that production of Subtilisin Carlsberg is specific for the species *B. licheniformis*.

One strain that has enjoyed particular attention is the bacitracin-producing strain Oxford A5 (N.C.T.C. 8720) which has been studied by Bernlohr and others (Bernlohr, 1964; Hall *et al.*, 1966). These authors did not make a direct comparison with Subtilisin Carlsberg, but unpublished work from Novo has shown that the main protease component of strain A5 is immunoelectrophoretically identical with Subtilisin Carlsberg.

Bacillus licheniformis is a common saprophytic *Bacillus* species which is closely related to *B. subtilis*. It is characteristically facultative in being able to reduce nitrate to nitrogen under anaerobic conditions. It is worth noting that there is no production of Subtilisin Carlsberg under anaerobic conditions. The organism may be used to produce several metabolites of industrial interest (Table 6). Its α-amylase differs from that of *B. amyloliquefaciens* in being more resistant to high temperatures and high pH values (Madsen *et al.*, 1973). In protease production, the enzyme is formed in a concentration that is very low, but sufficient to allow the organism to grow well on hydrolysed starch. This α-amylase,

Table 6

Metabolites of industrial interest
produced by *Bacillus licheniformis*

Enzymes	protease
	α-amylase
	penicillinase
Antibiotics	bacitracin
Solvents	butylene glycol glycerol
Polypeptide	D-glutamylpolypeptide

in combination with a cell-bound α-glucosidase (Hoegh, 1978), is able to hydrolyse starch completely to glucose. Commercial production of the α-amylase is performed with highly mutated strains (Outtrup and Aunstrup, 1975). The penicillinase that is produced is, under the conditions of protease production, cell bound and can be removed from the fermentation broth with the cells, so that it is not present in the protease product. It is usually requested that enzyme products are free from antibiotic activity, so strains unable to produce bacitracin must be used in protease production. It is of interest to note that bacitracin is a competitive inhibitor of Subtilisin Carlsberg (Mäkinen, 1970). D-Glutamylpolypeptide makes the medium viscous and slimey, and its formation in protease production should be avoided either by the use of strains that are unable to produce this polymer, or by proper composition of the medium.

Since protease production is always performed under aerobic conditions, there will only be insignificant formation of butylene glycol and glycerol which are anaerobic fermentation products.

b. *Regulation of protease synthesis.* Güntelberg (1954) reported production of protease corresponding to about 2.5 g of pure enzyme protein per litre of medium. No reports have been made on present commercial yields, but it is safe to assume that they are considerably higher. Güntelberg (1954) also reported that protease production starts towards the end of the logarithmic growth phase, and he also referred to the absence of spores from the cultures. Production of protease in the early stationary phase has been observed by others (Keay et al., 1972) and is common in production of extracellular enzymes.

The absence of spores is an advantage in industrial practice since complete sporulation stops growth and enzyme formation and spores are difficult to remove from the enzyme preparation. Asporogenic mutants have been shown to give improved protease yields (Aunstrup and Outtrup, 1973), probably because they permit a longer production phase. The mechanism of regulation of protease synthesis and the possible importance of the protease in sporulation have been the subject of extensive investigations which are excellently summarized by Priest (1977). However, despite our extensive knowledge of the mechanism of enzyme synthesis, no better way of improving yields in industrial processes other than empirical optimization of strain and medium has yet been described.

c. *The proteases produced.* Güntelberg and Ottesen (1952) prepared crystalline Subtilisin Carlsberg from the culture broth of *B. licheniformis* and they found only one proteolytic component. Hall *et al.* (1966) found evidence for two different proteases and a peptidase. Zuidweg *et al.* (1972) separated commercial protease products from *B. licheniformis* into three components by electrophoresis and showed that they were not formed by autodigestion. Verbruggen (1975) made a very thorough study of pure and industrial-grade Subtilisin Carlsberg using crossed immunoelectrophoresis. He was able to separate the protease into two antigenically different populations within each of which there were multiple components that were antigenically identical but had different electrophoretic mobilities. He concluded that the organism produces two primary protease components which then, by autodigestion, are transformed into isoenzymes. The major component is Subtilisin Carlsberg and the minor unidentified component constitutes only a small fraction of the enzyme preparations.

```
           Thr              Ile Pro Leu         Asp Lys Val Gln Ala        Phe Lys    Ala
NH₂ -Ala-Gln-Ser-Val-Pro-Tyr-Gly-Val-Ser-Gln-Ile-Lys-Ala-Pro-Ala-Leu-His-Ser-Gln-Gly-Tyr-Thr-Gly-Ser-Asn-Val-Lys-Val-Ala-Val-
                              10                              20                              30

Leu    Thr    Gln Ala                     Asn    Val               Phe    Ala Gly     Ala ━ Tyr Asn Thr
Ile-Asp-Ser-Gly-Ile-Asp-Ser-Ser-His-Pro-Asp-Leu-Lys-Val-Ala-Gly-Gly-Ala-Ser-Met-Val-Pro-Ser-Glu-Thr-Pro-Asn-Phe-Gln-Asp-
                        40                              50                              60

Gly    Gly                              Asp    Thr Thr                        Val Ser
Asp-Asn-Ser-His-Gly-Thr-His-Val-Ala-Gly-Thr-Val-Ala-Ala-Leu-Asn-Asn-Ser-Ile-Gly-Val-Leu-Gly-Val-Ala-Pro-Ser-Ser-Ala-Leu-
                  70                              80                              90

            Asn Ser Ser         Ser     Gly   Val Ser               Thr Thr    Gly
Tyr-Ala-Val-Lys-Val-Leu-Gly-Asp-Ala-Gly-Ser-Gly-Gln-Tyr-Ser-Trp-Ile-Ile-Asn-Gly-Ile-Glu-Trp-Ala-Ile-Ala-Asn-Asn-Met-Asp-
                  100                             110                             120

                    Ala              Thr    Met    Gln           Asn    Tyr    Arg
Val-Ile-Asn-Met-Ser-Leu-Gly-Gly-Pro-Ser-Gly-Ser-Ala-Ala-Leu-Lys-Ala-Ala-Val-Asp-Lys-Ala-Val-Ala-Ser-Gly-Val-Val-Val-Val-
              130                             140                             150

          Ser    Asn Ser         Thr Asn    Ile         Ala         Asp
Ala-Ala-Ala-Gly-Asn-Glu-Gly-Ser-Thr-Gly-Ser-Ser-Ser-Thr-Val-Gly-Tyr-Pro-Gly-Lys-Tyr-Pro-Ser-Val-Ile-Ala-Val-Gly-Ala-Val-
              160                             170                             180

        Asn Ser Asn                         Ala    Glu              Ala Gly Val Tyr       Tyr
Asp-Ser-Ser-Asn-Gln-Arg-Ala-Ser-Phe-Ser-Ser-Val-Gly-Pro-Glu-Leu-Asp-Val-Met-Ala-Pro-Gly-Val-Ser-Ile-Gln-Ser-Thr-Leu-Pro-
              190                             200                             210

Thr    Thr    Ala Thr Leu               *
Gly-Asn-Lys-Tyr-Gly-Ala-Tyr-Asn-Gly-Thr-Ser-Met-Ala-Ser-Pro-His-Val-Ala-Gly-Ala-Ala-Ala-Leu-Ile-Leu-Ser-Lys-His-Pro-Asn-
              220                             230                             240

Leu Ser Ala  Ser         Asn Arg     Ser Ser    Ala    Tyr        Ser
Trp-Thr-Asn-Thr-Gln-Val-Arg-Ser-Ser-Leu-Gln-Asn-Thr-Thr-Thr-Lys-Leu-Gly-Asp-Ser-Phe-Tyr-Tyr-Gly-Lys-Gly-Leu-Ile-Asn-Val-
              250                             260                             270

Glu
Gln-Ala-Ala-Ala-GlnCOOH
275
```

Fig. 1. Comparison of the amino-acid sequence of Subtilisins Novo and Carlsberg. The continuous sequence is that of Subtilisin Novo. The residues that differ in Subtilisin Carlsberg are given above the corresponding residue. The dash at residue 56 indicates that the corresponding residue is lacking in the Carlsberg enzyme. The active Ser 221 residue is denoted by an asterisk. From Markland and Smith (1971).

Bernlohr and Clark (1971) have described an intracellular protease of broad specificity, and Aiyappa and Lampen (1977) described a cell-bound protease of narrow specificity which is thought to release membrane-bound penicillinase. None of these proteases has been observed in commercial enzyme preparations, but their possible relationship with the above-mentioned unidentified protease has not been elucidated.

d. *Basic properties of Subtilisin Carlsberg.* Subtilisin Carlsberg was the first microbial protease that was prepared in crystalline form (in 1952), and its basic properties have been extensively studied. The amino-acid sequence was determined by Smith *et al.* (1968; Fig. 1). The enzyme consists of a single peptide chain with 274 amino-acid residues and, since cystine or cysteine residues are absent, no disulphide bridges are formed. The tertiary structure is spherical with a diameter of about 4.2 nm. The active site is formed by the residues: Ser 221, His 64 and Asp 32. The enzyme activity is inhibited by reagents that react with the serine group, such as di-isopropylfluorophosphonate (DFP) and phenylmethylsulphonyl fluoride (PMSF).

The determination of physicochemical constants of proteases is complicated by the presence of peptides formed by autodigestion. Some data are listed in Table 7.

Table 7

Some physical properties and esterase activities of Carlsberg and Novo Subtilisins From Ottesen and Svendsen (1970)

	Carlsberg	Novo
Number of amino-acid residues in the molecule	274	275
Molecular weight from amino-acid sequence	27,277	27,537
Isoelectric point	9.4	9.1
S_{201w} (Svedberg units) value	2.85	2.81
Intrinsic viscosity	3.2	3.6
$E_{280nm}^{1mgml^{-1}}$ value	0.95	1.17

Esterase activity with the following substrates:	Values for			
	K_m	V_{max}	K_m	V_{max}
N-Acetyltyrosine methyl ester	0.07	1,930	0.09	1,560
N-Acetylphenylalanine methyl ester	0.03	765	0.06	415
Toluene sulphonyl arginine methyl ester	0.04	117	0.03	7
Benzoylarginine ethyl ester	0.007	16	0.01	4.6

Subtilisin Carlsberg has a broad specificity and will hydrolyse most types of peptide bond and some ester bonds. It is also effective in transpeptidation and transesterification. Of the synthetic amino-acid esters, those containing aromatic amino-acid residues, e.g. N-acetyl-tyrosine methyl ester, are preferentially attacked. Some triglycerides, such as tripropionin and tributyrin, are also hydrolysed. In hydrolysis of proteins, activators are not required, and Ca^{2+} is not necessary for stability as is the case for many of the serine proteases. This means that sequestering agents like EDTA or tripolyphosphate do not inactivate the enzyme. The enzyme is stable over a broad pH range. The optimum pH value depends on the substrate and reaction conditions. A typical pH-stability curve is shown in Figure 2. It is seen that the enzyme is inactivated rapidly below pH 5 and above pH 11, presumably because of autodigestion caused by unfolding of the protein molecule. The enzyme is stable up to a temperature of about 50°C as seen in the stability curves in Figure 3. These stability tests were made in dilute buffer systems and may vary greatly from the stabilities observed under industrial conditions, where protein hydrolysates will often be present to stabilize the enzyme.

Subtilisin Carlsberg has good activity for hydrolysis of urea-denatured haemoglobin between pH 6 and 11 as shown in Figure 4,

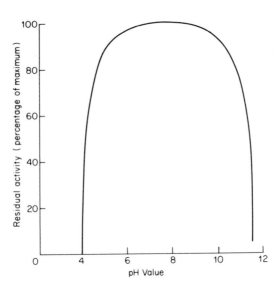

Fig. 2. Stability of Subtilisin Carlsberg at different pH values. The enzyme (0.2 Anson unit/ml) was incubated for 24 hours in phosphate buffer at 25°C and at the pH values indicated. Residual activity was assayed by the method of Anson (1939).

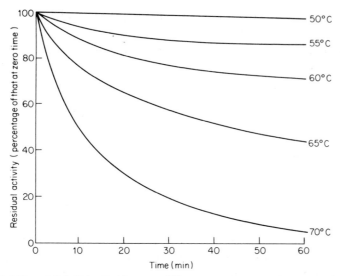

Fig. 3. Stability of Subtilisin Carlsberg at different temperatures. The enzyme (2,3 Anson units/ml) was incubated in tris–maleate buffer, at pH 8.5, for the times indicated. Residual activity was assayed by the method of Anson (1939).

and at temperatures as high as 60°C as seen in Figure 5. With soybean protein as the substrate (Fig. 6), the activity varies with pH value in much the same way as with haemoglobin. Again, it must be borne in

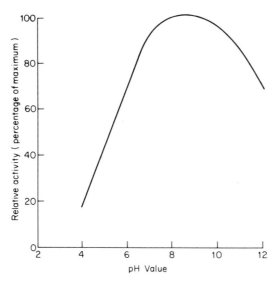

Fig. 4. Activity of Subtilisin Carlsberg at different pH values. The enzyme (0.02–0.2 Anson unit/l) was assayed at 25°C by a modified Anson (1939) method using urea-denaturated haemoglobin substrate and 10 min reaction time at the pH values indicated.

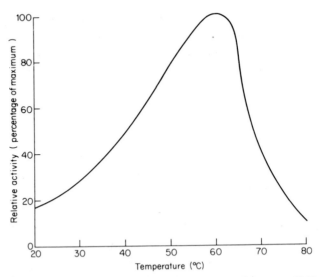

Fig. 5. Activity of Subtilisin Carlsberg at different temperatures. The enzyme (0.03–0.3 Anson unit/l) was assayed at pH 8.5 by a modified Anson method (1939) using urea-denatured haemoglobin substrate and 10 min reaction time at the temperatures indicated.

mind that experiments like this were performed under laboratory conditions and that, in most instances, experiments conducted under industrial conditions will give a slightly different result.

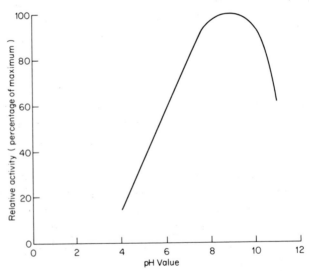

Fig. 6. Activity of Subtilisin Carlsberg at different pH values. The enzyme (1 Anson unit/l) was assayed at 50°C and a reaction time of 10 min at the pH values indicated using 8% soybean-protein as a substrate and determining the free amino groups formed with tri-nitrobenzene sulphonic acid (Lin et al., 1969).

Subtilisin Carlsberg will not hydrolyse proteins completely to amino acids, and the extent of hydrolysis will depend on the protein in question, the reaction conditions and the amount of enzyme used. The limit for casein is hydrolysis of 30–40% of the peptide bonds in the protein.

2. Subtilisin Novo

This alkaline serine protease is usually present as a side activity in commercial preparations of *Bacillus* α-amylase. The enzyme was first purified and crystallized in 1954 by Hagihara who used the commercial preparation Bacterial Protease Nagarse (B.P.N.) as a source of the protease. He therefore named the enzyme Subtilisin B.P.N. In 1960, Ottesen and Spector isolated a similar enzyme from Bacterial Proteinase Novo and called this enzyme Subtilisin Novo. Studies by Smith and his coworkers in 1968 revealed that the two enzymes had the same amino-acid sequence, thus they are identical.

Since protease is an unwanted component of α-amylase preparations, these have been developed in the last few years so that the protease content is minimized, and today most α-amylase preparations are practically free from alkaline protease. At the same time, preparations optimized for the alkaline protease have been developed, but they have limited use. The applications are detergents, dehairing and in brewing, but the market is insignificant compared to the market for Subtilisin Carlsberg.

a. *Microbial source.* Subtilisin Novo is produced by the *B. subtilis* variety that also produces liquefying α-amylase. This variety is now called *B. amyloliquefaciens* according to the Eighth Edition of Bergey's Manual (1974). The organism is present in most culture collections, and it is a characteristic of the species that a good yield of alkaline protease is produced. *Bacillus amyloliquefaciens* forms a number of other enzymes of commercial interest, including α-amylase, β-glucanase, hemicellulase and neutral protease. Commercial preparations will usually contain all of these enzymes, but in different amounts. The β-glucanase is always present in high concentration, and it is useful in brewing by hydrolysing barley gum.

b. *Preparation.* Regulation of enzyme synthesis in the closely related *B. subtilis* Marburg strain has been extensively studied, but little has been done to investigate protease synthesis in *B. amyloliquefaciens*. Protease synthesis starts after completion of the logarithmic growth phase, and it is closely related to the sporulation process. Mutants unable to produce alkaline protease are asporogenic, but asporogenic mutants may well produce alkaline protease in high yields. The productivity of the available strains is about 10% of that reported for *B. licheniformis*. The protease is produced by growing the organism on a rich medium under good aeration at a temperature between 30°C and 40°C and a pH between 6 and 7. Incremental addition of carbohydrate during the fermentation may be advantageous.

c. *Properties.* Subtilisin Novo contains 275 amino-acid residues in the sequence shown in Figure 1 (p. 60). There is an extensive homology with Subtilisin Carlsberg. Only 58 of the 275 residues differ. There is no cysteine residue in the molecule, hence no disulphide bridges. The *N*-terminal is alanine and the *C*-terminal is glutamine. The active site comprises Ser (221), His (54) and Asp (32). The enzyme is inhibited by serine reagents. Some of the basic properties are summarized in Table 7 (p. 61). Subtilisin Novo is more dependent on Ca^{2+} for stability than Subtilisin Carlsberg so, at high temperatures or extreme pH values, it is advisable to add Ca^{2+} to the reaction mixture.

The specificity for peptide bonds is broad, and the enzyme has esterase activity, but the specificity is different from that of Subtilisin Carlsberg. Activity and stability at different pH values and temperatures are similar to those of Subtilisin Carlsberg.

From the properties just described it is understood that Subtilisin Novo as a protease has little advantage over Subtilisin Carlsberg. This, combined with the higher costs of production caused by the relatively low yield, makes it understandable that the application of this protease is limited. In fact, it is primarily used in applications where the carbohydrases or the neutral protease produced by the *B. amyloliquefaciens* are useful.

3. Proteases from Alkalophilic Bacillus Species

The commercial success of the detergent proteases in the 1960s initiated a search for alkaline proteases that would act as better detergent proteases than Subtilisin Carlsberg. The properties that were looked for were primarily good stability under washing conditions, i.e.

pH 9–10, temperatures above 50°C, and the presence of surfactants and sequestering agents. The search was also extended to micro-organisms living under extreme conditions and, in 1967, it was found that especially stable proteases were formed by some members of the genus *Bacillus* which could grow at high pH values (pH 7–11). These organisms were first described in 1934 by Vedder, but were almost forgotten until their rediscovery.

These so-called alkalophilic bacilli are very common in nature (Aunstrup *et al.*, 1972) and may be isolated from most soil samples. They appear to be alkalophilic counterparts to the known *Bacillus* species, and they produce a number of different proteolytic enzymes. Some of these enzymes have properties close to those of the known subtilisins, but others are characterized by having maximum activity up to pH 12. These proteases are valuable enzymes for incorporating into detergents, and because of their good stability at high pH values they have made a new process for enzymic dehairing possible.

a. *Preparation.* The organisms used in the preparation of the new proteases may be characterized as alkalophilic counterparts to *B. subtilis* or *B. licheniformis.* These organisms are easy to grow, provided that the pH value is maintained above 7.5, e.g. by the addition of sodium carbonate or by including salts of metabolizable acids in the medium.

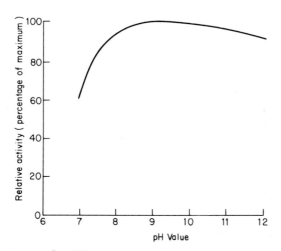

Fig. 7. Activity of Esperase[R] at different pH values. The enzyme (150 mg/l) was assayed at 25°C by a modified Anson (1939) method using urea-denatured haemoglobin substrate and 10 min reaction time at the pH values indicated.

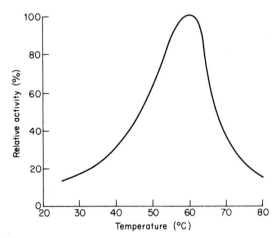

Fig. 8. Activity of Esperase® at different temperatures. The enzyme (150 mg/l) was assayed at pH 10.1 by a modified Anson (1939) method using urea-denatured haemoglobin substrate and 10 min reaction time at the temperatures indicated.

Strains and methods for production of the enzymes are described in several patents (Aunstrup *et al.*, 1973, Horikoshi and Ikeda, 1977; Nijenhuis, 1977). The strains secrete small amounts of other alkaline-tolerant enzymes, such as amylase, but over 90% of the enzyme content of the broth is alkaline protease.

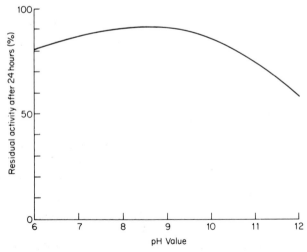

Fig. 9. Stability of Esperase® at different pH values. The enzyme (150 mg/l) was stored for 24 hours in phosphate buffer at the indicated pH values at 25°C. Residual activity was assayed by the method of Anson (1939).

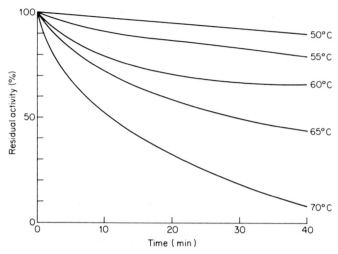

Fig. 10. Stability of Esperase® at different temperatures. The enzyme (150 mg/l) was incubated at pH 9 for the times indicated. Residual activity was assayed by the method of Anson (1939).

b. *Properties of the proteases.* These alkaline proteases are serine proteases like the subtilisins. They are single peptide chains free from disulphide bridges and carbohydrates. The N-terminal residue is alanine as in the subtilisins. The specificity for peptide-bond hydrolysis is broad, and the enzymes have esterase activity. The molecular weight is 20,000–30,000 and the iso-electric point about 11. Stability and activity curves for a typical enzyme of this type, namely Esperase®, are shown in Figures 7 to 10. It can be seen that the enzymes are stable and active in the pH range 6–12 and that they may be used at temperatures over 60°C.

B. Metalloproteases

These enzymes are endopeptidases that contain an essential metal atom, usually Zn. Their optimum pH value is around pH 7 so they are also called neutral proteases. Metalloproteases are found alone or together with other proteases in many micro-organisms. They are formed by several *Bacillus* species, for example *B. subtilis*, *B. cereus*, *B. megatherium* and *B. stearothermophilus* (Keay, 1972). They are also found in protease preparations from fungi, such as *Aspergillus oryzae*, where the metalloprotease occurs together with an acid and an alkaline protease.

The metalloproteases from *B. amyloliquefaciens* and *B. thermoproteo-lyticus* (thermolysin) are used for industrial purposes. The market for these enzymes is not large, but they find some applications in general protein hydrolysis and under circumstances where their special properties are advantageous. One example is hydrolysis of barley protein in the brewing process where an inhibitor in the barley inhibits serine proteases, but not the metalloproteases. Metalloproteases are more specific than serine proteases and this has led to the belief that they could be used as microbial rennet. Metalloproteases from *B. subtilis* and *B. cereus* have been developed for this purpose, but the application was unsuccessful because the non-specific proteolysis is so high that the cheese prepared became bitter and unpalatable. Thermolysin is the most heat-stable commercial protease available. It loses only 50% of its activity after one hour at 80°C (Endo, 1962), so that it is the enzyme of choice whenever a high-temperature proteo-lysis is necessary. Commercial preparations of neutral protease from *B. amyloliquefaciens* usually contain a substantial amount of serine protease and α-amylase. Only one commercial product, namely Neutrase® from Novo, is free from these two contaminating activities.

1. Microbial Sources

Metalloprotease from *B. amyloliquefaciens* may be produced by the same organisms used in the production of Subtilisin Novo and α-amylase. The strains are often named *B. subtilis* in the literature or in brochures, but the organism used commercially for both protease and amylase production is most likely *B. amyloliquefaciens*. Murray and Prince (1970) described isolation of a suitable production strain that is asporogenic, lacks a serine protease and, at the same time, produces a better yield of neutral protease than the parent strain. The ratio between neutral and alkaline proteases depends also on the compo-sition of the medium and fermentation conditions (Markkanen and Bailey, 1974). Since the neutral protease is less stable in the fermenta-tion medium than the alkaline protease, a short fermentation time favours production of the neutral protease.

The source of thermolysin, *B. thermoproteolyticus* Rokko, was named by its discoverer Endo (1962), but the species is not generally recognized and the organism should be classified as a variety of *B. stearothermophilus* (Walker and Wolf, 1971). Sidler and Zuber (1977) prepared a similar neutral protease by cultivating *B. stearothermophilus* N.C.I.B. 8924. Thermolysin is produced by growing the organism for about 24 hours at 55°C and pH 6–8 in submerged culture (Endo, 1962). It has not been published whether the organism produces other proteases or amylase.

2. Properties of Metalloproteases

The amino-acid sequence of thermolysin has been determined (Titani *et al.*, 1972). It contains 316 amino acids (Table 8) in a single peptide chain without disulphide bridges. The molecule is folded to form two lobes separated by a deep cleft in the bottom of which the essential zinc atom is found bound to two histidine residues and a glutamate residue. The zinc atom is essential for activity and the active site comprises six amino acids close to the zinc atom. Four calcium atoms are bound to the molecule, and they are important for the heat-stability of the enzyme.

The properties of the neutral protease from *B. amyloliquefaciens* are very close to those of thermolysin. The molecule consists of one peptide chain with 326 amino-acid residues and there is an extensive homology with thermolysin. The important difference between the two enzymes is in thermostability. When heated for 15 minutes at pH 7.2, thermolysin loses 50% of its activity at 84°C, whereas the corresponding temperature for the enzyme from *B. amyloliquefaciens* is 59°C. The reason for this difference in thermostability is not known, but it is assumed that the higher percentage of hydrophobic amino-acid

Table 8

Properties of the metalloprotease from *Bacillus amyloliquefaciens* (BA) and *Bacillus thermoproteolyticus* (thermolysin)

	Enzyme	
	BA	Thermolysin
Number of amino-acid residues in the molecule	326	316
Molecular weight	37,000	34,400
N-Terminal amino acid	Ala	Ile
C-Terminal amino acid	Leu	Lys
Number of zinc atoms per molecule	1	1
Number of calcium atoms per molecule	2	4

residues and the presence of four calcium atoms in thermolysin instead of two may be of importance.

Metalloproteases have optimum activity for hydrolysis of casein in the pH range 7–8 (Endo, 1962), and they are stable in the pH range 5–10. The stability is increased considerably by addition of Ca^{2+} to the reaction mixture and it is lowered by addition of sequestering agents. Strong sequestering agents, like EDTA, inhibit the enzyme by removing the zinc atom, whereas removal of the calcium atoms only affects the stability of the enzyme. Metalloproteases hydrolyse preferentially peptides with hydrophobic side chains such as phenylalanine and leucine. They have very weak esterase activity.

C. Acid Proteases

Many fungi produce proteases with pH optima between pH 3 and 5. These acid proteases are not inhibited by serine protease-inhibiting reagents, sequestering agents or by thiol-group reagents. Their properties are closely related to the animal digestive enzymes pepsin and rennin. These enzymes are of industrial interest in all areas where protein hydrolysis at low pH values is desired. Two applications are of major interest, namely hydrolysis of soybean protein to make soy sauce and milk coagulation in the preparation of cheese. Other applications involve improvement of the baking properties of flour, application as a digestive aid and prevention of chill haze in beer.

A number of acid proteases are commercially available, but most of the preparations are only used for laboratory purposes. Industrial use is practically limited to the enzymes from *A. niger*, *A. oryzae*, *Endothia parasitica*, *Mucor pusillus* and *M. miehei*. Sales of these proteases are at the rate of U.S. $5–10 million annually. The value of proteases from aspergilli is rather difficult to evaluate because a large part of it is used directly by the manufacturer in the production of soy sauce.

1. Protease from Aspergillus niger

The best known protease is the acid protease Aspergillopeptidase A produced by *A. saitoi* (synonymous with *A. phoenicis*). The enzyme was isolated in 1956 by Yoshida and it has been commercially available for over 20 years, but industrial use of the enzyme is limited. It is primarily used in digestive-aid preparations. Small amounts of acid protease are, however, found in other enzyme preparations from *A. niger* such as glucoamylase and pectinase. In both of these preparations, the protease component is undesired and is often removed. In use of glucoamylase in the distilling industry, acid protease is useful because it hydrolyses proteins in the mash thus providing nutrients for the yeast.

a. *Preparation.* Commercial production is performed in semisolid culture on a wheat-bran medium with addition of inorganic nitrogen. The methods used are not published. Ichishima (1970) described laboratory methods for both semisolid and submerged cultivation of *Aspergillus saitoi* A.T.C.C. 14332. Cultivation is performed at 30–35°C and at low pH values for 60 to 96 hours. The enzyme is recovered by precipitation with ethanol. Usually high activity of an acid carboxy-peptidase is also found in the precipitate.

Methods for preparation of similar enzymes have been described by Koaze *et al.* (1970). They used an organism called *A. niger* var. *macrosporus* (A.T.C.C. 16513) and they found two acid proteases, one with a pH optimum of 2.0 and another with a pH optimum 2.6.

b. *Properties.* Aspergillopeptidase A is a typical acid protease. It has a pH optimum for protein hydrolysis at 30°C between pH 2.5 and 3.0, and it is stable in the pH range 2–5. It has a relatively narrow specificity which resembles that of pepsin, peptide bonds with hydrophobic side chains being preferentially attacked. It has very little peptidase, amidase and esterase activity. Trypsinogen and chymotrypsinogen are activated by the enzyme. The molecule contains 283–289 amino-acid residues (molecular weight 34,000–35,000) organized in a single peptide chain with serine as the *N*-terminal and alanine as the *C*-terminal residues. The molecule contains two cysteine residues which form a disulphide-bridge. Methionine is not present in the molecule. There is no report on carbohydrate content of the molecule.

2. *Proteases from* Aspergillus oryzae

Aspergillus oryzae and the closely related *A. sojae* are important sources of industrial enzymes. *Aspergillus oryzae* is used for producing the traditional Takadiastase preparations which are multi-enzyme preparations prepared by semisolid fermentation of the fungus on wheat bran. The most important enzymes of Takadiastase are α-amylase and proteases, and commercial preparations will usually contain both enzymes in good quantities so that it is often difficult to know whether a preparation is an α-amylase or a protease. The protease components are acid, neutral and alkaline so that the preparation will show good proteolytic activity over a very broad pH range. Takadiastase is mainly used to make digestive-aid preparations and in the baking industry. Proteases from *A. oryzae* have also been applied in medicine for debridement and dissolution of blood clots (Bergkvist, 1966). Proteases from *A. sojae* are not commercially available. Enzymes prepared by this organism are exclusively used in the manufacture of soy sauce.

a. *Preparation.* Commercial manufacture of protease from *A. oryzae* is by semisolid cultivation on wheat bran. Commercial strains are not available, but the strain *A. oryzae* N.R.R.L 2160 has been used for laboratory production of the protease. When it is grown in submerged culture, the only enzyme produced is a serine protease (Klapper *et al.*, 1973). The ratio between the proteases varies with the conditions of the cultivation, and production of the acid protease is favoured by growth at low pH values.

b. *Properties.* *Aspergillus oryzae* produces acid, neutral and alkaline proteases, and it has been possible to isolate several isoenzymes of each type. The acid protease has an optimum pH value for protein hydrolysis at pH 4–4.5. It is stable between pH 2.5 and 6 and 50% is inactivated after 10 minutes at 60°C. The enzyme has limited esterase activity. It activates trypsinogen, and its active centre is similar to that of pepsin (Matsubara and Feder, 1971). When the enzyme is produced in semisolid culture, two components of the enzyme can be isolated. One is a glycoprotein with about 50% carbohydrate, whereas the other is carbohydrate-free. When the enzyme is isolated from submerged culture, all components contain carbohydrate (Tsujita and Endo, 1977).

Two neutral components have been isolated. They are typical metalloproteases and are inhibited by sequestering agents. One has an optimum pH value of 7, is stable in the pH range 5.5–12, and is rapidly inactivated above 50°C, whereas the other has optimum pH value at 5.5–6.0 and retains more than 70% activity after heating at 90°C for 10 minutes (Nakadai *et al.*, 1973).

The alkaline protease has an optimum pH value between 7 and 8.5; it is stable between pH 4.5 and 9, and is rapidly destroyed at 60°C. It is a serine protease and has properties similar to the serine proteases of *Bacillus* spp. The enzyme is stabilized by Ca^{2+}, and inhibited by serine reagents and by potato inhibitor (Klapper *et al.*, 1973).

Protease preparation from *A. oryzae* usually contain substantial amounts of peptidase which are important for production of free amino acids in the hydrolysate (Nakadai and Nasuno, 1977). The proteolytic components of *A. sojae* are parallel to those of *A. oryzae* although it is possible to differentiate the enzymes from the two species by electrophoresis (Nasuno, 1972).

3. Proteases from Mucor *spp.*

The two thermophilic *Mucor* species *M. pusillus* and *M. miehei* are important for their production of microbial rennet. The *M. pusillus* enzyme was introduced in 1969 and the *M. miehei* enzyme in 1972. The present value of the annual sales is over U.S. $5 million.

Both micro-organisms and enzymes are closely related, yet there are important differences in the two enzymes which are decisive for production of the enzymes as well as for their application. The proteases are solely used for milk coagulation.

a. *Protease from* Mucor pusilus. (i) *Preparation.* The strain found by Arima (1964), when he discovered the enzyme, is not publicly available, but the strains A.T.C.C. 16458 and N.R.R.L. 2543 are suited for production of the enzyme. The strain N.R.R.L. 2543 has been investigated thoroughly by Somkuti and Baloil (1968) who found that they could produce the enzyme in submerged culture in a medium containing a 5% suspension of wheat bran. The yields obtained are, however, extremely low when compared to the yields obtained in semisolid culture, so this method is always used in commercial practice. Details of the method used at present are not known, but it can be assumed that the fungus is grown on moistened wheat bran with various additives at a temperature of about 30°C, as described by Arima (1964). After cultivation for about 3 days, the enzyme is extracted from the bran with water and is recovered by precipitation with ethanol. During the cultivation, some unwanted side activities may be formed, e.g. cellulase, lipase and an unspecific protease. They may be removed by acid treatment, adsorption on silicium dioxide or other adsorbents. The commercial preparation contains only one protease component.

(ii) *Properties*. The enzyme is a typical acid protease with an aspartate residue in the active centre. It consists of a single peptide chain and the molecular weight is about 30,000. Although the molecule contains two cysteine groups, no disulphide bridge has been observed, and there is no carbohydrate attached to the molecule. The optimum pH value for hydrolysis of haemoglobin is pH 4 and for casein is pH 4.5. The enzyme is specific for peptide bonds involving aromatic side chains. It is stable in the pH range 3 to 6, and loses 90% of its activity after heating for 15 minutes at 65°C and pH 5.5. The properties are similar to pepsin and rennin, but there are differences in their milk-coagulating activity (see Section VI, pp. 98–100).

b. *Protease from* Mucor miehei. (i) *Preparation*. Several strains have been mentioned in connection with commercial production of the enzyme, namely C.B.S. 370.65 (Aunstrup, 1968), N.R.R.L. 3420 and N.R.R.L. 13042. These strains all produce the protease in approximately the same yield.

Mucor miehei is well suited for submerged cultivation. Good yields may be obtained in a medium containing 4% potato starch, 3% soybean meal and 10% barley (Aunstrup, 1968). The organism is cultivated at 30–35°C and the pH value is maintained between 5 and 7. During cultivation, lipase is formed, and this must be removed before the enzyme is used for cheese making. This may be achieved by adsorption, heat treatment or acid treatment at pH 2–4. Formation of non-specific protease is negligible in submerged culture, so special steps to remove this activity are unnecessary. The enzyme is usually sold in liquid form, and the recovery process consists simply in removing the mycelium from the broth and concentrating the enzyme-containing liquid by evaporation or ultrafiltration.

(ii) *Properties*. The protease from *Mucor miehei* is an acid-aspartate protease with a molecular weight of about 38,000. The molecule consists of a single peptide chain, and it contains about 6% carbohydrate. The enzyme is stable at 38°C between pH 3 and 6 with an optimum at pH 4.5 for both stability and activity against denatured haemoglobin.

The enzyme specifically hydrolyses peptide bonds involving aromatic and hydrophobic side chains. In its action on oxidized insulin B chain, it shows similarities with pepsin, rennin and the protease from *M. pusillus*. Its milk-coagulating property is also similar to that of these enzymes, but there are differences in the dependency on temperature and Ca^{2+} concentration (see Section VI, pp. 98–100).

4. *Protease from* Endothia parasitica

This enzyme was found in 1965 by Sardinas (1972) and was commercialized as a microbial rennet in 1969. It has a higher proteolytic activity than the rennets from *Mucor* spp., and therefore it is not so widely used. In Emmental cheese, it has shown good results.

a. *Preparation*. A suitable strain (e.g. A.T.C.C. 14729) is grown in submerged culture in a medium containing 3% soybean meal, 1% glucose, 0.3%, sodium nitrate, 1% skim milk, 0.05% potassium dihydrogen phosphate, and 0.025% magnesium sulphate. Cultivation continues for 48 hours at pH 6–7 and 28°C. Owing to the instability of the enzyme, recovery must be effected quickly at low temperature and preferably in the absence of oxygen. The enzyme is always marketed as a solid preparation.

b. *Properties*. The enzyme from *E. parasitica* has a molecular weight in the range 34,000–39,000. It is a single peptide chain lacking carbohydrate. It is stable in the pH range 4.0–4.5 with an optimum at about 4.5. It is inactivated in less than 5 minutes at 60°C. The isoelectric point is at 5.5, and the optimum for hydrolysis of denatured haemoglobin is at pH 2.0. Hydrolysis of oxidized insulin B chain shows a broader specificity than that of pepsin, rennet or the Mucor proteases. The milk-coagulating activity is less dependent on temperature and Ca^{2+} than that of rennin and the Mucor proteases.

V. METHODS OF PRODUCTION

A. Introduction

The fermentation methods used for production of microbial proteases are in principle similar to those used in production of other fermentation products. The general methods used are described elsewhere and will not be dealt with here. Some details are, however, peculiar to the production of protease and will be described.

The methods used in recovery and purification of the enzymes are more specialized, particularly in production of detergent enzymes where strict requirements have been raised both with regard to physical appearance and to stability. Present production methods have been developed with three objectives, namely to minimize production costs, to satisfy market requirements with regard to catalytic properties, purity and stability, and last but not least to satisfy the increasing requirements from the regulatory authorities. Protease production is not a large-scale operation, the yearly World-wide production being in the range 1,000–2,000 tonnes of pure enzyme protein. Some steps in the production process require sophistication but, in general, protease production is not complicated. Nevertheless, proteases and their formation by micro-organisms have enjoyed close attention from the scientific world, an interest that has resulted in an extensive literature on the subject, much more than its economic importance justifies. Most of the scientific literature on microbial proteases deals with subjects and methods which are far from those that are of direct interest to the industry. This presentation will attempt to describe industrial methods and points of view. The interesting theoretical aspects of enzyme production are therefore not discussed. The reader is referred to various reviews and books on the subject, including Boyer *et al.* (1960, 1976), Cohen (1977), Erickson (1976), Lorand (1976) and Priest (1977).

B. Fermentation Methods

1. Semisolid fermentation

Proteases from *Aspergillus* and *M. pusillus* are produced by semisolid fermentation. The most popular medium is wheat bran, which is moistened to 50% dry substance, inoculated with spores and left at 30–40°C under aeration. The reason for establishing semisolid fermentation for these products is that higher yields and a more suitable composition are obtained compared to submerged culture. The biochemical differences between semisolid and submerged culture have not been studied in detail, but various effects indicate that these differences are deep and fundamental. Basically, there is a difference in water concentration, as in semisolid fermentations one usually operates with about 50% dry substance. But another important difference is the fact that the mycelium in the semisolid fermentation is exposed to atmospheric air and is allowed to grow undisturbed in its natural form, a condition that permits development of spores and fruiting bodies to a much greater extent than the submerged fermentation. It is a general experience that, in semisolid fermentation, the range of enzymes produced at the same time from a single fungus is much broader than the range of enzymes produced under submerged cultivation. For example, in the production of α-amylase by *A. oryzae* in the bran process, there is a substantial concomitant production of protease and aminopeptidase. This production is almost completely absent from the submerged culture of the same strain.

Many of the extracellular enzymes of *A. oryzae* are glucoproteins and it has been shown for the acid protease that the protein : carbohydrate ratio is dependent on the growth form so that components with less carbohydrate or without carbohydrate are formed in semisolid culture. This interesting phenomenon is currently being investigated. It is of practical importance because the carbohydrate component usually gives the enzyme molecule added stability. In this connection, it is interesting to note that the protease of *M. pusillus*, which is formed in best yields in semisolid culture, is carbohydrate free, whereas the corresponding enzyme from *M. miehei*, which is a glucoprotein, is formed in best yields in submerged culture.

Semisolid cultivation has one disadvantage; it is almost impossible to feed ingredients to the medium during cultivation, which means that the changes of pH value and carbohydrate concentration are predetermined when the medium is inoculated. On the other hand, it appears that the fermentation runs more steadily if only the important problem of controlling the right humidity in the medium is solved.

Recovery of the enzymes prepared by semisolid fermentation is usually simple. The fermented mash is extracted with water or buffer at a suitable pH value, preferably in a countercurrent system so that the enzyme concentration in the extract becomes as high as possible. The extract is filtered and concentrated by ultrafiltration or evaporation, and this concentrate is used as an enzyme preparation after standardization and addition of preservatives, or the enzyme is precipitated with solvent or more rarely with salts.

2. Submerged fermentation

The proteases of *M. miehei*, *E. parasitica* and all species of *Bacillus* are produced in submerged fermentation. A list of suitable media and fermentation conditions is given in Table 9.

Many of the medium components are common inexpensive agricultural products which can be supplied in reliable, uniform quality. It is characteristic of the media that they are relatively concentrated; a dry-substance concentration of 10–15% is not unusual. The protein content of the media is also high. Usually, free amino acids in high concentrations are avoided, since they often inhibit protease formation.

Carbohydrate consumption during the fermentation is high, and it is a general practice to add carbohydrates during the fermentation in incremental amounts in order to keep the concentration low at all times. A similar technique was used by Güntelberg as early as 1954, and it has been described for industrial protease production by Kalabokias (1971). There is no report on an inhibition of protease production by high concentrations of carbohydrate, but it is a general experience that carbohydrate utilization becomes better by using this so-called fed-batch principle.

The carbohydrate may be fed as glucose, sucrose or starch hydrolysate. When starch hydrolysate is used, utilization by *B. amyloliquefaciens* is incomplete because this organism lacks the enzyme

Table 9

Composition of typical fermentation media used for microbial protease production (in g per litre)

Bacillus protease
  Starch hydrolystate, 50; soybean meal, 20; casein, 20; disodium hydrogen phosphate, 3.3 (Churchill *et al.*, 1973).
  Starch hydrolysate, 150; lactose, 4.3; cottonseed meal, 30; brewer's yeast, 7.2; soybean protein, 3.65; dipotassium hydrogen phosphate, 4.3; magnesium sulphate monohydrate, 1.25; trace metals (Feldman, 1971).
Fungal protease
  Corn starch, 30; cornsteep liquor, 5; soybean meal, 20; casein, 12; gelatin, 5; distiller's dried solubles, 5; potassium dihydrogen phosphate, 2.4; sodium nitrate, 1; ammonium chloride, 1; ferrous sulphate, 0.01 (Lehmann *et al.*, 1977).
Mucor rennet
  Potato starch, 40; soybean meal, 30; ground barley, 100; calcium carbonate, 5 (Aunstrup, 1968).

---

system necessary for a complete starch hydrolysis. Lactose has been advocated as a favourable carbon source by some authors. If it is supplied in the form of whey powder, it has the additional advantage of being a valuable protein supplement. Owing to its low solubility, lactose is not fed to the medium during the fermentation.

Some years ago, fish meal was highly recommended as a protein source for these fermentations. The use was, however, soon abandoned because the odour inherent to fish protein was transferred to the final product, and this created problems in some applications, e.g. in detergent enzymes.

In commercial practice, medium composition is optimized continuously over the years so that the balance between the various components is maintained, thus maximizing productivity of the micro-organism and minimizing the amount of unutilized components at the end of the fermentation. This balance is highly dependent on the interaction between the particular strain of micro-organism, the type of medium components, the fermentation equipment, and the process conditions. Fermentation time is also of importance, the medium designed for a long fermentation time with a high enzyme concentration being different from that designed for a short fermentation. It must be borne in mind that the goal of the industrial microbiologist is not to obtain the highest possible enzyme yield per unit of cell mass or the highest possible enzyme concentration in the final broth, but to minimize the total costs of the production process while the quality of the product is kept high.

Continuous or semicontinuous processes for protease production are not known to be used in commercial practice, although it has been shown by several authors that it is possible to maintain a high productivity of protease production for long periods of time (Jensen, 1972). The reason is that the economy of a fed-batch process is more attractive because of the relatively inefficient utilization of medium in the continuous process. Strain improvement is an important part of the work done in industry, but little is published about it. It is known that asporogenic mutants of Bacillus spp. are used to a large extent. They are advantageous because they often result in improved yields when compared with their sporogenic parent strains, while they also eliminate the risk of sudden sporulation with a consequent stop in enzyme production. Bacteriophage problems have not been reported in protease production. It is known that Bacillus spp. almost always harbour some sort of phage, and it has been claimed that this may improve protease production, but no thorough study of this problem has been made.

Formation of side products often becomes a problem in the fermentation of enzymes, and proteases are no exception, although many proteases are suprisingly pure. An example is Subtilisin Carlsberg, where over 60% of the protein component of crude commercial preparations is protease protein and no other enzymes are present in significant amounts. With other *Bacillus* species, several enzymes are produced which often will be left in the preparation. However, if they are wanted, they may be removed by mutation of the producing organism. A successful example of this kind of mutation is removal of the serine protease produced by the strain of *B. amylo-liquefaciens* used in production of neutral protease. In other cases, it has been impossible to remove the undesired component by mutation. An example of this is production of Mucor rennet, where secretion of protease is followed by lipase which must be removed before the preparation can be used in cheese manufacture. In this case, it appears that the lipase is so important to the organism's lipid metabolism that mutants without this enzyme will not survive. The enzyme therefore has to be removed in the recovery process.

Ideally the fermentation broth at the end of the fermentation should be a clear solution of the protease with a precipitate of the producing organism. In practice, the broth will often contain approximately 1% cell dry mass, 0.1–0.5% enzyme protein, 1–2% residues of medium components and fermentation products. The yield of enzyme protein based on added protein may be as high as 10%.

C. Recovery and Finishing

The purpose of the processes leading from the fermentation broth to the finished product is to remove impurities and to bring the enzyme into a stable form that is suitable for handling. Moreover, it is important that the processes are inexpensive, that the enzyme losses are small, and that the operations are performed in such a way that no impurities or harmful micro-organisms are introduced.

1. Pretreatment

The aseptic conditions of the fermentor cannot be maintained during the recovery process. All possible precautions are therefore taken to prevent infection during this process. The most important factor is the quality of the equipment used. It should be constructed of stainless steel or suitable plastic materials and designed in such a way that cleaning of all parts in contact with the process fluid is possible. It is especially important to avoid pockets in the apparatus where fermentation fluid may remain from batch to batch. It is also important to avoid construction materials that may leak toxic substances, such as heavy metals or plasticizers, into the fermentation fluid. The requirements for the equipment used are similar to, or higher than those for, equipment used in the dairy industry and much equipment for the dairy industry is well suited for use in an enzyme factory, e.g. piping, valves and heat exchangers. The plant must be kept scrupulously clean at all times. This is obtained by manual cleaning, by the use of high-pressure cleaning equipment, and by circulating disinfecting fluids through those parts of the equipment that cannot be reached. All residues of such cleaning fluids must be carefully removed before use of the equipment, and it is customary to steam-sterilize as much of the equipment as possible before use.

The fermentation broth will typically have a temperature of between 30°C and 40°C when the fermentation is finished and, in order to minimize further microbial growth and enzyme loss, the first step in the recovery process is a quick cooling to about 5°C in a heat exchanger. This low temperature is as far as possible maintained in the following recovery steps.

Solid materials in the broth must now be removed; these are mycelium or bacterial cells, coarse particles from the medium components and colloidal particles from cell debris or medium components. This separation may be done in several steps. First, the coarse particles are removed in a continuous decanter centrifuge, then the fine particles and colloidal material are removed in high-speed disc-type centrifuges. Alternatively, the separation may be done on continuous filters, such as drum filters or leaf filters. In most cases, this process is difficult to perform directly with a satisfactory result. The colloids will clog the filters or the resulting liquid is not clear. It is therefore customary to treat the broth so that the colloids are flocculated and become easier to remove. Numerous methods are possible and the method selected will, to a large extent, depend on the nature of the broth. Precipitation of salts, such as calcium phosphate, in the broth will often be a useful method for removing colloids. However, the method has the drawback in that the precipitate is bulky and its disposal difficult. It is better to use some of the flocculating agents that have been developed for water treatment. They are polyelectrolytes, e.g. polyamines, and the treatment may involve reaction with electrolytes of opposite charge, such as aluminium salts. Filter aids, such as diatomaceous earth, are often used to clarify the broth. They can for example be used as a coating on a rotary-drum filter.

A simple, but unfortunately rather unpredictable way, of aiding purification of the broth is to let it mature by standing for some time without agitation. This method sometimes gives a surprisingly good separation of the colloids, but there is a high risk of contamination and loss of enzyme activity. Pretreatment may also involve steps designed to remove undesired enzyme activities. An example of this is the removal of lipase from Mucor rennet by treatment at low pH values. In the same way, α-amylase activity may be removed. A heat treatment is often used in the preparation of enzymes to remove undesired components, but it is rarely used in the preparation of proteases since these enzymes usually have the lowest heat stability of the enzyme components in the broth.

After pretreatment, the broth ideally is a sparkling clear liquid with a low content of micro-organisms. It contains the protease in soluble form, together with metabolites formed during fermentation and residues from the medium. In a protease preparation, the only proteins present in the broth in significant amounts will be those of the active enzymes.

2. Purification

A number of methods can be used to convert the pretreated broth into a commercial enzyme product. The methods commonly used are shown in Figure 11.

Firstly, the pretreated broth is concentrated by evaporation or ultrafiltration. Evaporation is the traditional process. It is performed

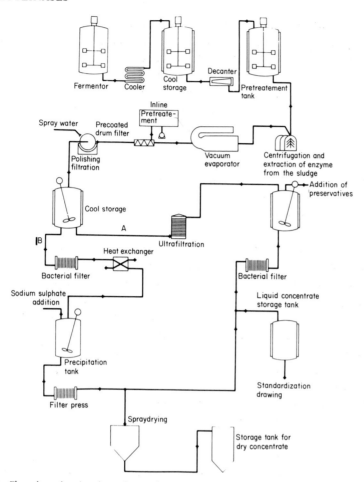

Fig. 11. Flow sheet showing the pathway of enzyme recovery in production of microbial protease.

in multistage vacuum evaporators which often must be custom built. The temperature is kept as low as possible, usually 30–50°C, and the operation should be as rapid as possible. The equipment is operating continuously and great care is necessary in order to avoid infection. Another problem is that the content of dry substance increases with enzyme concentration, which in some cases causes high viscosity and handling problems.

Ultrafiltration is often a preferred alternative to evaporation. The technique has been developed to a high degree of perfection in recent years. Plant-scale equipment is readily available. The process is inexpensive and it has the advantage that substances of low molecular-weight (below 10,000) are removed from the concentrate. Further, the process may be performed at low temperature (5°C), thus keeping the activity loss and contamination risk at a minimum. The method has the disadvantage that colloids sometimes clog membranes, and the cellulose type of membrane is attacked by cellulases in the broth. In such instances, more expensive types of membranes have to be used, for example composite membranes based on polysulphones on a carrier fabric. Owing to the increase in dry substance content during the concentration process, precipitates are usually formed. It is also impossible to avoid some contamination with micro-organisms. These particles are removed by a polishing filtration followed by a germ filtration.

We now have a concentrated solution of the enzyme which could be used as a commercial product were it not for the fact that it would soon be infected and deteriorate during storage. Therefore, the microbiological stability must be ensured by addition of preservatives. Very few preservatives can be used in enzyme products. The reason is that the preservative in most cases must be approved for use in food additives, it must not react with the enzyme protein, and usually the pH value of the solution is close to neutrality, which is an unsuitable pH value for most food preservatives such as benzoates or sorbates. The preservative of choice is often sodium chloride at a concentration of 18–20%. The salt is added directly to the concentrate in powder form and must be of high quality. After addition of preservatives, enzyme activity is adjusted to the specified value by suitable dilution with water containing preservatives. In some instances, it is necessary to add enzyme stabilizers to the preparation. This may be in the form of salts such as calcium salts for the metalloproteases or protein, such as casein, which on hydrolysis acts as a general protease stabilizer.

Protease preparations in liquid form offer a number of advantages. Thus, manufacturing costs are relatively low, handling is easy, and the risks involved in dust formation are avoided. In some applications, however, solid enzyme preparations have to be used, for example for detergents or in flour treatment. Solid enzyme preparations may be prepared by simply spray-drying a clear concentrate. Because of the high cost of removing water by spray-drying, the concentrate should have as high a dry-substance content as possible. Stabilizers or carriers may be added to the concentrate before drying. Usually, the fermentation broth will contain low molecular-weight substances, which are hygroscopic. Their presence in a spray-dried product is not acceptable. Therefore, as a rule, concentrates prepared by ultrafiltration are used for spray drying. The spray-drying process has some disadvantages. The enzymes are subjected to fairly high temperatures and oxidizing conditions, the process does not result in any purification of the enzyme, and finally the product has a low bulk density. It is an advantage that spray-dried products usually are easily soluble.

Salt precipitation may be carried out with ammonium sulphate or with sodium sulphate. Ammonium sulphate can only be used at acid to neutral pH values and, since it will develop ammonia under alkaline conditions, it cannot be used in detergent enzymes. Here sodium sulphate or an organic solvent must be used. The process is very simple in principle. Dry salt is added to the purified enzyme broth until a suitable concentration is reached and the precipitate is removed by filtration and dried. The salt in the filtrate may be regenerated and re-used. The wet salt cake present is dried in suitable equipment. The advantages of the salt-precipitation process are the relatively low costs of investment and the high solubility of the product. Disadvantages are the high salt concentration in the product, and the fact that aseptic conditions during the process are difficult to maintain.

Solvent precipitation may be performed with most water-miscible solvents such as ethanol, acetone and propan-2-ol (isopropanol). The solvent is mixed with the concentrate in the proper concentration and the precipitate is isolated by filtration and dried. The solvent is purified and recovered by distillation. The advantages of this process are the greater purity of the product and the higher specific activity which usually is obtained. Furthermore, it is no problem to prevent microbial growth during the process.

The costs of the investments in explosion-proof equipment in the recovery plant are high, and the loss of solvent must be kept at a very low level in order to operate economically. A number of variations in the precipitation processes have been described in the literature, for example the use of other precipitants like tannin, lignin or ligno-sulphonic acid. These methods have found limited use. It is possible to improve the precipitation methods by fractional precipitations or multiple fractional precipitations with intermediate purification steps. Commercial use of such methods has not been described. For most applications, such methods would probably be too expensive because it is difficult to avoid substantial loss of enzyme activity in a fractional precipitation.

3. Finishing of Solid Enzyme Preparations

A few years ago, the final step in preparation of solid enzyme products was grinding in a mill of the large lumps of the dried precipitate. The resulting fine powder was standardized by mixing with inert materials like salt, lactose or other suitable substances, such as flour in the case of enzyme for flour treatment. Such a preparation is in many ways satisfactory, but the small particles result in high dust levels when handling the product. Since exposure to enzyme dust may cause allergic reactions, it is preferable to make the enzyme preparations so that dust formation is minimized. A number of such methods have been developed in the last few years or have been adopted from other industries. The simplest method is to mix the enzyme powder with dust-preventing compounds like poly(ethylene glycol); another method is granulation with an inorganic salt, for example by fluid-bed granulation. The methods in use at present consist of granulation and coating with a water-soluble inert wax.

In the so-called prilling process, the powdered enzyme preparation is mixed directly with melted wax and subsequently spray-cooled in a cooling tower. In this way, granules of about 0.3–0.8 mm diameter are formed in which the enzyme particles are evenly distributed in the wax. A better solution would be to make a solid core of enzyme powder and cover it with wax. This can be achieved in the so-called marumerizer process (Fig. 12). In this process, the enzyme is mixed with an inert filler such as salt, a binder such as dextrin, and water. The resulting plastic mass is extruded and shaped into spheres in a so-called marumerizer. After drying, the spheres are covered with a water-soluble inert wax. Particles can be made with a diameter of 0.5 mm and upwards, and variation in size is small. Dust formation is practically eliminated. As an example, a detergent enzyme preparation made according to this method will produce less than 0.5 μg of pure enzyme dust per kilogram when tested as described by Harris and Rose (1972).

All detergent enzymes are now produced in the form of wax-covered granules, partly because of the elimination of dust, but also because these granules blend well with other components of the detergent. In applications where the wax cannot be used, such as in the brewery, the

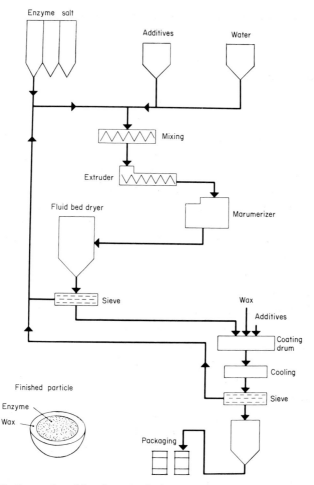

Fig. 12. Preparation of dust-free microbial protease by the marumerizer process.

enzyme may be dedusted by agglomeration but, in all applications where it is possible, liquid preparations are preferred, both from the point of view of the customer and of the manufacturer.

4. Immobilization

Proteolytic enzymes may be immobilized by most of the methods commonly used. No specific method will be described here, because there is not yet any significant production of immobilized proteolytic enzymes. It should be borne in mind, however, that immobilized proteases, which are to be used to prepare food components, must be made with materials that are approved for use in food, a fact that restricts the number of possible immobilization methods.

D. Standardization and Control

Final steps in the manufacturing process are standardization of enzyme activity and ensuring that the properties of the preparation meet the specifications. This procedure is important both from an economical, technical and safety point of view, and the number of tests performed, especially to ensure the safety of the products, is steadily increasing.

1. Standardization

Enzymes are always sold on an activity basis, so that the method for determining the activity is of primary economical importance. Unfortunately, protease activity is not easy to determine accurately. Protein substrates are not well defined. Several different peptide bonds are hydrolysed, and there is no way to determine directly which bonds and how many are hydrolysed. For some enzymes, simple and well-defined substrates may be used, for example esters like benzoyl-arginine ethyl ester for determination of the activity of Subtilisin Carlsberg or simple peptides with a chromophore group such as Bz-Phe-Vol-Arg-p-nitroaniline (Pozsgay et al., 1977). The use of such systems is not advisable since contamination with other enzymes, such as peptidases or esterases, may influence the result and thus give a false activity.

In general, the principle used for determining the activity of a commercial preparation consists of treating a relatively well-defined substrate with enzyme for a given time, inactivating the enzyme and determining the reaction products by a suitable method. The unit of activity is as a rule defined in a semi-arbitrary way although it may relate to the number of bonds hydrolysed. In order to avoid drift in the unit definitions, it is customary to use a standard preparation of the enzyme in question as a reference. The standard is analysed with every batch of enzymes, and the results are adjusted with a factor so that the correct result for the standard is obtained. The substrate may be casein, haemoglobin or other proteins that are readily available in pure form. Derivatives of proteins may also be used with advantage, for example nitrocasein or dimethylcasein. With these substrates, nitration makes it easier to quantify the hydrolysis products, while methylation reduces the colour formed by reaction of the unchanged substrate with the amino-group reagent.

The hydrolysis products after protease action may be determined by precipitating the hydrolysed protein with trichloracetic acid and measuring the ultraviolet absorption at 280 nm as in the Anson (1939) method. Since aromatic amino acids are responsible for the major part of the ultraviolet absorption, the result is often related to amounts of tyrosine dissolved.

Microbial rennets have to be standardized by their milk-coagulating activity and, in the methods used, the principle is a determination of the reaction time necessary for a given amount of enzyme to start coagulating a standard sample of milk. The method is very sensitive to reaction conditions, the quality of the milk, the concentration of Ca^{2+}, and even to the degree of agitation during the reaction. Various methods are used, and they differ in reaction conditions and in the way the starting point of coagulation is observed. The different methods are difficult to compare, especially if enzymes from different sources are used. A standard must always be applied, preferably of the same source as the enzyme in question.

In some instances, several proteolytic activities are present and must be tested in the same preparation. This is, for example, the case with microbial rennet, where it is necessary to have a test for non-specific protease activity apart from the test for milk-coagulating activity. Likewise, in assaying the protease from *B. amyloliquefaciens*, tests must be made both for the alkaline serine protease and the neutral metalloprotease. This may be done by determining the protease activity in Anson (1939) units before and after treatment with EDTA which will inactivate the metalloenzyme.

A number of the more common methods of analysis are shown in Table 10. There is no universal method, and each manufacturer relies on his own methods. This makes a comparison of enzymes from different suppliers difficult. It would be an ideal situation if everybody used the same analytical methods, but this would raise a number of problems. Firstly, it is difficult to reproduce the same method

Table 10

Methods for determination of protease activity

Name	Substrate	Method for detecting hydrolysis	Reference
Anson	Haemoglobin	Photometry of aromatic amino acids	Anson (1939)
Kumitz	Casein	Photometry of aromatic amino acids	Detmar and Vogels (1971)
Löhlein Volhard	Casein	Titration with hydrochloric acid	Küntzel (1955)
Dimethylcasein	Dimethylcasein	Colorimetry of reaction product of primary amino groups and trinitrobenzene sulphonic acid	Lin et al. (1969)
Azocasein	Diazotized casein	Colorimetry of hydrolysis products	Charney and Tomarelly (1947)
Berridge	Milk	Milk coagulation	British Standard 3624 (1963)

accurately in different laboratories, and secondly because all of the methods used at present have some advantages. Furthermore, small differences in enzyme properties might result in unrealistic differences in the results when a general method not designed for that particular enzyme is used. Enzyme activity is standardized as accurately as possible. Usually the standard deviation on the result for a single batch is less than 10%.

2. Control

Control measures can be divided into those important for application of the enzyme and those necessary to ensure the safety of the product. Before a new enzyme product is introduced on the market, extensive and lengthy tests are made with regard to stability during storage and efficiency in use. Furthermore, the toxicological and allergenic properties are investigated thoroughly in animal-feeding studies and other tests. The results of the tests are submitted to the authorities in the various countries where they are evaluated; further tests may be suggested. When the product is approved, it is usually under the condition that a certain number of analyses must be made on all batches of enzyme to be sold. Such analyses usually include: tests for heavy metals, mycotoxins and antibiotic activity; determination of the dust level in granulated enzyme preparations; microbiological tests such as total viable count, coliform count; tests for absence of *Escherichia coli*, *Salmonella* spp. or other pathogens. Usually a test for the absence of production strain is also included although the requirement usually is that the production strain is a harmless saprophyte.

Tests relevant to application of the procedure may include assay for undesired side activities, analysis of the concentration of preservatives and stabilizers such as Ca^{2+}, and appearance tests, i.e. colour, odour, density and viscosity in liquid preparations, and particle size and bulk density of solid preparations.

VI. INDUSTRIAL APPLICATIONS OF MICROBIAL PROTEASES

A. Detergents

Laundering processes are designed to remove dirt from clothes in the simplest possible way. In most processes, a washing agent is used. This may consist of one or more surface-active agents (surfactants), sequestering agents, alkali and other building materials. In Europe, it is common to include perborate, which acts as a bleaching agent by releasing hydrogen peroxide at temperatures above 50°C. In other areas, such as the U.S.A., bleaching is obtained by adding the bleaching agent separately in the form of hypochlorite after the washing process.

Clothes get dirty by adsorbing particles or solutions of the materials surrounding us and excreted by the body. The nature of the dirt is, of course, as varied as our surroundings, but may in general be classified as soil, small inert particles, lipids, carbohydrates, proteins and dyes. In the washing process, heat, alkaline conditions and the presence of surfactant and sequestering agents will dissolve or disperse most of these components, and the bleaching agent will decompose the undissolved dye. Proteinaceous dirt will not alway be removed in this process, and protein remaining on the fabric will be able to retain other dirt components. The reason is that proteins may coagulate on the fabric in such a way that they are not dissolved at the pH value (9–10) of the washing process. This coagulation may be caused by the alkaline pH value, temperature, enzymic processes such as the coagulation of blood, and by the action of oxidizing agents such as the hydrogen peroxide that develop in perborate-containing detergents above 50°C. This latter action is of particular importance since proteinaceous dirt not removed from the fabric when the temperature is raised to 50°C in the washing process will be precipitated on the fabric and then be difficult to remove.

The use of proteolytic enzymes in detergents can solve this problem. The enzymes hydrolyse dissolved or coagulated proteins and the soluble degradation products are easily rinsed off the farbic. It is easy to understand that enzymes can help remove blood stains, but it is often argued that blood stains are rare and that addition of enzymes to detergents for this reason is superfluous. The enzymes do, however, dissolve a number of other stains that are not so obvious, but which after several washings will give the clothing a general grey and unclean appearance. Those stains include proteins from body secretions and skin particles, foods such as milk, egg, meat and fish, and plant material such as grass. The value of proteases in removing such stains

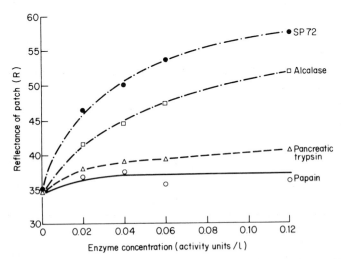

Fig. 13. Effect of including enzymes at different concentrations in detergent (composition given in Table 11) on the efficiency of washing. Enzymes used are indicated on the figure. SP 72 is a highly alkaline protease from an alkalophilic *Bacillus* sp. Alcalase® is a commercial preparation of Subtilisin Carlsberg. Each experiment involved five washes of E.M.P.A. 116-stained patches (3.7 × 7.3 cm). Washing was carried out in a Launder-O-Meter. The volume of washing solution was 300 ml, wash time 33 min, temperature of 25–90°C, and the water had a hardness of 10° German. Washing efficiency was assessed by the reflectance (R) of test patches after washing. From Dambmann *et al.* (1971).

has been proved in numerous tests involving both ordinary clothing and standard test patches soiled with defined compositions of proteins. An example of such standard patches is the E.M.P.A. 116 patch which is cotton fabric impregnated with blood, milk and carbon black. It is marketed by Eidgenössische Material Prüfungsanstalt, St. Gallen, Switzerland. The test may be performed in standard washing model equipment, such as the Terg-O-Tometer or the Launder-O-Meter, or together with ordinary clothing in a household washing machine. After washing and drying, the cleanliness of the test patch is estimated by measuring its light reflectance in suitable equipment. In this way, it is possible to obtain a quantitative estimate of washing efficiency of the enzyme. An example of the use of this method to compare different enzymes is shown in Figure 13 and Table 11 (Dambmann et al., 1971). By using these methods, valuable information can be obtained about the efficiency of proteases against different types of stains and with different washing conditions and detergent compositions. Tests of this nature are, however, artificial in nature and the ultimate test for the enzymes is always washing of ordinary clothing preferably in large-scale consumer tests. Extensive tests of this nature have been performed by the detergent manufacturers, and they have shown that enzymes are valuable detergent components in most washing processes.

The ideal application of enzymes is in the presoaking process where the clothing is soaked in the detergent solution at low temperature for a considerable time before the final washing. In this way, the enzyme has ample time to work, and a most satisfactory result is obtained. The

Table 11

Basic detergent composition for evaluating detergent action by the European method

Sodium LAS (40%)	25.0%
Alkyl phenol ethoxylate (11 EO)	3.0%
Soap, high titre (80%)	3.0%
Sodium carboxymethylcellulose (59%)	1.6%
Sodium ethylenediamine tetraacetic acid (tetrasodium salt)	0.18%
Sodium tripolyphosphate	38.0%
Sodium perborate tetrahydrate	25.0%
Sodium metasilicate	1.0%
Sodium sulphate and water; balance to	100.0%

disadvantage of the method is the long time required and the risk of microbial growth resulting from the many bacteria present in soiled clothing. This may give the clothing a rather unpleasant odour. In a common European machine-washing procedure, the washing process consists of prewash at 20–30°C for 10–20 min followed by a main wash where the suds are heated to the desired temperature (40, 60 or 95°C depending on the fabric) during 30–60 min. In this process, the enzyme will work well because it can exert its function during the prewash and the main wash up to about 60°C. In the U.S.A., washing is often performed by adding hot water at 50–60°C to the clothing and washing while the suds are cooling for 10–15 min. In this way the enzyme has less favourable conditions for its action, but nevertheless it has a significant effect on proteinaceous stains.

Enzyme action is dependent on the other components of the detergent. Oxidizing agents will inactivate the enzyme rapidly. Some surface-active agents also denature enzymes. This is especially true for cationic surfactants, whereas anionic surfactants are less harmful and nonionic surfactants usually have no influence on the stability of the enzymes. Most detergents contain tripolyphosphate as a sequestering agent. This means that the detergent proteases should not be dependent on Ca^{2+} for stability. Because of the risk of phosphate pollution, other sequestering agents have been proposed such as nitrilotriacetic acid or citric acid. The washing properties of these agents are generally not as good as those of tripolyphosphate, but it has been shown that, with these sequestrants, use of enzymes improves the washing efficiency significantly more than when tripolyphosphate is used as a sequestering agent.

Typically the enzyme content of a detergent composition is 0.5% of a preparation containing 3% active enzyme protein. If 5 g of detergent per litre of washing suds is used, the enzyme concentration of the washing sud will be 0.75 p.p.m. This concentration would seem to be very low were it not for the fact that the affinity between enzyme and protein makes the concentration around the proteinaceous stain higher. Highly alkaline proteases appear to have better washing properties than the enzyme from B. *licheniformis* as is shown in Figure 13 (p. 94). The reason is not their activity at pH values above 10 since the pH value of the washing suds rarely exceeds 10. Rather it is to their better stability at high pH values and towards certain surfactants, combined with their higher affinity towards proteinaceous dirts, which in turn may be explained by their larger ionic charge (the pI value for Subtilisin Carlsberg is 9.4, whereas for the highly alkaline proteases, the value is approx. 11).

The newly developed liquid detergents may also benefit from combination with enzymes because they allow for a so-called pre-spotting technique in which the stains are treated for a short time with the concentrated detergents before washing. In this way an efficient removal of proteinaceous stains can be obtained. Application of concentrated enzyme solutions in this way has been used for many years in the dry-cleaning industry. Since this operation does not include washing at extreme conditions, there have been no special requirements for the stability of the enzymes used, but it is obvious that the method described will be advantageous here also.

⌐Proteolytic enzymes have now been an important component in detergents for about 15 years. The washing habits of the future will determine whether this will continue. It is expected that the use of phosphate will decrease and the washing temperature will be lowered in order to save resources. Under these circumstances, it is to be expected that there will be an increasing demand for enzymes in the detergent compositions. ⌐

B. Microbial Rennet

Since the origin of cheese production, the manufacturing process has been adapted to the properties of calf rennet, which is an ideal enzyme for the purpose. The enzyme is readily available in an agricultural community, and it coagulates milk quickly at its natural pH value with little further degradation of the milk proteins. The disadvantage of the enzyme is that it can only be extracted from the fourth stomach of the unweaned calf, a source that has become in short supply in the last years because slaughtering of these young animals is uneconomical. Microbial enzymes have therefore been introduced as alternatives to calf rennet.

The properties of the three microbial enzymes that have obtained commercial success are close to those of rennin, but there are differences which become important in the cheese-making process. The ratio between milk-coagulating activity and proteolytic activity is lower for microbial enzymes than for pure calf rennet (Table 12).

Table 12

Ratio of milk-coagulating activity to proteolytic activity for three
microbial rennets

	Ratio of activities (arbitrary units)
Calf rennet	500–1,000
Mucor rennet	270–300
Enzyme from *Endothia parasitica*	80

Further, the variation of the milk-coagulating activity with temperature, pH value and Ca^{2+} concentration is different from that of calf rennet (Figs. 14, 15 and 16). Of special interest is the fact that the enzymes from species of *Mucor* are more dependent on temperature than calf rennet and that the enzyme from *M. pusillus* is more dependent on Ca^{2+} concentration.

If one looks at the so-called tertiary proteolysis, i.e. formation of non-protein nitrogen after coagulation (Fig. 17), it is seen that the protease from *M. miehei* gives a slightly higher value than calf rennet, but that the enzyme from *E. parasitica* gives a somewhat larger value. This means that use of microbial enzymes may result in shorter ripening time and slightly softer curd than that produced using calf

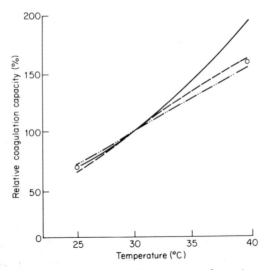

Fig. 14. Variation of coagulation capacity with temperature for various microbial rennets, compared with calf rennet (O); —— indicates the behaviour of rennet from *Mucor miehei*, – – – – of rennet from *Mucor pusillus*, and —··— of rennet from *Endothia parasitica*.

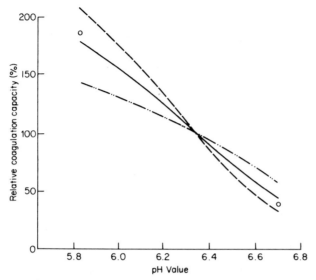

Fig. 15. Variation of coagulation capacity with pH value for various microbial rennets, compared with calf rennet (O); —— indicates the behaviour of rennet from *Mucor miehei*, ––––– of rennet from *Mucor pusillus*, and —··— of rennet from *Endothia parasitica*.

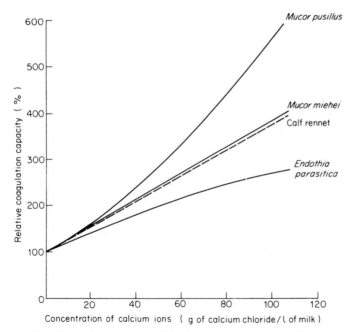

Fig. 16. Variation of coagulation capacity with calcium ion concentration for various microbial rennets, compared with calf rennet.

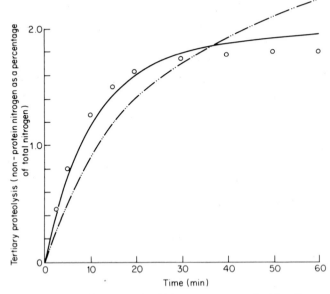

Fig. 17. Tertiary proteolysis of various microbial rennets compared with calf rennet (O); ——— indicates the behaviour of rennet from *Mucor miehei*, and —··— of rennet from *Endothia parasitica*.

rennet, and furthermore that the Endothia enzyme is less suitable for ripened cheese. The problem can be overcome if high scalding temperatures (50°C) are used. In this way, the Endothia enzyme is inactivated and further proteolysis prevented. A cheese such as Emmenthal is scalded at 50–55°C, and it has been shown that the Endothia enzyme is well suited for this application. The use of a high scalding temperature is a disadvantage with the Mucor enzymes since they require a temperature of 75–80°C for inactivation. Here, the high scalding temperature may increase the formation of non-protein nitrogen.

The Mucor proteases have been successfully used for preparation of most types of cheese by appropriate adjustment of the cheese-making technique. Since it may be important to minimize enzyme concentration in the curd, it is, especially when the enzyme from *M. pusillus* is used, often recommended to increase the Ca^{2+} content of the milk. Other modifications of the process technique include adjustment of stirring, cutting time and scalding temperature. Furthermore, the type and dosage of starter culture may be adapted to the microbial rennet. The main disadvantage with the Mucor proteases is the high temperature necessary for inactivation. This means that it is difficult to inactivate the enzyme present in the whey, especially with that from *M. miehei*, a fact that precludes use of the whey for milk products if it is not pasteurized at a suficiently high temperature.

The microbial rennets, primarily of the Mucor type, covered in 1978 about 10% of the total rennet market. They are competing with animal-rennet substitutes such as pig and beef pepsin, which are available in large quantities and have properties comparable to those of microbial rennets. Pepsins, especially pig pepsin, are often used in combination with calf rennet, and most so-called calf rennet sold today contains beef pepsin in different concentrations.

Microbial rennets have obtained an important position as a rennet substitute and may be used successfully, but for many applications rennin will still be the enzyme of choice. Indeed, there is no reason to believe that microbial rennets will take over the position of rennin in the future, unless there is an improvement in their properties. On the other hand, their lower price, consistent quality and unlimited availability will undoubtedly mean that microbial rennets will gain an increasing proportion of the market.

C. The Tanning Industry

In the tanning industry, proteases may be used for two purposes, namely unhairing and bating.

1. Unhairing

The use of enzymes for unhairing hides has been discussed since the beginning of this century (Green, 1952) and many microbial sources and methods of application have been suggested. Up to now, only limited use has been made of enzymes for this purpose. The reason is that enzymic methods are more expensive and more difficult to use than the chemical methods.

The unhairing method most common to-day consists basically of treatment with lime containing sodium sulphide. The unhairing action is rapid and efficient and, at the same time, the hide is swollen due to the alkali, an effect that makes subsequent handling of the hide easier. The hair is more or less damaged by the process, but this is of little importance since it usually has little or no commercial value. If the hair must be collected, such as is the case with sheep skin, lime and sulphide are painted on the flesh side of the hide and the hair will then loosen without damage.

The disadvantages of this type of process are primarily concerned with labour safety and environment. Accidents, sometimes fatal, due to the development of hydrogen sulphide are not uncommon, and under all circumstances the presence of sulphide makes work in the beam house unpleasant. Treatment of the resulting waste water, which contains lime, sulphide and decomposed hair, is a serious problem for tanneries situated in areas where waste-water treatment is required. Consequently there is a motivation in tanneries for a switch to an enzymic method.

Proteolytic enzymes of all types will loosen hair and wool under conditions where they are active. The hides may be treated with an enzyme solution in a drum or the enzyme solution may be painted or sprayed on the fleshy side of the skin. Since the enzyme reaction is highly dependent on the temperature, it is general practice in the fellmongering industry to heat the conditioning chambers to 30–35°C in order to obtain a satisfactory action. Because of the relatively high cost of enzymes, a minimum amount is used, which means that there may be problems with hair that is difficult to remove, such as the small hair. In order to save enzyme, sharpening agents such as sulphites may be added. These agents will intensify action of the enzymes, but at the same time add to the sewage problems. The enzyme acts well at moderately alkaline pH values using, for instance, protease from *B. licheniformis*. In this case, the hide becomes soft and flat, and it must undergo an alkali treatment before further processing in the tannery. An improved process has therefore been developed which makes use of a highly alkaline protease from an alkalophilic *Bacillus* species in combination with lime. By this treatment, which takes place at pH 12–12.5, a satisfactory unhairing and swelling of the hide can be obtained.

In a typical process of this type, the hides are treated with 1.5–10 times their own weight of a liquid containing 8% lime and 0.1% dehairing enzyme. The unhairing process is completed in 18–24 hours at 20–30°C.

Apart from the ecological advantages, the process destroys undesired pigments in the skin and an area increase of up to 6% has been observed. Furthermore, the subsequent bating operation can be shortened or avoided. Problems with fine hair are much less with this method than with other methods. The problem can be completely overcome if the hide undergoes a defatting process before unhairing. An improved method, which makes use of a combination of several proteolytic enzymes and salts of thioglycollic acid, has recently been described (Monsheimer and Pfleiderer, 1976). If enzymic unhairing is not properly performed and enzymic attack goes too far, the resulting leather will suffer from loose grain.

2. Bating

After unhairing, the hides are often defleshed and subsequently bated and delimed. Bating is a well-established process, the purpose of which is to make the leather soft and elastic. Hard leather, e.g. that used for soles, is usually not bated. Initially, the enzyme source for this purpose was bacterial enzymes from dung of dogs and birds but, as early as in 1910, Rohm discovered that proteolytic enzymes from pancreas could be used. The chemistry behind the process is not completely known, and the ultimate test for a successful result is still the appearance and feel or grip of the leather. It has, however, been shown that proteolytic enzymes alone can be used. Details of the bating operation, including pH value, temperature, enzyme concentration and duration, are a matter of experience and are developed individually by the tannery operator.

Bating preparations may contain pancreatic trypsin or proteases from *A. oryzae* or *B. amyloliquefaciens* or *B. licheniformis*, or it may contain mixtures of these enzymes. A very popular ingredient in these preparations has been wood flour, which facilitates spreading of the enzyme on the hide, but at the same time adsorbs some enzymes. It is therefore more economic to spread the enzyme directly on the hide. Proteolytic enzymes in the washing water accelerate water uptake. This is especially advantageous when dried hides are used as raw material.

Future developments in application of enzymes in the tanning industry are difficult to evaluate. The enzymic unhairing process is ready for use and may even be improved, but its widespread use is at present barred for economic reasons. Use of enzymes for bating will probably more and more involve microbial proteases in preference to pancreatic trypsin because of the rising cost of the latter.

D. The Baking Industry

Amylolytic enzymes have been used to adjust the properties of flour for well over 50 years. The source was initially malt, but later enzyme from *A. oryzae* was introduced. The purpose is to aid formation of fermentable sugars thereby giving the yeast better possibilities for leavening.

The first fungal preparations used were relatively weak in amylolytic activity, but rather high in protease activity. This caused an unpleasant side effect because the protease would attack the gluten of the flour and destroy its ability to retain carbon dioxide formed by the yeast. The resulting bread thus became flat and sticky (Reed, 1966). This problem has since been solved by the development of fungal amylase containing little or no protease. However, the fungal protease has now found an application of greater importance in baking, namely in adjustment of the properties of gluten.

In the preparation of dough, mechanical mixing combined with resting periods are necessary to prepare the dough for baking. This is due to the visco-elastic behaviour of gluten. The properties of gluten vary considerably, and therefore the dough preparation should be varied accordingly. This is inconvenient in an industrial bakery, and it is therefore preferred to adjust the properties of the flour instead. This may be done with oxidizing agents which will increase the strength of the gluten, by reacting with the thiol groups of gluten so creating disulphide bridges, or by reducing agents which will have the opposite effect. The agents used may be bromate, peroxide, sulphite or cysteine; they are undesirable in food and are therefore not permitted in several countries.

Another way is to weaken the gluten to a constant level by addition of protease and adjust the preparation of the dough accordingly. This means shortened mixing and resting times, an advantage in an industrial bakery. Protease from *A. oryzae* is preferred for this purpose. It has good activity at the pH value of the dough (5–6) and it is rapidly inactivated during the baking process. Neutral protease from *B. amyloliquefaciens* may be used just as well if the preparation is free from α-amylase. But, so far, it has not been widely accepted because of conservatism and because only the protease from *A. oryzae* is approved for this purpose in the U.S.A. where there is the largest application.

In the manufacture of biscuits and crackers, the dough should be very soft in order to prevent bending and wrinkling of the biscuit in the oven. Also, the precise impression of letters and decoration, which is common in biscuits, requires a very soft and plastic dough. In this case, bacterial proteases have shown excellent properties, especially the neutral protease. The amount of enzyme added to the flour is extremely small, of the order of magnitude 0.01 p.p.m. active enzyme protein. Most enzyme preparations for this purpose are therefore standardized to a suitable low activity with flour.

E. The Brewing Industry

The traditional application of proteases in the brewing industry is in preventing chill haze, a precipitate of protein and tannin formed by cold storage of beer. Although both fungal and bacterial enzymes may be used for this purpose, preference is given to the plant enzymes papain and bromelain. To the knowledge of the author, microbial proteases have not been used commercially to any significant extent. The reason for this is partly tradition and partly the fact that, in many countries, microbial proteases have not been approved for the purpose. Furthermore, they do not offer any advantages over plant enzymes.

In the last ten years, there has been a growing interest in substituting malt with unmalted barley or an increasing amount of brewing adjuncts. If this is done, it is necessary to add microbial α-amylase to obtain sufficient fermentability and microbial β-glucanase to lower the viscosity of the mash caused by barley β-glucans. Lowering the malt concentration has, however, a negative effect on the amount of free amino acids formed during mashing. It therefore becomes necessary to supplement with a protease. The best commercial enzyme for this purpose is the neutral metalloprotease from *B. amyloliquefaciens* since it has a pH optimum close to the pH value of wort (5–6), and it is not inhibited by the protease inhibitor of barley, which inhibits serine proteases. The protease has the added advantage that it activates the β-amylase of barley and thus increases the amylolytic activity of the mash. Protease addition must always be used where a large proportion of barley is used in the mashing, but it can also be applied to compensate for a malt with low proteolytic activity or to increase the concentration of amino acids in the wort when a high percentage of brewing adjuncts is used. The amount of proteinase used is small; usually less than 0.001% of the barley as active enzyme protein or less than 1 p.p.m. in the wort.

F. Meat Tenderization

Treatment of meat with proteases to improve its quality is an ancient craft in tropical countries where fresh pineapple was served with meat, or meat was packed in papaya leaves so that the proteolytic enzymes of those plants could tenderize the meat. The desired effect is primarily degradation of connective tissue collagen and elastin, and the main part of the enzyme action takes place in the temperature range 40–60°C.

Papain has been extensively used for this purpose. It may be applied to the meat immediately before preparation or the enzyme may be injected into the animal just before slaughtering. This is possible because the enzyme is injected in its inactive oxidized form, which then becomes active by reduction in the animal body. The ideal meat-tenderizing enzyme should attack connective tissue at low temperatures and be inactivated at about 50°C. It should also have good activity at the pH value of meat (4–5).

Several attempts have been made to introduce microbial enzymes for this purpose, both of fungal and bacterial origin. So far none of the enzymes has been successfully commercialized. Two problems seem difficult to overcome. The activity towards connective tissue is too low, added to which it has not been possible to commercialize a microbial enzyme, which like papain can be inactivated by oxidation and re-activated by reduction.

G. Protein Hydrolysis

The most important application of microbial proteases for production of protein hydrolysates is in the manufacture of soy sauce. Since this process does not involve use of commercial enzyme preparations, but has the protease production integrated in the production process, the method will not be described in detail here. The proteases are formed by *A. sojae*, and the resulting very extensive hydrolysis is obtained by a combined effect of several proteases and peptidases.

The use of commercial enzyme preparations for protein hydrolysis is small in size. It is most frequently used to change the functional properties of the protein and not to effect an extensive hydrolysis to free amino acids. For this purpose, acid hydrolysis is still the method of choice despite its inherent disadvantages (use of high-salt content in product, loss of tryptophan). The aim of the hydrolysis may be to lower viscosity, for example, in treatment of fish solubles, or to make the protein soluble at isoelectric pH values. An example is the treatment of soybean protein or casein to make protein products suitable for use in soft drinks and liquid food.

The problem is to avoid the formation of bitter-tasting peptides. These peptides usually contain hydrophobic amino-acid residues. The amount formed depends on the protein source and the degree of hydrolysis. Usually, formation of bitter peptides can be avoided if the degree of hydrolysis is kept sufficiently low, e.g. for soy protein below 10%.

The course of the hydrolysis depends on protein, enzyme and reaction conditions. In Figures 18 and 19, hydrolysis curves are shown for Alcalase®, a Subtilisin Carlsberg preparation, and Neutrase®, a preparation of the metalloprotease from *B. amyloliquefaciens*. It is seen that hydrolysis rate and the maximum degree of hydrolysis vary greatly, and that the maximum degree of hydrolysis obtained with the

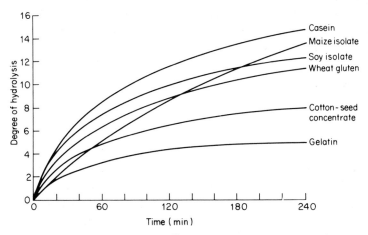

Fig. 18. Hydrolysis of proteins with Alcalase® using the pH-stat method. The substrate indicated (80%, w/v) was treated in a pH-stat with Alcalase (0.16%, w/v; activity 0.6 Anson unit/g) at pH 8.0 and 50°C for the time indicated. The degree of hydrolysis was determined by measuring alkali consumption.

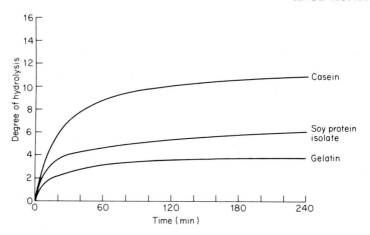

Fig. 19. Hydrolysis of proteins with Neutrase® using the pH-stat method. Substrate (8%, w/v) was treated in a pH-stat with Neutrase (0.16%, w/v; activity 0.5 Anson unit/g) at pH 7.0 and 50°C for the time indicated. The degree of hydrolysis was determined by measuring alkali consumption.

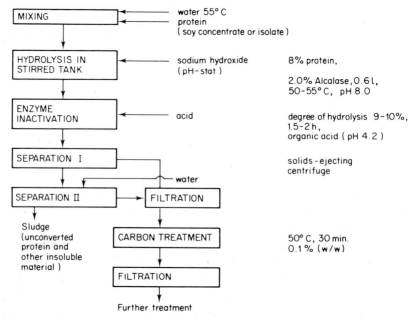

Fig. 20. Flow sheet for production of a non-bitter soluble soy protein hydrolysate suitable for incorporation into soft drinks and low-pH foods. The process parameters indicated are standard. For production purposes, the optimal parameters are established by experiment.

neutral protease is much less than that obtained with Subtilisin Carlsberg.

A method for preparation of a soluble hydrolysate from soybean protein is shown in Figure 20. The preparation obtained has no bitter taste, and it is completely soluble at pH 4–5. The yield of the product will depend on the raw material used, but it is usually close to 65%. The protein-containing residue may be hydrolysed again or used as feed. The amount of enzyme used is relatively high, about 300 p.p.m., calculated as parts of pure enzyme protein per part hydrolysed protein.

Another example of the use of enzyme hydrolysis is treatment of corn gluten, a protein with very little water retention. By treatment with proteases, it is possible to increase the water absorption of this protein significantly, so that a homogeneous suspension can be formed (Adler-Nissen, 1978).

H. Miscellaneous Applications

Throughout the ages, microbial proteases have been used for a variety of purposes of minor economic importance, such as recovery of photographic silver by hydrolysis of gelatin, degumming of natural silk, improvement of rubber recovery by digestion of undesired protein in latex skin, and various special cleaning processes such as cleaning of membranes for ultrafiltration or vital medical instruments such as dialysis equipment. One obvious application has never enjoyed commercial success, that is the application of proteases as a feed supplement. Extensive work has been done to test the value of enzymes as feed supplements, but there have not been any successful results with proteases, apparently because the proteolytic enzyme system of the digestive tract is so powerful that the relatively small amounts one can afford to add in the feed are insignificant. However, in the manufacture of feed, proteolytic enzymes have found some application in the preparation of milk substitutes for young animals. A useful, but yet macabre, application is the use of proteases to dissolve tissue in the coroner's laboratory in order to improve the possibilities for detection of intoxication or poisoning of dead bodies. Numerous other small applications of proteases could be mentioned. The microbial proteases are a very convenient tool whenever removal or degradation of proteins is needed.

I. Medical Applications

Although it is not intended that this chapter should cover the proteases used in medicine, it seems appropriate to mention the most important enzymes and their application. Digestive-aid preparations are widely used. They contain pancreatic enzymes, often supplemented by proteases and amylases from *Aspergillus* spp. Streptokinase produced from *Streptococcus pyogenes* is widely used in the treatment of thrombosis. The enzyme functions by activating plasminogen of the blood. Similar preparations have been developed from *A. oryzae*. Proteolytic enzymes have been widely studied for use as debridement agents, and a number of micro-organisms have been used as enzyme source, e.g. *Serratia marcescens*. There is some debate about the value of proteolytic enzymes for this purpose.

VII. SAFETY ASPECTS

Proteolytic enzymes are natural catalysts, the commercial products are nontoxic, and they are produced by harmless micro-organisms. When handled properly, they are without risk to user or consumer. Yet their use in detergents did arouse serious public concern in 1969–70, when it was reported that some workers in a detergent factory had contracted a pulmonary disease caused by proteolytic enzymes. The incidence was regarded of national significance in the U.S.A., and a committee under the auspices of the National Research Council studied all possible hazards derived from the use of proteases in detergents. From the report, which was published in 1971, it appears that the risk involved in use of detergent enzymes is insignificant (PB 204 118, 1971). The disease was caused by inhalation of high concentrations of enzyme dust in combination with detergent, which intensified the activity of enzyme on lung tissue. The manufacturers took immediate steps to lower the dust level. The Soap and Detergent Industry Association issued 'Recommended Operating Procedures for U.K. Factories Handling Enzyme Materials' (1971), and the enzyme manufacturers developed encapsulated dust-free enzymes. As a consequence, the health risks caused by enzymes in the factories were eliminated.

From this, it can be deduced that proteases in large concentrations, especially in the form of dust, may present a risk. It is obvious that the microbial proteases will attack moist skins and mucous membranes if present in high concentration and for a sufficient time. This will, however, only cause a passing discomfort and may easily be avoided by proper hygiene. The allergic reactions also caused by enzymes are more unpleasant and difficult to foresee. That the enzymes are allergens is not surprising in view of their proteinaceous nature. All proteins have allergenic properties and are especially potent when inhaled. Bacillus proteases are not particularly potent antigens when compared with many other proteins. Fungal proteases are somewhat more powerful antigens but, since they are solely used in industries where proper precautions may be taken, no allergy problems appear to be connected with their use.

Since the commercial enzymes are produced by organisms that are common inhabitants of our surroundings, a few people may become sensitive to enzymes without having been in contact with them. Farm workers may, for example, become sensitive to species of *Aspergillus* or *Bacillus* and their metabolic products through their work in the stable or with hay, and they may react when confronted with enzymes produced by micro-organisms to which they have been sensitized. Such incidences are, however, rare exceptions and their effects will probably not be noticed unless the individuals happen to come in contact with concentrated enzyme preparations. In order to minimize any risk for allergic reactions, formation of enzyme dust should be avoided by using liquid or encapsulated enzyme concentrates whenever possible, and good personal hygiene should be exercised when protease concentrates are handled.

VIII. CONCLUSION

Proteolytic enzymes find many useful applications, and for the enzyme industry they are of great economic importance. Yet when one looks at the variation in the proteolytic enzymes formed by micro-organisms and the amount of work devoted to the study of these enzymes, it is surprising that virtually no important new applications for proteolytic enzymes have been developed in well over 50 years. It is true that there have been improvements in methods and sources of enzymes, but the basic ideas in the application of these enzymes are exactly the same as they were. Does this mean that the brilliant results of modern biochemistry and microbiology only lead to academic satisfaction and not to technical advances or can we expect new exciting developments? Time will show.

X. ACKNOWLEDGEMENTS

C. Dambmann, F. S. Markland, T. K. Nielsen and Novo Industri A/S are thanked for permission to reproduce figures.

REFERENCES

Adler-Nissen, J. (1978). United States Patent 4,100,151.
Aiyappa, P. S. and Lampen, J. O. (1977). *Journal of Biological Chemistry* **252**, 1745.
Anson, M. L. (1939). *Journal of General Physiology* **22**, 79.
Arima, K. (1964). United States Patent 3,151,039.
Aunstrup, K. (1968). British Patent 1,108,287.
Aunstrup, K. and Outtrup, H. (1973). British Patent 1,303, 633.
Aunstrup, K., Outtrup, H., Andresen, O. and Dambmann, C. (1972). 'Fermentation Technology Today' (G. Terui, ed.), p. 299. Kyoto, Japan.
Aunstrup, K., Outtrup, H. and Andresen, O. (1973). United States Patent 3,723, 250.
Bergkvist, R. (1966). United States Patent 3,281,331.
Bergey's Manual of Determinative Bacteriology (1974). (R. E. Buchanen and N. E. Gibbons, eds.), 8th edition. Williams and Wilkins, Baltimore.
Bernlohr, R. W. (1964). *Journal of Biological Chemistry* **239**, 538.
Bernlohr, R. W. and Novelli, G. D. (1963). *Archives of Biochemistry and Biophysics* **103**, 94.
Bernlohr, R. W. and Clark, V. (1971). *Journal of Bacteriology* **105**, 276.
Boyer, P. D. ed. (1976). 'The Enzymes', vol. 3, 3rd edition. Academic Press, New York.
Boyer, P. D., Lardy, H. and Myrbäck, K. eds. (1960). 'The Enzymes', vol. 4, 2nd edition. Academic Press, New York.
British Standard (1963). Number 3624.
Boidin, A. and Effront, J. (1917). United States Patent 1,227,525.
Charney, J. and Tomarelly, R. M. (1947). *Journal of Biological Chemistry* **177**, 501.
Churchill, B. W., Steel, D. R. and Buss, D. R. (1973). United States Patent 3,740,318.
Cohen, R. L. (1977). *In* 'Genetics and Physiology of Aspergillus' (J. E. Smith and J. A. Pateman, eds.), p. 280. Academic Press, London.
Dambmann, C., Holm, P., Jensen, V. and Nielsen, M. H. (1971). *Developments in Industrial Microbiology* **12**, 11.
Damodaran, M., Govindarajan, V. S. and Subramanian, S. S. (1955). *Biochimica et Biophysica Acta* **17**, 99.
Detmar, D. A. and Vogels, R. J. (1971). *Journal of the American Oil Chemist's Society* **48**, 77.
Endo, S. (1962). *Journal of Fermentation Technology* **40**, 346.
Erickson, R. J. (1976). *In* 'Microbiology' (D. Schlessinger, ed.), p. 406, American Society for Microbiology, Washington.
Feldman, L. I. (1971). United States Patent 3,623,957.
Green, G. H. (1952). *Journal of the Society of the Leather Trade Chemists* **36**, 127.

Güntelberg, A. V. (1954). *Comptes Rendus des Traveaux du Laboratoire. Carlsberg, Series Chimie* **29**, 27.

Güntelberg, A. V. and Ottesen, M. (1952). *Nature, London* **170**, 802.

Hagihara, B. (1954). *Annual Reports of the Scientific Works of the Faculty of Science of Osaka University* **2**, 35.

Hall, F. F., Kunkel, H. O. and Prescott, J. M. (1966). *Archives of Biochemistry and Biophysics* **114**, 145.

Harris, R. G. and Rose, T. J. (1972). French Patent 2,182,638.

Hoegh, S. (1978). M.Sc. Thesis: Technical University of Denmark.

Horikoshi, K. and Ikeda, Y. (1977). United States Patent 4,052,262.

Ichishima, E. (1970). *Methods in Enzymology* **19**, 397.

Jensen, D. E. (1972). *Biotechnology and Bioengineering* **14**, 647.

Kalabokias, G. (1971). United States Patent 3,623,956.

Keay, L. (1972). *In* 'Fermentation Technology Today' (G. Terui, ed.). p. 289. Kyoto, Japan.

Keay, L., Moser, P. W. and Bernard, S. W. (1970). *Biotechnology and Bioengineering* **12**, 213.

Keay, L., Moseley, M. H., Anderson, R. G., O'Connor, R. J. and Wildi, B.S. (1972). *Biotechnology and Bioengineering Symposium No. 3*, p. 63.

Klapper, B. F., Jameson, D. M. and Mayer, R. M. (1973). *Biochimica et Biophysica Acta* **304**, 505.

Koaze, Y., Goi, H. and Hara, T. (1970). United States Patent 3,492, 204.

Küntzel, A. (1955). *In* 'Gerbereichemisches Taschenbuch', 6 edn., p. 85. Dresden.

Lehmann, R., Pfeiffer, H. F., Schindler, J. and Schreiber, W. (1977). United States Patent 4,062,732.

Lin, Y., Means, G. E. and Teeney, R. E. (1969). *Journal of Biological Chemistry* **244**, 789.

Lorand, L. ed. (1976). *Methods in Enzymology*, **45**, 1.

Madsen, G. B., Norman, B. E. and Slott, S. (1973). *Die Stärke* **25**, 304.

Mäkinen, K. K. (1970). *International Journal of Protein Research* **4**, 21.

Markkanen, P. H. and Bailey, M. J. (1974). *Journal of Applied Chemistry and Biotechnology* **24**, 93.

Markland, F. S. and Smith, E. L. (1971). *In* 'The Enzymes' (P. D. Boyer, ed.), volume 3, p. 568. Academic Press, New York.

Matsubara, H. and Feder, J. (1971). *In* 'The Enzymes' (P. D. Boyer, ed.), volume 3, p. 721. Academic Press, New York.

Monsheimer, R. and Pfleiderer, E. (1976). United States Patent 3,966,551.

Murray, E. D. and Prince, M. P. (1970). United States Patent 3,507,750.

Nakadai, T. and Nasuno, S. (1977). *Journal of Fermentation Technology* **55**, 273.

Nakadai, T., Nasuno, S. and Iguchi, N. (1973). *Agricultural and Biological Chemistry* **37**, 2703.

Nasuno, S. (1972). *Journal of General Microbiology* **70**, 29.

Nijenhuis, B.te (1977). United States Patent 4,002,572.

Ottesen, M. and Spector, A. (1960). *Comptes Rendus des Traveaux du Laboratoire Carlsberg* **32**, 63.

Ottesen, M. and Svendsen, I. (1970). *Methods in Enzymology* **19**, 199.

Outtrup, H. and Aunstrup, K. (1975). *Proceedings of the First International Congress of International Association of Microbiological Societies, Tokyo 1974* **5**, 204.

PB 204118 (1971). *Report of the* ad hoc *Committee on Enzyme Detergents. Enzyme-containing Laundering Compounds and Consumer Health.*

Pozsgay, M., Gaspar, R. and Elödi, P. (1977). *Federation of European Biochemical Societies Letters* **74**, 67.

Priest, F. G. 1977). *Bacteriological Reviews* **41**, 711.

Raper, K. B. and Fennel, D. I. (1965). 'The Genus Aspergillus.' Williams and Wilkins, Baltimore.

Reed, G. (1966). *In* 'Enzymes in Food Processing' (G. Reed, ed.), p. 238. Academic Press, New York.

Sardinas, J. L. (1972). *Advances in Applied Microbiology* **15**, 39.

Sidler, W. and Zuber, H. (1977). *European Journal of Applied Microbiology* **4**, 255.

Smith, E. L., De Lange, R. J., Evans, W. H., Landon, M. and Markland, F. S. (1968). *Journal of Biological Chemistry* **243**, 2184.

The Soap and Detergent Industry Association. (1971). *Annals Occupational Hygiene* **14**, 71.

Somkuti, G. A. and Baloil, F. J. (1968). *Journal of Bacteriology* **95**, 1407.

Takamine, J. (1894). United States Patent 525,820.

Titani, K., Hermodson, M. A., Ericsson, L. H., Walsh, K. A. and Neurath, H. (1972). *Nature, London* **238**, 35.

Tsujita, S. and Endo, A. (1977). *Journal of Bacteriology* **130**, 48.

Vedder, A. (1934). *Antonie van Leeuwenhoek* **1**, 141.

Verbruggen, R. (1975). *Biochemical Journal* **151**, 149.

Viccaro, J. P. (1973). United States Patent 3,748,233.

Walker, P. D. and Wolf, J. (1971). *In* 'Spore Research' (A. N. Barker, G. W. Gould and J. Wolf, eds.), p. 247. Academic Press, London.

Wouters, J. T. M. and Buysman, P. J. (1977). *Federation of European Microbiological Societies Microbiology Letters* **1**, 109.

Yoshida, F. (1956). *Bulletin of the Agricultural Chemical Society of Japan* **20**, 252.

Zuidweg, M. H. J., Bos, C. J. K. and van Welzen, H. (1972). *Biotechnology and Bioengineering* **14**, 685.

3. Amylases, Amyloglucosidases and Related Glucanases

WILLIAM M. FOGARTY AND CATHERINE T. KELLY

Department of Industrial Microbiology, University College, Dublin 4, Ireland

I. INTRODUCTION

Starch-degrading enzymes are distributed widely in the microbial, plant and animal kingdoms. They degrade starch and related polymers to yield products characteristic of individual amylolytic enzymes. Initially, the term amylase was used to describe enzymes hydrolysing α-1,4-glucosidic bonds of amylose, amylopectin, glycogen and related products (Greenwood and Milne, 1968b; Fischer and Stein, 1960; Meyer, 1952). In recent years, a number of new starch-degrading enzymes have been detected in micro-organisms (Boyer and Ingle, 1972; Buonocore *et al.*, 1976; W. M. Fogarty, E, Bourke and M. M. Dixon, unpublished observations; Fogarty and Griffin, 1975; Hasegawa *et al.*, 1976; Kainuma *et al.*, 1975; Kitahata and Okada, 1974; McWethy and Hartman, 1977; Nakamura and Horikoshi, 1976; Robyt and Ackerman, 1971; Saito, 1973; Suzuki *et al.*, 1976; Takasaki, 1976a; Taylor *et al.*, 1978; Urlaub and Wöber, 1978; Wang and Hartman, 1976; Yuhki *et al.*, 1977). The commercially significant amylolytic enzymes of microbial origin may be divided into the following classes.

1. Exo-Acting Amylases

a. *Amyloglucosidases (glucoamylases)*. These enzymes hydrolyse α-1,4 and α-1,6 linkages, and produce glucose as the sole end-product from starch and related polymers.

b. *β-Amylases*. These enzymes hydrolyse α-1,4 bonds but cannot bypass α-1,6 linkages in amylopectin and glycogen. They produce maltose from amylose, and maltose and a β-limit dextrin from amylopectin and glycogen.

c. *Other exo-acting enzymes*. These enzymes hydrolyse α-1,4 bonds and cannot bypass α-1,6 linkages and produce end-products other than maltose from starch substrates, e.g. the enzymes from *Pseudomonas stutzeri* (Robyt and Ackerman, 1971) and *Aerobacter aerogenes* (Kainuma *et al.*, 1972, 1975).

2. Endo-Acting Amylases

α-Amylases; these enzymes hydrolyse α-1,4 bonds and bypass α-1,6 linkages in amylopectin and glycogen.

3. Debranching Enzymes

These enzymes hydrolyse only α-1,6 linkages.

4. Cyclodextrin-Producing Enzymes

These enzymes hydrolyse starch to a series of non-reducing cyclic D-glucosyl polymers called cyclodextrins or Schardinger dextrins, e.g. the amylase from *Bacillus macerans* (Lane and Pirt, 1971; Schwimmer and Garibaldi, 1952; Thoma *et al.*, 1970).

II. THE SUBSTRATE: STARCH

A. Occurrence

Detailed reports on the properties and chemistry of starch have been published in a number of reviews (Banks and Greenwood, 1975; Greenwood, 1976; Manners, 1974; Radley, 1976). Starch occurs in the form of water-insoluble granules as the major reserve carbohydrate in all higher plants. It is produced commercially from the seeds, tubers and roots of plants. The major source of starch is corn (known as maize in Europe), from which it is extracted by a wet-milling process (Berkhout, 1976).

B. Composition

Starch is a heterogeneous polysaccharide composed of two high molecular-weight components, called amylose and amylopectin. These two polymers differ significantly in many physical properties (Table 1), notably molecular size, solubility in water, iodine-staining capacity

Table 1

Comparison of some properties of amylose and amylopectin

Property	Amylose	Amylopectin
Basic structure	Linear	Branched
Stability in aqueous solution	Retrogrades	Stable
Degree of polymerization	About 10^3	10^4–10^5
Average chain length	About 10^3	20–25
Iodine complex (λ_{max} nm)	650	550

and susceptibility to enzymic hydrolysis. Amylose is a linear molecule comprised of chains of α-1,4-linked D-glucose residues, whereas amylopectin is a highly branched polymer. Branching occurs through α-1,6 linkages. Amylose generally accounts for 20–25% of starch, depending on the source of the polymer. Starch may be separated into its two components by addition of a polar solvent, e.g. butan-1-ol, to a dispersion of starch (Greenwood, 1956). Methods for the detection and

Table 2

Sources of amyloglucosidases

Organism	Reference
Aspergillus awamori	Attia and Ali (1977); Hayashida et al. (1976); Qadeer and Kausar (1971); Watanabe and Fukimbara (1965); Yamasaki et al. (1977c).
Aspergillus batatae	Bendetskii et al. (1974).
Aspergillus cinnamomeus	Kurushima et al. (1974).
Aspergillus clavatus	Voitkova-Lepshikova and Kotskova-Kratokhvilova (1966).
Aspergillus foetidus	Bode (1966b).
Aspergillus niger	Barton et al. (1972); Freedberg et al. (1975); Lineback et al. (1969); Pazur and Ando (1959); Pazur et al. (1971).
Aspergillus oryzae	Miah and Ueda (1977); Morita et al. (1966).
Aspergillus phoenicis	Lineback and Baumann (1970).
Cephalosporium charticola Lindau	Krzechowsk and Urbanek (1975).
Cephalosporium eichhorniae	Day (1978).
Coniophora cerebella	King (1967).
Endomyces sp.	Fukui and Nikuni (1969); Hattori and Takeuchi (1962).
Endomycopsis capsularis	Ebertová (1966).
Humicola lanuginosa	Taylor et al. (1978).
Mucor rouxianus	Yamasaki et al. (1977d).
Neurospora crassa	Murayama and Ishikawa (1973).
Neurospora sitophila	W. M. Fogarty and M. M. Dixon (unpublished observations).
Penicillium oxalicum	Yamasaki et al. (1977a, b).
Piricularia oryzae	Yuhki et al. (1977).
Rhizopus delemar	Hamauzu et al. (1965); Pazur and Okada (1967).
Rhizopus javanicus	Watanabe and Fukimbara (1973).
Rhizopus niveus	Adachi et al. (1977); Ohnishi and Hiromi (1978); Ohnishi et al. (1976).
Saccharomyces diastaticus	Hopkins and Kulka (1957).
Torula thermophilia	Subrahmanyam et al. (1977).
Trichoderma viride	Schellart et al. (1976a, b).

assay of amylolytic enzymes are based on the use of starch or a starch derivative as a substrate. These methods have been outlined recently (Fogarty and Kelly, 1979a).

III. DISTRIBUTION, PROPERTIES AND MODE OF ACTION OF AMYLASES

A. Amyloglucosidase

1. Distribution and Substrate Specificity

Amyloglucosidase (EC 3.2.1.3, glucoamylase, γ-amylase) is an exo-acting enzyme that yields β-D-glucose (Ono *et al.*, 1965) by hydrolysing α-1,4 linkages consecutively from the non-reducing chain ends of amylose, amylopectin and glycogen. It also hydrolyses α-1,6 and α-1,3 linkages (Pazur and Kleppe, 1962). Amyloglucosidases occur almost exclusively in fungi (Table 2). Rates of substrate hydrolysis are affected by molecular size and structure, and by the next bond in sequence (Abdullah *et al.*, 1963; Fleming, 1968). Thus, the rate of hydrolysis of α-1,4 bonds increases with the molecular weight of the substrate (Fukui and Nikuni, 1969; Krzechowsk and Urbanek, 1975; Smiley *et al.*, 1971; Yamasaki *et al.*, 1977b, c). Similarly, in hydrolysis of α-1,6 linkages, the size of the substrate and the position of the linkage play a significant role in their susceptibility to hydrolysis. Thus, Phillips and Caldwell (1951) observed a 95% and a 92% conversion of corn starch and glycogen, respectively, and no hydrolysis of dextran or isomaltose by the enzyme from *Rhizopus delemar*. The optimum pH value curves of the enzymes from *A. niger* hydrolysing maltose and isomaltose were similar (Pazur and Ando, 1960; Pazur and Kleppe, 1962). Amyloglucosidases of *A. awamori* var. *kawachi* (Ueda *et al.*, 1974) and of *Rhizopus* sp. (Ueda and Kano, 1975) hydrolysed soluble waxy corn starch completely. Amylopectin, glycogen and soluble starch were completely hydrolysed by the enzymes from *M. rouxianus* (Yamasaki *et al.*, 1977d), *P. oxalicum* (Yamasaki *et al.*, 1977b) and *A. awamori* (Yamasaki *et al.*, 1977c) but, in no case, was isomaltose degraded. The enzyme from *A. niger* hydrolysed isomaltose at a much slower rate than maltose, but panose was hydrolysed at 70% of the rate for maltose (Abdullah *et al.*, 1963). Amyloglucosidase hydrolysed linear dextrans containing about 95% α-1,6 linkages, yet highly branched dextrans linked through α-1,2 or α-1,3 bonds were scarcely attacked. The linkages at such branch points in dextran may act as a barrier to the enzyme (Kobayashi and Matsuda, 1978).

Observing digestion of starch granules was facilitated by the use of electron-microscopic techniques (Shetty *et al.*, 1974; Smith and Lineback, 1976). Digestion of raw starch has been reported with a number of amyloglucosidases (Evers *et al.*, 1971; Hayashida, 1975; Manners, 1971a; Miah and Ueda, 1977; Shetty *et al.*, 1974; Smith and Lineback, 1976; Ueda *et al.*, 1974; Yamasaki *et al.*, 1977d). In certain instances, two or more forms of amyloglucosidase have been detected. Strong debranching activity and the ability to degrade raw starch are properties shared by amyloglucosidases I from *A. awamori* var. *kawachi* (Ueda *et al.*, 1974) and *Rhizopus* sp. (Ueda and Kano, 1975), whereas weak debranching activity and weak ability to hydrolyse raw starch are properties common to amyloglucosidases II from the same organisms.

Amyloglucosidases, I, I′ and II are produced in submerged culture by *A. awamori* var. *kawachi* (Hayashida, 1975). Type I hydrolysed raw starch and limit dextrin almost completely. Types I′ and II did not degrade raw starch, and hydrolysed limit dextrin 50% and less than 10%, respectively. Treating amyloglucosidase I with subtilisin (Hayashida *et al.*, 1976) or an acid protease (Hayashida and Yoshino, 1978) gave a modified enzyme unable to degrade raw starch. It was suggested that stepwise degradation of native amyloglucosidase I by proteases and glycosidases during the course of cultivation may account for multiple forms of the enzyme (Hayashida and Yoshino, 1978).

2. *Properties*

Structural characterization of amyloglucosidases I and II, from *A. niger* has been carried out (Fleming and Stone, 1965; Lineback and Aira, 1972; Lineback *et al.*, 1969; Pazur *et al.*, 1970, 1971; Russell and Lineback, 1970). Amyloglucosidases I and II have molecular weights of 74,900 and 54,300, and 13% and 18% carbohydrate, respectively (Lineback and Aira, 1972). The enzymes have different amino-acid compositions. Pazur *et al.* (1971) reported molecular weights of 99,000 and 112,000 for amyloglucosidases I and II from *A. niger*, respectively, and claimed that I and II possessed identical amino-acid compositions but that they differed in carbohydrate content. The enzymes from *A. phoenicis* (Lineback and Baumann, 1970), *R. delemar* (Pazur and Okada,

Table 3

Some physicochemical properties of amyloglucosidases

Organism	pH optimum	Temperature optimum (°C)	pH stability	Temperature stability (°C)	Molecular weight	Reference
Aspergillus awamori	4.5	60	5.0–9.0	Up to 50	83,700–88,000	Yamasaki et al. (1977c)
Aspergillus niger I	4.5–5.0			Up to 60	99,000	Lineback et al. (1969);
II	4.5–5.0			Up to 60	112,000	Pazur and Ando (1959)
Aspergillus oryzae I	4.5	60	4.0–7.0	Up to 40	76,000	
II	4.5	50	4.0–7.0	Up to 40	38,000	
III	4.5	40	4.0–7.0	Up to 40	38,000	Miah and Ueda (1977)
Cephalosporium charticola Lindau	5.4	60				
Endomycopsis capsularis	4.5	40–50	4.2–8.0		69,000	Krzechowsk and Urbanek (1975)
Mucor rouxianus I	4.6	55	4.0–8.0	Up to 50	59,000	Ebertová (1966)
II	5.0	55	4.0–7.5	Up to 50	49,000	Tsuboi et al. (1974);
Neurospora sitophila	5.0–6.5	50	5.0–9.0	Up to 40	23,000	Yamasaki et al. (1977d)
Penicillium oxalicum I	5.0	55–60	3.0–6.5	Up to 55	84,000	W. M. Fogarty and M. M. Dixon (unpublished observations)
II	4.5	60	3.0–6.5	Up to 55	86,000	Yamasaki et al. (1977a, b)
Piricularia oryzae	6.5	50–55	5.5–7.0	Up to 40	94,000	Yuhki et al. (1977)
Rhizopus delemar	4.5	40	4.5–6.3		100,000	Pazur and Okada (1967); Phillips and Caldwell (1951)

1967), *R. javanicus* (Watanabe, 1976; Watanabe and Fukimbara, 1973), *C. charticola* Lindau (Krzechowsk and Urbanek, 1975), *P. oxalicum* (Yamasaki *et al.*, 1977b) and *Endomycopsis* sp. 20–9 (Gracheva *et al.*, 1977) are reported to be glycoproteins. Amyloglucosidases of *A. phoenicis* (Lineback and Baumann, 1970) or *A. niger* (Lineback and Aira, 1972) do not appear to contain any amino sugars, whereas hexosamines were found in the enzymes from *R. javanicus* (Watanabe and Fukimbara, 1973) and *R. delemar* (Pazur *et al.*, 1971). The amino-acid composition of some amyloglucosidases has been determined (Fukui and Nikuni, 1969; Gracheva *et al.*, 1977; Pazur *et al.*, 1971; Lineback and Aira, 1972; Lineback and Baumann, 1970; Smiley *et al.*, 1971). They all contained methionine residues, except for the enzymes from *Endomyces* sp. (Fukui and Nikuni, 1969) and *Endomycopsis* sp. (Gracheva *et al.*, 1977). Tryptophan residues were present in all, except for the enzyme from *A. phoenicis* (Lineback and Baumann, 1970), but its presence was not determined in *A. niger* (Lineback and Aira, 1972). All enzymes contained half-cystine, except for that from *Endomycopsis* sp. (Gracheva *et al.*, 1977). The pH optima of these enzymes are generally in the range 4.5–5.0 (Ebertová, 1966; Kaji *et al.*, 1976; Lineback and Baumann, 1970; Miah and Ueda, 1977; Tsuboi *et al.*, 1974; Yamasaki *et al.*, 1977a, b, c, d; Table 3). Exceptions include amyloglucosidase II from the thermophilic fungus *Humicola lanuginosa* (Taylor *et al.*, 1978) which has an optimum pH value of 6.6. Usually, the enzymes are stable on the acid side of pH 7.0 (Hayashida *et al.*, 1976; Miah and Ueda, 1977; Ueda *et al.*, 1974). However, amyloglucosidase II from *H. lanuginosa* is stable up to pH 11.0 (Taylor *et al.*, 1978) and the enzymes from *C. rolfsii* (Kaji *et al.*, 1976) and *C. cerebella* (King, 1967) are stable up to 9.0. The temperature optima of most of these enzymes are in the range 40–60°C (Ebertová, 1966; Kaji *et al.*, 1976; Miah and Ueda, 1977; Tsuboi *et al.*, 1974), although amyloglucosidase II from *H. lanuginosa* has a temperature optimum between 65°C and 70°C (Taylor *et al.*, 1978).

Most amyloglucosidases are stable up to 40°C (Fukui and Nikuni, 1969; Miah and Ueda, 1977; Morita *et al.*, 1966; Ohga *et al.*, 1966). However, amyloglucosidase II from *H. lanuginosa* (Taylor *et al.*, 1978) appears to be more thermostable and retains 50% of its activity at 70°C, while the amyloglucosidase from another thermophilic fungus, *Torula thermophilia* (Subrahmanyam *et al.*, 1977), was stable for one hour at 60°C and retained 60% of its activity at 70°C. The iso-electric points range from 3.55 for *A. awamori* var. *kawachi* amyloglucosidase (Hayashida *et al.*, 1976; Ueda *et al.*, 1974) to 8.4 for the enzyme from *M. rouxianus* (Tsuboi *et al.*, 1974; Yamasaki *et al.*, 1977d).

Purified preparations of amyloglucosidase are capable of polymerizing glucose. This reverse reaction is accompanied by elimination of water (Underkofler *et al.*, 1965; Hehre *et al.*, 1969) and formation of maltose, isomaltose and other oligosaccharides. Although synthesis of maltose is very rapid, formation of the other saccharides occurs at a considerably lower rate. The amounts of these sugars formed are dependent on the substrate concentration.

The enzyme transglucosidase often occurs as an impurity in crude preparations of amyloglucosidase. It leads to formation of α-1,6-linked, non-fermentable oligosaccharides (Pazur and Ando, 1961). Thus, significant amounts of panose and isomaltose may be synthesized resulting in lower conversion of substrates into glucose.

B. β-Amylase

1. Distribution and Occurrence

β-Amylase (EC 3.2.1.2, α-1,4-glucan maltohydrolase) occurs widely in higher plants (Meyer *et al.*, 1953; Piguet and Fischer, 1952). Its occurrence as an extracellular enzyme in micro-organisms (Table 4) was established recently (W. M. Fogarty and E. Bourke, unpublished

Table 4

Distribution of microbial β-amylases

Organism	Reference
Bacillus cereus var. *mycoides*	Takasaki (1976a, b)
Bacillus megaterium	Higashihara and Okada (1974); Yamane and Tsukano (1977); W. M. Fogarty and M. T. O'Reilly (unpublished observations)
Bacillus polymyxa	Fogarty and Griffin (1973, 1975); Griffin and Fogarty (1973a, b, c)
Bacillus sp. BQ10	Shinke *et al.* (1974, 1975a, b)
Bacillus sp. IMD 198	W. M. Fogarty and E. Bourke (unpublished observations)
Pseudomonas sp. BQ6	Shinke *et al.* (1975a, c)
Streptomyces spp.	Shinke *et al.* (1974, 1975a); Koaze *et al.* (1974, 1975)

observations; Fogarty and Griffin, 1973, 1975; W. M. Fogarty and M. T. O'Reilly, unpublished work; Griffin and Fogarty, 1973a, b, c; Higashihara and Okada, 1974; Shinke *et al.*, 1974; Takasaki, 1976b). The enzyme degrades amylose, amylopectin and glycogen in an exo-fashion, from the non-reducing chain ends, by hydrolysing alternate glucosidic linkages. Thus, it produces the β-anomeric form of maltose. The enzyme is incapable of bypassing α-1,6-glucosidic linkages in branched polymers and degradation is therefore incomplete. Hydrolysis of amylopectin results in about 55% conversion into maltose with formation of a β-limit dextrin.

Plant β-amylases are thiol enzymes (Meyer *et al.*, 1953; Rowe and Weill, 1962) with no obvious requirement for metal ions. They are inactivated by thiol-reagents such as p-chloromercuribenzoate, and by oxidation. Oxidation occurs in dilute solution and is overcome by the presence of serum albumin and reduced glutathione (Walker and Whelan, 1960b). Schardinger dextrins competitively inhibit β-amylases (Thoma and Koshland, 1960).

2. Microbial β-Amylases

The amylase system produced by *Bacillus polymyxa* has been studied by a number of workers (Kneen and Beckord, 1946; Robyt and French, 1964; Rose, 1948; Tilden and Hudson, 1942). Originally it was considered to possess an endo-mechanism of attack in that it could

Table 5

Properties of some microbial β-amylases

Organism	Molecular weight	Optimum pH value	Temperature optimum (°C)	Reference
Bacillus cereus var.				
mycoides	35,000	7.0	50	Takasaki (1976b)
Bacillus megaterium	35,000	6.0	50	W. M. Fogarty and M. T. O'Reilly (unpublished observations)
Bacillus megaterium No. 32	—	6.5	—	Higashihara and Okada (1974)
Bacillus polymyxa	59,000	6.8	37	Fogarty and Griffin (1975)
Bacillus sp. BQ 10	160,000 80,000	6.0–7.0	45–55	Shinke *et al.* (1975b)
Bacillus sp. IMD 198	58,000	6.8	55	W. M. Fogarty and E. Bourke (unpublished observations)
Pseudomonas sp. BQ6	37,000	6.5–7.5	45–55	Shinke *et al.* (1975c)
Streptomyces sp.	—	6.7	40	Shinke *et al.* (1974)

either break or bypass α-1,6 linkages in amylopectin or starch and produce maltose in yields as high as 92–94% (Robyt and French, 1964). These data appeared to indicate that this enzyme contained properties of both α-amylase and β-amylase. Preliminary studies in this laboratory dealt with production of the enzyme (Fogarty and Griffin, 1973; Griffin and Fogarty, 1973a). Subsequently, the system was shown to consist of two components, a debranching enzyme specific for α-1,6 linkages of amylopectin and a β-amylase (Griffin and Fogarty, 1973b, c). The presence of these two activities explained the very high conversion of substrate into maltose with the original crude system and also established conclusively the existence of β-amylase in a micro-organism. Hydrolysis of amylopectin by this enzyme produced β-maltose and a limit dextrin similar to that produced by sweet-potato β-amylase. The enzyme hydrolysed maltotetraose but not maltotriose. Many of its properties are similar to typical plant β-amylases (Fogarty and Griffin, 1975). It was inhibited by Schardinger dextrins, p-chloromercuribenzoate and N-bromosuccinimide. The optimum pH value was 6.8 (Table 5) which is higher than that recorded for plant β-amylases (Balls et al., 1948; Piguet and Fischer, 1952). The molecular weight of the enzyme was 59,000 (W. M. Fogarty and E. Bourke, unpublished observations). Inhibition of the enzyme by p-chloromercuribenzoate could be reversed by addition of cysteine and mercaptoethanol. It was competitively inhibited by Schardinger dextrins (W. M. Fogarty and E. Bourke, unpublished observations). Although Marshall (1974) reported some of these observations, his suggestion that this bacterial β-amylase is not a thiol enzyme is inconsistent with observations from this laboratory.

β-Amylase was isolated recently from another source, namely Bacillus sp. IMD 198 (W. M. Fogarty and E. Bourke, unpublished observations). The enzyme had a molecular weight of 58,000 (Table 5). The limit dextrin produced by this enzyme from amylopectin was not further hydrolysed by sweet-potato β-amylase. It degraded maltotetraose, and was inhibited by thiol-reagents but, significantly, it was not inhibited by Schardinger dextrins.

The extracellular β-amylase of Bacillus megaterium No. 32 was purified (Higashihara and Okada, 1974) and the action pattern of the enzyme studied by an oligosaccharide-mapping procedure using maltodextrins as substrate (Pazur and Okada, 1966). The enzyme degraded substrates from the non-reducing ends and released maltose. Enzyme activity was effectively inhibited by *p*-chloromercuribenzoate but activity could be restored by addition of excess amounts of cysteine (Higashihara and Okada, 1974). *Bacillus cereus* var. *mycoides* β-amylase (Takasaki, 1976a, b) was purified to homogeneity by disc electrophoresis. Maltose produced from starch had the β-configuration and α-1,6 linkages in β-limit dextrin and pullulan were not hydrolysed. It was inhibited by heavy-metal ions and by *p*-chloromercuribenzoate. Inhibition by the latter compound could be reversed by cysteine. The molecular weight of the enzyme was 35,000 ± 5,000. *Pseudomonas* sp. BQ6 β-amylase was reported to have a molecular weight of 37,000 (Shinke *et al.*, 1975c) and many of its properties were similar to those of the enzyme from *B. cereus* var. *mycoides. Bacillus* sp. BQ 10 (Shinke *et al.*, 1975b) β-amylase was resolved into three fractions by gel filtration on Sephadex G-75. The temperature and pH optima of fractions I and II were 45–55°C and 6.0–7.0, respectively. It was reported that fraction I had slightly greater heat stability than fraction II and their molecular weights were 160,000 and 80,000, respectively. The activity of fraction III was too low to permit experimentation on its properties.

C. Other Microbial Exo-Acting Amylases

Recently, exo-acting amylases producing four- and six-carbon saccharides as end-products have been detected in microbial cultures. *Pseudomonas stutzeri* (Robyt and Ackerman, 1971) synthesizes an extracellular maltotetraose-producing amylase which releases large amounts of the tetrasaccharide from linear and branched substrates. Its action on glycogen and amylopectin gives maltotetraose and a high molecular-weight limit dextrin. It does not degrade Schardinger dextrins, has an exo-mechanism of substrate attack and hydrolyses the fourth α-1,4-glucosidic bond from the non-reducing end of susceptible substrates. The enzyme had a broad pH optimum in the range pH 6.5–10.5, and the optimum temperature for activity was 47°C. A number of different species of the enzyme were observed, with molecular weights ranging from 12,500 to 125,000. The two forms having molecular weights of 25,000 and 50,000 occurred in greatest amounts.

Aerobacter aerogenes synthesizes a maltohexaose-producing amylase (Kainuma *et al.*, 1972, 1975) which can be removed from cells with sodium dodecyl sulphate. This enzyme produces maltohexaose from

Table 6

Occurrence of some microbial α-amylases

Organism	Reference
Bacillus acidocaldarius	Buonocore *et al.* (1976)
Bacillus amyloliquefaciens	Welker and Campbell (1967c)
Bacillus caldolyticus	Grootegoed *et al.* (1973); Heinen and Heinen (1972)
Bacillus coagulans	Bliesmer and Hartman (1973)
Bacillus licheniformis	Meers (1972); Saito and Yamamoto (1975)
Bacillus stearothermophilus	Manning and Campbell (1961); Ogasahara *et al.* (1970)
Bacillus subtilis	Bailey and Markkanen (1975); Kokubu *et al.* (1978); Matsuzaki *et al.* (1974a, b)
Bacillus subtilis var. *amylosacchariticus*	Fujimori *et al.* (1978); Matsuno *et al.* (1978); Toda and Narita (1968)
Bacillus spp. (alkalophilic)	Boyer and Ingle (1972); Horikoshi (1971a); Yamamoto *et al.* (1972)
Bacteroides amylophilus	McWethy and Hartman (1977)
Clostridium acetobutylicum	Ensley *et al.* (1975)
Streptomyces aureofaciens	Hostinová and Zelinka (1978)
Thermoactinomyces vulgaris	Allen and Hartman (1972)
Thermomonospora curvata	Stutzenberger and Carnell (1977)
Thermomonospora vulgaris	Allam *et al.* (1977)
Thermomonospora viridis	Upton and Fogarty (1977)
Acinetobacter sp.	Onishi and Hidaka (1978)
Thermophile V-2	Hasegawa *et al.* (1976)
Pseudomonas saccharophila	Markowitz *et al.* (1956)
Aspergillus awamori	Watanabe and Fukimbara (1967)
Aspergillus batatae	Bendetskii *et al.* (1974)
Aspergillus foetidus	Hang and Woodams (1977)
Aspergillus niger	Aski *et al.* (1971)
Aspergillus oryzae	Vallier *et al.* (1977); Yabuki *et al.* (1977); Bata *et al.* (1978)
Aspergillus terricola	Aravina and Ponomarera (1977)
Mucor miehei	Adams and Deploey (1976)
Mucor pusillus	Adams and Deploey (1976)
Neurospora crassa	Gratzner (1972)
Penicillium expansum	Belloc *et al.* (1975)

starch, amylose and amylopectin. It was not active on Schardinger dextrins, pullulan or maltohexaitol and had an exo-mechanism of substrate attack. An interesting property of this enzyme is that β-limit dextrins are degraded to yield branched oligosaccharides. However, it does not hydrolyse amylopectin totally to low molecular-weight oligosaccharides. The optimum pH value for activity was 6.8 and the temperature optimum was 50°C. The enzyme was greatly inhibited by heavy metals. Only slight inactivation was produced by *p*-chloromercuribenzoate.

D. α-Amylase

1. *Distribution and Action on Substrates*

α-Amylase (α-1,4-glucan 4-glucanohydrolase, EC 3.2.1.1, endo-amylase) occurs widely in micro-organisms (Table 6). It hydrolyses α-1,4-glucosidic bonds in amylose, amylopectin and glycogen in an endo-fashion, but the α-1,6-glucosidic linkages in the branched polymers are not hydrolysed. The properties and mechanisms of action depend on the source of the enzyme. They are all endo-acting enzymes, and this characteristic causes a rapid decrease in iodine-staining power together with a rapid decrease in the viscosity of starch solutions. Hydrolysis of amylose by α-amylase effects its complete conversion into maltose and maltotriose, initially. Hydrolysis of maltotriose, which is a poor substrate, follows (Walker and Whelan, 1960a). Hydrolysis of amylopectin by α-amylase also yields glucose and maltose in addition to a series of branched α-limit dextrins. These dextrins of four or more glucose residues contain all of the α-1,6-glucosidic linkages of the original structure. With amylopectin or glycogen, the second stage of α-amylase degradation involves slow hydrolysis of maltotriose as well as slow hydrolysis of specific bonds near the branch points of the α-limit dextrins. Different α-amylases produce different α-limit dextrins (Umeki and Yamamoto, 1975, 1977; Whelan, 1960). In addition to singly branched α-limit dextrins, it is highly probable that multiple-branched dextrins are produced by some α-amylases (French, 1960; Roberts and Whelan, 1960). Iso-maltose is not formed in these reactions (Manners and Marshall, 1971) because α-1,6 linkages are resistant to α-amylases and they confer some stability on certain α-1,4 linkages near the branch points.

α-Amylases from different sources have been purified and many have been crystallized. Examples of purified preparations include those of *B. subtilis* (Moseley and Keay, 1970; Yamamoto, 1955), *B. coagulans* (Campbell, 1954), *B. stearothermophilus* (Manning and Campbell, 1961; Ogasahara *et al.*, 1970; Pfueller and Elliott, 1969), *B. amyloliquefaciens* (Welker and Campbell, 1967c), *B. licheniformis* (Saito, 1973), *Pseudomonas saccharophila* (Markowitz *et al.*, 1956), *Aspergillus oryzae* (Akabori *et al.*, 1954; Fischer and De Montmollin, 1951a, b; Toda and Akabori, 1963; Underkofler and Roy, 1951) and *A. niger* (De Song Chong and Tsujisaka, 1976; Minoda and Yamada, 1963). The homogeneity of most crystalline α-amylases has been confirmed by various techniques. For example, polyacrylamide-gel electrophoresis has been used to demonstrate homogeneity of the crystalline α-amylases of *B. amyloliquefaciens* (Welker and Campbell, 1967c), *B. stearothermophilus* (Pfueller and Elliott, 1969), and *A. oryzae* (McKelvy and Lee, 1969). Electrophoresis of purified α-amylase of *B. licheniformis* (Saito, 1973) indicated that it contained four active protein bands although the enzyme behaved as a single peak during ultracentrifugation.

The bacilli produce saccharifying and liquefying α-amylases (Fukumoto, 1963; Welker and Campbell, 1967a, b). In this review, we have retained the terminology used in original publications and the following are therefore synonymous: (1). *Bacillus subtilis* liquefying α-amylase, *Bacillus subtilis* var. *amyloliquefaciens* liquefying α-amylase and *Bacillus amyloliquefaciens* liquefying α-amylase. (2). *Bacillus subtilis* saccharifying α-amylase and *Bacillus subtilis* var. *amylosacchariticus* α-amylase. (3). *Bacillus subtilis* NRRL B3411 liquefying α-amylase is indistinguishable immunologically from *B. amyloliquefaciens* liquefying α-amylase. Saccharifying and liquefying amylases are distinguishable by their mechanisms of starch degradation by the fact that the saccharifying α-amylase produces an increase in reducing power about twice that of the liquefying enzyme (Fukumoto, 1963; Pazur and Okada, 1966). *Bacillus subtilis* var. *amylosacchariticus, B. subtilis* Marburg and *B. natto* all produce saccharifying α-amylase (Matsuzaki *et al.*, 1974b; Yoneda *et al.*, 1974). *Bacillus amyloliquefaciens* produces large quantities of liquefying α-amylase (Welker and Campbell, 1967a, b). On the basis of physiological and biochemical properties, Welker and Campbell (1967a) showed that highly amylolytic strains of *B. subtilis* were, in fact, strains of *B. amyloliquefaciens*. These findings confirmed the observation of Fukumoto (1943a) who classified the strains as *B. amyloliquefaciens* n. sp. Fukumoto (Fukumoto, 1943a, b). This organism was described subsequently as *B. amyloliquefaciens* (Welker and Campbell, 1967a, b, c), a strain of *B. subtilis* (Hagihara, 1960) or *B. subtilis* var. *amyloliquefaciens* (Tsuru, 1962; Tsuru and Fukumoto, 1963). The bacterium producing the saccharifying enzyme is referred to as *B. subtilis* or *B. subtilis* var. *amylosacchariticus* (Fukumoto, 1963). Authentic strains of *B. subtilis* produce α-amylases that differ immunologically

Table 7

Some properties of microbial α-amylases

	Molecular weight	Optimum pH value	Temperature optimum (°C)	Reference
Bacillus acidocaldarius	68,000	3.5	75	Buonocore *et al.* (1976)
Bacillus amyloliquefaciens	49,000	5.9	65	Borgia and Campbell (1978); Welker and Campbell (1967b)
Bacillus caldolyticus	—	5.4	70	Heinen and Heinen (1972)
Bacillus coagulans	—	5.2	57	Joyce (1977)
Bacillus licheniformis	22,500	5.0–8.0	76	Saito (1973)
Bacillus stearothermophilus	49,000	5.4–6.1	70	Pfueller and Elliott (1969)
Bacillus sp. NRRL B-3881	—	9.2	50	Boyer and Ingle (1972)
Bacillus subtilis	47,000 100,000	5.3–6.4	50	Menzi *et al.* (1957); Meyer (1952)
Bacillus subtilis NRRL B3411	48,000	6.0	60	Moseley and Keay (1970)
Bacillus saccharophila	—	5.25–5.75	40	Markowitz *et al.* (1956)
Thermophile, V-2	50,000	6.0–7.0	70	Hasagawa *et al.* (1976)
Streptomyces aureofaciens	40,000	4.6–5.3	40	Hostinová and Zelinka (1978)
Thermoactinomyces vulgaris	—	5.9–7.0	60	Kuo and Hartman (1967)
Thermomonospora curvata	62,000	5.5–6.0	65	Glymph and Stutzenberger (1977)
Thermomonospora viridis	—	6.5	55	Upton and Fogarty (1977)
Aspergillus niger (a) acid-unstable enzyme	61,000	5.0–6.0	35	Arai *et al.* (1968); Minoda *et al.* (1968)
(b) acid-stable enzyme	58,000	4.0–5.0	50	
Aspergillus oryzae	52,600	5.5–5.9	40	Fischer and De Montmollin (1951c); McKelvy and Lee (1969)

and electrophoretically from strains of *B. amyloliquefaciens* (Welker and Campbell, 1967b). In addition, the latter are highly proteolytic. The amino-acid composition of the α-amylase of *B. amyloliquefaciens* (Borgia and Campbell, 1978) is somewhat similar to that of the α-amylase of *B. subtilis* (*B. amyloliquefaciens* strain T) (Junge *et al.*, 1959) but differs from that of the α-amylases of authentic strains of *B. subtilis* (Yamane *et al.*, 1973). Furthermore, antiserum prepared against *B. amyloliquefaciens* α-amylase (Welker and Campbell, 1967b) does not cross-react with the α-amylase of *B. subtilis*.

2. Properties

The molecular weights of α-amylases are about 50,000 (Table 7). Monomer–dimer transformations were reported in the case of *B. subtilis* α-amylase (Menzi *et al.*, 1957). The dimer had a molecular weight of 100,000. Formation of higher aggregates than the dimer occurs in the presence of higher concentrations of zinc (Isemura and Kakiuchi, 1962). From the data of Mitchell *et al.* (1973) and Robyt and Ackerman (1973), it would appear that the α-amylase of *B. subtilis* has a basic subunit of 24,000 molecular weight. A molecular weight of 22,500 (Saito, 1973) has been reported for the α-amylase of *B. licheniformis*. The acid-unstable and acid-stable enzymes of *A. niger* have molecular weights of 61,000 and 58,000, respectively (Arai *et al.*, 1968). These values are similar to the value of 62,000 reported for the enzyme from *Thermomonospora curvata* (Glymph and Stutzenberger, 1977). The α-amylase of *B. acidocaldarius* is composed of one peptide chain having a molecular weight of 68,000 (Buonocore *et al.*, 1976) while the enzyme from *Bacteroides amylophilus* has a value of 92,000 although, in this case, the likelihood of the existence of a dimer was not excluded (McWethy and Hartman, 1977).

α-Amylases are usually stable in the pH range 5.5–8.0 and to extremes of pH value when supplied with a full complement of calcium ions. Optimum pH values for α-amylases generally occur between pH 4.8 and 6.5 (Manning and Campbell, 1961; Markowitz *et al.*, 1956; Menzi *et al.*, 1957) but with significant differences in the shapes of the pH–activity curves for different enzymes. *Bacillus acidocaldarius* produces a thermoacidophilic α-amylase having an optimum pH value at 3.5 and a temperature optimum at 75°C (Buonocore *et al.*, 1976), whereas the enzyme from *B. licheniformis* (Saito, 1973) has a broad range of activity between pH 5.0 and 9.0 and a temperature optimum of 76°C. In contrast, the amylases produced by alkaline *Bacillus* spp. (Boyer and Ingle, 1972; Horikoshi, 1971b) are not thermostable, although they have alkaline pH optima, whereas the enzyme produced by *B. stearothermophilus* (Ogasahara *et al.*, 1970) is thermostable and has little activity under alkaline conditions. The enzyme from *B. licheniformis*, as already indicated, is stable to both temperature and pH value in the alkaline range. The activity of most highly purified α-amylases is lost quickly above 50°C, but the presence of calcium ions may slow down this inactivation. All α-amylases are calcium metallo-enzymes (Fischer and Stein, 1960) having a minimum of one atom of this metal per molecule of enzyme, and the strength of binding of metal to protein depends on the origin of the enzyme. The metal is required for catalytic activity. In the presence of calcium ions, α-amylases are quite resistant to extremes of pH value, temperature, treatment with urea or exposure to some proteases (Stein and Fischer, 1958; Yamamoto, 1955). However, calcium-free amylases are highly susceptible to denaturation by similar treatments (Stein and Fischer, 1958). Once calcium-containing α-amylases have been denatured (Hagihara *et al.*, 1956; Yamanaka and Higashi, 1957), they are susceptible to protease degradation. Calcium was totally removed from the α-amylase of *B. subtilis* by chelation with ethylenediaminetetra-acetic acid (EDTA) or electrodialysis. Re-activation of the enzyme was achieved by addition of the metal. To obtain full activity, the enzyme required at least four gram-atoms of calcium per mole of enzyme (Hsiu *et al.*, 1964). The acid-stable and acid-unstable α-amylases of *A. niger* contain one gram-atom of calcium (Arai *et al.*, 1969) and the α-amylase from *A. oryzae* binds 10 gram-atoms of calcium per mole of enzyme. Nine of these atoms are bound very loosely and do not affect catalytic activity (Oikawa and Maeda, 1957). Removing the strongly bound calcium results in loss of activity (Kato *et al.*, 1967). Other divalent metals, such as strontium, barium and magnesium, could replace calcium without appreciable loss in activity (Toda and Narita, 1967), but it appears that these metals can replace calcium only when the original conformation of the protein is retained and the active site has not been altered. The single thiol function in Taka-amylase A (*A. oryzae*) may be involved in binding the strongly bound calcium atom (Toda *et al.*, 1968). The α-amylase of *B. licheniformis* is stable to EDTA treatment (Saito, 1973). The alkaline α-amylase of *Bacillus* sp. NRRL B-3881 was only

12% inactivated by EDTA (Boyer and Ingle, 1972), whereas the corresponding enzyme of *B. subtilis* var. *amylosacchariticus* was only 10% inactivated when treated with EDTA under similar conditions (Toda and Narita, 1968). These α-amylases would appear, therefore, to be somewhat atypical by virtue of their relative stability to EDTA. Heavy metals, e.g. silver, lead, copper and mercury, inhibit α-amylases (Greenwood and Milne, 1968a; Urata, 1957).

E. Amylopectin-Debranching Enzymes

1. Classification and Occurrence

Debranching enzymes hydrolyse α-1,6 linkages in amylopectin and/or glycogen and related polymers. They are classified into two groups. (1) Direct debranching enzymes that will degrade unmodified amylopectin and/or glycogen, and (2) indirect debranching enzymes, in which the substrate must first be modified by another enzyme or enzymes before the indirect debrancher will act on the polymer. There

Table 8

Occurrence of some microbial debranching enzymes

Organism	Type of enzyme	Reference
Aerobacter aerogenes	Pullulanase	Abdullah *et al.* (1966); Bender and Wellenfels (1961); Kainuma *et al.* (1978); Ohba and Ueda (1973); Ueda and Nanri (1967)
Bacillus cereus var. *mycoides*	Pullulanase	Takasaki (1976b)
Bacillus macerans	Pullulanase	Adams and Priest (1977)
Bacillus polymyxa	Pullulanase	W. M. Fogarty (unpublished observations)
Bacillus sp. RK00105	Pullulanase	W. M. Fogarty (unpublished observations)
Bacillus sp. No. 202-1	Pullulanase	Nakamura *et al.* (1975)
Pseudomonas saccharophila	Pullulanase	Norrman and Wöber (1975)
Pseudomonas stutzeri	Pullulanase	Wöber (1973)
Streptococcus mitis	Pullulanase	Walker (1968)
Streptomyces sp. No. 280	Pullulanase	Yagisawa *et al.* (1972)
Bacillus amyloliquefaciens	Isoamylase	Urlaub and Wöber (1975)
Cytophaga sp.	Isoamylase	Gunja-Smith *et al.* (1970)
Escherichia coli	Isoamylase	Jeanningros *et al.* (1976)
Escherichia intermedia	Isoamylase	Ueda and Nanri (1967)
Pseudomonas sp. SB15	Isoamylase	Harada (1977); Harada *et al.* (1968); Kuswanto *et al.* (1976); Yokoboyashi *et al.* (1970)
Streptomyces sp. No. 28	Isoamylase	Ueda *et al.* (1971)

are two types of direct debranching enzymes, namely pullulanase (EC 3.2.1.41) and isoamylase (EC 3.2.1.68). Although both enzymes attack the branch points in amylopectin, pullulanase will degrade the α-glucan pullulan, whereas isoamylase will not (Lee and Whelan, 1971). These enzymes are produced by a number of micro-organisms (Table 8).

2. Microbial Pullulanase

Pullulanase was first reported in *Aerobacter aerogenes* by Bender and Wallenfels (1961). It hydrolysed pullulan almost entirely to malto-triose. Kainuma *et al.* (1978) reported that the lowest molecular-weight substrates for pullulanase are 6^1- and 6^2-O-α-maltosylmaltose. Pullulanase does not hydrolyse isomaltose. The enzyme from *A. aerogenes* almost totally debranches amylopectin and its β-limit dextrin, but it has only slight or no activity on rabbit-liver glycogen and debranches its β-limit dextrin only to a limited extent (Lee and Whelan, 1971).

The pullulanases of *Bacillus macerans* (Adams and Priest, 1977) and *B. cereus* var. *mycoides* (Takasaki, 1976b) had very little, if any, activity on amylopectin but degraded its β-limit dextrin. Pullulanase may thus either have a lower affinity for the larger branches or its access to the α-1,6 linkages may be prevented by the highly branched structures (Adams and Priest, 1977). The precise location of α-1,6 linkages in a substrate is an important factor with almost all debranching enzymes (Manners, 1971b). However, the pullulanase from *Bacillus* sp. No. 202-1 (Nakamura *et al.*, 1975) hydrolyses all of the α-1,6 linkages in amylopectin, glycogen and pullulan. The enzyme from *B. polymyxa* (W. M. Fogarty, unpublished observations) also hydrolyses amylopectin, glycogen and pullulan, and the enzyme from *Streptococcus mitis* (Walker, 1968) gave an increase in iodine-staining power with amylopectin and glycogen of 50% and 30%, respectively.

Palmer *et al.* (1973) reported the occurrence of a pullulanase in *E. coli*, but Dessein and Schwartz (1974) were unable to detect any pullulanase activity in a number of strains of the same species. Enzymes produced by *Streptomyces* sp. No. 28 (Ueda *et al.*, 1971) and *E. intermedia* (Ueda and Nanri, 1967) hydrolysed pullulan but, in both cases, were called isoamylase. It is not correct to use the term isoamylase to describe an enzyme that attacks pullulan.

Table 9

Some properties of microbial debranching enzymes

	Optimum pH value	Temperature optimum (°C)	pH stability	Molecular weight	Reference
Pullulanase					
Aerobacter aerogenes	6.5	50	5.0–11.5	143,000	Eisele et al. (1972); Ohba and Ueda (1973); Ueda and Ohba (1972)
Bacillus cereus var. mycoides	6.0–6.5	50	7.0–8.75	112,000 ± 20,000	Takasaki (1976b)
Bacillus polymyxa	6.0–7.0	45	5.5–7.0	48,000	Fogarty and Griffin (1975); W. M. Fogarty (unpublished observations)
Bacillus sp. No. 202–1	8.5–9.0	55	6.5–11.0	92,000	Nakamura et al. (1975)
Streptomyces sp. No. 280	5.0–6.0	50	—	—	Yagisawa et al. (1972)
Isoamylase					
Cytophaga sp.	5.5	40	—	120,000	Gunja-Smith et al. (1970)
Escherichia coli	5.6–6.4	45–50	—	—	Jeanningros et al. (1976)
Escherichia intermedia	6.0	47	—	—	Ueda and Nanri (1967)
Pseudomonas sp. SB15	3.0–4.0	52	3.5–5.5	94,000	Harada (1977); Harada et al. (1968); Kuswanto et al. (1976); Yokobayashi et al. (1970)
Saccharomyces cerevisiae	6.0	25	—	—	Gunja et al. (1961)
Streptomyces sp. No. 28	5.0	60	—	—	Ueda et al. (1971)

3. Properties of Pullulanase

The pH optima of pullulanases (Table 9) occur between pH 5.0 and 7.0 (W. M. Fogarty, unpublished observations; Ohba and Ueda, 1973; Takasaki, 1976b; Ueda and Ohba, 1972; Walker, 1968; Yagisawa *et al.*, 1972) and have temperature optima of 45–50°C (W. M. Fogarty, unpublished observations; Ohba and Ueda, 1973; Takasaki, 1976b; Ueda and Ohba 1972; Yagisawa *et al.*, 1972), except for the enzymes from *Bacillus* sp. No. 202-1 and *S. mitis* which have optima at 55°C (Nakamura *et al.*, 1975) and 30°C (Walker, 1968), respectively. Molecular weights ranging from 56,000 to 150,000 are estimated for the pullulanase of *A. aerogenes* (Ohba and Ueda, 1973; Ueda and Ohba, 1972; Mercier *et al.*, 1972; Frantz *et al.*, 1966). In a more recent study involving a variety of techniques, it was indicated that the pullulanase of *A. aerogenes* is a single polypeptide chain with a molecular weight of 143,000 (Eisele *et al.*, 1972). Schardinger dextrins competitively inhibit the pullulanase from *A. aerogenes* (Marshall, 1973) but the *B. polymyxa* enzyme is not inhibited (W. M. Fogarty, unpublished observations).

4. Microbial Isoamylases

It is recommended (Lee and Whelan, 1971) that the name isoamylase be used to describe debranching enzymes of the type isolated from *Pseudomonas* sp. SB15 (Harada *et al.*, 1968; Yokobayashi *et al.*, 1970). The specificity of this enzyme has been described (Harada *et al.*, 1968; Yokobayashi *et al.*, 1969, 1970). It cleaves all of the α-1,6 interchain linkages in amylopectin and glycogen and has no action on pullulan (Harada *et al.*, 1968; Yokobayashi *et al.*, 1970). However, β-limit dextrin is not fully debranched, and this appears to be due to the specificity of the enzyme towards the shortened side chains, which consist only of 2–3 glucose residues (Yokobayashi *et al.*, 1970). A similar enzyme was produced by *Cytophaga* sp. (Gunja-Smith *et al.*, 1970). Because of the inability of bacterial isoamylases to attack pullulan, it is concluded that they degrade α-1,6 bonds at branch points in amylaceous substrates (Lee and Whelan, 1971). Using a high concentration of purified isoamylase from *Pseudomonas* sp. SB15 (Yokobayashi *et al.*, 1973), slight activity towards pullulan was demonstrated and it was therefore suggested that the definition of isoamylase be altered to describe an enzyme causing little or no cleavage of pullulan. Kainuma *et al.* (1978) observed that the best substrates for the isoamylase from *Pseudomonas* sp. SB15 were high molecular-weight polysaccharides. The lowest molecular-weight substrate degraded by this enzyme is the pentasaccharide 6^3-O-α-maltosyl-maltotriose. It appears that a minimum of three α-D-glucose residues are required by this enzyme in the B- or C-chains of the substrate (Kainuma *et al.*, 1978). Urlaub and Wöber (1975) reported the occurrence of an isoamylase in *B. amyloliquefaciens*: it was more active against glycogen than amylopectin and did not hydrolyse pullulan.

Yeast isoamylase partially debranched glycogen, amylopectin and their β-limit dextrins, but was inactive against α-limit dextrins (Lee and Whelan, 1971). Evans and Manners (1971) observed that yeast isoamylase hydrolysed the outermost interchain linkages in amylopectin.

The pH optima of isoamylases (Table 9) are in the range pH 5.0–6.4 (Gunja *et al.*, 1961; Gunja-Smith *et al.*, 1970; Jeanningros *et al.*, 1976; Ueda *et al.*, 1971). The temperature optima of isoamylases generally lie between 40°C and 52°C (Gunja-Smith *et al.*, 1970; Harada *et al.*, 1968; Jeanningros *et al.*, 1976; Ueda and Nanri, 1967; Yokobayashi *et al.*, 1970).

IV. PRODUCTION OF AMYLASES

A. Culture Maintenance and Preservation

Strains of micro-organisms used in commercial production of enzymes are maintained as pure cultures by standard microbiological techniques. Purity of culture implies exclusion of contaminating micro-organisms and, equally important, avoidance of variants developing in the strain itself. Single-spore or cell isolates should be examined for morphological and biochemical characteristics at regular intervals.

The main methods of preserving strains include: (1) transfer and storage at 5°C; (2) storage under mineral oil; (3) storage in soil, sand or silica gel; (4) lyophilization; (5) deep-freeze storage; (6) liquid-nitrogen storage; (7) storage in sterile distilled water.

The most fundamental method is use of agar slopes and storage of these at refrigeration temperatures. This is a simple but laborious method and there is risk of contamination and selection. Cultures prepared for storage on agar slopes may be covered with sterile mineral oil and stored at room or refrigeration temperatures (Dade, 1960; Hesseltine *et al.*, 1960; Little and Gordon, 1967; Tatarenko *et al.*, 1976; Krizková and Balan, 1975; Elliott, 1975; Braverman and Crosier, 1966; Fennell, 1960; Onions, 1971a, b; Weiss, 1957; Lapage *et al.*, 1970). Subculturing does not have to be carried out as frequently in this case, and so the amount of work involved and the risks of selection are decreased.

Fungal spores have been successfully stored in soil (Bakerspigel, 1953, 1954; Onions, 1971a; Pridham *et al.*, 1973), sand (Fennell, 1960) and silica gel (Perkins, 1962; Barratt *et al.*, 1965; Mayne *et al.*, 1971; Trollope, 1975). The survival period is greater in soil than in agar, but changes cannot be detected until recultivation. One advantage of dry storage is that many subcultures may be made from each dried sample. Bacterial cultures have also been successfully stored by drying (Iijima and Sakane, 1973; Trollope, 1975).

The advantages of lyophilization include the prolonged period of survival with no possibility of contamination during storage while the chances of strain variation are minimized because of infrequent subculturing. Viability reports have been published on survival of lyophilized cultures after different periods of time and using various protective agents (Mehrota and Hesseltine, 1958; Hesseltine *et al.*, 1960; Ellis and Roberson, 1968; Bosmans, 1974; Sarbhoy *et al.*, 1974; Belyakova *et al.*, 1975; Schipper and Bekker-Holtman, 1976; Semenov, 1973, 1975; Hall and Webb, 1975; Marshall *et al.*, 1973; Redway and Lapage, 1974). A viability test on 17–18-year-old fungal cultures showed that, of the 100 cultures tested, 91 were viable (Schipper and Bekker-Holtman, 1976). Not all fungi can be successfully lyophilized, but the majority of the sporing fungi, including *Aspergillus* spp. and *Penicillium* spp., survive well (Mehrota and Hesseltine, 1958).

Deep freeze is suitable for storage of certain types of organisms (Carmichael, 1956, 1962; Kramer and Mix, 1957; Webster *et al.*, 1958; Yamasato *et al.*, 1973). Storage in liquid nitrogen appears to be a satisfactory method of long-term storage (Hwang, 1966, 1968; Onions, 1971b; MacDonald, 1972; McDaniel and Bailey, 1968; Daily and Higgens, 1973; Moore and Carlson, 1975; Manchee, 1975; Wellman and Stewart, 1973). *Lactobacillus bulgaricus* did not remain viable after storage in liquid nitrogen (Smittle *et al.*, 1972). A very simple technique which is successful for storage of fungal cultures involves use of sterile distilled water (McGinnis *et al.*, 1974; Marx and Daniel 1975; Boeswinkel, 1976).

B. Screening and Selection of Micro-Organisms

Selection of an organism for production of an enzyme requires consideration of the following criteria: (1) extensive screening must be undertaken to select the most suitable organism; (2) the organism must produce the enzyme in good yield in a relatively short time and, ideally, in submerged culture; (3) the organism must grow and produce the enzyme on inexpensive readily available nutrients; (4) the organism should be easily removed from the fermentation liquor; (5) the enzyme should preferably be produced extracellularly and readily isolated from the fermentation liquor; (6) the organism should be non-pathogenic and unrelated phylogenically to a pathogen; (7) ideally, the organism should not produce toxins or other biologically active materials; (8) the organism should be genetically stable and not susceptible to bacteriophages.

1. Bacteria

Boidin and Effront (1917a, b) reported the first commercial method for production of amylases by strains of *B. subtilis* or *B. mesentericus*. Janke and Schaefer (1940) listed 37 species of bacteria that hydrolyse starch. The most active of these included strains of *B. mesentericus*, *B. subtilis*, *B. mycoides*, *B. macerans*, *B. albus*, *B. megatherium*, *B. vulgatus* and *B. ovalecticus*, while Kneen and Beckord (1946) studied the amylolytic systems produced by 41 strains of *B. subtilis*, seven of *B. polymyxa* and three of *B. macerans*.

Until quite recently, the organisms employed in commercial production of bacterial α-amylase have been described as *B. subtilis*. Different strains of *B. subtilis* produce either saccharifying or liquefying α-amylase. They may be distinguished from each other by the fact that, in hydrolysis of starch, increase in reducing power produced by the saccharifying α-amylase is approximately twice that of the liquefying enzyme. Highly amylolytic strains of *B. subtilis* producing liquefying α-amylase have been reclassified as *B. amyloliquefaciens* (Fukumoto, 1943a, b; Welker and Campbell, 1967a, b, c).

Bacillus subtilis Marburg, *B. amylosacchariticus* and *B. natto* produce similar saccharifying α-amylases. They have limited commercial uses because of the similarity in their properties to those of *Aspergillus* spp. while the enzymes of the latter organisms can be produced in much greater yields. *Bacillus amyloliquefaciens* is the bacterium most widely used in commercial production of α-amylase. Recently, *B. licheniformis* was shown to produce an α-amylase with better thermostability properties than *B. amyloliquefaciens*, which allows its use at temperatures in excess of 100°C (Madsen *et al.*, 1973). A mutant of this organism with high yields of α-amylase has been isolated, and is now used in commercial production (Outtrup *et al.*, 1970; Outtrup and Aunstrup, 1975). *Aerobacter aerogenes* is used for production of the starch-debranching enzyme pullulanase (Corn Products Company, 1971; Heady, 1974).

2. Fungi

There are a number of reports of strains of fungi yielding different amounts of amylolytic enzymes (Le Mense *et al.*, 1947; Dingle and Solomons, 1951; Kamachi and Tanaka, 1957; Corman and Langlykke, 1948). Pool and Underkofler (1953) observed that the proportions of α-amylase and amyloglucosidase produced by *A. oryzae* and *A. niger* varied, depending on whether the organism was cultivated in submerged or surface culture. *Aspergillus niger* produced considerably more amyloglucosidase than α-amylase, and the reverse was true for *A. oryzae*. Consequently, *A. niger* is one of the organisms used for commercial production of amyloglucosidase, and fungal α-amylase is produced commercially by *A. oryzae*. Commercial production of amylolytic enzymes from *A. oryzae* was originally developed by Takamine (1894, 1914) under the name Takadiastase, which is still sometimes used to describe fungal α-amylase. Descriptions of preparations of *A. oryzae* α-amylase and their properties have been reported by Wallerstein (1939), Sherman and Tanberg (1916), Oshima and Church (1923), Leopold and Germann (1940) and Harada (1931). Takadiastase, although manufactured and sold on the basis of its α-amylase content, contains various other enzymes (Tauber, 1949). The proportion of enzymes in the *A. oryzae* preparation is very important for many of its commercial applications. The proportion produced in surface culture is not successfully achieved by submerged culture of the organism (Lockwood, 1952). Processes for production of α-amylase have been patented by Kvesitadze *et al.* (1974) for the enzyme from *A. oryzae*, and by Belloc *et al.* (1975) for the enzyme from *Penicillium expansum*.

A number of species in the genus *Rhizopus*, including *R. niveus* (Komaki *et al.*, 1956), *R. delemar* (Phillips and Caldwell, 1951), *R. oryzae* (Lenney, 1962) and *R. javanicus* (Yamada, 1961), have been used for production of amylolytic enzymes including amyloglucosidase. Other organisms used for production of amyloglucosidase in submerged culture include *A. niger* (Corn Products Company, 1960, 1962; Armbruster, 1961; Aunstrup, 1967), *A. phoenicis* (Liggett *et al.*, 1959; Langlois and Turner, 1959; Baribo, 1967), *A. awamori* (Smiley, 1967; Dworschack and Nelson, 1972) and *A. foetidus* (Bode, 1966b).

Transglucosidase synthesizes α-1,6-linked saccharides, e.g. isomaltose and panose from glucose and maltose, respectively. Its presence in an amyloglucosidase preparation limits the saccharifying power of this enzyme. Species of the genus *Rhizopus* are reported not to produce transglucosidase (Lenney, 1962) but this enzyme is produced by species of *Aspergillus*. It must therefore be removed or inactivated before such preparations are used in commercial production of dextrose. Methods of removing or inactivating this enzyme will be dealt with later. Comparable yields of dextrose are obtainable with the amyloglucosidase of species of *Rhizopus* and *Aspergillus* when transglucosidase has been removed from the latter. Disadvantages of *Rhizopus* spp. are that they do not give satisfactory amyloglucosidase yields in submerged culture (Fukumoto, 1967).

Attempts to increase enzyme yields have been directed: (a) towards increasing biomass yield in a culture and thus enzyme yield; or (b) more specifically to produce higher levels of an enzyme irrespective of biomass yield. Four approaches have been used in this context: (1) selection of strains that give high yields of enzyme; (2) modification of environment, medium or conditions of growth to increase biomass and/or enzyme yield per unit of biomass; (3) induction and/or derepression of enzyme synthesis by alteration of the growth environment; and (4) genetic improvement to yield a strain having enhanced enzyme yields.

As most if not all micro-organisms used in commercial production of enzymes have not been genetically characterized, production of a high-yielding mutant is unpredictable.

C. Production of Mutants

The mutation process involves treatment of organisms with a mutagenic agent or agents and isolating mutants for subsequent testing and selection. The dose of mutagen used and the degree of kill achieved affect the efficiency of mutation but the optimal level of each has to be determined experimentally (Calam, 1970). Mutagenic agents most frequently used include ultraviolet radiation (Gratzner and Sheehan, 1969), chemical mutagens such as ethylene imine (Alikhanian, 1962), ethylmethane sulphonate (Kihlman, 1966), N-methyl-N'-nitro-N-nitrosoguanidine (NTG) (Adelburg et al., 1965; Yoneda et al., 1973; Yoneda and Maruo, 1975), β-propiolactone (Kihlman, 1966), X-rays (Minagawa and Hamaishi, 1962) and gamma rays (Outtrup and Aunstrup, 1975).

Artificial mutation of an organism involves treatment of a cell or spore suspension with a mutagenic agent until a 95–99.9% mortality is achieved. Surviving cells are cultured on nutritionally complete media to give colonies. Isolates are subsequently screened for enzyme production. Those that give increased yields compared with untreated cells are then re-examined in order to differentiate between stable and reversible mutants. By using different mutagenic agents, one increases the probability of obtaining a greater variety of mutants (Markkanen, 1975). Mutagens have different effects on the cell and, therefore, use of a variety of mutagens in succession allows different effects to be brought into play (Calam, 1964). Examples of the use of different mutagenic agents to obtain high-yielding mutants include those of Markkanen (1975) and Bailey and Markkanen (1975) who used ultraviolet radiation, ethylene imine and nitrosoguanidine (NTG) to obtain a mutant of B. subtilis producing double the yield of α-amylase obtained with the parent strain. Outtrup and Aunstrup (1975) treated B. licheniformis with gamma rays, ultraviolet radiation and NTG. The mutation programme included seven steps, and the survival rate after the initial treatment with gamma rays was 0.01%. An eventual improvement in yield of α-amylase of about 25-fold was achieved.

Tunicamycin-resistant mutants of B. subtilis NA64 (Sasaki et al., 1976) were obtained after treatment of the cells with NTG. One mutant yielded a five-fold increase in α-amylase production. Genetic analyses indicated that resistance to tunicamycin and hyperproductivity of α-amylase were induced by a single mutation. *Bacillus subtilis* NA64 was originally derived from *B. subtilis* Marburg 6160 by transformation with DNA from *B. natto* (Yamaguchi *et al.*, 1969). In *B. subtilis* 6160, the regulator gene *amy* R1 was replaced by *amy* R2 (from *B. natto*) and a three- to four-fold increase in amylase production over *B. subtilis* 6160 was obtained (Yamaguchi *et al.*, 1974). Hybrid α-amylases were produced by transformants of *B. subtilis*, and it was thought that DNA recombination events may have taken place in the α-amylase structural gene (Yamane and Maruo, 1975). Some mutants of *B. subtilis* produced enzymes with different physicochemical and immunological properties compared with the enzyme of the parent strain (Yamane and Maruo, 1974). Very high producers of α-amylase were also obtained by crossing *B. subtilis* var. *amylosacchariticus* as donor with *B. subtilis* NA20 as recipient (Yoneda *et al.*, 1974).

Aunstrup (1977) outlined a method for obtaining transglucosidase-free mutants of an amyloglucosidase-producing strain of *A. niger*. Fungal spores were treated with ultraviolet radiation or NTG. Other reports also describe use of ultraviolet radiation as a mutagenic agent in obtaining decreased levels of transglucosidase from amyloglucosidase-producing fungi (Corn Products Company, 1962). Ultraviolet radiation was also used to produce a mutant giving high yields of amyloglucosidase in *A. phoenicis* (Baribo, 1967) and in *A. foetidus* (Bode, 1966b). In the latter case, levels of transglucosidase were substantially decreased. Similar results have also been obtained with mutants of *A. niger* (Aunstrup, 1967; Armbruster, 1961). Lin (1972) produced a mutant of *Rhizopus formosaensis* by ultraviolet radiation and NTG treatment which produced 15 times as much amyloglucosidase as the parent strain in submerged culture.

Enzyme production can be improved by genetic recombination by both an increase in the rate of production and by transferring advantageous characters to high-yielding strains. The overall economics of the process may be improved either during the fermentation or at the post-fermentation stage. Improvement in amyloglucosidase production following genetic recombination of *A. niger* strains was reported by Ball *et al.* (1978). A high-yielding strain was crossed parasexually with a low-yielding strain, and the recombinants had the efficient broth-filtration characteristics of the latter.

D. Environmental and Other Factors

Scientific, including patent, literature contains innumerable formulations for media suited to enzyme production. The basic composition of all of these media includes a carbohydrate source, a nitrogen source and nutrient salts. Commercial media are invariably complex (Table 10). These complex ingredients or nutrient sources have the advantage of supplying a range of materials including carbohydrates, proteins, fats, amino acids, salts and trace elements, vitamins and various inducers. It is not just the presence of a given nutrient in a medium that is important but its availability to the particular organism in the correct concentrations and at the appropriate time in the fermentation cycle. The presence of large concentrations of low molecular-weight, readily metabolizable sugars or of amino acids can be detrimental. However, if these compounds become available by slow hydrolysis, they can be in a form and at a concentration required by the organism for maximum production of the enzyme. It is well established that free amino acids repress protease synthesis (Chaloupka and Kreckova, 1966), whereas free amino acids stimulate α-amylase synthesis in *B. amyloliquefaciens* (C. T. Kelly and W. M. Fogarty, unpublished observations). This obviously presents a problem if it is required to produce both enzymes simultaneously. Thus, a cheap form of protein must be provided for production of biomass and α-amylase that will not repress protease synthesis. Crude forms of protein used for fermentations involving *Bacillus* spp. include soyabean meal, distiller's solubles, fish meal and peanut meal. Similarly, in selecting carbohydrate for a production

Table 10

Ingredients of Fermentation Media

Material	Per cent (w/v) of:				Major growth factors
	Protein	Carbohydrate	Fat	Ash	
Fish meal	72.0	5.0	1.5	18.1	—
Soya meal	50.0	28.0	0.8	6.0	Niacin
Torula yeast	50.0	32.0	2.5	8.0	Inositol, niacin, thiamin, pyridoxine, riboflavin
Distiller's solubles	26.0	45.0	9.0	8.0	Pyridoxine, niacin
Corn-steep liquor	24.0	6.0	1.0	8.8	Niacin, inositol, pantothenic acid
Dried whey	13.0	72.0	1.0	8.0	Riboflavin, pantothenic acid
Corn meal	9.0	69.0	4.0	1.3	—

medium, it is highly beneficial to understand the mechanism regulating synthesis of the particular enzyme.

Investigations on amylase synthesis have been made with complex growth materials such as corn meal (Van Lanen and Le Mense, 1946) and defatted peanut meal and ammonium sulphate (Babber *et al.*, 1960). In other cases, synthetic or semisynthetic media, containing starch as carbon source and various nitrogen compounds, have been examined, e.g. ammonium sulphate (Minoda and Asai, 1960), ammonium tartarate (Oikawa, 1959), ammonium phosphate (Nomura *et al.*, 1956), potassium nitrate (Goering and Bruski, 1954) and casein hydrolysate (Shu and Blackwood, 1951). It is not possible to generalize as far as carbon:nitrogen ratios and extracellular enzyme formation are concerned. Optimum values varying from 20 (Minoda and Asai, 1960) to 1.2 (Shu and Blackwood, 1951) have been reported for α-amylase production by *Aspergillus* spp., and from 50 (Lulla, 1951) to 9 (Yoshida and Yamasaki, 1959) for α-amylase production by *B. subtilis*. A complication concerning the effect of complex nitrogenous sources is the likelihood of enzyme destruction by inducible proteases. Such a type of degradation might be involved in those cases where production of an extracellular enzyme reaches a peak and then disappears rapidly, as for example with α-amylase of *A. niger* PRL 558 (Shu and Blackwood, 1951) and α-amylase of various *Streptomyces* spp. (Simpson and McCoy, 1953).

Design of the fermentation medium for a process must take into consideration factors beyond simple nutrition. In submerged culture, viscosity, solubility and foaming characteristics are important, whereas ease of handling, free-flowing properties and surface area are of major significance in surface culture. The composition of the medium will have a major effect on the ease and method of control of pH value, e.g. a medium containing a high concentration of protein lessens the need for inorganic buffering systems. In some instances, selection of the proper ratio of carbohydrate to protein may be sufficient to control the pH value in the required range.

1. Exo-Acting Amylases

In *Aspergillus niger*, amyloglucosidase production is affected by both the nitrogen source and its concentration, carbohydrate concentration, the concentration of trace elements and pH value of the medium (Lineback *et al.*, 1966). The nitrogen source may affect both the quantity of the enzyme and the time of production (Barton *et al.*, 1969). Two periods of production occurred with an organic nitrogen source. The first of these took place after three days and the other after six days. Two forms, amyloglucosidase I and II, were produced by *A. niger* in a medium containing yeast extract, but only type I appears to be formed when ammonium chloride is used as a nitrogen source (Barton *et al.*, 1972). *Aspergillus awamori* var. *kawachi* produces different proportions of α-amylase and amyloglucosidase and different types of amyloglucosidase, depending on the environmental conditions (Hayashida, 1975). Amyloglucosidase I was produced when the organism was cultured in a synthetic medium with ammonium citrate as the nitrogen source, deficient in zinc and sufficient in phosphate, for 30 h at 35°C. Type I' was produced on cultivation in a synthetic medium containing ammonium acetate and sufficient in zinc, for 50 h at 30°C. Type II was obtained in a synthetic medium containing glucose, casein-peptone and adequate concentrations of inorganic elements, especially zinc and calcium, for 96 h at 25°C. An organic nitrogen source, e.g. penicillin waste mycelium, gave maximum enzyme formation by *A. awamori* (Qadeer and Kausar, 1971). Lower yields were obtained with corn-steep liquor or inorganic sources, such as sodium nitrate or ammonium chloride or sulphate. Maximum levels of amyloglucosidase were produced by *Torula thermophilia* with soybean as a nitrogen source (Subrahmanyam *et al.*, 1977). Thiamin was effective for mycelial growth of *Corticium rolfsii* but repressed production of the enzyme (Kaji *et al.*, 1976).

The nature of the carbohydrate in the culture medium also affects amyloglucosidase production (Attia and Ali, 1977; Kaji *et al.*, 1976; Qadeer and Kausar, 1971). Maltose, starch, dextrin and glucose gave satisfactory yields of amyloglucosidase with *A. awamori* (Attia and Ali, 1977). Very little enzyme was produced with sucrose, lactose, fructose or galactose in the medium. Agricultural byproducts, such as rice bran, glucose syrup and corn bran, supported good production of amyloglucosidase. *Corticium rolfsii* produces amyloglucosidase with starch, sucrose, glucose and galactose as the carbon source (Kaji *et al.*, 1976), although the last three compounds were utilized to a much lesser extent. Processes for producing high levels of amyloglucosidase with different media have been patented (Aunstrup, 1967; Langlois and Turner, 1959; Liggett *et al.*, 1959; Smiley, 1967; Armbruster, 1961).

Corn starch, ground corn, or ground yellow-dent corn were the carbon sources used. Nitrogen sources included potassium nitrate (Aunstrup, 1967), ammonium sulphate (Langlois and Turner, 1959) and corn-steep liquor solids (Liggett *et al.*, 1959; Baribo, 1967). Van Lanen and Smith (1968) used a medium consisting of cereal grain in high concentration to which a supplement of grain stillage or distiller's solubles was added for amyloglucosidase production by *A. niger*. Production of high yields of the enzyme by *A. phoenicis* involved preparation of an active inoculum by serially transplanting and cultivating the culture on four to five consecutive days. The initial pH value was 5.8–6.2 and, when the pH value fell to 3.8–4.2, it was maintained in that range. Dworschack and Nelson (1972) incorporated ammonia gas or ammonium hydroxide in the culture medium to control the pH value.

Two types of amyloglucosidase, which differ in acid stability, were produced by *A. awamori* var. *fumeus* (Watanabe and Fukimbara, 1965). At pH 2.0–2.5, the more acid-stable enzyme is produced and the less acid-stable species at pH 6.0-6.5. Amyloglucosidase production by the thermophilic fungus *Torula thermophilia* (Subrahmanyam *et al.*, 1977) was maximum at 50°C and pH 7.0, while production of the enzyme by *A. awamori* was carried out for four days at 35°C and aerated at a rate of 1.5 v/v/min (Smiley, 1967).

Starch and complex nitrogenous sources (polypeptone, casein, corn-steep liquor) are the carbon and nitrogen requirements of a range of bacteria that produce β-amylase (Griffin and Fogarty, 1973a; Okada and Higashibara, 1974; Shinke *et al.*, 1974, 1975a, b; Yamane and Tsukano, 1977). Glucose, in one instance, was shown to be more effective than starch as an inducer (Takasaki, 1976a, b). In all cases, a pH value of about 7.0 was most conducive to high yields of enzyme. Calcium and barium stimulated production of β-amylase by *B. cereus* var. *mycoides* (Takasaki, 1976a, b).

2. α-Amylases

Starch (or in the crude form as ground corn) is widely used as a carbon source in production of α-amylases (Allam *et al.*, 1977; Ensley *et al.*, 1975; Upton and Fogarty, 1977; Wojskowicz, 1977; Andrzejczuk-Hybel and Kaczkowski, 1971; Horikoshi *et al.*, 1974; Kvesitadze *et al.*, 1974; Fukumoto and Negoro, 1966; Yamada and Tomoda, 1966;

Outtrup and Aunstrup, 1975; Onishi, 1973). Wheat bran has been used with a number of strains of *B. subtilis* (Taha *et al.*, 1968; Minagawa and Hamaishi, 1962), and *A. foetidus* produces very high levels of α-amylase and amyloglucosidase when grown on baked-bean waste as compared to other media (Hang and Woodams, 1977).

A variety of nitrogen sources have been used for α-amylase production, including ammonium citrate (Allam *et al.*, 1977), mycological peptone (Upton and Fogarty, 1977), corn-steep liquor (Abd El-Akher *et al.*, 1973; Outtrup and Aunstrup, 1975; Belloc *et al.*, 1975; Yamada and Tomoda, 1966), calcium nitrate (Andrzejczuk-Hybel and Kaczkowski, 1971) and ammonium sulphate (Ingle and Boyer, 1972). Feniksova (1957) showed that, when ammonium sulphate was used as nitrogen source in production of α-amylase by *A. oryzae*, 95% of the enzyme was extracellular, whereas when the nitrogen source was potassium nitrate, the yield was decreased by 50%, and 82% of the enzyme was intracellular. With the latter salt, the pH value of the culture medium drifted with time to the alkaline side but, when a buffering system was used, the activity of the enzyme was equivalent to that obtained with ammonium sulphate, and 85% of the activity was detected extracellularly.

Formation of α-amylase by *B. subtilis* was stimulated by carbon dioxide (Gandhi and Kjaergaard, 1975). High concentrations of phosphate (0.1 M) stimulated α-amylase formation by *B. amyloliquefaciens*, whereas 0.04 M was adequate for maximum growth (Fukumoto *et al.*, 1957). Phytic acid, at concentrations of 0.05–0.1%, has been reported to be a specific stimulant of α-amylase formation by *A. niger*, *A. oryzae*, *A. awamori* and *B. subtilis*, but not by *R. javanicus* (Dunn *et al.*, 1959; Yamada, 1961).

Calcium and manganous ions were required for amylase production by *B. thuringiensis* (Tobey and Yousten, 1976). Calcium ions are especially significant in α-amylase synthesis and production, and are usually incorporated in the production media, e.g. for enzymes of *Penicillium expansum* (Belloc *et al.*, 1975), *B. subtilis* (Minagawa and Hamaishi, 1962), *Aspergillus* sp. (Yamada and Tomoda, 1966) and *Bacillus* sp. NRRL B-3881 (Ingle and Boyer, 1972). The α-amylase from *Thermomonospora vulgaris* was extracted from static cultures of the organism. No growth was obtained when the medium was devoid of calcium carbonate (Allam *et al.*, 1977). Increased yields of α-amylase are produced by *B. subtilis* in a medium containing starch, corn-steep water, glucose, dried brewer's yeast and diammonium phosphate w en an organosilicone copolymer is added to the medium (Wymes and Lloyd, 1968). The organosilicone copolymer also acts as an antifoam during production.

In a study of the effect of pH value and aeration on amylase production by *B. subtilis*, it was observed that, if the pH value is controlled during fermentation, aeration of the culture can be discontinued at the same time as cessation of growth, and α-amylase yields will be similar to those of an aerated culture (Mazza and Ertola, 1970). Markkanen and Bailey (1975) showed that α-amylase production by *B. subtilis* was notably increased by decreasing aeration at the end of the exponential phase of growth: α-amylase was produced in the pH range 6.5–8.0. Production of amylase by *Streptomyces olivaceus* was similar at pH values 5.8, 6.7 and 7.5, and growth of the culture was always accompanied by a shift in pH value towards the alkaline side (Wojskowicz, 1977). The optimum initial pH value for enzyme production by *A. foetidus* was 7.0, but maximal fungal growth occurred at pH 4.0 (Hang and Woodams, 1977). The pH values of the media for production of alkaline amylases were 9.0–9.4 (Ingle and Boyer, 1972) and 10.0 (Horikoshi *et al.*, 1974). Temperatures used for α-amylase production were between 37°C and 40°C for *B. subtilis* (Markkanen and Bailey, 1974), 30°C for another strain of *B. subtilis* (Abd El-Akher *et al.*, 1973), 37°C for *B. licheniformis* (Outtrup and Aunstrup, 1975), *Bacillus* sp. No. 38-2 (Horikoshi *et al.*, 1974) and *B. subtilis* (Fukumoto and Negoro, 1966) and 30–37°C for *Micrococcus halibius* (Onishi, 1973).

3. Debranching Enzymes

Starch, dextrins and maltose are all inducers of the debranching enzymes pullulanase and isoamylase (Harada *et al.*, 1968; Fujio *et al.*, 1970a; Ueda *et al.*, 1971; Yagisawa *et al.*, 1972; Hope and Dean, 1974; Urlaub and Wöber, 1975). The isoamylase of *E. intermedia* (Ueda and Nanri, 1967) was not detected when starch was used as carbon source but its synthesis was induced both by dextrin and maltose. In contrast, when maltose was supplied as carbon source, *A. aerogenes* produced only 5% of the debranching activity obtained with soluble starch (Fujio *et al.*, 1970b). Increased amounts of isoamylase were obtained from a number of species by use of maltitol (Horwath and Rotheim, 1976). A mutant of *Pseudomonas* sp. SB15 produces isoamylase constitutively

(Sugimoto et al., 1974a), whereas another mutant, strain S5, produces isoamylase in much greater yields than the original SB15 (Harada, 1977). In practically all cases, complex nitrogenous nutrients were required for enzyme production. Fujio et al. (1970a) reported that, with A. aerogenes, the level of activity with ammonium sulphate was only 24% of that obtained with ammonium acetate, and almost all of the enzyme was cell-bound. The Hayashibara Company (1973a) patented a process describing production of cell-bound α-1,6-glucosidases in media containing ammonium sulphate, but extracellular enzymes were produced by a number of organisms in media containing acetate. Although a considerable variety of organisms have been used for production of debranching enzymes, there appears to be a marked similarity in the composition of the media used (Corn Products Company, 1971; Hayashibara Company, 1971, 1972, 1973b; Yokobayashi et al., 1971; Masuda and Sugimoto, 1973; Gunja-Smith, 1974a, b; Sugimoto et al., 1974b; Bulich, 1976; Horwath et al., 1976, 1977; Horwath and Rotheim, 1976).

E. Synthesis and Regulation

The primary reason for the suitability of micro-organisms as sources of enzymes lies in the fact that their production may be increased by environmental and genetic manipulations. Factors that involve genetic expression include induction, catabolite repression and end-product inhibition (Enatsu and Shinmyo, 1978; Priest, 1977).

1. Induction

Synthesis of extracellular enzymes may be inducible, partially constitutive or totally constitutive depending on the micro-organism and the enzyme involved. Extracellular enzymes synthesized by members of the genus Bacillus would appear to be partially inducible. Many catabolic enzymes are in the inducible category, and are induced by the enzyme substrate. Substrates that are of high molecular weight cannot enter the cell and, therefore, it appears feasible to suggest that, in polymeric form, they are not directly involved in the induction process. It is thought that a small level of constitutive extracellular enzyme degrades the high molecular-weight substrate, and that the low molecular-weight products induce further enzyme synthesis on entering the cell.

Synthesis of α-amylase is controlled in some instances by this mechanism. For example, maltotetraose is the most powerful inducer of α-amylase in *B. stearothermophilus* (Welker and Campbell, 1963b, c) and *B. licheniformis* (Saito and Yamamoto, 1975). Other oligo-saccharides are less efficient inducers than maltotetraose. Synthesis of α-amylase by *Streptomyces olivaceus* is induced only in media containing maltose or starch. It is thought that induction of the enzyme by maltose was due to the presence of a contaminating oligosaccharide (Wojskowicz, 1977). Induction of α-amylase by non-growing mycelia of *Aspergillus oryzae* was best achieved by maltose (Yabuki *et al.*, 1977). Higher yields of amylase are known to be produced in media containing complex starch-containing materials, e.g. ground corn, rather than on artificially defined media (Adams and Deploey, 1976; Nyiri, 1971; Kvesitadze *et al.*, 1974) because these carbon compounds do not exert catabolite repression, and it would appear that they permit maximal induction of α-amylases.

In general, compounds containing α-1,4-glucosidic linkages are effective inducers of amylases. In contrast to this general observation, Tomomura *et al.* (1961) reported that isomaltose and panose were most effective inducers of α-amylase in *A. oryzae*. Suzuki and Tanabe (1962) observed that highest levels of α-amylase were produced when high concentrations of oligosaccharides, identified as isomaltose and maltose, were present. Furthermore, isomaltose was utilized more slowly than maltose, and this observation was offered as an explanation for the effectiveness of the former as an inducer of α-amylase.

Meers (1972) claims that no evidence was obtained to infer that an inducer was necessary for α-amylase production in *B. licheniformis*. Similarly, the α-amylases of *B. subtilis* (*B. amyloliquefaciens*) (Coleman, 1967) and *B. subtilis* (Sekigiuchi and Okada, 1972) were produced constitutively.

Bender and Wallenfels (1961) quoted maltose, maltotriose and pullulan as suitable inducing agents for pullulanase, whereas the Hayashibara Company (1971) and Corn Products Company (1971) reported that a starch hydrolysate (DE 5-10) was particularly suited to induction of this enzyme in *A. aerogenes*. Maltose, maltotriose, oligosaccharide mixtures and pullulan were equally effective in inducing the pullulanase of *K. aerogenes* in carbon-limited conditions (Hope and Dean, 1974).

Synthesis of β-amylase by *B. polymyxa* occurs only in the presence of starch or similar polymers (Griffin and Fogarty, 1973a). Maltose, one of two end-products of the action of this enzyme on amylopectin, induces only 50% of the activity obtained with starch as a carbon source. It is not inconceivable that the other product, β-limit dextrin, or an oligosaccharide derived from the combined action of β-amylase and the debranching enzyme (Griffin and Fogarty 1973a, b, c; Fogarty and Griffin, 1975) may have some inductive effect.

The cyclodextrin glucosyltransferase of *B. macerans* is produced in highest levels in starch-containing media, and accumulation of the enzyme increases as the carbon source declines (Lane and Pirt, 1973).

2. Catabolite Repression

Repression of constitutive or inducible enzyme synthesis in the presence of glucose or other readily utilizable carbon sources is known as catabolite repression. It accounts for or explains many aspects of extracellular enzyme synthesis (Schaeffer, 1969) and contributes to regulation of these enzymes. Most of the evidence substantiating the role of catabolite repression in controlling expression of extracellular enzyme genes has been derived from an understanding of β-galactosidase synthesis in *E. coli*. The rate of synthesis of this enzyme in constitutive mutants or after induction was inversely proportional to the growth rate of the organism. Thus, it was higher in cells when the growth rate was slow, as, for example, with lactate or succinate, and low in cells growing rapidly with glucose or galactose (Moses and Sharp, 1972). By comparison, α-amylase synthesis in *B. subtilis* 168 is related to the nature of the carbon source, being highest in media containing lactate or glutamate but lowest with glucose, which gave the fastest growth rate (Sekiguchi and Okada, 1972). In both *B. stearothermophilus* (Welker and Campbell, 1963a) and *B. subtilis* (*B. amyloliquefaciens*) (Coleman and Grant, 1966), rate of synthesis of α-amylase is inversely related to the growth rate of the organism when starch, glycerol or glucose are used as carbon sources. Meers (1972) reported that catabolite repression and growth rate were the main factors controlling α-amylase synthesis in *B. licheniformis*. In chemostat studies with carbon-limited conditions, high levels of α-amylase were produced with glucose, glutamate or alanine as the sole energy source. On the other hand, no enzyme was detected in nitrogen-limited cultures because of the presence of excess glucose. Under carbon-limited conditions, the rate of amylase production decreased with increased growth rate. Continuous-culture studies on *B. subtilis* (Heineken and O'Connor, 1972) showed the α-amylase was catabolite-repressed.

Bacillus licheniformis α-amylase is sensitive to catabolite repression (Saito and Yamamoto, 1975). Formation of the enzyme was suppressed by 0.5% glucose. Cyclic AMP did not alleviate the repressive effect of glucose, but it did stimulate α-amylase formation. Mutants of *B. licheniformis* were not sensitive to glucose, and were not stimulated by addition of cyclic AMP.

Glucose inhibited amylase production in *A. oryzae* (Nikolov *et al.*, 1970; Andrzejczuk-Hybel and Kaczkowski, 1971), *Streptomyces olivaceus* (Wojskowicz, 1977) and *B. amyloliquefaciens* (Ingle and Boyer, 1976). Glucose represses α-amylase synthesis in *B. subtilis* at the level of transcription (Priest, 1975).

V. APPLICATIONS

Conversion of starch into sugars, syrups and dextrins forms the major part of the starch-processing industry (Marshall, 1975). These hydrolysates are used as carbon sources in fermentation, as sources of sweetness and also, because of their physical characteristics, in a range of manufactured food products and beverages. Hydrolysis of starch to produce these products (Fig. 1), containing glucose, maltose and other oligosaccharides, is brought about by controlled degradation (Underkofler *et al.*, 1965; Norman, 1978; Barfoed, 1976; Hurst, 1975; Armbruster, 1974; Slott and Madsen, 1975). The degree of hydrolysis may be controlled so that end-products with the desired physical properties are obtained. Thus, in certain instances, such properties as osmotic pressure, sweetness, or resistance to crystallization may be important, whereas in other applications, maintenance of viscosity may be the important attribute (Palmer, 1975).

Preparation of starch hydrolysates may be effected using either acid or enzymes as catalysts. The traditional acid-catalysis methods are now being replaced by enzymic processes (Barfoed, 1976; Lenney, 1962). Enzymic degradation of starch has several advantages over acid processes. These are: (1) fewer byproducts are produced (Birch and Schallenberger, 1973; Greenshields and Macgillivray, 1972); (2) it is more specific; and (3) higher yields are obtained.

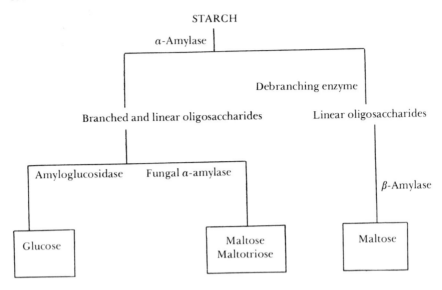

Fig. 1. Scheme showing enzymic hydrolysis of starch.

Hydrolysates are classified according to their content of reducing sugars, represented as D-glucose and described as DE or dextrose equivalent. Careful selection of enzyme, or combinations of enzymes, permits production of a range of products with well-defined chemical and physical properties. Thus, bacterial amylases can be used to produce modified starches with a low DE value as well as a range of maltodextrins with different DE values. When bacterial amylases are used in combination with amyloglucosidase, glucose is the sole end-product while, in association with fungal amylase, maltose syrups with high DE values are obtained. Each of the processes in production of hydrolysates involves firstly liquefaction of gelatinized starch and secondly saccharification of the thinned or liquefied product with the desired enzyme. The liquefaction step produces dextrins and is sometimes referred to as the dextrinizing step. Bacterial thermostable α-amylases are used for liquefaction. The reaction is generally terminated before significant hydrolysis takes place and when the average degree of polymerization is about 10–12. Two types of thermostable α-amylase are commercially available, one from *B. amyloliquefaciens* (Welker and Campbell, 1967c) and the other from *B. licheniformis* (Saito, 1973; Madsen *et al.*, 1973). Along with the differences in the physicochemical properties of these enzymes (which have been dealt with in Section B, p. 123), a significant difference also exists in the composition of the end-products produced by their action. Thus α-amylase from *B. amyloliquefaciens* produces mainly maltohexaose, whereas the enzyme from *B. licheniformis* produces chiefly maltose, maltotriose and maltopentaose (Norman, 1978).

Saccharification is the second stage in production of starch hydrolysates and utilizes thermolabile amylases which are not sufficiently heat-stable to be used in the liquefaction step. Their suitability is further dependent on the spectrum of end-products they produce. Sources of these enzymes include α-amylases from *Aspergillus oryzae* (Barfoed, 1976; Fischer and De Montmollin, 1951a, b, c), *Bacillus* spp. (Armbruster, 1974; Slott and Madsen, 1975), amyloglucosidases from *Aspergillus* spp. (Corman, 1967; Armbruster, 1961; Van Lanen and Smith, 1968) and *Rhizopus* spp. (Watanabe and Fukimbara, 1973; Ohnishi and Hiromi, 1978), and debranching enzymes from *A. aerogenes* (Hurst, 1975; Bender and Wallenfels, 1961).

A. Starch-Processing Industries

1. Liquefaction or Thinning of Starch Pastes

Liquefaction involves dispersion or gelatinization of starch into aqueous solution followed by partial hydrolysis. This process is the major outlet for thermostable α-amylases in the starch-processing industry (Armbruster, 1974; High and Rogols, 1966; Kooi *et al.*, 1966; Krebs, 1962; Land and Barton, 1966; Slott and Madsen, 1975; Table 11). The thinning of starches with enzymes may be improved by addition of quaternary amines to the starch–enzyme slurry (Speakman,

Table 11

Applications of amylases

Enzyme	Applications
α-Amylase	Starch syrups; adhesives; sizings; paper coating; brewing; baking industry; textiles; pharmaceuticals; animal feed; sewage treatment; digestive aid; detergents.
β-Amylase	Maltose syrups; brewing; distilling; baking.
Amyloglucosidase	Dextrose production; confectionery; baking; pharmaceuticals.

1969). For example, addition of dodecyl trimethyl ammonium chloride increases the rate of conversion of substrate, produces a product with a lower viscosity, and the conversion may be carried out at a lower temperature.

2. Saccharification

Saccharifying enzymes are selected on the basis of the end-products they produce. Thus, syrups containing a high concentration of maltose and maltotriose (Barfoed, 1976) are prepared industrially using α-amylase from *A. oryzae*. The temperature of conversion is usually about 55°C, which is considerably lower than that used with liquefaction enzymes. High DE-value maltose syrups have many applications in the food industry as they do not crystallize and are relatively non-hygroscopic.

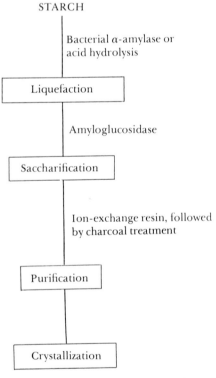

Fig. 2. Scheme showing production of crystalline D-glucose from starch.

Amyloglucosidase is used in the production of syrups containing 94–97% D-glucose (Bode, 1966a; Hayes, 1974; Hurst, 1975; Kerr, 1962; Langlois and Turner, 1959; Rentshler et al., 1962). These syrups are used for production of crystalline D-glucose (Fig. 2.; Knight, 1969) or fructose syrups (Zittan et al., 1975). The yield of glucose obtained with amyloglucosidase is dependent on the liquefaction process (Underkofler et al., 1965). It is important that, after liquefaction, the thinned starch should have a DE value of 10–12 in order to prevent retrogradation on cooling to 55–60°C to initiate the saccharification step. The quality of the saccharifying enzyme has a significant effect on the yield of glucose. Contaminating activities, in particular trans-glucosidase, should not be present as they result in production of α-1,6-linked oligosaccharides (Okazaki, 1956; Pazur and Ando, 1960; Thompson et al., 1954). Many methods have been described for the removal or inactivation of transglucosidase (Maher, 1968; White and Dworschack, 1973; Armbruster and Bruner, 1967; Croxall, 1966; Kooi et al., 1962).

Glucose may also be prepared using two saccharifying enzymes, namely amyloglucosidase and amylo-1,6-glucosidase, from a starch hydrolysate prepared with a bacterial α-amylase. The function of the debranching enzyme is to inhibit the reversionary action of amylo-glucosidase (Hurst, 1975). Amyloglucosidases hydrolyse α-1,6-gluco-sidic linkages but at a relatively slow rate (Pazur and Ando, 1960). Use of a debranching enzyme would therefore appear to have obvious advantages in production of D-glucose in the presence of amylo-glucosidases (Hurst, 1975). Debranching enzymes are produced by a number of micro-organisms (Fogarty and Kelly, 1979a, b; Wöber, 1976; Ohba and Ueda, 1975). Production of amyloses having excellent solubility and other properties has been described using debranching enzymes (Kurimoto and Sugimoto, 1975; Yoshida and Hirao, 1974).

B. Other Applications

1. Bread Making

Along with the gradual replacement of acid by thermostable enzymes in the starch-processing industry (Section V, p. 153), probably the most extensive use of fungal α-amylase has been in the supplementation of flour. A number of advantages are claimed for supplementing diastatically-deficient flour with fungal α-amylase (Harada, 1951; Hirchberg, 1957; Johnson and Miller, 1951).

2. Alcohol Production and Brewing

In alcohol production and brewing, the position of mould amylolytic enzymes (Ueda, 1957; Okazaki, 1957) in Japan, is similar to that held by malt in Europe and the U.S.A. However, supplementation of fungal enzymes in distilling mashes to augment cereal α- and β-amylases is being considered with increasing interest. The advantages claimed by such systems include uniform enzyme activity in mashes, increased rates of starch saccharification, alcohol yields and yeast growth (Van Lanen and Smith, 1968).

VI. CONCLUSION

World-wide output of enzymes has been estimated to value £82,000,000 (Bronn, 1976). World production of amylolytic enzymes in 1977 was estimated at 610 tonnes of pure enzyme protein (Aunstrup, 1977). This figure may be broken down into the following quantities.

Amyloglucosidase	300 tonnes
Bacterial amylase	300 tonnes
Fungal amylase	10 tonnes

The bulk of these enzymes are used in the starch-processing industry, where they have gradually been replacing acid as a hydrolytic agent. Of particular interest at this time is glucose isomerase which transforms the ultimate product of starch degradation, D-glucose, into fructose (Barfoed, 1976). Other potential new areas of application are in the alcoholic beverage and industrial alcohol industries, which have a World-wide output of £26,000,000,000 per annum (Bronn, 1976).

REFERENCES

Abdullah, M., Catley, B. J., Lee, E. Y. C., Robyt, J., Wallenfels, K. and Whelan, W. J. (1966). *Cereal Chemistry* **43**, 111.

Abd El-Akher, M., El-Leithy, M. A., El-Marsafy, M. K. and Kassim, S. A. (1973). *Zentralblatt für Bakteriologie, Parasitenkunde Infektionskrankheiten und Hygiene,* Abteilung II **128**, 48.

Abdullah, M., Fleming, I. D., Taylor, P. M. and Whelan, W. J. (1963). *Biochemical Journal* **89**, 35P.

Adachi, S., Nakanishi, K., Matsuno, R. and Kamikubo, T. (1977). *Agricultural and Biological Chemistry* **41**, 1673.

Adams, P. R. and Deploey, J. J. (1976). *Mycologia* **68**, 934.

Adams, K. R. and Priest, F. G. (1977). *Federation of European Microbiological Societies Microbiology Letters* **1**, 269.

Adelburg, E. A., Mandel, M. and Chen, C. C. G. (1965). *Biochemical and Biophysical Research Communications* **18**, 788.

Akabori, S., Hagihara, B. and Ikenaka, T. (1954). *Journal of Biochemistry, Tokyo* **41**, 577.

Alikhanian, S. I. (1962). *Advances in Applied Microbiology* **4**, 1.

Allam, A. M., Hussein, A. M. and Ragab, A. M. (1977). *Zentralblatt für Bakteriologie, Parasitenkunde Infektionskrankheiten und Hygiene*, Abteilung II **132**, 143.

Allen, M. J. and Hartman, P. A. (1972). *Journal of Bacteriology* **109**, 452.

Andrzejczuk-Hybel, J. and Kaczkowski, J. (1971). *Bulletin de L'Academie Polonaise des Sciences* **XIX** No. 5, 313.

Arai, M., Koyano, T., Ozana, H., Minoda, T. and Yamada, K. (1968). *Agricultural and Biological Chemistry* **32**, 507.

Arai, M., Minoda, Y. and Yamada, K. (1969). *Agricultural and Biological Chemistry* **33**, 922.

Aravina, L. A. and Ponomarera, V. D. (1977). *Microbiologiya* **46**, 379.

Armbruster, F. C. (1961). United States Patent 3,012,944.

Armbruster, F. C. (1974). United States Patent 3,853,706.

Armbruster, F. C. and Bruner, R. L. (1967). United States Patent 3,303,102.

Aski, K., Arai, M., Minoda, Y. and Yamada, K. (1971). *Agricultural and Biological Chemistry* **35**, 1913.

Attia, R. M. and Ali, S. A. (1977). *Zentralblatt für Bakteriologie, Parasitenkunde Infektionskrankheiten und Hygiene*, Abteilung II **132**, 322.

Aunstrup, K. (1967). British Patent 1,092,775.

Aunstrup, K. (1977). *In* 'Biotechnology and Fungal Differentiation' (J. Meyrath and J. D. Bu'lock, eds.), p. 157. Academic Press, London.

Babber, I., Behki, R. M. and Srinivasan, M. C. (1960). Indian Patent 66,096.

Bailey, M. J. and Markkanen, P. H. (1975). *Journal of Applied Chemistry and Biotechnology* **25**, 73.

Bakerspigel, A. (1953). *Mycologia* **45**, 596.

Bakerspigel, A. (1954). *Mycologia* **46**, 680.

Ball, C., Lawrence, A. J., Butler, J. M. and Morrison, K. B. (1978). *European Journal of Applied Microbiology and Biotechnology* **5**, 95.

Balls, A. K., Walden, M. K. and Thompson, R. R. (1948). *Journal of Biological Chemistry* **173**, 9.

Banks, W. and Greenwood, C. T. (1975). 'Starch and its Components' Edinburgh University Press.

Barfoed, H. C. (1976). *Cereal Foods World* **21**, 588.

Baribo, L. E. (1967). United States Patent 3,298,926.

Barratt, R. W., Johnson, G. B. and Ogata, W. N. (1965). *Genetics, Princeton* **52**, 233.

Barton, L. L., Lineback, D. R. and Georgi, C. E. (1969). *Journal of General and Applied Microbiology* **15**, 327.

Barton, L. L., Georgi, C. E. and Lineback, D. R. (1972). *Journal of Bacteriology* **111**, 771.

Bata, J., Vallier, P. and Colobert, L. (1978). *Experientia* **34**, 572.

Belloc, A., Florent, J., Mancy, D. and Verrier, J. (1975). United States Patent 3,906,113.

Belyakova, L. A., Lebedeva, M. K. and Napelina, L. N. (1975). *Mikologiya 1 Fitopatologia* **9**, 153.

Bender, H. and Wallenfels, K. (1961). *Biochemische Zeitschrift* **334**, 79.
Bendetskii, K. M., Tarovenko, V. L., Korchagina, G. T., Senatovora, T. P. and Khakhanova, T. S. (1974). *Biokhimiya* **39**, 802.
Berkhout, F. (1976). *In* 'Starch Production Technology' (J. A. Radley, ed.), p. 109. Applied Science Publishers Ltd., London.
Birch, G. G. and Schallenberger, R. S. (1973). *In* 'Molecular Structure and Function of Food Carbohydrates' (G. G. Birch and L. F. Green, eds.), p. 9. Applied Science Publishers, London.
Bliesmer, B. O. and Hartman, P. A. (1973). *Journal of Bacteriology* **113**, 526.
Bode, H. E. (1966a). United States Patent 3,249,512.
Bode, H. E. (1966b). United States Patent 3,249,514.
Boeswinkel, H. J. (1976). *Transactions of the British Mycological Society* **66**, 183.
Bosmans, J. (1974). *Mycopathologia et Mycologia Applicata* **53**, 1.
Boidin, A. R. and Effront, J. (1917a). United States Patent 1,227,374.
Boidin, A. R. and Effront, J. (1917b). United States Patent 1,227,525.
Borgia, P. T. and Campbell, L. L. (1978). *Journal of Bacteriology* **134**, 389.
Boyer, E. W. and Ingle, M. B. (1972). *Journal of Bacteriology* **110**, 992.
Braverman, S. W. and Crosier, W. F. (1966). *Plant Disease Reporter* **50**, 321.
Bronn, W. K. (1976). *Die Branntwein Wirtschaft* **116** (12), 216.
Bulich, A. A. (1976). United States Patent 3,963,575.
Buonocore, V., Caporale, C., de Rosa, M. and Gambacorta, A. (1976). *Journal of Bacteriology* **128**, 515.
Calam, C. T. (1964). *Progress in Industrial Microbiology* **5**, 1.
Calam, C. T. (1970). *Methods in Microbiology* **3A**, 435.
Campbell, L. L. (1954). *Journal of the American Chemical Society* **76**, 5256.
Carmichael, J. W. (1956). *Mycologia* **48**, 378.
Carmichael, J. W. (1962). *Mycologia* **54**, 432.
Chaloupka, J. and Kreckova, P. (1966). *Folia Microbiologica* **11**, 82.
Coleman, G. (1967). *Journal of General Microbiology* **49**, 421.
Coleman, G. and Grant, M. A. (1966). *Nature, London* **211**, 306.
Corman, J. (1967). United States Patent 3,296,092.
Corman, J. and Langlykke, A. F. (1948). *Cereal Chemistry* **25**, 190.
Corn Products Company (1960). British Patent 849,508.
Corn Products Company (1962). British Patent 904,423.
Corn Products Company (1971). British Patent 1,232,130.
Croxall, W. J. (1966). United States Patent 3,254,003.
Dade, H. A. (1960). *In* 'Herb', I.M.I. Handbook, p. 40. Commonwealth Mycological Institute, Kew.
Daily, W. A. and Higgens, C. E. (1973). *Cyrobiology* **10**, 364.
Day, D. F. (1978). *Current Microbiology* **1**, 181.
De Song Chong and Tsujisaka, Y. (1976). *Journal of Fermentation Technology* **54**, 264.
Dessein, A. and Schwartz, M. (1974). *European Journal of Biochemistry* **45**, 363.
Dingle, J. and Solomons, G. L. (1951). *Nature, London* **168**, 425.
Dunn, C. G., Fuld, G. J., Yamada, K., Urioste, J. M. and Casey, P. R. (1959). *Applied Microbiology* **7**, 212.
Dworschack, R. G. and Nelson, C. A. (1972). United States Patent 3,660,236.
Ebertová, H. (1966). *Folia Microbiologica* **11**, 422.
Eisele, B., Rasched, I. R. and Wallenfels, K. (1972). *European Journal of Biochemistry* **26**, 62.

3. AMYLASES, AMYLOGLUCOSIDASES AND RELATED GLUCANASES   161

Elliott, R. F. (1975). *Laboratory Practice* **24**, 751.

Ellis, J. J. and Roberson, J. A. (1968). *Mycologia* **60**, 399.

Enatsu, T. and Shinmyo, A. (1978). *Advances in Biochemical Engineering* **9**, 112.

Ensley, B., McHugh, J. J. and Barton, L. L. (1975). *Journal of General and Applied Microbiology* **21**, 51.

Evans, R. B. and Manners, D. J. (1971). *Carbohydrate Research* **20**, 339.

Evers, A. D., Gough, B. M. and Pybus, J. N. (1971). *Die Stärke* **23**, 16.

Feniksova, R. V. (1957). *Proceedings of the International Symposium on Enzyme Chemistry,* Tokyo-Kyoto, p. 482.

Fennell, D. I. (1960). *Botanical Review* **26**, 79.

Fischer, E. H. and De Montmollin, R. (1951a). *Nature, London* **160**, 606.

Fischer, E. H. and De Montmollin, R. (1951b). *Helvetica Chimica Acta* **34**, 1987.

Fischer, E. H. and De Montmollin, R. (1951c). *Helvetica Chimica Acta* **34**, 1994.

Fischer, E. H. and Stein, E. A. (1960). *In* 'The Enzymes' (P. D. Boyer, H. Lardy and K. Myrbäck, eds.), vol. 4, p. 313. Academic Press, New York.

Fleming, I. D. (1968). *In* 'Starch and Its Derivatives' (J. A. Radley, ed.), p. 498. Chapman and Hall Ltd., London.

Fleming, I. D. and Stone, B. A. (1965). *Biochemical Journal* **97**,13P.

Fogarty, W. M. and Griffin, P. J. (1973). *Journal of Applied Chemistry and Biotechnology* **23**, 166.

Fogarty, W. M. and Griffin, P. J. (1975). *Journal of Applied Chemistry and Biotechnology* **25**, 229.

Fogarty, W. M. and Kelly, C. T. (1979a). *Progress in Industrial Microbiology* **15**, 87.

Fogarty, W. M. and Kelly, C. T. (1979b). *In* 'Topics in Enzyme and Fermentation Biotechnology' (A. Wiseman, ed.), vol. 3, p. 45. Ellis Horwood, Chichester.

Frantz, B. M., Lee, E. Y. C. and Whelan, W. J. (1966). *Biochemical Journal* **100**, 7.

Freedberg, I. M., Levin, Y., Kay, C. M., McCubbin, W. D. and Katchalski-Katzir, E. (1975). *Biochimica et Biophysica Acta* **391**, 361.

French, D. (1960). *Bulletin of the Society of Chemistry and Biology* **42**, 1677.

Fujimori, H., Ohnishi, M. and Hiromi, K. (1978). *Journal of Biochemistry, Tokyo* **83**, 1503.

Fujio, Y., Shiosaka, M. and Ueda, S. (1970a). *Journal of Fermentation Technology* **48**, 8.

Fujio, Y., Sambuichi, M. and Ueda, S. (1970b). *Journal of Fermentation Technology* **48**, 601.

Fukui, T. and Nikuni, Z. (1969). *Agricultural and Biological Chemistry* **33**, 884.

Fukumoto, J. (1943a). *Journal of the Agricultural Chemical Society of Japan* **19**, 487.

Fukumoto, J. (1943b). *Journal of the Agricultural Chemical Society of Japan* **19**, 634.

Fukumoto, J. (1963). *Journal of Fermentation Technology* **41**, 427.

Fukumoto, J. (1967). Quoted in 'Starch Production Technology' (J. A. Radley, ed.), p. 302. Applied Science Publications.

Fukumoto, J. and Negoro, H. (1966). United States Patent 3,272,717.

Fukumoto, J., Yamamoto, T., Tsuru, D. and Ichikawa, K. (1957). *Proceedings of the International Symposium on Enzyme Chemistry* Tokyo-Kyoto, p. 479.

Gandhi, A. P. and Kjaergaard, L. (1975). *Biotechnology and Bioengineering* **17**, 1109.

Glymph, J. L. and Stutzenberger, F. J. (1977). *Applied and Environmental Microbiology* **34**, 391.

Goering, K. J. and Bruski, V. C. (1954). *Cereal Chemistry* **31**, 7.

Gracheva, I. M., Lushchik, T. A., Tyrsin, Yu. A. and Pinchukova, E. E. (1977). *Biokhimiya* **42**, 1603.

Gratzner, H. G. (1972). *Journal of Bacteriology* **111**, 443.

Gratzner, H. and Sheehan, D. N. (1969). *Journal of Bacteriology* **97**, 544.

Greenshields, R. N. and Macgillivray, A. W. (1972). *Process Biochemistry* **7** (12), 11.

Greenwood, C. T. (1956). *Advances in Carbohydrate Chemistry* **11**, 335.

Greenwood, C. T. (1976). *Advances in Cereal Science Technology* **1**, 119.

Greenwood, C. T. and Milne, E. A. (1968a). *Die Stärke* **20**, 139.

Greenwood, C. T. and Milne, E. A. (1968b). *Advances in Carbohydrate Chemistry* **23**, 281.

Griffin, P. J. and Fogarty, W. M. (1973a). *Journal of Applied Chemistry and Biotechnology* **23**, 301.

Griffin, P. J. and Fogarty, W. M. (1973b). *Biochemical Society Transactions* **1**, 397.

Griffin, P. J. and Fogarty, W. M. (1973c). *Biochemical Society Transactions* **1**, 1097.

Grootegoed, J. A., Lauwers, A. M. and Heinen, W. (1973). *Archives of Microbiology* **90**, 223.

Gunja, Z. H., Manners, D. J. and Maung, K. (1961). *Biochemical Journal* **81**, 392.

Gunja-Smith, Z. (1974a). United States Patent 3,790,446.

Gunja-Smith, Z. (1974b). British Patent 1,377,064.

Gunja-Smith, Z., Marshall, J. J., Smith, E. E. and Whelan, W. J. (1970). *Federation of European Biochemical Societies Letters* **12**, 96.

Hagihara, B. (1960). *In* 'Isolation of Biologically Active Proteins' (P. Alexander and R. J. Block, eds.), vol. 1, p. 32. Pergamon Press, New York.

Hagihara, B., Matsubara, H. and Okunuki, K. (1956). *Journal of Biochemistry, Tokyo* **43**, 469.

Hall, J. F. and Webb, T. J. B. (1975). *Journal of the Institute of Brewing* **81**, 471.

Hamauzu, Z.-I., Hiromi, K. and Ono, S. (1965). *Journal of Biochemistry, Tokyo* **57**, 39.

Hang, T. D. and Woodams, E. E. (1977). *Applied and Environmental Microbiology* **33**, 1293.

Harada, T. (1931). *Industrial and Engineering Chemistry* **23**, 1424.

Harada, T. (1951). *Bulletin of the Chemical Society of Japan* **24**, 105.

Harada, T. (1977). *Memoirs of the Institute of Scientific and Industrial Research, Osaka University* **34**, 49.

Harada, T., Yokobayashi, K. and Misaki, A. (1968). *Applied Microbiology* **16**, 1439.

Hasegawa, A., Miwa, M., Oshima, T. and Imahori, K. (1976). *Journal of Biochemistry, Tokyo* **79**, 35.

Hattori, Y. and Takeuchi, I. (1962). *Agricultural and Biological Chemistry* **26**, 316.

Hayashibara Company (1971). British Patent 1,233,743.

Hayashibara Company (1972). British Patent 1,285,204.

Hayashibara Company (1973a). British Patent 1,310,261.

Hayashibara Company (1973b). British Patent 1,336,599.

Hayashida, S. (1975). *Agricultural and Biological Chemistry* **39**, 2093.

Hayashida, S. and Yoshino, E. (1978). *Agricultural and Biological Chemistry* **42**, 927.

Hayashida, S., Nomura, T., Yoshino, E. and Hongo, M. (1976). *Agricultural and Biological Chemistry* **40**, 141.

Hayes, L. P. (1974). United States Patent 3,806,415.

Heady, R. E. (1974). United States Patent 3,806,419.

Hehre, E. J., Okada, G. and Genghof, D. S. (1969). *Archives of Biochemistry and Biophysics* **135**, 75.

Heineken, F. G. and O'Connor, R. J. (1972). *Journal of General Microbiology* **73**, 35.

Heinen, U. J. and Heinen, W. (1972). *Archives of Microbiology* **82**, 1.

Hesseltine, C. W., Bradle, B. J. and Benjamin, C. R. (1960). *Mycologia* **52**, 762.

Higashihara, M. and Okada, S. (1974). *Agricultural and Biological Chemistry* **38**, 1023.

High, R. L. and Rogols, S. (1966). United States Patent 3,251,748.

Hirchberg, L. M. (1957). *Food* **26**, 130 and 176.

Hope, G. C. and Dean, A. C. R. (1974). *Biochemical Journal* **144**, 403.

Hopkins, R. H. and Kulka, D. (1957). *Archives of Biochemistry and Biophysics* **69**, 45.

Horikoshi, K. (1971a). *Agricultural and Biological Chemistry* **35**, 1407.

Horikoshi, K. (1971b). *Agricultural and Biological Chemistry* **35**, 1783.

Horikoshi, K., Ikeda, Y. and Tanaka, Y. (1974). United States Patent 3,826,715.

Horwath, R. O. and Rotheim, P. (1976). United States Patent Application B566,585.

Horwath, R. O., Cole, G. W. and Lally, J. A. (1976). United States Patent Application B592, 146.

Horwath, R. O., Lally, J. A. and Rotheim, P. (1977). United States Patent 4,011,139.

Hostinová, E. and Zelinka, J. (1978). *Die Stärke* **30**, 338.

Hsiu, J., Fischer, E. H. and Stein, E. A. (1964). *Biochemistry, New York* **3**, 61.

Hurst, T. L. (1975). United States Patent 3,897,305.

Hwang, S. W. (1966). *Applied Microbiology* **14**, 784.

Hwang, S. W. (1968). *Mycologia* **60**, 613.

Iijima, T. and Sakane, T. (1973). *Cryobiology* **10**, 379.

Ingle, M. B. and Boyer, E. W. (1972). *Developments in Industrial Microbiology* **13**, 421.

Ingle, M. B. and Boyer, E. W. (1976). *In* 'Microbiology' (D. Schlessinger, ed.), p. 420. American Society for Microbiology, Washington D.C.

Isemura, T. and Kakiuchi, K. (1962). *Journal of Biochemistry, Tokyo* **51**, 385.

Janke, A. and Schaefer, B. (1940). *Zentralblatt für Bakteriologie Parasitenkunde Infektionskrankheiten und Hygiene*, Abteilung II **102**, 241.

Jeanningros, R., Creuzet-Sigal, N., Frixon, C. and Cattaneo, J. (1976). *Biochimica et Biophysica Acta* **438**, 186.

Johnson, J. A. and Miller, B. S. (1951). *Food Industry* **23**, 80.

Joyce, A. M. (1977). Ph.D. Thesis: National University of Ireland.

Junge, J. M., Stein, E. A., Neurath, H. and Fischer, E. H. (1959). *Journal of Biological Chemistry* **234**, 556.

Kainuma, K., Kobayashi, S., Ito, T. and Suzuki, S. (1972). *Federation of European Biochemical Societies Letters* **26**, 281.

Kainuma, K., Wako, K., Kobayashi, S., Nogami, A. and Suzuki, S. (1975). *Biochimica et Biophysica Acta* **410**, 333.

Kainuma, K., Kobayashi, S. and Harada, T. (1978). *Carbohydrate Research* **61**, 345.

Kaji, A., Sato, M., Kobayashi, M. and Murao, T. (1976). *Journal of the Agricultural Chemical Society of Japan* **50**, 509.

Kamachi, T. and Tanaka, S. (1957). *Nippon Jozo Kyokai Zasshi* **52**, 811.

Kato, I., Toda, H. and Narita, K. (1967). *Proceedings of the Japanese Academy* **43**, 38.

Kerr, R. W. (1962). United States Patent 3,017,330.

Kihlman, B. A. (1966). 'Actions of Chemicals on Dividing Cells'. Prentice-Hall, Englewood Cliffs, N.J., U.S.A.

King, N. J. (1967). *Biochemical Journal* **105**, 577.

Kitahata, S. and Okada, S. (1974). *Agricultural and Biological Chemistry* **38**, 2413.

Kneen, E. and Beckord, L. D. (1946). *Archives of Biochemistry* **10**, 41.

Knight, J. W. (1969). 'The Starch Industry'. Pergamon Press, Oxford.

Koaze, Y., Nakajima, Y., Hidaka, H., Niwa, T., Adachi, T., Yoshida, K., Ito, J., Niida, T., Shomura, T. and Ueda, M. (1974). United States Patent 3,804,717.

Koaze, Y., Nakajima, Y., Hidaka, H., Niwa, T., Adachi, T., Yoshida, K., Ito, J., Niida, T., Shomura, T. and Ueda, M. (1975). United States Patent 3,868,464.

Kobayashi, M. and Matsuda, K. (1978). *Agricultural and Biological Chemistry* **42**, 181.

Kokubu, T., Karube, I. and Suzuki, S. (1978). *European Journal of Applied Microbiology and Biotechnology* **5**, 233.

Komaki, T., Matsuba, Y., Okamoto, N., Yamada, T. and Sawada, K. (1956). *Denpun Kogyo, Gakkaishi* **4**, 89.

Kooi, E. R., Harjes, C. F. and Gilkison, J. S. (1962). United States Patent 3,042,584.

Kooi, E. R., Bruner, R. L. and Newkirk, T. H. (1966). United States Patent 3,264,193.

Kramer, C. L. and Mix, A. J. (1957). Transactions of the Kansas Academy of Science 60, 58.

Krebs, J. (1962). United States Patent 3,029,192.

Krizková, L. and Balan, J. (1975). Folia Microbiologica 20, 351.

Krzechowsk, M. and Urbanek, H. (1975). Applied Microbiology 30, 163.

Kuo, M. J. and Hartman, P. A. (1967). Canadian Journal of Microbiology 13, 1157.

Kurimoto, M. and Sugimoto, K. (1975). United States Patent 3,881, 991.

Kurushima, M., Sato, J. and Kitahara, K. (1974). Journal of the Agricultural Chemical Society of Japan 48, 665.

Kuswanto, K. P., Kato, K., Amemura, A. and Harada, T. (1976). Journal of Fermentation Technology 54, 192.

Kvesitadze, G. I., Kokonashvili, G. N. and Fenixova, R. V. (1974). United States Patent 3,826,716.

Land, C. E. and Barton, R. R. (1966). United States Patent 3,265, 586.

Lane, A. G. and Pirt, S. J. (1971). Journal of Applied Chemistry and Biotechnology 21, 330.

Lane, A. G. and Pirt, S. J. (1973). Journal of Applied Chemistry 23, 309.

Langlois, D. P. and Turner, W. (1959). United States Patent 2,893,921.

Lapage, S. P., Shelton, J. E., Mitchell, T. G. and Mackenzie, A. R. (1970). Methods in Microbiology 3A, 135.

Lee, E. Y. C. and Whelan, W. J. (1971). In 'The Enzymes' (P. D. Boyer, ed.), vol. 5, p. 91. Academic Press, New York.

Le Mense, E. H., Corman, J., Van Lanen, J. M. and Langlykke, A. F. (1947). Journal of Bacteriology 54, 149.

Lenney, J. F. (1962). United States Patent 3,039,936.

Leopold, H. and Germann, H. G. (1940). Zentralblatt für Bakteriologie Parasitenkunde Infektionskrankheiten und Hygiene, Abteilung II 102, 65.

Liggett, R. W., Mussulman, W. C., Rentshler, D. F. and Ziffa, J. (1959). United States Patent 2,881,115.

Lin, C.-F. (1972). In 'Fermentation Technology Today' (G. Terui, ed.), p. 327.

Lineback, D. R. and Aira, L. A. (1972). Cereal Chemistry 49, 283.

Lineback, D. R. and Baumann, W. E. (1970). Carbohydrate Research 14, 341.

Lineback, D. R., Georgi, C. E. and Doty, R. L. (1966). Journal of General and Applied Microbiology 12, 27.

Lineback, D. R., Russell, I. J. and Rasmussen, C. (1969). Archives of Biochemistry and Biophysics 134, 539.

Little, G. N. and Gordon, M. A. (1967). Mycologia 59, 733.

Lockwood, L. B. (1952). Transactions of the New York Academy of Sciences, Series II 15, 2.

Lulla, B. S. (1951). Archives of Biochemistry 7, 244.

MacDonald, K. D. (1972). Applied Microbiology 23, 990.

Madsen, G. B., Norman, B. E. and Slott, S. (1973). Die Stärke 25, 304.

Maher, G. G. (1968). Die Stärke 20, 228.

Manchee, R. J. (1975). Journal of Applied Bacteriology 38, 191.

Manners, D. J. (1971a). Biochemical Journal 123, 1P.

Manners, D. J. (1971b). Nature New Biology 234, 150.

Manners, D. J. (1974). In 'Essays in Biochemistry' (P. N. Campbell and F. Dickens, eds.), vol. 10, p. 37. Academic Press, London.

Manners, D. J. and Marshall, J. J. (1971). Carbohydrate Research 18, 203.

Manning, G. B. and Campbell, L. L. (1961). Journal of Biological Chemistry 236, 2952.

Markkanen, P. (1975). Technical Research Centre of Finland. Materials and Processing Technology Publication 12.

Markkanen, P. H. and Bailey, M. J. (1974). *Journal of Applied Microbiology* 19, 535.

Markkanen, P. H. and Bailey, M. J. (1975). *Journal of Applied Chemistry and Biotechnology* 25, 863.

Markowitz, A., Klein, H. P. and Fischer, E. H. (1956). *Biochimica et Biophysica Acta* 19, 267.

Marshall, J. J. (1973). *Federation of European Biochemical Societies Letters* 37, 269.

Marshall, J. J. (1974). *Federation of European Biochemical Societies Letters* 46, 1.

Marshall, J. J. (1975). *Die Stärke* 27, 377.

Marshall, B. J., Coote, G. G. and Scott, W. J. (1973). *Applied Microbiology* 26, 206.

Marx, D. H. and Daniel, W. J. (1975). *Canadian Journal of Microbiology* 22, 338.

Masuda, K. and Sugimoto, K. (1973). United States Patent 3,766,014.

Matsuno, R., Nakanishi, K., Ohnishi, M., Hiromi, K. and Kamikubo, T. (1978). *Journal of Biochemistry, Tokyo* 83, 859.

Matsuzaki, H., Yamane, K. and Maruo, B. (1974a). *Biochimica et Biophysica Acta* 365, 248.

Matsuzaki, H., Yamane, K., Yamaguchi, K., Nagata, Y. and Maruo, B. (1974b). *Biochimica et Biophysica Acta* 365, 235.

Mayne, R. Y., Bennett, J. N. and Tallant, J. (1971). *Mycologia* 63, 644.

Mazza, L. A. and Ertola, R. J. (1970). *Applied Microbiology* 19, 535.

McDaniel, L. E. and Bailey, E. G. (1968). *Applied Microbiology* 16, 912.

McGinnis, M. R., Padhye, A. A. and Ajello, L. (1974). *Applied Microbiology* 28, 218.

McKelvy, J. F. and Lee, Y. C. (1969). *Archives of Biochemistry and Biophysics* 132, 99.

McWethy, S. J. and Hartman, P. A. (1977). *Journal of Bacteriology* 129, 1537.

Meers, J. L. (1972). *Antonie van Leeuwenhoek* 38, 585.

Mehrota, B. S. and Hesseltine, C. W. (1958). *Applied Microbiology* 6, 179.

Menzi, R., Stein, E. A. and Fischer, E. H. (1957). *Helvetica Chimica Acta* 40, 534.

Mercier, C., Frantz, B. M. and Whelan, W. J. (1972). *European Journal of Biochemistry* 26, 1.

Meyer, K. H. (1952). *Experientia* 8, 405.

Meyer, K. H., Spahr, P. F. and Fischer, E. H. (1953). *Helvetica Chimica Acta* 36, 1924.

Miah, M. N. N. and Ueda, S. (1977). *Die Stärke* 29, 191 and 235.

Minagawa, T. and Hamaishi, T. (1962). United States Patent 3,031,380.

Minoda, Y. and Asai, T. (1960). Japan Patent 5046.

Minoda, Y. and Yamada, K. (1963). *Agricultural and Biological Chemistry* 27, 806.

Minoda, Y., Arai, M., Torigoe, Y. and Yamada, K. (1968). *Agricultural and Biological Chemistry* 32, 110.

Mitchell, E. D., Riquetti, P., Loring, R. H. and Carraway, K. L. (1973). *Biochimica et Biophysica Acta* 295, 314.

Moore, L. W. and Carlson, R. V. (1975). *Phytopathology* 65, 246.

Morita, Y., Shimizu, K., Ohga, M. and Korenaga, T. (1966). *Agricultural and Biological Chemistry* 30, 114.

Moseley, M. H. and Keay, L. (1970). *Biotechnology and Bioengineering* 12, 251.

Moses, V. and Sharp, P. B. (1972). *Journal of General Microbiology* 71, 181.

Murayama, T. and Ishikawa, T. (1973). *Journal of Bacteriology* 115, 796.

Nakamura, N. and Horikoshi, K. (1976). *Agricultural and Biological Chemistry* 40, 753.

Nakamura, N., Watanabe, K. and Horikoshi, K. (1975). *Biochimica et Biophysica Acta* 397, 188.

Nikolov, T., Kibarska, T. and Nedeva, D. (1970). *Comptes Rendus de L'Academie des Sciences Agricoles en Bulgarie* 3, 145.

Nomura, M., Maruo, B. and Akabori, S. (1956). *Journal of Biochemistry, Tokyo* **43**, 143.

Norman, B. E. (1978). *Proceedings of the 83rd Meeting of the Society for General Microbiology, Aberdeen.*

Norrman, J. and Wöber, G. (1975). *Archives of Microbiology* **102**, 253.

Nyiri, L. (1971). *International Chemical Engineering* **11**, 447.

Ogasahara, K., Imanishi, A. and Isemura, T. (1970). *Journal of Biochemistry, Tokyo* **67**, 65.

Ohba, R. and Ueda, S. (1973). *Agricultural and Biological Chemistry* **37**, 2821.

Ohba, R. and Ueda, S. (1975). *Agricultural and Biological Chemistry* **39**, 967.

Ohga, M., Shimizu, K. and Morita, Y. (1966). *Agricultural and Biological Chemistry* **30**, 967.

Ohnishi, M. and Hiromi, K. (1978). *Carbohydrate Research* **61**, 335.

Ohnishi, M., Yamashita, T. and Hiromi, K. (1976). *Journal of Biochemistry, Tokyo* **79**, 1007.

Oikawa, A. (1959). *Journal of Biochemistry, Tokyo* **46**, 463.

Oikawa, A. and Maeda, A. (1957). *Journal of Biochemistry, Tokyo* **44**, 745.

Okada, S. and Higashibara, M. (1974). United States Patent 3,804,718.

Okazaki, H. (1956). *Archives of Biochemistry and Biophysics* **63**, 322.

Okazaki, H. (1957). *Proceedings of the International Symposium on Enzyme Chemistry, Tokyo and Kyoto* **2**, 494.

Onions, A. H. S. (1971a). *Methods in Microbiology* **4**, 113.

Onions, A. H. S. (1971b). *Biological Journal of the Linnean Society* **3**, 189.

Onishi, H. (1973). United States Patent 3,767,530.

Onishi, H. and Hidaka, O. (1978). *Canadian Journal of Microbiology* **24**, 1017.

Ono, S., Hiromi, K. and Hamauzu, Z.-I. (1965). *Journal of Biochemistry, Tokyo* **57**, 34.

Oshima, K. and Church, M. B. (1923). *Industrial and Engineering Chemistry* **15**, 67.

Outtrup, H. and Aunstrup, K. (1975). *Proceedings of the First International Congress, International Association of Microbiological Societies* **5**, 204.

Outtrup, H., Andresen, O. and Aunstrup, K. (1970). German Patent Application 2,025,748.

Palmer, T. J. (1975). *Process Biochemistry* **10** (10), 19.

Palmer, T. N., Wöber, G. and Whelan, W. J. (1973). *European Journal of Biochemistry* **39**, 601.

Pazur, J. H. and Ando, T. (1959). *Journal of Biological Chemistry* **234**, 1966.

Pazur, J. H. and Ando, T. (1960). *Journal of Biological Chemistry* **235**, 297.

Pazur, J. H. and Ando, T. (1961). *Archives of Biochemistry and Biophysics* **93**, 43.

Pazur, J. H. and Kleppe, K. (1962). *Journal of Biological Chemistry* **237**, 1002.

Pazur, J. H. and Okada, S. (1966). *Journal of Biological Chemistry* **241**, 4146.

Pazur, J. H. and Okada, S. (1967). *Carbohydrate Research* **4**, 371.

Pazur, J. H., Knull, H. R. and Simpson, D. L. (1970). *Biochemical and Biophysical Research Communications* **40**, 110.

Pazur, J. H., Knull, H. R. and Cepure, A. (1971). *Carbohydrate Resarch* **20**, 83.

Perkins, D. D. (1962). *Canadian Journal of Microbiology* **8**, 591.

Pfueller, S. L. and Elliott, W. H. (1969). *Journal of Biological Chemistry* **244**, 48.

Phillips, L. L. and Caldwell, M. L. (1951). *Journal of the American Chemical Society* **73**, 3559 and 3563.

Piguet, A. and Fischer, E. H. (1952). *Helvetica Chimica Acta* **35**, 257.

Pool, E. L. and Underkofler, L. A. (1953). *Journal of Agricultural and Food Chemistry* **1**, 20 and 87.

Pridham, T. G., Lyons, A. J. and Phrompatima, B. (1973). *Applied Microbiology* **26**, 441.

Priest, F. G. (1975). *Biochemical and Biophysical Research Communications* **63**, 606.

3. AMYLASES, AMYLOGLUCOSIDASES AND RELATED GLUCANASES 167

Priest, F. G. (1977). *Bacteriological Reviews* **41**, 711.

Qadeer, M. A. and Kausar, T. (1971). *Pakistan Journal of Scientific and Industrial Research* **14**, 247.

Radley, J. A. (1976). 'Starch Production Technology'. Applied Science Publishers, London.

Redway, K. F. and Lapage, S. P. (1974). *Cryobiology* **11**, 73.

Rentshler, D. F., Langlois, D. P., Larson, R. F., Alverson, L. H. and Liggett, R. W. (1962). United States Patent 3,039,935.

Roberts, P. J. P. and Whelan, W. J. (1960). *Biochemical Journal* **76**, 246.

Robyt, J. F. and Ackerman, R. J. (1971). *Archives of Biochemistry and Biophysics* **145**, 105.

Robyt, J. F. and Ackerman, R. J. (1973). *Archives of Biochemistry and Biophysics* **155**, 445.

Robyt, J. and French, D. (1964). *Archives of Biochemistry and Biophysics* **104**, 338.

Rose, D. (1948). *Archives of Biochemistry and Biophysics* **16**, 349.

Rowe, A. W. and Weill, C. E. (1962). *Biochimica et Biophysica Acta* **65**, 245.

Russell, I. J. and Lineback, D. R. (1970). *Carbohydrate Research* **15**, 123.

Saito, N. (1973). *Archives of Biochemistry and Biophysics* **155**, 290.

Saito, N. and Yamamoto, K. (1975). *Journal of Bacteriology* **121**, 848.

Sarbhoy, A. K., Ghosh, S. K., Lal, S. P. and Lall, G. (1974). *Indian Phytopathology* **27**, 361.

Sasaki, T., Yamasaki, M., Maruo, B., Yoneda, Y., Yamane, K., Takatsuki, A. and Tamura, G. (1976). *Biochemical and Biophysical Research Communications* **70**, 125.

Schaeffer, P. (1969). *Bacteriological Reviews* **33**, 48.

Schellart, J. A., van Arem, E. J. F., van Boekel, M. A. J. S. and Middlehoven, W. J. (1976a). *Antonie van Leeuwenhoek* **42**, 239.

Schellart, J. A., Visser, F. M. W., Zandstra, T. and Middlehoven, W. J. (1976b). *Antonie van Leeuwenhoek* **42**, 229.

Schipper, M. A. A. and Bekker-Holtman, J. (1976). *Antonie van Leeuwenhoek* **42**, 325.

Schwimmer, S. and Garibaldi, J. A. (1952). *Cereal Chemistry* **29**, 108.

Sekiguchi, J. and Okada, H. (1972). *Journal of Fermentation Technology* **50**, 801.

Semenov, S. M. (1973). *Antibiotiki* **18**, 1026.

Semenov, S. M. (1975). *Antibiotiki* **20**, 779.

Sherman, H. C. and Tanberg, A. P. (1916). *Journal of the American Chemical Society* **38**, 1638.

Shetty, R. M., Lineback, D. R. and Seib, P. A. (1974). *Cereal Chemistry* **51**, 364.

Shinke, R., Nishira, H. and Mugibayashi, N. (1974). *Agricultural and Biological Chemistry* **38**, 665.

Shinke, R., Kunimi, Y. and Nishira, H. (1975a). *Journal of Fermentation Technology* **53**, 687.

Shinke, R., Kunimi, Y. and Nishira, H. (1975b). *Journal of Fermentation Technology* **53**, 693.

Shinke, R., Kunimi, Y. and Nishira, H. (1975c). *Journal of Fermentation Technology* **53**, 698.

Shu, P. and Blackwood, A. C. (1951). *Canadian Journal of Botany* **29**, 113.

Simpson, F. J. and McCoy, E. (1953). *Applied Microbiology* **1**, 228.

Slott, S. and Madsen, G. B. (1975). United States Patent 3,912,590.

Smiley, K. L. (1967). United States Patent 3,301,768.

Smiley, K. L., Hensley, D. E., Smiley, M. J. and Gasdorf, H. J. (1971). *Archives of Biochemistry and Biophysics* **144**, 694.

Smith, J. S. and Lineback, D. R. (1976). *Die Stärke* **28**, 243.

Smittle, R. B., Gilliland, S. E. and Speck, M. L. (1972). *Applied Microbiology* **24**, 551.

Speakman, E. L. (1969). United States Patent 3,425,909.

Stein, E. A. and Fischer, E. H. (1958). *Journal of Biological Chemistry* **232**, 867.

Stutzenberger, F. and Carnell, R. (1977). *Applied and Environmental Microbiology* **34**, 234.

Subrahmanyam, A., Mangallam, S. and Gopalkirshnan, K. S. (1977). *Indian Journal of Experimental Biology* **15**, 495.

Sugimoto, T., Amemura, A. and Harada, T. (1974a). *Applied Microbiology* **28**, 536.

Sugimoto, K., Hirao, M. and Masuda, K. (1974b). United States Patent 3,827,940.

Suzuki, H. and Tanabe, O. (1962). *Reports of the Fermentation Research Institute, Japan* p. 57.

Suzuki, Y., Yuki, T., Kishigami, T. and Abe, S. (1976). *Biochimica et Biophysica Acta* **445**, 386.

Taha, S. M., Mahmoud, S. A. Z. and Attia, R. M. (1968). *Journal of Botany of the United Arab Republic* **11**, 49.

Takamine, J. (1894). United States Patent 525,823.

Takamine, J. (1914). *Journal of Industrial and Engineering Chemistry* **6**, 824.

Takasaki, Y. (1976a). *Agricultural and Biological Chemistry* **40**, 1515.

Takasaki, Y. (1976b). *Agricultural and Biological Chemistry* **40**, 1523.

Tatarenko, E. S., Igolkina, E. V. and Man'ko, V. G. (1976). *Mikrobiologicheskii Zhurnal* **38**, 176.

Tauber, H. (1949). 'The Chemistry and Technology of Enzymes'. Wiley, New York.

Taylor, P. M., Napier, E. J. and Fleming, I. D. (1978). *Carbohydrate Research* **61**, 301.

Thoma, J. A. and Koshland, D. E. (1960). *Journal of the American Chemical Society* **82**, 3329.

Thoma, J. A., Brothers, C. and Spradlin, J. (1970). *Biochemistry, New York* **8**, 1768.

Thompson, A., Kimiko, A., Wolfrom, M. L. and Inatome, M. (1954). *Journal of the American Chemical Society* **76**, 1309.

Tilden, E. B. and Hudson, C. S. (1942). *Journal of Bacteriology* **43**, 527.

Tobey, J. F. and Yousten, A. A. (1976). *Developments in Industrial Microbiology* **18**, 499.

Toda, H. and Akabori, S. (1963). *Journal of Biochemistry, Tokyo* **53**, 102.

Toda, H. and Narita, K. (1967). *Journal of Biochemistry, Tokyo* **62**, 767.

Toda, H. and Narita, K. (1968). *Journal of Biochemistry, Tokyo* **63**, 302.

Toda, H., Kato, I. and Narita, K. (1968). *Journal of Biochemistry, Tokyo* **63**, 295.

Tomomura, K., Suzuki, H., Nakamura, N., Kuraya, K. and Tanabe, O. (1961). *Agricultural and Biological Chemistry* **25**, 1.

Trollope, D. R. (1975). *Journal of Applied Bacteriology* **38**, 115.

Tsuboi, A., Yamasaki, Y. and Suzuki, Y. (1974). *Agricultural and Biological Chemistry* **38**, 543.

Tsuru, D. (1962). *Agricultural and Biological Chemistry* **26**, 288.

Tsuru, D. and Fukumoto, J. (1963). *Agricultural and Biological Chemistry* **27**, 279.

Ueda, S. (1957). *Proceedings of the International Symposium on Enzyme Chemistry, Tokyo and Kyoto* **2**, 491.

Ueda, S. and Kano, S. (1975). *Die Stärke* **27**, 123.

Ueda, S. and Nanri, N. (1967). *Applied Microbiology* **15**, 492.

Ueda, S. and Ohba, R. (1972). *Agricultural and Biological Chemistry* **36**, 2381.

Ueda, S., Yagisawa, M. and Sato, Y. (1971). *Journal of Fermentation Technology* **49**, 552.

Ueda, S., Ohba, R. and Kano, S. (1974). *Die Stärke* **26**, 374.

Umeki, K. and Yamamoto, T. (1975). *Journal of Biochemistry, Tokyo* **78**, 889.

Umeki, K. and Yamamoto, T. (1977). *Journal of Biochemistry, Tokyo* **82**, 101.

Underkofler, L. A. and Roy, D. K. (1951). *Cereal Chemistry* **28**, 18.

Underkofler, L. A., Denault, L. J. and Hou, E. F. (1965). *Die Stärke* **6**, 179.
Upton, M. E. and Fogarty, W. M. (1977). *Applied and Environmental Microbiology* **33**, 59.
Urata, G. (1957). *Journal of Biochemistry, Tokyo* **44**, 359.
Urlaub, H. and Wöber, G. (1975). *Federation of European Biochemical Societies Letters* **57**, 1.
Urlaub, H. and Wöber, G. (1978). *Biochimica et Biophysica Acta* **522**, 161.
Vallier, P., Bata, J. and Colobert, L. (1977). *Annales de Microbiologie* **128B**, 359.
Van Lanen, J. M. and Le Mense, E. H. (1946). *Journal of Bacteriology* **51**, 595.
Van Lanen, J. M. and Smith, M. B. (1968). United States Patent 3,418,211.
Voitkova-Lepshikova, A. and Kotskova-Kratokhvilova, A. (1966). *Microbiologiya* **35**, 773.
Walker, G. T. (1968). *Biochemical Journal* **108**, 33.
Walker, G. T. and Whelan, W. J. (1960a). *Biochemical Journal* **76**, 257.
Walker, G. T. and Whelan, W. J. (1960b). *Biochemical Journal* **76**, 264.
Wallerstein, L. (1939). *Industrial and Engineering Chemistry* **31**, 1218.
Wang, L.-H. and Hartman, P. A. (1976). *Applied and Environmental Microbiology* **31**, 108.
Watanabe, K. (1976). *Journal of Biochemistry, Tokyo* **80**, 379.
Watanabe, K. and Fukimbara, T. (1965). *Journal of Fermentation Technology* **43**, 690.
Watanabe, K. and Fukimbara, T. (1967). *Journal of Fermentation Technology* **45**, 226.
Watanabe, K. and Fukimbara, T. (1973). *Agricultural and Biological Chemistry* **37**, 2755.
Webster, R. E., Drechsler, C. and Jorgenson, H. (1958). *Plant Disease Reporter* **42**, 233.
Weiss, F. A. (1957). *In* 'Society of American Bacteriologists Manual of Microbiological Methods', p. 99. McGraw-Hill Book Co., N.Y.
Welker, N. E. and Campbell, L. L. (1963a). *Journal of Bacteriology* **86**, 681.
Welker, N. E. and Campbell, L. L. (1963b). *Journal of Bacteriology* **86**, 687.
Welker, N. E. and Campbell, L. L. (1963c). *Journal of Bacteriology* **86**, 1202.
Welker, N. E. and Campbell, L. L. (1967a). *Journal of Bacteriology* **94**, 1124.
Welker, N. E. and Campbell, L. L. (1967b). *Journal of Bacteriology* **94**, 1131.
Welker, N. E. and Campbell, L. L. (1967c). *Biochemistry, New York* **6**, 3681.
Wellman, A. M. and Stewart, G. G. (1973). *Applied Microbiology* **26**, 577.
Whelan, W. J. (1960). *Die Stärke* **12**, 358.
White, W. H. and Dworschack, R. G. (1973). United States Patent 3,725,202.
Wöber, G. (1973). *Hoppe Seyler's Zeitschrift für Physiologische Chemie* **354**, 75.
Wöber, G. (1976). *European Journal of Applied Microbiology* **3**, 71.
Wojskowicz, J. (1977). *Acta Microbiologica Polonica* **26**, 149 and 171.
Wymes, R. A. and Lloyd, N. E. (1968). United States Patent 3,414,479.
Yabuki, M., Ono, N., Hoshino, K. and Fukui, S. (1977). *Applied and Environmental Microbiology* **34**, 1.
Yagisawa, M., Kato, K., Koba, Y. and Ueda, S. (1972). *Journal of Fermentation Technology* **50**, 572.
Yamada, K. (1961). United States Patent 2,976,219.
Yamada, N. and Tomoda, K. (1966). United States Patent 3,293,142.
Yamaguchi, K., Matsuzaki, H. and Maruo, B. (1969). *Journal of General and Applied Microbiology* **15**, 97.
Yamaguchi, K., Nagata, Y. and Maruo, B. (1974). *Journal of Bacteriology* **119**, 416.
Yamamoto, T. (1955). *Bulletin of the Agricultural and Chemical Society of Japan* **19**, 12.
Yamamoto, M., Tanaka, Y. and Horikoshi, K. (1972). *Agricultural and Biological Chemistry* **36**, 1819.
Yamanaka, T. and Higashi, T. (1957). *Journal of Biochemistry, Tokyo* **44**, 637.
Yamane, K. and Maruo, B. (1974). *Journal of Bacteriology* **120**, 792.
Yamane, K. and Maruo, B. (1975). *Biochimica et Biophysica Acta* **393**, 571.
Yamane, T. and Tsukano, M. (1977). *Journal of Fermentation Technology* **55**, 233.

Yamane, K., Yamaguchi, K. and Maruo, B. (1973). *Biochimica et Biophysica Acta* **295**, 323.

Yamasaki, Y., Suzuki, Y. and Ozawa, J. (1977a). *Agricultural and Biological Chemistry* **41**, 755.

Yamasaki, Y., Suzuki, Y. and Ozawa, J. (1977b). *Agricultural and Biological Chemistry* **41**, 1443.

Yamasaki, Y., Suzuki, Y. and Ozawa, J. (1977c). *Agricultural and Biological Chemistry* **41**, 2149.

Yamasaki, Y., Tsuboi, A. and Suzuki, Y. (1977d). *Agricultural and Biological Chemistry* **41**, 2189.

Yamasato, K., Okuno, D. and Ohtomo, T. (1973). *Cryobiology* **10**, 453.

Yokobayashi, K., Misaki, A. and Harada, T. (1969). *Agricultural and Biological Chemistry* **33**, 625.

Yokobayashi, K., Misaki, A. and Harada, T. (1970). *Biochimica et Biophysica Acta* **212**, 458.

Yokobayashi, Y., Sugimoto, K. and Sato, Y. (1971). United States Patent 3,560,345.

Yokobayashi, K., Akai, H., Sugimoto, T., Hirao, M., Sugimoto, K. and Harada, T. (1973). *Biochimica et Biophysica Acta* **293**, 197.

Yoneda, Y. and Maruo, B. (1975). *Journal of Bacteriology* **124**, 48.

Yoneda, Y., Yamane, K. and Maruo, B. (1973). *Biochemical and Biophysical Research Communications* **50**, 765.

Yoneda, Y., Yamane, K., Yamaguchi, K., Nagata, Y. and Maruo, B. (1974). *Journal of Bacteriology* **120**, 1144.

Yoshida, M. and Hirao, M. (1974). United States Patent 3,830,697.

Yoshida, A. and Yamasaki, M. (1959). *Biochimica et Biophysica Acta* **34**, 158.

Yuhki, A., Watanabe, T. and Matsuda, K. (1977). *Die Stärke* **29**, 265.

Zittan, L., Ponlsen, P. B. and Hemmirigsen, S. H. (1975). *Die Stärke* **27**, 236.

4. Glucose Oxidase, Glucose Dehydrogenase, Glucose Isomerase, β-Galactosidase and Invertase

S. A. BARKER AND J. A. SHIRLEY

Departments of Chemistry and Biological Sciences, University of Birmingham, Birmingham, England

I. GLUCOSE OXIDASE

A. Occurrence

Since its discovery in *Aspergillus niger* (Müller, 1928), glucose oxidase (EC 1.1.3.4; β-D-glucose:oxygen 1-oxidoreductase) has been reported in *Penicillium glaucum* (Franke, 1944), *Penicillium notatum* (Keilen and Hartree, 1948), *Penicillium amagasakiense* (Kusai *et al.*, 1960; Nakamura and Ogura, 1962) and the red alga *Iridophycus flaccidum* (Bean and Hassid, 1956).

These and other sources, including honey, have been reviewed by Schepartz and Subers (1964). In the honeybee, the pharyngeal gland is the source of the glucose oxidase.

B. Structure

O'Malley and Weaver (1972) have shown that glucose oxidase from *A. niger* contains 16% carbohydrate and two flavindinucleotide cofactors per molecule. The native enzyme has a molecular weight of 160,000, but reduction of the enzyme's two disulphide bonds with β-mercaptoethanol gives two molecular species with molecular weights of 80,000. By contrast, the glucose oxidase isolated from *Penicillium amagasakiense* consists of four polypeptide chains equal in molecular size (45,000). Two of these polypeptide chains are held together by a disulphide bond to form a dimer (81,000) and two dimeric units associate non-covalently to form a tetramer (160,000) (Yoshimura and Isemura, 1971).

C. Mechanism

Glucose oxidase catalyses oxidation of β-D-glucose to D-glucono-δ-lactone, which subsequently hydrolyses rapidly to D-gluconic acid:

$$\text{Glucose} + O_2 \rightleftharpoons \text{Glucono-}\delta\text{-lactone} + H_2O_2$$

$$\text{Glucono-}\delta\text{-lactone} + H_2O \rightleftharpoons \text{Gluconic acid}$$

It exhibits maximum activity at pH 5.6 and 35–40°C. Bright and Gibson (1967) suggest the reaction proceeds as follows:

$$E\text{–}Fad + Glucose \rightleftharpoons E\text{–}FAD\text{–}glucose$$

$$E\text{–}FAD\text{–}glucose \longrightarrow E\text{–}FADH_2\text{–}gluconolactone \longrightarrow E\text{–}FADH_2 + Gluconolactone$$

$$E\text{–}FADH_2 + O_2 \longrightarrow E\text{–}FAD\text{–}H_2O_2 \longrightarrow E\text{–}FAD + H_2O_2$$

Using deazaflavin instead of flavin in glucose oxidase, it has been shown that the deazaflavin–glucose oxidase is reactive towards substrate reduction but the reduced enzyme is not re-oxidized by oxygen. This provides direct evidence for transfer of substrate hydrogen to flavin (Fisher *et al.*, 1976). Deaza-FMN–glucose oxidase forms reversible 1:1 complexes with sulphite and cyanide. Hydroxylamine also forms a complex with glucose oxidase. These results were consistent with formation of covalent adducts via attack of the various nucleophiles at position 5 of (deaza)-flavin (Jorns and Hersh, 1976).

D. Specificity

Nuclear-magnetic-resonance spectroscopy has been used to study the anomeric specificity of glucose oxidase with 2-deoxy-D-glucose and D-glucosone (Feather, 1970). In both cases, the preferred substrate is β-pyranose so these reactions are stereochemically equivalent to that of β-D-glucose itself. Pazur and Kleppe (1964) showed that glucose oxidase from *A. niger* was capable of oxidizing D-aldohexoses, monodeoxy D-glucoses and O-methyl D-glucoses at different rates. Thus, the relative rate compared with D-glucose at 100 was: L-glucose, 0; 1,5-anhydro-D-glucitol, 0; 2-deoxy-D-glucose, 20; D-mannose, 1; 3-deoxy-D-glucose, 1; D-allose, 0.02; 3-O-methyl-D-glucose, 0.02; 4-deoxy-D-glucose, 2; D-galactose, 0.5; 4-O-methyl-D-glucose, 15; 5-deoxy-D-glucose, 0.05; 6-deoxy-D-glucose, 10; 6-O-methyl-D-glucose, 1. The structural features associated with particular importance to function as a substrate were a pyranose ring in the chain or C-1 conformation, an equatorially orientated hydroxyl group at position 1 and similarly at position 3.

E. Inhibitors

Rogers and Brandt (1971) showed that a structural analogue of D-glucose, namely D-glucal, was an inhibitor of glucose oxidase. This inhibition was competitive with respect to D-glucose and non-competitive with respect to oxygen. It was postulated that D-glucal binds to the active site of the oxidized form of *A. niger* glucose oxidase since the visible absorption spectrum of enzyme-bound FAD is perturbed in the presence of D-glucal. The inhibition constant had an average value of 0.13 M over the pH range 3.8–7.5. Chloride is also an inhibitor competitive with respect to the substrate D-glucose. Rogers and Brandt (1971) presented evidence that both D-glucal and chloride can bind simultaneously to the enzyme. The chloride ion is effective at an ionic strength of between 0.5 and 0.6.

Penicillium notatum glucose oxidase is inhibited by sodium nitrite, β-hydroxyquinoline and semicarbazide (Bean and Hassid, 1956). The enzyme from *A. niger* is markedly inhibited by Ag^+, Hg^{2+}, Cu^{2+}, *p*-chloromercuribenzoate and phenylmercuric acetate (Nakamura and Ogura, 1968). The first two metal ions inhibit oxidation of the reduced FAD moiety competing with molecular oxygen used as a hydrogen acceptor. Agents such as *N*-ethylmaleimide, mono-iodoacetate and mono-iodoacetamide had no effect glucose oxidase activity.

At very low oxygen levels, the substrate glucose inhibits the glucose oxidase reaction (Nicol and Duke, 1966). This behaviour with the enzyme from *A. niger* resembled that of another flavin enzyme, L-amino acid oxidase. The inhibition by glucose becomes apparent at an oxygen concentration below 20 μM (2% saturated oxygen).

F. Properties

Glucose oxidase as purchased often contains other enzymes. That from *A. niger*, supplied by Sigma Chemicals, had the following activities: glucose oxidase (56.4 mmol/min/mg of protein), gluconolactonase (625 mmol/min/mg of protein) and catalase (5170 mmol/min/mg of protein) (Cho and Bailey, 1977). The enzyme had an optimal pH value of 5.5 and exhibited 95% of its maximal activity at pH 5. The glucose oxidase in buffer solution was stable up to six months at 3°C but the half-life at 30°C was 11 days, substantially less than the half-life of 90 days at 27°C reported previously. Correcting for gluconolactonase contamination, the glucose oxidase activity obeys Michaelis–Menten kinetics with values for V_{m^3} of 16.95 mM/min and K_{m^3} of 192.3 mM. The only substrate for the enzyme action is β-D-glucose in the pyranose ring form. To protect the enzyme, 0.5 mM EDTA can be added to cope with de-activating metal ions. The catalase impurity can be suppressed with 0.1 mM KCN. According to Weibel and Bright (1971), $K_{catalysis}/K_{oxidation} \times 10^3 = 0.51$ mol/litre, and $K_{catalysis}/K_{reduced} = 0.071$ mol/litre.

G. Stability

Stabilization of glucose oxidase has been achieved by grafting polymers onto the enzyme. Thus, Hixson (1973) used a carbodiimide at 0°C and pH 4.75 to graft on the p-aminophenoxyacetyl derivative of polyvinyl alcohol (5% substitution). After separation of the grafted enzyme on Agarose A, the modified enzyme was appreciably more stable at 50°C retaining 77% of its original activity after four hours compared to 52% shown by the native enzyme. Grafts prepared contained 29–45% enzyme by weight, and enzyme molecules were 45–55% active. This corresponds to 25–30 polymer chains per enzyme molecule. Simple admixture of glucose oxidase with certain synthetic polymers also enhances the thermal stability of glucose oxidase. Thus, O'Malley and Ulmer (1973) found that copolymers of vinyl acetate with either vinyl pyrrolidone or vinyl alcohol were particularly effective. With the latter after four hours at 50°C, there was about twice as much enzymic activity remaining as with the native enzyme. By contrast, polyelectrolytes such as poly(methacrylic acid) and quaternized poly(vinyl pyridine) were found to destabilize glucose oxidase. Variable results were obtained with non-ionic polymers but, in general, they either behaved as inert additives or moderately enhanced the thermal stability of the enzyme. The presence of 12% substitution of acetate groups in polyvinyl alcohol was essential since the original polyvinyl alcohol showed no effect. The minimum concentration for this to be effective was 1%.

The glucose oxidase of honey and from the honeybee is quite different in properties (Schepartz and Subers, 1964) from the microbial type. They require extremely high substrate concentrations (1.5 M and

2 M respectively for the honey sources) and have pH optima at 6.1 and 6.7 respectively. By contrast, the mould enzyme requires only 0.1 M concentration of substrate. Another odd species is the glucose oxidase of the red alga (Bean and Hassid, 1956). It can oxidize not only glucose but D-galactose to D-galactonic acid and several reducing disaccharides, namely maltose, lactose and cellobiose. Its optimum pH is 5.0.

H. Microbiology

Mueller (1977) identified at least three gluconic acid-forming enzymes in cell-free extracts of *A. niger*. In addition to glucose oxidase there was a glucose dehydrogenase (EC 1.1.99.10) and an enzyme or a mixture of enzymes catalysing cleavage of 6-phosphogluconate into gluconate and inorganic phosphate. 2,6-Dichlorophenol-indophenol was a hydrogen acceptor for the glucose dehydrogenase *in vitro*.

Glucose oxidases of fungal origins are at present commercially available from several companies, and a useful comparison of those from *A. niger* and *Penicillium amagasakiense* has been published by Hayashi and Nakamura (1976). The enzyme from *P. amagasakiense* is excreted into the culture medium, whereas that from *A. niger* remains in the mycelia during submerged culture. The enzymes from the three commercial *A. niger* sources differed from that of the Penicillium source although they all shared much of a common structure. The authors suggested that the enzymes might have been evolved from a common ancestral precursor. Both are glycoproteins with mannose as the dominant sugar. However, the turnover number (V/e value) of the Penicillium enzyme was about twice as large as those of the Aspergillus enzymes and the value for the Michaelis constant for glucose was considerably lower for the Penicillium enzyme.

I. Enzyme Production

Zetelaki (1970) studied the effectiveness of aeration and agitation in production of glucose oxidase by *A. niger* on a semi-industrial scale (500 l). The rate of growth of the mould and activity of glucose oxidase per gram of mycelium rose with increasing agitation. The concentration of dissolved oxygen in the fermentation broth as well as the rate of respiration increased in direct proportion to the increased speed of agitation, and assimilation of sugars was accelerated. Earlier work (Zetelaki and Vas, 1968), in 10-litre glass fermenters had shown that glucose oxidase activity from an oxygenated culture was twice as potent as that from an aerated one.

Zetelaki (1970) employed *A. niger* 1026/5, a strain from the collection of the Central Food Research Institute, Budapest. The medium contained 5% sucrose, 0.2% $Ca(NO_3)_2,2H_2O$, 0.75% citric acid, 0.025% KH_2PO_4, 0.025% KCl, 0.025% $MgSO_4$, $7H_2O$, 0.001% $FeCl_3$, $6H_2O$ and 2% corn-steep liquor. The rate of aeration was one vol/vol/min in media (400 l) at 28°C contained in stainless-steel fermenters (500 l). Speeds of agitation were varied between 153, 230 and 306 rev./min corresponding to oxygen solution rates of 48, 115 and 185 mmol of $O_2/l/h$, respectively. The glucose oxidase concentrations in the mycelia after 24 hours were 17%, 10.3% and 12.3% at the agitation speeds and 101 kPa (one atmosphere) and 10.4% at 306 rev./min at 202 kPa (two atmospheres). Mycelial dry weights of the 24-hour cultures were 0.98%, 1.33% and 1.38% at the three speeds, and the increase in pressure at the highest speed resulted in a 60% increase in mycelial weight. Thus, 72 hours were required at 153 rev./min to achieve maximum mycelial weight compared to 28 and 44 hours earlier at 230 rev./min and 306 rev./min, respectively. Highest total enzyme activity 2370 Sarrette units/100 ml was attainable after 16 hours at 306 rev./min/202 kPa (two atmospheres) followed by 2025 at 20 hours/306 rev./min/101 kPa (one atmosphere) and 1925 at 48 hours/230 rev./min/101 kPa (one atmosphere). Activity of glucose oxidase per gram of mycelium was highest (1340 standard units/g) in cultures agitated at 306 rev./min. The workers issued a warning that, after depletion of the sugar, a sudden activation of the enzyme was found accompanied by a partial autolysis of the mycelia. In their earlier work, Zetelaki and Vas (1968) had shown that glucose oxidase activity was highest in a medium containing 5% sugar and did not get higher when it was increased to 7%. The strain, *A. niger* 1026/5, was selected by Proszt (1963).

In *Penicillium vitale*, maximum glucose oxidase production (70–80 units/ml) was observed in submerged cultures when the inoculum was grown for 36–43 hours at 26°C with aeration at one l of air/l of medium/minute. This inoculum was used at 4–6% concentration with a fermentation temperature of 28°C for 72–78 hours (Vakulenko *et al.*,

1976). When the inoculum volume was increased to 10–20%, the fermentation temperature decreased to 25°C and a lower aeration rate employed, catalase was mainly observed.

Maximum yield of glucose oxidase with minimum catalase activity was obtained with an *A. niger* mutant culture after 72 hours. The inoculum (7 ml grown on millet for 30 days at 28°C) was added to 50 ml of a medium containing 10% sucrose and 0.036 g of KH_2PO_4/l, pH 6.5–7.0 (Dzherova *et al.*, 1974). Extraction of glucose oxidase from *A. niger* is claimed to be effected by treatment with enzyme A from *Achromobacter lunatus* (Hirago and Hirano, 1975). Thus *A. niger* I.A.M. 2020 was cultured aerobically at 30°C for 24 hours in a medium containing 3% sucrose, 0.2% yeast extract, 0.2% $NaNO_3$, 0.1% $CaCl_2$, and 0.05% $MgSO_4$. Addition of the enzyme A (2 g) at pH 6.0 followed by incubation for eight hours produced a reaction mixture that, when filtered, contained 2.1 units of glucose oxidase/ml compared with 2.4 units/ml for a sonicated preparation.

Japanese workers (Nakamatsu *et al.*, 1975) reported that the best glucose oxidase producer was *P. purpurogenum* No. 778. Some 32,000 units of activity/ml of broth were obtained in a simple medium (pH 6.0) of 10% beet molasses, 0.7% $NaNO_3$ and 0.2% KH_2PO_4 in submerged culture of three days at 30°C. The purified enzyme had an optimum pH value of 5.0 and an optimum temperature of 35°C. It was stable at pH 5–7.0 at 40°C for two hours and at temperatures below 50°C at pH 5.6 for 15 minutes. The apparent K_m value was 12.5 mM for its specific substrate glucose.

Gorniak and Kaczkowski (1973) showed that sodium nitrate was the most effective nitrogen source for production of glucose oxidase by *P. notatum* and its subsequent excretion into the medium. Ammonium salts inhibited oxidase activity, and urea showed different dynamics of enzyme accumulation and changes in pH value compared with ammonium carbonate.

J. Purification

Glucose oxidase from *A. niger* was purified to homogeneity by Tiselius and starch-gel electrophoresis as well as by ultracentrifugation (Swoboda and Massey, 1965). The essential steps from the crude commercially available extract were: (1) dialysis against 0.1 M sodium acetate buffer (pH 4.5), with removal of a brown precipitate; (2) absorption of the enzyme from the supernatant onto Amberlite CG-50 in the same buffer with its subsequent elution by 0.1 M sodium acetate buffer (pH 5.0); (3) precipitation of the enzyme at 90% saturation with respect to ammonium sulphate; (4) dialysis against 0.1 M phosphate buffer (pH 6.0); (5) absorption of the enzyme from this solution on DEAE-cellulose and subsequent elution with 0.2 M phosphate buffer (pH 6.0); (6) repetition of step (3) and dialysis of the product against water. Some 300–400 mg of enzyme were obtained from 50 g of *A. niger* extract. In this state, it could be stored for a number of years in the deep freeze at 10 mg/ml without loss of activity.

Kusai *et al.* (1960) crystallized the glucose oxidase of *Penicillium amagasakiense*. Again, the starting point was a commercially available extract 'Deoxine' from Nagase and Co. Ltd. The enzyme is secreted into the medium from this strain in submerged cultures. Steps (1) and (2) were as above; step (3) gave a precipitate with 60–80% ammonium sulphate. In step (4), the product was dissolved in 0.05 M sodium acetate (pH 5.0) and dialysed against the same buffer; (5) involved fractionation on Amberlite CG-50 Type II with the same buffer, and (6) a repeat of step (4). In step (7), the product was dissolved in water and induced to crystallize by addition of ammonium sulphate at room temperature. Some 272 mg of crystals were recovered from 100 g of 'Deoxine'. The crystalline enzyme showed maxima at 278 nm, 380 nm and 460 nm in the oxidized form, and at 273 nm in the presence of glucose. The peaks at 380 nm and 400 nm due to FAD disappeared on reduction of the enzyme with glucose or sodium dithionite.

Yoshimura and Isemura (1971) purified glucose oxidase from *Penicillium amagasakiense* to homogeneity as judged by ultracentrifugation and electrophoretic studies. Their purification was again starting from Nagase Sangyo Co. Ltd. product. Steps were as follows: (1) dialysis against 0.01 M acetate buffer (pH 5.6); (2) absorption of the enzyme on a DEAE-cellulose column with the same buffer and elution with 0.1 M buffer (pH 5.6); (3) again the active fraction was passed down a DEAE-cellulose column with 0.045M buffer (pH 5.6). The first of two fractions was retrieved and; (4) re-precipitated with 65% saturated ammonium sulphate.

Tsuge *et al.* (1975) reported an improved method of purification based on Sigma Type II *A. niger* glucose oxidase. The steps were: (1) the commercial preparation (2 g) was dialysed against 1 mM potassium–

sodium phosphate buffer (pH 7); (2) absorbed from the same buffer on DEAE-cellulose from which it was eluted with 100 mM phosphate buffer (pH 7.0); (3) recovery by ammonium sulphate precipitation; (4) dialysis against 10 mM phosphate buffer (pH 7.0); (5) gradient elution from DEAE-cellulose column with 10 going to 200 mM buffer (pH 7.0); (6) the fraction having the highest activity was recovered by ammonium sulphate precipitation and; (7) fractionated on a Sephadex G-200 column.

The product was pure by disc-gel electrophoresis at two pH values. Analysis showed 74% protein, 16.4% carbohydrate, 2.4% amino sugar, about 0.4% metal ash and 0.9% FAD, equivalent to a $94.1 \pm 3.6\%$ total. The pure enzyme showed maximum absorptions at λ 278, 383 and 454 nm. The enzyme had a specific activity of 172 μmol of oxygen consumed/min/mg of protein with only 0.00013 litre g^{-1} s^{-1} catalase activity, indicating that it was essentially catalase free.

K. Uses

Glucose oxidase is widely used in clinical diagnosis as a key indicator of carbohydrate metabolism. Thus, dip tests for detection of glucose in urine use a combination of glucose oxidase and peroxidase. The latter enzyme catalyses the reaction of hydrogen peroxide with a potassium iodide chromogen to oxidize the chromogen to a green-to-brown colour. Approximately 5.5 mmol of glucose/litre is detectable. Glucose response can be read after 30 seconds.

Manufacturers nowadays report faithfully on known impurities in enzymes. Thus, that from *A. niger*, sold in the crude state by one company, contained amylase, maltase, catalase, glycogenase, invertase and galactose oxidase, but best grades contain only the merest traces of catalase, amylase and invertase. The top grades are recommended for determination of glucose in blood. In the enzyme reaction, the product, hydrogen peroxide, reacts with a leucodye (e.g. O-dianisidine) converting it into a dye. An alternative is ABTS (2,2'-azino-di-(3-ethyl-benzthiazoline sulphonic acid-6) ammonium salt). One manufacturer sells immobilized glucose oxidase under the name 'Enzygel' as a lyophilizate containing sucrose. Another company sells nylon tube-immobilized glucose oxidase under the name Catalinks®

for use in continuous-flow systems of glucose analysis in conjunction with 4-aminophenazone. Immobilized glucose oxidase is used in the Y.S.I. 23AM glucose Analyser for assay of glucose in whole blood, plasma and serum.

II. GLUCOSE DEHYDROGENASE

A. Occurrence

Oxidation of glucose to gluconic acid is carried out by many bacteria including species of *Acetobacter*, *Aerobacter*, *Azotobacter* and *Pseudomonas*, and can be linked to structural elements that carry cytochromes reducible by glucose oxidation. Bacteria that have been shown to possess particulate glucose dehydrogenases are listed in Table 1. In some instances, bacteria contain a soluble glucose dehydrogenase (EC 1.1.99.10) that is independent of added NAD^+ or $NADP^+$ for its action. Hauge (1960) established that the glucose dehydrogenase of *Bacterium antitratum* was associated with cytochrome b, both probably soluble precursors of the particulate system.

B. Structure

Soluble glucose dehydrogenase from *Bacillus antitratum* has a molecular weight of $86,000 \pm 4\%$ and shows a $S_{20,\omega}$ value of 6.2. Its turnover number is $320,000$ min^{-1}. Its prosthetic group is visible in the broad band at 347 nm which, on reduction with glucose, changes to a sharper band at 337 nm (Hauge, 1964). When the enzyme is boiled, a factor that will re-activate enzyme preparations inactivated by gel filtration at pH 2.5 is liberated. This factor can be purified on DEAE-Sephadex. The maximum of the oxidized form is shifted to 330 nm. The liberated prosthetic group in its oxidized form is also characterized by a maximum at 248 nm with a shoulder at 270–280 nm. It appears to have an acid dissociation constant with a pK value of 1.7.

Glucose dehydrogenase from *B. megaterium* is a stable tetramer at pH 6.5 but is readily dissociated into four inactive protomers at pH 8.5 (Pauly and Pfleiderer, 1977). Rapid re-association and re-activation is achieved by dropping the pH value to 6.5. The quaternary structure is

Table 1

List of some bacteria that possess particulate glucose dehydrogenases

Bacteria	Reference
Pseudomonas fluorescens	Wood and Schwerdt (1953)
Pseudomonas pseudomallei	Dowling and Levine (1956)
Pseudomonas quercito-pyrogallica	Bentley and Slechta (1960)
Pseudomonas fragi	Weimberg (1963)
Pseudomonas graveolens	Nishizuka and Hayaishi (1962)
Rhodopseudomonas spheroides	Niederpruem and Doudoroff (1965)
Acetobacter sp. e.g. suboxydans	King and Cheldelin (1957)
Xanthomonas phaseoli	Hochster and Katznelson (1958)
Bacterium anitratum	Hauge (1960)

stabilized by sodium chloride and the coenzyme NAD^+. Higher concentrations of salt inhibit unfolding of the enzyme by 8 M urea. Inactivity of the subunit is due to the loss of its coenzyme-binding capacity. The apparent dissociation constants (mM) for various ligands were: NAD^+, 0.6; NADH, 7.1; $NADP^+$, 1.4; adenosine 5'-diphosphoribose, 4.9; adenosine, 4.4; adenine, 14.0; AMP, 55; ADP, 52; NMN, 44; ribose 5'-phosphate, 60.

C. Mechanism

Glucose dehydrogenase from *Ps. aeruginosa* strain O.S.U. 64 can oxidize 2-deoxy-D-glucose to 2-deoxygluconic acid. Cell-free extracts were used to demonstrate this reaction (Williams and Eagon, 1959). Eagon (1971) showed that 2-deoxyglucose was taken up and oxidized by glucose-grown cells of *Ps. aeruginosa* at a rate (85%) approaching that for uptake and oxidation of glucose. 2-Deoxyglucose was transported by the cells via passive diffusion and sufficient oxygen was utilized to permit complete conversion. A concentration of 2-deoxyglucose, 1000-fold that of glucose, inhibited the rate of glucose uptake by about 50%. Glucose on the other hand, strongly inhibited uptake of 2-deoxyglucose. This was attributed to the enzyme having a greater affinity for glucose than 2-deoxyglucose. Glucose dehydrogenase is an inducible enzyme in this organism.

D. Specificity

Hauge (1964) showed that the glucose dehydrogenase from *B. anitratum* attacked primarily the β-glucose form, although he did not regard the specificity as absolute. Hauge (1960) reported that the relative reaction rates with 0.02 M solutions of the following aldoses were: D-glucose, 100: D-galactose, 66; L-arabinose, 51; D-mannose, 30; D-xylose, 27; D-ribose, 7.5; 2-deoxy-D-glucose, 4.5; D-arabinose, 1.6; DL-glyceraldehyde, 1.0; L-xylose, 0. D-Fructose, sorbitol, mannitol and D-glucose 6-phosphate were not attacked. More precise evaluation was evident from a study of the Michaelis constants (K_m values, M $\times$ 10^3) which were 1.1 for D-glucose, 22 for L-arabinose and 51 for D-xylose. The reducing disaccharides maltose, lactose and cellobiose were attacked almost as readily as D-glucose, their K_m values being 4.7, 3.3 and 2.2 mM, respectively. With melibiose, a K_m value of 0.2 M was found.

In early experiments, Wood and Schwerdt (1953) used sonicated cell extracts of *Ps. fluorescens* to demonstrate that they were able to oxidize glucose without addition of accessory hydrogen carriers such as phenazine dyes or methylene blue. Two atoms of oxygen per mol of glucose were taken up by glucose if oxidation of glucose 6-phosphate and 6-phosphogluconate was abolished by carrying out the reaction in the presence of phosphate buffer (pH 7). Alternatively, the enzyme fraction could be freed from these activities by precipitation with ammonium sulphate. Cytochrome carriers were suggested. Neither NAD^+ nor $NADP^+$ was reduced by the enzyme in the presence of glucose. In this respect, it therefore differed from the NAD^+-specific glucose dehydrogenase in liver and the flavoprotein glucose oxidase from moulds.

Dowling and Levine (1956) showed that, in *Malleomyces pseudomallei*, oxidation of glucose to gluconolactone was not stimulated by ATP, NAD^+ or $NADP^+$ but oxygen was consumed. The reaction was performed on a cell extract that could also oxidize galactose in the same way. Bentley and Slechta (1960) also described a cell-free particulate preparation from *Ps. quercito-pyrogallica* which directly oxidized aldoses, lactose and maltose to the corresponding lactones with consumption of oxygen but without production of hydrogen peroxide.

A glucose dehydrogenase was obtained from *Ps. graveolens* A.T.C.C.

4683 grown in the presence of lactose (Nishizuka and Hayaishi, 1962). The enzyme was solubilized by deoxycholate–butan-1-ol extraction and was then purified 40–70-fold by chloroform treatment, followed by fractionation with protamine sulphate, ammonium sulphate and ethanol. However, the authors reported that 2,6-dichlorophenol-indophenol, methylene blue and ferricyanide served as electron acceptors for the soluble enzyme, but not oxygen or nicotinamide nucleotides. The prosthetic group appeared to contain FAD and was closely associated with a haemoprotein electron-transfer system. The optimum pH value was 5.6. It reacted with D-glucose (reactivity, 100); D-galactose, 97; D-mannose, 19; D-talose, 54; D-ribose, 97; D-xylose, 95; L-arabinose, 106; 3-deoxy-D-glucose, 113; D-frucose, but not with L-xylose, L-lyxose, L-rhamnose, L-fucose, D-fructose, L-sorbose, D-glucosamine or D-galactosamine. The disaccharides lactose (85), maltose (95) and cellobiose (96) were good substrates. The particulate fraction from *Ps. graveolens* reacted directly with oxygen but otherwise had a similar spectrum of activity against the sugars already listed, apart from a weak lactose dehydrogenase activity. Two different dehydrogenases are therefore suspected one of which, lactose dehydrogenase, is induced by D-ribose, L-arabinose or lactose.

The activities for lactose, maltose, cellobiose and D-ribose were constant during enzyme purification. Further, when cells were grown with D-glucose or D-galactose, lactose and D-ribose were not oxidized to a measurable extent. The enzyme activities with D-glucose, D-galactose, L-arabinose and D-fucose paralleled each other during purification, and appeared to form a separate group.

Relative substrate activities of the enzyme from *Rhodopseudomonas spheroides* were: D-glucose, 73; D-mannose, 50; D-galactose, 63; L-arabinose, 68; D-xylose, 68; D-ribose, 58; D-lyxose, 20; D-arabinose, 5 (Niederpruem and Doudoroff, 1965).

King and Cheldelin (1957) used deoxycholate to solubilize the particulate enzyme responsible for glucose oxidation in *Acetobacter suboxydans* A.T.C.C. 621. The preparation did not then need exogenous cofactors or inorganic ions. The reduced cytochrome of the solubilized particulate preparation showed absorption maxima at 426–428, 528–530 and 558 nm. The bacterium had been grown in a glycerol–yeast extract and disintegrated by grinding or sonication. Deoxycholate extracts of the particulate fraction consumed oxygen rapidly up to a limit of one atom per molecule of glucose, although extensive dialysis was carried out before use because of the inhibitory effects of deoxycholate. 2,6-Dichlorophenol-indophenol, but not methylene blue, could replace oxygen as an electron acceptor. Only glucose and galactose were appreciably oxidized. Maltose, lactose, fructose, sucrose, cellobiose raffinose were not attacked. Since non-particulate cell-free extracts were unable to oxidize galactose, the enzyme exhibiting this activity occurs only in the particulate fraction.

E. Inhibitors

The glucose dehydrogenase of *A. suboxydans* (King and Cheldelin, 1957) had optimum activity at pH 5.5. Cyanide at 0.5 mM, pH 7.0, and azide at 0.5 mM, pH 6.0 or 7.0, inhibited about 90% of the oxidation effected by the enzyme.

F. Properties

The glucose dehydrogenase from *A. suboxydans*, when solubilized with deoxycholate, had an optimum pH of 5.5 but could not tolerate temperatures above 40°C (King, 1966). The primary product of oxidation of glucose is gluconolactone, as indicated by the hydrox-amate test. Its K_m value for glucose is 85 mM, and it contains cytochromes reducible by glucose. Oxygen is a good receptor, as is 2,6-dichlorophenol-indophenol. Triphenyltetrazolium salt and methylene blue are inactive.

G. Stability

Particulate glucose dehydrogenase from *Ps. quercito-pyrogallica* loses its ability to use oxygen as acceptor when treated with EDTA, and the presence of 1 mM EDTA abolished oxygen uptake by that of *Ps. fluorescens*. This property, however, may reflect the nature of the electron-transport chain since dehydrogenases from *B. antitratum* and *A. suboxydans* are not affected by EDTA. The soluble glucose dehydro-genase of *B. anitratum* has a variable tendency to lose activity particularly during the last stages of its purification where its prosthetic group may be lost.

H. Microbiology

Niederpruem and Doudoroff (1965) described a particulate enzyme preparation from *Rhodopseudomonas spheroides* which had aldose dehydrogenase activity towards glucose and mannose. The activity was dependent on a soluble cofactor which, while present in their particulate fraction, was unavailable to the enzyme. It could, however, be liberated by boiling or treatment with salts at high concentrations. The cofactor also appears in the soluble fraction of aerobic cells after exponential growth has ceased. Extracts of cells grown anerobically in light possessed the apoenzyme but not the cofactor. Similar cofactor activity was found in *B. anitratum* but not in *E. coli* or *Ps. fluorescens*.

Lynch *et al.* (1975) studied the effect of temperature on activity and synthesis of glucose-catabolizing enzymes in *Ps. fluorescens*. In particular, the glucose dehydrogenase was assayed at 30°C by incubation with 2.5 mM glucose and 0.14 mM 2,4-dichlorophenol-indophenol in 20 mM phosphate buffer (pH 6.0). The rate of decrease in absorbance at 600 nm was then measured. Cell-free extracts were prepared by resuspending cell pellets in 0.1 M Tris buffer (pH 7.4), disrupting in a French pressure cell at 117.3 MPa (17,000 lbf/in^2) and using the supernatant after centrifugation at 10,000 g for 10 min at 5°C.

Synthesis of glucose dehydrogenase was induced during growth with glucose at 5°C and 20°C and reached a maximum level as the stationary phase was reached at both temperatures. The level of induction was up to 2.3-fold at 5°C and up to 4.5-fold at 20°C. During growth with glucose at 30°C, glucose dehydrogenase was present at low or undetectable levels at all stages of growth from the early logarithmic to the stationary phase.

Glucose dehydrogenase was present at low levels during growth on gluconate or 2-oxogluconate at 30°C. However, there was some induction of glucose dehydrogenase during growth with 2-oxogluconate at 20°C representing a 2–5-fold increase over that with the same inducer at 30°C.

Lee *et al.* (1977) in a taxonomic study of some pseudomonad-like marine bacteria, assayed 2,6-dichlorophenol-indophenol-linked dehydrogenases on the particulate fractions by the method of Kersters (1967). With glucose as a substrate, activity was found in seven strains having Phenom A but not strains B, C, D or E. Phenom A includes *Ps. fluorescens*, *Ps. fragi*, *Ps. putida* and *Ps. cepacia* (N.C.M.B.I. 129, 320, 401 and N.C.I.B. 10649, respectively).

I. Enzyme Production

Hauge (1964) favours his original strain of *B. anitratum* for glucose dehydrogenase production finding it superior to strain N.C.T.C. 7844. However, experience showed that the best growth medium contained, per litre of tap water: 15 g of sodium succinate,$6H_2O$, 4 g of NH_4Cl, 2 g of K_2HPO_4 and 0.06 g of $MgSO_4,7H_2O$. This bacterium grew well under vigorous aeration at 28°C either in a 20-litre or 170-litre fermenter particularly in the latter where the generation time was 100 minutes. Addition of glucose to the medium was avoided since it inhibited synthesis of glucose dehydrogenase.

J. Purification

With *B. antitratum* (Hauge, 1964), the frozen cell paste was treated as follows at 2–6°C: (1) passed through a pressure cell (9–12 tonnes pressure) as a paste (137 g) diluted with 130 ml of 0.1 M phosphate buffer (pH 6); (2) after dilution with 45 ml of buffer and 200 ml of water, cells and debris were removed by centrifugation at 20,000 g; (3) the supernatant and washings were made 1% with respect to protamine sulphate and the precipitate was discarded; (4) particulate dehydrogenase was precipitated from the supernatant at 45–55% saturation with ammonium sulphate; (5) soluble dehydrogenase was precipitated between 58 and 70% saturation with ammonium sulphate; (6) soluble enzyme was dialysed against 5 mM phosphate (pH 7); (7) the basic protein constituting the enzyme was purified on a DEAE-cellulose column in 5 mM phosphate (pH 7); (8) further purification was achieved on a carboxymethyl-cellulose column in 0.01 M phosphate (pH 6); (9) active enzyme was fractionated on Sephadex G-25 in 0.01 M phosphate (pH 6); and (10) final purification was achieved on carboxymethyl-cellulose eluted with 0.1 M phosphate (pH 7).

K. Uses

Like the other glucose dehydrogenase that requires NAD^+, this enzyme should also be favoured for blood glucose determinations using reduction in absorbance at 600 nm of 2,6-dichlorophenol-indophenol. An early review on the principles and advantages of using a glucose dehydrogenase is that of Gerbig (1976). It should have application also in manufacture of gluconic acid and its lactone.

III. GLUCOSE DEHYDROGENASE

A. Occurrence

In some instances, bacteria or their spores contain a soluble glucose dehydrogenase (β-D-glucose: NAD(P) 1-oxidoreductase: EC 1.1.1.47) that is dependent on added NAD$^+$ or NADP$^+$ for its action. These include *Bacillus cereus* (Bach and Sadoff, 1962), *B. megaterium* (Garvard and Combre, 1959), *B. subtilis* (Kunita and Fukumara, 1956) and *Pediococcus pentosaceous* (Lee and Dobrogosz, 1965).

B. Structure

Pauly and Pfleiderer (1975) have shown that the glucose dehydrogenase from *B. megaterium* M1286 has a molecular weight of 116,000 by gel-permeation chromatography, in agreement with the reports of other workers (Hedrick and Smith, 1968) who cited 120,000 and 118,000 for molecular-weight measurements by electrophoresis and density-gradient centrifugation respectively. The enzyme dissociates into subunits of 30,000, indicative of four polypeptide chains which each carry a coenzyme-binding site as shown from fluorescent titration of the NADH-binding sites. This glucose dehydrogenase is highly specific for β-D-glucose and is capable of using either NAD$^+$ or NADP$^+$. It is insensitive to thiol-group inhibitors, heavy metal ions and chelating agents. Dissociation into subunits of 30,000 takes place in 0.1% sodium dodecyl sulphate and 8 M urea, as shown by dodecyl sulphate–polyacrylamide-gel electrophoresis. By contrast, in polyacrylamide disc electrophoresis, the purified enzyme exhibits three active bands, representing the native tetrameric form, and two monomeric forms (mol. wt. 30,000), arising under the conditions of pH value and ionic strength of this method.

C. Mechanism

Lee and Dobrogosz (1965) have purified a glucose dehydrogenase from *Pediococcus pentosaceous* that is specific for $NADP^+$ as a cofactor and oxidizes only D-glucose and its analogue 2-deoxy-D-glucose. The reaction is:

β-D-Glucose + $NADP^+$ $\rightleftharpoons$ D-Glucono-δ-lactone + NADPH + H^+

Values for K_m were 23 mM for glucose and 0.2 mM for $NADP^+$. The optimum pH value was 7.0 and the equilibrium constant 13.4×10^{-7} at pH 6.4.

D. Specificity

The relative rates of attack of the glucose dehydrogenase from *B. megaterium* on equilibrated aldoses were as follows, compared with D-glucose at 100%: 2-deoxy-D-glucose, 114; D-xylose, 6; D-mannose, 0.8; 2-deoxy-2-amino-D-glucose, 14; no action was found on D-galactose, D-arabinose, D-ribose, D-fructose, L-rhamnose, *N*-acetyl-2-deoxy-2-amino-D-glucose, *myo*-inositol and D-glucose 6-phosphate.

In the presence of NAD^+, the *B. subtilis* glucose dehydrogenase dehydrogenated D-glucose and 2-deoxy-D-glucose at about the same rates. It showed no activity with a large number of other compounds including *myo*-inositol and D-xylose. With either of its substrates, the enzyme reduced NAD^+ more efficiently than $NADP^+$. Thus its K_m value (mM) was 0.23 with 0.1 M glucose and NAD^+, 0.26 with 1 M glucose and $NADP^+$, 0.24 with 0.1 M 2-deoxyglucose and NAD^+, where the sugar concentration was kept constant and the coenzyme varied. Under the reverse set of conditions, the K_m value (mM) was 55.9 with 0.5 mM NAD^+ and glucose, 47.8 with 0.5 mM NAD^+ and 2-deoxyglucose and 42.9 with 0.5 mM $NADP^+$ and glucose (Fujita *et al.*, 1977).

E. Inhibitors

The enzyme from *B. megaterium* is inhibited by *N*-acetyl-2-deoxy-2-amino-D-glucose and myo-inositol but is unaffected by a great number of metabolic intermediates including fructose and glucose 1- and 6-phosphates which inhibit the enzyme from mammalian liver.

F. Properties

The enzyme from *B. megaterium* shows a sharp pH optimum at pH 8 in Tris–HCl buffer that shifts to pH 9 in acetate–borate buffer. The limiting Michaelis constant at pH is 4.5 mM for NAD^+ and 47.5 mM for glucose. The dissociation constant for NAD^+ is 0.69 mM.

G. Stability

The enzyme from *B. megaterium* is stable for long periods when frozen or in solution at pH 6.5 containing 3 M sodium chloride and a protein concentration of 0.5 mg/ml. Diluted enzyme can be stabilized by adding 0.5% polyvinyl pyrollidone. Irreversible inactivation of the glucose dehydrogenase follows lowering of the ionic strength of the enzyme solution.

Remarkable effects were noted for reactivation of glucose dehydrogenase of *B. subtilis* when it was inactivated at 87°C for 30 min. Spores treated in this way yielded inactive enzyme, reactivated by 300 mM EDTA, 100 mM dipicolinic acid and some salts (100 mM NaCl, KCl, LiCl or $(NH_4)_2SO_4$). Other compounds, like lutadinic acid and quinolinic acid, showed the same reactivating effect. It was noted that the molecular weight of the active glucose dehydrogenase was about 100,000, whereas that of the inactive glucose dehydrogenase in heated spores was about 40,000.

H. Microbiology

Glucose dehydrogenase was formed in *Pediococcus* sp. growing on a variety of carbon sources (glucose, fructose, glycerol, arabinose and gluconate) under aerobic conditions, and with glucose under anaerobic conditions. It appeared to be a constitutive enzyme, although decreased levels were evident when cells were grown anaerobically or when gluconate was the growth substrate. The organism was grown in the manner described by Dobrogosz and Stone (1962) at 37°C in shake culture. Strains showing positive indications of the enzyme were *P. pentosaceus* AZ-25-5, 65, TJ-10-1, 4, AZ-37-4, 64, with lower levels in *P. cerevisiae* and very low levels in *Leuconostoc*

mesenteroides and *Streptococcus durans* but not many other species of
Streptococcus or *Lactobacillus*.

I. Enzyme Production

Details for cultivation of *B. megaterium* M 1286 were given by
Banauch *et al.* (1975). Growth for glucose dehydrogenase production
was carried out at 28°C for 20–24 hours in a medium containing
mineral salts, yeast extract, peptone and glucose. Initial purification
involved ammonium sulphate precipitation of the smashed cell extract.
The fraction at 0.6 to 0.9 saturation was fractionated on poly(ethylene
glycol)dimethacrylate PGM-2000.

Fujita *et al.* (1977) found that *B. subtilis* produced glucose dehydrogenase late on during sporulation. This enzyme could react with NAD^+
or $NADP^+$ as a cofactor in its oxidation of glucose. Strangely, another
enzyme detected had a higher affinity for *myo*-inositol than glucose,
and was therefore designated *myo*-inositol: NAD^+ 2-oxidoreductase
(EC 1.1.1.18). This latter enzyme, although having glucose dehydrogenase activity, was a catabolite-repressible inositol dehydrogenase
which, though it reacted with glucose, required NAD^+ as a specific
cofactor and had a different mobility from the true glucose dehydrogenase. The true glucose dehydrogenase appeared to be made in the
forespore compartment and is present in spores. The American
workers defined T_x as the sporulating stage at x hours after the end
of exponential growth. At least 80% of cells had sporulated by T_8.
Glucose dehydrogenase was produced only after $T_{2.5}$ and its specific
activity increased up to T_5. At T_5, five to six times more enzyme
activity was observed in the forespore fraction than in a fraction
containing the extract of the mother cells, although the latter contained seven times more protein. By contrast, the inositol dehydrogenase showing glucose dehydrogenase activity was produced during
vegetative growth and development period.

Fujita *et al.* (1977) used transformable *B. subtilis* 60015 which
requires L-tryptophan and L-methionine for growth. Medium for
glucose dehydrogenase production comprised Difco nutrient broth (8 g)
in 950 ml of water to which was added the following salts: 5 ml each
of 0.14 M $CaCl_2$, 0.01 M $MnCl_2$, 0.2 M $MgCl_2$ and 2 M potassium
phosphate (pH 6.5), 0.5 ml of $FeCl_3$ (2 mM in 0.01 M HCl), 5 ml of
L-tryptophan (5 mg/ml) and 1 ml of L-methionine (10 mg/ml). It was

found that addition of 5 or 10 mM potassium acetate (pH 6.8) increased the extent of sporulation. Large quantities of spores were made from cells grown in 300 litres of this medium to which 1 mM α-methylanthranilate was added to retard spore germination. Aeration was 1 l of air/l of medium/min, and the rotation speed 120 rev./min. After spores had formed in almost all cells, the remaining vegetative cells were lysed with lysozyme.

J. Purification

Partly purified glucose dehydrogenase, obtained from B. *megaterium*, was the starting product for purification of glucose dehydrogenase to homogeneity used by Pauly and Pfleiderer (1975). The procedure comprised: (a) fractionation on hydroxylapatite using a gradient of sodium dihydrogen phosphate from 33 mM to 300 mM containing 0.5 M NaCl; (b) fractionation of the concentrated enzyme on Sephadex QAEA-50 with a gradient of sodium chloride from 0.1 M to 0.5 M in a 33 mM phosphate buffer (pH 6.5) containing 0.05 M EDTA. The purest fraction was eluted at 0.25 M sodium chloride.

In members of the Lactobacillaceae, synthesis of glucose dehydrogenase was confined to species in the genus *Pediococcus*. The procedure for enzyme purification used cell-free extracts of the organism obtained by shaking cells with glass beads in 0.02 M sodium phosphate buffer (pH 7.0) containing 0.02 M sodium chloride and 1 mM EDTA. The steps in purification were: (1) removal of material precipitated with protamine sulphate: (2) precipitation of a fraction from the supernatant by 55% saturation with ammonium sulphate, heating at 60°C for five minutes and discarding the precipitate; (3) recovery of activity by increasing the degree of saturation to 75%; (4) dialysis against the 0.02 M sodium phosphate buffer; (5) sequential absorption of the activity on calcium phosphate, solubilization with the buffer and reprecipitation with ammonium sulphate; and (6) fractionation on DEAE-Sephadex. The presence of salt (NaCl, KCl, NH_4Cl or $(NH_4)_2SO_4$) in the buffer was essential for stabilizing the enzyme. Except for ammonium sulphate, these salts stimulated activity of the enzyme with optimal activity at 0.3 M. The behaviour of this type of glucose dehydrogenase conforms with that reported earlier by Strecker and Korkes (1952) and Metzger et al. (1964).

A heat-resistant glucose dehydrogenase is found in B. *cereus* cultures

(Bach and Sadoff, 1962) which are in the initial stages of sporulation. It is identical but more heat resistant than the spore-free enzyme. A labile glucose dehydrogenase can be extracted from germinated spores of *B. cereus*; it has a higher pH optimum and differs from its stable counterparts. Throughout, glucose dehydrogenase was assayed using glucose and NAD$^+$ as substrates. The bulk of the NADH oxidase impurity could be deactivated by prior heating at 65°C. Purification involved: (a) precipitation of the active fraction with ammonium sulphate; (b) removal of nucleic acid impurities with 2% protamine sulphate; (c) dialysis against 0.1 M Tris buffer (pH 7.6); and (d) final fractionation on a DEAE-cellulose column with removal of the active fraction in a sodium chloride gradient. The vegetative cell and spore glucose dehydrogenases showed identity in the agar gel-diffusion technique against antibody prepared in rabbits. The Michaelis constants were 7 mM for glucose and 0.2 mM for NAD$^+$. The enzyme will not oxidize fructose or galactose.

Hachisuka and Tochikubo (1971) isolated a soluble glucose dehydrogenase from the spores of *B. subtilis* PC1 219 grown on meat extract agar at 37°C for five days. This enzyme acted on glucose in the presence of NAD$^+$ and phosphate buffer (pH 6.8) containing EDTA. Under these conditions, NAD$^+$ was reduced. To isolate the enzyme, the spores were disrupted in phosphate buffer (pH 6.8; 100 mM) using glass beads in a sonic oscillator. Centrifugation was effected at 100,000 g, to remove particulate NADH oxidase. Partially purified enzyme could be obtained by fractionation on Sephadex G-200 in 50 mM phosphate buffer (pH 6.8) and dialysis of the active fraction.

Purification of the glucose dehydrogenase from *B. subtilis* (Fujita *et al.*, 1977) comprised: (a) repeated washing of the spores; (b) breaking the spores with glass beads, centrifugation at 7,000 g and extraction of the pellet with buffer followed by high-speed centrifugation at 67,000 g; (c) precipitation of activity from the supernatant with ammonium sulphate; (d) dialysis of the precipitate against phosphate buffer; (e) fractionation on Ultragel with phosphate buffer containing glycerol; (f) purification on DEAE-cellulose with gradient of potassium chloride and dialysis; (g) fractionation on hydroxylapatite and specific elution with 0.23 M phosphate buffer; and (h) final fractionation on ω-aminohexylagarose.

K. Uses

Lutz and Flückiger (1975) have used glucose dehydrogenase supplied by Merck and Co. in a Gemsaec Centrifugal Analyzer. They advocate its use for analysis of glucose in biological fluids since compounds like uric acid, a natural constituent of serum and plasma, interfere with glucose oxidase determinations. Again they chose this in preference to the hexokinase reaction where two enzymic steps are required, the first of which is non-specific. The glucose dehydrogenase of *B. cereus* was preferred.

IV. GLUCOSE ISOMERASE

A. Occurrence

Glucose isomerase (EC 5.3.1.5; D-xylose ketol-isomerase; xylose isomerase) has been reported from the micro-organisms listed in Table 2.

Table 2

List of some bacteria that synthesize glucose isomerase

Bacteria	Reference
Pseudomonas hydrophila	Marshall and Kooi (1957).
Aerobacter aerogenes	Natake and Yoshimura (1963)
Aerobacter cloacae	Tsumira and Sato (1965)
Bacillus coagulans	Danno (1971)
Bacillus megaterium	Takasaki and Tanabe (1964)
Escherichia intermedia	Natake (1966)
Brevibacterium pentoaminoacidum	Ichimura *et al.* (1965)
Leuconostoc mesenteroides	Yamanaka (1963)
Micrococcus sp.	Yamanaka (1965)
Paracolobacterium aerogenoides	Takasaki and Tanabe (1967)
Streptomyces albus	Takasaki *et al.* (1969)
Streptomyces phaeochromogenus	Tsumira and Sato (1965)
Lactobacillus brevis	Kent and Emery (1973)

B. Structure

An initial report of an X-ray crystallographic study of D-xylose isomerase has been presented (Berman et al., 1974) based on two crystal forms derived from a Streptomyces species. Form A, crystallized as hexagonal prisms in the presence of Co^{2+} and Mg^{2+} ions, showed a = 93.9, b = 99.4, c = 102.7 (estimated error 0.05) A; its density was 1.21 g per cc. Form B, obtained in a vapour diffusion chamber from 60% 2-methyl-2,4-pentanediol, showed a = 94.5, b = 98.9, c = 87.0 (estimated error 0.15) A; its density was 1.27 g per cc. At low resolution, the extinctions were consistent with the space group value of I222 for both forms. Since there are eight equivalent entities per unit cell, these must comprise two tetrameric molecules having four identical subunits. This deduction agrees with data from sodium dodecyl sulphate–polyacrylamide-gel electrophoresis of this enzyme.

The subunit structure and amino-acid composition of xylose isomerase from Streptomyces albus, which is manganese-dependent, has been investigated by Hogue-Angeletti (1975). She maintained that the hydrogen-transfer reaction of the enzyme was intramolecular. The strain used for production of the enzyme was the A.T.C.C. 21175 employed by Clinton Corn Processing Co. The molecular weight of the native enzyme was 165,000 which, in 6 M guanidine hydrochloride, dissociated to give subunits with a molecular weight of 43,000, similar to the 40,000 molecular weight obtained from the S-carboxymethyl-xylose isomerase in sodium dodecyl sulphate. Again, these data suggested a tetrameric structure for the protein composed of four polypeptide chains of equal, or nearly equal, molecular weight. The number of residues in each 41,500 g of protein was: aspartic acid, 45.1; threonine, 14; serine, 9; glutamic acid, 38.0; proline, 17.9; glycine, 35.3; alanine, 45.2; valine, 18.3; methionine, 8.0; isoleucine, 9.8; leucine, 35.9; tyrosine, 8.6; phenylalanine, 23.3; histidine, 9.6; lysine, 9.9; arginine, 33.2; half-cystine and tryptophan (approx. 8).

C. Mechanism

Rose (1975) has reviewed the general mechanism of aldose-ketose isomerase reactions. Xylose isomerase was then the exception of all the isomerases tested, and it showed a degree of hydrogen exchange with water in the course of catalysis. In this respect, he related it to other Mn^{2+}-dependent isomerases which also show low degrees of exchange. Prolonged incubation revealed some enzyme-dependent exchange suggesting that it, too, was a base-catalysed enediol mechanism. Hydrogen transfer was thought to abstract the C-2 proton of the aldose or the C-1 proton of the ketose. Stereochemical evidence was presented for a cis-enediol intermediate, and that, with the 2R aldose (D-xylose) as substrate, a 1R–1H* ketose product was produced in *H_2O. As predicted, the enzyme prefers the α-D-xylose anomer.

Proton-magnetic relaxation studies of interaction of D-xylose and xylitol with D-xylose isomerase have been reported (Young et al., 1975). A greater paramagnetic effect of enzyme-bound Mn^{2+} on the α-anomer of D-xylose was observed, supporting the specificity of D-xylose isomerase. The amount of xylitol necessary to displace α-D-xylose from the substrate–enzyme–Mn^{2+} complex was consistent with the K_m value of α-D-xylose and the K_i value for xylitol. It was suggested that two small ligands, such as water molecules, intervene between the bound metal ion and the bound substrate in the active ternary complex.

D. Specificity

Some problems are encountered in removing other aldose isomerases from xylose isomerase. This is illustrated in its isolation from *Lactobacillus plantarum*, strain 124-2 (A.T.C.C. 8041). Even the best preparations from this source still have a L-arabinose isomerase impurity (Horecker, 1974). The enzyme-catalysed reaction is easily reversible and is specific for D-xylose and D-xylulose. The forward assay to ketoses can be helped by introduction of borate buffer (pH 8.2) that complexes selectively with the ketose and shifts the position of equilibrium. Normal non-complexing buffers are used for the reverse reaction. Thus, some 82% of the D-xylulose is converted into D-xylose under these conditions. The enzyme has no action on D-ribulose or L-xylulose (Burma and Horecker, 1958). Earlier, Mitsuhashi and Lampen (1953) reported that the enzyme had no action on D-glucose or D-xylose 5-phosphate.

Hochster and Watson (1954) reported that xylose isomerase from *Ps. hydrophilia* (N.R.C. 492) preferred D-xylose and not D-ribose, D-arabinose, L-arabinose, D-lyxose or L-rhamnose. Again, D-glucose was not reported to be a substrate. The enzyme required Mn^{2+} and Mg^{2+} ions to be fully operative at its optimum pH value of 7.5. However, Marshall and Kooi (1957), also using *Ps. hydrophila* grown on D-xylose, found that interconversion of D-glucose to D-fructose did occur but required the presence of arsenate. Arsenate was not required with the xylose isomerase from *L. brevis* which catalysed the same conversion (Yamanaka, 1963). He concluded that D-xylose isomerase and D-glucose isomerase were the same enzyme. Some further light was shed on the problem when it was discovered that an enzyme from *Escherichia intermedia* H.N. 500, grown on glucose, could also catalyse conversion of glucose into fructose in the presence of arsenate. In fact, the true substrate of the enzyme was glucose 6-phosphate and it was envisaged that the non-enzymically formed glucose–arsenate complex posed as substrate analogue (Natake, 1968).

The current position is therefore that, although there have been reports of D-xylose isomerases not acting on D-glucose, those that are known to act on D-glucose also act on D-xylose. Thus, the xylose isomerases of *L. brevis*, *S. albus* and *B. coagulans* show K_m values for D-xylose and D-glucose of 5 mM and 0.92 M, 32 mM and 16 M, and 1.1 mM and 90 mM, respectively. The first and the third of these enzymes were also stated to act on D-ribose with K_m values of 0.67 M and 83 mM, respectively (Noltmann, 1972). All three of these enzymes have been crystallized and they can be further characterized by their sensitivity to polyhydric alcohols and metal-ion requirements (Mn^{2+}, Co^{2+} and Mg^{2+} and Co^{2+}, Mn^{2+} and Mg^{2+}, respectively).

E. Inhibitors

Xylitol is one of the best inhibitors of glucose isomerase having a K_i value of 5 mM for production of glucose from fructose at 60°C and pH 6.8 in 0.01 M succinate containing cobalt and magnesium (Lee *et al.*, 1976).

F. Properties

Glucose isomerase from *L. brevis* had an apparent K_m value for glucose conversion into fructose of 47.8% (w/v) in buffer (0.02 M Tris, pH 7.0) and 25.6% (w/v) in water calculated from a Lineweaver–Burk plot with purified enzyme. The corresponding values for V_{max} (mg of fructose per hour per ml of enzyme) were 20, 78 and 1350, respectively. The percentage of fructose in the equilibrium mixture at various temperatures was 58% (35°C), 64% (50°C) and 63% (60°C) attained under the same conditions of enzyme concentration and initial concentration of sugar (12%, w/v) in 100 hours, 40 hours and 20 hours, respectively (Kent and Emery, 1973). Reaction mixtures always contained both manganese sulphate and cobaltous chloride.

Streptomyces albus glucose isomerase was stated, when crystalline, to contain both cobalt and magnesium in the proportion of about 1.4 atoms and 0.3 atom per mol of enzyme respectively (Takasaki *et al.*, 1969). It has a high optimum pH value of 8–8.5 and an optimum temperature of 80°C. It was stable over a wide range of pH (4.5–11) but was inhibited by heavy metals such as Ag^+, Hg^{2+} and Cu^{2+}. It was estimated that the enzyme had a molecular weight of 187,000.

G. Stabilization

Zittan *et al.* (1975) investigated the relationship between various process parameters and the stability of their enzyme from *B. coagulans*. They emphasized the importance of using adequate enzyme in short process times, since the fructose produced gives rise slowly to byproduct and coloured products that might decrease the stability of the enzyme. This is particularly true under alkaline conditions where colouration calls for increased use of decolourizing carbon. They showed that, while the enzyme was active up to 90°C, the stability at that temperature was too low for industrial use. In terms of productivity per unit of enzyme, they preferred 65°C with a lower limit of 60°C to minimize the risk of microbial infection. Below pH 5, they found that the enzyme was irreversibly denatured. Although the optimum was at pH 8.0–8.5, it exhibited 70% of optimum activity at pH 6.0. Cobalt ions could be omitted from the glucose feed at pH values above 8. Thus, at pH 8.5, it retained 52% of its activity after 450 hours in the absence of cobalt and 42% in its presence. In the presence of cobalt ions at pH 7.6, the retention was virtually the same (41%) but there was a severe loss in its absence to 28% residual activity. A somewhat similar behaviour was evident in the requirement for magnesium ions, but this was in the concentrations required above pH 8 (0.4 mM) and below pH 8 (8 mM). Their enzyme was inhibited by calcium ions, so a molar ratio of Mg : Ca of at least 10 was required. The enzyme was sensitive to oxygen and the purity of the glucose feed, particularly with respect to ions. The enzyme is sold by Novo Ltd. under the name Sweetzyme®.

Gist Brocades in Holland claim a half-life of about 500 hours for their Maxazyme® operating at 65°C under nitrogen gas at pH 7.7 and a 40% glucose feed containing 3 mM Mg^{2+}. Their glucose isomerase was derived from a strain of *Actinoplanes missouriensis* under an agreement with Anheuser-Busch (1974). The enzyme is strongly inhibited by copper, nickel or zinc ions. Other strains of this genus, e.g. *A. philippinsis* and *A. armeniacus*, are producers of glucose isomerase with no need for xylose or xylan in the culture medium comprising corn-steep liquor, salts and tap water.

H. Microbiology

Kent and Emery (1973) increased the glucose isomerase yield some four times over that previously reported for *Lactobacillus brevis* using strain N.C.D.O.-474 and a medium containing 2% xylose, 0.5% glucose, 5% bacteriological peptone, 3% yeast extract, 1% sodium acetate, 0.04% $MnSO_4,4H_2O$, 0.01% $MgSO_4,7H_2O$, and 0.01% $CoCl_2,6H_2O$. Growth was at 30°C without aeration using mild agitation only. Use of xylose for industrial production of glucose isomerase would be prohibitive so, earlier, a search had been made and some *Streptomyces* species were found to produce the enzyme in a medium containing xylan or xylan-containing materials such as wheat-bran, corn cob and corn hull (Takasaki *et al.*, 1969). Typically, *Strep. albus* grew in such a medium of 3% wheat bran, 2% corn-steep liquor and 0.024% $CoCl_2,6H_2O$ at 30°C and pH 7. The enzyme was also inducible by xylose but not by glucose. With both of these bacteria, the enzyme was present in cells. In a British patent (Agency of Industrial Science and Technology, 1968) assigned to a Japanese Agency, Disclosure is made that several species of *Streptomyces* will produce similar glucose isomerase (*Strep. albus* strain Y.T.-No. 4; *Strep.* species *flavovirens*, *echinatur* and *achromogenus*).

Lee *et al.* (1972) discovered that micro-organisms of the diphtheroidic genus *Arthrobacter* nov. sp. N.R.R.L. B-3726, N.R.R.L. B-3727 and N.R.R.L. B-3728 were capable of producing glucose isomerase in the absence of both xylose and xylan. This patent was assigned to the R. J. Reynolds Tobacco Company (1977) and is operated by ICI Ltd. The medium contained 2% glucose, 0.01% $MgSO_4,7H_2O$, 0.2% KH_2PO_4, 0.6% $(NH_4)_2HPO_4$, 0.15% yeast extract and 0.5% meat protein, at 30°C. The glucose isomerase can act on both glucose and xylose, and it can be produced in the presence of either sugar. Only Mg^{2+} ions were cited as being required for maximum activity. The enzyme showed good activity between 50–90°C, preferably 60–75°C and between pH 6 and 10, preferably pH 8.

I. Enzyme Production

Burma and Horecker (1958) used xylose-grown cells of *L. plantarum* a mixture of 1% xylose and 0.1% glucose autoclaved separately and added to sterile media. Cells were harvested from the media after 24 hours at 37°C. In addition to sugars, this contained 0.4% yeast extract, 1% nutrient broth, 1% sodium acetate, 0.02% $MgSO_4,7H_2O$, 0.001% NaCl, 0.001% $FeSO_4,7H_2O$ and 0.001% $MnSO_4,4H_2O$. Cells were centrifuged and washed with 20 mM sodium bicarbonate solution before disruption with glass beads. High-speed centrifugation at 25,000 *g* gave an active supernatant. Treatment of this with manganese sulphate and removal of the precipitate enabled the purified enzyme to be recovered from the supernatant by ammonium sulphate precipitation.

Shieh (1976) described a medium for producing glucose isomerase which was based on molasses, corn-steep liquor and an inorganic nitrogenous salt. *Actinoplanes missouriensio* was grown on 3% beet molasses, 1% corn-steep liquor, 0.2% $NaNO_2$, 0.001% $MgSO_4,7H_2O$, and 0.06% Dow Corning Antifoam. This medium overcame the expense of growing the micro-organism in a xylose- or xylan-containing medium. A further improvement was later reported (Shieh, 1976), by substituting 1.5% soyflour for the corn-steep liquor in this medium, which increased enzyme activity from 42.0 units/ml to 67.5 units/ml after 72 hours of fermentation with the N.R.R.L.-B3342 strain. Soyflour had the added advantage that it did not need to be filtered as did the corn-steep liquor-based medium.

The medium favoured for *Streptomyces* A.T.C.C. 21175 was a

cotton-seed hull hydrolysate sufficient to provide 1% xylose (Cotter *et al.*, 1977). After 44 hours, the broth was filtered and the filter-cake washed with water. Glucose isomerase was then extracted from the cake (160 g) with 180 ml of water containing 1% octadecyltrimethyl-ammonium chloride at 58°C. After an hour, all of the glucose isomerase was in the aqueous phase.

Production of glucose isomerase in *Strep. albus* Y.T. No. 5 was enhanced by culturing it in 50 ml of medium (pH 7.5) containing (w/v) 2% corn-steep liquor, 3% wheat bran, 0.024% $CoCl_2,6H_2O$ and 0.5% rapeseed cake. When grown at 30°C for 40 hours with shaking, addition of the rapeseed cake increased the enzyme level from 20.7 units/ml to 33.3 units/ml (Takasaki, 1977).

Streptomyces albus was one of a number of species (others included *Strep. gracilis*, *Strep. matensis*, *Strep. niveus* and *Strep. platensis*) used for production of glucose isomerase by Buki *et al.* (1976). The medium contained (g/l): wheat bran, 30; corn-steep liquor (50% dry material content), 40; $MgSO_4,7H_2O$, 1; and $CoCl_2,6H_2O$, 0.24 (pH 7). After inoculation with 1 ml of a spore suspension, growth was continued for 48 hours at 28°C with aeration. A medium (100 ml), containing (g/l): corn-steep liquor, 40; $(NH_4)_2HPO_4$, 4; sorbitol, 4; glucose, 8; and $CoCl_2$, 0.24 (pH 7.5), was inoculated with 5 ml of the previous culture medium and then shaken (with 1 g of xylose after 28 hours) for 40 hours at 28°C. The filtrate contained 40 Takasaki units of glucose isomerase/ml. Eventual conversion of 1.67 M glucose into 0.8 M fructose could be achieved at 70°C and pH 6.5 in the presence of magnesium sulphate and cobaltous chloride. Takasaki (1975) reported that cobalt ions, added at a concentration that did not inhibit optimal cell growth, markedly increased glucose isomerase production if added after the beginning of the fermentation. Thus *Strep. albus* Y.T. No. 4 was aerobically cultured at 30°C in a medium containing 3% (w/v) wheat flour, 2% (w/v) corn-steep liquor and 1 mM cobaltous chloride added after 30 hours. After 43 hours, the glucose isomerase content was 27.8 units/ml compared to 20 units/ml in the absence of additional cobaltous salt. The importance of xylose in the culture medium was also stressed by the same worker (Takasaki, 1976). Thus, when the Y.T. No. 4 strain was cultured with shaking at 30°C for 35 hours on a medium containing (w/v) 3% wheat bran (with a xylan content of about 20 g), 2% corn-steep liquor, 0.024% $CoCl_2,6H_2O$ and 0.1% $MgSO_4$, addition of an acid hydrolysate of cotton-seed cake

Table 3

Different types of glucose isomerase activity detected in various bacteria. From Coker and Gardner (1976). Enzyme activities are expressed as the amount of enzyme that will convert one μmol of glucose into fructose in one minute at 60°C

Activity	Species of bacterium			
	Brevibacterium incertum N.R.R.L. B.5383	Flavobacterium deverans N.R.R.L. B-5384	A.T.C.C. 10,829	Streptomyces phaeochromogens A.T.C.C. 15,486
Cellular activity	4.91	5.63	0.829	4.93
Cell-bound enzyme	23.65	20.1	0.89	17.18
Soluble enzyme yield	8.88	10.25	4.48	6.87
Total enzyme yield	32.53	30.35	11.75	24.05

equivalent to 0.5% xylose improved the glucose isomerase level from 20.8 units/ml to 40.5 units/ml.

Various types of glucose isomerase activity were measured by Coker and Gardner (1976) in a group of bacteria (Table 3).

Meers (1976) grew *Arthrobacter* strain N.R.R.L. B-3728 in a culture medium in which either the carbohydrate or nitrogen source was the growth-limiting substrate. The other culture constituents were: 40% $MgSO_4$, 2 ml/l; PO_4^{3-}, 1.596 g/l; Na_2SO_4, 0.075 g/l; K_2SO_4, 0.45 g/l; yeast extract, 0.05 g/l; Fe^{2+}, 3 p.p.m.; Cu^{2+}, 0.075 p.p.m.; Mn^{2+}, 0.375 p.p.m.; Zn^{2+}, 0.345 p.p.m.; Ca^{2+}, 0.075 p.p.m.; H_3PO_3, 0.384 p.p.m.; and Na_2MoO_4, 0.135 p.p.m. The temperature was maintained at 30°C and the pH value at 6.9. In nitrogen-limitation experiments ammonium sulphate was employed. In carbohydrate-limitation experiments, the pH value was maintained with ammonia gas. The best carbohydrate conversion (34 and 46%) and enzyme activities (1,150 and 1,150 units/g dry wt of cells) were obtained with either xylose or glucose, respectively, as the limiting sources.

Kelly and Meers (1975) also investigated glucose isomerase production by various *Curtobacterium* species. Thus *C. helvolum* N.C.L.B. 10,352 was grown for 24 hours at 30°C in 10 litres of a medium containing (g/l): meat extract, 5; yeast extract, 2.5; peptone, 10; NaCl, 5; $MgSO_4$, 0.5; $CoCl_2$, 0.05; and D-xylose, 10. The cells effected a conversion of glucose into fructose to the extent of 42% fructose when incubated with glucose and magnesium sulphate in a phosphate buffer (pH 7.5) at 65°C for 48 hours.

J. Enzyme Purification

Lee *et al.* (1976) developed an affinity chromatography system for purification of glucose isomerase. Xylitol was coupled to Sepharose 4B activated by cyanogen bromide followed by reaction, successively, with 1,6-hexanediamine and succinic anhydride. The glucose isomerase adhered strongly to the column, but specific elution with xylitol was impractical because of the slow elution rate. However, the enzyme can be eluted with 0.4 M sodium chloride. An affinity column, based on xylitol attached directly to the Sepharose via hexanoic acid, showed pronounced leakage of enzyme activity. Economic use of such a procedure was only considered viable for recovery of enzyme from solutions containing low concentrations.

K. Uses

The realization that only a single process step via immobilized glucose isomerase would convert non-sweet glucose into a glucose–fructose syrup at high substrate concentrations, near neutral pH values and with a yield of 42% in terms of fructose, created a tremendous growing market for the product syrup variously called Isomerose®, Isosyrup®, Corn Sweet™ and Isosweet® by different companies. Since the syrup was effectively as sweet as sucrose, there was no need initially for a product-separation step. More attractive still was the opportunity afforded to manufacturers of producing the syrup at a price that would compete or undercut the price of sucrose. Thus, in 1972, an equivalent product, invert sugar, was selling at $0.12/lb compared with $0.11/lb for Isomerose® 100. These products were sold as syrups which markedly lowered the price calculated on a dry basis. Thus, in 1973, crystalline sucrose, and glucose cost $0.127 and $0.135/lb, respectively. Low-cost production of syrups with a much higher content of fructose is only just becoming feasible. The firm of A.D.M. Corn Sweeteners already sell Corn Sweet® 55 and 90 with these contents of fructose sugar. A. F. Staley Mfg. Co., Decatur, Illinois, U.S.A. announced (February, 1978) the availability, later in the year, of second generation high-fructose corn syrups containing 55% fructose forecasted as a replacement for medium invert sugar and sucrose in soft drinks. Immense potential to produce fructose syrups is building up in the

U.S.A. where, at the Staley plant alone, the capacity is for 500 million lb. Hubinger, a subsidiary of Heinz Co., advertised in January, 1978 that they are awaiting business for their 550 million lb plant. Even small countries like Yugoslavia are building such plants in this case via collaboration with Miles Laboratories Inc., Elkhart, Indiana, U.S.A.

V. *BETA*-GALACTOSIDASE

A. Occurrence

The enzyme β-galactosidase (galactohydrolase, lactase, EC 3.2.1.23) has been detected in *E. coli* (Rickenberg, 1974), *B. megaterium* (Yeung *et al.*, 1976) and *L. plantarum* (Sahyoun and Durr, 1972).

B. Structure

The molecular weight of β-galactosidase from *B. stearothermophilus* was estimated to be 215,000 by disc-gel electrophoresis (Goodman and Pederson, 1976).

C. Mechanism

Huber *et al.* (1975) have shown that one of the products of *E. coli* β-galactosidase acting on lactose is allolactose (6-O-β-D-galacto-pyranosyl-β-D-glucopyranose). Surprisingly, allolactose, not lactose, is the natural inducer of the *lac* operon in *E. coli*. The only products of the action of β-galactosidase on allolactose were galactose and glucose at low concentrations ($\leqslant 0.05$ M). Above this concentration of substrate, more glucose was produced than galactose because of synthesis of oligosaccharides. With both lactose and allolactose, the optimum pH value was 7.8–7.9: α-allolactose is initially acted on about twice as rapidly as β-allolactose. With allolactose at 30°C, the V_{max} value was 49.6 units/mg (compare with lactose, 32.6 units/mg) and the K_m value 1.2 mM (lactose, 2.53 mM).

D. Specificity

In the search for an ideal β-galactosidase, many sources of enzyme have been investigated since those of selected *E. coli* strains did not possess as great an activity towards lactose as towards synthetic nitrophenyl-galactosidase substrates. Wondolowski and Woychik (1974) found that one of the best was *E. coli* A.T.C.C.-26 galactosidase which had an activity towards lactose equal to 40% of that toward synthetic substrates.

E. Inhibitors

Bacillus stearothermophilus β-galactosidase is strongly inhibited by galactose (K_i value 2.5 mM; Goodman and Pederson, 1976). Under comparable ˙ conditions acting on *o*-nitrophenyl-β-D-galacto-pyranoside, the activity exhibited in the presence of various sugars was: lactose, 88%; glucose, 80%; xylose, 91%; cellobiose, 67%. Other sugars, like fructose, mannose, mannitol, sucrose and raffinose, had no effect. *Bacillus megaterium* β-galactosidase is stimulated by glucose (Rohlfing and Crawford, 1966).

F. Properties

The β-galactosidase from *B. stearothermophilus* (Goodman and Pederson, 1976) has a pH optimum of 6–6.4 and a very sharp dependence on pH value on the acid side (5–6) of its optimum. The optimum temperature was 65°C and, at 55°C, the K_m value was 0.11 M for lactose and 9.8 mM for *o*-nitrophenyl-β-D-galactopyranoside. The K_m value for *B. subtilis* β-galactosidase reported by Anema (1964) was 42 mM for *o*-nitrophenyl-β-D-galactopyranoside, while the K_i values were 0.71 M for lactose and 42 mM for galactose.

Homogeneous β-galactosidase from an *Arthrobacter* sp. had a molecular weight of 500,000, a pH optimum of 6.3–6.4 and K_m values of 0.21 mM and 10.5 mM for *o*-nitrophenyl-β-D-galactopyranoside and lactose, respectively (Donnelly *et al.*, 1977). The enzyme resembled β-galactosidase from *E. coli* in its metal-dependency, but it was less stable to heat and disruption by urea than the latter. Inhibition of the enzyme by zinc ions was suppressed in the presence of 20 mM inorganic phosphate. Crude extracts of the *Arthrobacter* sp. contained several electrophoretically distinct forms of the enzyme.

Electrophoretically homogeneous β-galactosidase (EC 3.2.1.23) isolated from *Aspergillus oryzae* RT-102 (F.E.R.M.-P1680) was free from α-galactosidase, α- and β-mannosidase, α- and β-*N*-acetylhexosaminidase and proteolytic activities. The enzyme hydrolysed links in aryl-β-galactosides, lactose and lactosides as well as β-galactosyl links in urinary glyco-asparagines and asialo-α-acid glycoprotein. In aqueous solution, it retained full activity at 4°C for several months. At pH 4.5, the optimum pH value, and 37°C full activity was maintained for several days (Akasaki *et al.*, 1976). Another strain of *A. oryzae* (F.E.R.M.-P3065) had a pH optimum of 4.5 and temperature optimum of 50°C. It was stable at pH 4.0–7.5 and 45°C (Kobayashi and Kasuga, 1976).

G. Stability

Bacillus stearothermophilus β-galactosidase is relatively stable at 50°C losing only half of its activity after 20 days at this temperature. At 60°C, more than 60% of the activity was lost in 10 min (Goodman and Pederson, 1976). Addition of 4 mg of bovine serum albumin/ml provided some protection at this temperature; the enzyme lost only 18% of its activity under the same conditions. After a variable lag period, samples of β-galactosidase exhibited an exponential decay in activity with half-lives of 250 days at room temperature, 75 days at 0°C, and 60 days at −20°C.

H. Microbiology

About 95% of 0.11 serotypes of *Ps. aeruginosa* are able to hydrolyse *o*-nitrophenyl-β-D-galactopyranoside but not lactose (Poindron *et al.*, 1977). In all, some 647 strains of this type were examined. The β-galactosidase was located in the periplasm.

Goodman and Pederson (1976) found that several strains of thermophilic aerobic spore-forming bacilli synthesized a β-galactosidase active against both the galactoside and lactose. Some five of eight strains of *B. stearothermophilus* produced the enzyme constitutively. This constitutivity was not the result of a temperature-sensitive repressor. These observations are in marked contrast to the normally inducible synthesis of β-galactosidase by other members of the genus *Bacillus*. Anema (1964) has, however, isolated a constitutive mutant of *B. subtilis*.

Okon *et al.* (1975) found that addition of 0.5% lactose to a glucose–mineral medium induced *Sclerotium rolfsii* to produce a β-galactosidase. This effect was inversely related to the glucose concentration in the test range of 0.5–2.5%. *Vibrio* El Tor contains a constitutive β-galactosidase but it cannot use lactose as it lacks β-galactoside permease (Bhattacharya *et al.*, 1974). The β-galactosidase from *E. coli* is sensitive to glucose repression, and this repression is overcome by cAMP (Perlman and Pastan, 1968). The β-galactosidase from this vibrio is sensitive to glucose repression and this can be partially overcome by a similar addition of cAMP.

Yeung *et al.* (1976) showed that, when *B. megaterium* was grown on D-galactose as the sole carbon source, the cells actively synthesized β-galactosidase. However, D-galactose, when added to a glucose-grown culture, did not induce β-galactosidase apparently because of the glucose inhibition of transport of galactose. On the other hand, when glucose was added to a galactose-grown culture, transport of galactose continued at a lower but significant rate, whereas further synthesis of β-galactosidase was halted. Glucose inhibition of β-galactosidase synthesis in the pre-induced culture was not overcome by addition of cAMP or cGMP.

Aspergillus nidulans mutants, altered in carbon catabolite regulation, were isolated by selecting for mutants of the *cre* A 217 strain capable of using acetamide as the sole nitrogen source in the presence of sucrose. In addition to *cre* A mutants, strains with mutations in two new genes, *cre* B and *cre* C, were found. The *cre* B and *cre* C mutants grow poorly on some carbon sources and have low levels of some enzymes of carbon catabolism, e.g. β-galactosidase (Hynes and Kelly, 1977).

Successful induction of β-galactosidase in *E. coli* has been achieved with methyl-β-D-glucoside, 3-nitrophenyl-β-D-glucoside, picein, cellobiose and gentiobiose at 1 mM. At 10 mM, only cellobiose and gentiobiose acted as inducers (Daoud *et al.*, 1977).

In *E. coli*, the *ebg* A gene, whose product does not hydrolyse lactose effectively *in vivo*, may evolve so that its product hydrolyses lactose *in vivo* well enough to permit utilization of lactose as a sole carbon source (Hall, 1976). At least three variants of the evolved gene product may be obtained, and these variants may be distinguished from each other and from the original gene product by kinetic studies.

Russian workers have shown that over 100 streptomycin-resistant mutants of *E. coli* strain CA 165 carrying the ochre suppressor sup B were able to suppress ochre mutation in the β-galactosidase gene of the same bacteria as mutation of phage T4 carrying no suppressor (Oganesyan *et al.*, 1977).

I. Enzyme Production

Higgins *et al.* (1978) used the following medium (g/l) for large-scale production of β-galactosidase from *E. coli*: $(NH_4)_2SO_4$, 8.0; NaCl, 3.0; Na_2HPO_4, 5.3; KH_2PO_4, 1.6; triammonium citrate, 0.5 and either glycerol, 15.0 or glycerol, 12.0 or glucose, 3.9. Seed cultures always employed glycerol as the sole carbon source together with double the concentrations of the two phosphates. It was noted that glucose caused repression of β-galactosidase synthesis and so glycerol was partially replaced with glucose in order to get a rapid transition from initial batch fermentation to a steady enzyme concentration in continuous culture. Previous work (Gray *et al.*, 1973) had shown that the glycerol and salts medium was superior to a medium containing glucose and yeast extract. The kinetics of β-galactosidase synthesis by bacteria growing on the later medium were reported by Gray *et al.* (1971).

Aspergillus oryzae microsporus WA-197 (F.E.R.M.-P 3065) was cultured on a wheat bran medium at 30°C for three days to effect β-galactosidase production (Kobayashi and Kasuga, 1976). Extraction of the mould and bran yielded an enzyme having 21 units/ml. After mixing with activated carbon and centrifuging, the soluble enzyme was purified on DEAE-cellulose equilibrated with 0.05 M phosphate buffer (pH 7.0), and the absorbed enzyme was eluted with 0.1 M phosphate buffer (pH 5) containing 0.4 M sodium chloride. After desalting with Sephadex and ethanol precipitation, the β-galactosidase had an activity of 12,000 units/g.

A similar bran-containing medium was used to grow *Penicillium citrinum* A.T.C.C. 9849 to yield a crude β-galactosidase precipitable by three volumes of alcohol added to five volumes of the water-extracted enzyme (Tsujisaka *et al.*, 1976). Purification of the enzyme involved ammonium sulphate fractionation and column chromatography on DEAE-Sephadex, SP (sulphopropyl)-Sephadex and Sephadex G-200. It had an optimum temperature of 50°C and pH value of 4.5, but was stable at pH 3.5–8 at 4°C for 24 hours and < 55°C at pH 6 for 15 minutes. It was inactivated by 1 mM Cu^{2+}.

Whey, diluted three-fold and containing 0.2% ammonium phosphate was suitable for growth of mycelia of *Trametes versicolor* and *Tyromyces albellus*. After 120 hours at 20°C and aeration with 0.08 mmol of oxygen/l/min, the supernatant was found to contain 100-fold higher β-galactosidase activity than that of the biomass (Lobarzewski, 1976).

Mucor pusillus I.F.O. 4578 has been cultured on a medium (pH 6.0) containing (w/v) 8% defatted soybean powder, 0.1% sorbitol, 0.5% KH_2PO_4, 0.02% $MgSO_4,7H_2O$ and 0.02% $CaCl_2,2H_2O$ (Tomoda *et al.*, 1976). After shaking at 28°C for seven days, β-galacosidase was recovered from the supernatant (700 ml) by addition of acetone, yielding 8.3 g of powder with an activity of 0.296 unit/mg. The optimum pH value and temperature were 2.5 and 60°C, respectively. Although the enzyme was generally stable between pH 3 and 9, it was inactivated at temperatures greater than 60°C at pH 7 for 30 minutes. It was inhibited by Fe^{3+}, Sn^{2+}, Pb^{2+} and Hg^{2+}.

Among *Kluyveromyces fragilis* strains (41), the ability to produce β-galactosidase varied 60-fold (Mahoney *et al.*, 1975). The four best strains were U.C.D. 55–31 (N.R.R.L. Y–1196), U.C.D. C21 (−), U.C.D. 72–297 (−) and U.C.D. 55–61 (N.R.R.L. Y–1109). The last named was monitored for biosynthesis of lactase on media containing lactose and sweet whey with 15% lactose. The best yield of lactase was obtained at an aeration rate of ≥ 0.2 mmol of oxygen/litre/min. Supplementary growth factors were unnecessary for good yields of lactase when the yeast was grown on whey-containing media. The enzyme was extractable by toluene autolysis (2% v/v) at 37°C in 0.1 M potassium phosphate buffer (pH 7.0) containing 0.1 mM manganese chloride and 0.5 mM magnesium sulphate. The lactase was precipitable by addition of acetone and, in the concentrated form, retained its activity for at least seven days at 22°C, for three weeks at 4°C and for six weeks at −20°C.

An acid-active and acid-stable β-galactosidase is claimed by Kiuchi and Tanaka (1976) from *Aspergillus oryzae* A.T.C.C. 20,423. The medium of 20 g of wheat bran in 20 ml of tap water in a 500 ml flask was cultivated for five days at 30°C after inoculation. The culture filtrate from 10 flasks was harvested (232 g) and extracted with 2 litres of water to yield an enzyme with an activity of 57.4 and 43.3 units/ml with o-nitrophenyl-β-D-galactopyranoside and lactose, respectively. The enzyme was decolourized with activated carbon and precipitated with propan-2-ol. The precipitate, after resolution and freeze-drying, yielded 4.3 g of white powdered β-galactosidase with activities of 16,300 and 12,400 units/g with o-nitrophenyl-β-D-galactopyranoside and lactose respectively.

The inoculum of A. oryzae was spores grown on potato–dextrose agar.

J. Enzyme Purification

Isolation of β-galactosidase from E. coli M.L.308 was examined for the feasibility of adapting it to a continuous-flow process (Higgins et al., 1978). It became possible to operate a semi-continuous process which, when fully continuous, would yield 100 g of protein/h comprising 23% β-galactosidase. The modified isolation sequence comprised: (a) fermentation at 37°C in a one m³ working volume vessel fitted with three six-bladed turbine impellers; (b) harvesting the bacterial slurry recovered by passage at 5°C, first through a plate heat-exchanger and then a disk-bowl centrifuge using a flow rate of 250 l/h to yield 37 l of bacterial slurry/h; (c) disruption in a high-pressure homogenizer wherein it was found that β-galactosidase was released faster than the remainder of the bacterial protein; (d) removal of cell debris followed by controlled heat treatment to remove RNA (a heat-exchanger with water at 65°C as heating medium was used with a holding time of 5 minutes); (e) precipitation of β-galactosidase by mixing the protein stream with a saturated ammonium sulphate solution in a stirred continuous-flow tank to a level of 40% saturation with respect to the salt. With the modified system, the time between removal from the fermentor to an ammonium sulphate paste was 4–5 hours.

Affinity chromatography columns have been developed by Robinson et al. (1974) for purification of β-galactosidase from E. coli. Biospecific absorption was achieved on agarose gel substituted with p-amino-phenyl-β-D-thiogalactopyranoside. On a 1.8-litre column, some 5 g of pure enzyme could be processed every two-hour cycle. The purified enzyme appeared to contain an inactive fragment of the enzyme in addition to the major enzyme component.

K. Uses

Ngo *et al.* (1976) used lactase in various lengths of nylon tubing to effect hydrolysis of lactose in skim milk and to catalyse hydrolysis of *o*-nitrophenyl-β-D-galactoside. The enzyme was attached to the nylon by glutaraldehyde after the tubes had been etched with calcium chloride in aqueous methanol and partially hydrolysed with 4.5 M hydrochloric acid. The industrial objective was to provide milk products in which the lactose is already hydrolysed (i.e. predigested) since many individuals are deficient in intestinal lactase and suffer from various abdominal disorders after drinking milk (Rosenweig, 1969; Bedine and Bayless, 1973).

Wadiak and Carbonell (1975) used *E. coli* lactase entrapped in cellulose-nitrate membrane microcapsules. A thin semipermeable membrane was formed around an aqueous haemoglobin solution containing the enzyme. The membrane stopped diffusion of the enzyme, but allowed substrate and products to permeate in order to carry out the chemical reaction. The enzyme in this form was an effective catalyst for conversion of lactose into glucose and galactose in pasteurized and homogenized whole milk. High enzyme loading could be achieved.

Many systems for conversion of whey lactose into glucose–galactose syrups have been described, and lactases with pH optima of 3.5–4.5 or 6.0–7.0 are available on many types of supports. Roland and Alm (1975) utilized such modified whey syrups with grape juice for wine alcohol fermentations giving palatable wines containing 10–12.5% alcohol. *Aspergillus niger* lactase was entrapped in a gel produced by irradiating poly(vinylpyrollidone) at 1 Mrad/h under nitrogen. This could effect hydrolysis of lactose in acidified whey at 50°C using a column packed with the gel and sawdust, without appreciable activity loss, for 12 days (Maeda, 1975).

Acidified whey is a product of cottage cheese manufacture and contains protein, lactic acid, lactose and various salts. Much protein can be recovered by ultrafiltration so that the target of the lactase is this acidified lactose generally containing 15–20% solids of which 70–80% is lactose. Hydrolysis would give a sweetened whey ultra-filtrate that could be recycled through the dairy by adding it to ice-cream, egg-nog, yogurt and other dairy products. Its sweetness could be further increased by treatment with glucose isomerase.

A novel use has been made of the β-galactosidase activity of *E. coli*, namely in effecting enrichment of *Shigella* spp. in competition with a 100-fold higher population of *E. coli*. The enzyme hydrolyses 4-chloro-2-cyclopentylphenyl-β-D-galactopyranoside liberating 4-chloro-2-cyclopentylphenol which is selectively toxic to *E. coli*, but not to *Shigella sonnei* or *S. flexneri* (Park *et al.*, 1976). It should be noted that the ability to utilize lactose is an important criterion for differentiating genera of the Enterobacteriaceae. Species of *Shigella*, by contrast, are incapable of utilizing lactose.

Ohmiya *et al.* (1977) have advocated the use of cells of *L. bulgaricus*, *E. coli* and *Kluyv. lactis* immobilized in polyacrylamide beads as catalysts for hydrolysing lactose, since they retained 27–61% of the β-galactosidase activity of intact cells. Stability, optimum temperature and pH value were unaffected by the immobilization. The K_m values of β-galactosidase under these conditions were 4.2, 5.4 and 30 mM, respectively: the first two systems were unaffected by galactose but the last one was. The systems effected high conversions when introduced into skim milk; its flavour was unaffected but its sweetness was considerably enhanced.

VI. INVERTASE

A. Occurrence

Invertase (β-D-fructofuranoside fructohydrolase: β-fructosidase; EC 3.2.1.26) has been isolated from *Saccharomyces cerevisiae* (Sutton and Lampen, 1962), *Sacch. carlsbergensis* (Eddy and Williamson, 1957), baker's yeast (Hoshino and Momose, 1966) and *Neurospora* sp. (Trevithick and Metzenberg, 1966).

B. Structure

Although iodine partially inhibited invertase action, reagents such as cysteine, glutathione, ascorbic acid and sodium amalgam did not re-activate the enzyme (Shall and Waheed, 1969). Maximal inhibition of 50% required 0.1 mM iodine with the enzyme at 8.15 mM. The reaction took only a few seconds, and mercaptoethanol and mercaptoethylamine re-activated the enzyme. From the variation of V_{max} value with pH value, the enzyme reaction required a single protonated acid with a pK_a value of about 6.8 and was presumed to be an imidazole. This group was not chemically modified by iodine, so it was suggested that the iodine oxidized cysteine residues to cystine and that cysteine, applied externally, was prevented from re-activating by charge repulsion from a carboxylate anion at or near the active centre.

C. Mechanism

In yeast β-fructosidase, rupture of the glycosidic link with sucrose as substrate occurs on the fructose side of the glycosidic oxygen (Koshland and Stein, 1954). Koshland (1953) reported that invertase liberates glucose from sucrose as α-D-glucose.

D. Specificity

In addition to sucrose, raffinose and stachyose are substrates for invertase (Adams et al., 1943). It is important to distinguish the enzyme under discussion from the sucrase–isomaltase complex reported by Quaroni and Semenza (1976) from rabbit small intestine which, within its glycoprotein structure, has two subunits, one splitting sucrose and maltose and the other isomaltose, maltose and palatinose. Maltose is not a substrate for invertase.

Barnett (1977), in a study of taxonony among nutritional yeasts, found only two exceptions (0.4%) to the rule that any yeast that utilized raffinose also used sucrose, but not conversely. As a double glycoside, sucrose is both an α-D-glucoside and a β-D-fructoside; hence it can be hydrolysed by an α-glucosidase or a β-fructofuranosidase. Nearly all yeasts that utilize raffinose do so by initially using a β-fructofuranosidase to give melibiose and fructose. Because it is a β-fructofuranosidase, invertase attacks molecules having unsubstituted terminal ends of β-D-fructofuranosyl units. Other furanosyl residues, even α-fructofuranosides, are not substrates. Minor impurities in commercial preparations may comprise α- and β-glucosidase and α- and β-galactosidase.

E. Inhibitors

Yeast β-fructosidase has been labelled by an active site-directed inhibitor of the enzyme [³H]conduritol-β-epoxide (Braun, 1977). This compound is a racemic mixture of 1-D-1,2-anhydro-*myo*-inositol and 1-L-1,2-anhydro-*myo*-inositol, but only the latter is the reactive compound with the enzyme. During inactivation of the enzyme, 1–2 mol of this inhibitor becomes bound covalently per mol of enzyme. The reaction is reversed at pH 9 by mild alkaline treatment whereby 1-D- or L-chiro-inositol is released. Presumably, this had been bound by an ester link to a carboxylic acid group at the active site of the enzyme. This agrees with the pH-dependence of the inactivation of this group (pK$_a$ 3.05). 1-L-1,2-Anhydro-*myo*-inositol has the structure:

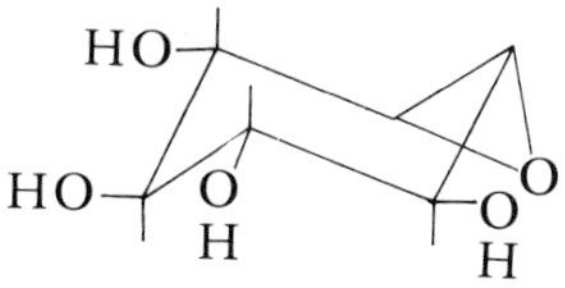

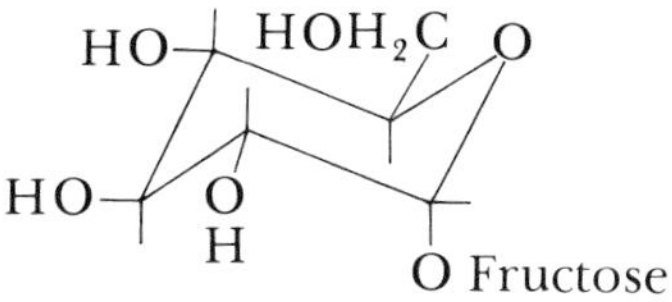

Sucrose

When raffinose or stachyose is added to the reaction mixture, the extent of inactivation by the epoxide is markedly lowered (Braun, 1976).

Yeast invertase is inactivated by iodine possibly due to reversible oxidation of a methionine sulphur atom (Waheed and Shall, 1971). Oxidation of some 5–6 tryptophan residues in the invertase molecule, if effected with N-bromosuccinimide, gave complete inactivation of the enzyme (Leskovac et al., 1975).

Invertase was not inhibited by iodoacetamide, iodoacetic acid or N-ethylmaleimide, but p-chloromercuribenzoate inhibited at high levels of inhibitor to enzyme. Thus, with 0.5 μM enzyme, inhibitor concentrations of 10 μM, 0.1 mM and 1 mM gave 0, 34% and 66% inhibition, respectively. Invertase is reversibly inhibited by heavy-metal ions such as Cu^{2+}, Hg^{2+} and Ag$^+$ (Myrbäck and Willstaedt, 1955).

F. Properties

Neumann and Lampen (1967) reported that yeast invertase from a
mutant (FH4C) of *Sacch. cerevisiae* purified to homogeneity on
polyacrylamide-gel electrophoresis had a molecular weight of
270,000 as determined by sedimentation equilibrium. Electrophoresis on cellulose acetate at pH values between 2 and 12 revealed a
single protein which also contained the enzymic activity. It had a
specific acitivity of 2,700–3,000 μmol of sucrose hydrolysed per minute
per mg of protein at 30°C. About 50% carbohydrate is present in the
invertase, most of it is mannan with some 3% glucosamine. Amino-acid
analysis revealed the following number of residues in each molecule:
glycine, 71; alanine, 68; serine, 114; threonine, 84; proline, 65;
valine, 69; isoleucine, 40; leucine, 83; phenylalanine, 80; tyrosine, 65;
tryptophan, 33; half-cystine, 5; methionine, 21; aspartic acid, 178;
glutamic acid, 115; arginine, 27; histidine, 16; lysine, 60; glucosamine, 38. Together with amide nitrogen, with 150 molecules, some
96% of the nitrogen in the enzyme was accounted for. Tracey (1963)
reported K_m values of 33 mM and 44 mM for cell-wall and soluble
invertase, respectively. The enzyme exhibits optimum activity between
pH 4.7 and 4.9.

Gascon *et al.* (1968) found that both internal and external invertases
from *Sacch. cerevisiae* have the same K_m values for sucrose and raffinose,
and neither showed an appreciable transferase activity (external
invertase K_m values was 26 mM with sucrose, 150 mM with raffinose;
internal invertase K_m was 25 mM with sucrose, 150 mM with raffinose).
Both enzymes exhibited maximum activity at pH 4.5. However, the
internal invertase showed reversible inactivation at acid pH values
unlike the external invertase. As expected, the internal invertase had
the higher mobility in polyacrylamide-gel electrophoresis at pH 6.8
and 8.6. Both enzymes were immunologically related. However, there
were wide disparities in their amino acid compositions, in particular
the absence of cysteine from the internal enzyme.

Neurospora crassa produces two forms of invertase both of which are
glycoproteins. The more abundant form ($S_{20\omega}$ = 10.3S) may be
dissociated into one with a $S_{20,\omega}$ value of 5.2S. Both forms have
identical Michaelis constants and enzymic properties (Tashiro and
Trevithick, 1977).

G. Stability

The pH stability of external and internal yeast invertases differ markedly (Gascon et al., 1968). The external glycoprotein-type invertase is stable at 30°C between pH 3 and 7.5. The internal protein invertase is stable from pH 6 to 9 and can be stored for at least one month at pH 7.5 at 4°C and for several months in the frozen state without loss of activity. At 30°C it loses activity slowly at pH 5.0, the rate of inactivation increasing towards lower pH values; the loss at 4°C is much slower. The activity was restored completely by incubating in 0.02 M Tris-HCl buffer, pH 7.5, for 5 min at 30°C. The enzyme is stable for several months in the dry state at 4°C.

H. Microbiology

Toda (1976) studied the effect of glucose on invertase formation in batch cultures of *Saccharomyces carlsbergensis* L.A.M.1068 (wild type), and the dilution rate effect on invertase activity per unit of yeast cell mass in a continuous culture. Experimental correlations between the dilution rate and invertase specific activity (E/X) in chemostat cultures led to recognition of an optimum for enzyme synthesis at a particular intermediate growth rate. The value of E/X increased from 1.1 units/mg of biomass in batch cultures to 13 units/mg of biomass in chemostat cultures. A mutant strain (A3) showed the highest value for E/X of 25 units/mg of biomass at high dilution rates where glucose repression was observed in the wild strain. Invertase is only one of a number of microbial enzymes whose synthesis is subjected to catabolite repression. Attempts to select mutants *Sacch. carlsbergensis* L.A.M.1068 (wild strain) resistant to glucose repression were unsucccessful. These attempts involved use of ultraviolet radiation and nitrosoguanidine. Some yeast isolates with high invertase activity were obtained when the wild strain was grown in the presence of 2-deoxyglucose.

In derepressed cells, most of the invertase is external (cell-bound but outside the cytoplasmic membrane where it occurs as a mannan protein). The invertase inside the cytoplasmic membrane is smaller and is not accessible to substrate. The glycoprotein external enzyme contains 50% mannose and 3% glucosamine. To examine the two types,

Gascon and Lampen (1968) used yeast strain 303–67 and the repression resistant mutant FH4C. The parent *Saccharomyces* strain 303–67 is diploid and homozygous for the Su_2 gene, and forms invertase in appreciable amounts only after glucose had disappeared from the medium. Strain FH4C was obtained by ultraviolet irradiation of strain 303–67. It produced high levels of external invertase even when growing in the presence of glucose, and this was accompanied by a corresponding increase in the levels of internal enzyme. The internal enzyme contains little or no carbohydrate.

Babczinski and Tanner (1978) described a glucose oxidase method for locating yeast invertase activity directly on sodium dodecyl sulphate–polyacrylamide gels. When crude membrane fractions were prepared from *Sacch. cerevisiae* actively synthesizing external invertase, these membranes showed an activity band on such gels additional to the external and internal invertases. In the soluble fraction, this form was absent. It had a molecular weight of about 190,000 and appeared to be a precursor of the external enzyme. This membrane-associated precursor amounted to less than 5% of the total yeast invertase activity. A picture is now emerging suggesting that the external glycoprotein invertase and its precursor have been built up by glycosylation reactions catalysed exclusively by membrane-bound enzymes and that dolichoyl phosphate-bound sugars are intermediates for some of these reactions. It was speculated that the precursor consisted of a protein core with a molecular weight of 135,000 and all of the carbohydrate core units of the external invertase. Addition of tunicamycin to yeast cells synthesizing external invertase inhibited further synthesis of the enzyme by 97% and strongly inhibited formation of the membrane-bound precursor.

I. Enzyme Production

Toda (1976) used a basal medium at 30°C composed of (g/l): $(NH_4)H_2PO_4$, 15.0; $(NH_4)_2HPO_4$, 1.5; KH_2PO_4, 2.0; $MgSO_4,7H_2O$, 0.2; NaCl, 0.01; $FeSO_4,7H_2O$, 0.005; $ZnSO_4,7H_2O$, 0.005; $MnSO_4,3H_2O$, 0.005; $CuSO_4,H_2O$, 0.0005; streptomycin sulphate, 0.1 at pH 5.0 for invertase production from *Sacch. carlsbergensis*. This basal medium was supplemented by (g/l): glucose, 5.0 and yeast extract, 0.4 for one of the media (YL). Invertase activity was assayed on cells suspended in 0.1 M acetate buffer (pH 5.0) containing sucrose (40 mg/ml) at 50°C. In batch culture using a jar fermenter (inoculum, 5%, v/v), invertase activity rose with increases in cell mass during the first 10 hours of growth, while glucose remained, but thereafter no significant increase of invertase was observed. Invertase specific activity in continuous culture showed that the highest value of E/X was obtained from cells harvested at a dilution rate of 0.15 h^{-1}.

Tashiro and Trevithick (1977) employed a wild-type strain of *Neurospora crassa* for invertase production. Conidia were used to inoculate Fries medium (Beadle and Tatum, 1945) to which was added sucrose at 0.3% (w/v). The culture was grown with gyratory shaking at 200 rev./min at 30°C and an initial pH value of 4. This was used to inoculate a jar fermenter containing 10 litres of medium grown for 25 h at 22°C and an initial pH value of 5.0 with agitation from two six-bladed turbine impellers at 400 rev./min. Thereafter, this was used as the inoculum of a tank fermenter (1,000 litres) containing 800 litres of medium. Growth continued for 24 hours at 30°C and an initial pH value of 5.0 with agitation (100 rev./min), vigorous aeration (1.8 m^3/min) and a head pressure of 34.5 kPa (5 lbf/in^2). Lactose was then added at 1.2% and the pH value adjusted to 7.0. After 42 hours at 30°C and pH 7.0, some 7.5 kg (wet weight) of mycelium was harvested. Invertase purification was carried out by the method of Metzenberg (1962).

Distinct improvements in the yield of invertase and its activity have been reported by Serova *et al.* (1976), when *Aspergillus awamori* strain 16 was grown in a culture medium containing molasses, NaNO$_3$, KCl, MgSO$_4$ and potassium phosphate to which soybean flour (2.5–3%) was added. The enzyme could be recovered by precipitation with a final concentration of 80% ethanol, 71% propan-2-ol or 75% acetone, preferably at pH 6.0–6.5 with ethanol. Optimum temperature and pH value of the β-fructofuranosidase activity were pH 4.6 and 50–55°C. In thermal stability, it was one to two orders of magnitude higher than that of the *Sacch. cerevisiae* particularly in the presence of sucrose.

It has been suggested that the catabolite repression of invertase synthesis by glucose operates at the level of transcription and translation, and produces an increase in the rate of RNA degradation (Elorza *et al.*, 1977). In the presence of glucose, synthesis of enzyme took place in *Sacch. cerevisiae* when the sugar concentration was < 1%. At higher concentrations, enzyme formation was repressed. Analysis of the glucose effect before RNA inhibition showed that the sugar interfered with transcription of DNA into invertase mRNA. Invertase activity was independent of glucose, suggesting that hexose does not produce catabolite inhibition of invertase activity. Inhibition of invertase translation by glucose turned out to be reversible, but the amount of enzyme produced was dependent on the duration of treatment. Yeast synthesizes invertase in media containing maltose and sucrose.

Invertase formation in *Sacch. cerevisiae* is inhibited by tunicamycin (Hašilik and Tanner, 1976). This appears to be associated with interference in synthesis of glycoproteins since invertase, which is a mannoprotein when externally localized, was only one of a number of such enzymes affected (e.g. carboxypeptidase Y and proteinase A).

Formation of invertase in a number of *Penicillium* species has been monitored (Nikolskaya *et al.*, 1975). Highest levels were observed in *P. expansum* (second day, surface growth) and *P. frequentas* (fourth day, surface growth). In deep cultivation, *P. vitale* produced extracellular invertase which was enhanced in media favouring catalase production but decreased in media favouring glucose oxidase production. This organism produced two invertases, an acid and an alkaline one.

Synthesis of invertase in *Pichia polymorpha* was strongly affected by the presence of 2-deoxyglycose and the degree of inhibition was directly related to the concentration of this analogue (Villa *et al.*, 1975). This was also true of another typical extracellular enzyme, acid phosphatase.

An intracellular invertase was induced in *Clostridium pasteurianum* growing on sucrose as its carbon source, but synthesis could be repressed by addition of fructose to the medium (Laishley, 1975). Addition of glucose under the same conditions did not affect invertase synthesis, which could be induced in a glucose-metabolizing culture by addition of sucrose. The purified enzyme had a pH optimum of 6.5 and an apparent K_m value of 79.5 mM for sucrose. It was completely inactivated by two minutes at 60°C. Thiol groups in the enzyme seemed essential for its activity.

Intracellular invertase is released into cultures of *Streptococcus mutans* strains SL-1 during the late logarithmic and stationary phases of growth (Osborne *et al.*, 1975). Both intracellular and extracellular invertases have the same optimum pH value of 6.2, optimum temperature of 38–39°C, molecular weight of 48,000, and the same electrophoretic mobility. *Actinomyces viscosus* cultured at 37°C for 48 hours on a complex medium containing sucrose produced extracellular invertase activity (Palenik and Miller, 1975) exhibited as β-fructofuranosidase against sucrose. Similar activity was induced in the presence of raffinose.

J. Enzyme Purification

Leskovac *et al.* (1975) used a commercially available (Boehringer) yeast invertase which had a λ_{max} value at 261 nm with E_{280}/E_{260} 0.58. The dialysed enzyme in 0.1 M sodium phosphate buffer (pH 6.0) was loaded onto a DEAE-Sephadex column similarly equilibrated. Some 2 g of enzyme was separated on a column (2 cm × 50 cm) using a 0–0.5 M sodium chloride gradient: the invertase was eluted in a single peak at 0.35–0.40 M sodium chloride and behaved chromatographically like the external isoenzyme of yeast invertase. After dialysis and freeze drying it had E_{280}/E_{260} value of 1.73.

K. Uses

Boudrant and Cheftel (1975) absorbed invertase on Amerlite IRA 93 resin and used it in a tubular reactor on a 50% (w/v) sucrose solution. The reaction was carried out for eight days at pH 4 and 30°C. At a selected flow rate, the degree of sucrose hydrolysis remained close to 76%. The enzymic activity seemed to be stable for an extended period of one month at 30°C and pH 3 or 4 when an industrial sucrose syrup was passed down a tubular reactor of the enzyme.

Sucrose inversion has also been effected by whole-cell invertase of *Sacch. pastorianus* entrapped in spherical agar pellets used in a continuously fed fluidized bed (Toda and Shoda, 1975). The enzyme showed good stability at 47.5°C and pH 5.0.

An attractive possibility in an enzyme reactor is to entrap invertase in cellulose triacetate fibres (Marconi *et al.*, 1974). A half-life of 5,300 days was calculated for a sample of invertase fibres which displayed 15–65% of that of the free enzyme.

REFERENCES

Adams, M., Richtmyer, N. K. and Hudson, C. S. (1943), *Journal of the American Chemical Society* **65**, 1369.
Agency of Industrial Science and Technology (1968). British Patent 1,103,394.
Akasaki, M., Suzuki, M., Funakoshi, I. and Yamashina, I. (1976). *Journal of Biochemistry, Tokyo* **80**, 1195.
Anema, P. J. (1964). *Biochimica et Biophysica Acta* **89**, 495.
Anheuser-Busch (1944). United States Patent 3,834,988.
Babczinski, P. and Tanner, W. (1978). *Biochimica et Biophysica Acta* **538**, 426.
Bach, J. A. and Sadoff, H. L. (1962). *Journal of Bacteriology* **83**, 699.
Banauch, D., Brümmer, W., Eberling, W., Metz, H., Rindfrey, H., Lang, H., Leybold, K. and Rick, W. (1975). *Zeitschrift für Klinika Chemie et Klinicka Biochemie* **13**, 101.
Barnett, J. (1977). *Journal of General Microbiology* **99**, 183.
Beadle, G. N. and Tatum, E. L. (1945). *American Journal of Botany* **32**, 678.
Bean, R. C. and Hassid, W. J. (1956). *Journal of Biological Chemistry* **218**, 425.
Bedine, M. S. and Bayless, T. M. (1973). *Gasteroenterology* **65**, 735.
Bentley, R. and Slechta, L. (1960). *Journal of Bacteriology* **79**, 346.
Berman, H. W., Rubin, B. H. Carrell, H. and Glusker, J. P. (1974). *Journal of Biological Chemistry* **249**, 3983.
Bhattacharya, G., Dastidar, P. G., Sinha, A. M., Majunder, P. K. and Chaterjee, G. C. (1974). *Canadian Journal of Microbiology* **20**, 897.
Boudrant, J. and Cheftel, C. (1975). *Biotechnology and Bioengineering* **17**, 827.
Braun, H. (1977). *Biochimica et Biophysica Acta* **485**, 141.
Braun, H. (1976). *Biochimica et Biophysica Acta* **452**, 452.
Bright, H. J. and Gibson, Q. H. (1967). *Journal of Biological Chemistry* **242**, 994.
Buki, K., Szentirmal, A., Tolgyesi, L., Hegyaljai, K. G. and Iia, L. (1976). Hungarian Patent 12,415.
Burma, D. P. and Horecker, B. L. (1958). *Journal of Biological Chemistry* **231**, 1053.
Cho, Y. K. and Bailey, J. E. (1977). *Biotechnology and Bioengineering* **19**, 185.
Coker, L. E. and Gardner, D. E. (1976). United States Patent 3,956,066.
Cotter, W. P., Lloyd, N. E. and Chen, J. C. (1977) Canadian Patent 1,004,613.
Danno, G. (1971) *Agricultural and Biological Chemistry* **35**, 997.
Daoud, M. S. Greenwood, I. J. and Morgan, M. R. J. (1977). *American Chemical Abstracts* **86**, 51032.
Dobrogosz, W. J. and Stone, R. W. (1962). *Journal of Bacteriology* **84**, 716, 724.
Donnelly, W. J., Fhaolain, I. N. and Patching, J. W. (1977). *International Journal of Biochemistry* **8**, 101.
Dowling, J. H. and Levine H. B. (1956). *Journal of Bacteriology* **72**, 555.
Dzherova, A., Grigorov, I. and Slokoska, L. (1974). *Prilozhna Microbiologia* **3**, 5.
Eagon, R. G. (1971). *Canadian Journal of Biochemistry* **49**, 606.
Eddy, A. A. and Williamson, D. H. (1957). *Nature, London* **179**, 1252.
Elorza, M. V., Villanueva, J. R. and Sentandreu, R. (1977). *Biochimica et Biophysica Acta* **475**, 103.
Feather, M. S. (1970). *Biochimica et Biophysica Acta* **220**, 127.
Fisher, J., Spencer, R. and Walsh, C. (1976). *Biochemistry, New York* **15**, 1054.
Franke, W. (1944). *Annalen* **555**, 111.
Fujita, Y., Ramaley, R. and Freese, E. (1977). *Journal of Bacteriology* **132**, 282.
Gascon, S. and Lampen, J. O. (1968). *Journal of Biological Chemistry* **243**, 1567.

Gascon, S., Neumann, N. P. and Lampen, J. O. (1968). *Journal of Biological Chemistry* **248**, 1573.

Gavard, R. and Combre, C. (1959). *Comptes Rendus de l'Institut Pasteur, Paris* **249**, 2245.

Gerbig, K. (1976). *Medical Laboratory* **29**, 1.

Goodman, R. E. and Pederson, D. M. (1976). *Canadian Journal of Microbiology* **23**, 798.

Gorniak, H. and Kaczkowski, J. (1973). *Bulletin of the Academy Polish Science, Series Biology* **21**, 576.

Gray, P. P., Dunnill, P. and Lilly, M. D. (1973). *Biotechnology and Bioengineering* **15**, 1179.

Gray, P. P., Dunnill, P. and Lilly, M. D. (1971). *Journal of Fermentation Technology* **50**, 381.

Hachisuka, Y. and Tochikubo, K. (1971). *Journal of Bacteriology* **107**, 442.

Hall, B. G. (1976). *Journal of Molecular Biology* **107**, 71.

Hašilik, A. and Tanner, W. (1976). *Antimicrobial Agents and Chemotherapy* **10**, 402.

Hauge, J. G. (1960). *Biochimica et Biophysica Acta* **45**, 250, 263.

Hauge, J. G. (1964). *Journal of Biological Chemistry* **239**, 3630.

Hayashi, S. and Nakamura, S. (1976). *Biochemica et Biophysica Acta* **438**, 37.

Hedrick, J. L. and Smith, A. J. (1968). *Archives of Biochemistry and Biophysics* **126**, 155.

Higgins, J. J., Lewis, D. J., Daly, W. H., Mosqueira, F. G., Dunnill, P. and Lilly, M. D. (1978). *Biotechnology and Bioengineering* **20**, 159.

Hirago, K. and Hirano, K. (1975). Japanese Patent 75,95,476.

Hixson, H. F. (1973). *Biotechnology and Bioengineering* **15**, 1011.

Hochster, R. M. and Katznelson, H. (1958). *Canadian Journal of Biochemistry and Physiology* **36**, 669.

Hochster, R. M. and Watson, R. W. (1954). *Archives of Biochemistry and Biophysics* **48**, 120.

Hogue-Angeletti, R. A. (1975). *Journal of Biological Chemistry* **250**, 7814.

Horecker, B. L. (1974). In 'Methods of Enzymatic Analysis' (H. U. Bergmeyer, ed.), p. 1371. Elsevier, Amsterdam.

Hoshino, J. and Momose, A. (1966). *Journal of General and Applied Microbiology* **12**, 163.

Huber, R. E., Wallenfels, K. and Kurz, G. (1975). *Canadian Journal of Biochemistry and Physiology* **53**, 1035.

Hynes, M. J. and Kelly, J. M. (1977). *Molecular and General Genitics* **150**, 193.

Ichimura, M., Hirose, Y., Katsuya, N. and Yamada, K. (1965). *Journal of the Agricultural Chemical Society, Japan* **39**, 291.

Jorns, M. S. and Hersh, L. B. (1976). *Journal of Biological Chemistry* **251**, 4872.

Keilen, D. and Hartree, E. F. (1948). *Biochemical Journal* **42**, 221.

Kelly, J. M. and Meers, J. L. (1975). German Patent 2,513,903.

Kent, C. A. and Emery, A. N. (1973). *Journal of Applied Chemistry and Biotechnology* **23**, 689.

Kersters, K. (1967). *Antonie van Leeuwenhoek* **33**, 63.

King, T. E. (1966). *Methods in Enzymology* **9**, 109.

King, T. E. and Cheldelin, V. H. (1957). *Journal of Biological Chemistry* **224**, 579.

Kiuchi, A. and Tanaka, Y. (1976). German Patent 2,431,297.

Kobayashi, T. and Kasuga, S. (1976). Japanese Patent 76,151,385.

Koshland, D. E. (1953). *Biological Reviews* **28**, 416.

Koshland, D. E. and Stein, S. S. (1954). *Journal of Biological Chemistry* **208**, 139.

Kunita, N. and Fukumara, T. (1956). *Medical Journal of Osaka University* **6**, 955.

Kusai, K., Sekuzu, I., Hagihara, B., Okunuki, K., Yamauchi, S. and Nakai, M. (1960). *Biochimica et Biophysica Acta* **40**, 555.

Laishley, E. J. (1975). *Canadian Journal of Microbiology* **21**, 1711.

Lee, C. K. and Dobrogosz, W. J. (1965). *Journal of Bacteriology* **90**, 653.

Lee, C. K., Hayes, L. E. and Long, M. E. (1972). United States Patent 3,645,848.

Lee, J. V., Gibson, D. M. and Shewan, J. M. (1977). *Journal of General Microbiology* **98**, 439.

Lee, Y. H., Wankat, P. C. and Emery, A. N. (1976). *Biotechnology and Bioengineering* **18**, 1639.

Leskovac, V., Trivić, S., Parkov-Pericin, D. and Prodanov, V. (1975). *Biochimica et Biophysica Acta* **405**, 482.

Lobarzewski, J. (1976). *Journal Przem. Fermentazione Rolny* **20**, 13.

Lutz, R. A. and Flückiger, J. (1975). *Clinical Chemistry* **21**, 1372.

Lynch, W. H., Macleod, J. and Franklin, M. (1975). *Canadian Journal of Microbiology* **21**, 1560, 1553.

Maeda, H. (1975). *Biotechnology and Bioengineering* **17**, 1571.

Mahoney, R. R., Nickerson, T. A. and Whitaker, J. R. (1975). *Journal of Dairy Science* **58**, 1620.

Marconi, W., Gulinelli, S. and Morisi, F. (1974) *Biotechnology and Bioengineering* **16**, 50.

Marshall, R. O. and Kooi, E. R. (1957). *Science, New York* **125**, 648.

Meers, J. L. (1976). German Patent 2,533,193.

Metzenberg, R. L. (1962). *Archives of Biochemistry and Biophysics* **96**, 469.

Metzger, R. P., Wilcox, S. S. and Wick, A. N. (1964). *Journal of Biological Chemistry* **239**, 1769.

Mitsuhashi, S. and Lampen, J. O. (1953). *Journal of Biological Chemistry* **204**, 1011.

Mueller, H. M. (1977). *Zentralblatt fur Bakteriologie, Parasitenkunde, Infektionskrankheiten und Hygiene, Abteilung 2* **132**, 14.

Müller, D. (1928). *Biochemische Zeitschrift* **199**, 136.

Myrbäck, K. and Willstaedt, E. (1955). *Akiv für Kemi* **8**, 367.

Nakamatsu, T., Akamatsu, T., Miyajima, R. and Shiio, I. (1975). *Agricultural Biological Chemistry* **39**, 1803.

Nakamura, T. and Ogura, Y. (1962). *Journal of Biochemistry, Tokyo* **52**, 214.

Nakamura, T. and Ogura, Y. (1968). *Journal of Biochemistry, Tokyo* **64**, 439.

Natake, M. (1966). *Agricultural Biological Chemistry* **30**, 887.

Natake, M. (1968). *Agricultural Biological Chemistry* **32**, 303.

Natake, M. and Yoshimura, S. (1963). *Agricultural Biological Chemistry* **27**, 342.

Neumann, N. P. and Lampen, J. O. (1968). *Biochemistry, New York* **6**, 468.

Ngo, T. T., Narinsingh, D. and Laidler, K. J. (1976). *Biotechnology and Bioengineering* **18**, 119.

Nicol, M. J. and Duke, F. R. (1966). *Journal of Biological Chemistry* **241**, 4292.

Niederpruem, D. J. and Doudoroff, M. (1965). *Journal of Bacteriology* **89**, 697.

Nikolskaya, E. A., Kirillova, L. M. and Sinyavskaya, O. I. (1975). *Mikrobiol. Zh.* (*Kiev*) **37**, 597.

Nishizuka, Y. and Hayaishi, O. (1962). *Journal of Biological Chemistry* **237**, 2721.

Noltmann, E. A. (1972). *In* 'The Enzymes' (P. D. Boyer, ed.), vol. 6, p. 271. Academic Press, New York.

Oganesyan, M. G., Mugnetsyan, E. G. and Dzhanpoladyn, L. O. (1977). *American Chemical Abstracts* **86**, 185723e.

Ohmiya, K., Ohashi, H., Kobayashi, T. and Schmizu, S. (1977). *Applied and Environmental Microbiology* **33**, 136.

Okon, Y., Chet, I. and Henis, Y. (1975). *Canadian Journal of Microbiology* **21**, 1123.

O'Malley, J. J. and Ulmer, R. W. (1973). *Biotechnology and Bioengineering* **15**, 917.

O'Malley, J. J. and Weaver, J. L. (1972). *Biochemistry, New York* **11**, 3527.

Osborne, R. M., Lamberts, B. L. and Roush, A. H. (1975). *Experientia* **31**, 1399.

Palenik, C. J. and Miller, C. H. (1975). *Journal of Dental Research* **54**, 186.

Park, C. E., Rayman, M. K., Szabo, R. and Stankiewicz, Z. K. (1976). *Canadian Journal of Microbiology*, **22**, 654.

Pauly, H. E. and Pfleiderer, G. (1975). *Hoppe-Seyler's Zeirschrift für Chemie* **356**, 1613.

Pauly, H. E. and Pfleiderer, G. (1977). *Biochemistry, New York* **16**, 4599.

Pazur, J. H. and Kleppe, K. (1964). *Biochemistry, New York* **3**, 578.

Perlman, R. L. and Pastan, I. (1968). *Journal of Biological Chemistry* **243**, 5420.

Poindron, P., Bourguignat, A., Scheid, F. and Lombard, Y. (1977). *Canadian Journal of Microbiology* **23**, 798.

Proszt, G. (1963). *Elelmezesi Ipar* **17**, 339.

Quarroni, A. and Semenza, G. (1976). *Journal of Biological Chemistry* **251**, 3250.

Reynolds, C. K., Tobacco Co. (1977). United States Patent 4,061,539.

Rickenberg, H. V. (1974). *Annual Review of Bacteriology* **28**, 353.

Robinson, P. J., Wheatley, M. A., Janson, J. C., Dunnill, P. and Lilly, M. D. (1974). *Biotechnology and Bioengineering* **16**, 1103.

Rogers, M. J. and Brandt, K. G. (1971). *Biochemistry, New York* **10**, 4624, 4636.

Rohlfing, S. R. and Crawford, I. P. (1966). *Journal of Bacteriology* **92**, 1258.

Roland, J. F. and Alm, W. L. (1975). *Biotechnology and Bioengineering* **17**, 144.

Rose, I. A. (1975). *Advances in Enzymology* **43**, 491.

Rosenweig, N. S. (1969). *Journal of Dairy Science* **52**, 585.

Sahyoun, N. and Durr, I. F. (1972). *Journal of Bacteriology* **112**, 421.

Schepartz, A. I. and Subers, H. H. (1964). *Biochimica et Biophysica Acta* **85**, 228.

Serova, Y. Z., Dobrolinskaya, G. M. and Yarovenko, V. L. (1976). U.S.S.R. Patent 533,632.

Shieh, K. K. (1976). United States Patent 3,992,262 and 4,003,793.

Shall, S. and Waheed, A. (1969). *Biochemical Journal* **111**, 33P.

Strecker, H. J. and Korkes, S. (1952). *Journal of Biological Chemistry* **196**, 769.

Sutton, D. D. and Lampen, J. O. (1962). *Biochimica et Biophysica Acta* **56**, 303.

Swoboda, B. E. P. and Massey, V. (1965). *Journal of Biological Chemistry* **240**, 2209.

Takasaki, Y. (1975). Japanese Patent 7532,317.

Takasaki, Y. (1976). Japanese Patent 7673,184.

Takaskai, Y. (1977). Japanese Patent 7707,480.

Takasaki, Y. and Tanabe, O. (1964). *Agricultural and Biological Chemistry* **28**, 740.

Takasaki, Y. and Tanabe, O. (1967). *Kogyo Gijutsuin, Hakko Kenkyusho Kenkyo Hokoku* **31**, 1.

Takasaki, Y., Kosugi, Y. and Kanbayashi, A. (1969) *Fermentation Advances* p. 561.

Tashiro, Y. and Trevithick, J. R. (1977). *Canadian Journal of Biochemistry* **55**, 249.

Toda, K. (1976). *Biotechnology and Bioengineering* **18**, 1103.

Toda, K. and Shoda, M. (1975). *Biotechnology and Bioengineering* **17**, 481.

Tomada, K., Miyata, K. and Yamazaki, O. (1976). Japanese Patent 7670,877.

Tracey, M. V. (1963). *Biochimica et Biophysica Acta* **77**, 149.

Trevithick, J. R. and Metzenberg, R. L. (1966). *Journal of Bacteriology* **92**, 1010.

Tsuge, H., Natsuaki, O. and Ohashi, K. (1975). *Journal of Biochemistry, Tokyo* **78**, 835.

Tsujisaka, Y., Watanabe, Y., Takenishi, S. and Kibesaki, Y. (1976). Japanese Patent 76,142,593.

Tsumira, N. and Sato, T. (1965). *Agricultural and Biological Chemistry* **29**, 1123 and 1129.

Vakulenko, V. A., Vitkovskaya, V. A. and Artyukhova, A. N. (1976). *American Chemical Abstracts* **85**, 190687e.

Villa, T. G., Notario, V., Benitez, T. and Villanueva, J. R. (1975). *Archives of Microbiology* **105**, 335.

Wadiak, D. T. and Carbonell, R. G. (1975). *Biotechnology and Bioengineering* **17**, 1157.

Waheed, A. and Shall, S. (1971). *Biochimica et Biophysica Acta* **242**, 172.

Weibel, M. K. and Bright, H. J. (1971). *Journal of Biological Chemistry* **246**, 2274.

Weimberg, R. (1963). *Biochimica et Biophysica Acta* **67**, 349.

Williams, A. K. and Eagon, R. G. (1959). *Journal of Bacteriology* **77**, 167.

Wondolowski, M. and Woychik, J. H. (1974). *Biotechnology and Bioengineering* **16**, 1633.

Wood, W. A. and Schwerdt, R. F. (1953). *Journal of Biological Chemistry* **201**, 501.

Yamanaka, K. (1963). *Agricultural and Biological Chemistry* **27**, 265 and 271.

Yamanaka, K. (1965). Japanese Patent 20, 230.

Yeung, K. H., Chaloner-Larsson, G. and Yamzaki, H. (1976). *Canadian Journal of Biochemistry* **54**, 854.

Young, J. M., Schray, K. J. and Mildyan, A. S. (1975). *Journal of Biological Chemistry* **250**, 9021.

Yoshimura, Y. and Isemura, Y. (1971). *Journal of Biochemistry, Tokyo* **69**, 839.

Zetelaki, K. Z. (1970). *Biotechnology and Bioengineering* **12**, 379.

Zetelaki, K. Z. and Vas, K. (1968). *Biotechnology and Bioengineering* **10**, 45.

Zittan, L., Poulsen, P. B. and Hemmingsen, St. H. (1975). *Die Stärke* **27**, 236.

5. Pectic Enzymes

FRANK M. ROMBOUTS AND WALTER PILNIK

Department of Food Science, Agricultural University, De Dreijen 12, 6703 BC Wageningen, The Netherlands

I. INTRODUCTION

Pectic substances are structural polysaccharides, occurring mainly in the middle lamella and primary cell wall of higher plants. Although they are present in quantities generally not exceeding 1% of the weight of fresh plant material (Doesburg, 1965; Kawabata, 1977), they are largely responsible for the integrity and coherence of plant tissues. The middle lamella is the cementing layer between cells in plant tissues. It consists principally of pectic substances and hemicelluloses, but its detailed composition is as yet largely unknown (Ishii, 1978). Degradation of middle-lamella pectin results in tissue disintegration by cell separation (maceration). In the primary cell wall, cellulose microfibrils are bound essentially through a hemicellulose–pectin network into a more or less rigid matrix (Keegstra et al., 1973). Its degradation requires pectic as well as other enzymes.

Pectolysis is an important phenomenon associated with many biological processes in which plant material is involved, including elongation of cells and growth of plants, ripening of fruits and abscission of leaves. Pectic enzymes are also produced by bacteria, fungi, yeasts, insects, nematodes and protozoa. Microbial pectolysis plays an important role in plant pathogenesis, symbiosis, decomposition of plant deposits, digestion of plant foods, retting processes, certain fermentations and fruit and vegetable spoilage. Moreover, pectic enzymes from fungi are produced industrially to be used as processing aids for the extraction, clarification and depectinization of fruit juices, for the maceration of fruits and vegetables and for the extraction of vegetable oil.

In this article, we try to cover the many aspects of pectolysis in a number of sections dealing with classification and properties of pectic enzymes, microbial polygalacturonate metabolism, regulatory aspects of microbial pectic-enzyme synthesis, the role of pectic enzymes in the above-mentioned processes and industrial production and application of pectinases. For additional reading, we refer to reviews by Macmillan and Sheiman (1974), Fogarty and Ward (1974) and Rexová-Benková and Markovič (1976).

II. PECTIC ENZYMES

A. Pectic Substances

Pectic substances, as they are extracted from various plant sources, are heteropolysaccharides (molecular weight 30,000 to 300,000) consisting

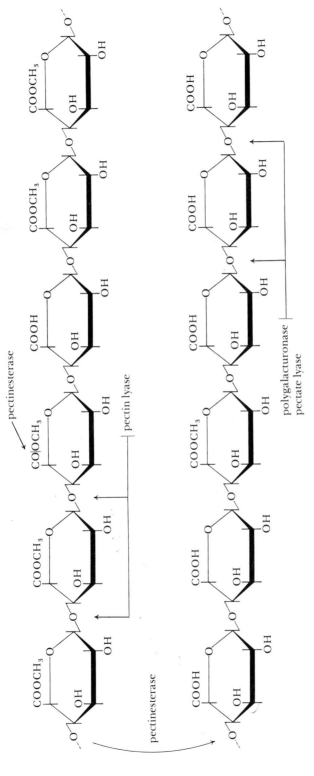

Fig. 1. Fragment of a pectin molecule and points of attack of pectic enzymes.

of a main chain of partially methyl-esterified $(1 \rightarrow 4)$-α-D-galacturonan (Fig. 1). The demethylated compound is known as pectic acid or polygalacturonic acid. Various numbers of L-rhamnopyranose residues are known to be linked, through their C-1 and C-2 atoms in the main galacturonan chain. Depending on the plant source of the pectin, more or less of the galacturonate residues may be acetylated at C-2 and C-3. Side chains of neutral-sugar residues, mainly galactose, arabinose and xylose, are covalently linked to C-2 or C-3 of the galacturonate residues, or to C-4 of the rhamnose residues (Aspinall, 1970; Pilnik and Voragen, 1970). The galacturonan part by itself may account for 50% (potato, sugar beet) to 90% (apple, citrus) of the dry weight of pectin preparations from various plant materials.

The activity of the various groups of pectic enzymes on galacturonan-rich preparations has been described in considerable detail, but little is known about the effects of side chains, acetyl groups and rhamnose residues on enzyme activity. Accordingly, classification of pectic enzymes is based on their attack on the galacturonan backbone of the pectin molecule (Fig. 1).

B. Classification of Pectic Enzymes

Pectic enzymes are classified into two main groups, namely de-esterifying enzymes (pectinesterases) and chain-splitting enzymes (depolymerases). Pectinesterase (pectin methylesterase, pectase) is classified as a carboxyl ester hydrolase, its systematic name being pectin pectyl-hydrolase, EC 3.1.1.11 (Enzyme Nomenclature, 1973). It de-esterifies pectin, producing methanol and pectic acid.

The depolymerases (Table 1, Fig. 2) split the glycosidic bonds of their preferred substrates either by hydrolysis (hydrolases) or by β-elimination (lyases). The best substrate for endopolygalacturonase (poly(1,4-α-D-galacturonide) glycanohydrolase, EC 3.2.1.15) action is pectate which it hydrolyses in a more or less random fashion, producing a series of oligogalacturonates, of which mono-galacturonate, digalacturonate and, sometimes, trigalacturonate accumulate. Its activity decreases with increasing degree of esterification of the substrate. Exopolygalacturonase (poly(1,4-α-D-galacturonide) galacturonohydrolase, EC 3.2.1.67) liberates galacturonate residues by a terminal attack on the pectate polymer.

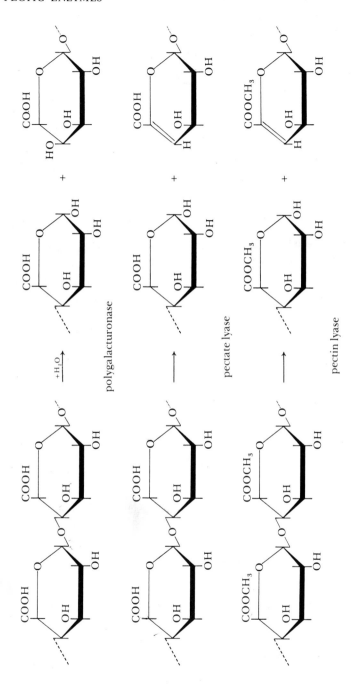

Fig. 2. Splitting of glycosidic bonds in pectin by hydrolysis (polygalacturonase) and by β-elimination (pectate lyase and pectin lyase).

Table 1

Properties of pectate and pectin depolymerases. From Enzyme Nomenclature
(1973)

Name	EC number	Preferred substrate	Mode of attack
Hydrolases			
Endopolygalacturonase	3.2.1.15	Pectate	Random
Exopolygalacturonase	3.2.1.67	Pectate	Terminal
Lyases			
Endopectate lyase	4.2.2.2	Pectate	Random
Exopectate lyase	4.2.2.9	Pectate	Terminal
Endopectin lyase	4.2.2.10	Pectin	Random

The β-eliminative attack of the lyases on their substrates results in
formation of a double bond between C-4 and C-5 in the monomer at
the newly formed nonreducing end (Fig. 2). Pectates and low-methoxyl
pectins are the best substrates for endopectate lyase (poly(1,4-α-
D-galacturonide) lyase, EC 4.2.2.2) which it attacks more or less at
random, producing a series of unsaturated oligogalacturonates of
which the unsaturated dimer and the unsaturated trimer accumulate.
Exopectate lyase (poly(1,4-α-D-galacturonide) exolyase, EC 4.2.2.9)
liberates unsaturated dimers from the reducing end of pectate. Pectin
lyase (poly(methoxygalacturonide) lyase, EC 4.2.2.10) is the only
depolymerase specific for high-methoxyl pectin. It attacks this substrate at random, ultimately accumulating the lower unsaturated
oligogalacturonate methyl esters, which resist further attack. The
activity of all these enzymes is commonly expressed in units, one unit
being the amount of enzyme that splits one microequivalent of ester
bonds or glycosidic bonds per minute, under specified assay
conditions.

The exopolygalacturonases and exopectate lyases already mentioned
either prefer high molecular-weight substrates or are indifferent
towards the chain length of the substrate. In additon to these enzymes,
oligogalacturonases (oligogalacturonate hydrolases and oligogalacturonate lyases) have been isolated from various micro-
organisms. These enzymes split galacturonate oligomers in preference
to long-chain substrates, with a rate approximately inversely proportional to chain length. As these cell-bound enzymes seem to be

important primarily in microbial catabolism of galacturonates, they are dealt with in the section on polygalacturonate metabolism.

C. Pectinesterases

Pectinesterases are formed by higher plants, numerous fungi and some yeasts and bacteria. Particularly well studied are some plant pectinesterases, but knowledge about fungal, bacterial and yeast pectinesterases is rather limited. Multiple molecular forms and isoenzymes of pectinesterase have been found in a number of fruits, as well as in certain fungi. The molecular and kinetic characteristics of the better known enzymes are summarized in Table 2.

Pectinesterases may be assayed by different methods (Rexová-Benková and Markovič, 1976). Usually, activity is measured by continuously recorded titration of the free carboxyl groups which are produced from the pectin substrate. Methods to determine the enzymically released methanol are very useful when enzyme activity has to be determined at lower pH values, in buffered systems, or in plant tissues. A sensitive and accurate method consists of conversion of methanol into volatile methyl nitrite which is then sampled by the head-space technique and analysed by gas–liquid chromatography (Bartolome and Hoff, 1972; Baron et al., 1978).

Pectinesterases from plants and fungi have a high specificity towards the methyl ester of pectic acid. The methyl ester of alginic acid (polymannuronic acid) is not attacked (MacDonnell et al., 1950), and reduction of some of the methylgalacturonate residues in a pectin chain to galactose residues causes a marked inhibition of enzyme activity (Solms and Deuel, 1955). The rates of hydrolysis of the ethyl ester of pectic acid by various plant and fungal enzymes is 6 to 16% of the rates of hydrolysis of the methyl ester (MacDonnell et al., 1950). Orange pectinesterase hydrolyses propyl and allyl esters of pectic acid even more slowly, and ethylene glycyl and glyceryl esters not at all (Manabe, 1973b).

Below a degree of polymerization of about 10, the rate of hydrolysis by orange pectinesterase decreases with decreasing chain length of the substrate and drops to zero for the trimethyl ester of the trimer (McCready and Seegmiller, 1954). The Michaelis constant (K_m value) of orange pectinesterase decreases markedly with the degree of esterifi-

Table 2

Properties of some plant and microbial pectinesterases

Source of enzyme	Molecular weight	Iso-electric point	Specific activity (units/mg of protein)	Optimum pH value	K_m Value for pectin (mg/ml)	Reference
Fruits						
Banana I[a]	30,000	8.9	457	6.0		Brady (1976)
II[a]	30,000	9.4	529	6.0		
Orange (*Citrus natsudaidai*)			2,200	8.0	2.3	Manabe (1973a)
Orange (*Citrus sinensis*) I[a]	36,200	10.0	694	7.6	0.083	Versteeg et al. (1978)
II[a]	36,200	11	762	8.0	0.0046	
Plums (*Prunus salicina*)			25	7.5	0.1	Theron et al. (1977)
Tomato[a]	27,500		1,150	6–9	0.74	Lee and Macmillan (1968, 1970); Lee et al. (1970)
Tomato[a]	26,300	8.4				Delincée and Radola (1970); Delincée (1976)
Tomato			724	8.0	2.40	Nakagawa et al. (1970a, b)
Tomato[a]	27,800					Marković (1974)
Fungi						
Acrocylindrium sp.				7.5	0.7	Kimura et al. (1973)
Coniothyrium diplodiella I[a]				4.8		Endo (1964a)
II[a]				4.8		
Corticium rolfsii[b]	37,000			3.5		Yoshihara et al. (1977)
Fusarium oxysporum	35,000		203	7.0		Miller and Macmillan (1971)
Bacteria						
Clostridium multifermentans[c]	400,000		48	9.0	0.74	Miller and Macmillan (1970, 1971); Sheiman et al. (1976)

[a] (One of) multiple molecular forms or isoenzymes.
[b] This enzyme is active at low pH values; stable in the pH range 1–11.
[c] This enzyme is complexed with exopectate lyase.

cation of the substrate (Versteeg, 1979). The preferred points of attack on a pectin molecule are probably methyl ester groups adjacent to free carboxyl groups (Solms and Deuel, 1955). However, on very highly methylated pectin, about half of the pectinesterase activity may be initiated at the reducing end of the molecules (Lee and Macmillan, 1970; Miller and Macmillan, 1971). The pectinesterase of *Clostridium multifermentans*, which is complexed with exopectate lyase, attacks its substrate from the reducing end only (Sheiman *et al.*, 1976).

Enzymic de-esterification of pectin proceeds linearly along the molecule, by a so-called single-chain mechanism (Heri *et al.*, 1961). In this way, blocks of free carboxyl groups are formed which make the pectin extremely calcium-sensitive (Kohn *et al.*, 1968). Usually enzymic de-esterification of pectin does not go to completion, but levels off at about 10% esterification (Solms and Deuel, 1955).

The activity of plant pectinesterases is markedly influenced by divalent and monovalent cations (MacDonnell *et al.*, 1945). They are also competitively inhibited by pectate (end-product inhibition). Oligogalacturonates with a degree of polymerization of 8 and higher are effective inhibitors (Termote *et al.*, 1977). According to a theory of Lineweaver and Ballou (1945), plant pectinesterase would form an inactive ionic complex with pectate. The stimulatory effect of cations on pectinesterase activity would be due to removal of inhibition by liberating the enzyme from its inactive ionic complex, rather than to specific activation of the enzyme.

D. Endopolygalacturonases

Endopolygalacturonases are produced by numerous plant-pathogenic and saprophytic fungi and bacteria, and by some yeasts. Also, they are formed in higher plants, especially in soft fruits. Recently, quite a number of endopolygalacturonases, mainly from fungi, have been thoroughly purified. Their molecular and kinetic properties are compiled in Table 3. As with pectinesterases, multiple molecular forms and isoenzymes of endopolygalacturonases appear to be produced by many of the organisms examined. Most of these enzymes have a molecular weight of 30,000 to 35,000, and a glycoprotein nature has been established for several of them. With the rather narrow range of molecular weights and K_m values, striking differences exist in specific

Table 3

Properties of some plant and microbial endopolygalacturonases

Source of enzyme	Molecular weight	Iso-electric point	Specific activity (units/mg of protein)	Optimum pH value	K_m Value for pectate (mg/ml)	Reference
Fruits						
Tomato	52,000		47	4.5	2.7	Takehana et al. (1977)
Tomato I[a]	84,000			4.5		Pressey and Avants (1973a)
II[a]	44,000			5.0		
Fungi						
Aspergillus niger I[a]		3.8		4.0		Koller (1966)
II[a]		4.5		4.5	1.7	Koller and Neukom (1969)
III[a]				5.5		
Aspergillus niger I[a]	35,000		81	4.1		Cooke et al. (1976)
II[a]	85,000		44	3.8		
Aspergillus niger	46,000		75	5.0	0.54	Heinrichová and Rexová-Benková (1977)
Aspergillus japonicus	35,500		1,362	4.5		Ishii and Yokotsuka (1972a)
Botrytis cinerea	69,000		2,049	4.0	1.2	Urbanek and Zalewska-Sobczak (1975)
Fusarium oxysporum I[b]	37,000	7.0	194	5.0	0.54	Strand et al. (1976)
II[b]	37,000	7.0	148	5.0	0.80	Cervone et al. (1977, 1978)
Rhizoctonia fragariae I[b]	36,000	6.8	1,866	5.0	0.75	
II[b]	36,000	7.1	1,845	5.0	0.54	
Rhizopus arrhizus	30,300		92	5.0	0.54	Liu and Luh (1978)
Trichoderma koningii I[b]	32,000	6.41		5.0	0.80	Fanelli et al. (1978)
II[b]	32,000	6.57		5.0	0.85	
Verticillium albo-atrum	30,000		2,075	6.5	1.5	Wang and Keen (1970)
Yeasts						
Kluyveromyces fragilis			168	4.4		Phaff (1966)
Bacteria						
Erwinia carotovora			362	5.3		Nasuno and Starr (1966)
Pseudomonas cepacia			125	4.5		Ulrich (1975)

[a] This enzyme exists in multiple forms.

activities, which reflect a great variation in turnover numbers (molecular activities) with these enzymes.

Endopolygalacturonases are specific for polygalacturonate; the rate and degree of hydrolysis of pectins decrease rapidly with increasing degree of esterification. The same is true for glycyl esters of pectate (Pilnik et al., 1973). The activity on oligogalacturonates decreases with decreasing degree of polymerization. Digalacturonate is not hydrolysed, and some of the enzymes do not attack the trimer (Rexová-Benková and Markovič, 1976). Pectates of which the secondary hydroxyl groups at C-2 and C-3 are partially acetylated are still degraded with the same maximum velocity, but the values of the apparent Michaelis constant are increased and the degradation limits are decreased (Rexová-Benková et al., 1977).

Endopolygalacturonases may be assayed by measuring the rate of increase of reducing groups or the decrease in viscosity of the substrate solution. A number of methods are available for measuring the increase in reducing groups (Rexová-Benková and Markovič, 1976), but the colorimetric test of Nelson-Somogyi (Spiro, 1966) or that of Milner and Avigad (1967) are preferred. Both methods are relatively convenient and avoid strong alkaline reaction conditions, under which substrates having a higher degree of esterification are split by β-elimination (Albersheim et al., 1960), so that unrealistically high amounts of reducing groups would be estimated.

Viscosimetry is a very sensitive assay method for endopolygalacturonases. With these enzymes, a 50% drop in specific viscosity of a pectate solution corresponds to hydrolysis of only a few per cent of the glycosidic bonds (Rombouts and Pilnik, 1972). For different enzymes, this percentage may vary, even under carefully standardized test conditions (Endo, 1964b, c, d; Pressey and Avants, 1973a). It is likely that these variations reflect differences in action pattern of the enzymes, comparable to those reported for α-amylases (Robyt and French, 1967). An endo-enzyme may hydrolyse (randomly) one bond in a single enzyme–substrate encounter, followed by complete dissociation of enzyme and products (multichain attack). In the case of single-chain, multiple attack, a single random hydrolytic scission is followed by a number of non-random attacks on one of the products, resulting in liberation of oligogalacturonates. The endopolygalacturonase of *Kluyveromyces fragilis* shows the first type of action pattern. This enzyme hydrolyses pectate through a series of higher oligogalacturonates, which it subsequently hydrolyses until finally digalacturonate and monogalacturonate accumulate (Phaff, 1966). An example of an enzyme showing the latter action pattern is the polygalacturonase of *Colletotrichum lindemuthianum*. The enzyme is truly an endopolygalacturonase, but the initial products, as well as the final products of pectate hydrolysis, are predominantly trigalacturonate and digalacturonate (English *et al.*, 1972).

Endopolygalacturonases also show considerable differences in action patterns on oligogalacturonates. These differences are determined by the nature of the active site of the enzymes, but more specifically by the size of the substrate-binding site and the position of the catalytic groups. For example, Rexová-Benková (1973) assumes the active site of an endopolygalacturonase from *A. niger* to be composed of four subsites and the catalytic groups to be situated between subsites 1 and 2 (Fig. 3). In her review, she distinguishes two more types of active site, namely for the endopolygalacturonases of tomato and of *Erwinia carotovora* (Rexová-Benková and Markovič, 1976). These active sites are composed of a rather limited number of subsites, and other loci may take part in interactions of the enzyme with substrates of higher degree of polymerization (Koller and Neukom, 1969).

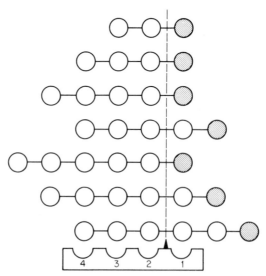

Fig. 3. Preferred cleavage patterns of oligogalacturonates and schematic model for the active site of *Aspergillus niger* endopolygalacturonase. O Indicates a galacturonate residue; ⊘ a reducing galacturonate residue; —— a glycosidic bond; 1, 2, 3 and 4 are binding subsites; and ▲ a catalytic site. From Rexová-Benková (1973).

The catalytically reactive groups of *Aspergillus niger* endopoly-galacturonases are a histidine group and a carboxylate group (Rexová-Benková and Slezarik, 1970; Cooke *et al.*, 1976; Rexová-Benková and Mračková, 1978). In the substrate molecule, carboxyl groups and some secondary alcohol groups are presumed to be involved in interaction with the enzyme (Koller and Neukom, 1969; Rexová-Benková *et al.*, 1977).

E. Exopolygalacturonases

Exopolygalacturonases are found in higher plants, in the intestinal tracts of a number of insects, in fungi and in some bacteria. They are usually assayed by measuring reducing groups of liberated galacturonate monomers or, more specifically, by enzymic determination of galacturonate (Nagel and Hasegawa, 1967a; Bateman *et al.*, 1970; Wagner and Hollmann, 1976).

Recently, a number of reports appeared on exopoly-galacturonases of plants including carrots (Pressey and Avants, 1975a; Heinrichová, 1977), peaches (Pressey and Avants, 1975a), citrus fruits (Riov, 1975), cucumbers (Pressey and Avants, 1975b), pears (Pressey and Avants, 1976), apples (Bartley, 1978) and oat seedlings (Pressey and Avants, 1977). These plant enzymes prefer moderately high to high molecular-weight pectates, which they attack at the non-reducing end liberating monogalacturonate. Oligogalacturonates and even digalacturonate are also degraded. The enzymes have optimum pH values of about 5.0; they are stimulated by calcium ions and hydrolyse their substrates by a multichain mechanism (Pressey and Avants, 1973b, 1975b). Hydrolysis of pectates does not proceed to completion, apparently because of structural irregularities of the substrates (Hatanaka and Ozawa, 1964).

The activity of insect exopolygalacturonases is also proportional to the chain length of the pectate substrate, from which they liberate galacturonate monomers. Interestingly, these enzymes attack at the reducing end (Courtois *et al.*, 1968; Foglietti *et al.*, 1971).

Many intracellular microbial exopolygalacturonases have a preference for oligogalacturonates, and these oligogalacturonate hydrolases are described in the section on polygalacturonate metabolism. In addition to these, an extracellular exopolygalacturonase of

A. niger (Heinrichová and Rexová-Benková, 1976) and a cell-bound enzyme of *Acrocylindrium* sp. (Kimura and Mizushima, 1973) have been described, and these have no clear preference with regard to substrate chain length.

Exopolygalacturonases hydrolysing digalacturonates from the non-reducing end of pectate were isolated from *Erwinia aroideae* (Hatanaka and Ozawa, 1971b) and the culture filtrate of a *Pseudomonas* sp. (Hatanaka and Imamura, 1974). These enzymes, which prefer high molecular-weight substrates, are to be named poly(1,4-α-D-galacturonide) digalacturonohydrolases.

F. Endopectate Lyases

Endopectate lyases are produced by various groups of bacteria and by some plant pathogenic fungi. Table 4 presents a list of recently described enzymes. All of these enzymes have very high optimum pH values and an absolute requirement of calcium ions for their activity. The β-eliminative attack of these enzymes on their substrates results in formation of products with a double bond between C-4 and C-5. Conjugation of the double bond with the carboxyl group at C-5 brings about absorption with a maximum at 235 nm. Pectate lyases are therefore assayed most conveniently with a recording ultraviolet spectrophotometer, the molar extinction coefficient being 4,600 $cm^2 \cdot mmol^{-1}$ (Macmillan and Phaff, 1966).

Pectates are very good substrates for endopectate lyases but, for enzymes from two *Arthrobacter* strains and from *Bacillus polymyxa*, the best substrates are pectins with degrees of esterification of 21, 44 and 26% esterification, respectively (Pilnik *et al.*, 1973). Highly substituted (1→4)-α-D-galacturonans, such as gum tragacanth and gum karaya, are not degraded (Rombouts *et al.*, 1978). However, the hydroxyl groups at C-2 and C-3 are not essential for the catalytic reaction, since the Vi antigen from an *Escherichia* sp., containing (1→4)-α-D-linked 2-deoxy, 2-acetamido, and 3-O-acetyl-galacturonate residues, is degraded by enzymes from *Bacillus sphaericus* and *B. polymyxa* (McNicol and Baker, 1970). Using tetragalacturonate, Atallah and Nagel (1977) showed that the true substrate for lyase activity is the calcium salt of the substrate, while the free tetragalacturonate anion is a competitive inhibitor.

With endopectate lyases, the viscosity of the substrate solution

Table 4

Properties of some endopectate lyases

Source of enzyme	Molecular weight	Iso-electric point	Specific activity (units/mg of protein)	Optimum pH value	K_m Value for pectate (mg/ml)	Reference
Bacteria						
Bacillus polymyxa[a]				8.3–9.6	0.056–0.0065	Nagel and Wilson (1970)
Bacillus subtilis	33,000	9.85		8.5		Chesson and Codner (1978)
Erwinia aroideae	37,000			9.1		Kamimiya et al. (1977)
Erwinia carotovora			90	8.5		Moran et al. (1968a)
Erwinia chrysanthemi[a]	30,000–36,000	9.4–4.6		9.8–8.2		Garibaldi and Bateman (1971); Pupillo et al. (1976)
Erwinia chrysanthemi		9.4	320	9.0		Basham (1974)
Erwinia rubrifaciens	41,000	6.25	450	9.5	5.0	Gardner and Kado (1976)
Pseudomonas fluorescens	42,300	10.3	956	9.4	0.10	Rombouts et al. (1978)
Streptomyces fradiae			176	9.1		Sato and Kaji (1975)
Xanthomonas campestris			1,050	9.5		Nasuno and Starr (1967)
Fungi						
Cephalosporium sp.			364	9.9	0.018	Atallah and Nagel (1977)
Hypomyces solani[a]	32,400–42,000	10.2–10.5		8.5		Hancock (1976)

[a] These enzymes have multiple molecular forms.

decreases rapidly with respect to the number of glycosidic bonds broken. Differences exist among various enzymes (Nagel and Wilson, 1970), and with a single enzyme under various experimental conditions such as temperature and pH value (Rombouts, 1972). These differences point towards a multiple attack mechanism, just as with endo-polygalacturonases.

Activity of endopectate lyases decreases with decreasing chain length of oligogalacturonate substrates. More specifically, the Michaelis constant increases with decreasing chain length (Nagel and Hasegawa, 1967b; Nagel and Wilson, 1970; Atallah and Nagel, 1977), whereas the maximum velocity, under conditions of substrate saturation, remains fairly constant (Nagel and Wilson, 1970; Atallah and Nagel, 1977). Most of the endopectate lyases described are capable of degrading trigalacturonate and unsaturated trigalacturonate as the lowest molecular-weight substrates, but the enzymes of an unidentified *Bacillus* sp. (Hasegawa and Nagel, 1966; Nagel and Hasegawa, 1967b) and of *B. pumilus* (Davé and Vaughn, 1971) do not attack these sub-strates. These properties, and the products formed from oligogalac-turonates (including the unsaturated ones), allowed Rexová-Benková and Markovič (1976) to distinguish two different action patterns for endopectate lyases.

G. Exopectate Lyases

Exopectate lyase is the only pectic depolymerase produced by *Clostridium multifermentans* (Macmillan and Vaughn, 1964; Macmillan *et al.*, 1964), *Erwinia dissolvens* (Castelein and Pilnik, 1976), another *Erwinia* sp. (Hatanaka and Ozawa, 1972, 1973) and *Streptomyces nitrosporeus* (Sato and Kaji, 1977a, b). This type of enzyme is also produced by *Erwinia aroideae* (Okamoto *et al.*, 1964a, b) and *Fusarium culmorum* (Urbanek *et al.*, 1976).

The enzymes prefer pectate over pectins, and polymethyl-galacturonate-methylglycoside is not attacked at all. Unsaturated digalacturonates (and exceptionally unsaturated trigalacturonates; Sato and Kaji, 1977b) are split off from the reducing end of the substrate. High optimum pH values (8.0 to 9.5) are reported. An absolute requirement for calcium ions exists for the enzymes of *Cl. multifermentans* and *Strep. nitrosporeus*, but the activity of those of the

Erwinia spp. is relatively indifferent towards calcium ions. In one case (Hatanaka and Ozawa, 1973), a strong activation is reported with sodium ions.

The reaction rates of exopectate lyases on polygalacturonate and oligogalacturonates are about the same, and trigalacturonate is the lowest molecular-weight substrate to be degraded (Macmillan *et al.*, 1964; Hatanaka and Ozawa, 1972). The enzyme from *Strep. nitrosporeus* has the highest affinity for tetragalacturonate, degrades hexa-galacturonate most rapidly, and trigalacturonate not at all (Sato and Kaji, 1977a). The extracellular exopectate lyase of *Cl. multifermentans* occurs together with pectinesterase in a complex with a molecular weight of 400,000 (Miller and Macmillan, 1971). The complex is active on highly esterified pectin, producing free carboxyl groups and unsaturated digalacturonates in a ratio of $2:1$. Apparently, the two types of activity operate in a co-ordinated manner as a 'molecular disassembly line' in which a pectin chain passes from the esterase site to the lyase site without intermediate dissociation and rebinding (Sheiman *et al.*, 1976).

H. Endopectin Lyases

Endopectin lyases are produced almost exclusively by fungi, the only exceptions known so far being a soft-rot pseudomonad (Ohuchi and Tominaga, 1974), a strain of *E. carotovora* (Almengor-Hecht and Bull, 1978) and one of *E. aroideae* (Kamimiya *et al.*, 1972, 1974). Some properties of the pectin lyases, which have been thoroughly purified, are summarized in Table 5.

The method of choice for assaying these enzymes is again measure-ment of the increase in absorbance at 235 nm, the molar extinction coefficient for unsaturated methylated oligogalacturonates being 5,500 $cm^2 \cdot mmol^{-1}$ (Edstrom and Phaff, 1964a). Highly esterified pectins are the best substrates, while pectate, pectic-acid amide and the glycyl ester of pectate are not degraded (Voragen *et al.*, 1971; Voragen, 1972; Pilnik *et al.*, 1973, 1974). Most pectin lyases are markedly stimulated by calcium and other cations, the stimulation being dependent on pH value and the degree of esterification of the substrate (Edstrom and Phaff, 1964a; Voragen, 1972; Ishii and Yokotsuka, 1972b, 1975).

Pectin lyases are endo-enzymes that cause a rapid drop in viscosity of

Table 5

Properties of some endopectin lyases

Source of enzyme	Molecular weight	Iso-electric point	Specific activity (units/mg of protein)	Optimum pH value	K_m Value for pectin (mg/ml)	Reference
Fungi						
Alternaria mali I	28,000		176	8.7		Hasui *et al.* (1976)
II	31,000		577	8.2		
Aspergillus fonsecaeus			19	5.2		Edstrom and Phaff (1964a, b)
Aspergillus japonicus	32,000	7.7	355	6.0		Ishii and Yokotsuka (1975)
Aspergillus niger		3.5	24	5.2		Albersheim and Killias (1962)
Aspergillus niger		3.5		5.9	2.2	Amadò (1970)
Aspergillus niger I[a]	35,400	3.65	17	6.0	5.0	Van Houdenhoven (1975)
II[a]	33,100	3.75	44	6.0	0.9	
Aspergillus sojae	32,000		77	5.5		Ishii and Yokotsuka (1972b)
Dothidea ribesia	31,200	8.9		8.4	3.2	Knobel and Neukom (1974)
Bacteria						
Erwinia aroideae	30,000		400	8.1		Kamimiya *et al.* (1972, 1974)

[a] These enzymes are glycoproteins, with single peptides.

the substrate solution with respect to the number of glycosidic bonds broken. Activity decreases rapidly with the chain length of methyl oligogalacturonates (Edstrom and Phaff, 1964b; Voragen, 1972; Van Houdenhoven, 1975). The lowest molecular-weight substrates to be degraded are fully methylated tetragalacturonate and fully methylated trigalacturonate for enzymes from *Aspergillus fonsecaeus* (Edstrom and Phaff, 1964b) and *A. niger* (Voragen, 1972; Van Houdenhoven, 1975), respectively. Kinetic studies with two molecular forms of pectin lyase on fully methylated oligogalacturonates made it possible to show that the number of subsites in the substrate-binding site of these two enzymes are 8 and 9 or 10. Evidence was also obtained for the involvement of carboxyl groups and a tyrosine residue in binding of the substrate (Van Houdenhoven, 1975).

III. MICROBIAL POLYGALACTURONATE METABOLISM

Attack of extracellular microbial enzymes on pectic substances results in accumulation of a variety of products. Methanol is formed, due to pectinesterase activity. D-Galacturonate and oligogalacturonates are the final products of exo- and endopolygalacturonase activity. Pectate lyase produces a series of unsaturated oligomers, namely oligo-galacturonates of which the terminal residue at the non-reducing end is a 4,5-dehydrogalacturonate. Pectin lyases produce methyl esters of unsaturated oligomers. The subject of this section is the further breakdown and utilization of these products by micro-organisms.

Methanol serves as a carbon and energy source for a number of micro-organisms that use this compound (and methane) either obligatory or facultatively. For its metabolism, we refer to Schlegel (1972) and Kosaric and Zajic (1974). Pectin is a major natural source of methanol, but it is not known whether any methanol-utilizing micro-organisms produce pectinesterase.

Very little research has been done on pectin metabolism in fungi, and nothing is known, for instance, about the fate of methyl esters of unsaturated oligogalacturonates, which are products of fungal pectin lyase. That is why most of this chapter deals with polygalacturonate and uronate metabolism in bacteria, a subject that has been studied in considerable detail in the last two decades.

A. Oligogalacturonases

Several micro-organisms capable of utilizing polygalacturonate appear to produce oligogalacturonases. These are cell-bound enzymes that convert oligogalacturonates or unsaturated oligogalacturonates into monomeric units. As these enzymes attack their substrates from either the reducing end or the non-reducing end, they could be called exo-enzymes. However, the essential difference between these cell-bound enzymes and exopolygalacturonase and exopectate lyase is that they preferentially attack short-chain substrates. The rate of attack of these enzymes decreases rapidly with increasing degree of polymerization of the substrate (Table 6).

Both oligogalacturonate hydrolases and oligogalacturonate lyases have been described. Oligogalacturonate hydrolyses attack their substrates from the non-reducing end, whereas oligogalacturonate lyases operate from the reducing end. Two different types of oligogalacturonate hydrolases have been isolated from *Bacillus* sp. (Table 6). The one described by Hasegawa and Nagel (1968) specifically hydrolyses oligogalacturonates, ultimately to galacturonic acid only. The other one, which is called an unsaturated oligogalacturonate hydrolase (Nagel and Hasegawa, 1968), is specific for unsaturated oligogalacturonates, of which it only hydrolyses the unsaturated monomeric unit, leaving the glycosidic bonds between 'normal' galacturonate units unattacked. Preiss and Ashwell (1963a) showed that the unsaturated monomeric product isomerizes rapidly into 4-deoxy-L-*threo*-5-hexoseulose uronate (4-deoxy-5-oxo-D-glucuronate). Oligogalacturonate hydrolase prepared from the mycelium of *A. niger* (Hatanaka and Ozawa, 1969a, b) was found to exert both activities of oligogalacturonate hydrolases from *Bacillus* sp., but the activity on unsaturated dimer was only about one-sixtieth of that on the dimer. Two so-called exopolygalacturonases purified by Mill (1966a, b) from the mycelium of *A. niger* may now also be classified as oligogalacturonate hydrolases. At an optimum pH value of 4.5 to 5.0, these enzymes attacked digalacturonate and trigalacturonate much more rapidly than pectic acid, producing only galacturonate.

Oligogalacturonate lyases (oligogalacturonide lyases, EC 4.2.2.6) readily attack oligogalacturonates and unsaturated oligogalacturonates. By transelimination, they remove unsaturated monomers from the reducing end of their substrates, thus converting unsaturated

Properties of some microbial oligogalacturonases

Micro-organism	Substrates[a]	Products from		Optimum pH value	Reference
		Dimer[b]	u-Dimer		
Oligogalacturonate hydrolases; attack of substrate from non-reducing end					
Bacillus sp.	trimer > tetramer > pentamer > dimer > acid-soluble pectic acid	monomer	none	6.5	Hasegawa and Nagel (1968)
Bacillus sp.[c]	u-dimer > u-trimer > u-tetramer > u-pentamer	none	u-monomer monomer	6.5	Nagel and Hasegawa (1968)
Aspergillus niger	dimer > pectic acid > u-dimer	monomer	u-monomer monomer	—	Hatanaka and Ozawa (1969a, b)
Oligogalacturonate lyases; attack of substrate from reducing end					
Erwinia carotovora	u-dimer > dimer > trimer > tetramer, pectate	u-monomer monomer	u-monomer	7.2	Moran et al. (1968b)
Erwinia aroideae	u-dimer > u-trimer > dimer > trimer > tetramer > acid-soluble pectic acid	u-monomer[d] monomer	u-monomer[d]	7.0	Hatanaka and Ozawa (1970)
Pseudomonas sp.[e]	tetramer > trimer, u-trimer > dimer, u-dimer > acid-soluble pectic acid > pectate	u-monomer monomer	u-monomer	7.0	Hatanaka and Ozawa (1971a)

[a] These are listed in order of rate of degradation.

[b] Terminology and abbreviations: monomer, D-galacturonic acid; dimer, digalacturonic acid; u-oligomers (unsaturated oligomers), oligo-galacturonates of which the terminal unit at the non-reducing end is a 4,5-dehydrogalacturonate; u-monomer, 4-deoxy-5-oxo-D-glucuronic acid (4-deoxy-L-threo-5-hexoseulose uronic acid) the degree of polymerization of acid-soluble pectic acid is 15–20.

[c] This enzyme, which attacks the bond adjacent to a terminal unsaturated residue only, is called an unsaturated oligogalacturonate hydrolase.

[d] The product formed in the reaction mixture was identified as 4-deoxy-5-oxo-D-fructuronic acid.

[e] This oligogalacturonate lyase, unlike those of Erwinia spp., is stimulated by calcium ions.

oligogalacturonates into only unsaturated monomers. The products from oligogalacturonates are unsaturated monomers and a single D-galacturonate monomer from each oligomer. Production of 4-deoxy-5-oxo-fructuronate by the oligogalacturonate lyase from *Erwinia aroideae* (Table 6) suggests the presence of an isomerase in this enzyme preparation.

B. Uronate Metabolism

Apparently, bacterial polygalacturonate degradation leads to production of two different types of monomers, namely D-galacturonate and 4-deoxy-L-*threo*-5-hexoseulose uronate (4-deoxy-5-oxo-D-glucuronate). These two monomers are further metabolized along at least three different pathways (Fig. 4). 4-Deoxy-L-*threo*-5-hexoseulose uronate is an end-product of bacterial degradation of polymers containing D-glucuronate, L-iduronate and D-galacturonate. Preiss and Ashwell (1963a, b) studied metabolism of this compound by a pectolytic species of *Pseudomonas*. Following isomerization into 3-deoxy-D-*glycero*-2,5-hexodiulosonate (4-deoxy-5-oxo-D-fructuronate), the compound is reduced by an oxodeoxygluconate dehydrogenase into 2-oxo-3-deoxygluconate which is phosphorylated by oxodeoxygluconate kinase and further metabolized along the Entner–Doudoroff pathway (Fig. 5). D-Galacturonate is metabolized by a number of pectolytic as well as non-pectolytic bacteria, along at least two different pathways (Fig. 4). In *E. carotovora* (Kilgore and Starr, 1959a), *E. coli* (Ashwell, 1962), *B. polymyxa* (Nagel and Hasegawa, 1967a) and *Aeromonas* sp. (Farmer and Eagon, 1969), D-galacturonate is isomerized by uronate isomerase (D-glucuronate ketol-isomerase, EC 5.3.1.12) to D-tagaturonate. Following reduction by D-altronate dehydrogenase (D-altronate: NAD$^+$ 3-oxidoreductase, EC 1.1.1.58) to D-altronate, and dehydration by D-altronate dehydrase (D-mannonate hydrolyase, EC 4.2.1.8), 2-oxo-3-deoxygluconate is formed (Fig. 5). This compound is phosphorylated and further metabolized by the Entner-Doudoroff pathway. Similarly, in *E. carotovora*, *E. coli* and *Aeromonas* sp., D-glucuronate is isomerized to D-fructuronate which, after reduction by D-mannonate dehydrogenase (D-mannonate: NAD$^+$ 5-oxidoreductase, EC 1.1.1.57) to D-mannonate and dehydration by D-mannonate dehydrase (D-mannonate hydrolyase, EC 4.2.1.8), also yields 2-oxo-3-deoxygluconate.

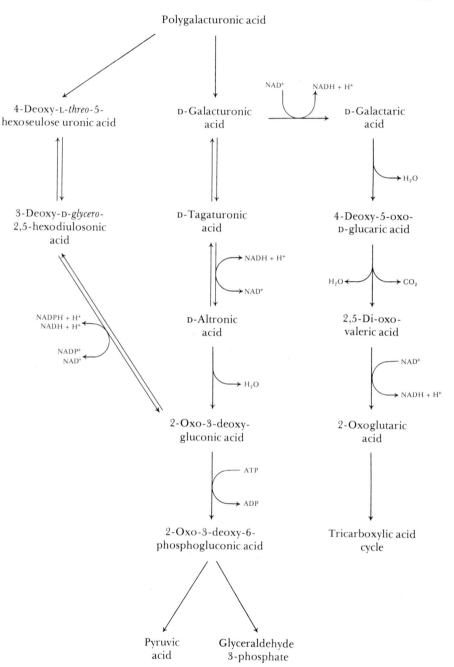

Fig. 4. Pathways of polygalacturonic acid metabolism in bacteria.

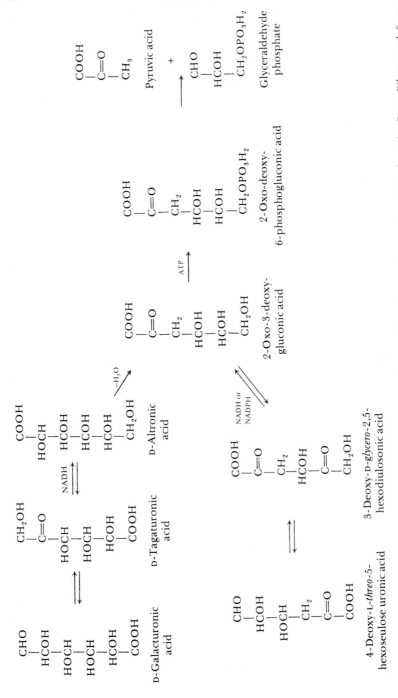

Fig. 5. Pathways for metabolism of D-galacturonic acid and 4-deoxy-L-*threo*-5-hexoseulose uronic acid in bacteria. From Kilgore and Starr (1959a) and Preiss and Ashwell (1963a, b).

Synthesis of uronate isomerase is induced by either D-galacturonate or D-glucuronate in the culture medium, and appears to be a single enzyme which isomerizes both compounds (Kilgore and Starr, 1959a; Ashwell, 1962). In *E. coli*, D-tagaturonate and not D-galacturonate appeared to be the true inducer for the three galacturonate degradative enzymes, namely uronate isomerase, altronate dehydrogenase and altronate dehydrase. D-Glucuronate and D-fructuronate induced synthesis of all five enzymes necessary for conversion of D-galacturonate and D-glucuronate into 2-oxo-3-deoxygluconate. Also, D-mannonic amide served as gratuitous inducer of all five enzymes (Robert-Baudouy *et al.*, 1974; Stoeber *et al.*, 1974). The extensive studies of this group on the physiological and genetic aspects of hexuronate metabolism in *E. coli* have shown that the structural genes of the five enzymes referred to are located on the chromosome in three presumed distinct operons belonging to two regulatory units. More recent studies were concerned with the inducible aldo-hexuronate transport system, which is specific for both galacturonate and glucuronate (Nemoz *et al.*, 1976).

When grown on pectin, *B. polymyxa* produces an isomerase-reductase system specific for D-galacturonate. The purified system does not show activity on D-glucuronate, 4-deoxy-L-*threo*-5-hexoseulose uronate, oligogalacturonates, unsaturated oligogalacturonates or on a variety of reducing sugars. It is therefore very useful for enzymic determination of D-galacturonate (Nagel and Hasegawa, 1967a).

An alternative pathway for D-galacturonate metabolism is its oxidation to D-galactarate which, after several conversions, enters the tricarboxylic acid cycle as 2-oxoglutaric acid (Fig. 4). Oxidation of D-galacturonate with NAD$^+$-dependent uronate dehydrogenase (uronate: NAD$^+$ 1-oxidoreductase, EC 1.2.1.35) was discovered by Kilgore and Starr (1959b) in *Ps. syringae*. The enzyme was studied in detail by Bateman *et al.* (1970) and by Wagner and Hollmann (1976). The enzyme is induced by the presence of either galacturonate or glucuronate in the culture medium. It has a molecular weight of 60,000 and is composed of two subunits of molecular weight 30,000. The subunits are inactive. The enzyme is specific for D-galacturonate and D-glucuronate, which it converts into D-galactarate (D-mucate) and D-glucarate, respectively, but its affinity for galacturonate is much higher than that for glucuronate. Since complete oxidation of these uronates occurs, with stoicheiometric yields of NADH, the enzyme is very suitable for quantitative determination of these uronates (Bateman *et al.*, 1970).

D-Galacturonate catabolism by the 2-oxoglutarate pathway as depicted in Figure 4 was shown to be common to many species of *Pseudomonas* (Trudgill and Widdus, 1966). The enzymes involved in the different reactions of this pathway were studied in considerable detail by Jeffcoat *et al.* (1969a, b) and by Jeffcoat (1974). Variations in the 2-oxoglutarate pathway have been described, for instance, in *Agrobacterium tumefaciens* (Chang and Feingold, 1970). A number of species of the Enterobacteriaceae, which can utilize D-glucarate as the sole source of carbon and energy, metabolize this compound by a different pathway, leading to glycerate and pyruvate (Blumenthal and Fish, 1963).

It is difficult to estimate the relative importance of the three pathways of polygalacturonate metabolism in bacteria. However, since pectate lyase is the predominant polygalacturonate depolymerase in bacteria and oligogalacturonate lyases have been demonstrated in a number of bacteria, the pathway of 4-deoxy-L-*threo*-5-hexoseulose uronate may be more important than the tagaturonate pathway. At present, catabolism of galacturonate via galactarate seems to be confined to species of *Pseudomonas* and *Agrobacterium*.

IV. REGULATORY ASPECTS OF PECTIC ENZYME SYNTHESIS

A. Bacteria

Pectic enzymes are produced by a great variety of bacteria, important genera being *Pseudomonas*, *Xanthomonas*, *Erwinia*, *Bacillus* and *Clostridium*. Various plant-pathogenic and saprophytic *Pseudomonas* spp. produce one or more pectic enzymes (Hildebrand, 1972). *Pseudomonas fluorescens* elaborates an inducible, extracellular endopectate lyase (Fuchs, 1965; Zucker and Hankin, 1970, 1971). Enzyme formation in a pectin-containing medium is greatly stimulated by a heat-labile factor (not pectinesterase) present in fresh potato tissue or acetone-dried powder of potato tissue. However, maximum enzyme production may require 10 to 20 generations of growth in inducing medium (Zucker and Hankin, 1970, 1971). Constitutive synthesis of great amounts of endopectate lyase is common with *Ps. marginalis* and *Ps. fluorescens* strains that cause soft rots (Zucker *et al.*, 1972; Rombouts *et al.*, 1978).

Xanthomonas spp. synthesize pectinesterase and endopectate lyase. While synthesis of pectinesterase is inducible, pectate lyase is either inducible or constitutive, and constitutive enzymes are apparently not necessary for phytopathogenicity (Starr and Nasuno, 1967; Nasuno and Starr, 1967). Synthesis of constitutive extracellular endopectate lyase of *Aeromonas liquefaciens* is subject to catabolite repression (Hsu and Vaughn, 1969). Under conditions of restricted growth (slow feeding of glucose, glycerol or pectate to carbon-limited cultures), the enzyme is formed at a maximum differential rate 500 times greater than with excess of substrate present. Also, with unrestricted growth in media containing less readily utilizable carbon sources, such as citrate and aspartate, a certain relief from catabolite repression is observed. A mutant strain was isolated which could readily synthesize the enzyme while growing in the presence of excess glucose or pectate.

Within the genus *Erwinia*, producers of all of the known pectic enzymes are to be found. Some of them, for instance *E. carotovora*, *E. chrysanthemi* and *E. aroideae*, all members of the soft-rot section of the genus, produce large amounts of endopectate lyase, and also endopolygalacturonase. Only part of the pectate lyase activity is excreted into the culture medium, but Moran *et al.* (1968a) found the cell-bound and extracellular endopectate lyase of *E. carotovora* to be identical. Recently, Chatterjee and Starr (1977) obtained evidence that, in *E. chrysanthemi*, pectate lyase synthesis and excretion are controlled by one single gene, while endopolygalacturonase production is controlled by another gene.

Representative strains of *E. carotovora* and *E. aroideae* were found to be constitutive with respect to endopectate lyase synthesis (Moran and Starr, 1969). In batch cultures, the differential rate of enzyme formation was low on glucose, high on glycerol and even higher on pectate as the sole carbon and energy source, and it was concluded that enzyme production is under catabolite repression. Hubbard *et al.* (1978) confirmed this catabolite repression in *E. carotovora*. They also showed that an exogenous supply of adenosine 3':5'-monophosphate (cAMP, the effector of catabolite activator protein which is responsible for regulation of all enzymes under catabolite repression) could relieve catabolite repression, and increase the differential rate of enzyme production in glucose-containing medium above that in pectate-containing medium without added cAMP. A decrease in intracellular cAMP could be correlated with catabolite repression of endopectate lyase synthesis. In addition to repression, induction of pectate lyase synthesis also occurs in *E. carotovora*. Both Tsuyumu (1977) and Hubbard (1978) showed that products of pectate degraded by pectate lyase (unsaturated digalacturonate and unsaturated trigalacturonate) are more effective than pectate as inducers of pectate lyase synthesis. Whether or not these products are true inducers remains to be established. Apparently, with *E. carotovora*, constitutively formed pectate lyase when degrading pectate produces unsaturated oligogalacturonates. These are taken up by the cell and, in the absence of conditions leading to catabolite repression, directly or indirectly induce the cell to increase pectate lyase synthesis.

Stimulation of pectate lyase production by certain plant extracts is sometimes observed. A combination of pectin and a heat-labile factor from potato tissue was the best medium for pectate lyase production by *E. carotovora* (Zucker and Hankin, 1970). Other vegetable extracts also had pronounced repressive or stimulating (depending on the concentration) effects on pectate lyase production (Zucker *et al.*, 1972). Pectate lyase production in *E. aroideae* was greatly stimulated, amongst others, by an alkali-labile factor (molecular weight 400) isolated from carrot extracts (Tsuchida *et al.*, 1968; Tomizawa *et al.*, 1970).

A pectin lyase-producing strain of *E. carotovora* was isolated by means of a chemostat enrichment procedure (Almengor-Hecht and Bull, 1978). In a pectin-limited chemostat culture, the differential rate of pectin lyase production passed through a maximum at a dilution rate of 0.04 h^{-1}, an observation that probably reflects a balance between enzyme induction and catabolite repression. A pectin lyase was also produced extracellularly by a strain of *E. aroideae* growing in a pectin-containing medium in the presence of nalidixic acid, an inhibitor of DNA biosynthesis (Kamimiya *et al.*, 1972, 1974). Enzyme production apparently involved protein synthesis *de novo*, as it required casein hydrolysate in the culture medium and was susceptible to chloramphenicol inhibition. The precise mechanism of stimulation of pectin lyase formation by nalidixic acid was not clarified.

Certain other genera within the Enterobacteriaceae, notably *Yersinia* and *Klebsiella*, produce polygalacturonase and pectate lyase (Starr *et al.*, 1977). However, the amounts of enzymes elaborated by these bacteria are lower than those produced by soft-rot *Erwinia* spp.; furthermore, the enzymes are not excreted rapidly and massively into the growth medium.

Three species of the genus *Bacillus* are reputable endopectate lyase producers; these are *B. subtilis* (Nortje and Vaughn, 1953; Kurowski and Dunleavy, 1976a; Chesson and Codner, 1978), *B. polymyxa* (Nagel and Vaughn, 1962; Nagel and Wilson, 1970; Fogarty and Griffin, 1973), and *B. pumilus* (Davé and Vaughn, 1971; Davé *et al.*, 1976). A few other species also make this enzyme. These include *B. sphaericus* (McNicol and Baker, 1970), *B. circulans* (Joyce and Fogarty, 1976) and *B. stearothermophilus* (Karbassi, 1973).

The pectate lyase produced by *B. subtilis* is a true extracellular enzyme which is produced constitutively during growth on a wide range of carbon and nitrogen sources. Kurowski and Dunleavy (1976a) found that the differential rate of enzyme synthesis is maximum on glucose, indicating a complete lack of catabolite repression. The growth-linked nature of pectate lyase synthesis was confirmed by chemostat experiments, which showed that the differential rate of enzyme synthesis was constant over a wide range of dilution rates. Their strain produced no pectinesterase, and pectin was a poor substrate for growth and enzyme formation. However, both pectate lyase and pectinesterase were produced by a strain isolated from carrots (Chesson and Codner, 1978).

Bacillus polymyxa produces pectate lyase constitutively, but the differential rate of enzyme production is increased in media containing D-galacturonate, pectate and pectin. With pectin as the substrate, pectinesterase is also induced (Nagel and Vaughn, 1961). Enzyme production and excretion occur in the growth phase, but a small amount of cell-associated enzyme is liberated only during the stationary phase of growth and subsequent lysis of the cells (Nagel and Vaughn, 1962). Four different endopectate lyases were isolated from this organism (Nagel and Wilson, 1970), but there are at present no genetic or physiological data about their origin.

The diversity of adaptation within the genus *Bacillus* referred to by Priest (1977) may be demonstrated by pectate lyase synthesis at 65°C by a thermophilic strain (Karbassi, 1973), and at an optimum pH value of about 9 by an alkalophilic strain (Kelly and Fogarty, 1978).

Although a number of *Clostridium* species are known to produce various pectic enzymes (Ng and Vaughn, 1963; Lund and Brockle-hurst, 1978), very little is known about their synthesis. The interesting observation that *Cl. multifermentans* and *Cl. butyricum* rapidly fermented both pectin and pectate, but not galacturonate, lead to the discovery of the inducible exopectate lyase–pectinesterase complex produced by the former species (Macmillan and Vaughn, 1964; Sheiman *et al.*, 1976).

Recently, Kurowski and Dunleavy (1976b) reported on pectate lyase, the only pectic enzyme produced by *Cytophaga johnsonii*. Enzyme synthesis was inducible and also subject to catabolite repression. The best inducer was galacturonate, followed by pectate and pectin. Catabolite repression was not alleviated by an exogenous supply of cAMP. Enzyme production was growth-linked. The enzyme occurred partly in the medium and was partly cell-bound. The bound enzyme was probably superficially located, as it displayed its full activity from intact cells.

No literature is available on the mechanism of secretion of extracellular pectic enzymes produced by bacteria. We therefore refer to the general reviews by Glenn (1976) and Priest (1977).

B. Fungi

Early studies by Phaff (1947) revealed that polygalacturonase and pectinesterase are strongly induced in *Penicillium chrysogenum* by pectin, pectic acid, gum tragacanth, D-galacturonate, mucate and L-galactonate. One of his speculations, namely that mucate (D-galactarate) is possibly the only stimulatory compound (the true inducer), is still very attractive, and should be checked with mutants. *Penicillium expansum* produces polygalacturonase and pectin lyase in apple tissue (Swinburne and Corden, 1969; Spalding and Abdul-Baki, 1973). Formation of both enzymes is repressed by a variety of sugars, and also by galacturonate and glutamate. Dialysis of apple medium results in greatly increased levels of enzyme production (Spalding *et al.*, 1973). *Penicillium digitatum* produces pectin lyase constitutively at a relatively constant rate, independent of culture conditions (Lobanok *et al.*, 1977).

Aspergillus niger is an effective producer of pectinesterase, polygalacturonase and pectin lyase. Tuttobello and Mill (1961) reported very good enzyme production in a liquid medium of groundnut flour extract with 2% sucrose and 2% pectin, the enzymes being produced in a stage when the sucrose concentration in the medium dropped rapidly. A synergistic effect of pectin and lactose on synthesis of pectic enzymes by *A. niger* was also described by Feniksova and Moldabaeva (1967). Likewise, Nyeste and Holló (1963) had found that the best media were those with a mixed carbohydrate composition (pectin and either glucose or starch). Vasu (1967) showed that the pectic enzymes produced by *A. niger* remain partly associated with the mycelium, and that secretion of these enzymes is favoured by the presence of pectic substances in the culture medium. Multiple molecular forms of different pectic enzymes occurred, some of them being produced inducibly, others constitutively (Vasu, 1974). Extensive studies were made on the kinetics of pectic enzyme formation by *A. awamori* and *A. foetidus* in media containing sucrose or a mixed carbon source (Zetelaki-Horváth and Békassy-Molnár, 1973; Zetelaki, 1976; Zetelaki-Horváth, 1978).

Polygalacturonase production by an adenine-requiring mutant of *A. niger* was inducible and subject to catabolite repression (Shinmyo *et al.*, 1978). Glucose, fructose, or intermediates of glycolysis, but not tricarboxylic acid-cycle intermediates, repressed enzyme synthesis. Catabolite repression by glucose occurred quickly and almost completely. As the corresponding messenger RNA was rather stable, it was suggested that catabolite repression of polygalacturonase formation occurred at the translation level. Little is known about enzyme synthesis in the many other pectolytic *Aspergillus* species (Edstrom and Phaff, 1964a; Sreekantiah *et al.*, 1975; Abdel-Fattah and Mabrouk, 1977).

Various plant-pathogenic *Fusarium* (*Hypomyces*) species produce pectic hydrolases, lyases and pectinesterase (Bateman, 1966; Papavizas and Ayers, 1966; Hancock, 1968; Patil and Dimond, 1968; Miller and Macmillan, 1971). Pectate lyase production by *H. solani* f. sp. *cucurbitae* is growth-related, induced by pectate, but not by pectin, and glucose-repressed (Hancock *et al.*, 1970). Differences in molecular and catalytic properties were found between pectate lyases produced by this organism in pectate-containing medium and in infected squash (*Cucurbita maxima*) plants (Hancock, 1976). Endopolygalacturonase synthesis by *F. roseum* is induced by mucate and by polygalacturonate at pH 6.5, and also by pectin at pH 3.5. Pectin depolymerase and pectinesterase are induced on pectin at pH 3.5 and 6.5 (Perley and Page, 1971). Extracellular polygalacturonase synthesis by *F. oxysporum* f. sp. *lycopersici* is induced in a pectin-containing medium, and stimulated by addition of small concentrations of sugars or galacturonate, but repressed by higher concentrations of these compounds (Patil and Dimond, 1968).

Sequential induction of various pectic enzymes, glycosidases, hemicellulases and cellulases was reported to occur in *F. oxysporum* f. sp.

lycopersici (Jones *et al.*, 1972; Cooper and Wood, 1973, 1975), *Colletotrichum lindemuthianum* (English *et al.*, 1971), *Verticillium albo-atrum* (Cooper and Wood, 1973, 1975) and in *Rhizoctonia solani* (Lisker *et al.*, 1975). For instance, in *Verticillium albo-atrum* cultures growing on tomato stem cell walls, the enzymes appear sequentially over two to nine days in the order: endopolygalacturonase, exo-arabanase, endopectin lyase, endoxylanase and cellulase. Synthesis of each enzyme is usually induced specifically by growth-limiting concentrations of the breakdown products of its substrate, and strongly repressed by concentrations that allow unrestricted growth (Cooper and Wood, 1975). Apparently a number of plant-pathogenic fungi produce small amounts of various extracellular enzymes constitutively, increased enzyme synthesis being inducible and subject to catabolite repression. This conclusion may explain, at least in part, the poor or even negative correlation between enzyme production in culture media and virulence of plant pathogens which is sometimes found, for instance, with *Verticillium* species (Wiese *et al.*, 1970).

Pyrenochaeta terrestris produces small amounts of extracellular endopolygalacturonase constitutively. Enzyme synthesis is maximally induced by pectin, followed by galacturonate, mucate, pectate and dulcitol, in that order (Keen and Horton, 1966). Synthesis of polygalacturonase in the presence of pectin is stimulated by low concentrations of sugars, but enzyme formation is repressed in media with high concentrations of sugar (Keen and Horton, 1966; Goodenough and Kempton, 1974).

In cultures of *Acrocylindrium* sp., various pectic enzymes, and enzymes of galacturonate metabolism, are rapidly induced by galacturonate and, with a lag period, also by oligogalacturonates and pectate. The existence of a co-ordinated regulatory mechanism for synthesis of all of these enzymes was considered (Kimura and Mizushima, 1974).

Production and secretion of pectic enzymes may start at an early stage in the development of a fungus. Barash (1968) studied endopolygalacturonase liberation during germination of spores of *Geotrichum candidum*, and observed that release of enzyme occurred even before germ-tube appearance. Also, conidia of *Botrytis cinerea* contain endopolygalacturonase which is readily released on wetting while, after the onset of germination, more enzyme is being synthesized (Verhoeff and Liem, 1978).

Calonge *et al.* (1969) observed multivesicular bodies in the hyphae of

Sclerotinia fructigena induced to produce extracellular enzymes. They found evidence that these bodies are cytoplasmic constituents, involved in extracellular enzyme secretion, by a type of reverse pinocytosis.

C. Yeasts

Yeasts are generally not known as being powerfully pectolytic. Groups of yeasts have been surveyed for pectolytic properties by Luh and Phaff (1951), Roelofsen (1953), Bilimoria and Bhat (1962), Kotomina and Pisarnitskii (1974), Call and Emeis (1978a, b) and Wimborne and Rickard (1978). Although quite a number of yeast species give positive results in tests screening for pectolytic properties, only one species, namely *Kluyveromyces fragilis* (*Saccharomyces fragilis*) and its imperfect form *Candida pseudotropicalis*, secretes large amounts of a pectic enzyme, namely endopolygalacturonase, into the culture medium. High levels of endopolygalacturonase are produced constitutively by those strains which slowly ferment lactose (Phaff, 1966). Interestingly, enzyme production with this yeast is completely repressed in media at or above 60% oxygen saturation, and derepressed by replacement of oxygen with carbon dioxide or nitrogen (Wimborne and Rickard, 1978).

V. THE ROLE OF PECTOLYTIC MICRO-ORGANISMS

This section of the chapter deals with the many ways in which micro-organisms, through their pectic enzymes, interfere with plants and plant material. Pectolysis is by no means limited to this field, as it is also part of many physiological processes occurring in plants. Pectinesterase may be isolated from any plant source. Also, exo-polygalacturonase is being found in a rapidly increasing number of plants and fruits, whereas the occurrence of endopolygalacturonase seems to be restricted to soft fruits only. The activity of one or more of these enzymes has been correlated with growth of plants and elongation of plant cell walls (Bryan and Newcomb, 1954; Glasziou and Inglis, 1958; Pressey and Avants, 1977), fertility and fertilization (Kroh and Loewus, 1978), consistency changes of ripening fruits (Hobson, 1963, 1964; Nakagawa *et al.*, 1971; Markovič *et al.*, 1975; Pressey and Avants, 1976; Bartley, 1978; Sawamura *et al.*, 1978), abscission of leaves and fruits (Greenberg *et al.*, 1975) and succulence (Binet, 1976). The occurrence and significance of multiple molecular forms of some of these plant enzymes (Hultin and Levine, 1963, 1965; Hultin *et al.*, 1966; Pressey and Avants, 1972; Roeb and Stegemann, 1975; Brady, 1976; Evans and McHale, 1978; Versteeg *et al.*, 1978), as well as regulation of their synthesis and activity (Goren *et al.*, 1976; Sawamura *et al.*, 1978), are important topics being studied by plant physiologists and food biochemists.

A. Plant Pathogenesis and Symbiosis

The involvement of pectic enzymes in plant pathogenesis has been reviewed regularly, for example by Bateman and Millar (1966), Starr and Chatterjee (1972), Hall and Wood (1973) and Mussell and Strand (1977). Also, rather lengthy lists of plant-pathogenic and other micro-organisms, and their pectic enzymes, were compiled by Hildebrand (1972), Rombouts and Pilnik (1972) and Fogarty and Ward (1974).

It has been known for some time that a single, well-purified endopectic enzyme, namely endopectate lyase, can cause plant-tissue maceration, electrolyte loss and cell death (Dean and Wood, 1967; Mount *et al.*, 1970; Garibaldi and Bateman, 1971). It is thought that the killing of cells in plant tissue treated with this enzyme results from a loss in the ability of enzymically degraded cell walls to support the protoplast, which becomes leaky (Basham and Bateman, 1975a, b; Gardner and Kado, 1976).

Pectic enzymes are not the only polysaccharide-degrading enzymes involved in plant pathogenesis. Often, a pathogen's attack on plant tissue is initiated by pectic enzymes, perhaps because the pectic substances are most readily accessible. But as soon as other polysaccharides in the plant cell become exposed by degradation of pectic material, the pathogen may be induced to produce high levels of other enzymes, including glycosidases, hemicellulases and cellulases. This may lead to a sequence of enzyme induction, as was described for a number of fungi (see Section IV, p. 257). In such systems, exo-enzymes and glycosidases may play an important role. They may provide the monomeric or oligomeric compounds necessary for specific induction of certain endo-enzymes (Cooper and Wood, 1975). In addition to these enzymes, other enzymic factors, for which substrates are as yet unknown, may be involved in tissue maceration and plant pathogenesis. Such a factor was purified from *A. japonicus* (Ishii and Kiho, 1976; Ishii, 1977). This factor stimulates tissue maceration by endo-pectin lyase and endopolygalacturonase, although it does not macerate by itself or stimulate the enzymes in their activity on pectin and pectate.

One of the general defence mechanisms of plants against invading pathogens is through endopolygalacturonase inhibitors which they have associated with their cell walls (Bock *et al.*, 1970, 1972; Albersheim and Anderson, 1971; Jones *et al.*, 1972; Anderson and Albersheim, 1972; Fisher *et al.*, 1973). These inhibitor proteins, found in many plants, can completely inhibit microbial endopolygalacturon-ases but do not interfere with the activity of plant endopoly-galacturonases or other microbial polysaccharide-degrading enzymes.

The primary cell walls of a variety of dicotyledonous plants appear very similar. The same is true for cell walls in monocotyledons, although these differ from those of dicotyledons. Also, many of the plant pathogens can produce large and similar patterns of poly-saccharide-degrading enzymes, which may degrade almost every glycosidic linkage known to occur in plant cell-wall polymers. These considerations led Albersheim and Anderson-Prouty (1975) to the conclusion that, although these enzymes are essential in general pathogenesis, they are not the determinants of specificity in host–pathogen systems. While this is probably true, we consider a discussion about host–pathogen specificity as being beyond the scope of this article and, for that reason, refer to Albersheim and Anderson-Prouty (1975), Cooper (1977), Deverall (1977) and Mussell and Strand (1977).

Low levels of pectic enzyme(s) are produced by *Rhizobium* spp. (Hubbell *et al.*, 1978), and perhaps by other symbiotic micro-organisms (Lindeberg and Lindeberg, 1977). The role of the *Rhizobium* enzyme(s) in the infection and nodulation process in leguminous plants is not clear. However, it is thought that they do not play a direct role in the determination of host specificity.

B. Microbial Pectolysis in Fresh, Processed and Fermented Fruits and Vegetables

Considerable losses of fresh fruits and vegetables, during transport or in store, may be caused by fungal and bacterial infections (Lund, 1971). A number of examples, compiled from the literature, are given in

Table 7. Spoilage of acidic fruits is usually caused by fungi, and characteristic infections are known for every kind of fruit. Vegetables, or fruits having a pH value around 6.5, are most commonly spoiled by bacteria, notably soft-rot *Erwinia* and *Pseudomonas* spp.

Pectic enzymes play an essential role in rotting of fruits and vegetables and, in many cases, the symptoms have been correlated with a particular enzyme, usually an endopectic enzyme (Table 7). However, this has not always been possible. For instance, Byrde *et al.* (1973) could not completely correlate symptoms of leakage from apple tissue

Table 7

Examples of microbial spoilage of fresh fruits and vegetables

Product	Micro-organism	Enzymes involved	Reference
Apples	*Gilbertella persicaria*	Endopolygalacturonase	Mehrotra *et al.* (1971)
	Penicillium expansum	Endopolygalacturonase	Swinburne and Corden (1969)
	Penicillium expansum	Endopectin lyase	Spalding and Abdul-Baki (1973)
	Pezicula malicorticis	Various pectic enzymes	Koch and Schulz (1974)
	Sclerotinia fructigena	Endopectin lyases	Byrde and Fielding (1968)
	Sclerotinia fructigena	Endopolygalacturonase	Fielding and Byrde (1969)
Apricots	*Rhizopus arrhizus*	Endopolygalacturonase	Luh *et al.* (1978)
	Rhizopus stolonifer	Polygalacturonase	Harper *et al.* (1973)
Avocados	*Colletotrichum gloeosporioides*	Endopolygalacturonase	Barash and Khazzam (1970)
Carrots	*Clostridium* spp.		Perry and Harrison (1977)
Citrus	*Geotrichum candidum*	Endopolygalacturonase	Barash and Eyal (1970)
	Penicillium digitatum	Endopectin lyase	Bush and Codner (1968)
	Penicillium italicum	Endopectin lyase	Bush and Codner (1970)
Grapes	*Botrytis cinerea*[a]	Various pectic enzymes	Drawert and Krefft (1978a, b)
Papayas	*Corynespora cassiicola*	Various pectic enzymes	Agarwal and Gupta (1978)
Peaches	*Monilia fructicola*	Various pectic enzymes	Paynter (1975)
Potatoes	*Clostridium* spp.	Various pectic enzymes	Lund and Brocklehurst (1978)
	Erwinia carotovora		Lund and Nicolls (1970)
	Fusarium roseum	Various pectic enzymes	Perley and Page (1971)
	Pseudomonas spp.	Endopectate lyase	Zucker and Hankin (1970, 1971)
Strawberries	*Rhizoctonia fragariae*	Endopolygalacturonases	Cervone *et al.* (1977, 1978)
Tomatoes	*Botrytis cinerea*	Pectic and other enzymes	Verhoeff and Warren (1972)

[a] This fungus causes "Edelfäule" or noble rot.

and cell death with any single enzyme characterized from *Sclerotinia fructigena*. Possibly, in this case, a lytic factor or a factor that stimulates maceration by other enzymes was involved (Ishii, 1977).

Fresh fruits may be particularly rich in pectic enzymes, for instance pectinesterase (citrus fruits, tomato) and endopolygalacturonase (tomato, avocado). These enzymes will also play a role in the infection process, as was demonstrated by Barash and Khazzam (1970) in avocado. In that particular case, the fruit's own endopoly-galacturonase activity was substantially induced by infection with *Colletotrichum gloeosporioides*, thus rendering the fruit even more predisposed for further invasion by the pathogen. This increase in polygalacturonase activity is perhaps correlated with increased ethylene production which is often observed as a response of fruits and vegetables to fungal and bacterial infection (Lund, 1973).

Some natural defence mechanisms against microbial infection exist in fruits and vegetables. For instance, they contain inhibitors of fungal polygalacturonase (see Section V, p. 261). Also, in many fruits and vegetables, polyphenols and polyphenol oxidase occur. In healthy tissues, substrate and enzyme cannot interact. However, in rotting tissue, enzymic oxidation of polyphenols leads to the formation of tannins (black rots, brown rots) which are powerful inhibitors of enzymes, including pectic enzymes of the invading parasite (Hathway and Seakins, 1958; Hunter, 1974, 1978).

Texture losses of canned or bottled fruit products may be caused by thermostable fungal endopolygalacturonases, which were produced before canning and processing (Harper, 1971; Harper *et al.*, 1973; Paynter, 1975; Luh *et al.*, 1978). It is also possible that the fungus itself survives the heat treatment, as is the case with *Byssochlamys* spp. which have heat-resistant ascospores (Put and Kruiswijk, 1968). Even at the low oxygen tensions and high concentrations of sugar and acid in canned fruits, these fungi may grow, produce pectic enzymes (and patulin, a mycotoxin) and cause softening, or even complete disinte-gration of fruit tissues (Put and Kruiswijk, 1968; Rice, 1977; Rice and Beuchat, 1978).

Certain spoilage phenomena in processed and fermented olives are attributable to pectolytic and cellulolytic micro-organisms. The processing of ripe olives includes lye-treatments and oxidation steps to produce the dark colour. During subsequent washing to leach out the alkali, a softening may occur. This softening, commonly called 'sloughing', is caused by a variety of pectolytic Gram-negative bacteria, including species of *Aeromonas, Achromobacter, Escherichia* and *Enterobacter* (Vaughn *et al.*, 1969b) as well as by certain cellulolytic bacteria (Patel and Vaughn, 1973). Softening in fermenting and fermented olives is ascribed to certain yeasts (*Rhodotorula* spp.) developing in a film on the surface of the brine (Vaughn *et al.*, 1969a). Particularly serious problems of softening and gas-pocket formation may be caused by various pectolytic yeasts, when low concentrations of salt are used in the brine (Vaughn *et al.*, 1972). Softening of brined and fermented cucumbers is probably attributable to pectolytic fungi which are introduced into curing brines chiefly by way of the fungus-laden flowers that remain attached to the cucumbers (Etchells *et al.*, 1958) and less to pectolytic yeasts known to occur in pickle brines (Bell and Etchells, 1956). However, re-use of pickle brines is possible if pectic enzymes are inactivated by an adequate heat treatment (Chavana and McFeeters, 1977).

Microbial pectolysis also occurs in fermenting cacao and coffee beans. During fermentation, the mucilage layer surrounding these beans is decomposed. Washing of these beans before drying is thereby greatly facilitated. Micro-organisms involved in maceration of the juicy tissue of cacao beans are various yeasts, including species of *Candida, Pichia, Saccharomyces* and *Zygosaccharomyces* (Roelofsen, 1953). The yeasts involved in coffee fermentation are species of *Saccharomyces, Schizosaccharomyces* and *Candida* (Agate and Bhat, 1966; Van Pee and Castelein, 1971). Pectolytic representatives of the Enterobacteriaceae, notably *Erwinia dissolvens*, were also isolated from fermenting coffee (Frank and Dela Cruz, 1964; Van Pee and Castelein, 1972).

C. Flax Retting and Water Logging

Production of textile fibres from flax, hemp and jute is an old but ailing branch of agricultural industry. It involves a retting process, which is based on the ability of certain pectolytic micro-organisms to liberate cellulose fibres from stems of fibre plants. The process is carried out either aerobically or anaerobically. The micro-organisms involved in the aerobic 'dew retting' of flax are *Cladosporium herbarium, Pullularia pullulans*, and species of *Rhodotorula, Cryptococcus* and *Pseudomonas* (Wieringa, 1956). *Bacillus polymyxa* and *Pseudomonas* sp. are important in the aerobic retting process of jute and hemp (Ali, 1958; Betrabet and Bhat, 1958). In the anaerobic flax-retting process, usually carried out in concrete tanks, various *Clostridium* spp. are involved, *Cl. felsineum* being the main retting agent (Avrova, 1977). Active strains of this organism are being selected for production of pectolytin, a preparation of *Cl. felsineum* spores to be used for accelerating the flax-retting process (Avrova, 1978).

Another long-standing practice is the water-storage of soft wood. The desirable effect is to render the sapwood more permeable for preservatives. An increase in permeability of the wood is associated with bacterial destruction of pit membranes (Fogarty *et al.*, 1974). It appears that, during water storage, pectolytic bacteria (*Bacillus subtilis*, *Cytophaga johnsonii*) effect this increase in permeability of the wood (Ward and Fogarty, 1973; Kurowski and Dunleavy, 1976a, b).

D. Pectolytic Micro-organisms in Soils and on Plants

Soil is a large reservoir of pectolytic micro-organisms. Wieringa (1949) counted up to 3 millions of these (mostly actinomycetes) per g of soil. They contribute to the decomposition and mineralization of plant deposits. In his extensive studies on pectolytic micro-organisms from soils, Kaiser (1961) surveyed plant deposits, arable land, garden soils, geologically different types of soil and mud from ponds and lakes, in France and Zaïre, as well as the rhizosphere of various plants. Three large groups of pectolytic microbes are encountered in French soils, namely fungi, actinomycetes and (other) bacteria. Most of the actinomycetes belong to the genus *Streptomyces*. Important groups of other bacteria are species of *Pseudomonas*, *Bacillus* and *Clostridium*. In the volcanic lakes of Zaïre there is a preponderance of *Clostridium*, *Bacillus*, *Pseudomonas*, *Achromobacter* and *Acinetobacter* species. The rhizosphere is relatively rich in numbers of aerobic pectolytic micro-organisms.

Most of the soil isolates of *Arthrobacter* spp. are pectolytic (Rombouts and Pilnik, 1971). Other groups of pectolytic micro-organisms encountered in soils are members of the Myxobacteriales and species of *Flavobacterium* (Dorey, 1959; Kaiser, 1961), *Micromonospora*, *Microbispora*, *Actinoplanes* and *Streptosporangium* (all actinomycetes; Kaiser, 1971), *Azotobacter* (Monzon de Asconegui and Kaiser, 1972) and *Rhizobium* (Hubbell *et al.*, 1978). The phyllosphere of healthy plants harbours a variety of pectolytic bacteria, mainly Gram-negative, such as species of *Pseudomonas* (Rombouts, 1972) and *Flavobacterium* (Lund, 1969).

E. Pectolytic Micro-Organisms in Other Habitats

A wide array of pectin-fermenting bacteria and protozoa occur in the rumen. These micro-organisms, living under the strictly anaerobic conditions that prevail in the rumen, are largely different from those referred to so far. The following pectolytic bacteria have been isolated: *Bacteroides ruminicola*, *Bacteroides succinogenes*, *Butyrivibrio fibrisolvens*, *Lachnospira multiparus*, *Peptostreptococcus* sp., *Ruminococcus flavefaciens* (Dehority, 1969; Gradel and Dehority, 1972) and recently also large *Treponema* species (Ziołecki, 1979). Many of these, although fermenting pectin vigorously, do not utilize D-galacturonic acid which, apparently, is not an important intermediate in pectin fermentation by these bacteria (Gradel and Dehority, 1972). *Ruminococcus flavefaciens* produces a number of polysaccharide-degrading enzymes, including pectin-esterase and pectin lyase (possibly pectate lyase), the latter enzyme being primarily cell-associated during exponential growth (Pettipher and Latham, 1979a, b). The *Treponema* spp. are outspoken pectin fermenters, accumulating acetate, formate and some succinate as fermentation products. Their pectic enzymes are pectate lyase (optimum pH 8.0 to 9.0), polygalacturonase and pectinesterase (Wojciechowicz and Ziołecki, 1979).

Pectin-fermenting rumen protozoa are *Dasytrychum ruminantium*, *Isotricha intestinalis*, *Isotricha prostoma* and *Ophryoscolex purkinei* (Abou Akkada and Howard, 1961; Mah and Hungate, 1965; Prins and Van Hoven, 1977; Van Hoven and Prins, 1977). All of these produce pectinesterase and pectate lyase.

Pectin is not attacked by the digestive juices in the intestinal tract of higher animals and humans (Chenoweth and Leveille, 1975). It is passed on to the colon, where it may be partially degraded by a number of pectolytic bacteria (Werch *et al.*, 1942).

VI. INDUSTRIAL PRODUCTION AND UTILIZATION OF PECTIC ENZYMES

A. Selection of Micro-Organisms

A first step in the development and improvement of an industrial production process for pectic enzymes is the search for productive strains. Screening for production of specific pectic enzymes may be done with the 'cup-plate' technique, originally described for pectic enzymes and other polysaccharide-degrading enzymes by Dingle *et al.* (1953). An attractive alternative is to rely on simple performance tests, for instance clarification of apple juice in a test tube, a filter test with homogenized fruit tissue, or a maceration test with potato discs (Endo and Miura, 1961; Arima *et al.*, 1964).

A serious restriction in the search for productive strains stems from the fact that the enzymes of only a few micro-organisms have the stamp of approval for use in the food industry. That is probably the main reason why commercial 'pectinases' (the name commonly used for industrial pectic enzyme preparations) are all derived from *Aspergillus* species, mainly from *A. niger*. The pectic enzyme preparations from this fungus (containing polygalacturonases, pectinesterases and pectin lyases) generally perform well in fruit- and vegetable-juice technology, the main field of application of pectinases (Rombouts and Pilnik, 1978). Nevertheless, the economy of the enzyme-production process may be improved through selection of mutants that are more productive (Zetelaki-Horváth and Dobra-Seres, 1972), resist catabolite repression, or form large amounts of enzyme without an inducer (Faith *et al.*, 1971).

Bacterial pectinases are not produced industrially, although certain bacteria, such as *B. subtilis* (Chesson and Codner, 1978) and *Ps. fluorescens* (Rombouts *et al.*, 1978), synthesize large amounts of pectate lyase constitutively. This enzyme should have a potential for maceration of chopped vegetables, and for application in manufacture of potato puree.

B. Pectinase Production

Microbial enzymes are produced industrially essentially by three different methods, namely the surface-bran culture method, the deep-tank (submerged) process, and the two-stage submerged process. Because of its easier control, the deep-tank method is more widely used now (Faith *et al.*, 1971), but it is our impression that most pectinases are still produced by the surface method, carried out in rotating drums. This may be for reasons of enzyme yield, or for obtaining the desired composition of the enzyme mixture (Charley, 1969; Meyrath and Volavsek, 1975).

A crucial factor in pectinase production is the composition of the medium. Laboratory experiments have shown that very high yields of enzyme are obtained in media with well-balanced concentrations of mixed carbon sources, such as sucrose and pectin, lactose and pectin or glucose and pectin, the enzymes being produced at a stage when the sugar concentration in the medium drops rapidly (Tuttobello and Mill, 1961; Nyeste and Holló, 1963; Vasu, 1967; Feniksova and Moldabaeva, 1967). In addition to its inducing effect, the presence of pectin in the production medium favours release of pectic enzymes from the mycelium (Vasu, 1967; Nyiri, 1968). However, for reasons of economy, pectin is not used very much in production media. It is substituted by dried sugar-beet cossettes (Zetelaki, 1976; Szajer, 1978), citrus peel, or apple pomace which, together with wheat bran, are the main carbon and energy sources.

Control of pH value is another important factor in pectinase production. Greatest enzyme production is achieved when the pH value drops from an initial value of about 4.5 to a more or less constant value of 3.5 in the course of the fermentation, which usually takes three to six days (Nyiri, 1968; Zetelaki-Horváth, 1978). A marked inactivation of enzymes occurs at extremes of pH value (below pH 3 and above pH 7; Schröder and Müller-Stoll, 1962).

With the use of complex media, a variety of enzymes other than pectic enzymes are induced. These include hemicellulases, cellulases, glycosidases, proteinases, esterases and oxidoreductases. They are normally present in commercial pectinases, and some of them may adversly affect the flavour of the fruit products in which they are applied (Jakob et al., 1973; Burkhardt, 1976, 1978; Schreier et al., 1978).

At the end of the fermentation, the enzymes are extracted from the semisolid medium and mycelium, the dilute enzyme solution is concentrated and the enzymes are then precipitated with organic solvents or inorganic salts. The dried precipitate is milled, mixed with stabilizers and inert ingredients (gelatin, glucose, diatomaceous earth) for standardization, packaged and stored until delivery. Some of the products are sold as liquid concentrates. As these steps are not essentially different from those in the preparation of other microbial extracellular enzymes, we feel no need for further description and refer to Nyiri (1969), Faith et al. (1971), Melling and Phillips (1975) and other articles in this volume.

Table 8

Pectinase manufactures

Manufacturer	Location	Brand name
C. H. Boehringer Sohn	Ingelheim, Federal Republic of West Germany	Panzym products
G. B. Fermentation Ind. Inc.	Kingstree, South Carolina, U.S.A.	Klerzyme
Grindstedvaerket A/S	Brabrand, Denmark	Pektolase products
Kikkoman Shoyu Co., Ltd.	Tokyo, Japan	Pectolyase
Miles Laboratories, Inc.	Elkhart, Indiana, U.S.A.	Spark-L
Miles Kali-Chemie GmbH	Hannover-Kleefeld, Federal Republic of West Germany	MKC-pectinases
Novo Industri A/S	Laufen, Switzerland	Ultrazym, Irgazym
Röhm GmbH	Darmstadt, Federal Republic of West Germany	Pectinol products, Rohament P
Rohm & Haas Co.	Philadelphia, Pennsylvania, U.S.A.	Pectinol products
Société Rapidase	Séclin N, France	Rapidase products, Clarizyme
Swiss Ferment Co. Ltd.	Basle, Switzerland	Pectinex products

So far, pectinases are not sold as immobilized products, but performance studies have been done with an immobilized multipectic enzyme system (Young, 1976) and with an immobilized pectin lyase (Hanisch et al., 1978). With food regulations becoming ever more restrictive, there is a need for further purification of pectinases. Therefore, it seems worthwhile to study the possibilities of purification of pectinases by affinity chromatography, for instance on cross-linked pectate and derivatives (Rexová-Benková and Tibenský, 1972; Vijayalakshmi et al., 1978; Rombouts et al., 1979).

It is estimated that the World-wide food-enzyme production represents a value of about U.S.$45 million, of which perhaps one-quarter, or less, relates to pectinases, manufactured by various companies in Europe, U.S.A. and Japan (Table 8).

C. Application of Pectinases

1. Fruit-Juice Extraction

Fruit juices are generally extracted by pressing, but this is by no means easy with soft fruits, like blackcurrants, strawberries and raspberries.

Partial destruction of pectin, through enzymic treatment of the pulp ('Maischefermentierung'), facilitates pressing and ensures high yields of juice and anthocyanin pigments (Charley, 1969; Delecourt, 1972). The same is true for the Concord variety of *Vitis labrusca*, which, in U.S.A., is used for extraction of red grape juice (Neubeck, 1975). In France, a heat treatment (55–77°C) of crushed red grapes ('thermovinification') is recommended for rapid solubilization of anthocyanins. This process is followed by enzymic extraction of juice, pressing and fermentation of the clear juice, an attractive alternative to the traditional fermentation on the skins (Terrier and Blouin, 1974; Cordonnier and Marteau, 1976).

In the European apple-juice industry, the labour situation necessitated introduction of mechanized and automated pressing equipment at a time when the pressing quality of the apples decreased because of the variety available (Golden Delicious) and because of prolonged storage. In this situation, enzymic treatment of pulp became a necessity and was shown also to increase the juice yield from apples of good pressing quality (Pilnik and De Vos, 1970; Bielig *et al.*, 1971; De Vos and Pilnik, 1973). Enzyme preparations capable of rapidly degrading the highly esterified apple pectin are used most successfully (Voordouw *et al.*, 1974).

Inhibition of pectic enzymes by polyphenols from fruits presents a problem which, however, for apples was overcome by addition of insoluble polyvinylpyrrolidone, or by aeration, leading to enzymic oxidation and subsequent condensation of polyphenols to insoluble dark pigments (De Vos and Pilnik, 1973). Juices obtained from pulps treated enzymically have a higher methanol content than those obtained only by mechanical pressing. It is difficult to say whether this constitutes a health hazard, as little is known about the subacute toxicity of methanol. At any rate, methanol may be lost in the vapour during concentration, and aroma stripping may be done before enzyme treatment (Kwasniewski, 1975).

2. Fruit-Juice Clarification

This is the oldest and still largest use of pectinases, applied mainly to apple juice and further to pear and grape juices. Freshly-pressed juices are rather viscous and have a persistent turbidity. Addition of pectinase results in a rapid drop in viscosity and a 'break' of the turbidity; cloudy particles agglomerate, forming flocs which settle out. A clear juice is then obtained either by filtration or by centrifugation. In the case of apple juice, the drop in viscosity may be caused by the combined activity of pectinesterase and endopolygalacturonase on the highly esterified pectin in solution. The enzymes also solubilize part of the negatively charged pectin coating of the cloud particles, to an extent that sites of the positively charged protein nucleus are exposed. The destabilized cloud particles then coagulate and precipitate through electrostatic interaction (Endo, 1965; Yamasaki et al., 1967). Apple-juice clarification is possible also with pectin lyase (Ishii and Yokotsuka, 1972c) but this enzyme is less effective in clarification of grape juice which contains pectin with a degree of esterification of only 45 to 60% (Ishii and Yokotsuka, 1973). Clarification and complete depectinization are prerequisites for juices that are to be concentrated. It can be done at raised temperatures (45–50°C), saving both enzyme and time, and avoiding problems of infection and unwanted fermentation (Grampp, 1977).

3. Maceration and Liquefaction

In enzymic maceration, suspensions of loose cells are produced from fruit and vegetable tissues (Grampp, 1969). The process may be applied in the manufacture of fruit nectars (comminuted fruit juices). These are pulpy drinks prepared from a variety of fruits (apples, pears, peaches, apricots, berries, guava, papaya and passion fruit) by grinding the fruits, adding water, acids and sugar, homogenization and pasteurization. Especially when firm-flesh fruits, like apples , are used, the cloud stability of the product may be increased by using a pectinase in conjunction with the mechanical dispersion process (Strübi et al., 1975). Also, enzymes can be used to lower the viscosity of the products, making it possible to prepare concentrated nectar bases (Šulc and Ciric, 1968). The process is further recommended for producing finely dispersed constituents for baby foods from carrots and other vegetables (Grampp, 1972).

Maceration can be achieved by a pectinase (Rohament P®) relatively rich in endopolygalacturonase and almost free from pectinesterase and pectin lyase. It is claimed that this preparation selectively degrades middle-lamella pectin, which is believed to have a low degree of esterification (Grampp, 1969, 1972). However, maceration can also be done with a pure pectin lyase (Ishii and Yokotsuka, 1975).

Enzymic treatment of pulp can be carried through to effect almost complete liquefaction of fruits and vegetables, if cellulase is used in conjunction with pectic enzymes (Pilnik et al., 1975). In this process, cell walls are largely digested. A slightly viscous, clear or turbid suspension of enzyme-resistant tissue, skin particles and pits is obtained, which can be centrifuged, or filtered to give juice in 90 to 98% yield. The process may be suitable for products for which no juice-extraction equipment exists (tropical fruits).

4. Application in Citrus Technology

In the citrus industry, pectinases are used mainly for valorization of byproducts. The course pulp that is sieved out of freshly-pressed citrus juice contains considerable concentrations of juice and soluble solids. These are recovered in a counter-current washing operation, in which pectinases are used to obtain so-called pulp wash with maximum concentrations of soluble solids and low viscosity, so that it may be concentrated (Braddock and Kesterson, 1976). Another important application is with the peels and rags left after juice extraction. These products contain large amounts of substances capable of providing desirable cloudiness to beverages, products for which a great demand exists in the soft-drink industry. Strongly turbid, low-viscosity cloudy preparations, suitable for concentration, may be obtained by thermal–mechanical comminution, followed by a pectinase treatment (Larsen, 1969). Finally, pectinases are also useful in isolation of essential oils and carotenoid pigments from citrus peels (Platt and Poston, 1962).

Pectinases have been used in fruit and vegetable technology for almost 50 years. Their application will continue to expand, as they are an indispensible tool in the further development of this branch of technology, with its trends towards the use of a greater variety of raw materials, more complete utilization of raw materials, speeding-up of processes and production of an increased variety of bases and finished products.

REFERENCES

Abdel-Fattah, A. F. and Mabrouk, S. S. (1977). *Chemie Mikrobiologie und Technologie der Lebensmittel* 5, 38.

Abou Akkada, A. R. and Howard, B. H. (1961). *Biochemical Journal* **78**, 512.

Agarwal, G. P. and Gupta, S. (1978). *Current Science (India)* **47**, 161.

Agate, A. D. and Bhat, J. V. (1966). *Applied Microbiology* **14**, 256.

Albersheim, P. and Anderson, A. J. (1971). *Proceedings of the National Academy of Sciences of the United States of America* **68**, 1815.

Albersheim, P. and Anderson-Prouty, A. J. (1975). *Annual Review of Plant Physiology* **26**, 31.

Albersheim, P. and Killias, U. (1962). *Archives of Biochemistry and Biophysics* **97**, 107.

Albersheim, P., Neukom, H. and Deuel, H. (1960). *Archives of Biochemistry and Biophysics* **90**, 46.

Ali, M. M. (1958). *Applied Microbiology* **6**, 87.

Almengor-Hecht, M. L. and Bull, A. T. (1978). *Archives of Microbiology* **119**, 163.

Amadò, R. (1970). Dissertation No. 4536: Eidgenoessische Technische Hochschule, Zürich.

Anderson, A. J. and Albersheim, P. (1972). *Physiological Plant Pathology* **2**, 339.

Arima, K., Yamasaki, M. and Yasui, T. (1964). *Agricultural and Biological Chemistry* **28**, 248.

Ashwell, G. (1962). *Methods in Enzymology* **5**, 190.

Aspinall, G. (1970). 'Polysaccharides', p. 116. Pergamon Press, Oxford.

Atallah, M. T. and Nagel, C. W. (1977). *Journal of Food Biochemistry* **1**, 185.

Avrova, N. P. (1977). *Mikrobiologiya* **46**, 346.

Avrova, N. P. (1978). *Applied Biochemistry and Microbiology* **13**, 547.

Barash, I. (1968). *Phytopathology* **58**, 1364.

Barash, I. and Eyal, Z. (1970). *Phytopathology* **60**, 27.

Barash, I. and Khazzam, S. (1970). *Phytochemistry* **9**, 1189.

Baron, A., Calvez, J. and Drilleau, J.-F. (1978). *Annales de Falsifications et de l'Expertise Chimique* **71**, 29.

Bartley, I. M. (1978). *Phytochemistry* **17**, 213.

Bartolome, L. G. and Hoff, J. E. (1972). *Journal of Agricultural and Food Chemistry* **20**, 262.

Basham, H. (1974). Ph.D. Thesis: Cornell University.

Basham, H. G. and Bateman, D. F. (1975a). *Physiological Plant Pathology* **5**, 249.

Basham, H. G. and Bateman, D. F. (1975b). *Phytopathology* **65**, 141.

Bateman, D. F. (1966). *Phytopathology* **56**, 238.

Bateman, D. F. and Millar, R. L. (1966). *Annual Review of Phytopathology* **4**, 119.

Bateman, D. F., Kosuge, T. and Kilgore, W. W. (1970). *Archives of Biochemistry and Biophysics* **136**, 97.

Bell, T. A. and Etchells, J. L. (1956). *Applied Microbiology* **4**, 196.

Betrabet, S. M. and Bhat, J. V. (1958). *Applied Microbiology* **6**, 89.

Bielig, H. J., Wolff, J. and Balcke, K. J. (1971). *Flüssiges Obst* **38**, 408.

Bilimoria, M. H. and Bhat, J. V. (1962). *Journal of the Indian Institute of Science* **44**, 15.

Binet, P. (1976). *Physiologie Végétale* **14**, 283.

Blumenthal, H. J. and Fish, D. C. (1963). *Biochemical and Biophysical Research Communications* **11**, 239.

Bock, W., Krause, M. and Dongowski, G. (1970). *Die Nahrung* **14**, 375.

Bock, W., Dongowski, G. and Krause, M. (1972). *Die Nahrung* **16**, 787.

Braddock, R. J. and Kesterson, J. W. (1976). *Journal of Food Science* **41**, 82.

Brady, C. J. (1976). *Australian Journal of Plant Physiology* **3**, 163.

Bryan, W. H. and Newcomb, E. H. (1954). *Physiologia Plantarum* **7**, 290.

Burkhardt, R. (1976). *Deutsche Lebensmittel-Rundschau* **72**, 417.

Burkhardt, R. (1978). *Deutsche Lebensmittel-Rundschau* **74**, 205.

Bush, D. A. and Codner, R. C. (1968). *Phytochemistry* **7**, 863.

Bush, D. A. and Codner, R. C. (1970). *Phytochemistry* **9**, 87.

Byrde, R. J. W. and Fielding, A. H. (1968). *Journal of General Microbiology* **52**, 287.

Byrde, R. J. W., Fielding, A. H. and Archer, S. A. (1973). *In* 'Fungal Pathogenicity and the Plant's Response' (R. J. W. Byrde and C. V. Cutting, eds.), p. 39. Academic Press, London.

Call, H. P. and Emeis, C. C. (1978a). *Chemie, Mikrobiologie und Technologie der Lebensmittel* **5**, 137.

Call, H. P. and Emeis, C. C. (1978b). *Chemie, Mikrobiologie und Technologie der Lebensmittel* **5**, 143.

Calonge, F. D., Fielding, A. H. and Byrde, R. J. W. (1969). *Journal of General Microbiology* **55**, 177.

Castelein, J. M. and Pilnik, W. (1976). *Lebensmittel-Wissenschaft und Technologie* **9**, 277.

Cervone, F., Scala, A., Foresti, M., Cacace, M. G. and Noviello, C. (1977). *Biochimica et Biophysica Acta* **482**, 379.

Cervone, F., Scala, A. and Scala, F. (1978). *Physiological Plant Pathology* **12**, 19.

Chang, Y. F. and Feingold, D. S. (1970). *Journal of Bacteriology* **102**, 85.

Charley, V. L. S. (1969). *Chemistry and Industry*, 635.

Chatterjee, A. K. and Starr, M. P. (1977). *Journal of Bacteriology* **132**, 862.

Chavana, S. and McFeeters, R. F. (1977). *Lebensmittel-Wissenschaft und Technologie* **10**, 290.

Chenoweth, W. L. and Leveille, G. A. (1975). *In* 'Physiological Effects of Food Carbohydrates' (A. Jeanes and J. Hodge, eds.), p. 312. American Chemical Society, Washington, D. C.

Chesson, A. and Codner, R. C. (1978). *Journal of Applied Bacteriology* **44**, 347.

Cooke, R. D., Ferber, E. M. and Kanagasabapathy, L. (1976). *Biochimica et Biophysica Acta* **452**, 440.

Cooper, R. M. (1977). *In* 'Cell Wall Biochemistry Related to Specific Host Plant–Pathogen Interactions' (B. Solheim and J. Raa, eds.), p. 163. Universitetsforlaget, Oslo.

Cooper, R. M. and Wood, R. K. S. (1973). *Nature, London* **246**, 309.

Cooper, R. M. and Wood, R. K. S. (1975). *Physiological Plant Pathology* **5**, 135.

Cordonnier, R. and Marteau, G. (1976). *Bulletin de l'Office International du Vin* **49**, 490.

Courtois, J.-E., Percheron, F. and Foglietti, M.-J. (1968). *Comptes Rendus de l'Académie des Sciences, Paris, Série D* **266**, 164.

Davé, B. A. and Vaughn, R. H. (1971). *Journal of Bacteriology* **108**, 166.

Davé, B. A., Vaughn, R. H. and Patel, I. B. 1976). *Journal of Chromatography* **116**, 395.

Dean, M. and Wood, R. K. S. (1967). *Nature, London* **214**, 408.

Dehority, B. A. (1969). *Journal of Bacteriology* **99**, 189.

Delecourt, R. (1972). *Bulletin de l'Association des Anciens Elèves de l'Institut des Industries de Fermentation de Bruxelles* **15**, 82.

Delincée, H. (1976). *Phytochemistry* **15**, 903.

Delincée, H. and Radola, B. J. (1970). *Biochimica et Biophysica Acta* **214**, 178.

Deverall, B. J. (1977). 'Defence Mechanisms of Plants'. Cambridge University Press, Cambridge.

De Vos, L. and Pilnik, W. (1973). *Process Biochemistry* **8** (8), 18.

Dingle, J., Reid, W. W. and Solomons, G. L. (1953). *Journal of the Science of Food and Agriculture* **4**, 149.

Doesburg, J. J. (1965). Pectic substances in fresh and preserved fruits and vegetables. IBVT Communication No. 25. Institute for Research on Storage and Processing of Horticultural Produce, Wageningen, The Netherlands.

Dorey, M. J. (1959). *Journal of General Microbiology* **20**, 91.

Drawert, F. and Krefft, M. (1978a). *Lebensmittel-Wissenschaft und Technologie* **11**, 79.

Drawert, F. and Krefft, M. (1978b). *Phytochemistry* **17**, 887.

Edstrom, R. D. and Phaff, H. J. (1964a). *Journal of Biological Chemistry* **239**, 2403.

Edstrom, R. D. and Phaff, H. J. (1964b). *Journal of Biological Chemistry* **239**, 2409.

Endo, A. (1964a). *Agricultural and Biological Chemistry* **28**, 757.

Endo, A. (1964b). *Agricultural and Biological Chemistry* **28**, 535.

Endo, A. (1964c). *Agricultural and Biological Chemistry* **28**, 543.

Endo, A. (1964d). *Agricultural and Biological Chemistry* **28**, 551.

Endo, A. (1965). *Agricultural and Biological Chemistry* **29**, 229.

Endo, A. and Miura, Y. (1961). *Agricultural and Biological Chemistry* **25**, 382.

English, P. D., Jurale, J. B. and Albersheim, P. (1971). *Plant Physiology* **47**, 1.

English, P. D., Maglothin, A., Keegstra, K. and Albersheim, P. (1972). *Plant Physiology* **49**, 293.

Enzyme Nomenclature (1973). International Union of Pure and Applied Chemistry–International Union of Biochemistry Commission on Enzyme Nomenclature. *In* 'Comprehensive Biochemistry' (M. Florkin and E. H. Stotz, eds.), vol. 3, 3rd edition. Elsevier Publishing Company, Amsterdam.

Etchells, J. L., Bell, T. A., Monroe, R. J., Masley, P. M. and Demain, A. L. (1958). *Applied Microbiology* **6**, 427.

Evans, R. and McHale, D. (1978). *Phytochemistry* **17**, 1073.

Faith, W. T., Neubeck, C. E. and Reese, E. T. (1971). *In* 'Advances in Biochemical Engineering' (T. K. Ghose and A. Fiechter, eds.), p. 77. Springer–Verlag, New York.

Fanelli, C., Cacace, M. G. and Cervone, F. (1978). *Journal of General Microbiology* **104**, 305.

Farmer III, J. J. and Eagon, R. G. (1969). *Journal of Bacteriology* **97**, 97.

Feniksova, R. V. and Moldabaeva, R. K. (1967). *Applied Microbiology and Biochemistry* **3**, 283.

Fielding, A. H. and Byrde, R. J. W. (1969). *Journal of General Microbiology* **58**, 73.

Fisher, M. L., Anderson, A. J. and Albersheim, P. (1973). *Plant Physiology* **51**, 489.

Fogarty, W. M. and Griffin, P. J. (1973). *Biochemical Society Transactions* **1**, 263.

Fogarty, W. M. and Ward, O. P. (1974). *In* 'Progress in Industrial Microbiology (D. J. D. Hockenhull, ed.), vol. 13, p. 59. Churchill Livingstone, Edinburgh.

Fogarty, W. M., Griffin, P. J. and Joyce, A. M. (1974). *Process Biochemistry* **9** (7), 11.

Foglietti, M.-J., Levaditou, V., Courtois, J.-E. and Chararas, C. (1971). *Comptes Rendus de la Société Biologique* **165**, 1019.

Frank, A. H. and Dela Cruz, A. S. (1964). *Journal of Food Science* **29**, 850.

Fuchs, A. (1965). *Antonie van Leeuwenhoek* **31**, 323.

Gardner, J. M. and Kado, C. I. (1976). *Journal of Bacteriology* **127**, 451.

Garibaldi, A. and Bateman, D. F. (1971). *Physiological Plant Pathology* **1**, 25.

Glasziou, K. T. and Inglis, S. D. (1958). *Australian Journal of Biological Sciences* **11**, 127.

Glenn, A. R. (1976). *Annual Review of Microbiology* **30**, 41.

Goodenough, P. W. and Kempton, R. J. (1974). *Phytopathologische Zeitschrift* **81**, 78.

Goren, R., Riov, J. and Greenberg, J. (1976). *Acta Universita Nicolai Copernici* **37**, 215.

Gradel, C. M. and Dehority, B. A. (1972). *Applied Microbiology* **23**, 332.

Grampp, E. (1969). *Deutsche Lebensmittel Rundschau* **65**, 343.

Grampp, E. (1972). *Dechema Monographien* **70**, 178.

Grampp, E. (1977). *Food Technology* **31** (11), 38.

Greenberg, J., Goren, R. and Birk, Y. (1975). *Physiologia Plantarum* **34**, 1.

Hall, J. A. and Wood, R. K. S. (1973). *In* 'Fungal Pathogenicity and the Plant's Response' (R. J. W. Byrde and C. V. Cutting, eds.), p. 19. Academic Press, London.

Hancock, J. G. (1968). *Phytopathology* **58**, 62.

Hancock, J. G. (1976). *Phytopathology* **66**, 40.

Hancock, J. G., Eldridge, C. and Alexander, M. (1970). *Canadian Journal of Microbiology* **16**, 69.

Hanisch, W. H., Rickard, A. D. and Nyo, S. (1978). *Biotechnology and Bioengineering* **20**, 95.

Harper, K. A. (1971). *Chemistry and Industry*, 462.

Harper, K. A., Beattie, B. B. and Best, D. J. (1973). *Journal of the Science of Food and Agriculture* **24**, 527.

Hasegawa, S. and Nagel, C. W. (1966). *Journal of Food Science* **31**, 838.

Hasegawa, S. and Nagel, C. W. (1968). *Archives of Biochemistry and Biophysics* **124**, 513.

Hasui, Y., Miyairi, K., Okuno, T., Sawai, K. and Sawamura, K. (1976). *Annales of the Phytopathological Society of Japan* **42**, 228.

Hatanaka, C. and Imamura, T. (1974). *Agricultural and Biological Chemistry* **38**, 2267.

Hatanaka, C. and Ozawa, J. (1964). *Agricultural and Biological Chemistry* **28**, 627.

Hatanaka, C. and Ozawa, J. (1969a). *Berichte des Ohara Instituts für Landwirtschaftliche Biologie* **15**, 15.

Hatanaka, C. and Ozawa, J. (1969b). *Berichte des Ohara Instituts für Landwirtschaftliche Biologie* **15**, 25.

Hatanaka, C. and Ozawa, J. (1970). *Agricultural and Biological Chemistry* **34**, 1618.

Hatanaka, C. and Ozawa, J. (1971a). *Agricultural and Biological Chemistry* **35**, 1617.

Hatanaka, C. and Ozawa, J. (1971b). *Berichte des Ohara Instituts für Landwirtschaftliche Biologie* **15**, 47.

Hatanaka, C. and Ozawa, J. (1972). *Agricultural and Biological Chemistry* **36**, 2307.

Hatanaka, C. and Ozawa, J. (1973). *Agricultural and Biological Chemistry* **37**, 593.

Hathway, D. E. and Seakins, J. W. T. (1958). *Biochemical Journal* **70**, 158.

Heinrichová, K. (1977). *Collection of Czechoslovak Chemical Communications* **42**, 3214.

Heinrichová, K. and Rexová-Benková, L'. (1976). *Biochimica et Biophysica Acta* **422**, 349.

Heinrichová, K. and Rexová-Benková, L'. (1977). *Collection of Czechoslovak Chemical Communications* **42**, 2569.

Heri, W., Neukom, H. and Deuel, H. (1961). *Helvetica Chimica Acta* **44**, 1945.

Hildebrand, D. C. (1972). *In* 'Proceedings of the Third International Congress on Plant Pathogenic Bacteria' (H. P. Maas Geesteranus, ed.), p. 331. Centre for Agricultural Publishing and Documentation, Wageningen, The Netherlands.

Hobson, G. E. (1963). *Biochemical Journal* **86**, 358.

Hobson, G. E. (1964). *Biochemical Journal* **92**, 324.

Hsu, E. J. and Vaughn, R. H. (1969). *Journal of Bacteriology* **98**, 172.

Hubbard, J. P. (1978). PhD. Thesis: University of Massachusetts.

Hubbard, J. P., Williams, J. D., Niles, R. M. and Mount, M. S. (1978). *Phytopathology* **68**, 95.

Hubbell, D. H., Morales, V. M. and Umali-Carcia, M. (1978). *Applied and Environmental Microbiology* **35**, 210.

Hultin, H. O. and Levine, A. S. (1963). *Achives of Biochemistry and Biophysics* **101**, 396.

Hultin, H. O. and Levine, A. S. (1965). *Journal of Food Science* **30**, 917.

Hultin, H. O., Sun, B. and Bulger, J. (1966). *Journal of Food Science* **31**, 320.

Hunter, R. E. (1974). *Physiological Plant Pathology* **4**, 151.

Hunter, R. E. (1978). *Phytopathology* **68**, 1032.

Ishii, S. (1977). *Phytopathology* **67**, 994.

Ishii, S. (1978). *Plant Physiology* **62**, 586.

Ishii, S. and Kiho, K. (1976). *Phytopathology* **66**, 1077.

Ishii, S. and Yokotsuka, T. (1972a). *Agricultural and Biological Chemistry* **36**, 146.

Ishii, S. and Yokotsuka, T. (1972b). *Agricultural and Biological Chemistry* **36**, 1885.

Ishii, S. and Yokotsuka, T. (1972c). *Journal of Agricultural and Food Chemistry* **20**, 787.

Ishii, S. and Yokotsuka, T. (1973). *Journal of Agricultural and Food Chemistry* **21**, 269.

Ishii, S. and Yokotsuka, T. (1975). *Agricultural and Biological Chemistry* **39**, 313.

Jakob, M., Hippler, R. and Lüthi, H. R. (1973). *Lebensmittel-Wissenschaft und Technologie* **6**, 138.

Jeffcoat, R. (1974). *Biochemical Journal* **139**, 477.

Jeffcoat, R., Hassall, H. and Dagley, S. (1969a). *Biochemical Journal* **115**, 969.

Jeffcoat, R., Hassall, H. and Dagley, S. (1969b). *Biochemical Journal* **115**, 977.

Jones, T. M., Anderson, A. J. and Albersheim, P. (1972). *Physiological Plant Pathology* **2**, 153.

Joyce, A. M. and Fogarty, W. M. (1976). *Proceedings of the Society for General Microbiology* **3**, 126.

Kaiser, P. (1961). D.Sc. Thesis: Institut Pasteur, Paris.

Kaiser, P. (1971). *Annales de l'Institut Pasteur, Paris* **121**, 389.

Kamimiya, S., Izaki, K. and Takahashi, H. (1972). *Agricultural and Biological Chemistry* **36**, 2367.

Kamimiya, S., Nishiya, T., Izaki, K. and Takahashi, H. (1974). *Agricultural and Biological Chemistry* **38**, 1071.

Kamimiya, S., Itoh, Y., Izaki, K. and Takahashi, H. (1977). *Agricultural and Biological Chemistry* **41**, 975.

Karbassi, A. (1973). Ph.D. Thesis: University of California, Davis.

Kawabata, A. (1977). *Memoirs of the Tokyo University of Agriculture* **19**, 115.

Keegstra, K., Talmadge, K. W., Bauer, W. D. and Albersheim, P. (1973). *Plant Physiology* **51**, 188.

Keen, N. T. and Horton, J. C. (1966). *Canadian Journal of Microbiology* **12**, 443.

Kelly, C. T. and Fogarty, W. M. (1978). *Canadian Journal of Microbiology* **24**, 1164.

Kilgore, W. W. and Starr, M. P. (1959a). *Journal of Biological Chemistry* **234**, 2227.

Kilgore, W. W. and Starr, M. P. (1959b). *Nature, London* **183**, 1412.

Kimura, H. and Mizushima, S. (1973). *Agricultural and Biological Chemistry* **37**, 2589.

Kimura, H. and Mizushima, S. (1974). *Journal of General and Applied Microbiology, Tokyo* **20**, 33.

Kimura, H., Uchino, F. and Mizushima, S. (1973). *Agricultural and Biological Chemistry* **37**, 1209.

Knobel, H. R. and Neukom, H. (1974). *Phytopathologische Zeitschrift* **80**, 244.

Koch, O. and Schulz, F. A. (1974). *Phytopathologische Zeitschrift* **81**, 184.

Kohn, R., Furda, I. and Kopec, Z. (1968). *Collection of Czechoslovak Chemical Communications* **33**, 264.

Koller, A. (1966). Dissertation No. 3774. Eidgenoessische Technische Hochschule, Zürich.

Koller, A. and Neukom, H. (1969). *European Journal of Biochemistry* **7**, 485.

Kosaric, N. and Zajic, J. E. (1974). *In* Advances in Biochemical Engineering', (T. K. Ghose, A. Fiechter and N. Blakebrough, eds.), vol. 3, p. 89. Springer-Verlag, Berlin.

Kotomina, E. M. and Pisarnitskii (1974). *Applied Biochemistry and Microbiology* **10**, 623.

Kroh, M. and Loewus, M. W. (1978). *Phytochemistry* **17**, 797.

Kurowski, W. M. and Dunleavy, J. A. (1976a). *European Journal of Applied Microbiology* **2**, 103.

Kurowski, W. M. and Dunleavy, J. A. (1976b). *Journal of Applied Bacteriology* **41**, 119.

Kwasniewski, R. (1975). *Industries Alimentaires et Agricoles* **92**, 225.

Larsen, S. (1969). *International Fruit Juice Union, Technical Reports* **9**, 109.

Lee, M. and Macmillan, J. D. (1968). *Biochemistry, New York* **7**, 4005.

Lee, M. and Macmillan, J. D. (1970). *Biochemistry, New York* **9**, 1930.

Lee, M., Miller, L. and Macmillan, J. D. (1970). *Journal of Bacteriology* **103**, 595.

Lindeberg, G. and Lindeberg, M. (1977). *Archives of Microbiology* **115**, 9.

Lineweaver, H. and Ballou, G. A. (1945). *Archives of Biochemistry* **6**, 373.

Lisker, N., Katan, J. and Henis, Y. (1975). *Canadian Journal of Microbiology* **21**, 1298.

Liu, Y. K. and Luh, B. S. (1978). *Journal of Food Science* **43**, 721.

Lobanok, A. G., Mikhailova, R. V. and Sapunova, L. I. (1977). *Mikrobiologiya* **46**, 747.

Luh, B. S. and Phaff, H. J. (1951). *Archives of Biochemistry and Biophysics* **33**, 212.

Luh, B. S., Ozbilgin, S. and Liu, Y. K. (1978). *Journal of Food Science* **43**, 713.

Lund, B. M. (1969). *Journal of Applied Bacteriology* **32**, 60.

Lund, B. M. (1971). *Journal of Applied Bacteriology* **34**, 9.

Lund, B. M. (1973). *In* 'Fungal Pathogenicity and the Plant's Response' (R. J. W. Byrde and C. V. Cutting, eds.), p. 69. Academic Press, London.

Lund, B. M. and Brocklehurst, T. F. (1978). *Journal of General Microbiology* **104**, 59.

Lund, B. M. and Nicolls, J. C. (1970). *Potato Research* **13**, 210.

MacDonnell, L. R., Jansen, E. F. and Lineweaver, H. (1945). *Archives of Biochemistry* **6**, 389.

MacDonnell, L. R., Jang, R., Jansen, E. F. and Lineweaver, H. (1950). *Archives of Biochemistry and Biophysics* **28**, 260.

Macmillan, J. D. and Phaff, H. J. (1966). *Methods in Enzymology* **8**, 632.

Macmillan, J. D. and Sheiman, M. I. (1974). *In* 'Food Related Enzymes' (J. R. Whitaker, ed.), p. 101. American Chemical Society, Washington, D.C.

Macmillan, J. D. and Vaughn, R. H. (1964). *Biochemistry, New York* **3**, 564.

Macmillan, J. D., Phaff, H. J. and Vaughn, R. H. (1964). *Biochemistry, New York* **3**, 572.

Mah, R. A. and Hungate, R. E. (1965). *Journal of Protozoology* **12**, 131.

Manabe, M. (1973a). *Agricultural and Biological Chemistry* **37**, 1487.

Manabe, M. (1973b). *Journal of the Agricultural Chemical Society of Japan* **47**, 385.

Marković, O. (1974). *Collection of Czechoslovak Chemical Communications* **39**, 908.

Marković, O., Heinrichová, K. and Lenkey, B. (1975). *Collection of Czechoslovak Chemical Communications* **40**, 769.

McCready, R. M. and Seegmiller, C. G. (1954). *Archives of Biochemistry and Biophysics* **50**, 440.

McNicol, L. N. and Baker, E. E. (1970). *Biochemistry, New York* **9**, 1017.

Mehrotra, M. D., Kaiser, P. and Reynaud, C. (1971). *Annales de l'Institut Pasteur, Paris* **120**, 81.

Melling, J. and Phillips, B. W. (1975). *In* 'Handbook of Enzyme Biotechnology' (A. Wiseman, ed.), p. 58. John Wiley and Sons, Inc., New York.

Meyrath, J. and Volavsek, G. (1975). *In* 'Enzymes in Food Processing' (G. Reed, ed.), 2nd ed., p. 255. Academic Press, New York.

Mill, P. J. (1966a). *Biochemical Journal* **99**, 557.

Mill, P. J. (1966b). *Biochemical Journal* **99**, 562.

Miller, L. and Macmillan, J. D. (1970). *Journal of Bacteriology* **102**, 72.

Miller, L. and Macmillan, J. D. (1971). *Biochemistry, New York* **10**, 570.

Milner, Y. and Avigad, G. (1967). *Carbohydrate Research* **4**, 359.

Monzon de Asconegui, M. A. and Kaiser, P. (1972). *Annales de l'Institut Pasteur, Paris* **122**, 1009.

Moran, F. and Starr, M. P. (1969). *European Journal of Biochemistry* **11**, 291.

Moran, F., Nasuno, S. and Starr, M. P. (1968a). *Archives of Biochemistry and Biophysics* **123**, 298.

Moran, F., Nasuno, S. and Starr, M. P. (1968b). *Archives of Biochemistry and Biophysics* **125**, 734.

Mount, M. S,, Bateman, D. F. and Basham, H. G. (1970). *Phytopathology* **60**, 924.

Mussell, H. and Strand, L. L. (1977). *In* 'Cell Wall Biochemistry Related to Specific Host Plant-Pathogen Interactions' (B. Solheim and J. Raa, eds.), p. 31. Universitetsforlaget, Oslo.

Nagel, C. W. and Hasegawa, S. (1967a). *Analytical Biochemistry* **21**, 411.

Nagel, C. W. and Hasegawa, S. (1967b). *Archives of Biochemistry and Biophysics* **118**, 590.

Nagel, C. W. and Hasegawa, S. (1968). *Journal of Food Science* **33**, 378.

Nagel, C. W. and Vaughn, R. H. (1961). *Archives of Biochemistry and Biophysics* **93**, 344.

Nagel, C. W. and Vaughn, R. H. (1962). *Journal of Bacteriology* **83**, 1.

Nagel, C. W. and Wilson, T. M. (1970). *Applied Microbiology* **20**, 374.

Nakagawa, H., Yanagawa, Y. and Takehana, H. (1970a). *Agricultural and Biological Chemistry* **34**, 991.

Nakagawa, H., Yanagawa, Y. and Takehana, H. (1970b). *Agricultural and Biological Chemistry* **34**, 998.

Nakagawa, H., Seikiguchi, K., Ogura, N. and Takehana, H. (1971). *Agricultural and Biological Chemistry* **35**, 301.

Nasuno, S. and Starr, M. P. (1966). *Journal of Biological Chemistry* **241**, 5298.

Nasuno, S. and Starr, M. P. (1967). *Biochemical Journal* **104**, 178.

Nemoz, G., Robert-Baudouy, J. and Stoeber, F. (1976). *Journal of Bacteriology* **127**, 706.

Neubeck, C. E. (1975). *In* 'Enzymes in Food Processing' (G. Reed, ed.), 2nd ed., p. 397. Academic Press, New York.

Ng, H. and Vaughn, R. H. (1963). *Journal of Bacteriology* **85**, 1104.

Nortje, B. K. and Vaughn, R. H. (1953). *Food Research* **18**, 57.

Nyeste, L. and Holló, J. (1963). *Zeitschrift für Allgemeine Mikrobiologie* **3**, 37.

Nyiri, L. (1968). *Process Biochemistry* **3** (8), 27.

Nyiri, L. (1969). *Process Biochemistry* **4** (8), 27.

Ohuchi, A. and Tominaga, T. (1974). *Annals of the Phytopathological Society of Japan* **40**, 22.

Okamoto, K., Hatanaka, C. and Ozawa, J. (1964a). *Agricultural and Biological Chemistry* **28**, 331.

Okamoto, K., Hatanaka, C. and Ozawa, J. (1964b). *Berichte des Ohara Instituts für Landwirtschaftliche Biologie* **12**, 115.

Papavizas, G. C. and Ayers, W. A. (1966). *Phytopathology* **56**, 1269.

Patel, I. B. and Vaughn, R. H. (1973). *Applied Microbiology* **25**, 62.

Patil, S. S. and Dimond, A. E. (1968). *Phytopathology* **58**, 676.

Paynter, V. (1975). Ph.D.-Thesis: Clemson University, Clemson, South Carolina, U.S.A.

Perley, A. F. and Page, O. T. (1971). *Canadian Journal of Microbiology* **17**, 415.

Perry, D. A. and Harrison, J. G. (1977). *Nature, London* **269**, 509.

Pettipher, G. L. and Latham, M. J. (1979a). *Journal of General Microbiology* **110**, 21.

Pettipher, G. L. and Latham, M. J. (1979b). *Journal of General Microbiology* **110**, 29.

Phaff, H. J. (1947). *Archives of Biochemistry* **13**, 67.

Phaff, H. J. (1966). *Methods in Enzymology* **8**, 636.

Pilnik, W. and De Vos, L. (1970). *Flüssiges Obst* **37**, 430.

Pilnik, W. and Voragen, A. G. J. (1970). *In* 'The Biochemistry of Fruits and Their Products' (A. C. Hulme, ed.), vol. 1, p. 53. Academic Press, London.

Pilnik, W., Rombouts, F. M. and Voragen, A. G. J. (1973). *Chemie, Mikrobiologie, Technologie der Lebensmittel* **2**, 122.

Pilnik, W., Voragen, A. G. J. and Rombouts, F. M. (1974). *Lebensmittel-Wissenschaft und Technologie* **7**, 353.

Pilnik, W.,Voragen, A. G. J. and De Vos, L. (1975). *Flüssiges Obst* **42**, 448.

Platt, W. C. and Poston, A. L. (1962). United States Patent 3,058,887.

Preiss, J. and Ashwell, G. (1963a). *Journal of Biological Chemistry* **238**, 1571.

Preiss, J. and Ashwell, G. (1963b). *Journal of Biological Chemistry* **238**, 1577.

Pressey, R. and Avants, J. K. (1972). *Phytochemistry* **11**, 3139.

Pressey, R. and Avants, J. K. (1973a). *Biochimica et Biophysica Acta* **309**, 363.

Pressey, R. and Avants, J. K. (1973b). *Plant Physiology* **52**, 252.

Pressey, R. and Avants, J. K. (1975a). *Phytochemistry* **14**, 957.

Pressey, R. and Avants, J. K. (1975b). *Journal of Food Science* **40**, 937.

Pressey, R. and Avants, J. K. (1976). *Phytochemistry* **15**, 1349.

Pressey, R. and Avants, J. K. (1977). *Plant Physiology* **60**, 548.

Priest, F. G. (1977). *Bacteriological Reviews* **41**, 711.

Prins, R. A. and Van Hoven, W. (1977). *Protistologica* **13**, 549.

Pupillo, P., Mazzuchi, U. and Pierini, G. (1976). *Physiological Plant Pathology* **9**, 113.

Put, H. M. C. and Kruiswijk, J. Th. (1968). *Annales de l'Institut Pasteur, Lille* **19**, 171.

Rexová-Benková, L'. (1973). *European Journal of Biochemistry* **39**, 109.

Rexová-Benková, L'. and Markovič, O. (1976). In 'Advances in Carbohydrate Chemistry and Biochemistry' (R. S. Tipson and D. Horton, eds.), vol. 33, p. 323. Academic Press, New York.

Rexová-Benková, L'. and Mračková, M. (1978). *Biochimica et Biophysica Acta* **523**, 162.

Rexová-Benková, L'. and Slezárik, A. (1970). *Collection of Czechoslovak Chemical Communications* **35**, 1255.

Rexová-Benková, L'. and Tibenský, V. (1972). *Biochimica et Biophysica Acta* **268**, 187.

Rexová-Benková, L'., Mračková, M., Luknár, O. and Kohn, R. (1977). *Collection of Czechoslovak Chemical Communications* **42**, 3204.

Rice, S. L. (1977). Ph.D. Thesis: University of Georgia, Athens, Georgia, U.S.A.

Rice, S. L. and Beuchat, R. (1978). *Mycopathologia* **63**, 29.

Riov, J. (1975). *Journal of Food Science* **40**, 201.

Robert-Baudouy, J. M., Portalier, R. C. and Stoeber, R. (1974). *European Journal of Biochemistry* **43**, 1.

Robyt, J. F. and French, D. (1967). *Archives of Biochemistry and Biophysics* **122**, 8.

Roeb, L. and Stegemann, H. (1975). *Biochemie und Physiologie der Pflanzen* **168**, 607.

Roelofsen, P. A. (1953). *Biochimica et Biophysica Acta* **10**, 410.

Rombouts, F. M. (1972). Ph.D. Thesis: Agricultural University, Wageningen, The Netherlands.

Rombouts, F. M. and Pilnik, W. (1971). *Antonie van Leeuwenhoek* **37**, 247.

Rombouts, F. M. and Pilnik, W. (1972). *Chemical Rubber Company Critical Reviews in Food Technology* **3**, 1.

Rombouts, F. M. and Pilnik, W. (1978). *Process Biochemistry* **13**, (8), 9.

Rombouts, F. M., Spaansen, C. H., Visser, J. and Pilnik, W. (1978). *Journal of Food Biochemistry* **2**, 1.

Rombouts, F. M., Wissenburg, A. K. and Pilnik, W. (1979). *Journal of Chromatography* **168**, 151.

Sato, M. and Kaji, A. (1975). *Agricultural and Biological Chemistry* **39**, 819.

Sato, M. and Kaji, A. (1977a). *Agricultural and Biological Chemistry* **41**, 2193.

Sato, M. and Kaji, A. (1977b). *Agricultural and Biological Chemistry* **41**, 2199.

Sawamura, M., Knegt, E. and Bruinsma, J. (1978). *Plant and Cell Physiology* **19**, 1061.

Schlegel, H. G. (1972). 'Allgemeine Mikrobiologie', 2nd edition, p. 340. George Thieme Verlag, Stuttgart.

Schreier, P., Drawert, F., Steiger, G. and Mick, W. (1978). *Journal of Food Science* **43**, 1797.

Schröder, H. and Müller-Stoll, W. R. (1962). *Zeitschrift für Allgemeine Mikrobiologie* **2**, 109.

Sheiman, M. I., Macmillan, J. D., Miller, L. and Chase, T. (1976). *European Journal of Biochemistry* **64**, 565.

Shinmyo, A., Davis, I. K., Nomoto, F., Tahara, T. and Enatsu, T. (1978). *European Journal of Applied Microbiology* **5**, 59.

Solms, J. and Deuel, H. (1955). *Helvetica Chimica Acta* **38**, 321.

Spalding, D. H. and Abdul-Baki, A. A. (1973). *Phytopathology* **63**, 231.

Spalding, D. H., Wells, J. M. and Allison, D. W. (1973). *Phytopathology* **63**, 840.

Spiro, R. G. (1966). *Methods in Enzymology* **8**, 3.

Sreekantiah, K. R., Jaleel, S. A., Narajana Rao, D. and Raghavendra Rao, M. R. (1975). *European Journal of Applied Microbiology* **1**, 173.

Starr, M. P. and Chatterjee, A. K. (1972). *Annual Review of Microbiology* **26**, 389.

Starr, M. P. and Nasuno, S. (1967). *Journal of General Microbiology* **46**, 425.

Starr, M. P., Chatterjee, A. K., Starr, P. B. and Buchanan, G. E. (1977). *Journal of Clinical Microbiology* **6**, 379.

Stoeber, F., Lagarde, A., Nemoz, G., Novel, G., Novel, M., Portalier, R., Pouyssegur, J. and Robert-Baudouy, J. (1974). *Biochimie* **56**, 199.

Strand, L. L., Corden, M. E. and MacDonald, D. L. (1976). *Biochimica et Biophysica Acta* **429**, 870.

Strübi, P., Escher, F. and Neukom, H. (1975). *Die Industrielle Obst- und Gemüseverwertung* **60**, 349.

Šulc, D. and Ciric, D. (1968). *Flüssiges Obst* **35**, 230.

Swinburne, T. R. and Corden, M. E. (1969). *Journal of General Microbiology* **55**, 75.

Szajer, I. (1978). *Acta Microbiologica Polonica* **27**, 237.

Takehana, H., Shibuya, T., Nakagawa, H. and Ogura, N. (1977). *Technical Bulletin of the Faculty of Horticulture, Chiba University* **25**, 29.

Termote, F., Rombouts, F. M. and Pilnik, W. (1977). *Journal of Food Biochemistry* **1**, 15.

Terrier, A. and Blouin, J. (1974). *Le Progrès de l'Agricole et Vinicole* **91**, 604.

Theron, T., De Villiers, O. T. and Schmidt, A. A. (1977). *Agrochemophysica* **9**, 7.

Tomizawa, H., Izaki, K. and Takahashi, H. (1970). *Agricultural and Biological Chemistry* **34**, 1064.

Trudgill, P. W. and Widdus, R. (1966). *Nature, London* **211**, 1097.

Tsuchida, T., Nakamura, K., Fujii, Y. and Takahashi, H. (1968). *Agricultural and Biological Chemistry* **32**, 1355.

Tsuyumu, S. (1977). *Nature, London* **269**, 237.

Tuttobello, R. and Mill, P. J. (1961). *Biochemical Journal* **79**, 51.

Ulrich, J. M. (1975). *Physiological Plant Pathology* **5**, 37.

Urbanek, H. and Zalewska-Sobczak, J. (1975). *Biochimica et Biophysica Acta* **377**, 402.

Urbanek, H. and Zalewska-Sobczak, J. and Krzechowska, M. (1976). *Bulletin de l'Academie Polonaise des Sciences, Série des Sciences Biologiques*, Cl II, **24**, 635.

Van Houdenhoven, F. E. A. (1975). Ph.D. Thesis: Agricultural University, Wageningen, The Netherlands.

Van Hoven, W. and Prins, R. A. (1977). *Protistologica* **13**, 599.

Van Pee, W. and Castelein, J. M. (1971). *East African Agricultural and Forestry Journal* **36**, 308.

Van Pee, W. and Castelein, J. M. (1972). *Journal of Food Science* **37**, 171.

Vasu, S. (1967). *Revue Roumaine de Biochimie* **4**, 67.

Vasu, S. (1974). *Revue Roumaine de Biochimie* **11**, 211.

Vaughn, R. H., Jakubczyk, T., Macmillan, J. D., Higgins, T. E., Davé, B. A. and Crampton, V. M. (1969a). *Applied Microbiology* **18**, 1969.

Vaughn, R. H., King, A. D., Nagel, C. W., Ng, H., Levin, R. E., Macmillan, J. D. and York, G. K. (1969b). *Journal of Food Science* **34**, 224.

Vaughn, R. H., Stevenson, K. E., Davé, B. A. and Hyan, C. (1972). *Applied Microbiology* **23**, 316.

Verhoeff, K. and Liem, J. I. (1978). *Phytopathologische Zeitschrift* **91**, 110.

Verhoeff, K. and Warren, J. M. (1972). *Netherlands Journal of Plant Pathology* **78**, 179.

Versteeg, C. (1979). Ph.D. Thesis: Agricultural University, Wageningen, The Netherlands.

Versteeg, C., Rombouts, F. M. and Pilnik, W. (1978). *Lebensmittel-Wissenschaft und Technologie* **11**, 267.

Vijayalakshmi, M. A., Segard, E., Rombouts, F. M. and Pilnik, W. (1978). *Lebensmittel-Wissenschaft und Technologie* **11**, 288.

Voordouw, G., Voragen, A. G. J. and Pilnik, W. (1974). *Flüssiges Obst* **41**, 282.

Voragen, A. G. J. (1972). Ph.D. Thesis: Agricultural University, Wageningen, The Netherlands.

Voragen, A. G. J., Rombouts, F. M. and Pilnik, W. (1971). *Lebensmittel-Wissenschaft und Technologie* **4**, 126.

Wagner, G. and Hollmann, S. (1976). *European Journal of Biochemistry* **61**, 589.

Wang, M. C. and Keen, N. T. (1970). *Archives of Biochemistry and Biophysics* **141**, 749.

Ward, O. P. and Fogarty, W. M. (1973). *Journal of the Institute of Wood Science* **6**, 8.

Werch, S. C., Jung, R. W., Day, A. A., Friedemann, T. E. and Ivy, A. C. (1942). *Journal of Infectious Diseases* **70**, 231.

Wieringa, K. T. (1949). Proceedings of the Fourth International Congress for Microbiology, p. 482. Rosenkilde and Bagger, Copenhagen.

Wieringa, K. T. (1956). *Netherlands Journal of Agricultural Science* **4**, 204.

Wiese, M. V., DeVay, J. E. and Ravenscroft, A. V. (1970). *Phytopathology* **60**, 641.

Wimborne, M. P. and Rickard, P. A. D. (1978). *Biotechnology and Bioengineering* **20**, 231.

Wojciechowicz, M. and Ziołecki, A. (1979). *Applied and Environmental Microbiology* **37**, 136.

Yamasaki, M., Kato, A., Chu, S.-Y. and Arima, K. (1967). *Agricultural and Biological Chemistry* **31**, 552.

Yoshihara, O., Matsuo, T. and Kaji, A. (1977). *Agricultural and Biological Chemistry* **41**, 2335.

Young, L. (1976). Ph.D. Thesis: Cornell University, Ithaca, New York.

Zetalaki, K. (1976). *Process Biochemistry* **11** (7), 11.

Zetelaki-Horváth, K. (1978). *Acta Alimentaria* **7**, 209.

Zetelaki-Horváth, K. and Békássy-Molnár, E. (1973). *Biotechnology and Bioengineering* **15**, 163.

Zetelaki-Horváth, K. and Dobra-Seres, I. (1972). *Acta Alimentaria* **1**, 139.

Ziołecki, A. (1979). *Applied and Environmental Microbiology* **37**, 131.

Zucker, M. and Hankin, L. (1970). *Journal of Bacteriology* **104**, 13.

Zucker, M. and Hankin, L. (1971). *Canadian Journal of Microbiology* **17**, 1313.

Zucker, M., Hankin, L. and Sands, D. (1972). *Physiological Plant Pathology* **2**, 59.

6. Cellulases

JOSTEIN GOKSØYR AND JONNY ERIKSEN

Department of Microbiology and Plant Physiology, University of Bergen, Allégt. 70, 5014 Bergen-Universitetet, Norway

I. INTRODUCTION

Cellulose is probably the most abundant biological compound on terrestrial Earth. Cellulosic materials have a correspondingly wide use, as lignified cellulose (wood) for construction purposes, and as more or less pure cellulose in paper, fibres and textiles. As a consequence of this, and since in addition cellulose is the dominating waste material from agriculture, in the form of stalks, stems and husks, cellulose is also one of the main waste products both in nature and from man-made activities. From this, one should expect a considerable interest in cellulolytic organisms and in their cellulases. Until now, it may be fair to say that industrial interest in production and use of cellulolytic enzymes has not been excessive. Some of the reasons for this we hope to clarify in this chapter. Briefly, it may be stated that the cellulolytic enzyme system is quite complex, and the structure of cellulose as it occurs in nature makes it hard to obtain rapid and efficient hydrolysis by enzymes. Further research ought to make it possible to overcome at least some of these difficulties.

II. CELLULOSE AND CELLULOSIC RAW MATERIALS

A. Natural Sources

Cellulose is the principal constituent of the cell wall of most terrestrial plants. Another important source is cotton, which are the hairs in the fruit of the cotton plant. Here, cellulose is present in a relatively pure state, whereas in cell walls it is found in close association with other polysaccharides (hemicelluloses, pectin) and lignin. The cellulose content in some materials is given in Table 1.

In the technical exploitation of natural cellulosic sources, it is often necessary to remove non-cellulosic components. This can be done by various methods, giving more suitable cellulose fibres for further refinement. The current utilization of some main natural sources of cellulose is shown in Table 2.

Despite an enormous and World-wide utilization of natural cellulosic sources, there are still abundant quantities of cellulose-containing raw materials or waste products that are not exploited, or which could be used more efficiently. The main problem in this respect is, however, to develop processes that are economically profitable.

In recent years, considerable attention has been given to the possibility of creating an enzymic-hydrolysis industry, and many pilot investigations have been started. As stated by Ghose et al. (1975), any further economic utilization of cellulosic materials should be based on

Table 1

The cellulose content of some natural, industrial and agricultural materials

Material	Percent cellulose	Reference
Cotton	91	Gascoigne and Gascoigne (1960)
Wood		
Spruce and pine	41	Virkola (1975)
Birch	40	Virkola (1975)
Pulp		
Mechanical spruce	41	Virkola (1975)
Unbleached sulphite	79	Virkola (1975)
Unbleached sulphate	77	Virkola (1975)
Neutral sulphite–semichemical	56	Virkola (1975)
Dissolving pulp	92.5	Virkola (1975)
Newspaper	40–80	Updegraff (1971)
Newspaper	61	Grethlein (1978)
Furfural process waste	40	Markkanen and Eklund (1975)
Straw		
Wheat	30.5	Virkola (1975)
Rice	32.1	Virkola (1975)
Rye	34.0	Pigden and Heaney (1969)
Barley	40	Peiterson (1975a)
Oat	42.8	Donefer et al. (1969)
Bagasse		
Entire bagasse	46	Srinivasan and Han (1969)
Fibre bagasse	56.6	Srinivasan and Han (1969)
Pith bagasse	55.4	Srinivasan and Han (1969)

their current exploitation patterns. With this in mind, the most useful raw materials for an enzymic-hydrolysis industry are: unutilized forest resources, waste wood left in forests, barking wastes and waste fibres of forest industries, waste paper and board in communal wastes, agricultural wastes, unutilized non-wood plants and wastes of food industries. All of these potential raw materials contain, in addition to cellulose, various amounts of hemicellulose, lignin, extractives and inorganic compounds (see the tables in Virkola, 1975). For this reason, most of them are not suitable for enzymic hydrolysis without modification. Mechanical treatments to reduce particle size and induce swelling are frequently necessary, and also in many cases chemical treatments to remove non-cellulosic compounds.

The raw materials mentioned above are those that can be exploited in a large-scale enzymic hydrolysis of cellulose. The final aim for the process may vary, and include sugar production, protein production

Table 2

Cellulose-containing materials, and their current use. From Virkola (1975)

Raw material	Current use
Non-wood plants, bagasse, straw	Pulp and paper industry Fuel Acid-hydrolysis industry
Brush wood, short rotation wood	Pulp and paper industry Chip and fibre-board industry
Forest wood residuals	Pulp and paper industry Chip and fibre-board industry
Sawood and plywood mill wastes	Chip and fibre-board industry Pulp and paper industry Fuel
Peat	Fuel As a soil-improving agent
Waste fibres (pulp and paper mills)	Re-use in pulp, paper-board industry Fuel
Waste paper (community)	Re-use in paper and board industry Acid-hydrolysis industry
Barking waste	Fuel As a soil-improving agent

and production of biologically active substances. Such raw materials can obviously also be used for cellulase production, and the cellulase used for other purposes as in the food industry and removal of fibres from waste waters.

B. Structure of Cellulose

Cellulose is a linear glucose polymer (polyalcohol), composed of anhydroglucose units coupled to each other by β-1,4-glucosidic bonds (Fig. 1). The anhydroglucose units adopt the chair configuration shown in the figure, with the units rotated 180° about the main axis in proportion to each other. The result is an unstrained linear configuration with minimum steric hindrance. The number of glucose units in the cellulose molecule varies. In a survey given by Cowling and Brown (1969), D.P. (degree of polymerization) values between 15 and 14,000 are reported, with a mean value about 3,000. However, even higher D.P. values have been found (Emert et al., 1974). Sihtola and Neimo (1975) assumed a D.P. value of about 10,000 in native cellulose,

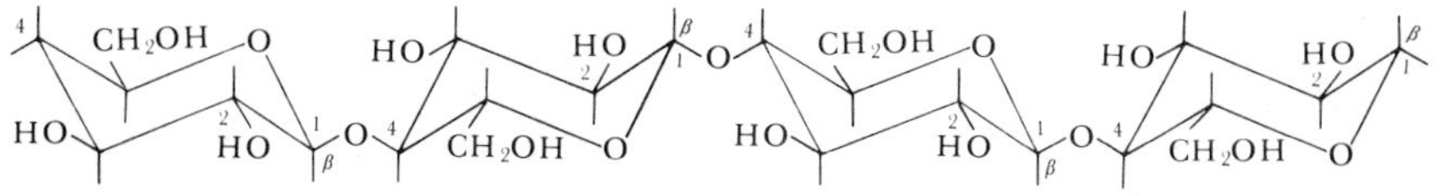

Fig. 1. Conformational structure for cellulose.

which corresponds to a molecular weight of 1.5 million. The length of the anhydroglucose unit is 0.515 nm, and the total length of a cellulose molecule may thus be above 5 μm.

Cellulose molecules are coupled together by hydrogen bonds to give larger units, which are ordered in a crystalline manner. There are different opinions about the number of cellulose molecules in such units and how they are organized. There seems, however, to be a general agreement about the structure of the unit cell of the cellulose crystallite. This is shown in Figure 2, as proposed by Meyer and Misch (1937), but with the axes changed in accordance with the current practice in polymer crystallography, where the c axis is the fibre axis and the monoclinic angle (γ) is obtuse (Gardner and Blackwell, 1974). The model contains four glucose residues in each unit cell. These

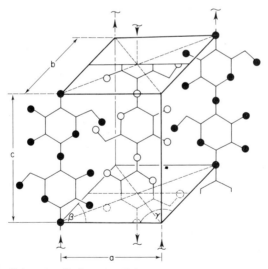

Fig. 2. The monoclinic unit cell of native cellulose, according to Meyer and Misch (1937), but which the fibre axis designated as the c axis. The chains lying in the front right and hind left corners have not been drawn. Note that the central chain is in an antiparallel position. According to Gardner and Blackwell (1974), all chains are probably parallel. The dimensions are: 1 = 0.835 nm, b = 0.79 nm, c = 0.103 nm, $\gamma = 90°$.

include two in the centre and one-fourth of each of the eight residues which are placed at the corners of the monoclinic cell, and which are shared by each of the four unit cells which meet at the corners. The dimensions of this unit cell were shown to be: a = 0.835 nm, b = 0.79 nm, c = 0.103 nm and $\gamma = 96°$.

This model was supported by the work of Gardner and Blackwell (1974) using X-ray crystallography on highly crystalline cellulose from the cell walls of the algae *Valonia ventricosa*. In addition, they proposed a parallel-chain model for cellulose, with all molecules lying in the same direction, and a hydrogen-bonding network with one intermolecular and two intramolecular bonds per glucose residue (Fig. 3). The two intramolecular bonds (O3–O5' and O6–O2') will support the glucosidic bond in maintaining the rigidity of the cellulose molecule. Hydrogen bonding is not possible along the unit cell diagonals or along the b axis. In the b direction, the molecules are attached to each other by van der Waals forces only (Sihtola and Neimo, 1975).

As already mentioned, there is some disagreement as to how cellulose molecules are organized in larger units. A fibrous structure can normally be seen in the microscope. These fibres are generally assumed to be composed of microfibrils, 5–30 nm wide. Some authors, such as Preston (1971), think that the microfibrils are also the basic

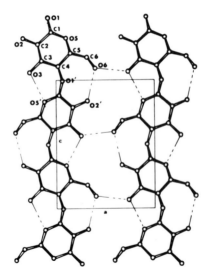

Fig. 3. The hydrogen-bonding network in cellulose. Each glucose residue forms two intra-molecular bonds (O3–H · · · O5' and O6 · · · H–O2') and one intermolecular bond (O6–H · · · O3). From Gardner and Blackwell (1974).

structural units, varying in transverse dimension from about 10 nm × 5 nm to 30 nm × 15 nm, and with a central crystalline region with lateral dimensions varying with those of the microfibril. Others claim that the microfibrils are in turn built up of elementary fibrils, attached to each other by hydrogen bonds (Frey-Wyssling and Mühletahler, 1963; Heyn, 1965; Hepler et al., 1970). In considering these opinions, one should bear in mind the extreme length and relative rigidity of the cellulose molecules. One molecule will certainly pass through areas with a very high order, the crystallites, and other areas with tangles and therefore less ordered and amorphous structure. This is visualized in a model by Sihtola and Neimo (1975) (Fig. 4).

The chemical and physical properties of cellulose are to a large extent related to its tertiary structure. The solubility of poly-β-1,4-glucosides in water decrease rapidly with a D.P. above 4–5. Cellulose fibres absorb water readily and swell, but the swelling is limited to amorphous regions of the fibre, as it is counteracted by the strong hydrogen-bond network of the crystalline regions. An average cellulase molecule has a diameter of 5 nm (Cowling and Brown,, 1969), assuming that it is globular. This means that cellulases or other molecules similar in size will not be able to penetrate the interior structure of the fibrils in native cellulose. They can only act on the surface of the microfibrils or elementary fibrils (crystalline regions). The number of glucosidic bonds available for enzyme action will thus to a large extent depend on the degree of swelling of the cellulose.

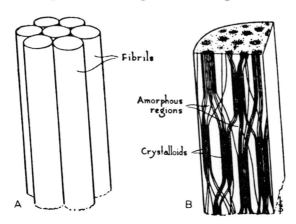

Fig. 4. Alignment and composition of elementary fibrils of cellulose. A: Bundle of parallel fibrils (held together crosswise by hydrogen bonds). B: Lateral section view of one fibril. From Sihtola and Neimo (1975).

A swelling in excess of that obtained by mere hydration can be obtained by suitable mechanical or physical treatment, such as steam treatment, milling and ultrasonic treatment. Also, mineral acids and alkalis in high concentrations will induce swelling of the whole fibre, since these agents are able to break the hydrogen-bond network and penetrate into the crystalline areas. For efficient hydrolysis of cellulose by cellulases *in vitro*, a pretreatment to cause swelling is frequently necessary. The complexity and cost of this pretreatment may be a crucial factor in a technical exploitation of enzymes for hydrolysis of cellulose.

III. THE CELLULASE COMPLEX

A. History and Earlier Concepts

Since the first observations of the action of cellulolytic enzymes *in vitro* (probably Seilliere, 1906), an extensive research has been carried out on cellulolytic enzymes from a variety of micro-organisms. Reviews of these earlier works are given by Siu (1951), Gascoigne and Gascoigne (1960), Reese (1963), Norkrans (1963, 1967), Whitaker (1971) and Emert *et al.* (1974).

A major advance in the understanding of cellulase activity was made by introduction of the $C_1–C_x$ concept by Reese *et al.* (1950). This concept was based on the observation that culture filtrates of some cellulolytic micro-organisms hydrolysed native cellulose, whereas those from other strains were able to hydrolyse only soluble cellulose derivatives such as carboxymethylcellulose. Reese *et al.* (1950) suggested that C_1 was a component capable of forming shorter, linear cellulose chains from native cellulose. These chains could then be hydrolysed by C_x, which also hydrolysed soluble cellulose derivatives, and which was the cellulase proper. Following this concept, C_1 was considered a 'prehydrolytic factor', not necessarily enzymic. A protein corresponding to this factor was first isolated by Selby and Maitland (1967). In accordance with Siu (1951), Selby (1968) suggested that C_1 could be a 'hydrogen bondase', acting by removing hydrogen bonds between adjacent cellulose chains and thereby making them available for attack by the C_x enzyme. An argument in favour of this was given by King and Vessal (1969), on the basis of the very low activation energies found for C_1 action on native cellulose (Rautela and King, 1968). However, as pointed out by Eriksen and Goksøyr (1977), this argument is not valid in a kinetically complex system like the cellulase system.

Eriksson and Pettersson (1972) isolated a 'C_1-type' protein from *Sporotrichium pulverulentum,* and showed this to be an enzyme catalysing splitting of cellobiose and glucose units from the ends of cellulose chains. Similar exoglucanases were later isolated from a number of fungi. The finding that cellobiose is an important end-product of cellulolytic activity also gives a rational explanation of the fact that cellobiase activity is needed to obtain maximal enzyme hydrolysis of native cellulose.

B. Present Status

The development in techniques for separation and identification of proteins during the last ten years has led to isolation of several cellulase proteins with distinct properties. This has created some confusion in terminology. Several authors maintain use of the Reese terminology, and assume that C_1 is synonymous with the exo-glucanase, whereas others state that use of C_1 for the exoglucanase is unfortunate since it probably does not act as the first step in cellulose hydrolysis. Some authors even maintain the C_1 concept besides the exoglucanase.

Reese *et al.* (1950) proposed the C_1-C_x concept as a hypothetical model but with very little, if any, evidence for a prehydrolytic step. In much later work, it seems to have obtained a degree of inner truth. The important conclusion in their work, which is still valid, is that native cellulose is frequently hydrolysed by a two-step or three-step reaction sequence. If C_1 is defined as causing a process that occurs before action of C_x (the endoglucanase), it is wrong to use the term as a synonym for exoglucanases. If, on the other hand, C_1 is defined as an enzyme present in 'true cellulolytic' fungi that acts together with endo-glucanases to hydrolyse native cellulose, C_1 may be used as a substitute for exoglucanase. A thorough discussion of the problem related to the connection between C_1 and exoglucanase is given by Wood and McCrae (1977b).

In agreement with most authors during the last 2–3 years, we will avoid the use of the terms C_x and C_1 and, instead, use the abbreviated names endoglucanase for 1,4-β-D-glucan 4-glucan-hydrolase (EC 3.2.1.4) and exoglucanase for 1,4-β-D-glucan cellobiohydrolase (EC 3.2.1.91). With present knowledge of cellulolytic enzyme systems, it is still hard to make generalizations valid for all types of micro-organisms. In this survey, we have therefore found it useful to present information about cellulases in a restricted number of micro-organisms, which might be considered typical for the main groups. Within each group, generalizations may to some extent be made. The first two groups represent micro-organisms where only endo-glucanase activity has been found with certainty, while the three later groups show both exo- and endoglucanase activity.

1. Sporocytophaga spp. (bacteria)

Sporocytophaga and Cytophaga spp. are gliding bacteria which grow in close contact with cellulose fibrils that are their natural substrate, and which they hydrolyse efficiently for use as carbon and energy source. Their cellulolytic ability has been studied by several authors, and the cellulases of S. myxococcoides were isolated and characterized by Osmundsvåg and Goksøyr (1975). Only endoglucanase activity is found in this bacterium, and the major proportion of this was in the growth medium (about 90%). By gel filtration and ion-exchange chromatography, two major cellulase components were isolated and purified. Cellulase II was present in the greatest amount. It had a molecular weight of 52,000, a pI value of 4.75, and a relatively broad pH optimum (5.5–7.5). Cellulase I had a molecular weight of 46,000, a pI value of 7.5, and a pH optimum of 6.5–7.5. These cellulases were probably not glycoproteins, since only trace amounts of carbo-hydrate could be detected. The amino-acid compositions were fairly similar, but cellulase I had a specific activity (measured on carboxy-methylcellulose) about six times higher than cellulase II. A cell-associated cellulase was partly purified. It showed similarities with cellulase II.

These cellulases were typical random-acting endoglucanases, causing a rapid decrease of viscosity in carboxymethylcellulose solutions compared to the rate of formation of reducing end groups. No effect could be found on native cellulose with purified enzymes, concentrated cell-free supernatant, extracts from broken cells or with supernatant from Triton X-100-treated cells. These results seem to be fairly typical for cellulolytic bacteria. Those studied most extensively are *Cellvibrio (Pseudomonas) fulvus* (Berg *et al.*, 1972a, b; Berg, 1975) and a cellulolytic *Pseudomonas fluorescens* strain (Suzuki *et al.*, 1969; Yamane *et al.*, 1970a, b, 1971; Suzuki, 1975). The common feature of all cellulolytic bacteria seems to be that *in vitro* (cell-free) systems have a high activity on soluble cellulose derivatives, but very low if any activity towards native, highly ordered cellulose.

2. Coniophora cerebella

This is a basidiomycete causing brown rot in wood. The cellulases in this organism were studied by Eidså (1972). The cellulases are predominantly extracellular. The culture filtrate showed high endoglucanase activity, but very low activity on native cellulose. By gel filtration and ion-exchange chromatography, two cellulases (A and B) were isolated and purified. Cellulase A had a molecular weight of 42,000; the pI was 3.55 and the pH optimum was around 4.7. Cellulase B had a molecular weight of 38,500, a pI value of 2.2 and a pH optimum of 4.2. Both contained less than 5% carbohydrate. The very low isoelectric point, in particular of cellulase B, is interesting. The cellulolytic activity of the culture filtrate was also stable at low pH values (between 2 and 7), while the purified cellulases were stable only at pH values between 5.5 and 7.5 (cellulase A) and 4 and 6.5 (cellulase B). It should be noted that this fungus is a strong acid producer, rapidly lowering the pH value in its growth medium to 2–2.5. This is certainly also the case in its natural habitat, wood.

A very low activity could be detected on cotton by the culture filtrate; this amounted to 10% hydrolysis in five days by a 30–40 times-concentrated culture filtrate.

By comparison of viscosimetric measurements and rate of formation of reducing end groups with carboxymethylcellulose as substrate, it was found that *C. cerebella* cellulases had a less random action than the endoglucanases from *Sporocytophaga* sp. and *Trichoderma koningii*. In particular, this was the case with *C. cerebella* cellulase B. Of interest also is the observation that, when mixed with exoglucanase from *Trichoderma koningii*, the *C. cerebella* cellulases did not show any enhanced activity towards native cellulose. Thus, these endoglucanases do not act in quite the same manner as the endoglucanases in fungi with the complete cellulase system (Section III.B.3, p. 294).

Few other enzymic studies have been made on brown-rot basidiomycetes. A comparison between brown- and white-rot fungi by
Highley (1975) indicated, however, that brown-rot fungi only produce
endoglucanases, in contrast to white-rot fungi. This means that, in
culture filtrates of brown-rot fungi, the endoglucanase activity is high,
but the ability to attack native cellulose is very low.

3. Trichoderma viride (Ascomycetes and Deuteromycetes)

Within these groups of fungi, there are several highly cellulolytic
species. They differ from the micro-organisms already described in
that their culture filtrates are able to accomplish extensive hydrolysis of
native cellulose. An exception to this is *Myrothecium verrucaria*, which
grows profusely on cellulose, but where the culture filtrate seems to
contain only endoglucanases (Halliwell, 1961, 1966; Selby *et al.*, 1963;
Selby and Maitland, 1965). A number of these fungi have been
extensively studied. We choose here for reference *Trichoderma viride*
since, within this species, strains have been described with the highest
cellulolytic activity known to date. Strains of *T. viride* (together with
Aspergillus niger) are at present probably the most important cellulase
producers for commercial purposes.

It should be noted that, within the genus *Trichoderma*, species
definition is very difficult (Rifai, 1969). Simmons (1977) found that *T.
viride* QM6a, on which much of the work by different groups has been
carried out, belongs to the *T. longibrachiatum* aggregate, but is
sufficiently different morphologically to warrant a separate species
name. He has thus proposed that *T. viride* QM6a should be renamed *T.
reesei* Simmons. To avoid confusion with the existing literature, we have
retained the name *T. viride* on this strain and the mutants derived from
it.

In a series of papers (Berghem and Pettersson, 1973, 1974; Berghem
et al., 1975, 1976), the cellulase system in a commercial cellulase
product of *T. viride* (Onozuka SS, All Japan Biochemicals Co. Ltd.,
Japan) was studied. Three different enzymes related to cellulase activity
are present in this product, namely exoglucanase (β-1,4-glucan
cellobiohydrolase), endoglucanase (β-1,4-glucan glucanhydrolase) and
β-glucosidase or cellobiase. By fractionation of the exoglucanase, four
different isoenzymes were found. One of these (cellulase IV:D) was
purified extensively. It was found to have a molecular weight of 42,000,

and contained 9.2% carbohydrate; it is thus a glycoprotein. Mannose was found to be the dominant saccharide. The isoelectric point was 3.79, consistent with the high content of acidic amino acids found, and the pH optimum was about 4.8. The enzyme was completely inactivated by exposure for three minutes to a temperature of 78°C. The enzyme acted as a typical exoglucanase, giving practically no decrease in the degree of polymerization of amorphous cellulose, and giving cellobiose as the predominant product of hydrolysis with Avicel, phosphoric acid-swollen Avicel, and cellotetraose as substrate. For efficient hydrolysis to occur, cellobiose must be removed, since it had an inhibitory action on the enzyme. This had previously been shown by Halliwell and Griffin (1973) for the exoglucanase from *Trichoderma koningii*.

Gum and Brown (1976) have studied another of the exoglucanases from *T. viride*, namely cellobiohydrolase C. This was purified from another commercial preparation, Meicelase P, produced by Meiji Seika Kaisha. Ltd., Tokyo, Japan. This cellulase could be obtained in gram quantities by a simple chromatographic separation, based on its high affinity for cellulose at pH 5.0 in the presence of acetate buffer. It had a molecular weight of 48,000 and was also a glycoprotein, containing 26.4, 4.8, 2.4 and 3.4 moles of mannose, glucose, galactose and glucosamine, respectively, per mole of enzyme. The carbohydrates were predominantly bound as monosaccharides, but chains with up to five saccharide units could be detected.

Trichoderma viride also contains several endoglucanases. Two of these were purified extensively by Berghem *et al.* (1976). One had a low molecular weight, about 12,500, whereas the other was in the normal range for cellulases, about 50,000. The low molecular-weight enzyme had an iso-electric point at 4.60 and contained 21% carbohydrate, while the other was iso-electric at 3.39 and contained 12% carbo-hydrate. The yields from 75 g of commercial cellulase were about 60 mg of the low molecular-weight type and 90 mg of the other type. The total amount of endoglucanases present in the commercial extract was estimated to be about 12% of the protein content. Both enzymes acted as typical endoglucanases with a very low ability to attack native cellulose. The low molecular-weight type was, however, able to form free fibres from filter paper. The third enzyme with cellulolytic activity, namely cellobiase, was also isolated by Berghem and Pettersson (1974). The purified enzyme had a molecular weight of 47,000, and was not a glycoprotein. Its isoelectric point was 5.74. The yield of this particular enzyme was 53 mg per 100 g of commercial cellulase. The enzyme had no activity towards Avicel or carboxymethylcellulose. It was active towards aryl-β-glucosides, cellobiose and cellotetraose. With the last substrate, a strong substrate inhibition was found. They concluded that the substrate specificity of this enzyme was more similar to that of an aryl-β-glucosidase than to that of a cellobiase.

The closely related *T. koningii* cellulases have been studied by British groups (Halliwell, 1965a, b; Halliwell and Griffin, 1973; Wood, 1968; Wood and McCrae, 1972, 1975, 1978). In this fungus, one exoglucanase and at least four endoglucanases have been isolated and characterized, one of which was of the low molecular-weight type. Other fungi whose cellulases have been extensively purified are *Penicillium funiculosum* (Selby, 1968; Wood and McCrae, 1977b) and *Fusarium solani* (Wood and Phillips, 1969; Wood, 1969, 1971; Wood and McCrae, 1977a). The three enzymes, exoglucanase, endoglucanase and cellobiase, act in a synergistic or co-operative manner on native cellulose. This has been shown by several authors on different fungi within this group (Selby and Maitland, 1967; Selby 1968; Wood, 1968, 1975; Wood and McCrae, 1972). A thorough study

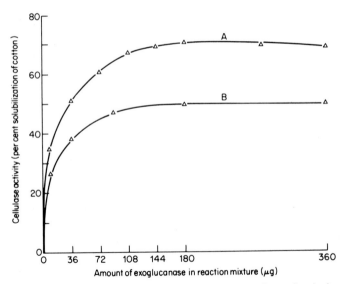

Fig. 5. Cellulase (cotton-solubilizing) activity of different proportions of endoglucanase and exoglucanase from *Trichoderma koningii*. Curve A, 1,000 units endoglucanase; curve B, 250 units endoglucanase, with 2 mg of cotton and seven days incubation at 30°C. From Wood (1975).

by Wood (1975) revealed that there is a stoicheiometric relationship between the amount of endo- and exoglucanase which gives highest cellulolytic activity (Fig. 5). In *T. koningii* this corresponded to the ratio between exo- and endoglucanase in the culture filtrate. With filtrates from cultures of other fungi, addition of extra exoglucanase from *T. koningii* enhanced the cellulolytic ability. These fungi thus seemed to be more or less deficient in exoglucanase activity.

Careful experiments (Wood, 1968; Wood and McCrae, 1977b) have shown that a mixture of purified exoglucanase, endoglucanase and cellobiase, in the same concentrations as in the culture filtrate, gave very much the same activity towards highly crystalline cellulose as the culture filtrate. It is thus not necessary to assume the presence of hypothetical factors in culture filtrates of these fungi. Cellulases within this group are not very species specific. Thus, for the fungi that produce free exoglucanase, this enzyme from one fungus will act synergistically with endoglucanase from another. Such a synergistic effect is, however, more or less absent with endoglucanases from *Myrothecium verrucaria* or other fungi, such as *Stachybotrys atra*, *Gliocladium roseum* and *Memnionella echinata* (Wood, 1975), that seemingly produce only endoglucanases. The synergistic action is not fully understood. The effect of cellobiase is most easy to explain, since cellobiose acts as a product inhibitor on exoglucanase. One possible explanation for the activity of a mixture of exo- and endoglucanases on native, highly ordered cellulose is that the attack is started by the endoglucanase on the surface of fibrils. However, glycosidic bonds that become broken will have a high chance of being re-established because cellulose chains are lying in a very ordered manner, supported by a large number of hydrogen bonds. In the presence of exoglucanase, cellobiose is removed from free ends, leaving a gap where the glucosidic bond was first broken. Thus, the hydrolytic process when first started, will become irreversible. This explanation is rather opposite from that presented by Reese *et al.* (1950). Wood and McCrae (1978) demonstrated that, in *T. koningii*, all four endoglucanases gave about the same synergistic effect with the exoglucanase on dewaxed cotton, when the incubation time was 18 hours. Only two of them could, however, produce extensive solubilization of cotton in combination with the exoglucanase (seven days incubation time). They suggest that, for attack on the highly crystalline parts of cellulose in cotton, formation of a loose complex between exo- and endoglucanase is necessary to secure an instantaneous attack by the exoglucanase after the endoglucanase has broken the first glycosidic bond. This hypothesis presumes a 1 : 1 molar ratio between the two glucanases for maximal synergism to occur. Unfortunately, the saturation data presented by Wood (1975) do not permit calculation of molecular ratios between exo- and endoglucanase.

4. Chaetomium thermophile *var*. dissitum (*a thermophilic fungus*)

The cellulases from this fungus, which is an ascomycete, were studied by Eriksen and Goksøyr (1976, 1977). The cellulases are extracellular, and like other ascomycetes, both exoglucanase and endoglucanase are found, together with β-glucosidase. One exoglucanase and one endoglucanase were isolated and purified by ion-exchange chromatography and gel filtration. There seemed to be only one enzyme of each type in this fungus. The exoglucanase had a molecular weight of 67,000, and the endoglucanase 41,000: their isoelectric points were 4.55. The pH optimum of the combined system (culture filtrate) on cotton was about 5.0. The fungus itself had a minimum temperature for growth at 27°C, and a maximum temperature at 57°C. Optimum growth was at 45–50°C. In the culture filtrate, the optimum temperature for cellulase activity with carboxymethylcellulose as substrate varied between 77°C with 0.5 hours incubation and 58°C with 10 hours incubation time. With cotton as substrate, the optimum temperature was, however, found to be 58°C regardless of incubation time. The enzyme system was to some extent stabilized by its substrate, but carboxymethylcellulose had a higher stabilizing effect than cotton. The temperature stability of the enzyme was highest at pH 6.0.

With the purified enzymes, the effect of temperature on activity was studied with carboxymethylcellulose and cotton as substrates. Activation energies, Q_{10} values and optimum temperatures (one hour incubation time) were determined from Arrhenius plots. The results are summarized in Table 3. Low Q_{10} values shown by the complete system are noteworthy. They demonstrate that, in fact, little is gained in reaction rate by increasing the reaction temperature. Also, the optimum and maximum temperatures of cellulases from thermophilic fungi do not seem to lie much higher than those from mesophilic strains. This may not be so surprising, since mesophilic cellulases are also able to tolerate temperatures at which thermophilic fungi can

Table 3

Effects of temperature on cellulases from the thermophilic fungus *Chaetomium thermophile* var. *dissitum*. From Eriksen and Goksøyr (1977)

Fraction	Activation energy (kJ)	Q_{10} Value	Temperature optimum (°C)
Substrate: Carboxymethylcellulose			
Exoglucanase	52.0	1.9	71
Endoglucanase	26.1	1.4	66
Exoglucanase with endoglucanase	26.9	1.3	69
Culture filtrate	27.3	1.4	71
Substrate: Cotton			
Exoglucanase	14.7	1.2	53
Endoglucanase	47.2	1.8	55
Exoglucanase with endoglucanase	18.9	1.3	57
Culture filtrate	22.3	1.3	59

Incubation time was one hour in all assays.

grow. The other group of thermophilic cellulolytic micro-organisms, the actinomycetes, may be able to grow at higher temperatures than the fungi. However, as far as we know, no systematic studies of their cellulases have been carried out.

5. Sporotrichium pulverulentum

This is a basidiomycete causing white rot in wood. This fungus has been studied extensively by Eriksson's group (Eriksson and Pettersson, 1975a, b; Almin *et al.*, 1975). A survey of the results are presented by Eriksson (1978). These fungi are characterized by simultaneous attack on lignin, cellulose and other polysaccharides in wood. They have a cellulase system similar to that in the ascomycete–deuteromycete group. Thus, in *S. pulverulentum*, five endoglucanase isoenzymes were found, with molecular weights ranging between 28,300 and 37,500. With the exception of one, all of these were glycoproteins, with a carbohydrate content ranging between 10.5 and 2.2%. With carboxymethylcellulose as substrate, both the molecular activities and the Michaelis constants were determined. The molecular activities varied between 710 and 50 glucosidic bonds broken per second, and from these data it was calculated that, in the culture filtrate, the relative activity of the five endoglucanases was $4:1:1:1:1$. The K_m values varied between 9 and 2 g of carboxymethylcellulose per l. One exoglucanase was found. This enzyme formed a mixture of glucose and cellobiose as products, and is thus somewhat different from the exoglucanase in *Trichoderma viride*. It had a molecular weight of 48,600 and lacked carbohydrate. It was calculated that the weight ratio of endoglucanase protein to the exoglucanase protein in the culture filtrate of this fungus was approximately 1:1. One or several β-glucosidases were also found. These were to some extent bound to the cell wall.

The enzymes showed the normal synergistic effect on native cellulose when mixed. No such effect was found, however, when acid-swollen cellulose (Walseth-cellulose) was used as substrate. One interesting observation with the cellulase system of *S. pulverulentum* was that cellulose degradation in the culture filtrate was dependent on oxygen. In the absence of oxygen, degradation (measured as weight loss of cotton) was about the same as that from a mixture of endo- and exoglucanases, but only about half of that from a culture filtrate incubated in an oxygen-containing atmosphere. This indicates that oxidative processes also take part in cellulose degradation in this fungus. Two oxidizing systems have been purified. One is a haem-protein which can oxidize cellobiose to cellobionolactone. The cellobionic acid eventually formed might prevent transglycosylation reactions from taking place, as they would if high concentrations of cellobiose were built up. The other system is a cellobiose–quinone oxidoreductase (Westermark and Eriksson, 1974a, b, 1975). This enzyme seems to couple together degradation of cellulose and lignin. One of the three isoenzymes occurring in this fungus was isolated and purified. It is a flavoprotein with FAD as prosthetic group, and it produced cellobionolactone as the oxidation product of cellobiose. Cellopentaose was also oxidized, but not cellulose. It also oxidized lactose and β-4-glucosylmannose, but not β-4-mannosylglucose. This implies that the C-2 hydroxyl residue of the non-reducing end of the disaccharide is necessary for specificity. The quinone requirements were less specific, and the enzyme was able to reduce both *ortho*- and *para*quinones. This oxidizing enzyme worked together with a lactonase, also found in the culture medium.

Eriksson (1978) has presented a comprehensive scheme for cellulose and lignin degradation in *S. pulverulentum* (Fig. 6). It is presumed that there is an extracellular regulation of substrate breakdown, in which cellobiose transformations take an important part. Thus, the cellulo-

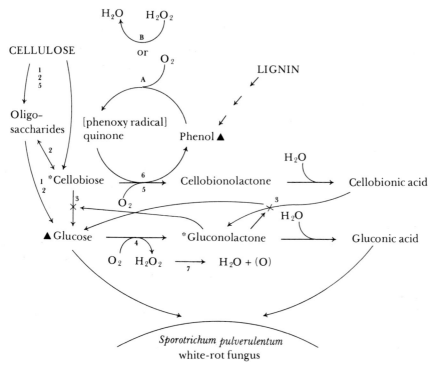

Fig. 6. Proposed enzyme mechanisms for degradation of cellulose and lignin and their extracellular regulation in *Sporotrichium pulverulentum* from Eriksson (1978). Enzymes involved in cellulose degradation are: 1, endo-1,4-β-glucanase; 2, exo-1,4-β-glucanase; 3, β-glucosidase; 4, glucose oxidase; 5, cellobiose oxidase; 6, cellobiase; 7, catalase. Enzymes involved in lignin oxidation are: A, laccase; B, peroxidase. An asterisk denotes products that regulate enzyme activity; gluconolactone inhibits 3; cellobiose increases transglycosylations. A triangle denotes products that regulate enzyme synthesis; glucose and gluconic acid → catabolite repression; phenols → repression of glucanases.

lytic system in this white-rot fungus seems to be the most complex of those described.

IV. DETERMINATION OF CELLULASE ACTIVITY

A multitude of assay systems for cellulase have been flourishing in the literature. This is natural, since the cellulase complex is a multi-enzyme system acting on a substrate where neither the real substrate concentration ('concentration' of available glycosidic bonds) nor the rate of reaction is easy to determine. We have found it convenient to distinguish between biochemical assay methods, suited for determination of the activity of separate components, and 'practical' methods, perhaps better suited for determination of the capability of a culture filtrate or a commercial cellulase product to degrade native cellulose.

A. Biochemical Assay Methods

1. Endoglucanases

For these, soluble cellulose derivatives are convenient substrates. Carboxymethylcellulose has been most frequently used, but it should be noted that the viscosity of carboxymethylcellulose solutions is affected by pH value and ionic strength of the solvent, as demonstrated by Iwasaki et al. (1964). In addition, such solutions change phase from sol at temperatures above 37°C to gel at lower temperatures (Li et al., 1965). Hydroxyethylcellulose might be preferable, since it is a neutral compound (Child et al., 1973). With both of these substrates, the decrease in viscosity is most frequently measured. For several endoglucanases, it has been found that the change in inverse specific viscosity is proportional to incubation time and enzyme concentration, at least within certain limits (Thomas, 1956; Osmundsvåg and Goksøyr, 1975; Eriksen and Goskøyr, 1976). The enzyme unit for endoglucanase activity has been defined as the amount of enzyme required to change the inverse specific viscosity by 0.001 min^{-1} under the described conditions (Osmundsvåg and Goksøyr, 1975). This unit meets one of the requirements by the International Union of Biochemistry, in that initial reaction velocities should be measured. However, the change in viscosity gives no information about the molecular reaction velocity of the enzyme. Almin and Eriksson (1967) and Almin et al. (1967) have developed a method by which viscosity changes can be recalculated to absolute units. The method is, however, quite cumbersome since it requires careful analysis of the carboxymethylcellulose used. It has thus not become widely used. What seems to be desired is that some biochemical firm should offer a standardized carboxymethyl- or hydroxyethylcellulose, with a known conversion factor from viscosity changes to absolute units.

Also, to some extent, in use for endoglucanases is the determination of the reducing end groups formed by hydrolysis. Normally, the amount of reducing end groups is expressed as glucose equivalents, determined either by the dinitrosalicylic acid method, the Nelson-Somogyi method or preferably by the ferricyanide method, which has the greatest sensitivity (Wood and McCrae, 1972). These analyses give the rate of hydrolysis directly in absolute values, but they are less sensitive than the viscosimetric technique, requiring longer incubation times. It is the only technique applicable when using insoluble cellulose as substrate. On the other hand, when using this technique, it must be borne in mind that the other enzymes in the cellulase system also form reducing end groups.

2. Exoglucanases

Direct measurements of exoglucanase activity can only be made with purified enzymes, since no available substrate is specific for exo-glucanase activity. Preferably, a cellulose with a low degree of polymerization should be used, such as Avicel or phosphoric acid-swollen cellulose. Activity must be measured as the rate of formation of reducing end groups but, for a complete analysis, the hydrolysis products should also be determined (cellobiose and/or glucose). Product inhibition of the enzyme requires that short incubation times are used, if hydrolysis products are not removed. Frequently, exo-glucanase activity is not measured directly, but only expressed in a semiquantitative manner as the synergistic effect obtained in a mixture with endoglucanase and cellobiase.

3. β-Glucosidase (Cellobiase)

For this enzyme, standard assay methods are available, such as using a β-glucoside with a chromophore (e.g. o-nitrophenyl-β-D-gluco-pyranoside; Wood, 1968) or specific determination of the glucose formed, by coupling to a glucose oxidase system.

4. Synergism (Coupled Cellulases)

For determination of the activity of a coupled cellulase system, the standard method is to measure formation of reducing end groups, with a moderately to highly polymerized cellulose (Avicel, filter paper, cotton). The degree of synergism is taken as the amount of reducing end groups formed above that of the separate components when used alone (with a highly polymerized cellulose, this is close to zero). The incubation time is primarily dependent on the sensitivity of the assay method, and lies normally between several hours and several days. The observation by Wood and McCrae (1978) that synergism experiments may give different results whether short (18 hour) or long (seven days) incubation times are used, should be noted. Only in the latter case may the action of the combined enzymes on the highly crystalline, recalcitrant parts of the cellulose be determined. This demonstrates that synergism experiments should be carried out in a manner that gives extensive hydrolysis of the substrate.

B. Practical Assay Methods

Under this heading, we will consider methods that are used for measuring the cellulolytic capability of an impure cellulase product, like those that are at present commercially available. We would like to point out however that, with increasing knowledge about the mechanisms of cellulase activity, a product declaration of commercial cellulases ought to contain information about the absolute amounts of the different cellulase components present. The practical methods can be divided into two main groups, depending on the purpose for which the cellulase will be used.

1. Saccharolytic methods

The standard method is based on saccharification of filter paper. *Filter paper activity* was defined by Mandels and Weber (1969) as the amount (in mg) of reducing sugars formed in one hour, when 50 mg of Whatman No 1 filter paper (1 cm × 6 cm) was incubated with one ml of enzyme solution and one ml of 0.05 M sodium citrate buffer (pH 4.8) at 50°C. Reducing sugars were determined by the dinitrosalicylic acid method with glucose as a standard. Since the relationship between incubation time, enzyme concentration and amount of reducing sugar formed is not linear, Mandels and Weber (1969) also introduced the term *Filter paper unit* (F.P.U.). This equals the inverse of the dilution required to give 0.50 mg of reducing sugars (as glucose) under the above-mentioned conditions. This enzyme unit is thus dimensionless. Mandels *et al.* (1975) have changed the definition of F.P.U. so that 0.5 ml of culture filtrate or enzyme solution was used for incubation (total volume 1.5 ml), and the unit is determined by diluting to a filter-paper activity of 2.0 mg of reducing sugars per hour.

Yet another definition is used by Mandels *et al.* (1976). In accordance with the recommendations from the International Union of Biochemistry, the unit is expressed as μM of glucose per min, but is still based on the release of 2 mg of reducing sugars per hour under the defined conditions. Thus, when comparing cellulase activities presented by different authors, great care must be taken to check that a comparison is at all possible.

The so-called C_1 activity is still in some use (e.g. Ghose *et al.*, 1975). This is an expression of the ability to hydrolyse recalcitrant celluloses like cotton. The incubation time normally is 24 hours or more, otherwise the conditions are the same as for determination of filter-paper activity.

Leisola and Linko (1976) have suggested the use of Avicel dyed with Remazol brilliant blue R, and measuring formation of soluble dyed products. This can be done very rapidly, after filtration of the insoluble unchanged dyed Avicel. The reaction rate is not linear with time, since rather few sites on the substituted cellulose chains will fulfil the claims for enzymic attack (unsubstituted neighbouring groups), and, at the same time, result in release of dyed fragments small enough to be soluble. With a standardized dyed Avicel, the method may, however, give useful information about the cellulolytic capabilities of a cellulase product.

2. Measurement of the degradation of organized cellulose structures

There are instances where cellulase is not used to accomplish complete saccharification, but for treatment of fibrous structures for various purposes. Several assay methods, more or less quantitative, have been used to characterize this property. The most quantitative one is probably measurement of loss of tensile strength of cotton fibres. Carefully standardized conditions must be applied (see Selby *et al.*, 1963, for a description of a macromethod with cotton yarn). Formation of free fibres from filter paper is another characteristic that seems to be mostly related to endoglucanase activity. The method, which is semiquantitative, is described by Berghem et al. (1975).

V. BIOSYNTHESIS OF CELLULASE, AND RELEASE FROM CELLS

Our discussion is based on generally accepted concepts of protein synthesis and regulation mechanisms (Lehninger, 1977). Thus, cellulase proteins are assumed to be synthesized on ribosomes in the cytoplasm with specific mRNA templates transcribed from 'cellulase genes', one gene for each cellulase isoenzyme. They are then in some manner passed through the cellular membrane and, to a greater or lesser degree, bound to cell-wall material before or simultaneously with release into the external medium. At all of these stages, reactions may take place which affect the yield of cellulase in the medium.

A. Regulation of Cellulase Protein Synthesis

Cellulases are generally assumed to be adaptive enzymes, i.e. subject to some kind of genetic regulation. As a rule, high cellulase activity is found with cellulose as the carbon source, but with no or very low activity with glucose, as long as the glucose concentration is not growth-limiting. Genetic control of enzyme synthesis is generally thought to be mediated by three different mechanisms, which may operate independently of each other, or in combination, namely induction, repression and catabolite repression.

If cellulase synthesis is regulated by an induction mechanism, the apparent difficulty is that cellulose itself cannot act as inducer, because it cannot enter the cell. Inducible enzymes are, however, normally produced in small amounts also in the absence of inducer, because the repressor system will not function as an absolute block. Thus, low levels of cellulase will be formed constitutively, and may give soluble hydrolysis products of cellulose which can enter cells and function as inducer. There has been a search for inducers of this kind, and the best candidate is probably cellobiose (Mandels and Reese, 1960). It is, however, difficult to prove this conclusively, because cellobiose and similar hydrolysis products are rapidly hydrolysed to give glucose, which undoubtedly acts as a catabolite repressor. The induction effect can thus only be obtained at low concentrations of carbohydrate, and then it is again difficult to separate from a catabolite-repression effect. In a more recent paper, Mandels *et al.* (1975) were in doubt whether cellobiose is a true inducer. The reason for this is that *Trichoderma viride*, in their experiments, always gave very low yields of cellulase after slow addition of cellobiose. High yields could be obtained by using high concentrations of cellobiose and slowing down metabolism by decreased aeration, suboptimal temperature, a marginal nutrient deficiency or a marginally toxic excess of trace metals (Mandels and Reese, 1960).

A seemingly specific cellulase inducer was discovered by Mandels *et al.* (1962) in the disaccharide sophorose (2-O-β-D-glucopyranosyl-D-glucose). This sugar can occur as an impurity in commercial glucose, and it was found to have a potent inductive activity for cellulase in *T. viride*. This was more closely studied by Nisizawa *et al.* (1971a, b). They followed extracellular cellulase (endoglucanase) formation in washed mycelia, and found that, among a number of compounds tested, only sophorose and gentiobiose enhanced extracellular cellulase formation. The effect of sophorose was about seven times that of gentiobiose. The otpimum concentration of sophorose was 1 mM. At higher concentrations, much less cellulase was formed. Cellulase was produced at a maximal rate 6–8 hours after addition of sophorose. With cellulose present, cellulase synthesis started much later. Cellulase formation was completely inhibited by puromycin and actinomycin D. In a tracer experiment, using radioactively-labelled amino acids to follow protein synthesis *de novo*, it was shown that addition of sophorose caused synthesis of proteins with chromatographic characteristics similar to those of cellulase. Thus, it seems as if addition of sophorose results in synthesis of cellulase proteins. It is noteworthy that formation of β-glucosidase but not α-amylase was also stimulated. The latter is a strong argument in favour of sophorose acting as a true inducer for the cellulase system.

Loewenberg and Chapman (1977) used tritiated sophorose to study its fate during the induction process. They found that it was rapidly metabolized to carbon dioxide and water, and only small amounts were used to induce synthesis of cellulase. Uptake of sophorose started after a lag of one hour, and reached a half-maximum value at 14 hours. The half-life of sophorose in the medium was less than five hours, but the presence of free sophorose was required for continuous cellulase production. Thus, several small additions of sophorose induced much more cellulase synthesis than an equivalent single dose.

Sophorose is efficient as an inducer in only a limited number of cellulolytic fungi. Not even all *T. viride* strains responded (Mandels *et al.*, 1962). There may be many reasons for this, such as the inability of sophorose to enter cells, inability of the fungus to convert sophorose into the inducer (if this is not sophorose), or a too rapid catabolism of the disaccharide.

Cellulase regulation has also been studied in the bacterium *Pseudomonas fluorescens* (var. *cellulose*) by Suzuki and his group. In batch culture, this bacterium produced cellulase in significant amounts on sophorose and cellulose, but not on glucose and cellobiose (0.5% sugar concentration). With slow feeding of sugars, cellulase was also produced on glucose and cellobiose (Yamane *et al.*, 1970a). The rate of bacterial growth was the same on sophorose as on glucose. From this, it was tentatively concluded that formation of cellulase was induced by sophorose and suppressed by catabolite repression under conditions of excess supply of easily metabolizable sugars. More careful studies (Suzuki, 1975) have, however, made it questionable whether cellulases really are inducible. With washed-cell suspensions, harvested from the mid-phase of exponential growth, short-time incubation experiments were carried out. With glucose, a very sharp concentration range was found where cellulase synthesis took place. Maximum cellulase formation was found at 4 mM, and this was apparently turned towards repression at 6 mM glucose. When diglucosides were added, there was a marked cellulase formation with laminaribiose and sophorose. The former gave a peak at about the same concentration (glucose equivalents) as with glucose but, with sophorose, maximum cellulase activity was found at about 8 mequiv. glucose, and no repression was seen up to 20 mequiv. This might support the hypothesis that sophorose acted as an inducer. However, when α-amylase activity was measured, very much the same behaviour was found as with cellulase, only with maxima at lower concentrations. This is in contrast to the results with *T. viride*, and may be taken to indicate that the effect of sophorose on *Ps. fluorescens* is not specific on cellulases; it does not act as an inducer.

Adenosine 3':5'-monophosphate (cAMP) is believed to play a central role in the mechanism of catabolic repression, both in prokaryote and eukaryote organisms. As long as glucose is available as an energy source, the cAMP concentration is low, thus preventing (via a specific protein) synthesis of enzymes capable of utilizing other energy sources (Lehninger, 1977). To a washed-cell suspension of *Ps. fluorescens*, cAMP as well as other nucleotides (AMP, ATP) were added together with either glucose or sophorose. In all cases, except with AMP, a decrease in relative cellulase activity was found (Suzuki, 1975). This is contrary to what should be expected if cellulase synthesis was controlled by a cAMP-mediated catabolite repression. Recently, Kight-Olliff and Fitzgerald (1978) have obtained similar effects with induction of alkylsulphatase in *Ps. aeruginosa*. They discuss the possibility that, in *Pseudomonas* sp., catabolic repression is regulated by ATP concentration, and not by cAMP, as in *E. coli*. Considering the possible regulatory mechanisms for cellulase synthesis, it should be noted that a possible alternative to catabolite repression is glucose repression, i.e. that glucose or an intermediate on the glycolytic pathway acts as an end-product repressor according to the Jacob–Monod model.

B. Release in the Medium

Not many experiments have been carried out to study this phenomenon but, with *Trichoderma viride*, Chapman and Loewenberg (1976) claim to have demonstrated that the enzyme is formed on rough endoplasmic reticulum, which is incorporated into complex Golgi equivalents that bud off cellulase-containing vesicles. These vesicles, initially found in the cytoplasm, later fuse with the plasma membrane and release their contents into the surrounding medium by exocytosis. When cellulase formation is induced, the appearance of the cytoplasmic cellulase-containing vesicles precedes the appearance of cellulase in the medium by five hours.

In *Ps. fluorescens*, Suzuki (1975) has shown that the extracellular enzymes, cellulase and α-amylase, were found predominantly in the membrane-bound polyribosome fraction, whereas the intracellular enzymes, cellobiase and β-glucosidase, were found almost equally in the free and membrane-bound polyribosome fractions. The same kind of distribution was observed when enzyme synthesis was studied in systems *in vitro*. He concluded that distinctive polyribosomes participating in formation of secretory proteins seem to exist in bacterial cells as well as in eukaryotic cells.

Berg and Pettersson (1977) have, with *T. viride*, forwarded a somewhat different view about release of cellulases into the medium. They pointed out that carbohydrates in cellulases (which as a rule are glycoproteins) resemble those in the cell wall. When cellulase production was followed in a growing culture, they found that cellulases were first bound to particles (cellulose or mycelium; this was difficult to decide) but, when growth stopped some time after cellulose had become depleted, there was a marked increase in free cellulases and a concomitant decrease in the particle-bound activity. They took this as an argument that cellulase was first bound to the cell wall, and was released together with some carbohydrate in an autolytic process. An observation that may indicate that release of cellulase is in some degree rate-limiting for formation is that addition of Tween 80 to the medium enhances cellulase formation (Mandels *et al.*, 1975; Ghose *et al.*, 1975).

As a final comment on this section, it may safely be stated that present knowledge about the biochemistry and molecular biology of cellulase synthesis is scanty. Enzyme protein synthesis is evidently regulated in some manner, but there are few, if any, stringent proofs that this is by induction, repression or catabolite repression. For the white-rot fungus *Polyporus adustus*, Eriksson and Goodell (1974) proposed that all cellulase genes are under control of a single regulator gene. But one does not know whether this is generally the case since, in most experiments with other organisms, only bulk endoglucanase activity has been measured. It might even be that regulation of cellulase synthesis is by the ribosomes, if extracellular enzymes are synthesized on special ribosomes either connected with the Golgi system or by membranes (in bacteria). Two points should be considered in this connection. One is that 'induction' of cellulase takes a very long time. The other is that cellulases are relatively inefficient enzymes, and the yield of extracellular proteins (mainly cellulases) in a cellulose-producing culture is often 20% of the weight of the cellulose consumed (Mandels *et al.*, 1975). This means that, under these conditions, cellulases are by far the dominating proteins synthesized.

VI. GROWTH CONDITIONS AND CELLULASE YIELD

A. Choice of Organism and Strain Selection

Although cellulases are produced by a number of micro-organisms, only a few appear to be good candidates for high yields of the enzyme in the culture medium. In most cases, the practical interest in the use of cellulases is connected with their saccharifying capabilities. This means that the cellulase product should contain all three components (endoglucanase, exoglucanase and β-glucosidase) in appropriate amounts. As a consequence, fungi within the ascomycete and deutero-mycete groups, possibly together with some white-rot basidiomycetes, are the most likely candidates. Within these groups, there is a considerable diversity in cellulase production, as shown in Table 4.

Mainly through the work of Reese and Mandels and their collaborators, *T. viride* has been singled out as the most potent cellulase producer, and a majority of the studies on the physiology of cellulase

Table 4

Production of cellulase by various fungi. From Mandels and Weber (1969)

Strain number	Organism	Cellulase activity		Hydrolysis of cotton (per cent)	
		C_1 Units	C_x Units	Sugar	Weight loss
Buffer alone (0.05 M acetate)		0	0	0	1
6a	*Trichoderma viride*	50.0	50.0	58	53
826	*Chrysosphorium pruinosum*	30.0	70.0	11	19
137g	*Penicillium pusillum*	27.0	110.0	22	23
1224	*Fusarium moniliforme*	NT	3.5	39	39
72f	*Aspergillus terreus*	5.0	36.0	28	24
806	Basidiomycete	5.0	75.0	15	23
94d	*Stachybotrys atra*	1.0	8.0	5	6
B814	*Streptomyces* sp.	0.7	40.0	9	12
38g	*Fusarium roseum*	0.7	10.0	9	10
381	*Pestalotiopsis westerdijkii*	0.7	60.0	4	8
460	*Myrothecium verrucaria*	0.4	28.0	2	2
459	*Chaetomium globosum*	0.2	0.5	NT	NT

Cultures were grown on Solka Floc, except *Penicillium pusillum* 137g which was grown on cotton duck. C_1 units indicate action on cotton sliver for 24 hours at 40°C. C_x units indicate action on carboxymethylcellulose for 30 minutes at 50°C. NT indicates that the value was not tested. Hydrolysis of cotton was measured by acting on 1% cotton sliver for 35 days at 29°C (four changes of enzyme).

formation has been carried out on this organism. As has been emphasized by several authors, interest in *T. viride* should not lead one to ignore other micro-organisms. Among such should be mentioned *T. koningii* and probably other *Trichoderma* species, *Fusarium solani* (Wood and Phillips, 1969), *Penicillium funiculosum* (Wood and McCrae, 1977b) and *Sporotrichium pulverulentum* (Eriksson, 1978).

Even within one species, there may be considerable strain differences in cellulase production. This is shown for *T. viride* in Table 5 (Mandels and Weber, 1969) where cellulase activity for a number of wild-type strains is quoted. By selection of mutants, strains with even higher cellulase activity have been obtained. Thus, Mandels *et al.* (1971) have isolated mutants of *T. viride* strain QM6a which gave between two and three times higher cellulase activities, measured as filter-paper units. These strains, designated QM9123 and QM9414, are probably among the most potent cellulase producers extensively studied. They are available from the American Type Culture Collection as A.T.C.C. numbers 24449 and 26921, respectively. Morozova (1975) has discussed possible genetic approaches to the improvement of cellulase-producing strains, and also isolated mutants with enhanced cellulase activity.

Recently, Montenecourt and Eveleigh (1977) have prepared mutants from *T. viride* QM6a (by them called *T. reesei*). One of these, named NG-14, is claimed to produce five times the filter paper-degrading

Table 5

Cellulase activity of some *Trichoderma viride* strains. From Mandels and Weber (1969)

Strain number	Cellulase activity		Hydrolysis of Solka Floc (per cent)
	C_1 Units/ml	C_x Units/ml	
6a	33	44	57
317	31	30	80
3161	28	19	49
2098	25	18	65
7733	25	17	47
4093	19	7	23
2647	8	16	36

Cultures were grown on 1.0% Solka Floc. Hydrolysis of 6% Solka Floc was for 42 days at 29°C. For an explanation of enzyme units, see the caption for Table 4.

activity per ml of culture medium and twice the specific activity per mg of excreted protein, when compared with strain QM9414. One interesting property of stain NG-14 is that the cellulase complex is partially derepressed. Complete derepression was not obtained, although mutant selection took place under conditions of catabolic repression (5% glycerol as carbon source).

B. Medium Composition

No general medium composition can be given for growth and optimum cellulase production, since the medium must be adapted to the organism in use. In the many experiments with *T. viride*, a basal medium after Mandels and Reese (1957) has been most frequently used, either directly or with slight modifications. This is a mineral medium with nitrogen supplied as a mixture of ammonium sulphate and urea, and with trace elements added. The pH value of this medium is 5.3. The medium has often been modified by addition of proteose peptone and Tween 80. The carbon source used in connection with this basal medium has varied, according to the purpose of the investigation.

A philosophy frequently met with in cellulase studies is that it is desirable to produce the enzyme on the same substrate that is to be hydrolysed, in order to ensure the proper mix of enzymes (Mandels *et al.*, 1975). This is certainly true if the growing culture is used as the enzyme source directly. If, on the other hand, the goal is to produce cellulase in high yields for commercial purposes, the price of cellulase at present is so high compared with that of pure cellulose, that little is gained in not using pure cellulose if this increases yield or production efficiency.

C. Culture Techniques

Practically all reports on cellulase production are based on small-scale experiments, either with batch cultures, which may be stationary or submerged with aeration by shaking or by sparges, or continuous or semicontinuous cultures, with a more or less sophisticated control of growth parameters. In Japan, a *koji*-type process has been in use for commercial production of cellulase from *T. viride*. In this process, wheat bran is steamed, piled in trays and inoculated with a spore suspension. The trays are sprayed during growth with a water mist, and forced air is circulated among the trays. When growth is completed, the *koji* is extracted, the extract filtered, and the enzyme precipitated with ammonium sulphate. If further purification is desired, the enzyme is redissolved, passed through an ion-exchange resin, and precipitated with solvent (Toyama, 1963). The *koji*-process is quite effective in producing high yields of cellulase, but it requires considerable handling. Two Japanese patents have been issued for production of cellulase by submerged fermentation, on nutrient salts together with 12.5% wheat bran (Kawaji *et al.*, 1964) or 3% soybean cake (Maejima and Yoshino, 1964).

D. Culture Conditions

In a growing culture, cellulase yields appear to depend on a complex relationship involving a variety of factors, such as inoculum size, carbon source (cellulose quality), pH value, temperature, presence of inducers and/or inhibitors, medium additives (surface-active agents), batch size, aeration and growth time. From different experiments carried out in a number of laboratories, it is hard to make generalizations about the effects of each of these factors separately. We have therefore preferred to present a number of typical growth and cellulase-production experiments, together with the conclusions that may be drawn from each of these.

1. Experiments with Trichoderma viride

Mandels and Weber (1969) used a laboratory fermenter with a 10-litre culture volume in a semicontinuous system to produce relatively large quantities of cellulase from *T. viride*. The carbon source was Solka Floc cellulose, added to the standard mineral medium which had been supplemented with 0.05–0.2% peptone and 0.1% Tween 80. The initial pH value was 5.0 and the temperature 29°C. The culture was aerated with 3 litres/min, and stirred by an impeller at 120 rev./min. With this system, more than 100 l of active culture filtrate was harvested in 60 days. The maximum filter-paper activity was 2.26. This value was only half of that obtained with shake flasks. In shake flasks, however, 12–18 days of growth were necessary to obtain highest activity. The fermenter was run with a dilution rate of between 0.1 and 0.3 per day (0.0042–0.0125 h^{-1}). During growth, the pH value fell to about 3.0. From their different experiments, they concluded that *T. viride* produced cellulase when grown on cellulose together with nutrient salts alone, but higher and more stable yields were often obtained when a small quantity of a soluble carbohydrate (0.05–0.1% glucose) was added to the medium. The optimum cellulose concentration was between 0.5 and 1.0%, depending on the type of activity measured, and on the concentration of peptone in the medium. Optimum peptone concentration was 0.1 to 0.2%, depending on the cellulose concentration. Concentrations higher than 0.5% were strongly inhibitory to production of cellulase. Tween 80 was also found to stimulate cellulase production, particularly when peptone was present. Foam was a problem in some experiments, and it did not respond well to addition of antifoam agents. But, when Tween 80 was added, the foam became light and ceased to be a problem. Mandels and Weber (1969) concluded that the semicontinuous system appeared be a practical technique for producing cellulase on a large scale. *Trichoderma viride* grew vigorously and in a homogeneous manner, and the low pH values and use of cellulose as substrate did not encourage contaminants.

In a later publication, Mandels *et al.* (1975), using the same type of semicontinuous fermenter system, studied cellulase production in more detail. In this case, pH value was recorded continuously during the experiment. Two independently controlled peristaltic pumps were used for feeding and harvesting the culture. Every 15 minutes, the harvest pump was switched on for a 5–10 second pulse, depending on the dilution rate required, allowing a delivery of 25–50 ml of culture suspension. This was followed by an identical pulse for the nutrient or feed pump to deliver the replacement volume. This rapid-pulse feeding was found to function much better than a slow continuous feeding for suspensions of cellulose or mycelium. Growth was at 28°C with impeller speed at 100–200 rev./min and aeration at 0.2–0.4 litre per litre of culture volume per min.

In this experimental system, the two mutant strains of *T. viride*, QM9123 and QM9414, were found to grow readily on many substrates. A vigorous production of acid occurred when carbohydrate was consumed above a minimal rate and with ammonium as the nitrogen source. The rate of acid production was directly related to the rate of carbohydrate consumption. A significant feature of all pH value patterns was that, as soon as the carbohydrate was consumed, pH value started to rise. Thus, the organism had the ability to neutralize or reverse the acid condition it had produced, perhaps by secretion of ammonium ions. Under unfavourable conditions, *T. viride* will grow slowly on cellulose, producing low levels of cellulase. In such cases, there was only a slight fall in pH value. Under favourable conditions, large quantities of cellulase were produced and the rate of acid production was even higher than with glucose, indicating that, during early growth, the organism may be capable of consuming cellulose as rapidly as glucose. However, when grown on cellulose, much higher levels of extracellular proteins (including cellulases) were produced compared with on glucose.

During the rapid fall in pH value, cellulases were produced. This indicated that cellulase synthesis is not brought about by a slow feed of carbohydrate, but rather that cellulase 'induction' occurred during a period of high metabolic activity. The authors think that acid production during growth may have a regulatory function. Thus, rapid metabolism and the corresponding fall in pH value affected the activities of β-glucosidase and cellulase. When the pH value fell from 4.0 to 3.0, most of the β-glucosidase activity was lost. The saccharifying ability remained stable until a pH value of 2.5, where about 45% of the activity was lost. Endoglucanase appeared to be the most stable, decreasing by only about 20% at this pH value. By limiting β-glucosidase and cellulase activities, glucose uptake would also be limited. This again would result in a retarded metabolism and a rise in pH values, permitting further accumulation of enzyme. This system thus allows for a negative feedback control over enzymes that exist outside the cells. A practical consequence is that a careful control of pH value is necessary for obtaining maximum yields of cellulase.

Mandels *et al.* (1975) further confirmed the positive effects on cellulase production by adding peptone and Tween 80 to the medium. Another interesting observation is that, while the cultures grew most rapidly at 30°C, maximal enzyme production was obtained at 25°C. The inoculum had a marked effect on growth and cellulase production in cellulose-containing media. Use of mycelial inoculum shortened the lag that occurred when spores were used.

Ghose *et al.* (1975) studied cellulase formation by *T. viride* QM9123 in shake flasks, 5-litre stirred tanks and shallow fermenters. The standard mineral medium of Mandels and Weber (1969), to which 0.075% proteose peptone was added, was used. Highest yield of cellulase was found with 0.5% Solka Floc cellulose and with 0.7% VP cellulose (a purified powdered cellulose from native cotton). The latter substrate in shake flasks gave the highest yields so far recorded. In 5-litre stirred fermenters the yields were 2–3 times lower than in shake flasks. Peptone, at 0.075% concentration, improved the yield compared with other organic nitrogen sources. Tween 80, at 0.5% concentration, gave maximal cellulase yield when added to the culture medium after 24 hours of growth. Sodium oleate, optimal at 0.3%, was less efficient than Tween 80. A single feeding of cellobiose (0.01%) enhanced cellulase yield slightly compared with no feeding of cellobiose. Better yields were obtained using step feeding of cellobiose, with one 0.005% addition per day. In the fermenter, increasing the agitation above 100–200 rev./min had no positive effect on cellulase yield. The same was found for aeration at rates above one litre per litre of culture volume per min. Maximal rate of cellulase synthesis was found between the sixth and eighth day. Analysis of the growth data showed that cellulase biosynthesis was not growth associated.

Sternberg (1976) has shown that a doubling of cellulase production could be obtained in batch fermentations with 10-litre volume by increasing the cellulose concentration from 0.75 to 2% (Solka Floc cellulose), by increasing the nitrogen concentration in the medium correspondingly, and by controlling pH value. When left uncontrolled, pH value may drop to about 2.5 in a medium containing 2% cellulose. Growth of *T. viride* will then be extremely slow, and cellulase enzymes are inactivated. If the pH value was kept between 4.0 and 3.0, 2% cellulose gave higher yield than 0.75%, while, at pH 5.0, the yields were about the same. Production of β-glucosidase occurred later in the fermentation than that of the other cellulase enzymes. β-Glucosidase production occurred when the pH value was rising and the enzyme was protected from the inactivation which took place at pH values below 3.5. The final levels of β-glucosidase in cultures containing 2% cellulose were 4–5 times higher than in those containing 0.75% cellulose.

Peitersen (1977) used a continuous-culture system with ball-milled cellulose as the carbon source, to determine the growth characteristics and cellulase production of *T. viride* at pH 5.0 and 30°C. Steady-state values for cell protein, cellulose and cellulase for different substrate concentrations and dilution rates ($0.033–0.080$ h^{-1}) were obtained. Under steady-state conditions, 60–75% of the cellulose was consumed. Cellulase activity in the fermentation broth increased slightly with increasing substrate concentration, and decreased with increasing dilution rate, whereas the specific rate of cellulase production (enzyme units per mg of cell protein per hour) was fairly independent of dilution rate, with a maximum around a value of 0.05 h^{-1}. Following step changes in substrate concentration and dilution rate, new steady-state values were reached after 3 to 5 residence times for cell protein and cellulose, and 4 to 6 residence times for cellulase activity.

Brown *et al.* (1975) and Brown and Zainudeen (1977) have explored the possibilities for cellulase production by *T. viride* QM9123 with glucose as the single carbon source. Relatively high cellulase activities could be obtained in stationary-phase cultures in 70 hours (2.4 filter paper units), if the minimum pH value during growth was low (between 2.5 and 3.0) and there was some oxygen limitation. In continuous cultures (Brown and Zainudeen, 1977), growth characteristics were studied at three controlled pH values (2.5, 3.0 and 4.0) over a range of dilution rates. Steady-state conditions were obtained, and values for maximum specific growth rate, endogenous metabolism coefficient, yield and maintenance coefficients for glucose were derived and correlated to the effect of hydrogen-ion concentration. It was concluded that the effect of increased hydrogen-ion concentration was a slowing down of metabolic processes without greatly altering the stoicheiometry and without causing any deterioration in cell structure. Steady-state cellulase concentrations fell as dilution rates increased. They have presented a product-formation model based on the idea that the cells require a period of time to elapse before they are mature enough to produce enzyme. This model was found suitable for correlation of the results at pH 3.0.

One of the problems in large-scale production of cellulase is the long fermentation times required. In most of the laboratory experiments reported so far, this is partly due to the use of a small inoculum, causing lag and/or a long growth time before enough cellulase-producing mycelium is formed. Mitra and Wilke (1975) have suggested the use of a two-stage fermenter, growing *T. viride* on glucose in the first stage and on cellulose in the second. A significant increase in cellulase production over single-stage fermenters was found. In a pilot-plant investigation using a 400-litre fermenter, Nystrøm and

Kornuta (1975) have shown that fast and reproducible growth could be obtained by increasing the inoculum volume from 1% of the culture volume to 10%. The inoculum was pregrown on the same cellulose-containing medium, but the cellulase activity of the inoculum was of little importance for the final yield of cellulase.

Besides these studies on cellulase production from *T. viride* on cellulose, there are reports on growth and cellulase production on cellulosic waste materials. Peitersen (1975a) used barley straw, and showed that straw pretreated with sodium hydroxide under pressure was a satisfactory source of cellulose for *T. viride* QM9123. Peitersen (1975b) extended this by using a mixed culture of *T. viride* and a yeast, either *Saccharomyces cerevisiae* or *Candida utilis*. The yeast was inoculated 24–32 hours after the fungus. An aerated 5-litre fermenter was used, with barley straw that had been treated with 2–4% sodium hydroxide as the carbon source. In comparison with fermentations employing *T. viride* alone, production time for maximum yields of cellulase and cell protein was lowered by several days. Besides accelerating cellulase production, it was also found that the yeast had a positive effect on growth of the fungus.

Griffin *et al.* (1974, 1975) found that feedlot waste contained essentially all of the necessary nutrients for batch fermentation of *T. viride*. The organism could utilize two-thirds of the carbohydrate in the feedlot waste while producing cellulases in quantities comparable to the yields on cellulose. For good degradation of the fibres in the feedlot waste, it was found necessary to separate them from liquids. Fermentation of feedlot waste with *T. viride* appeared an interesting method for recycling this waste material.

2. Experiments with other fungi

Relatively few comprehensive papers have been published on the physiology of cellulase production in micro-organisms other than *T. viride*. Eriksson and Larson (1975) have studied fermentation of waste mechanical fibres from a newsprint mill by *Sporotrichium pulverulentum*. They also used cellulose from other sources with different degrees of polymerization and crystallinity, and concluded that the more complex the carbon source, the more difficult it was to digest and the more enzyme had to be produced for its degradation. Among the nitrogen sources tested, highest cellulase activity was obtained (after four days) with urea, and with the pH value adjusted to 4.6 before autoclaving. It is suggested that this is due to the relatively high final pH value in the medium with this nitrogen source. This is opposite to the results with *T. viride*, where a low pH value was desirable. In both cases, however, the argument is that a pH value different from the optimum for the cellulases stimulated the fungus to produce more cellulase. By using data from other research, where the endo- and exoglucanases of *S. pulverulentum* have been extensively purified, it was calculated that the extracellular enzymes could account for approximately 30% of the protein in the mycelium, while the endo- and exoglucanases accounted for up to 55% of the extracellular protein.

VII. APPLICATION OF CELLULASES

A. Present Uses

In Japan, cellulases (from *A. niger* and *T. viride*) are produced commercially at a rate of 45 tonnes per year, representing an economic output per year of 170×10^6 yen (Yamada, 1977). According to Toyama (1969), production at that time was somewhat higher, and about 6 tonnes were exported to Europe (West Germany) and 1.2 tonnes to Australia.

Commercial production of cellulases and an efficient utilization of these enzymes are infested with many problems. Both the economic cost and the limited efficiency of known cellulase preparations make many potential applications non-profitable at present. To increase process efficiency, one can search for more active cellulases, or develop cheap methods for pretreatment of the substrates, making them more accessible for enzymic degradation. Most of the potential raw materials for an enzymic saccharification industry contain, in addition to cellulose, various concentrations of other compounds which make them not directly suitable for enzymic hydrolysis. For most of these materials, chemical treatments of different kinds are necessary and, in many cases, also mechanical treatments to decrease particle size and induce swelling. Another factor one must take into consideration is the product inhibition found with cellulases. Thus, in order to obtain an efficient hydrolysis, the lower molecular-weight carbohydrates formed should be removed at rates proportional to their formation. This can be accomplished either mechanically, by dialysis through membranes, by enzyme immobilization, or by coupling directly to a fermentation process where glucose is consumed.

For these reasons, use of cellulases had been limited to a few specific applications, and probably mainly as components of digestive aids together with other hydrolytic enzymes. Such digestive aids are described as useful for extractions of green-tea components, soybean or coconut-cake protein, sweet potatoes or corn starch and agar; production of unicellular vegetables, vinegar from citrus pulp and seaweed jelly; for removing soybean seedcoats; modifying food tissues such as vegetables, rice and glutinous rice; isolation of plant unicells and protoplasts in an active state; decomposition of cooked soybean; enzyme digestion of excreta; and increasing tensile strength of paper (Toyama, 1969). The cellulase exported to Australia is reported to be used as a feed supplement for cattle.

B. Pilot Investigations

1. Cellulase-reactor systems

Katz and Reese (1968), using concentrated enzyme from *T. viride*, produced hydrolysates containing 30% glucose. The experiment was done on a very small scale (1 ml), and in larger scale experiments the yields were considerably lower. Ghose and Kostick (1969) studied enzymic saccharification of cellulose pulp (Solka Floc) in semi- and continuously agitated systems by using untreated enzymes from submerged fermentation of *T. viride*. Pretreatment of the cellulose by heating, followed by grinding, and grinding followed by heating, led to considerable differences in the rate of saccharification. It was possible to reactivate the residues from the digestion to a high degree of susceptibility to enzymic hydrolysis by heat treatment and milling. Saccharification of cellulose was shown to be possible by milling the substrate in contact with the enzyme. Semi-continuous hydrolysis of heated and milled Solka Floc (10%, w/v), with daily replacement of 40% of the reaction volume with fresh enzyme-substrate suspensions, maintained the sugar concentration in the effluent over a period of several days at 5.1–5.6%. Ghose and Kostik (1970) described a model for continuous enzymic saccharification of cellulose with simultaneous removal of glucose syrup. Cellulases from *T. viride* were concentrated in various cut-off membranes. The concentrated (5–8-fold) enzymes were used to saccharify finely ground cellulose (Solka Floc) in stirred tanks and membrane reactors. Nearly 14% glucose concentration was achieved in less than 50 hours in stirred tanks by digesting a 30% cellulose suspension. Based on the experimental data, a model system was proposed for continuous steady-state saccharification of finely ground cellulose with continuous removal of concentrated glucose syrup, and a feedback of enzyme.

Markkanen and Eklund (1975) studied hydrolysis of furfural-process waste which contained about 40% cellulose. The furfural process proved to be satisfactory pretreatment of cellulose before enzymic hydrolysis. With *T. viride* QM9414 as the enzyme source, a syrup containing about 5% glucose could be obtained in four days using high concentrations of enzyme. The enzyme converted α-cellulose into glucose almost completely. Careful control of pH value seemed to be important for increasing the efficiency of the enzyme. The hydrolysate was shown to be suitable for growth of micro-organisms. An advantage with this system was that the furfural-process waste could be enzymically hydrolysed directly, since the furfural process itself destroyed the crystalline structure of cellulose. The main problem seemed to be that the enzyme preparations should be more concentrated in order to get high concentrations of glucose in a short time.

Toyama and Ogawa (1975) studied saccharification of agricultural cellulosic wastes with cellulase from *T. viride*. Production of thick sugar solutions was attempted by saccharifying delignified rice straw and bagasse with commercial preparations of cellulase. Powdered wastes were delignified by boiling with 1% sodium hydroxide for 3 hours. After incubation with a 1% Meicelase CEP-233 solution at 10% substrate concentration, a pH value of 5.0 and a temperature of 45°C for 48 hours, solutions containing 8.5% sugar were obtained. With a concentration of 25% substrate, solutions containing 13% sugar were obtained under otherwise similar conditions. They found that the presence of lactose in the commercial enzyme preparation significantly inhibited cellulolytic activity. Inactivation of cellulolytic activity with xylose and glucose was also studied. Decomposition was not obtained if the alkali treatment of the substrates was omitted. This treatment, on the other hand, caused a considerable weight loss of about 50%.

Immobilized cellulase, in a collagen-fibril matrix was studied by Karube *et al.* (1977). This immobilized cellulase was more stable than

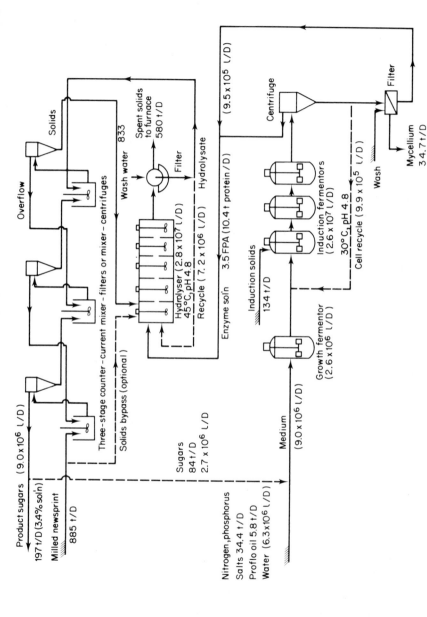

Fig. 7. Flow diagram for an enzymic hydrolysis plant utilizing 885 tonnes (t) newsprint per day (D). From Wilke and Yang (1975).

the native cellulase, but had also distinctly less activity during the first 40 hours of an incubation experiment. Later, the immobilized cellulase was more active than the native form. Cellulose was found to be hydrolysed quatitatively with immobilized cellulase, and the final reaction product was identified as glucose. It was used in a recycling fluidized-bed reactor with 100 ml total volume. The optimum flow velocity for enzyme hydrolysis was 1 cm/second.

As far as we know, large-scale saccharification reactor experiments have never been carried out. However, Wilke and his collaborators have made process-development studies and cost analysis for enzymic hydrolysis of newspaper waste. Wilke and Yang (1975) presented detailed plans for a plant that would hydrolyse 885 tonnes of newsprint per day. The plant (Fig. 7) consisted of a set of fermenters for growth and cellulase production by *T. viride*. The fungus was first grown on glucose, whereafter cellulase induction was carried out in three consecutive fermenters with newsprint cellulose (134 tonnes per day) as substrate. The culture supernatant was centrifuged off and used directly as a source of enzyme, while part of the mycelium was recycled to the first induction fermenter. The hydrolyser was fed with milled newsprint and enzyme solution, operated at 45°C, and had a retention time of 40 hours. The reaction product was contacted counter-currently with the feed solids in a series of three mixer-filter stages for enzyme recovery. By this process, a total enzyme recovery of 95% was predicted on the basis of enzyme-absorption studies carried out in the laboratory.

Cost analyses were carried out for a numbr of different modes of operation. The two dominating parameters in determing the cost per kg of glucose were the conversion efficiency of cellulose into glucose, and the enzyme activity of the culture broth. Assuming 50% conversion, the best situations gave a cost of between 0.12 and 0.08 U.S.$/kg of glucose in a 3.5–4% solution. Increasing the conversion efficiency to 69% gave a cost of 0.053 U.S.$/kg.

Grethlein (1978) has compared this system with the costs for acid hydrolysis of newsprint in a plant of similar size, and found that the cost for acid hydrolysis is in the range of 0.039 to 0.054 U.S.$/kg. Acid hydrolysis is thus cheaper (and probably also requires a less sophisticated plant to operate) than enzymic hydrolysis. The difference is not so great, however, that it may not change in favour of enzymic hydrolysis if more efficient strains for cellulase production were developed.

Wilke *et al.* (1976) made plans for an integrated processing scheme in which the sugar solution produced by enzymic hydrolysis was used directly in a fermentation plant for production of ethanol and yeast. Preliminary designs and cost studies were developed to provide a rough perspective on the potential economic feasibility of this technique for utilization of waste cellulose. Their analysis suggested that, as energy costs increase and petrochemical raw materials become scarce, production of ethanol from cellulosic waste materials could become a viable alternative. Menezes *et al.* (1978) investigated the effect of fungal cellulases as an acid for saccharification of cassava. Culture filtrates from *T. viride* and an unidentified soil basidiomycete, when used together with commercial amylases, enhanced both the rate of sugar formation and the degree of solubilization, at the same time as it increased the viscosity of hydrolysates. An increased ethanol yield would thus be expected from alcoholic fermentation of this hydrolysate. They expected this increased yield to compensate for the extra costs of the cellulase production.

2. *Fermentations with cellulolytic micro-organisms*

An alternative to using cellulases in saccharification systems is to use cellulolytic micro-organisms directly in fermenter operations, either singly or in mixture with non-cellulolytic micro-organisms. Especially for protein production, this would be an attractive possibility. Updegraff (1971) made extensive screening studies on celluloytic fungi and bacteria. He found that *Myrothecium verrucaria* gave the maximum rate of protein biosynthesis from ball-milled newspaper. The maximum rate of protein biosynthesis and the maximum protein yields were 0.3 g/l/day and 1.42 g/l, respectively, from medium containing 4% solid substrate. By evaporation of the culture, a final product was obtained which contained 3.3 g of protein (10% of the total solids) per litre of medium. Both the synthesis rate and the final yield were below those obtainable by growing Fungi Imperfecti, yeasts or bacteria on soluble substrates, and the process seemed unlikely to compete economically with other protein-producing processes.

Ek and Eriksson (1975) described fermentation experiments with the intention of developing processes based on solid lignocellulosic waste. The white-rot fungus *Sporotrichium pulverulentum* was used as the organism, and most of the work was concentrated on conversion of waste fibres from a sulphite mill into protein. A protein yield of

25–30% was obtained in batch cultures. Combined cultures with *S. pulverulentum* and *Candida utilis* were also tried. A similar mixed-culture system was tried by Peiterson (1975b), with *T. viride* and *Sacch. cerevisiae*. He found that addition of the yeast resulted in an increased overall rate of protein production, but concluded also that such a process cannot compete economically with other protein preparations like soyflour, fish protein concentrates or single-cell protein from more easily accessible substrates.

Single-cell protein production by the thermotolerant fungus *Chaetomium cellulolyticum* was studied by Moo-Ygung *et al.* (1977). This fungus showed a 50–100% faster growth and over 80% more final biomass protein formation than *T. viride*, when cultivated on Solka-Floc cellulose or partially delignified sawdust. Chahal and Wang (1978) found that *C. cellulolyticum*, growing on 1% Solka-Floc, grew with a protein-synthesis rate of 0.09 h^{-1}. The rate of protein synthesis increased more than three-fold when more substrate was added immediately after complete conversion of the first batch of substrate. These authors claim that *C. cellulolyticum* might be the most suitable organism for cyclic-batch or continuous fermentation of cellulose for protein production.

Only a few investigations have been done using cellulolytic bacteria. Srinivasan (1975) and Srinivasan *et al.* (1977) reported studies on growth and protein production with *Cellulomonas* sp., a cellulolytic coryneform bacterium, grown either alone or in combination with other micro-organisms. A considerable increase in yield was normally obtained in mixed cultures. Humphrey (1975) has likewise reported preliminary studies with *Thermoactinomyces* sp., a thermophilic cellulo-lytic actinomycete.

REFERENCES

Almin, K. E. and Eriksson, K.-E. (1967). *Biochimica et Biophysica Acta* **139**, 238.
Almin, K.-E., Eriksson, K.-E. and Pettersson, B. (1975). *European Journal of Biochemistry* **51**, 207.
Almin, K.-E., Eriksson, K.-E. and Jansson, C. (1967). *Biochimica et Biophysica Acta* **139**, 248.
Berg, B. (1975). *Canadian Journal of Microbiology* **21**, 51.
Berg, B. and Pettersson, G. (1977). *Journal of Applied Bacteriology* **42**, 65.
Berg, B., Hofsten, B. and Pettersson, G. (1972a). *Journal of Applied Bacteriology* **35**, 201.
Berg, B., Hofsten, B. and Pettersson, G. (1972b). *Journal of Applied Bacteriology* **35**, 215.
Berghem, L. E. R. and Pettersson, L. G. (1973). *European Journal of Biochemistry* **37**, 21.

Berghem, L. E. R. and Pettersson, L. G. (1974). *European Journal of Biochemistry* **46**, 295.

Berghem, L. E. R., Pettersson, L. G. and Axiö-Fredriksson, U.-B. (1975). *European Journal of Biochemistry* **53**, 55.

Berghem, L. E. R., Pettersson, L. G. and Axiö-Fredriksson, U.-B. (1976). *European Journal of Biochemistry* **61**, 621.

Brown, D. E. and Zainudeen, M. A. (1977). *Biotechnology and Bioengineering* **19**, 941.

Brown, D. E., Halsted, D. J. and Howard, P. (1975). In 'Symposium on Enzymatic Hydrolysis of Cellulose' (M. Bailey, T.-M. Enari and M. Linko, eds.), pp. 137. SITRA, Aulanko, Finland.

Chahal, D. S. and Wang, D. I. C. (1978). *Mycologia* **70**, 160.

Chapman, C. M. and Loewenberg, J. R. (1976). *Supplement to Plant Physiology* **57**, 70.

Child, J. J., Eveleigh, D. E. and Sieben, A. S. (1973). *Canadian Journal of Biochemistry* **51**, 39.

Cowling, E. B. and Brown, W. (1969). 'Cellulases and Their Applications'. Advances in Chemistry Series (R. F. Gould, ed.), vol. 95, p. 152. American Chemical Society Publications, Washington, D.C.

Donefer, E., Adeleye, I. O. A. and Jones, T. A. O. C. (1969). In 'Cellulases and Their Applications', Advances in Chemistry Series (R. F. Gould, ed.), vol. 95, p. 328. American Chemical Society Publications, Washington, D.C.

Eidså, G. (1972). Thesis: University of Bergen, Norway.

Ek, M. and Eriksson, K.-E. (1975). In 'Applied Polymer Symposium', no. 28, p. 197. John Wiley & Sons, Inc.

Emert, G. H., Gum, E. K., Lang, J. A., Liu, T. H. and Brown, R. D., Jr. (1974). In 'Food Related Enzymes', Advances in Chemistry Series, vol. 126, p. 79. American Chemical Society Publications, Washington, D.C.

Eriksen, J. and Goksøyr, J. (1976). *Archives of Microbiology* **110**, 233.

Eriksen, J. and Goksøyr, J. (1977). *European Journal of Biochemistry* **77**, 445.

Eriksson, K.-E. (1978). *Biotechnology and Bioengineering* **20**, 317.

Eriksson, K.-E. and Goodell, E. W. (1974). *Canadian Journal of Microbiology* **20**, 371.

Eriksson, K.-E. and Larson, K. (1975). *Biotechnology and Bioengineering* **17**, 327.

Eriksson, K.-E. and Pettersson, B. (1972). In 'Biodeterioration of Materials' (A. H. Walters and E. H. Hueck-Van Der Plas, eds.), vol. 2, p. 116. Applied Sciences Publishers Ltd., London.

Eriksson, K.-E. and Pettersson, B. (1975a). *European Journal of Biochemistry* **51**, 193.

Eriksson, K.-E. and Pettersson, B. (1975b). *European Journal of Biochemistry* **51**, 213.

Frey-Wyssling, A. and Mühlethaler, K. (1963). *Makromolekulare Chemie* **62**, 25.

Gardner, K. H. Blackwell, J. (1974). *Biochimica et Biophysica Acta* **343**, 232.

Gascoigne, J. A. and Gascoigne, M. M. (1960). *Biological Degradation of Cellulose*. Butterworth and Co. Limited, London.

Ghose, T. K. and Kostick, J. A. (1969). In 'Cellulases and Their Applications', Advances in Chemistry Series (R. F. Gould, ed.), vol. 95, p. 415. American Chemical Society Publications, Washington, D.C.

Ghose, T. K. and Kostick, J. A. (1970). *Biotechnology and Bioengineering* **12**, 921.

Ghose, T. K., Pathak, A. N. and Bisaria, V. S. (1975). In 'Symposium on Enzymatic Hydrolysis of Cellulose' (M. Bailey, T.-M. Enari, and M. Linko, eds.), p. 111. SITRA, Aulanko, Finland.

Grethlein, H. E. (1978). *Biotechnology and Bioengineering* **20**, 503.

Griffin, H. L., Sloneker, J. H. and Inglett, G. E. (1974). *Applied Microbiology* **27**, 1061.

Griffin, H. L., Kaneshiro, T., Kelson, B. F. and Sloneker, J. H. (1975). In 'Symposium on Enzymatic Hydrolysis of Cellulose' (M. Bailey, T.-M. Enari, and M. Linko, eds.), p. 419. SITRA, Aulanko, Finland.

Gum, E. K. and Brown, R. D. (1976). *Biochimica et Biophysica Acta* **446**, 371.

Halliwell, G. (1961). *Biochemical Journal* **79**, 185.

Halliwell, G. (1965a). *Biochemical Journal* **95**, 35.

Halliwell, G. (1965b). *Biochemical Journal* **95**, 270.

Halliwell, G. (1966). *Biochemical Journal* **100**, 315.

Halliwell, G. and Griffin, M. (1973). *Biochemical Journal* **135**, 587.

Hepler, P. K., Foskett, D. E. and Newcomb, E. H. (1970). *American Journal of Botany* **57**, 85.

Heyn, A. N. J. (1965). *Journal of Applied Physiology* **36**, 2088.

Highley, T. L. (1975). *Wood and Fiber* **6**(4), 275.

Humphrey, A. E. (1975). *In* 'Symposium on Enzymatic Hydrolysis of Cellulose' (M. Bailey, T.-M. Enari and M. Linko, eds.), p. 437. SITRA, Aulanko, Finland.

Iwasaki, T., Tokuyasu, K. and Funatsu, M. (1964). *Journal of Biochemistry, Tokyo* **55**, 30.

Karube, I., Tanaka, S., Shirai, T. and Suzuki, S. (1977). *Biotechnology and Bioengineering* **19**, 1183.

Katz, M. and Reese, E. T. (1968). *Applied Microbiology* **13**, 419.

Kawaji, S., Ishikawa, T., Saito, K., Inohara, T., Kubo, A. and Tejima, E. (1964). Japanese Patent 2986.

Kight-Olliff, L. C. and Fitzgerald, J. W. (1978). *Canadian Journal of Microbiology* **24**, 811.

King, K. W. and Vessal, M. I. (1969). *In* 'Cellulases and Their Applications', Advances in Chemistry Series (R. F. Gould, ed.), vol. 95, pp. 7. American Chemical Society Publications, Washington, D.C.

Lehninger, A. L. (1977). 'Biochemistry', Second Edition. Worth Publishers, Inc., New York.

Leisola, M. and Linko, M. (1976). *Analytical Biochemistry* **70**, 592.

Li, L. H., Flora, R. M. and King, K. W. (1965). *Archives of Biochemistry and Biophysis* **111**, 439.

Loewenberg, J. R. and Chapman, C. M. (1977). *Archives of Microbiology* **113**, 61.

Maejima, K. and Yoshina, H. (1964). Japanese Patent 2985.

Mandels, M. and Reese, E. T. (1957). *Journal of Bacteriology* **73**, 269.

Mandels, M. and Reese, E. T. (1960). *Journal of Bacteriology* **79**, 816.

Mandels, M. and Weber, J. (1969). *In* 'Cellulases and Their Application', Advances in Chemistry Series (R. F. Gould, ed.), vol. 95, p. 391. American Chemical Society Publications, Washington, D.C.

Mandels, M., Parrish, F. W. and Reese, E. T. (1962). *Journal of Bacteriology* **83**, 400.

Mandels, M., Weber, J. and Parizek, R. (1971). *Applied Microbiology* **21**, 152.

Mandels, M., Sternberg, D. and Andreotti, R. E. (1975). *In* 'Symposium on Enzymatic Hydrolysis of Cellulose' (M. Bailey, T.-M. Enari and M. Linko, eds.), p. 81. SITRA, Aulanko, Finland.

Mandels, M., Andreotti, R. and Roche, C. (1976). *In* 'Enzymatic Conversion of Cellulosic Materials: Technology and Applications' (E. L. Gaden, M. H. Mandels, E. T. Reese and L. A. Spano, eds.), p. 21. Wiley, New York.

Markkanen, P. and Eklund, F. (1975). *In* 'Symposium on Enzymatic Hydrolysis of Cellulose' (M. Bailey, T.-M. Enari and M. Linko, eds.), p. 337. SITRA, Aulanko, Finland.

Menezes, T. J. B., Arakaki, T., DeLamo, P. R. and Sales, A. M. (1978). *Biotechnology and Bioengineering* **20**, 555.

Meyer, K. H. and Misch, L. (1937). *Helvetica Chimica Acta* **20**, 232.

Mitra, G. and Wilke C. R. (1975). *Biotechnology and Bioengineering* **17**, 1.

Montenecourt, B. S. and Eveleigh, D. E. (1977). *Applied and Environmental Microbiology* **34**, 777.

Moo-Young, M., Chahal, D. S., Swan, J. E. and Robinson, C. W. (1977). *Biotechnology and Bioengineering* 19, 527.

Morozova, E. S. (1975). *In* 'Symposium on Enzymatic Hydrolysis of Cellulose' (M. Bailey, T.-M. Enari and M. Linko, eds.), p. 193. SITRA, Aulanko, Finland.

Nisizawa, T., Suzuki, H. Nakayama, M. and Nisizawa, K. (1971a). *Journal of Biochemistry, Tokyo* 70, 375.

Nisizawa, T., Suzuki, H. and Nisizawa, K. (1971b). *Journal of Biochemistry, Tokyo* 70, 387.

Norkrans, B. (1963). *Annual Revue of Phytopathology* 1, 325.

Norkrans, B. (1967). *Advances in Applied Microbiology* 9, 91.

Nystrøm, J. M. and Kornuta, K. A. (1975). *In* 'Symposium on Enzymatic Hydrolysis of Cellulose' (M. Bailey, T.-M. Enari and M. Linko, eds.), p. 181. SITRA, Aulanko, Finland.

Osmundsvåg, K. and Goksøyr, J. (1975). *European Journal of Biochemistry* 57, 405.

Peitersen, N. (1975a). *Biotechnology and Bioengineering* 17, 361.

Peitersen, N. (1975b). *Biotechnology and Bioengineering* 17, 1291.

Peitersen, N. (1977). *Biotechnology and Bioengineering* 19, 337.

Pigden, W. J. and Heaney, D. P. (1969). *In* 'Cellulases and Their Applications, Advances in Chemistry Series' (R. F. Gould, ed.), vol. 95, p. 245. American Chemical Society Publications, Washington, D.C.

Preston, R. D. (1971). *Journal of Microscopy* 93, 7.

Rautela, G. S. and King, K. W. (1968). *Archives of Biochemistry and Biophysics* 123, 589.

Reese, E. T. (1963). 'Advances in Enzymatic Hydrolysis of Cellulose and Related Materials'. Pergamon Press, London.

Reese, E. T., Siu, R. G. H. and Levinson, H. S. (1950). *Journal of Bacteriology* 59, 485.

Rifai, M. A. (1969). 'A revision of the genus Trichoderma, Mycological Papers, No. 116.

Seilliere, G. (1906). *Compte Rendu de la Société de Biologie* 61, 205.

Selby, K. (1968). *In* 'Biodeterioration of Materials' (A. H. Walters and J. J. Elphick, eds.), p. 62. Elsevier Publishing Company Ltd, Amsterdam–London–New York.

Selby, K. and Maitland, C. C. (1965). *Biochemical Journal* 94, 578.

Selby, K. and Maitland, C. C. (1967). *Biochemical Journal* 104, 716.

Selby, K., Maitland, C. C. and Thompson, K. V. A. (1963). *Biochemical Journal* 88, 288.

Sihtola, H. and Neimo, L. (1975). 'Symposium on Enzymatic Hydrolysis of Cellulose' (M. Bailey, T.-M. Enari and M. Linko, eds.), p. 9. SITRA, Aulanko, Finland.

Simmons, E. G. (1977). *Proceedings of the Second International Mycological Congress*, Tampa, Florida, p. 618.

Siu, R. G. H. (1951). 'Microbial Decomposition of Cellulose'. Reinhold, New York.

Srinivasan, V. R. (1975). *In* 'Symposium on Enzymatic Hydrolysis of Cellulose' (M. Bailey, T.-M. Enari and M. Linko, eds.), p. 393. SITRA, Aulanko, Finland.

Srinivasan, V. R. and Han, Y. W. (1969). *In* 'Cellulases and Their Applications', Advances in Chemistry Series (R. F. Gould, ed.), vol. 95, p. 447. American Chemical Society Publications, Washington, D.C.

Srinivasan, V. R., Fleenor, M. B. and Summers, R. J. (1977). *Biotechnology and Bioengineering* 19, 153.

Sternberg, D. (1976). *Biotechnology and Bioengineering* 18, 1751.

Suzuki, H. (1975). *In* 'Symposium on Enzymatic Hydrolysis of Cellulose' (M. Bailey, T.-M. Enari and M. Linko, eds.), p. 155. SITRA, Aulanko, Finland.

Suzuki, H., Yamane, K. and Nisizawa, K. (1969). *In* 'Cellulases and Their Applications', Advances in Chemistry Series (R. F. Gould, ed.), vol. 95, p. 60. American Chemical Society Publications, Washington, D.C.

Thomas, R. (1956). *Australian Journal of Biological Science* 9, 159.

Toyama, N. (1963). *In* 'Advances in Enzymic Hydrolysis of Cellulose and Related Materials' (E. T. Reese, ed.), p. 235. Pergamon Press, London.

Toyama, N. (1969). *In* 'Cellulases and Their Applications', Advances in Chemistry Series (R. F. Gould, ed.), vol. 95, p. 359. American Chemical Society Publications, Washington, D.C.

Toyama, N. and Ogawa, K. (1975) *In* 'Symposium on Enzymatic Hydrolysis of Cellulose' (M. Bailey, T.-M. Enari and M. Linko, eds.), p. 375. SITRA, Aulanko, Finland.

Updegraff, D. M. (1971). *Biotechnology and Bioenginering* **13**, 77.

Virkola, N.-E. (1975). *In* 'Symposium on Enzymatic Hydrolysis of Cellulose' (M. Bailey, T.-M. Enari and M. Linko, eds.), p. 23. SITRA, Aulanko, Finland.

Westermark, U. and Eriksson, K.-E. (1974a). *Acta Chemica Scandinavica* **B28**, 204.

Westermark, U. and Eriksson, E.-E. (1974b). *Acta Chemica Scandinavica* **B28**, 209.

Westermark, U. and Eriksson, K.-E. (1975). *Acta Chemica Scandinavica* **B29**, 419.

Whitaker, D. R. (1971). *In* 'Enzymes' (P. D. Boyer, ed.), vol. V, p. 273. Academic Press, New York.

Wilke, C. R. and Yang, R. D. (1975). *In* 'Symposium on Enzymatic Hydrolysis of Cellulose' (M. Bailey, T.-M. Enari and M. Linko, eds.), p. 485. SITRA, Aulanko, Finland.

Wilke, C. R., Cysewski, G. R., Yang, R. D. and Stockar, U. V. (1976). *Biotechnology and Bioengineering* **18**, 1315.

Wood, T. M. (1968). *Biochemical Journal* **109**, 217.

Wood, T. M. (1969). *Biochemical Journal* **115**, 457.

Wood, T. M. (1971). *Biochemical Journal* **121**, 353.

Wood, T. M. (1975). *Biotechnology and Bioengineering Symposium*, no. 5, p. 111.

Wood, T. M. and McCrae, S. I. (1972). *Biochemical Journal* **128**, 1183.

Wood, T. M. and McCrae, S. I. (1975). In 'Symposium on Enzymatic Hydrolysis of Cellulose' (M. Bailey, T.-M. Enari and M. Linko, eds.), p. 231. SITRA, Aulanko, Finland.

Wood, T. M. and McCrae, S. I. (1977a). *Carbohydrate Research* **57**, 117.

Wood, T. M. and McCrae, S. I. (1977b). *In* 'Proceedings of Bioconversion Symposium', p. 111. IIT Haus Khas, New Dehli-110029, India.

Wood, T. M. and McCrae, S. I. (1978). *Biochemical Journal* **171**, 61.

Wood, T. M. and Phillips, D. R. (1969). *Nature, London* **222**, 986.

Yamada, K. (1977). *Biotechnology and Bioengineering* **19**, 1563.

Yamane, K., Suzuki, H., Hirotani, M., Ozawa, H. and Nisizawa, K. (1970a). *Journal of Biochemistry, Tokyo* **67**, 9.

Yamane, K., Suzuki, H. and Nisizawa, K. (1970b). *Journal of Biochemistry, Tokyo* **67**, No. 1, 19.

Yamane, K., Yoshikawa, T., Suzuki, H. and Nisizawa, K. (1971). *Journal of Biochemistry, Tokyo* **69**, 771.

7. Immobilized Enzymes

S. A. BARKER

Department of Chemistry, University of Birmingham, Birmingham, England

1. INTRODUCTION

By 1968, the biotechnology of immobilized enzymes had reached the market place with one American (Miles Laboratories) and one South African (Serevac) company selling enzymes immobilized by covalently attaching them to an ethylene–maleic anhydride copolymer or carboxymethylcellulose hydrazide. In April 1969, products under the trade name Enzacryl were launched (Koch-Light Laboratories) based on specially synthesized copolymers of polyacrylamide. These and future ones were all developed at the University of Birmingham and later Wolverhampton Polytechnic by Dr. Epton. By late 1969, the potential importance of such products was realized, in particular for clinical analysis, then just starting its growth phase as pharmaceuticals became more expensive to launch and the interval became longer and longer before they were permitted to be sold. The technology of enzyme immobilization was to have its counterpart in immobilization of antibodies and antigens for the, then, rapidly expanding technique of radio-immuno-assay and later enzyme immuno-assay. The full potential of the technology has not yet been grasped by instrument companies, and enzyme electrodes are only just reaching the market with the enzyme thermistor still to come. However, by 1970, the Science Research Council had made sizeable grants to support the new technology particularly for developing design procedures for immobilized enzyme reactors (such as that at University College, Swansea in Wales) using enzymes immobilized in polyacrylamide gels. We (University of Birmingham, England) were supported for developing a combined enzyme reactor/separator device, and Strath-clyde University in Scotland for development of enzymes immobilized in non-aqueous systems. At the University of Liverpool in England, work went ahead on development of affinity chromatography for enzyme purification which was to offer later a great potential for simultaneous immobilization since, unexpectedly, the enzyme was often not eluted by its substrate. At the University of St. Andrews in Scotland, enzymes were being attached to the interior of plastic tubes as future components of autoanalysers. At University College in London, a massive grant marked their achievements both as engineers and as chemists. Already they had coupled an immobilized enzyme with a soluble immobilized cofactor and were developing magnetic supports for ease of recovery of the catalyst. By 1975, immobilized enzymes had achieved full status as industrial catalysts and were present in full-scale or pilot-plant industrial operation for hydrolysis of starch (Corning Research and Northern Regional Laboratory, U.S.A.) and for isomerization of glucose (Standard Brands Inc., and other companies in the U.S.A.). It is the last reaction, based on conversion of glucose into fructose with glucose isomerase, that has seen a massive growth rate and is now pre-eminent in the world of immobilized enzymes. Here was a cheap way to get a home-based industry to provide essential sweetener requirements for a nation. It took the barely profitable glucose, that had already started to plateau out in the U.K. and U.S.A., and produced a product that could always undercut imported sucrose. While this has boomed in free-enterprise United States, the U.K. has been tied down by E.E.C. regulations and prevented from developing this industry. Indeed, the massive plant specially built for the purpose still lies idle at the time I write, a monument to the inability of the scientist to convey to the politician the opportunities of the new technology.

II. DEVELOPMENT OF IMMOBILIZATION TECHNOLOGY

Early methodology depended mainly on adsorption or entrapment techniques. The enzymes used for treating hides before tanning had from the 14th Century been preserved by mixing them with sawdust and drying them. Indeed, this practice still persists in primitive tanneries in Africa where the enzymes of the pancreas are so preserved. Entrapment was typified by the ease with which an aqueous solution of enzyme and monomer acrylamide could be polymerized by γ-radiation or addition of a chemical initiator. Both these types, under certain circumstances, may be admirable but their deficiencies are not appreciated. Entrapment of an enzyme is quite good for enzymes acting on low molecular-weight substrates but useless for those acting on large polymers because of their inaccessibility to the enzyme catalyst. Their main disadvantage, like the adsorption technique, is leakage of enzyme. However, entrapment under rigidly controlled conditions can mimic a true cellular system, and offers a device for having intracellular enzymes as well as others attached to the surface of the sphere. Further, as enzyme biotechnology progresses to multi-enzyme reactors, each type of enzyme can be provided with its own micro-environment so important for their eventual co-existence in a medium that may well be hostile for one of the enzymes in a soluble form. Micro-environments provide a means of acting as an immobilized buffer system as well as providing safe anchorage for parts of the protein catalyst without undue disturbance of its favoured conformation. Three devices are available to upgrade adsorption technology. The simplest is to use ionic attraction to retain the enzyme, and here DEAE-cellulose and DEAE-Sephadex have found favour. An elegant but largely undeveloped technique is to use an affinity support. When the enzyme is not inhibited by such attachment, the results are excellent and so is the thermostability. A less reliable but simpler technique is the use of hydrophobic forces with supports such as the alkyl-Sepharose type.

In 1968, my colleagues and I applied for a patent to cover immobilization of enzymes by reaction with the 3-(p-diazophenoxy)-2-hydroxypropyl and 2-hydroxy-3-(p-isothiocyanatophenoxy)propyl ethers of cellulose (Barker *et al.*, 1971b). We claimed that previous supports had given low (< 3%) retention of activity when calculated as a percentage of the activity which that amount of enzyme protein bound to the cellulose derivative would display in its original soluble form. Our retentions were much higher but decreased from 60% at low loadings to 37% at higher loadings as the sites became more crowded for α-amylase immobilized on 3-(p-diazophenoxy)-2-hydroxypropyl cellulose. A second advantage was that Axen and Porath (1966) had previously shown that the β-amylase derivative of the 2-hydroxy-(p-isothiocyanatophenoxy)propyl-Sephadex was inactive, whereas our corresponding cellulose derivative had activity (16.6% retention). It was also evident that, depending on the enzyme and type of linking, enhanced thermostability was often achievable (e.g. β-amylase or α-amylase attached via the diazo derivative). Further investigation showed that the type of linking was crucial to the re-use properties of the immobilized enzyme (Barker *et al.*, 1970a). With α-amylase, the isothiocyanato-coupled derivative was superior to the diazo-coupled cellulose analogue. However, even better than these were analogous derivatives based on copolymers of polyacrylamide in which iso-thiocyanato-coupled α-amylase was better than diazo-coupled enzyme, and these were better than an acid azide-coupled α-amylase. We also learned that physically absorbed enzyme is often not removed with extensive washing, and that the polyacrylamide-based derivatives (Enzacryl AA and AH) were superior also in this respect. One procedure will invariably remove physically absorbed enzyme, i.e. incubation with substrate, and this was one purpose of submitting the samples to re-use procedures. With β-amylase, acrylamide derivatives were again superior to cellulose derivatives as regards stability to re-use, the diazo-coupled acrylamide being the best. One aspect of the technology neglected by many workers was highlighted by this work particularly with the diazo derivatives. What happens to the unreacted groups on the support polymer? Diazonium groups were found to be surprisingly stable and options exist for deliberately self-annealing them or coupling them with, for example, β-naphthol. It is a matter of trial and error as to which micro-environment suits the enzyme best. Certainly it may be crucial for heat stability. With the acrylamide series, acid hydrazide-coupled α-amylase was more stable than isothio-cyanato-coupled enzyme, which in turn was more stable than diazo-coupled α-amylase annealed with phenol. All were more stable than soluble α-amylase (Barker et al., 1970b).

Supports of the above type were suitable for chemical research purposes but not for either the industrial reactor market or, from some aspects, the clinical assay market. They were first-generation matrices and, for the clinical assay market, a support had to be provided to the medical scientist which, for activation, required only a suspension with an enzyme or antibody in a buffer solution. The consumer had to be able to do 'his own thing' and not be sold enzymes already coupled to matrices or matrices requiring some chemical manipulative sequence. For this purpose, cellulose carbonate was developed (Barker et al., 1971a) and patented (Barker et al., 1972). The optimum pH value for reactivity with proteins was between 6.5 and 8.5 below 4°C, i.e. extremely delicate conditions suitable for the preservation of essential biological and catalytic activity. It was in an extensive study of this product, with a whole range of enzymes, that it was noticed that enzymes like chymotrypsin, while showing good retention of estero-lytic activity against low molecular-weight substrates, had very poor activity against high molecular-weight substrates like casein. Macro-porous cellulose carbonate, instead of the type derived from micro-crystalline cellulose, greatly improved activity against casein (Kennedy et al., 1973). Another second-generation-type support was developed in the Enzacryl series (Koch-Light Laboratories, 1977), this time using a stabilized coating produced by diazotization of m-diaminobenzene which gives a precursor of Bismarck Brown and a reactive polymeric surface of diazonium groups that can be stabilized as the fluoroborate (Gray et al., 1974). Such a concept of coating a surface instead of making a specialized polymer brought development nearer to the industrial requirement, since the support could be now designed to the engineer's requirement as regards reactor type (e.g. tube, fixed bed or stirred tank) and the chemist's requirement as regards porosity, spurious adsorption properties and micro-environment. In the last decade, demands by the layman for 'safer' chemicals, not only in his environment but in the manufacture of anything he buys, particularly food, have imposed even greater restraints on the ideal industrial-type immobilized enzyme reactor.

One concept developed by us was that the coating for enzyme immobilization should be as near as possible to a permitted food additive, in our case titanium oxide. Through the chance use of titanium chloride for reduction of an aromatic nitro to an amino group prior to diazotization and enzyme coupling, we noticed that certain controls that had not been diazotized were also enzymically active. Such controls are often a useful guide to the adsorptive properties of the matrix itself, in this case cellulose. Indeed, when we repeated the titanium chloride treatment on cellulose, washed it and then offered it enzyme, we found good enzyme activity. It was quickly discovered that a host of support materials having hydroxyl groups, whether C–OH or Si–OH as well as –CO–NH–, would chelate with titanium chloride and other chlorides (including $FeCl_3$, $SnCl_4$, $ZrCl_4$ and VCl_3) and give a product that could be dried and stored for long periods (Emery et al., 1972). The nature of the drying step depended on the support. While temperature with inorganic supports like glass, silica and other oxides could rise to 100–140°C, with polysaccharides like cellulose a temperature of 45°C prevented unnecessary degradation. Thereafter, washing with water or suitable buffer served to replace stable chloride groups in the titanium chelate with unstable hydroxyl groups. At this stage, the material cannot be dried and must be contacted with enzyme to complete the other half of the chelate sandwich now only containing titanium and oxygen. Such chelates are suprisingly stable to different pH values and the properties of the immobilized enzyme are first class, exhibiting very high enzyme efficiences and, with an inorganic support, surprisingly resistant to bacterial contamination. After exhaustion of enzyme activity, such surfaces can be replated with fresh enzyme by a repetition of the above process. Surprisingly, this chelation can also be used to provide active immobilized antibiotics, so that further protection of the enzyme reactor is achieved (Kennedy et al., 1974; Barker and Kennedy, 1975).

An alternative but related support is simply freshly precipitated

Table 1

Some reactive groups for enzyme immobilization

Reagent	Reactive group	Type
Glutaraldehyde	$-CH=O$	Dialdehyde
m-Phenylene diamine	$-N\equiv NX^-$	Diazonium
Cyanogen bromide	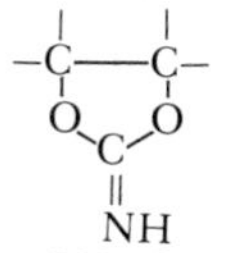 ![iminocarbonate structure]	Iminocarbonate
Sodium periodate	$-CH=O$	Dialdehyde
Carbodiimide-N-hydroxysuccinimide	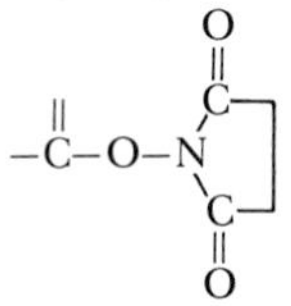 ![activated ester structure]	Activated ester
Titanium tetrachloride	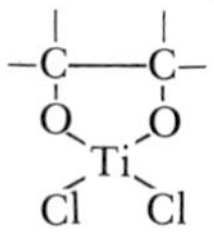 ![metal chelate structure]	Metal chelate
Succinic anhydride-thionylchloride	![acid chloride structure]	Acid chloride
Thiophosgene	$-N=C=S$	Isothiocyanate
N-Hydroxysuccinimido bromoacetate	![bromoacetamide structure]	Bromoacetamide
Hydrazine-nitrous acid	![acyl azide structure]	Acyl azide
Ethyl chloroformate	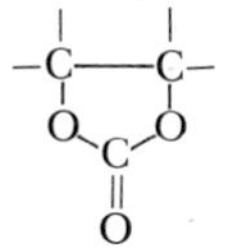 ![cyclic carbonate structure]	Cyclic carbonate
Acetyl-H$^+$	$-CH=O$	Aldehyde

hydrous titanium and other oxides (Barker *et al.*, 1975b). This has the immense advantage that enzymes and other biologically active compounds can be reversibly immobilized so that, in certain situations, it can be used as a recovery vehicle (e.g. for antibiotics). It also is completely re-usable. Various versions of this technique permitted coating of magnetic materials (Kennedy *et al.*, 1977b). One outstanding property of this immobilization procedure is that it is applicable to living micro-organisms and is so delicate that they can still respire (Kennedy *et al.*, 1976).

More and more workers are looking to use solid supports as blocking groups which can be subsequently removed. This aspect has become even more important with the emergence of enzyme immuno-assay as the deadly rival of radio-immune assay. Here the economics depend on the ease of producing the conjugate, often containing an enzyme. Newcomers to the Enzacryl series (Barker *et al.*, 1975a) are based on copolymers of *N*-acryloylcysteine and acrylamide. The reversibility of the SH → S–S → SH transformation makes these ideal supports for such syntheses, besides their advantages as enzyme supports. The support can be transformed to its cyclic thiolactone adduct which can home on different groups in an enzyme protein depending on the pH value of the environment ($-NH_2$ alkaline pH value; $-OH$ acid pH value; phenolic group, acid pH value after a longer time). Finally, the Enzacryl series is completed by a polymer that is based on the psychology that a customer will always do what he is already in the habit of doing. Basing this on the fact that chemists and others are used to regenerating ion-exchange resins with acid, an Enzacryl Acetal was offered where the aldehyde group is liberated by such treatment and then will couple merely by contacting with the enzyme.

III. RECENT IMMOBILIZATION TECHNOLOGY

A. Glucose Isomerase

No one method is the best to immobilize all enzymes. This is because enzymes are most prone to lose activity at their catalytic and binding sites. Since these differ in chemical nature, the mode of attachment of the enzyme to its support, whether it be reaction through an amino group or a carboxyl group of the enzyme, will be the most important factor if no other steps are taken. One obvious but rarely imple-mented approach is to protect the active site during enzyme immobilization either by use of the substrate that preferentially binds there or by a reversible chemical modification at the active site (e.g. $-SH$-active → $-S-S$-inactive → $-SH$ active after immobilization). Another classic example is protection of papain, a thiol-containing enzyme, using zinc or mercury ions followed by reactivation with cysteine.

The second step to observe is the nature of the support and its interaction with the immobilized enzyme. The tertiary structure or conformation of the enzyme is vital to its role as a catalyst. Immobilization of the enzyme to preserve activity necessarily implies that that tertiary structure is largely intact if considerably less mobile. What it should not be is abnormally distorted by hydrophobic bonding, hydrogen bonding or electron transition-complex formation with the support. This is the area of least knowledge where experimentalists should try all three types of support providing such interactions to determine that most suitable for the enzyme.

Often, immobilization of an enzyme within its natural environment, the bacterial cell, is most attractive. Then it is conversion of the starting material into a form suitable for an enzyme reactor that assumes greater importance. Thus Ito *et al.* (1977) treated cells of *Streptomyces* sp. containing glucose isomerase with $4MgCO_3 \cdot Mg(OH)_2 \cdot 5H_2O$ and 20% $Ca(OH)_2$ while bubbling with carbon dioxide at 55°C and pH 6.5–8.0. The resulting floc could then be extruded through 2 mm holes and dried at 30°C to yield immobilized cells which, in columns, continuously izomerized glucose for 14 days with conversions of 40.7–47.0% into fructose. Magnesium ions are essential for the stability of most glucose isomerase molecules, so that its use here was not fortuitous; on the other hand, calcium ions deplete activity but can be permitted in a small ratio to magnesium ions.

An attempt to retain more of the natural environment of the cell is illustrated by attachment of *Strep. phaeochromogenes* to chitosan (8% by weight of cells). Firstly, the glucose isomerase was fixed in cells by preheating for ten minutes at 80°C and then attached to soluble chitosan by reaction with 25% glutaraldehyde for one hour at room temperature (Maekawa *et al.*, 1977).

Immobilization on ion exchangers and ion-exchange resin, with retention of the enzyme by ionic forces, has long been a simple and favoured procedure. Fujita *et al.* (1977) immobilized soluble glucose isomerase on a cross-linked styrene–divinylbenzene copolymer with quaternary ammonium ion groups by suspending the swollen resin in a solution of the enzyme for six hours at 50°C. Isomerization could then be effected using a feed of glucose, magnesium sulphate and ferrous sulphate at pH 7.5 and 60°C. The ion-exchange resin could be re-used when the catalytic activity of the enzyme was exhausted.

Entrapment of enzymes and cells within gelatin has been presented under a number of disguises for the privilege of obtaining patent protection. Thus, glucose isomerase gel particles are claimed by Leroy *et al.* (1977) and were produced using a microbial cell sediment with gelatin and adding this to ice water through spin nozzles resulting in formation of gel filaments in which the enzyme was retained by cross-linking with glutaraldehyde. Such methods are very successful for enzymes acting on low molecular-weight substrates but pose problems when high molecular-weight substrates cannot find access to the enzyme.

Even now there are claims to the use of whole cells as resting cells for enzymic conversions. Lee (1977) stated that *Flavobacterium arborescens* containing glucose isomerase could be used to manufacture fructose by stirring whole cells with 50% glucose syrup containing 0.01 M magnesium chloride at pH 8.0; 45% conversion required 20 hours at 60°C. Such procedures are unlikely ever to be used nowadays by industry because of the prolonged reaction time.

Coated active cells of a *Streptomyces* sp. containing glucose isomerase are claimed by Miyoshi *et al.* (1977) using a method that superficially resembles active-site enzyme protection by substrate. Glucose and cells as powders were mixed and coated repeatedly (three times) with 2.5% ethyl cellulose in ethanol. After removal of ethanol, the preparation was soaked in water to dissolve glucose crystals and then dried to afford the encapsulated enzyme preparation.

Micro-encapsulation of cells containing glucose isomerase has been achieved at pH 8 using a quaternized methyl methacrylate–2-vinyl-pyridine-divinylbenzene random copolymer (Miyoshi *et al.*, 1977). Immobilization of glucose isomerase on colloidal silica (Ludox HS-30) was achieved by adsorption followed by freezing in the original state or after gelatinization for solidification of the resulting product (Ushiro, 1977).

One of the oldest methods of immobilizing enzymes is to polymerize the monomer in the presence of that enzyme. Two recent patents illustrate the old trick still being played. Kawashima and Umeda (1977) irradiated acrylic acid, *NN'*-methylenebisacrylamide and glucose isomerase together in frozen granules containing magnesium oxide. Better activity was attained in the presence of the magnesium ions. Similarly, films were made of microbial cells by irradiating them in xylylene diisocyanate (1 mole), polyoxyethyleneglycol (mol. wt. 1,500;

750 g) and 1.1 mole of 2-hydroxyethyl methacrylate and benzoin methyl ether. It was found imperative to cool the film to 5°C during irradiation by a low-voltage mercury lamp for 3 min (Fukui et al., 1977a). Further examples given by Iida and Hasegawa (1977) particularly applied to 1% glucose isomerase (150 parts) mixed with polypropylene glycol (mol. wt. 2,000), dimethylacrylate (100), and benzoin methyl ether (1). The hydrophilic photopolymerizing resin was kept in an atmosphere of 100% relative humidity without any crack forming in the film.

The use of enzymes in a sprayable paint is an attractive one. Thus, one version of this claimed by Kato et al. (1977) uses a spray containing 10% butyral resin solution followed by glucose isomerase in 5% carrageenan and then a spray of butyral resin again. The glucose isomerase retained some 34.5% of its original activity.

Hupkes (1978) has discussed the compressibility of glucose isomerase particles made by a gelatin-capsule entrapment process. Kent and Emery (1974) used the titanium-chelation technique to immobilize glucose isomerase (*Lactobacillus brevis*) on cellulose. In this procedure, the cellulose was contacted with 15% titanium chloride in 15% hydrochloric acid, thereafter dried at 45°C and washed with 0.02 M Tris buffer (pH 7) before coupling the enzyme at 4°C for 18 hours. Further washing with Tris buffer containing 0.5 M sodium chloride was carried out before assay.

Three main parameters in this procedure were studied. In the first, the ratio of protein available for coupling to activated solid was varied. In the range studied, the proportion of protein coupled was low (maximum around 10%) and this decreased as the coupling protein concentration in the solution increased. Specific activity of bound enzyme decreased with increasing content of bound protein. However, the method was outstanding in the considerable increase in the apparent specific activity usually obtained.

Time of coupling was varied, and it was found that almost 75% of the maximum activity was bound in 6 hours with completion being achieved in 18 hours. While most protein was coupled at pH 7, there was only a small variation in activity bound in the range pH 5 to 9. There was an apparent displacement of the pH optimum of cellulose-bound glucose isomerase by 0.6 unit to the acid side. Thus, the immobilized enzyme had a pH optimum of 6.0 and the activity–pH profile was slightly broader for the soluble enzyme. Ions (MN^{2+} and

Co^{2+}) were essential to the immobilized enzyme. The immobilized enzyme had a lower temperature optimum (50°C) than the soluble enzyme.

B. Invertase

A novel way of immobilizing invertase (β-fructofuranosidase) is to prepare an emulsion of the enzyme solution (five parts) with five parts of polystyrene dissolved in 94 parts of carbon tetrachloride. On lyophilization, a porous enzyme polymer was produced that inverted sucrose solutions (Nakamura and Ono, 1977b). Another version of this procedure was to mix the invertase solution (five parts) with five parts of diacetyl cellulose (degree of acetylation, 39.4%) dissolved in 94 parts of dioxane and lyophilize it (Nakamura and Ono, 1977a).

A whole range of enzymes, including invertase, have been immobilized in films. Thus, the mixture containing xylene diisocyanate, poly(ethylene glycol) and 2-hydroxyethyl methacrylate can be polymerized to a resin which is hardened by ultraviolet irradiation. A patent was granted for immobilization of enzymes by mixing an aqueous dispersion of enzymes with a photopolymerizable resin and then treating the mixture with ultraviolet irradiation. In one version, a buffered solution of the enzyme containing the resin and benzoin methyl ether was applied as a mixture to a glass plate and irradiated at 35°C for two minutes to give a transparent enzyme-containing film. The technique was applicable to glucose oxidase, glucoamylase, glucose isomerase and even cells of *Proteus vulgaris* (Fukui *et al.*, 1977b).

A related patent (Fukui *et al.*, 1977c) claims enzyme immobilization in photopolymerizable resins that contain $\geqslant 2$ ethylene unsaturated groups as well as ionic hydrophilic groups. Thus Epikote 1001, a polyglycidyl resin, was treated with adipic acid and the resulting product reacted with succinic anhydride. After addition of glycidyl methacrylate, the resulting resin was mixed with the enzyme solution and benzoin ethyl ether. Ultraviolet irradiation at 25°C for five minutes produced a polymer film.

In general, it is preferable to attach covalently the enzyme to any polymer, no matter in what form it is presented. Thus, the enzyme activity of invertase entrapped in cellulose triacetate fibres leaked over the first 3–4 days of operation, although this effect could be modified by addition of different amounts of poly(ethylene glycol) of different molecular weights (200–1,000) to the spun emulsion (Marani *et al.*, 1977).

Porosity and accessibility have always been major aims in devising ways of enzyme immobilization. The well-established method of using ionic attraction to hold the enzyme to a support was revamped by Brouillard (1977) by basing it on a bed of regenerated cellulose in sponge form that was chemically modified by treating it with sodium sulphate, sodium hydroxide and diethylaminoethylchloride hydrochloride. The resulting DEAE sponge cellulose could be used for a wide variety of enzymes including invertase, glucoamylase and glucose isomerase. The sponge material used in the column reactor had a density of or less than 1 g/18 cc and a water flow porosity such that > 0.5 gal/min/ft^2 flowed through the bed.

A related version is the use of polyurethane foams with isocyanate end groups to react chemically with the enzyme. For example, 1 g of poly(ethylene glycol) (mol. wt. 1,000) was mixed with 0.229 g of toluene diisocyanate with heating to form a polymer whose molecules contained NCO groups at either end, and was capable of forming a foam on agitation in which enzyme could be covalently bound (Hartdegan and Swann, 1976).

Periodate oxidation of cellulose affords dialdehyde cellulose which can be used directly for enzyme immobilization. A suitable reactor was devised by winding cotton yarn to a depth of 1.4 cm on a roll of stainless-steel screen to give a cartridge (6.25 cm diam; 10 cm long). The cartridge was placed in a stainless-steel holder and 0.11 M periodic acid (3.5 litres) was circulated through it for four days. After washing, the enzyme was immobilized at pH 8 (Wildi and Weeks, 1976).

C. *Alpha*-Amylase

Immobilization of α-amylase has been achieved by entrapment in a polymer produced by γ-radiation at less than 60°C of a mixture of methylolacrylamide/methanol/phosphate buffer (pH 7) containing 1% α-amylase (5:90:10, by vol.). The white opaque polymer was produced after irradiation with γ-rays for 15 hours at 1 × 10^5 R/h. The entrapped α-amylase showed 68% of its original activity (Kaetsu *et al.*, 1977a). In a related version by the same authors (Kaetsu *et al.*, 1977b),

10 parts of 1% α-amylase were first mixed with 40 parts of activated charcoal which was then dispersed in water and mixed with 50 parts of hexanediol methacrylate. Again, the enzyme was immobilized by irradiating with γ-rays at − 24°C to initiate polymerization.

The utility of α-amylase immobilized on Duolite DS-73141 has been demonstrated by Smiley et al. (1975) who indicated that approximately 910 kg of enzyme-resin would treat 4.5 million litres of paper-mill effluent/ day with a continuous fluidized bed-type reactor.

Acylated α-amylase was immobilized on a Millipore filter and used for continuous hydrolysis of soluble starch. The heavily palmitylated α-amylase-immobilized derivative lost activity only during the early stages, retaining constant activity later. The butyryl, caprylyl and palmityl α-amylase derivatives were characterized in this study (Okada and Urabe, 1976).

Senju and Tanaka (1976) used halogenamide residue-containing polymers to immobilize α-amylase, invertase, papain and other enzymes by covalent attachment. Cellulose and hydroxyethylcellulose were treated with acrylonitrile and then hydrogen peroxide to give their carbamoyl derivatives. These were converted into the chloramides by treatment with sodium hypochlorite. A graft polymer from gelatinized starch and saponified polyacrylonitrile in ratios from 1 : 1.5 to 1 : 1.9 was used to entrap α-amylase and other enzymes. Such polymers were capable of absorbing more than 300 parts of water by weight per part of the water-insoluble solids. This property enabled the enzyme solution to be first mixed with polymer which then swelled. After addition of sufficient water-soluble minerals salts, the polymer shrank again and entrapped the enzyme (Weaver et al., 1976).

D. Beta-Amylase

As β-amylase can be reversibly absorbed on starch, an enzyme from Bacillus cereus was absorbed on starch that had been heated at 60°C for one hour, and then cooled to 40°C. After being stirred for 30 minutes, some 95% of the β-amylase from a culture supernatant (500 ml) was absorbed on 80 g of starch presented as a suspension (200 ml). The enzyme had 85% of the activity of the original but was eluted completely by 10% maltose (Takasaki and Takahara, 1976a). A mixture of 1% gelatinized soluble starch and 40 g of active carbon was effective in absorbing 90.1% of the β-amylase and 100% of α-1,6-glucosidase from 1 litre of enzyme solution. Using this mixture (3,000 β-amylase and 310 α-1,6-glucosidase units), some 100 ml of a soluble starch solution (10%) was converted at 50°C and pH 6–6.5 into 90.5% maltose, 7.5% maltotriose and 2.0% other oligosaccharides (Takasaki and Takahara, 1976b).

Hydrophic interactions were utilized to immobilize β-amylase and amyloglucosidase on hexyl-Sepharose (Caldwell *et al.*, 1976b). Saturation occurred at 35 mg of β-amylase/ml of packed gel and activity leakage was low. The relative activity of the immobilized enzyme was inversely related to the amount of enzyme, to a given volume, having a maximum value of about 50%. Continuous maltose production from a soluble starch substrate was effected with column reactors based on such an immobilized β-amylase. Gels with a degree of substitution of hexyl groups from 0.02 to 0.70 mole of hexyl side chain/mole of galactose residue were studied. Adsorbate activity was a maximum at 0.51 (Caldwell *et al.*, 1976a).

Matensson (1975) has patented a process for maltose manufacture based on the use of an immobilized β-amylase–pullulanase mixture. Thus Bio-Gel CM100 (acrylamide–acrylic acid copolymer) was treated succcessively with β-amylase at pH 4, 1-cyclohexyl-3-(2-morpholine ethyl)carbodiimide p-toluenesulphonate and pullulanase at pH 4.2. This reactor was then fed α-amylase-hydrolysed potato starch which it converted into 70% maltose and 15.8% maltotriose. Entrapment of β-amylase, as might be expected with its requirement to act on a high molecular-weight substrate, was unsatisfactory. Thus, β-amylase and acrylamide irradiated in the frozen state yielded a spongy membrane, but the entrapped enzyme showed only 4.1% of its activity compared with 18.0% (α-amylase), 33.7% (glucose oxidase) and 69.2% (invertase) (Kawashima and Umeda, 1975).

E. Glucoamylase

Glucoamylase has been immobilized on imido esters, such as the imido ester of polyacrylonitrile, using the amidine reaction (Oosawa, 1977). The imido ester of polyacrylonitrile was prepared by treatment of polyacrylonitrile (13 g) with methanol (100 ml) through which dried hydrogen chloride gas was bubbled for three hours at 0°C. Some 200 mg of this carrier was mixed with glucoamylase in 0.1 M phosphate buffer (5 ml). Coupling was best effected at pH 6 below 50°C.

Porous glass is particularly suitable for immobilizing glucoamylase, although the size of the pores employed are critical. Typical of the patents assigned to the Corning Glass Co. is that of Tomb and Weetall (1977) in which the water durability of the glass was improved by coating it with zirconia or titania. The glass was silanized and the available amino groups were attached to the enzyme by glutaraldehyde or diazotization.

Many inorganic supports can be silanized. Thus, silicic acid grafted with $(EtO)_3-Si-(CH_2)_2-CH-(OEt)_3$ was treated with glucoamylase by stirring in phosphate buffer (pH 7) for 15 hours at 4°C to give an immobilized form of the enzyme (Meiller and Monsan, 1977). The same group (Mazarguil et al., 1976) treated silica microbeads, which had been dried at 150°C under vacuum for four hours, with a silane for six hours at 140°C. After washing with acetone, drying, suspending in 0.1 M hydrochloric acid and treating with sodium nitrite, the support, after further washing at pH 7, would couple easily with the enzyme shaken for 48 hours at 2°C. Remarkable stability was reported by Lee (1976) using glutaraldehyde-bound glucoamylase immobilized on silica. When the feed to the reactor was starch dextrin with a dextrose equivalent value (DE; a measure of reducing power) of 26, a maximum conversion of 94% into glucose was attainable with a DE of 96.4. Enzyme half-life was directly correlated with the reciprocal of the temperature.

Various methods of impregnating and coating titania beads were described by Rosevear (1976). Tannic acid in acetone/water (5 : 1, v/v) was used as a precipitating agent followed by treatment with a cross-linking agent such as glutaraldehyde. Freeze-drying in combination with the precipitating agent led to a 15% increase in the activity of the preparation. Earlier, Solomon and Levin (1975) had demonstrated that glucoamylase was adsorbed on an acid-activated molecular sieve and an alumina. The immobilized enzymes exhibited some 50–100% of their initial activity and possessed stability at high temperatures.

Glucamylase has been immobilized on a m-phenylenediamine–glutaraldehyde copolymer (Kraznobajev and Boeniger, 1975) which optionally contained a filler such as silica. Thus, 50 ml of glutaraldehyde (25%) and silica gel (30 g) were added to a solution of 5 g of

1,3-phenylenediamine in chloroform, to which 7 M hydrochloric acid and water were added. The coated silica particles were left for 10 minutes, then diazotized and washed. The enzyme could then be immobilized by contacting in buffer (pH 6.5).

Covalent bonding was responsible for immobilizing glucoamylase to carbonyl polymers based on cross-linked preparations of a sulphited aldehyde or ketone polymer produced from a vinyl monomer, especially polyacrolein-bisulphite adducts and an aliphatic or aryl dialdehyde (e.g. glutaraldehyde or benzene diacetaldehyde). Reaction of the enzyme occurred between 5 and 65°C under alkaline conditions (Forgione, 1974). Immobilization of glucamylase by γ-irradiation of a solution of the enzyme in N-vinylpyrrolidone monomer has also been reported (Maeda *et al.*, 1974).

F. Glucose Oxidase

Cyanogen bromide activation of polysaccharides is a well-established means of binding enzymes to the latter, although workers are sensitive about the toxiological implications of the process. Thus, glucose oxidase was coupled to cyanogen bromide-activated polyvinylidene fluoride (Kynar)-cellulose ultrafiltration membranes by passing the enzyme solution through the membrane at a pressure of ≥ 0.7 atmospheres (70.7 kPa), ideally 2.1–8.4 atmospheres (212–848 kPa). Reactions with the substrates were effected by passing through the substrate solutions at 0.35 atmospheres (35.4 kPa) (Gregor, 1977a).

Glucose oxidase can also be copolymerized with acrylamide and NN'-methylenebisacrylamide after first vinylating the protein with an alkylating or acylating monomer (Jaworek *et al.*, 1976). Nylon can be modified by treating its amino groups with chlorinating reagents followed by reaction with a bifunctional coupling reagent. Thus nylon-6 tubing (2 m) was achieved with a solution of phosphorus pentachloride in chloroform and then reacted with ethylenediamine. The resulting nylon derivative could be bound to an enzyme via reaction with ethyleneglycol-bis(propionic)bis(hydroxysuccinimide) ester (Horn *et al.*, 1977).

For automatic analysis of glucose, Miller *et al.* (1977) immobilized both glucose oxidase and horseradish peroxidase, either separately or together, on diazotized polyaminostyrene beads. The enzyme-containing beads were packed into a glass coil. In the automated assay, based on a fluorimetric determination with homovanillic acid, glucose could be determined at the p.p.m. concentrations. Experience showed that the beads containing the two enzymes immobilized simultaneously were much more active than beads that contained only one of the enzymes. After 40 days storage at 2°C, immobilized glucose oxidase retained more than 60% of the initial activity.

When a solution of polyvinyl chloride (solute, 14 parts) in tetrahydrofuran (solvent, 56 parts) and isopropyl alcohol (propan-2-ol; 30 parts) was mixed with 2 parts of glucose oxidase, coated onto a polyester monomer cloth to a depth of 0.06 mm and thereafter dried, an immobilized active glucose oxidase film was produced (Murata and Ishikawa, 1976).

It has been found that higher activities (catalase, 3–5-fold; glucose oxidase, 40-fold) were obtained by immobilizing preformed soluble conjugates of glucose oxidase and catalase cross-linked with glutardialdehyde than by immobilizing the untreated enzymes. The enzyme conjugate was immobilized on glass beads (non-porous) treated with γ-aminopropyltriethoxysilane and activated with glutardialdehyde (Atallah et al., 1976).

An alternative method of immobilizing glucose oxidase on nylon is the basis of the commercial product now sold by Miles Laboratories. Hornby and Morris (1976) alkylated nylon 6 and prepared the succinic acid hydrazide and adipic acid hydrazide derivatives. These were activated with glutaraldehyde before enzyme binding. Burrows and Lilly (1976) also used polyamized surfaces to immobilize glucose oxidase by creating either an excess of either amine or carboxyl end groups thereon of at least 100 equiv./10^6 g of polymer. Thus, 5,240 parts of hexamethylenediammonium adipate and 162.4 parts of hexamethylenediamine were autoclaved with 5,792 parts of water under nitrogen. The resultant polyamide was extruded into water, dried and ground, and contained 392.7 equiv. of amine end groups and 13.5 equiv. of carboxyl end groups/10^6 g of nylon polymer. Enzymes could be readily immobilized by stirring with the nylon polymer and a buffered aqueous solution of glutaraldehyde. A glucose oxidase–catalase mixture, β-galactosidase and alcohol dehydrogenase were so immobilized.

G. *Beta*-Galactosidase

The main purpose of immobilizing β-galactosidase (lactase) is to hydrolyse lactose in milk and milk products. Thus, Pastore (1977) immobilized the β-galactosidase of *Saccharomyces fragilis* in cellulose-acetate fibres and then used it after sterilizing the milk at 145°C for 2–3 hours and cooling to 4°C. After 20 hours of circulating the milk through the enzyme reactor, hydrolysis of lactose was 75–80% complete with an enzyme/milk ratio of 2 : 1,000. About 450 g of such a fibre reactor treated 10,000 kg of milk with a loss of activity of approximately 10%.

The β-galactosidases of *Macrophomina phaseoli* and *Sclerotium tuliparum* have been immobilized on porous glass beads (Sugiura *et al.*, 1977). Although K_m values for the immobilized enzymes were 2–5-fold higher than the soluble enzymes, other properties such as optimum temperature, pH stability and sensitivity to metal ions were little changed. However, a higher relative activity towards whey, lactose and skim milk was found for the immobilized enzyme from *M. phaseoli* compared with the native enzyme. The enzymes from *M. phaseoli* and *S. tuliparum* retained 100% and 78%, respectively, of their original activities in the immobilized form when stored for two months at pH 5.0 and 37°C compared with 62% and 6%, respectively, for the soluble enzymes. It was concluded that the enzyme from *M. phaseoli* was more effective for hydrolytic cleavage of lactose in dairy products.

A patent for lactase immobilized on glass has been issued to the Corning Company (Lynn, 1978). In a general procedure for inorganic supports, particularly porous glass, the surface is treated first with *p*-phenylenediamine so that the free amino groups left can be covalently bound to enzymes by diazotization or a Schiff-base reaction.

Controlled pore-size silica was used by colleagues at the Corning Glass Works (Messing *et al.*, 1976) for immobilizing lactase from *Aspergillus niger*. Here, the surface was treated with *o*-dianisidine followed by functionalization of the residual amine with glutaraldehyde or with nitrite to form the diazonium salt. The immobilized enzyme had a half-life of 52 days at 50°C with a 5% lactose feed at pH 3.5.

β-Galactosidase has been coupled under a pressure of 70 p.s.i.g.

(approx. 485 kPa) to a homoporous matrix membrane that had been activated with cyanogen bromide (Gregor, 1977b). Such membranes were made by casting from a 15% solution of cellulose acetate and, after enzyme coupling, could be used to treat lactose in skim milk.

Another patent from the United States Department of Agriculture (Meeri et al., 1976) concerned immobilization of lactase on a phenolic polymer modified by introduction of aldehyde or diazonium salt groups in the chain. One such commercial resin, Duolite S-30 which is a phenol formaldehyde resin, was aldehydically modified by reaction with phosphoryl chloride and dimethyl formamide. The immobilized enzyme exhibited an activity of 50 μmoles of lactose hydrolysed/min/g of moist resin.

Lactase immobilized on alumina particles, then cross-linked with glutaraldehyde and washed with citrate–phosphate buffer, has been used in a fluidized-bed enzyme reactor as part of a pilot plant to hydrolyse lactose in raw cottage-cheese whey (Charles et al., 1976, 1977). Kraznobajev (1977) immobilized lactase from A. niger on phenylene diamine–glutaraldehyde-coated pumis granulate and this system was evaluated for its ability to act on the lactose.

Masri et al. (1976) immobilized β-galactosidase on chitosan in a soluble form, transformed into a gel by addition of a polyfunctional cross-linking agent such as dialdehyde starch, glyoxal or glutaraldehyde. The gel was later treated with a reducing agent, such as sodium borohydride, to anneal residual aldehyde groups. Glyoxal was the preferred reagent for β-galactosidase. Suzuki (1977) produced enzyme gels of β-galactosidase by γ-irradiation of polyvinyl alcohol, polyvinylpyrrolidone and the enzyme. A relatively large dose of radiation (2–5 Mrad) was required. No enzyme leakage was evident and the polyvinylpyrrolidone–β-galactosidase gel was evaluated for lactose hydrolysis in acid whey using a packed column reactor.

β-Galactosidase was one of a number of enzymes cited that could be immobilized after reaction in the presence of a poly(ethylene glycol) and polyisocyanate to form an expanded polyurethane (Grace and Company, 1976). Typical reactants were ethylene glycol and toluene diisocyanate, mixed at 65°C, to which the enzyme was added on cooling. This patent corresponds to that issued to Hartdegan and Swann (1976).

Immobilization in the form of a soluble polymeric derivative, later recoverable by precipitation, is the idea incorporated in a patent issued to Schneider (1976). Thus, β-galactosidase was bound to N-(4-carboxy-phenyl)acrylamide polymer. Adjustment of the pH value to 4.5 after the reaction enabled the bound enzyme to be recovered.

In the U.S.S.R., workers have immobilized fungal β-galactosidase preparations on silaceous carriers using glutaraldehyde (Pappel et al., 1976). Highest activities were observed with the enzyme immobilized on polycarbamide-coated silochrome, but the β-galactosidase had higher stability on nylon-coated silochrome. Lactase has also been immobilized on resin triazine derivatives such as the one prepared by reaction of Duolite A7 (OH$^-$) with 5% sodium carbonate containing 2-carboxymethylamine-4,6-dichloro-s-triazine for 10 min at pH 11 (Marumo et al., 1976).

A review of the preparation, characterization and scale-up of a lactase system immobilized on inorganic supports has been written by Pitcher et al. (1976). Inorganic supports include oxides of zirconium, titanium and silicon.

H. Pronase from *Streptomyces griseus*

This enzyme has been covalently coupled to cyanogen bromide-activated Sepharose. In this procedure, it retained 22.6% of its original activity compared with almost complete retention on ω-amino-alkyl-Sepharose where the number of methylene groups was 8, 10 or 12 (Byun and Wold, 1976). In the latter procedure, the optimum pH value of the enzyme was unaltered, whereas direct attachment to Sepharose broadened it considerably. The K_m values were (on leucine-p-nitro-anilide) 0.83 mM for soluble pronase, 2.22 mM for the enzyme directly coupled to Sepharose and 0.91 mM for the ω-aminoalkyl-Sepharose version.

Pronase, as well as trypsin and subtilisin, has been attached to the polysaccharide derivative, chitosan, activated by treatment with glutaraldehyde. The enzyme was treated with the activated support in borate buffer (pH 8) containing 0.02 M Ca^{2+} and stirred for two hours at 2°C. It was then extensively washed including a treatment with sodium borohydride in borate buffer. Activity was preserved by storage in 0.05 M borate buffer (pH 7.5) containing sodium azide or by lyophilization in the presence of mannitol (Leuba, 1976).

The potentialities of various immobilized proteinases in determining the structure of proteins has been pointed out by Royer (1974). When attached to the arylamine derivative of porous glass, immobilized papain retained its activity against N-benzoylarginine ethyl ester, leucyl-p-nitroanilide, serum albumin and β-lactoglobulin. Carboxypeptidase C, when immobilized, had limited activity and stability but the corresponding carboxypeptidases A and B retained considerable activity against hippuryl-L-phenylanaline and L-arginine. Immobilized pronase caused 65% hydrolysis of β-lactoglobulin in six hours of digestion, a value that was raised to 93% by similarly immobilized aminopeptidase. Immobilized aminopeptidase also effected near total hydrolyses of the aminoethylated A and B chains of insulin, particularly the A chain (96%).

I. Papain

Papain has been attached to supports, such as titanium oxide and cellulose, via a precoat of diazotized 1,3-diaminobenzene which tends to polymerize during the diazotization process (Kennedy et al., 1977a). These derivatives possessed high esterolytic, but low proteolytic, activity. During its attachment to such supports, papain was protected by complexing it with zinc ions which later could be removed by washing the immobilized enzyme with EDTA. Enhanced proteolytic activity was attained by substituting macroporous cellulose for the microcrystalline cellulose previously used. In its potential application in improving the haze stability of beer, a vital question is the nature of the surface of the immobilized support surrounding the papain. Herbert and Scriban (1976) found supports such as bovine albumin, kieselguhr or porous glass in packed bed reactors quite unsatisfactory because of absorption of phenolic compounds on the enzyme beads.

Papain and other proteolytic enzymes (trypsin, chymotrypsin) have been immobilized on copolymers of ethylene and maleic anhydride cross-linked with divinylbenzene (optimum 5.8%). While papain in this covalent attachment showed an unchanged optimum pH value, that of trypsin was shifted 0.7 of a pH unit towards the acid region and that of chymotrypsin 1.5 pH units towards the alkaline region. While the thermostability of the papain was unchanged, that of the other two enzymes increased. Trypsin and chymotrypsin retained 35–40% of the esterase activity of the soluble enzymes when immobilized (Kurichev *et al.*, 1977).

A related copolymer, also based on ethylene-maleic acid anhydride but cross-linked with hydrazine and further reacted with excess hydrazine after enzyme immobilization, has been patented (Goldstein *et al.*, 1977). Post-treatment with hydrazine converts any unchanged anhydride residues into acylhydrazide groups. In the immobilization step, a buffer solution (pH 8.4) of the enzyme is incubated overnight with a buffered suspension of the polymeric hydrazide. Another version is to treat the ethylene–maleic anhydride copolymer with *p,p'*-diaminodiphenylmethane in dimethylsulphoxide. The resulting resin is then diazotized for attachment to the enzyme.

J. Neutral Proteinase from *Bacillus subtilis*

A close study has been made of this enzyme immobilized on Silichrome C-80. While the soluble proteinase was more stable at low temperatures (maximum activity, 40°C), the immobilized enzyme was less active at low temperatures and showed maximum activity at 60–70°C. Again, the soluble enzyme had an optimum pH value of 7–7.2, while the immobilized enzyme showed 96–100% activity at pH 5.5–10.0. Finally, the immobilized preparation was insensitive to changes in pH value between 4.5 to 9.2 and temperature changes from 10°C to 40°C for 1.5–2.0 hours. The soluble enzyme was especially sensitive at pH 4.5 (Kalashnikova *et al.*, 1975).

Proteinase from *Bacillus subtilis* can also be immobilized by procedures devised by CPC (United Kingdom) International (1976) which are also applicable to glucose isomerase. In these, the enzyme is attached to the inner surface of a porous material, such as silicate-based ceramic materials, with a mean pore diameter at least as large as that of the enzyme. Thus, titanium oxide or alumina is pretreated with 0.05 M magnesium acetate and 0.01 M cobaltous chloride (pH 7.5) for 15 min at 60°C and three hours at room temperature. The support is then suitable for enzyme adsorption, which took over four hours. After extensive washing, the enzyme showed good stability.

IV. IMMOBILIZED ENZYME REACTORS

A. Membrane Reactors

Part of the design specification for a membrane-immobilized enzyme is the essential requirement to heat-stabilize it for repeated use. Numerous authors have established that attachment to water-soluble polysaccharides has a high probability of accomplishing this, with the advantage that the enzyme conjugate has other improved properties such as resistance to proteolytic digestion. Thus, Marshall and Humphreys (1977) established that maximum stability of catalase attached to dextran, activated with cyanogen bromide, was obtained after reaction for 20 hours at 23°C and pH 8 followed by addition of glycine. The ratio of catalase:dextran during coupling was 1:10. Initial activation of the dextran was achieved at pH 10.7 and by adding it to the solution of solid cyanogen bromide with subsequent decreases in pH to 9 and 8 at which stage the catalase was added. Glycine addition to the conjugate prevented development of turbidity and eventual precipitation. It also permitted dialysis and freeze-drying to be applied with confidence. When heated at 52°C, the enzyme conjugate was much more stable than the native enzyme, which rapidly lost activity (Marshall and Rabinowitz, 1976). This closely paralleled

Table 2

Reactor types for immobilized enzymes

Batch reactors	
Stirred-tank reactor	— catalyst suspended by stirrer
Stirred-tank reactor	— catalyst on impeller blades
Stirred-tank reactor	— catalyst on basket stirrer wall
Air-stirred reactor	— catalyst suspended by air
Total-recycle reactor	— catalyst suspended by substrate movement up a column and down through a reservoir
Continuous reactors	
Tubular reactor	— substrate moves continuously through an open tube of catalyst
Fluidized-bed reactor	— substrate moves upwards continuously through air; suspended catalyst retained within a tube
Packed-bed reactor (plugged flow reactor)	— substrate moves upwards continuously through a packed immobilized bed of catalyst
Continuous stirred-tank reactor	— substrate fed continuously through a catalyst suspended by a stirrer

the results obtained by Marshall and his coworkers on the dextran conjugates of α-amylase and trypsin with respect to heat stability. Attachment of dextran also improved resistance to proteolytic degradation, inactivation by protein denaturants and other adverse conditions.

The nature of the polysaccharide stabilizing the enzyme against proteolytic digestion has been investigated (Vegarud and Christensen, 1975). Thus lysozyme, chymotrypsin, casein and serum albumin were stabilized by attachment to dextrans of different molecular weights (10,000 and 3,000). Starch dextrin (mol. wr. 6,000) was also assessed. Even low-level incorporation of dextrin into the enzyme yielded a significant decrease in proteolytic degradation. All of the workers emphasized the necessity for initial optimization of the coupling procedure. Christensen *et al.* (1977) showed the remarkable heat-stabilizing effect of dextran attachment to lysozyme and β-glucosidase tested at 100°C and 60°C, respectively.

A second design specification for the membrane reactor is provision of a soluble non-dialysable polymeric form of coenzyme which can be used over prolonged periods. Yamazaki *et al.* (1976) have provided such a derivative in polylysine-bound succinyl-NAD^+ by coupling succinyl-NAD^+ to polylysine using a carbodiimide reagent (ethyldimethyl-aminopropylcarbodiimide hydrochloride). The succinyl attachment was found to be surprisingly stable, and the polylysine to exert a protective function against decomposition by bacteria.

Immobilized NAD^+ responded well in a membrane reactor and retained at least 85% of the initial coenzymic activity even after dialysis for one week. It was applied in the membrane reactor with alcohol dehydrogenase and lactate dehydrogenase (equation 1) and lactate was continuously produced with a half-life of ten days. The main loss of activity was due to absorption of the enzymes on the membrane.

$$\text{Ethanol} + 2\,\text{Pyruvate} = \text{Acetic acid and 2 Lactate} \qquad (1)$$

However, there was an innate decrease in the ability of poly-lysine-NAD^+ to act as a cofactor. The V_{max} value for the modified coenzyme was 2.9 μmol/min for lactate dehydrogenase, and 1.8 μmol/min for alcohol dehydrogenase, compared with the corresponding values of 13.8 and 9.7, respectively, for the native NAD^+.

Oviously, the requirement for a coenzyme is not always imperative so that glucose oxidase, for example, can be substituted for glucose dehydrogenase which also requires NAD⁺. Most industrial enzymes in any case do not require coenzymes. Thus, starch hydrolysis was accomplished by a mixture of α- and β-amylases in a membrane reactor (Closset *et al.*, 1974).

A third design specification for a membrane reactor is preservation of the operational stability of the reactor against exoenzymes from micro-organisms which may degrade the enzyme catalyst, cause clogging of the membrane and/or attack on the membrane. While bacteriocides are very important, a novel suggestion is use of lysozyme which was found to have beneficial effects during attack of *Bacillus subtilis* on a immobilized enzyme. Mattiasson (1977) has pointed out that other enzymes from *Arthrobacter luteus* are available for a similar approach.

Hollow fibres offer an attractive way of immobilizing enzymes in a continuous-reactor form (Corus and Olson, 1977). Their large surface-to-volume ratio facilitates substrate transport. The American workers found that enzyme stabilities in hollow fibres of polysulphone which had been preconditioned with bovine albumin approached the stabilities of the free enzyme. Further advantages were the ease with which a hollow fibre cartridge could be loaded with enzyme, operated, cleaned and sterilized, and regenerated. Leakage of the enzyme was prevented by selecting a sufficiently tight fibre membrane and inactivation of enzyme on the fibres by conditioning the latter. Romicon P10 fibre cartridges, preconditioned with bovine albumin, were found to be excellent. In this system, hollow fibre-immobilized α-galactosidase offered a convenient system for hydrolysis of α-galactosides, such as raffinose, in sugar beets. Similarly, invertase loaded into fibres by ultrafiltration performed well on 0.5 M sucrose when recycled through the cartridge at 37°C using a peristaltic pump. The cartridge could be later freed from invertase by pumping 6 M urea into the fibre lumen, across the membrane and out on the shell side of the cartridge. When reloaded at three times the previous enzyme loading, the initial rate of glucose production was increased by a factor of 1.8. Since the rate of reaction increased as the square root of the enzyme concentration for complete diffusional control, this indicated that the rate of reaction was almost completely limited by diffusional effects at this higher enzyme loading.

B. Open-Tube Enzyme Reactors

Ngo *et al.* (1976) evaluated a β-galactosidase-containing tubular reactor in which the enzyme was covalently attached to the inner surface of nylon tubing. This was used to effect hydrolysis of lactose in skim milk. With a single 46-metre tube and continuous circulation, some 90% of the lactose was removed within 20 hours. A battery of ten such tubes, with a single passage, at a flow rate of 2 cm/sec would remove more than 90% of the lactose in less than 40 minutes. Immobilization was effected after pretreating the tubes with 18.6% (w/w) calcium chloride in 18.6% (v/v) water in methanol for 20 min at 50°C. The etched tubes were then partially hydrolysed with 4.5 M HCl and, after washing, reacted with 12.5% glutaraldehyde in 0.2 M sodium bicarbonate (pH 9.0) and later the enzyme. The rate of hydrolysis in the enzyme-containing tubes increased with increasing flow rate since the thickness of the Nernst diffusion layer in the liquid phase was lowered with the increasing flow rate, so that enzyme was more readily accessible to substrate.

One of the major problems with polysulphone membranes, which are replacing those of cellulose acetate, is that this type of ultra-filtration membrane suffers rapid loss of flux when the process fluids contain protein. Velicangil and Howell (1977) showed that this could be overcome by a self-cleansing device which involves attaching papain to the membrane so that the clogging proteins were hydrolysed as they were deposited. In effect, these workers made a membrane reactor by covalently attaching the enzyme to the membrane. To achieve this, membranes were hydrolysed in 3 M HCl at 45°C for 45 min and then perfused with 6.25% (w/v) glutaraldehyde in 0.1 M sodium borate buffer (pH 8.5) at 4°C for 50 minutes in the ultrafiltration cell. Papain was immobilized by continuing the perfusion for two hours with papain in phosphate buffer (pH 6.2) protected by EDTA, cysteine and mercaptoethanol in the solution. The bound enzyme retained 15% of its free-solution activity. The initial flux loss due to immobilization of papain was rapidly compensated for by the improved performance over a longer period, and this improved at higher operating temperatures.

C. Plug Flow-Type Continuous Enzyme Reactors

Ryu and Chung (1977) used upward flow through a column of immobilized glucose isomerase in their plug flow-type reactor. The enzyme was in the form of cells of *Streptomyces phaechromogenus* which had been heat-treated at 65°C for one hour to fix and thereby immobilize the intracellular glucose isomerase. The continuous tubular reactor consisted of a jacketed glass column, and the enzyme was retained by Millipore filter membranes top and bottom. Each membrane was sandwiched between nylon cloths and stainless-steel sieve plates (20 mesh) placed on both sides of the membrane. With a feed substrate of 2.3 M glucose and an enzyme loading of 7012 units/litre, the half-life of the enzyme was 94 hours. The activity decay followed first-order kinetics. The maximum attainable conversion of glucose into fructose was 57%, and the maximum reaction rate for this forward reaction was about 35% lower than that for the reverse reaction. The various kinetic constants were: $K_m(f)$ 0.24 M, $K_m(b)$ 0.49 M, $V_{max}(f)$ 11 μmol/min/mg of dry cell weight, $V_{max}(b)$ 16.6 μmol/min/mg and $K_{eq.}$ 1.35 at the optimal temperature of 70°C and pH 7.0. When operated in the continuous-reactor mode, it was found necessary to maintain the mean residence time in the reactor to less than three hours to avoid significant side-product and colour production. The heat of the reaction is only about 2.2 kcal/mol (9.2 J/mol) and so poses no problems of heat transfer to and from the reactor. The activation energy was 13.6 kcal/mol (57.1 J/mol). The up-flow mode was preferred in the context of problems with pressure drop, possible oxidation by air bubbles trapped in the reactor system, flow pattern and reactor performance. Computer simulation of the reactor system showed the productivity of the enzyme as a function of mean space time, substrate feed concentration and conversion. This enabled optimal operating conditions to be defined (residence time three hours and substrate concentration 2.3 M). One important feature found in this type of reactor was that the reaction rate decreased so rapidly that the rate is only about 10% of the capacity rate when the reaction mixture passes 40% of the reactor length or the time equivalent, i.e. the enzyme located in the latter half of the reactor was not utilized to its full capacity. This same phenomenon explains why, if enzyme loading is increased by 100%, the increase in productivity is only 60%. Despite these problems for which solutions were suggested, the plug flow-type of reactor was significantly better than a continuously stirred-tank reactor system in terms of productivity. Initially, this advantage was 26% and dropped to about 15% after five days of operation.

Marsh and Tsao (1976) studied glucamylase immobilized on porous glass in a packed-bed enzyme reactor again operated preferably by upward flow of the substrate, maltose. The packing could not be fluidized or expanded because of retention by 100 mesh screens. Because porous glass is a poor thermal conductor, temperature and concentration profiles in the column were examined. It was found preferable to feed the reactor slightly below the optimum temperature (58.5°C) and allow the heat of reaction to warm the reaction mixture to achieve maximum overall conversion in the 1.8 m-tall column. The heat of hydrolysis of maltose to glucose was -1220 cal/mol $(-5$ J/mol) at 55°C. The reactor was also operated using a feed of 27% 15 DE starch solution with an inlet temperature of 55.3°C. Conversion into glucose reached 80% some 160 cm along the column. The values mimicked those obtained with a 27% maltose feed.

D. Continuous Stirred-Tank Reactors

Choi and Fan (1973) developed mathematical models for the transient behaviour of encapsulated enzyme reactor systems, such as the continuous stirred-tank reactor. It was assumed that the membrane in which the enzyme was occluded was inert, and that there was no hindrance effect on substrate transport through the pore space of the membrane. Further, it was envisaged that no conformational change took place at the active site of the enzyme during encapsulation, and that all rate processes occurred under an isothermal condition. More unlikely, but included in their assumptions, was that the kinetic rate and the Michaelis constants of the encapsulated enzyme were identical with those of the free native enzyme. A material balance over the total volume of a continuous stirred-tank reactor system was deduced as:

$$V\phi_b \frac{DS}{dt} = FS_f - FS - V\varepsilon \frac{k[E]Si}{K_m + Si}$$

with the initial condition $S = 0$ at $t = 0$.

From this they derived:

$$K_f = \frac{\varepsilon K[E]V}{S_f F} \quad \text{and} \quad \tau = \frac{tF}{V\phi_b}$$

where:

V = volume of an immobilized continuous stirred-tank enzyme reactor
ϕ_b = volume fraction of bulk void space of bed
F = volumetric feed rate
S_f = substrate concentration of feed
S = substrate concentration of the bulk of the flow
V = volume of an immobilized continuous stirred-tank enzyme reactor
$[E]$ = enzyme concentration based on the void space of particle
Si = substrate concentration in the enzyme phase of encapsulated enzyme particle
ε = volume of enzyme phase per unit volume of bed
K_1 = dimensionless catalytic reaction constant
τ = dimensionless time
t = time

E. Continuous-Flow Bioreactors Containing Whole Cells

Saini and Vieth (1975) developed a simple inexpensive and efficient process of whole-cell immobilization based on the use of reconstituted hide collagen as a support material. This was used to immobilize cells of *Streptomyces venezuelae* containing glucose isomerase. The mechanical strength of the collagen membrane was enhanced by tanning, which also served to improve its hydrothermal durability and resistance to microbial attack, so permitting its use at 70°C in a continuous-flow bioreactor for several months. The reactor was a jacketed column provided with a bed support, feed distributor and preheat section. Tanned chips of the biocatalyst (60 mm × 60 mm) were packed into the column which was supplied with feed at the bottom by a variable-speed peristaltic pump. The feed comprised 0.1–3 M glucose containing 0.01 M Mg^{2+}, 0.1% (w/v) methyl paraben and 0.1% (w/v) sodium bicarbonate at pH 7.0 ± 0.2.

Cells were fixed by heating the aqueous suspension of cells (pH ~7) at 80°C for 1.5 hours with shaking. Any micro-organisms in the collagen were destroyed by suspending it (100 g/l) in 3% hydrogen peroxide at pH 7 followed by thorough washing to remove residual peroxide. During fixation, the enzyme becomes attached to the denatured intracellular mass. Denaturation by pH value accounts for about 64% of the loss of enzyme activity suffered, while tanning causes only 11%. However, a product is formed with a 50-day half-life. Cells are attached to the collagen by non-covalent bonds, while the tanning agent (formaldehyde or glutaraldehyde) cross-links the supranetwork of collagen. The optimum number of such cross-links are 20–25 per collagen molecule.

When in contact with substrate, the collagen membrane imbibes about twice its weight of solvent, reflecting the openness of the structure. In a continuous-flow reactor, one important factor for assessment is transport of glucose from the bulk fluid to the membrane fluid interface which requires a driving force, namely the concentration difference. This depends on the velocity pattern in the fluid near the surface, on the physical properties of the fluid and on the intrinsic rate of the biochemical reaction in the membrane. Further, there will be a temperature gradient between the bulk fluid and the membrane surface, the size of which will depend on the heat of reaction, the reaction rate constant and the heat-transfer coefficient between the fluid and the membrane surface.

The second assessment factor is intramembrane transport of substrate to the enzyme site, since intramembrane concentration and temperature gradients can cause the rate to vary with position. Saini and Vieth (1975) found that the external as well as the internal mass-transfer limitations for glucose-to-fructose conversion in the system under study were negligible.

Other important factors are formation of an enzyme–substrate complex, the bioreaction at the active site to form an enzyme–product complex, desorption of the product, transport of product from the interior to the outer surface of the membrane, and transport of product from the membrane-fluid interface to the bulk-fluid stream. A knowledge of the rate-limiting step under various conditions is essential for optimum reactor design. In their system, Saini and Vieth (1975) found that, somewhat below 0.4 M glucose, the process was controlled by adsorption on enzyme active sites whereas, above that concentration, the rate-determining step is the rate of reaction at the active site.

F. Microbial Catalysts in Reactors

Vandamme (1976) has pointed out that, heretofore, industrial exploitation of micro-organisms has been mainly restricted to use of intact growing cells, resting cells or their soluble cell-free enzyme preparations. While growing cells are used primarily for industrial microbiological syntheses, intact-cell or enzyme preparations serve as catalysts in bioconversion processes. As this review illustrates, immobilized enzymes are gradually entering the latter class. However, microbial cells offer particular challenges to extend their utility in more novel methods of presentation which take advantage of their pre-established structure.

One fact that has hardly been touched on is the deliberate attachment of additional enzymes to the cell surface. Examples in the literature (Hough and Lyons, 1972; Kosugi and Suzuki, 1973; Takasaki, 1974) illustrate that the titanium-chelation method can attach amyloglucosidase and α-amylase to Sacch. cerevisiae, B. subtilis, E. coli and A. niger. Further, α-amylase, invertase and amyloglucosidase have been coupled to Streptomyces strains that produce glucose isomerase, so that the cells can produce fructose directly from starch. Immobilization of cell components is an extension of this concept and already membrane-particles, mitochondria, chromatophores and yeast microbodies have been selectively immobilized (Vandamme, 1976).

The renewable property of living cells, or even a limited entity such as a particular metabolic sequence (e.g. photosynthesis) or a recycling capacity with respect to a certain coenzyme, is most attractive for exploitation even by limited immobilization (e.g. having a restricted freedom of movement). One example of this is the use of growing cells of Acetobacter sp. in a tower fermenter rendered sluggish in movement by aggregation to limit their rate of wash-out from the top of the tower by fresh substrate and air. Normally, in vinegar manufacture, they have been used unmodified in the tower where they convert up to 99% of the etahnol content of the charging wort into acetic acid. Barker et al. (1978) used, as aggregating agents, hydrous titanium oxide for Acetobacter strains that produced polysaccharide and a hydrous titanium-cellulose chelate for those strains that did not. These additions enabled the rate of production of acetic acid to be increased within the one fermenter. Further, it conferred greater resistance to disturbance by frothing, so that greater aeration rates could be employed, as well as giving better consistency of production. Conversion of alcohol into acetic acid proceeds at around pH 3. Below this pH value, the titanium chelating capacity decreases, while above this it increases. Hence with fluctuating pH value, due to inflow of the substrate raising it and the acetic-acid production lowering it, chelation with cells is variable so permitting cell renewal and disposal.

G. Future Developments

Past experience has shown that multi-enzyme systems in which previously isolated enzymes are subsequently put together to catalyse a sequence of three to four reactions in a reactor that uses them in an immobilized form, will rarely be economic industrially. Thus in a sequence:

$$\text{Starch} \underset{}{\overset{1}{\rightleftharpoons}} \text{Glucose 1-phosphate} \underset{}{\overset{2}{\rightleftharpoons}}$$

$$\text{Glucose 6-phosphate} \underset{}{\overset{3}{\rightleftharpoons}} \text{Fructose 6-phosphate} \underset{}{\overset{4}{\rightleftharpoons}} \text{Fructose}$$

most of the problems are in the architecture of the reactor and the method of presentation of immobilized enzymes. Reaction 1 requires excess phosphate to drive it, accessibility of the large starch molecule to the immobilized enzyme, and some means to separate the product glucose 1-phosphate to make it available to the enzyme in reaction 2, but not to the less specific phosphatase in reaction 4. A dialysis membrane solves some of these problems and poses the question as to whether an insoluble immobilized phosphorylase is really the best solution, or whether a reactor–separator in which a soluble but artificially stabilized phosphorylase (e.g. by linkage to a poly-saccharide) is confined by dialysis or even ultrafiltration membrane. Further consideration also suggests that another membrane, this time ionic, is required to confine the sugar phosphates, in particular fructose 6-phosphate for reaction 4, and separate it and them from the neutral fructose. With the small substrates in reactions 2, 3 and 4, immobilized enzymes are practical solutions to catalysis and where these are placed determines where the reactions will occur. They act as on–off switches and could, if required, be placed along the walls of a tubular reactor in their appropriate sequence. Most of these solutions would be quite practical for analytical procedures with dilute solutions, as we have shown. For preparative operations, other solutions would have to be sought that avoided membranes. In the medium range of concentrations, a pulsed reactor–separator, in which the reactor column is not only the enzyme support but a molecular sieve (e.g. Spheron) separating a large substrate molecule from the small product molecule as in reaction 1, is a realistic solution. For reaction 4, a pulsed reactor–separator, in which the reactor column carries the enzyme catalyst and separates by ion exclusion, provides the answer. The open tubular reactor with the catalyst on the walls might be replaced with segments of a fluidized bed of particles partly limited in movement by open networks that retain the particles but not the substrate. In addition to these problems are those associated with creating micro-environments for the four different enzymes to work at one harmonious pH value besides providing the glucose 1,6-diphosphate for reaction 2. While separation may help to displace the normally unfavourable equilibrium of reaction 3, an oxyanion to complex selectively with the product is a more practical solution.

This discussion points to some of the problems associated with future application of immobilized enzyme technology, and gives a warning that evolution took a long time to evolve a cellular structure for multi-enzyme systems and most of the suggested solutions of how to set up reactions 1–4 have a direct analogy in nature: (a) conversion of intracellular protein enzymes into an extracellular glycoprotein counterpart with enhanced thermostability (e.g. various forms of invertase); (b) placement of enzymes acting on high molecular-weight substrates at the outer surface or in the extracellular fluid (e.g. glucamylase of *A. niger*); (c) placement of enzymes acting on low molecular-weight substrates on the inner surface or in the intra-cellular fluid (e.g. phosphoglucomutase); (d) use of the ionic membranes to separate sugar phosphates and neutral carbohydrates, both large and small.

Future developments should therefore build on natural environments, and not seek to create completely new ones, except where a real industrial requirement exists, like coupling conversion of starch to fructose together in one simultaneous reaction sequence. It is striking that, with all of the new plant installed for glucose-to-fructose conversion using immobilized glucose isomerase, not one has taken up the challenge of immobilized amyloglucosidase, despite the major technological advantages this would confer. Perhaps they wait for the two enzymes immobilized in the one column and geared together at one pH value.

REFERENCES

Atallah, M. T., Wasserman, B. P. and Hultin, H. O. (1976). *Biotechnology and Bioengineering* **18**, 1833.
Axen, R. and Porath, J. (1966). *Nature, London* **210**, 367.
Barker, S. A. and Kennedy, J. F. (1975). British Patent 1,402,427.
Barker, S. A., Somers, P. J. and Epton, R. (1970a). *Carbohydrate Research* **14**, 323.

Barker, S. A., Somers, P. J. Epton, R. and Maclaren, J. V. (1970b). *Carbohydrate Research* **14**, 287.

Barker, S. A., Cho Tun, H., Doss, S. H., Gray, C. J. and Kennedy, J. F. (1971a). *Carbohydrate Research* **17**, 471.

Barker, S. A., Somers, P. J. and Epton, R. (1971b). British Patent 1,232,619.

Barker, S. A., Kennedy, J. F. and Gray, C. J. (1972). British Patents 1,289,548 and 1,289,549

Barker, S. A., Gray, C. J. and Cowling, D. T. (1975a). British Patent 1,380,898.

Barker, S. A., Kay, I. M. and Kennedy, J. F. (1975b). United States Patent 3,912,593.

Barker, S. A., Greenshields, R. M., Humphreys, J. D. and Kennedy, J. F. (1978). British Patent 1,514,425.

Brouillard, R. E. (1977). United States Patent 4,029,546.

Burrows, H. G. and Lilly, M. D. (1976). German Patent 2,450,132

Byun, S. M. and Wold, F. (1976). *Hanguk Sikpum Kwahakhoe Chi* **8**, 253.

Caldwell, K. D., Axen, R., Bergwall, M., Olsson, I. and Porath, J. (1976a). *Biotechnology and Bioengineering* **18**, 1605.

Caldwell, K. D., Axen, R., Bergwall, M. and Porath, J. (1976b). *Biotechnology and Bioengineering* **18**, 1573.

Charles, M., Coughlin, R. W. and Julkowski, K. (1976). *United States Environmental Protection Agency in Proceedings, National Symposium Food Process Wastes* 7th, p. 83.

Charles, M., Coughlin, R. W. and Julkowski, K. (1977). *American Chemical Abstracts* **87**, 66721.

Choi, P. S. K. and Fan, L. T. (1973). *Journal of Applied Chemistry and Biotechnology* **23**, 531.

Christensen, T. B., Vegarud, G. and Birkeland, A. J. (1977). *Process Biochemistry* **11**, 25.

Closset, C. P., Cobb, J. T. and Shah, Y. T. (1974). *Biotechnology and Bioengineering* **16**, 345.

Corus, R. A. and Olson, A. C. (1977). *Biotechnology and Bioengineering* **19**, 1.

CPC International (1976). Netherlands Patent Application 74 09,455.

Emery, A. N., Novais, J. M. and Barker, S. A. (1972). British Patent 1,346,631.

Forgione, P. S. (1974). United States Patent 3,847,743.

Fujita, Y., Matsumoto, A., Miyachi, I., Imai, N., Kawakami, I., Hishida, T. and Kamata, A. (1977). German Patent 2,720,538.

Fukui, S., Tanaka, A., Iida, T. and Hasegawa, E. (1977a). Japanese Patent 77,110,889.

Fukui, S., Yamamoto, T. and Iida, T. (1977b). German Patent 2,629,692.

Fukui, S., Yamamoto, T. and Iida, T. (1977c). German Patent 2,629,693.

Goldstein, L., Blumberg, S., Katchalski, E. and Levin, Y. (1977). Israel Patent 42,254.

Grace, W. and Co. (1976). Belgian Patent 833,191.

Gray, C. J., Livingstone, C. M., Jones, C. M. and Barker, S. A. (1974). *Biochimica et Biophysica Acta* **341**, 457.

Gregor, H. P. (1977a). United States Patent 4,033,822.

Gregor, H. P. (1977b). German Patent 2,650,920.

Hartdegan, F. J. and Swann, W. E. (1976). German Patent 2,612,138.

Herbert, J. P. and Scriban, R. (1976). *Brauwissenschaft* **29**, 365.

Horn, J., Batz, H. G. and Jaworek, D. (1977). German Patent 2,603,319.

Hornby, W. E. and Morris, D. L. (1976). German Patent 2,625,255.

Hough, J. S. and Lyons, T. P. (1972). *Nature, London* **235**, 389.

Hupkes, J. V. (1978). *Staerke* **30**, 24.

Iida, T. and Hasegawa, E. (1977). Japanese Patent 77,143,282.

Ito, H., Izuho, Y., Sugihara, S. and Ozawa, S. (1977). Japanese Patent 77,125,690.

Jaworek, D., Botsch, H. and Maier, J. (1976). *Methods in Enzymology* **44**, 195.

Kaetsu, I., Kumakura, M. and Yoshida, M. (1977a). Japanese Patent 77 47,981.

Kaetsu, I., Kumakura, M. and Yoshida, M. (1977b). German Patent 2,705,377.

Kalashnikova, A. M., Menyailova, I. I., Nakhapetyan, L. A., Gracheva, I. M. and Konurkina, A. A. (1975). *Prikl. Biokhim. Mikrobiol.* **11**, 842.

Kato, S., Matsuura, T. and Ozaki, H. (1977). Japanese Patent 77,151,789.

Kawashima, K. and Umaeda, K. (1975). *American Chemical Abstracts* **83**, 143667.

Kawashima, K. and Umeda, K. (1977). Japanese Patent 77,120,187.

Kennedy, J. F., Barker, S. A. and Rosevear, A. (1973). *Journal of the Chemical Society, Perkin I*, 2293.

Kennedy, J. F., Barker, S. A. and Zamir, A. (1974). *Antimicrobial Agents and Chemotherapy* **16**, 777.

Kennedy, J. F., Barker, S. A. and Humphreys, J. D. (1976). *Nature, London* **261**, 242.

Kennedy, J. F., Barker, S. A. and Pike, V. W. (1977a). *Biochimica et Biophysica Acta* **484**, 115.

Kennedy, J. F., Barker, S. A. and White, C. A. (1977b). *Die Stärke* **29**, 240.

Kent, C. A. and Emery, A. N. (1974). *Journal of Applied Chemistry and Biotechnology* **24**, 663.

Koch-Light Labaratories (1977). British Patents 1,484,564 and 1,484,565.

Kosugi, Y. and Suzuki, H. (1973). *Journal of Fermentation Technology* **51**, 895.

Kraznobajev, V. and Boeniger, R. (1975). *Chimia* **29**, 123.

Kraznobajev, V. (1977). *Chimia* **31**, 110.

Kurichev, V. A., Folomeeva, O. G., Terteryan, R. A., Khrapov, V. S., Koslova, T. P. and Lebedeva, O. G. (1977). *American Chemical Abstracts* **86**, 135582.

Lee, D. (1976). *American Chemical Abstracts* **85**, 175501.

Lee, C. K. (1977). United States Patent 4,061,539.

Leroy, P., Devos, F. and Huchette, M. (1977). German Patent 2,644,496.

Leuba, J. L. (1976). German Patent 2,522,484.

Lynn, M. (1978). United States Patent 4,072,566.

Maeda, H., Suzuki, H., Yamauchi, A. and Sakimae, A. (1974). *Biotechnology and Bioengineering* **16**, 1517.

Maekawa, Y., Inoue, H., Otaki, M. and Matsumoto, H. (1977). Japanese Patent 77,130,879.

Marani, H., Bartoli, F., Bendoricchio, G. and Morisi, F. (1977). *Chimica Industria, Milan* **59**, 243.

Marsh, D. R. and Tsao, G. T. (1976). *Biotechnology and Bioengineering* **18**, 349.

Marshall, J. J. and Humphreys, J. D. (1977). *Biotechnology and Bioengineering* **19**, 1739.

Marshall, J. J. and Rabinowitz, M. L. (1976). *Biotechnology and Bioengineering* **18**, 1325.

Marumo, H., Kimura, K. and Watanabe, K. (1976). *Japan Kokai* **76**, 125,790.

Masri, M. S., Randall, V. C. and Stanley, W. L. (19076). United States Patent Application 706,980.

Matensson, K. B. (1975). German Patent 2,411,255.

Mattiasson, B. (1977). *Biotechnology and Bioengineering* **19**, 777.

Mazarguil, H., Meiller, F. and Monsan, P. (1976). French Patent 2,306,213.

Meeri, M. S., Randall, V. G. and Stanley, W. L. (1976). United States Patent Application 712,295.

Meiller, F. and Monsan, P. (1977). German Patent 2,638,467.

Messing, R. A., Bialousz, L. and Lindner, R. E. (1976). *Journal of Solid-Phase Biochemistry* **1**, 151.

Miller, J. N., Rocks, B. F. and Burns, D. T. (1977). *Analytica Chimica Acta* **93**, 353.

Miyoshi, T., Ishimatsu, Y. and Kimura, S. (1977). Japanese Patent 77,120,185.

Murata, K. and Ishikawa, Y. (1976). Japanese Patent 76,115,990.

Nakamura, T. and Ono, Y. (1977a). Japanese Patent 77,105,281.

Nakamura, T. and Ono, Y. (1977b). Japanese Patent 77,105,282.

Ngo, T. T., Narinesingh, D. and Laidler, K. J. (1976). *Biotechnology and Bioengineering* **18**, 119.

Okada, H. and Urabe, I. (1976). *American Chemical Abstracts* **85**, 16336.

Oosawa, T. (1977). Japanese Patent 77,34,979.

Pappel, K., Kipper, H., Kostner, A., Kulikova, A. and Tikhonirova, A. S. (1976). *Transactions Tallin Politekh Institute* **402**, 3.

Pastore, M. (1977). *Latte* **2**, 456.

Pitcher, W. H., Ford, J. R. and Weetall, H. H. (1976). *Methods in Enzymology* **44**, 792.

Rosevear, A. (1976). German Patent 2,527,884.

Royer, G. P. (1974). *In* 'Immobilized Enzyme Technology' (H. H. Weetall and S. Suzuki, eds.). Plenum Press, New York.

Ryu, D. Y. and Chung, S. H. (1977). *Biotechnology and Bioengineering* **19**, 159.

Schneider, M. (1976). German Patent 2,624,002.

Saini, R. and Vieth, W. R. (1975). *Journal of Applied Chemistry and Biotechnology* **25**, 115.

Senju, R. and Tanaka, H. (1976). German Patent 2,603,599.

Smiley, K. L., Hofreiter, B. T., Boundy, J. A. and Rogovin, S. P. (1975). *Proceedings of the International Symposium on Biodegradation*, 1001. Applied Science.

Solomon, B. and Levin, Y. (1975). *Biotechnology and Bioengineering* **17**, 1323.

Sugiura, M., Suzuki, M. and Sasaki, M. (1977). *Journal of Fermentation Technology* **55**, 575.

Suzuki, H. (1977). *Process Biochemistry* **12**, 9.

Takasaki, Y. (1974). *Agricultural Biological Chemistry* **38**, 1081.

Takasaki, Y. and Takahara, Y. (1976a). Japanese Patent 76 70 873.

Takasaki, Y. and Takahara, Y. (1976b). Japanese Patent 767 875.

Tomb, W. H. and Weetall, H. H. (1977). United States 4,025,667.

Ushiro, S. (1977). German Patent 2,713,861.

Vandamme, E. J. (1976). *Chemistry and Industry* 1070.

Vegarud, G. and Christensen, T. B. (1975). *Biotechnology and Bioengineering* **17**, 1391.

Velicangil, O. and Howell, J. A. (1977). *Biotechnology and Bioengineering* **19**, 1891.

Weaver, M. O., Bagley, E. B., Fanta, G. F. and Doane, W. M. (1976). United States Patent 3,985,616.

Wildi, B. S. and Weeks, L. E. (1976). United States Patent Application B430,213.

Yamazaki, Y., Maeda, H. and Suzuki, H. (1976). *Biotechnology and Bioengineering* **18**, 1761.

8. Steroid Conversions

KLAUS KIESLICH

Gesellschaft für Biotechnologische Forschung mbH D-3300 Braunschweig-Stöckheim, West Germany

I. INTRODUCTION

Microbiological transformations have been used for some considerable time in the chemical synthesis of pharmaceutical products of various classes. The advantages of microbiological reactions as opposed to comparable chemical processes lie in the regiospecificity and stereospecificity of the enzymes and in the mild conditions employed. The objective of microbiological transformations may be a specific reaction, with the minimum of side reactions, or the linking of several reactions in a sequence as part of a transformation process.

Microbiological reactions have been most widely exploited in the field of steroids. The first processes, patented by Schering (Germany) in 1937, and involving reduction of the 17-keto group with yeast (Schoeller, 1937; Mamoli, 1937) and oxidation of the 3β-hydroxy-5-ene structure to the 3-oxo-4-ene system with bacteria (Koester, 1937) were not, in fact, of much practical use. Corresponding chemical reactions were still more satisfactory than the new, as yet novel, microbial methods using pure cultures. The significance of microbiological steroid transformation only became apparent with the important discovery made by the Upjohn Company in 1950 (Murray and Peterson, 1950), in which a hydroxyl group was introduced at the 11α-position in progesterone by means of a *Rhizopus* species, and a new shorter path to cortisone was opened up. The 11β-hydroxylation of Compound RS to hydrocortisone (Collingsworth et al., 1952), which was discovered shortly after, widened the path to further glucocorticoids. Finally, in 1955, Schering (U.S.A.) discovered that increased effectiveness could be achieved by using a microbiological trans-

Table 1

Types of reaction involved, and organisms used, in microbial conversions of steroids

Reaction type	Micro-organism	Reference
HYDROXYLATION		
1α-OH	*Acremonium strictum*	Ambrus *et al.* (1975)
	Acremonium kiliense	Ambrus *et al.* (1975)
	Aspergillus clavatus A.T.C.C. 9598	Petzoldt and Elger (1977)
	Calonectria decora A.T.C.C. 14767	Petzoldt and Elger (1976)
	Mucor spinosis C.B.S. 29563	Petzoldt and Elger (1977)
1β-OH	*Botryodiplodia malorum*	Greenspan *et al.* (1974), Greenspan and Rees
	C.B.S. 13450	(1975)
	Coniosporium rhizophilum Preuss	Schwartz and Protiva (1972)
1β,11α-(OH)₂	*Aspergillus ochraceus* C.B.S. 13752	Jones *et al.* (1972)
2α-OH	*Rhizopus nigricans*	Procházka *et al.* (1974)
2-OH, 4-OH (ring A aromatic)	*Aspergillus flavus*	Schubert *et al.* (1971)
6α-OH	*Absidia orchidis*	Joská *et al.* (1973)
	Aspergillus flavus	Schubert *et al.* (1971)
	Botryodiplodia malorum C.B.S. 13450	Petzoldt *et al.* (1977a)
	Penicillium chrysogenum	Schubert *et al.* (1975a)
6β-OH	*Absidia orchidis*	Engelfried *et al.* (1976)
	Aspergillus flavus	Schubert *et al.* (1972)
	Chaetomium cochliodes	Karim and Marsheck (1976), Marsheck and Karim (1973)
	Gilmaniella sp.	Čapek *et al.* (1975a)
	Mucor griseocyanus	Engelfried *et al.* (1976)
	Paecilomyces sp.	Čapek *et al.* (1976)
	Rhizopus arrhizus A.T.C.C. 11145	Schneider (1974), Holland and Auret (1975a)
6β,11α-(OH)₂	*Rhizopus arrhizus* Fischeri	Favero *et al.* (1977)
	Aspergillus ochraceus	Wagner *et al.* (1976)
	Epicoccum purpurascens	El-Monem *et al.* (1972)
	Rhizopus nigricans	Smith *et al.* (1973)
	Trichothecium roseum	El-Rafai *et al.* (1974a, b)
6β,14α-(OH)₂	*Mucor griseocyanus* A.T.C.C. 1207	Valcavi *et al.* (1974), Valcavi *et al.* (1973b)
7α-OH	*Absidia orchidis*	Engelfried *et al.* (1976)
	Beauveria bassiana	Sanda *et al.* (1977)
	Coriolus hirsutus I.F.O. 4917	Wada and Ishida (1974)
	Cunninghamella elegans C.B.S. 16753	Crabb *et al.* (1975)

Table 1 (contd.)

Reaction type	Micro-organism	Reference
	Diplodia natalensis A.T.C.C. 9055	Petzoldt et al. (1976)
	Mucor griseocyanus A.T.C.C. 1207a	Valcavi et al. (1973a, b)
	Mucor sp. (unidentified)	K. Kieslich (unpublished observations)
	Rhizopus nigricans	Procházka et al. (1974)
7β-OH	Aspergillus flavus	Schubert et al. (1972)
	Cladosporium sp.	Čapek and Fassitová (1974)
	Rhizopus arrhizus	Jones et al. (1976)
	Rhizopus circinans	Jones et al. (1976)
	Rhizopus nigricans	Procházka et al. (1974)
9α-OH	Mycobacterium fortuitum N.R.R.L.-B-8119	Wovcha (1977), Antosz et al. (1977)
	Nocardia corallina A.T.C.C. 4237	Görlich (1973d)
	Nocardia opaca A.T.C.C. 11996	Hörhold et al. (1976)
10-OH	Aspergillus flavus	Schubert et al. (1974)
11α-OH	Aspergillus ochraceus	W. Reusche and K. Kieslich (unpublished observation), Kerb et al. (1974, 1975a, 1976), Koninklijke Nederlansche Gist-en Spiritusfabriek (1973), Ripka (1972), Kieslich et al. (1971), Boswell and Ripka (1972, 1973), Wiechert et al. (1974), K. Kieslich (unpublished observation), Wagner et al. (1976)
	Aspergillus ustus	Sedlaczek et al. (1973a, b)
	Aspergillus sp.	Samanta et al. (1976, Zedan et al. (1976), Abdel-Fattah and Badawi (1974)
	Beauveria bassiana	Sanda et al. (1977)
	Beauveria sp.	Ordzhonikdze (1976)
	Fusarium graminearum	Sedlaczek et al. (1973a, b)
	Fusarium lini	Yamashita et al. (1976a, b)
	Giberella cingulata A.T.C.C. 10534	Engelfried et al. (1976)
	Monosporium olivaceum	Sedlaczek et al. (1973a, b)
	Pellicularia filamentosa I.F.O. 6675	Wiechert et al. (1974)
	Rhizopus arrhizus A.T.C.C. 11145	Holland and Auret (1975b)

Table 1 (*contd.*)

Reaction type	Micro-organism	Reference
	Rhizopus nigricans	Morozova *et al.* (1973), Belič *et al.* (1972), Procházka *et al.* (1974)
	Scopulariopsis sp.	Čapek *et al.* (1976)
	Tieghemella orchidis	All-Union Chemical and Pharmacological Research Institute (1973), Mogilnitsky (1973), Mogilnitsky *et al.* (1977), Garcia-Rodriguez *et al.* (1977a, b)
11β-OH	*Cunninghamella bainieri* A.T.C.C. 9244	Edwards *et al.* (1972)
	Cunninghamella blakesleeana	El-Monem *et al.* (1972)
	Curvularia lunata	El-Monem *et al.* (1972)
	Curvularia lunata N.R.R.L. 2174	Kerb *et al.* (1976)
	Curvularia lunata N.R.R.L. 2380	Petzoldt *et al.* (1972), Alig *et al.* (1976a, b), Kieslich *et al.* (1971, 1976, 1978), Kerb *et al.* (1975a), Boswell and Ripka (1972, 1973), Scribner (1973), Kulig and Smith (1974), Alig *et al.* (1977)
12α-OH	*Glomerella cingulata* A.T.C.C. 12097	Petzoldt *et al.* (1977a)
12β-OH	*Neurospora crassa*	Maugras *et al.* (1973)
12β,15α-(OH)$_2$	*Calonectria decora*	Vidic *et al.* (1974)
14α-OH	*Curvularia lunata*	Wagner *et al.* (1976), Lang (1975), Kreiser and Lang (1976)
	Helminthosporium buchloes C.B.S.	Dias and Pettit (1972)
	Mucor griseocyanus A.T.C.C. 1207	Valcavi and Innocenti (1973), Valcavi *et al.* (1973c, 1974)
	Stemphylium sp.	Čapek and Fassitová (1974)
14α,21-(OH)$_2$	*Epicoccum purpurascens*	El-Monem *et al.* (1972)
15α-OH	*Aspergillus flavus*	Balashova *et al.* (1975a)
	Beauveria bassiana	Sanda *et al.* (1977)
	Colletotrichum antirrhini A.T.C.C. 11598	Smithies and Eskins (1973)
	Giberella baccata	Smithies and Eskins (1973)
	Gilmanella sp.	Čapek *et al.* (1975b)
	Fusaria sp.	Petzoldt and Wiechert (1974)
	Penicillia sp.	Petzoldt and Wiechert (1974)

Table 1 *(contd.)*

Reaction type	Micro-organism	Reference
15β-OH	*Cunninghamella blakesleeana* *Curvularia lunata* *Paecilomyces* sp.	Micková *et al.* (1976) Kieslich *et al.* (1978) Čapek *et al.* (1976)
16α-OH	*Bacillus megaterium* N.R.R.L.-B-938	Buzby and Greenspan (1973)
	Streptomyces roseochromogenes N.R.R.L.-1233	Iida *et al.* (1975a), Permachem, (1976), Iida *et al.* (1975b), Yamashita and Kurosawa (1975), Iida and Iizuka (1976)
	Streptomyces roseochromogenes I.F.O. 1308	Anner and Wieland (1973), Cross (1972), Yamashita and Kurosawa (1975)
16β-OH	*Aspergillus niger* A.T.C.C. 9142	Yamashita *et al.* (1976a)
	Aspergillus niger N.R.R.L. 599	Yamashita *et al.* (1976b)
	Mucor sp.	K. Kieslich (unpublished observation)
17α-OH	*Actinomyces* sp. *Curvulaia lunata*	El-Kady *et al.* (1972) Wagner *et al.* (1976), Lang (1975), Kreiser and Lang (1976)
	Trichothecium roseum	El-Rafai *et al.* (1974a, b)
18-OH	*Streptomyces roseochromogenes* A.T.C.C. 13400	Braham *et al.* (1975)
21-OH	*Aspergillus niger* A.T.C.C. 9142	Holland and Auret (1975a)
Various mono- and dihydroxylations depending on the substrate structure	*Absidia orchidis* *Acromyx fungus* *Aspergillus ochraceus*	Bell *et al.* (1975a) Jones *et al.* (1975a, b) Blunt *et al.* (1971), Bell *et al.* (1972a, b, 1974), Clegg *et* *al.* (1973), Evans *et al.* (1975), Jones *et al.* (1975a, b)
	Calonectria decora	Jones *et al.* (1976), Bell *et al.* (1972b, c), Ashton *et al.* (1974), Evans *et al.* (1975), Chambers *et al.* (1975a, b), Jones *et al.* (1975a)
	Daedalea rufescens	Jones *et al.* (1976), Bell *et al.* (1975b)
	Diaporthe celastrina *Leptoporus fissilis* *Ophiobolus herpotrichus* *Penicillium urticae* *Rhizopus arrhizus*	Bell *et al.* (1975c) Denny *et al.* (1976) Chambers *et al.* (1975a, b) Blunt *et al.* (1971) Bell *et al.* (1973)

Table 1 (*contd.*)

Reaction type	Micro-organism	Reference
	Rhizopus circinans	Bell *et al.* (1973)
	Rhizopus nigricans	Jones *et al.* (1975a, 1976),
		Browne *et al.* (1973),
		Chambers *et al.* (1973),
		Ashton *et al.* (1974)
	Syncephalastrum racemosum	Bell *et al.* (1975a)
	Wojnowicia graminis	Chambers *et al.* (1975a, b)
EPOXIDATION		
5α,6α-epoxide	*Rhizopus nigricans*	Joská *et al.* (1973)
	Cunninghamella elegans	Crabb *et al.* (1975)
	C.B.S. 16753	
OXIDATION −OH → =O		
3α-OH → 3C=O	*Candida albicans*	Ghrai *et al.* (1975)
	Pseudomonas testosteroni	Skalhegg (1973, 1974)
	A.T.C.C. 11936	
3β-OH → 3C=O	*Arthrobacter simplex*	Görlich (1973a, b)
	A.T.C.C. 6946	
	Brevibacterium sterolicum	Terada *et al.* (1973)
	Nocardia asteroides	Görlich (1973a, b)
	A.T.C.C. 3308	
	Nocardia corallina	Görlich (1973c)
	A.T.C.C. 4237	
	Nocardia restrictus	Görlich (1973c, d)
	A.T.C.C. 14887	
	Nocardia restrictus C.B.S.	Belič and Sočič (1972), Belič
	15745	*et al.* (1973, 1977a, b)
7α-OH → 7-C=O	*Bacterioides fragilis*	Edenharder *et al.* (1976)
	Bacterioides thetalotaomicron	Edenharder *et al.* (1976)
17β-OH → 17-C=O	*Brevibacterium sterolicum*	Markert and Träger (1975)
	A.T.C.C. 21387	
	Candida albicans	Ghrai *et al.* (1975),
		Cremonesi *et al.* (1975)
	Cylindrocarpon radicicola	Al Rawi *et al.* (1974)
	A.T.C.C. 11011	
	Pseudomonas sp. A.T.C.C.	Kundsin *et al.* (1972)
	27330	
	Pseudomonas testosteroni	Antonini (1974)
	Streptomyces hydrogenans	Markert and Träger (1975),
		Markert *et al.* (1975)
20α-OH → 20-C=O	*Candida albicans*	Ghrai *et al.* (1975)
20β-OH → 20-C=O	*Candida albicans*	Ghrai *et al.* (1975)
OXIDATION 3β-OHΔ⁵ → 3-C=OΔ⁴		
	Arthrobacter simplex I.F.O.	Nagase Sangyo K.K. (1976)
	12069	
	Brevibacterium sterolicum	Uwajima *et al.* (1973), Terada
	A.T.C.C. 21397	*et al.* (1974)

Table 1 (*contd.*)

Reaction type	Micro-organism	Reference
	Corynebacterium cholesterolicum 3134	Kikkoman Shoyu K.K. (1977)
	Corynebacterium hydrocarboclastus	Iizuka (1974)
	Flavobacterium dehydrogenans A.T.C.C. 13930	Flines and Waard (1967), Kerb *et al.* (1975b), Wiechert *et al.* (1974)
	Flavobacterium dehydrogenans	Palmowski and Träger (1974), Kohler and Träger (1975)
	Flavobacterium luceoloratum	Udvardy *et al.* (1976)
	Flavobacterium peregrinum	Schwartz and Protiva (1974)
	Nocardia cholesterolicum N.R.R.L. 5762	Masureka and Goodhue (1976), Goodhue and Hugh (1975), Eastman Kodak (1977a)
	Nocardia sp. N.R.R.L. 5786	Mazureka and Goodhue (1976), Goodhue and Hugh (1975), Eastman Kodak (1977b)
	Nocardia erythropolis A.T.C.C. 17896	Bergmeyer *et al.* (1974a)
	Nocardia erythropolis A.T.C.C. 4277	Bergmeyer *et al.* (1974b)
	Nocardia formica A.T.C.C. 14811	Boehringer Mannheim (1975)
	Nocardia restrictus A.T.C.C. 14887	Soliman (1975)
	Nocardia rhodochrous N.C.I.B. 10554	Buckland *et al.* (1975)
	Nocardia rhodochrous N.C.I.B. 10555	Richmond (1973)
	Nocardia rhodochrous	Buckland *et al.* (1975, 1976)
	Proactinomyces N.C.T.B. 9158	Boehringer Mannheim (1975), Gruber *et al.* (1976)
	Proactinomyces A.T.C.C. 17895	Bergmeyer *et al.* (1974b), Whitehead and Flegg (1974)
	Schizophyllum commune I.F.O. 4928	Sugiura *et al.* (1976)
	Streptomyces griseocarnus	Szentirmai (1977), Kerényi *et al.* (1975)
	Streptomyces hydrogenans	Palmowski and Träger (1974), Kohler and Träger (1975)
	Streptomyces violascens	Fukuda *et al.* (1973)

Table 1 (contd.)

Reaction type	Micro-organism	Reference
DEHYDROGENATION		
Δ^1	Alternaria sp.	Čapek and Fassitová (1974)
Δ^1	Arthrobacter simplex A.T.C.C. 6946	Irmscher et al. (1973), Oxlea and Rosindale (1972), Ciba-Geigy (1970), Hoffmann-La Roche AG (1976), Wiechert et al. (1974), Alig et al. (1976a, 1977), Görlich (1973a, b)
Δ^1	Arthrobacter simplex N.R.R.L.-B-8055	Gallegra (1976)
Δ^1	Arthrobacter simplex	Scribner (1973), Marx et al. (1973), Sterkin et al. (1973), Udvardy (1974), Ryu et al. (1973), Penasse and Nomine (1974)
Δ^1	Bacillus lentus A.T.C.C. 13805	Kieslich et al. (1971, 1976, 1978), Kerb et al. (1974, 1975a, b), Wiechert et al. (1974)
Δ^1	Glomerella cingulata A.T.C.C. 10534	Kerb et al. (1974, 1975b), Costa Pla (1973)
Δ^1	Mycobacterium globiforme	Golovtzeva and Korovkina (1977), Borman et al. (1975), Union of Soviet Socialist Republic Academy of Sciences (1972, 1973a, b), Klimontovich et al. (1975), Koshcheenko et al. (1975), Skryabin et al. (1974a, b, 1976), Sterkin et al. (1973), El-Rafai et al. (1974a)
Δ^1	Nocardia asteroides A.T.C.C. 3308	Görlich (1973a, b)
Δ^1	Nocardia corallina A.T.C.C. 999	Riva and Toscano (1976)
Δ^1	Nocardia restrictus A.T.C.C. 14887	Nicholas (1975), Ikegawa and Nambara (1973), Görlich (1973c, d)
Δ^1	Septomyxa affinis	Kominek (1973)
Δ^4	Nocardia restrictus A.T.C.C. 14887	Nambara et al. (1973)
$\Delta^{8(9)}$	Clostridium paraputrificum	Babcock and Campbell (1972)

Table 1 (contd.)

Reaction type	Micro-organism	Reference

Δ^1-DEHYDROGENATION WITH AROMATIZATION OF RING A

Substrate

19-nor	Arthrobacter simplex A.T.C.C. 6946	Babcock and Campbell (1972)
19-nor	Septomyxa affinis A.T.C.C. 6737	Babcock and Campbell (1972)
19-OH	Arthrobacter ureafaciens	Shirasaka et al. (1972)
19-OH	Corynebacterium equi	Shirasaka et al. (1972)
19-OH	Proactinomyces globerula	Protiva and Schwartz (1974)
10-CHO	Nocardia asteroides A.T.C.C. 3308	Görlich (1973a)
10-CHO	Nocardia corallina A.T.C.C. 4237	Görlich (1973b)
10-CHO	Nocardia restrictus A.T.C.C. 14887	Görlich (1973c)
10-CH$_3$	Clostridia sp.	Goddard et al. (1975), Goddard and Hill (1972)
	Escherichia coli	Goddard and Hill (1972)

OXIDATION AND DEHYDROGENATION OF 3β-OH,Δ5 → 3C=O$\Delta^{1,4}$

	Fusarium solani	Kondo and Mitsugi (1973b)
	Mycobacterium globiforme	Irmscher et al. (1973)

OXIDATION AND DEHYDROGENATION OF 3β-OH,5αH → 3C=O^{1,4}

	Arthrobacter simplex	Belič et al. (1975)
	Nocardia restrictus C.B.S. 15745	Belič and Sočič (1972), Belič et al. (1973, 1977a)

BAYER-VILLIGER-OXIDATION TO 17-C=O OR TESTOLACTONE

	Cylindrocarpon radicicola	Carlström (1973), El-Tayeb et al. (1977)
	Fusarium javanicum var. ensiforme	David et al. (1973)
	Fusarium lini I.F.O. 7156	Abdel-Fattah and Badawi (1975a, b), Yamashita and Kurosawa (1975), Tom et al. (1975)
	Fusarium solani	Zedan and El-Tayeb (1977)
	Humicola sp.	Čapek et al. (1975a, b)
	Paecilomyces sp.	Čapek et al. (1976)
	Penicillium lilacinum N.R.R.L. 895	Carlström (1973, 1974a, b)

OXIDATIVE SIDE-CHAIN DEGRADATION

Substrate

Cholic acids → 17-C=O$\Delta^{1,4}$	Pseudomonas sp. N.C.I.B. 10590	Barnes et al. (1974, 1975, 1976)
Sterols → 17-C=O$\Delta^{1,4}$	Arthrobacter simplex I.A.M. 1660	Kanebo (1974), Mitsubishi Chemical Industries (1977a, b)

Table 1 (*contd.*)

Reaction type	Micro-organism	Reference
	Arthrobacter simplex I.F.O. 12069	Kanebo (1975a, b)
	Brevibacterium lipolyticum I.A.M. 1398	Nishikawa *et al.* (1975)
	Mycobacterium phlei I.F.O. 3158	Törmökény *et al.* (1975), Nishikawa *et al.* (1975)
	Mycobacterium phlei N.R.R.L. 3683	Krachy *et al.* (1972)
	Mycobacterium phlei N.R.R.L. 8153	Wovcha and Biggs (1977)
	Mycobacterium phlei N.R.R.L. 8154	Wovcha and Biggs (1977)
	Nocardia corallina I.F.O. 3338	Kanebo (1974a, b, c, d)
Sterols → 17-C=OΔ^4	*Mycobacterium* N.R.R.L. 3805	Marsheck and Krachy (1973), Weber *et al.* (1977a, b, c)
Sterol-3β-O-derivatives → 17-C=O, 3βORΔ^5	*Mycobacterium* sp. N.R.R.L. 3805	Eder *et al.* (1977), Weber *et al.* (1978)
19-Hydroxysterol → estrone	*Corynebacterium* sp. N.R.R.L.-B-3931	Mallett (1973)
Sterols → 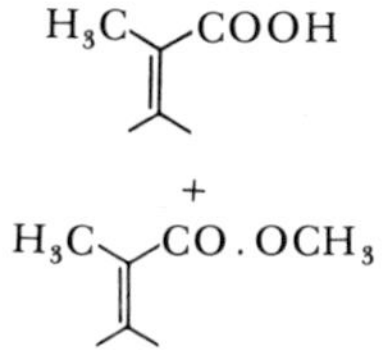	*Mycobacterium* sp.	Martin (1977)

$$H_3C \diagup COOH$$

$$+$$

$$H_3C \diagup CO.OCH_3$$

Sterols → 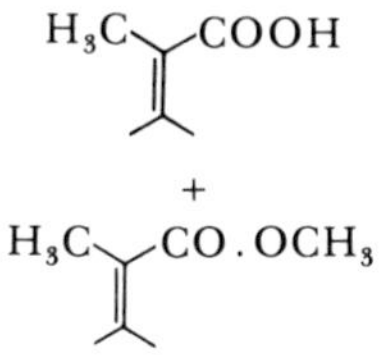	*Mycobacterium* sp. N.R.R.L.-B-8054	Jiu and Marsheck (1976, 1977)

$$H_3C \diagup COOH$$

$$+$$

$$H_3C \diagup CO.OCH_3$$

Sterols → 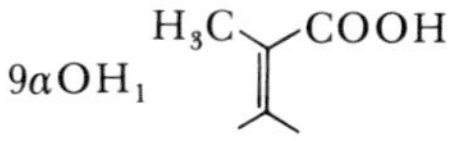	*Mycobacterium* sp. N.R.R.L.-B-8119	Antosz *et al.* (1977)

$$9\alpha OH_1 \quad H_3C \diagup COOH$$

Sterols → 9αOH,17-C=O,Δ^4	*Mycobacterium* sp. N.R.R.L.-B-8128	Knight and Wovcha (1977a, b)
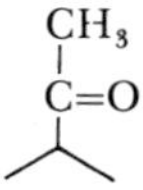 20-OH Sterols →	*Curvularia lunata* A.T.C.C. 12017	Canonica *et al.* (1974)
$\begin{array}{c} CH_3 \\ | \\ C=O \end{array}$ + 17-C=O	*Rhizopus arrhizus* A.T.C.C. 11145	Canonica *et al.* (1974)
	Rhizopus nigricans A.T.C.C. 6261	Canonica *et al.* (1974)

Table 1 (*contd.*)

Reaction type	Micro-organism	Reference
20-OH-Sterols → 17-C=O	*Fusarium lini* A.T.C.C. 9593	Tom *et al.* (1975)

OXIDATIVE CLEAVAGE OF RING B YIELDING SECO ACIDS

	Arthrobacter simplex A.T.C.C. 14887	Probst (1974)
	Arthrobacter simplex	Shionogi (1972), Hayakawa *et al.* (1977)
	Corynebacterium simplex A.T.C.C. 6997	Alig *et al.* (1976a), Hayakawa *et al.* (1972, 1976a), Hashimoto and Hayakawa (1977)
	Nocardia opaca A.T.C.C. 11996	Hörhold *et al.* (1976)
	Norcardia sp.	Tan *et al.* (1972), Probst *et al.* (1973)
	Streptomyces rubescens	Hayakawa *et al.* (1976b)
	Unidentified bacteria	Howe *et al.* (1973)

OXIDATIVE DEGRADATION OF RINGS E AND F

	Arthrobacter simplex	Belič *et al.* (1975)
	Fusarium solani	Kondo and Mitsugi (1973b)
	Stachylidium bicolor	Shionogi Co. Ltd. (1975), Kondo and Mitsugi (1973b)
	Verticillium strains	Shionogi Co. Ltd. (1975), Kondo and Mitsugi (1973b)

CLEAVAGE OF 3-ETHER AND 3-GLYCOSIDE BONDS

Substrate		
estradiol-3-OCH$_3$	*Bacterium mycoides*	Schubert *et al.* (1975a)
	Corynebacterium sp.	Schubert *et al.* (1975b)
	Nocardia erythropolis	Schubert *et al.* (1975c)
3β-*O*-glycosides	Various soil bacteria and fungi	Yosika *et al.* (1972a, b, 1973), Kitagawa *et al.* (1977), Yosika *et al.* (1974), Ford *et al.* (1977), Nogai *et al.* (1970)

FORMATION OF GLUCOSIDES

3-OH →3-*O*-β-glucoside	*Rhizopus* sp.	Petzoldt *et al.* (1974)
16α-OH → 16α-*O*-β- glucoside	*Aspergillus niger* N.R.R.L. 337	Pan and Lerner (1972)

REDUCTION C=O → -OH

3C=O → 3α-OH	*Saccharomyces cerevisiae*	Glaxo Laboratories Ltd. (1973)
3C=OΔ^4 → 3α-OH,5βH	*Clostridium paraputrificum*	Bokkenheuser *et al.* (1976), Schubert *et al.* (1972)
3C=OΔ^4 → 3β-OH,5-H	*Tetrahymena piriformis*	Lamontagne *et al.* (1977)
3C=OΔ^4 → 3β-OH,Δ^4	*Tetrahymena piriformis*	Lamontagne *et al.* (1976)

Table 1 *(contd.)*

Reaction type	Micro-organism	Reference
3C=OΔ⁴ → 3α- and 3β-OH,5α-H	*Rhodotorula glutinis*	Schubert *et al.* (1972)
8,14 Seco-14-C=O → 14α-OH	*Saccharomyces drosophilarum*	Törmörkény *et al.* (1976)
	Saccharomyces elongosporus	Gulaya *et al.* (1976)
	Penicillium citrinum	Törmörkény *et al.* (1976)
8,14-Seco-17-C=O → 17α-OH	*Kloeckera magna* a.o.	Takeda Chemical Industries (1972)
8,14-Seco-17-C=O → 17α-OH	*Pichia farinosa*	American Home Products (1972)
	Rhizopus arrhizus	Eignerová and Procházka (1974)
	Saccharomyces cerevisiae	Eignerová and Procházka (1974), Gulaya *et al.* (1975)
17-C=O → 17α-OH	*Neurospora crassa*	Maugras *et al.* (1973)
19 = O → 19-OH	*Arthrobacter simplex*	Görlich (1973d)
20 C=O → 20α-OH	*Fusarium solani*	Plourde *et al.* (1972)

REDUCTION OF PEROXIDES

10-O-OH → 10-OH	*Curvularia lunata*	Kulig and Smith (1974)
17α-O-OH → 17α-OH	*Aspergillus ochraeceus*	Tan (1972)

HYDROGENATION OF DOUBLE BONDS

Aromatic A-ring → 3C=OΔ⁴	*Aspergillus flavus*	Schubert *et al.* (1973, 1974)
3C=OΔ⁴ → 3C=O,5α-H	*Mycobacterium smegmatis*	Hörhold *et al.* (1975, 1977), Lestrovaja *et al.* (1977)
	Nocardia corallina A.T.C.C. 13259	Lefebvre *et al.* (1974a)
	Nocardia opaca A.T.C.C. 4176	Lefebvre *et al.* (1974b)
	Penicillium decumbens	Nambara *et al.* (1977)
3C=O,Δ⁴ → 3α- and 3β-OH,5αH	*Rhodotorula glutinis*	Schubert *et al.* (1972)
3C=O,Δ⁴ → 3-C=O,5β-H 3α-OH,5β-H	*Clostridium paraputrificum*	Bokkenheuser *et al.* (1976), Schubert *et al.* (1972)
3C=O,Δ⁴ → 3β-OH,5β-H	*Eubacterium* sp. A.T.C.C. 21408	Eyssen *et al.* (1973, 1974)
Δ⁴,¹⁶ → 5αH,16,17-H	*Nocardia corallina* A.T.C.C. 13259	Lefebvre *et al.* (1974a)

DEHYDROXYLATION

12αOH → 12α-H	*Clostridium perfringens*	Saltsman (1976)
4OH → 4-H (?)	*Cunninghamella elegans* N.R.R.L. 1393	Rosazza *et al.* (1976)

HYDROLYSIS OF ESTERS

3β-OHC → 3β-OH	*Candida rugosa* A.T.C.C. 14830	Boehringer Mannheim (1974), Eastman Kodak (1977)

Table 1 (contd.)

Reaction type	Micro-organism	Reference
	Flavobacterium esteroaromaticum A.T.C.C. 8091	Fishman (1960), Kieslich (1978)
	Flavobacterium dehydrogenans A.T.C.C. 13930	Kerb *et al.* (1975a, b), Wiechert *et al.* (1974)
	Nocardia cholesterolicum N.R.R.L. 5767	Masureka and Goodhue (1976)
	Pseudomonas fluorescens A.T.C.C. 31156	Sugiura *et al.* (1976), Terada and Uwajima (1976)
11β-OHC → 11β-OH	*Flavobacterium dehydrogenans* A.T.C.C. 13930	Rausser and Shapiro (1972), Teutsch *et al.* (1970)
16β-OHC → 16β-OH	*Flavobacterium esteroaromaticum* H.T.C.C. 8091	Fishman (1960), Kieslich (1978)
17α-OHC → 17α-OH	*Flavobacterium dehydrogenans* A.T.C.C. 13930	E. Merck AG (1972)
21-OHC → 21-OH	*Flavobacterium dehydrogenans*	Kieslich *et al.* (1978), de Flines and Waard (1967), Kerb *et al.* (1975), Wiechert *et al.* (1974), Rausser and Shapiro (1972), Teutsch *et al.* (1970), E.-Merck AG (1972)
	Bacillus megaterium	Ryzhkova *et al.* (1974)
	Corynebacterium sp. N.R.R.L.-3-3791	Grove *et al.* (1971)
REACTIONS WITH ISOMERIZATIONS		
3β-OH,5α,6α-epoxid → 6α-OH,3-C=OΔ⁴	*Flavobacterium dehydrogenans* A.T.C.C. 13930	Kieslich *et al.* (1978)
3β-OH,Δ^{1(10)5} → 3-OH,Δ^{1,3,5(10)}	*Proactinomyces globerula*	Protiva and Schwarz (1974)
3β-OH → 3α-OH	*Nocardia restrictus* C.B.S. 15745	Belič *et al.* (1977a)
	Mycobacterium phlei	Törmörkény *et al.* (1975)
12α-OH → 12β-OH	*Pseudomonas* sp. N.C.I.B. 10590	Barnes *et al.* (1974, 1976)

formation when introducing a double bond in the 1,2-position of hydrocortisone to make prednisolone (Nobile, 1958). The bacterium used, *Corynebacterium simplex* (*Arthrobacter* sp.), was employed later in the 1-dehydrogenation of a number of more advanced corticoid structures.

These historic discoveries are milestones in the development of this method, and results up to the mid-1960s are contained in four great monographs (Akhrem and Titow, 1965; Čapek et al., 1966; Charney and Herzog, 1967; Iizuka and Naito, 1967). The range of use of the various types of reaction in microbiologial transformations is described in other books dealing with steroid and non-steroid substrates (Sih and Rosazza, 1976; Wallen et al., 1959; Fonken and Johnson, 1972; Skryabin and Golovteva, 1976). The subsequent reports of microbial reactions on steroids or special classes of steroid structures up to 1972/73 were summarized in more recent reviews (Akhrem and Titow, 1970; Marsheck, 1971; Smith, 1974; Vezina et al., 1971; Hayakawa, 1973; Beukers et al., 1972; Vezina and Rakhit, 1974).

This article attempts to present the position reached in the last few years, up to 1978, as far as literature and patents are concerned. It deals with new processes already in use technically, the potential for their use in laboratory syntheses, and the more important trends in application and elucidation of enzyme–substrate structure dependences or enzyme properties essential for practical use.

Since the aim of microbiological transformations is ultimately to effect specific steroid combinations from prescribed starting structures, this review is subdivided on the basis of substrate structure. However, in pursuing the desired microbiological transformation of a steroid with a given micro-organism, different results are often achieved because of the organism's substrate specificity. In that case, either one can optimize that type of reaction, through selection of a more suitable micro-organism, or, within the limits of the proposed synthesis, a chemical change in the substrate structure. For this reason, a table of the types of reaction and of the micro-organisms used is supplied in Table 1.

II. PREGNANES

A. Progesterone

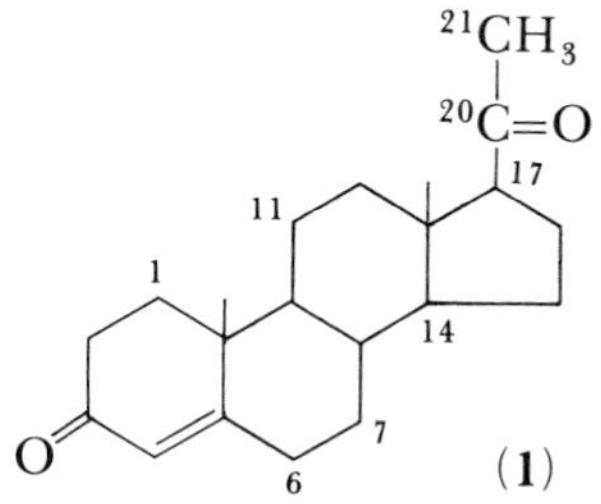

(1)

Progesterone (1), which is readily obtainable by chemical degradation of stigmasterol and from byproducts obtained from diosgenin, namely gallic acid and hecogenin, was the best starting material for synthesizing corticoids. The essential 11α-hydroxylation by *Rhizopus nigricans* A.T.C.C. 6227b or *Aspergillus ochraceus* was described as early as 1960; the yields were high and substrate concentrations were from 20 g to 50 g/litre (Weaver *et al.*, 1960). Although a great many other suitable fungi were already known as a result of widespread investigations (Dulaney *et al.*, 1955), new strains of *Aspergillus* with different degrees of direction specificity for these reactions were described as well (Samanta *et al.*, 1976; Zedan *et al.*, 1976; Abdel-Fattah and Badawi, 1974). Variations in medium composition (Abdel-Fattah and Badawi, 1974; Sallam *et al.*, 1973) and changes in pH value may partially influence the relationship between principal and subsidiary hydroxylations (El-Kady and Allam, 1973). It was possible to separate 11β- and 21-hydroxylating coenzymes from *A. niger* 12y (Abdel-Fattah and Badawi, 1975a, b). Degradation of the progesterone side chain by *Fusarium* spp., which was already known, must be considered an undesirable sequential reaction. Progesterone and 17α-hydroxyprogesterone yield up to 45% 11α-hydroxy-4-androstene-3,17-dione and 11α-hydroxy-testosterone with *Fusarium* strain IFO 7156 (Yamashita *et al.*, 1976a).

The 11β-hydroxylation, which is equally desirable for synthesis of corticoids, takes place without giving satisfactory yields, in contrast to the relatively high directional specificity of 11α-hydroxylation. *Curvularia lunata, Cunninghamella blakesleeana* and *Epicoccum purpurascens*, which were already known in this connection, have more recently been found to effect formation of 14α- and 21-, or 14α- and 11β, 14α-, or 11α-,21- and 6β,11α- mono- or dihydroxylated progesterone (El-Monem *et al.*, 1972).

For transformation of the progesterone side chain into a cortisol structure, microbiological hydroxylations in the 17α- and 21-positions, had already been investigated and patented. Unfortunately, the fungus *Tricothecium roseum*, already known for its ability to bring about 17α-hydroxylation, attacks the 6β- and 11α-positions as well, either in parallel or sequentially (El-Refai *et al.*, 1974a). Formation of byproducts can be influenced only to a slight extent by medium supplementation (Sallam *et al.*, 1974a). Some *Actinomyces* species show the same hydroxylation pattern with maximum yields of 18% 17α-hydroxyprogesterone (El-Kady *et al.*, 1972). Hydroxylation at position

21 by *A. niger* A.T.C.C. 9142, which was previously described as having a 50% yield, is surprisingly diverted when acting on the 20-keto-reduced substrates 20α- and 20β-hydroxy-4-pregnen-3-one, to attack exclusively the 11α- and 15β-positions (Holland and Auret, 1975a). As the neighbouring positions, activated by the 20-keto group, generally offer simple and good possibilities for chemical attack, it is unlikely, given the present state of the technique, that microbiological procedures will be found practical, even in the future, for introducing 17α- and 21-hydroxy groups.

In basic research into enzymic capabilities in fungi, progesterone transformations were used as a diagnostic feature of various fungus genera (Table 2). All of the species within a genus, for *Paecilomyces* spp., always showed the same transformation behaviour. This astonishingly unified picture could in practice be used with advantage to shorten screening processes for selecting suitable strains for certain reactions. The results described so far, however, can presumably only indicate certain trends in distribution of enzymic properties in various fungi and would have to be precisely defined by extending these systematic

Table 2

Progesterone conversions effected by fungal species

Species	Group introduced	Reference
Stemphylium (7 strains of 4 species)	14α-OH	Čapek and Fassitova (1974)
Cladosporium (18 strains of 6 species)	7β-OH	Čapek and Fassitova (1974)
Alternaria (8 strains of 4 species)	1-Dehydro	Čapek and Fassitova (1974)
Humicola (16 strains of 4 species)	Testolactone	Čapek *et al.* (1975a)
Gilmaniella (2 strains of 1 species)	15α-OH + 6β-OH	Čapek *et al.* (1975a)
Scopulariopsis (14 strains of 5 species)	11α-OH	Čapek *et al.* (1975b)
Paecilomyces (23 strains of 6 species)	Testolactone	Čapek *et al.* (1975b)
Paecilomyces (2 strains of 2 species)	6β-OH and 15β-OH	Čapek *et al.* (1976)

investigations, for practical experience has shown that relatively heterogeneous relationships can be found within any one genus.

The characteristic reactions of species in a genus, which have been brought to light in this way are only valid with identical substrates. *Rhizopus arrhizus* and *R. circinans* are known to hydroxylate progesterone in the 11α-position but, from 3β-hydroxy-5-pregnan-20-one, they form also the 7β-hydroxy derivative (Jones *et al.*, 1976).

A further instructive example of how the direction of hydroxylation depends on substrate structure can be seen in *Coniosporium rhizophilum*. Progesterone is hydroxylated by this fungus, as is desoxycortico-sterone simultaneously in the 1β- and 6β- positions. As in many other examples of hydroxylation of this substrate, dihydroxylation is not observed with compound S, and only the 1β- and 11α-mono-hydroxylated compounds are isolated (16% and 11%; Schwartz and Protiva, 1972).

B. 20-Oxo-5α-pregnanes

$$CH_3$$
$$C=O$$

H   (2)

Variations in substrate structure can affect the direction of this reaction to varying degrees, as can be seen from the results of comparative incubations of eight dioxygenated 5α-pregnanes (2) with *Calonectria decora, Daedalea refuscens* and *Rhizopus nigricans* (Jones *et al.*, 1976; Table 3). Only one 1β-11α-dihydroxylation of a 3β-hydroxy-5α-pregnan-20-one with *A. ochraceus* C.B.S. 13252 (Jones *et al.*, 1972) was patented because of good yields. A few substituted structures may be similarly transformed.

A 3β-acetoxy-5β-pregnan-20-one can be dihydroxylated with *C. decora* in the same way as the 5α-structure in the 12β- and 15α-positions. In this instance, the 3-acetate is still obtained, which is interesting for possible syntheses of digoxygenin (Vidic *et al.*, 1974).

Table 3

Yields in transformations by three fungal species of 20-oxo-5a-pregnanes.
Percentage yields are given in parentheses

	Organism		
Substrate	*Calonectria decora*	*Daedalea rufescens*	*Rhizopus nigricans*
3,20-(CO)$_2$	12β,15α-(OH)$_2$	3β,7α-(OH)$_2$ (4%)	11α-OH (22%)
		3β,7β-(OH)$_2$ (5%)	3β,7β,11α-(OH)$_3$ (7%)
3β-OH, 20-CO	12β,15α-(OH)$_2$ (24%)	7α-OH (9%)	7β,12β-(OH)$_2$ (22%)
		7β-OH (9%)	7β,11α-(OH)$_2$ (10%)
3β-OH Δ^{16} 20-CO	Not identified	14α,15β-(OH)$_2$ (21%)	7α-OH (13%)
			7α,15α-(OH)$_2$ (12%)
			7-CO,15β-OH (12%)
7,20-(CO)$_2$	12β-OH (18%) or	3β,7α-(OH)$_2$ (13%)	3α,11α-(OH)$_2$ (10%)
	1α,12β-(OH)$_2$	3β,7β-(OH)$_2$ (3%)	4α,11α-(OH)$_2$ (12%)
	(33%)		
			3β,11α-(OH)$_2$ (14%)
7α-OH,20-CO	12β-OH (17%)	3β-OH (27%)	3β-OH (24%)
11,20-(CO)$_2$	Not identified	3β-OH (35%)	Not identified
11α-OH 20-CO	Not identified	3β-OH (13%)	Not identified
11β-OH 20-CO	Not identified	Not identified	Not identified

C. Substituted Progesterones

An 11α-hydroxylation of progesterone might also be significant in production of anti-inflammatory 16α- and 17α-methyl-substituted corticoids. Methyl substitutions of progesterone in the 16α- and 16α,17α-positions, however, lower the yield of the 11α-hydroxylation product with *R. nigricans* when stimulated by dihydroxylations (Gabinskaya *et al.*, 1974). The compound 16α,17α-dimethylprogesterone (3) still gives a yield of 61% of the 11α-hydroxy compound (Morozova *et al.*, 1973). In the author's experience, 16α-methylprogesterone (4) also undergoes a normal 11α-hydroxylation with *A. ochraceus* (W. Reusche and K. Kieslich, unpublished observations). For synthesis of a *D*-homoanalogue of Alfaxalon®, which has anaesthetizing properties, it was possible to hydroxylate *D*-homoprogesterone (5) in the 11α-position with *A. ochraceus* A.T.C.C. 1008, whereas 11β-hydroxylation with *C. lunata* N.R.R.L. 2174 is less satisfactory because of undesirable side reactions (Kerb *et al.*, 1975a). A

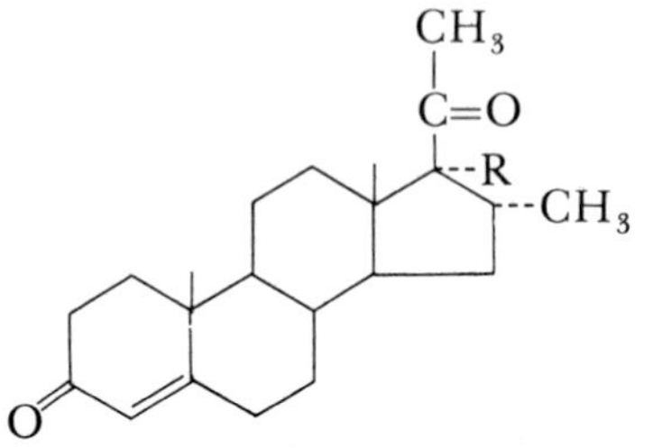

16β-methyl substituent (6) displayed a surprising influence on the known 14α-hydroxylation of unsubstituted progesterone with *Mucor griseocyanus* A.T.C.C. 1207a. The reaction was almost completely diverted to the otherwise secondary attack on the 7α-position (Valcavi *et al.*, 1973b). Transformation by *R. arrhizus* A.T.C.C. 11145 of 31 pregnane structures, with or without an oxygen function in the 11,17,20 and 21 positions, was investigated for 6β-hydroxylation. An 11-oxo or 11β-hydroxy function possesses a strong stimulatory effect, whereas an 11α-hydroxyl group should produce a slight inhibition (Schneider, 1974). On the other hand, 11α-hydroxylation by *R. nigricans* is not, as the principal reaction, affected by large-scale structural changes, as is shown by the substrate 16β-21-methyl-*D*-nor-5-pregnen-3β-ol-20,21-dion-21-ethyleneketal (7) (Belič *et al.*, 1972). Hydroxylation in the 15α position of a 6α-hydroxyprogesterone with *Gibberella baccata* or *Colletotrichum antirrhini* A.T.C.C. 11598 should also be seen as a normal course for the reaction (Smithies and Eskins, 1973). Hydroxylation in the 6- and 15-positions of products of this type of substrate have not yet found any practical importance.

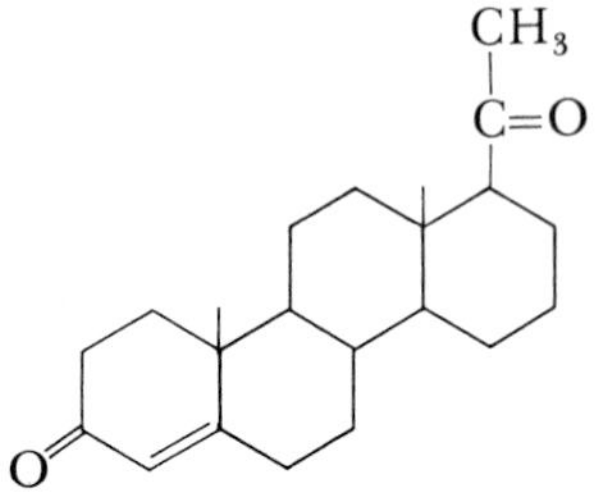

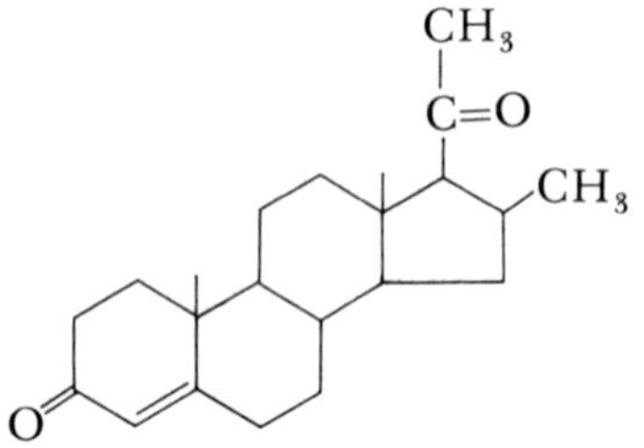

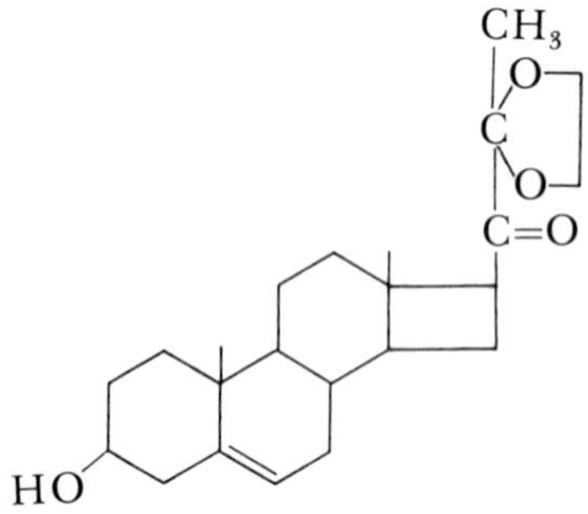

D. 17α-Hydroxyprogesterones

Since 21-desoxyprednisolones are also strongly anti-inflammatory, 11-hydroxylation of suitable 17α-hydroxyprogesterone bonds is important in preparation processes. Introduction of a 11β-hydroxy group by *Curvularia lunata* N.R.R.L. 2380 was described recently for the following substrates: 17α-hydroxy-1α,2α-methylene-4,6-pregnadiene-3,20-dione (**8**) and its 17-acetate (**9**) (Petzoldt *et al.*, 1972), 6-chloro-17α-hydroxy-1α,2α-methylene-4,6-pregnadiene-3,20-dione (**10**) and its 17-acetate (**11**) (Belič *et al.*, 1972), 17aα-acetoxy-*D*-homo-4-pregnene-3,20-dione (**12**) (Alig *et al.*, 1976b) and 17aα-hydroxy- 21-chloro-*D*-homo-4-pregnene-3,20-dione (**13**) (Alig *et al.*, 1976a). For 16-difluoromethylene-17α-hydroxy-4-pregnen-3,20-dione (**14**), *Cunninghamella bainieri* A.T.C.C. 9244 was used (Edwards *et al.*, 1972).

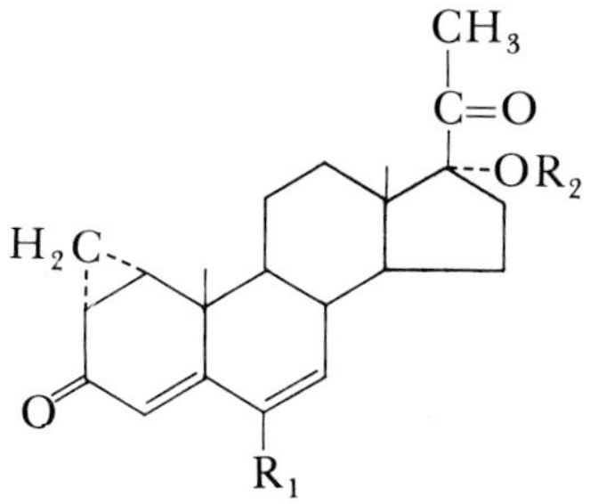

	R_1	R_2
(8)	H	H
(9)	H	Ac
(10)	Cl	H
(11)	Cl	Ac

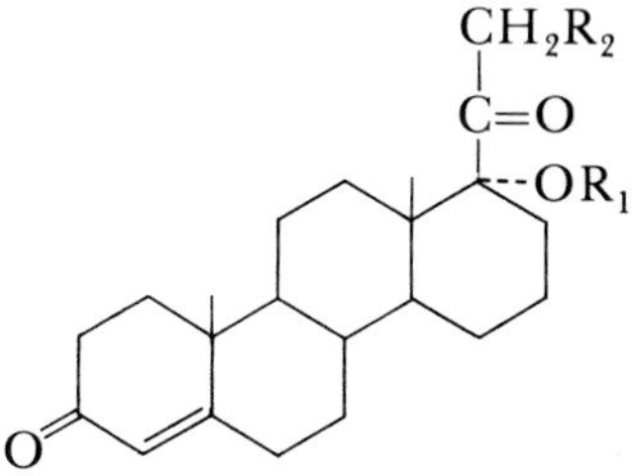

	R_1	R_2
(12)	Ac	H
(18)	H	Cl

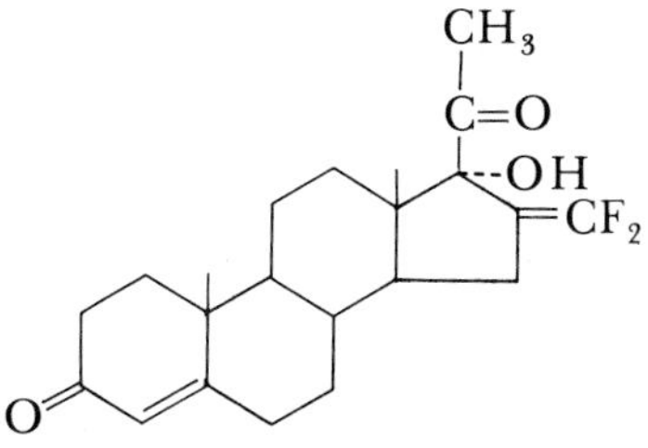

(14)

11α-Hydroxylations were also carried out with the efficient strain *Aspergillus ochraceus* using as substrates 14α,17α-dihydroxy-4-pregnene-3,20-dione (15) (Koninklijke Nederlandsche Gist-en Spiritusfabriek N. V., 1973), and 6,6-difluoro-16α,17α-dihydroxy-4-pregnene-3,20-dione (16) (Ripka, 1972). With inclusion of a 17α-hydroperoxy-progesterone (17), the expected 11α-hydroxy compound is obtained with loss of the peroxide structure (Tan, 1972). For 11α-hydroxylation of 6-chloro-17α-acetoxy-1α,2α-methylene-4,6-pregnadiene-3,20-dione (11), it is more advantageous to use *Gibberella cingulata* A.T.C.C. 10534 (Engelfried *et al.*, 1976).

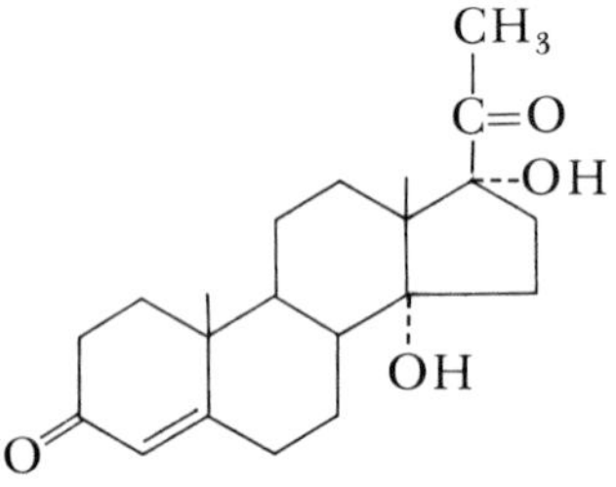

CH₃
C=O
---OH

ÓH

O

(15)

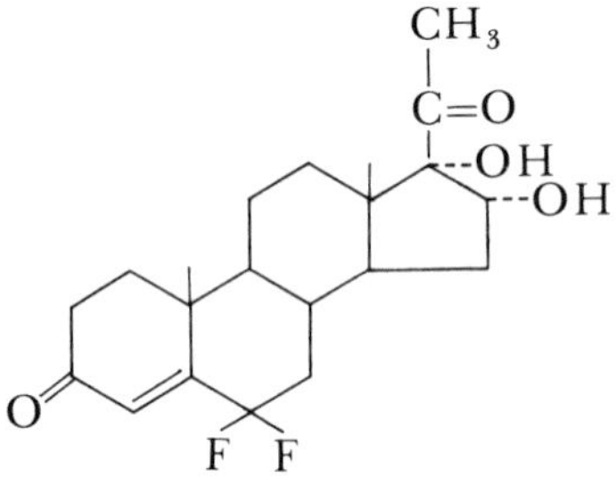

CH₃
C=O
---OH
---OH

O

F F

(16)

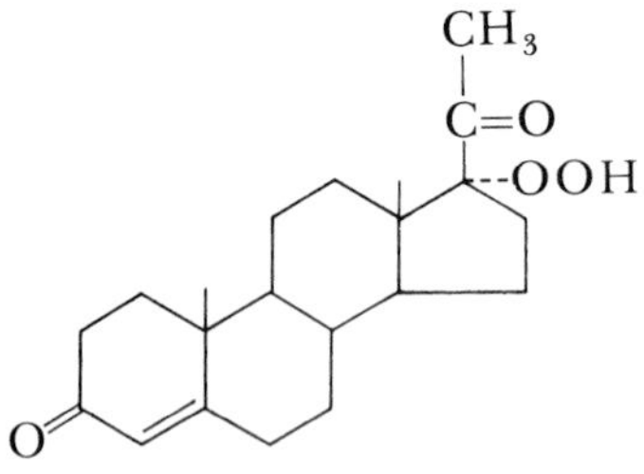

CH₃
C=O
---OOH

O

(17)

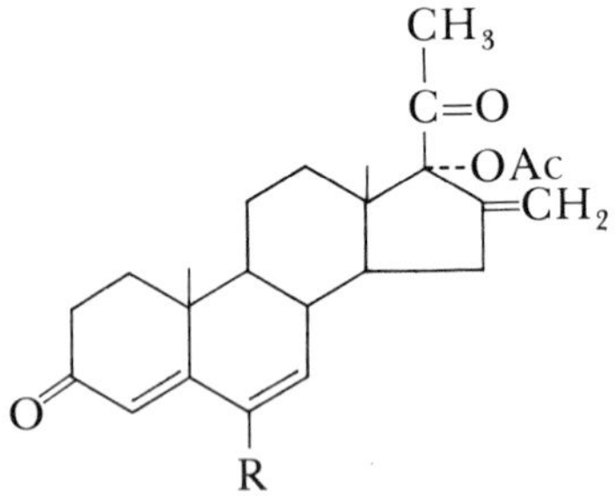

CH₃
C=O
---OAc
=CH₂

O

R

(18)   H
(19)   Cl

Cunninghamella blakesleeana, which is normally an 11β-hydroxylating agent, attacks 17α-acetoxy-16-methylene-4,6-pregnadiene-3,20-dione (18) and its 6-chloro-derivative (19) predominantly in the 15β-position (Micková *et al.*, 1976).

The presence of a 16α-hydroxy group was known to improve anti-inflammatory effectiveness and, in this connection, mention was made of using the well-known strain *Streptomyces roseochromogenes* to introduce this function on the following substrates: 6-chloro-21-fluoro-17α-hydroxy-4,6-pregnadiene-3,20-dione (**20**) (Anner and Wieland, 1973), 4-oxa-17α-hydroxy-5α-1-pregnene-3,20-dione (**21**) and its 5β-analogues (**22**) as well as the corresponding 19-nor structures (**23, 24**; Cross, 1972).

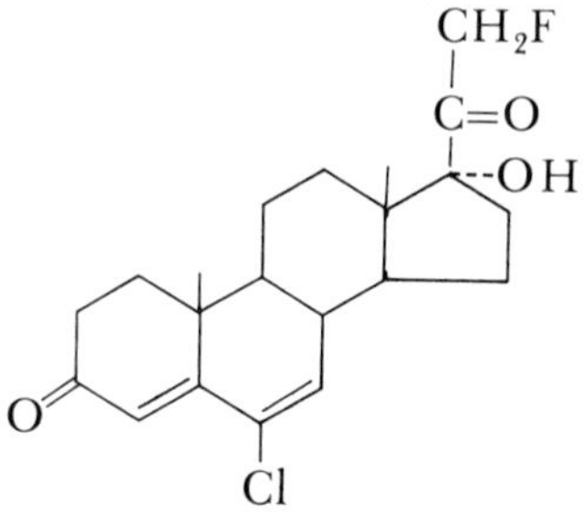

CH$_2$F
|
C=O
--OH

O

Cl

(**20**)

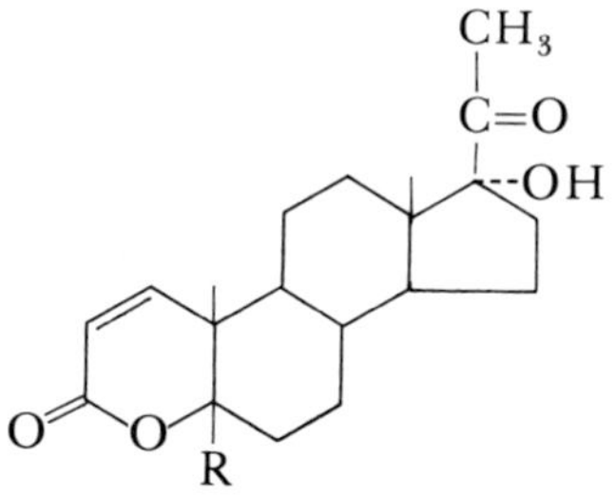

CH$_3$
|
C=O
--OH

O   O
    R

	R
(**21**)	αH
(**22**)	βH

CH$_3$
|
C=O
--OH

O   O
    R

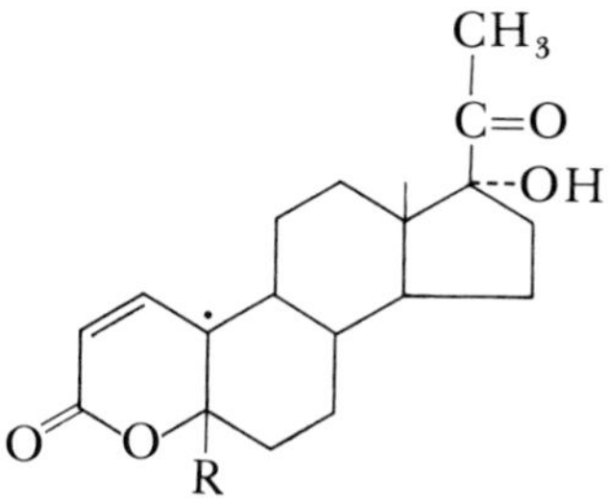

	R
(**23**)	αH
(**24**)	βH

The final transition to desoxyprednisolone structures is carried out by 1-dehydrogenation with the approved *Corynebacterium simplex* for which the following substrates may serve as examples: 9β,11β-epoxy-17α-hydroxy-16-methylene-4-pregnene-3,20-dione (**25**; Irmscher *et al.*, 1973); 17α-acyloxy-11β-hydroxy-4-pregnene-3,20-dione (**26**; Oxlea and Rosindale, 1972); 6α,9α-difluoro-16α-methyl-17α-acetoxy-4-pregnene-3,11,20-trione (**27**; Ciba-Geigy, 1970); 17aα-acetoxy-11β-hydroxy-D-homo-4-pregnene-3,20-dione (**28**; Alig *et al.*, 1976b; Hoffmann-La Roche A. G., 1976). 16α,17α-Isopropylidene-3β,16α,17α-trihydroxy-5-pregnen-20-one is dehydrogenated with

Mycobacterium globiforme, a reaction accompanied by transformation of the 3β-hydroxy-5-ene system to the 3-keto-Δ⁴-structure (Zelinski, 1975).

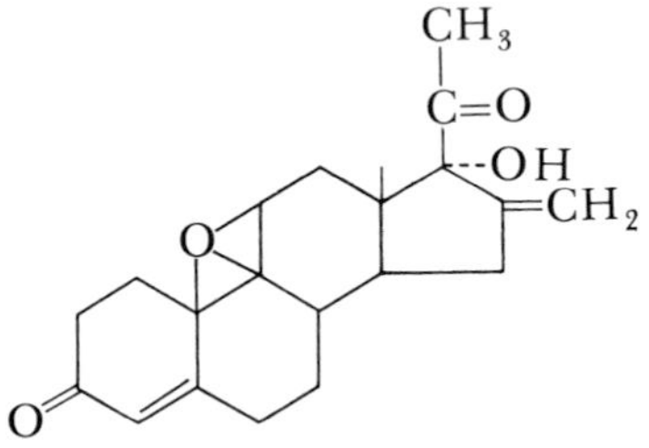

(25) 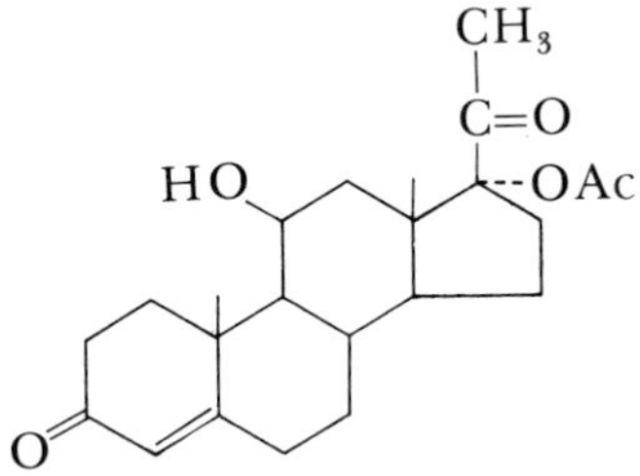   (26)

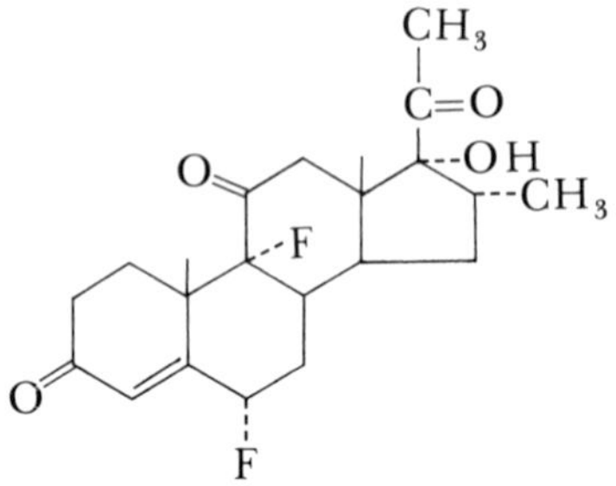

(27) 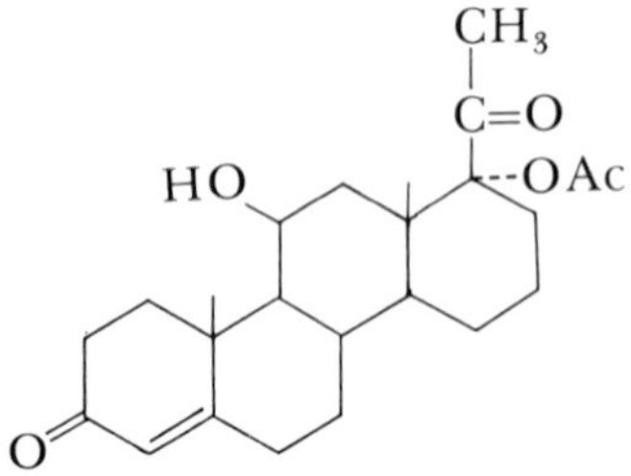   (28)

This Oppenauer oxidation can be achieved with cell pastes of *Nocardia rhodochrous* in a new process using different 3β-hydroxy-5-ene structures; this will be described in more detail in connection with cholesterol (Buckland *et al.*, 1975, 1976). In the same section, hydrogenation of the Δ⁴ double bond is reported for *Nocardia corallina* A.T.C.C. 13259. In the presence of a 4,16-pregnadiene-3,20-dione, this simultaneously hydrogenates the Δ¹⁶ double bond but, unfortunately, this has not yet been used in practice (Lefebvre *et al.*, 1974a, b).

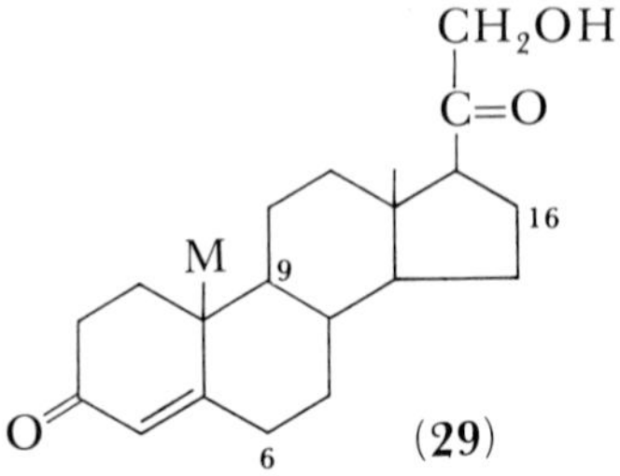

(29)

E. Desoxycorticosterone Structures

Since 1964 (Raspé *et al.*, 1964), 16α-methyl-corticosterone derivatives have been bonded, and shown to have a highly anti-inflammatory effect and, therefore, microbiological 11-hydroxylation and 1-dehydrogenation are of technical interest in these classes of structures. Recently, substrates were specifically synthesized for synthesis of 6α,16α-dimethyl-1-dehydrocorticosterone from the appropriate desoxycorticosterone structures (Table 4). The 21-acetates used were hydrolysed in the usual way by means of the esterase portion of the fungus before hydroxylation.

To produce a 6α,21-dihydroxy-6β,16α-dimethyl-4-pregnene-3,20-dione (**31**) (63% yield), which can easily be dehydrogenated chemically to the 6-methyl-3-oxo-4,6-diene structure, a reaction was used which has already been described and is typical of oxidation of a 3β-hydroxy-5α,6α-epoxide with *Flavobacterium dehydrogenans* A.T.C.C. 13936 (Kieslich *et al.*, 1978). 11β-Hydroxylation of this epoxide with

Table 4

Transformations of derivatives of 6,16α-dimethylprogesterone

Derivative	Organism	Group introduced	Reference
21-Acetoxy-6α,16α-dimethyl-4-pregnene-3,20-dione	*Curvularia lunata*	11β-OH	Kieslich *et al.* (1971)
	Aspergillus ochraceus	11α-OH	Kieslich *et al.* (1971)
21-Acetoxy-6β,16α-dimethyl-4-pregnene-3,20-dione	*Curvularia lunata*	11β-OH	Kieslich *et al.* (1971)
21-Acetoxy-6,16α-dimethyl-4,6-pregnadiene-3,20-dione	*Curvularia lunata* N.R.R.L. 2380	11β-OH	Kieslich *et al.* (1978)
11β,21-Dihydroxy-6α,16α-dimethyl-4-pregnene-3,20-dione	*Bacillus lentus*	Δ¹	Kieslich *et al.* (1971)
21-Acetoxy-9α-fluoro-11β-hydroxy-6α,16α-dimethyl-4-pregnene-3,20-dione	*Bacillus lentus*	Δ¹	Kieslich *et al.* (1972)
11β,21-Dihydroxy-6,16α-dimethyl-4,6-pregnadiene-3,20-dione	*Bacillus lentus*	Δ¹	Kieslich *et al.* (1978)
11α,21-Dihydroxy-6,16α-dimethyl-4,6-pregnadiene-3,20-dione	*Bacillus lentus*	Δ¹	Kieslich *et al.* (1978)

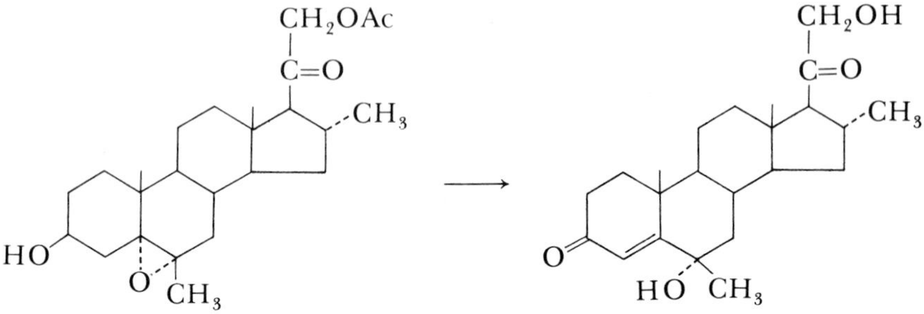

CH$_2$OAc
|
C=O
.-CH$_3$

HO

O̊ CH$_3$

(30)

⟶

CH$_2$OH
|
C=O
.-CH$_3$

O

HO̊ CH$_3$

(31)

CH$_2$OH
|
C=O
.-CH$_3$

O

F

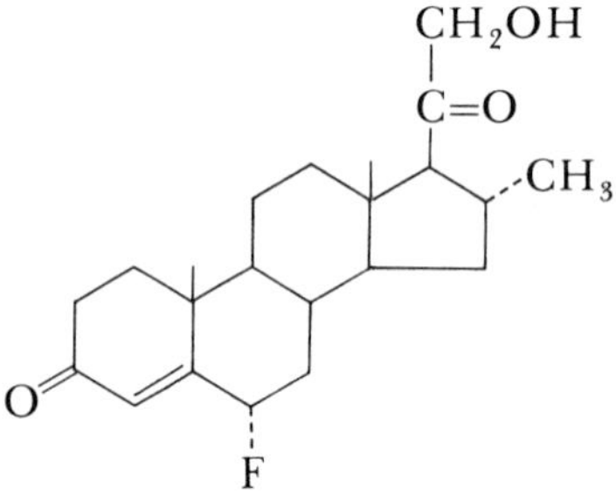

(32)

Curvularia lunata N.R.R.L. 2380 is completely suppressed; in a bad yield, only a 15β-hydroxy product can be isolated (Kieslich *et al.*, 1978). Such blocking of the 11β-hydroxylation may be ascribed to stereo-chemical effects of the methyl substituents; these are particularly evident with the substrate 6α-fluoro-21-hydroxy-16α,18-dimethyl-4-pregnene-3,20-dione (32) (Kerb *et al.*, 1975b). The 11-hydroxy function was successfully introduced here only by 11α-hydroxylation with *Aspergillus ochraceus* A.T.C.C. 1008 or *Glomerella cingulata* A.T.C.C. 10534 (Kerb *et al.*, 1974, 1975b). Conversely, a deliberate suppression of unwanted side hydroxylations is possible on the basis of steric substrate structure, and this was already known to be used in some 11β-hydroxylations on an industrial scale (Flines and Waard, 1967; Kieslich and Raspé, 1967; Lee *et al.*, 1969).

Results of hydroxylations of methyl-substituted desoxycortico-sterone structures with *Mucor griseocyanus* A.T.C.C. 1207, on the other hand, have been of no practical interest so far. It is known that desoxycorticosterone is hydroxylated by this fungus at the 14α-position (Valcavi and Innocenti, 1973b) and, with a longer fermentation period, at the 7α-position too. A 16α-methyl group

Table 5

Microbial transformation of derivatives of D-homocorticosterone

Derivative	Organism	Group introduced	Reference
21-Acetoxy-17α-methyl-D-homo-4-pregnene-3,20-dione	*Curvularia lunata* N.R.R.L. 2380	11β-OH	Kerb *et al.* (1975a)
21-Acetoxy-6α-fluoro-D-homo-4-pregnene-3,20-dione	*Curvularia lunata* N.R.R.L. 2380	11β-OH	Kerb *et al.* (1975a)
21-Acetoxy-6α-fluoro-17α-methyl-4-pregnene-3,20-dione	*Curvularia lunata* N.R.R.L. 2380	11β-OH	Kerb *et al.* (1975a)
11β,21-Dihydroxy-17α-methyl-D-homo-4-pregnene-3,20-dione	*Bacillus lentus* A.T.C.C. 13805	Δ¹	Kerb *et al.* (1975a)
11β,21-Dihydroxy-6α-fluoro-D-homo-4-pregnene-3,20-dione	*Bacillus lentus* A.T.C.C. 13805	Δ¹	Kerb *et al.* (1975a)
11β,21-Dihydroxy-6α-fluoro-D-homo-4-pregnene-3,20-dione	*Bacillus lentus* A.T.C.C. 13805	Δ¹	Kerb *et al.* (1975a)
6,6-Difluoro-17α,21-dihydroxy-4-pregnene-3,20-dione	*Curvularia lunata*	11β-OH	Boswell and Ripka (1972, 1973)
6,6-Difluoro-16α-methyl-17α-hydroxy-4-pregnene-3,20-dione	*Curvularia lunata* A.T.C.C. 12017 N.R.R.L. 2308	11β-OH	Scribner (1973)
17aα,21-Dihydroxy-D-homo-4-pregnene-3,20-dione	*Curvularia lunata* N.R.R.L. 2308	11β-OH (7α-OH)	Kieslich *et al.* (1976)
6α-Methyl-17aα,21-dihydroxy-D-homo-4-pregnene-3,20-dione	*Curvularia lunata* N.R.R.L. 2308	11β-OH (6β-OH)	Kieslich *et al.* (1976)
6α-Fluoro-17aα,21-dihydroxy-D-homo-4-pregnene-3,20-dione	*Curvularia lunata* N.R.R.L. 2308	11β-OH	Kieslich *et al.* (1976)
17aα-Acetoxy-17α-methyl-21-hydroxy-D-homo-4-pregnene-3,20-dione	*Curvularia lunata* N.R.R.L. 2308	11β-OH	Kieslich *et al.* (1976)
17aα,21-Dihydroxy-D-homo-4,16-pregnadiene-3,20-dione	*Curvularia lunata* N.R.R.L. 2308	11β-OH	Kieslich *et al.* (1976)
21-Acetoxy-17aα,17α-dihydroxy-D-homo-4-pregnene-3,20-dione	*Curvularia lunata* N.R.R.L. 2308	11β-OH	Kerb *et al.* (1975a)
17α,21-Dihydroxy-15α,16α-methylene-4-pregnene-3,20-dione	*Pellicularia filamentosa* I.F.O. 6675	11β-OH and 11α-OH	Wiechert *et al.* (1974)
17α,21-Dihydroxy-B-nor-4-pregnene-3,20-dione	*Beauveria bassiana*	11α-OH 7α-OH 15β-OH	Sanda *et al.* (1977)
17α,21-Dihydroxy-15α,16α-methylene-4-pregnene-3,20-dione	*Aspergillus ochraceus*	11α-OH	Wiechert *et al.* (1974)
6,6-Difluoro-16α,17α,21-trihydroxy-4-pregnene-3,20-dione	*Aspergillus ochraceus*	11α-OH	Boswell and Ripka (1972, 1973)

blocks attack at the 14α-position, favours 7α-hydroxylation and, with a longer fermentation period, leads to the 7α,12β-dihydroxy compound (Valcavi et al., 1973c). A methyl group in the 6α-position, on the other hand, does not affect 14α-hydroxylation (Valcavi et al., 1974), but diverts the secondary 7α-hydroxylation to attack the 6β-position (Valcavi et al., 1973a). Alteration of the side chain to form desoxycorticosterone-20-ethylene ketal, with its yield of 30% 14α-hydroxy compound, is presumably of little interest (Valcavi, 1971).

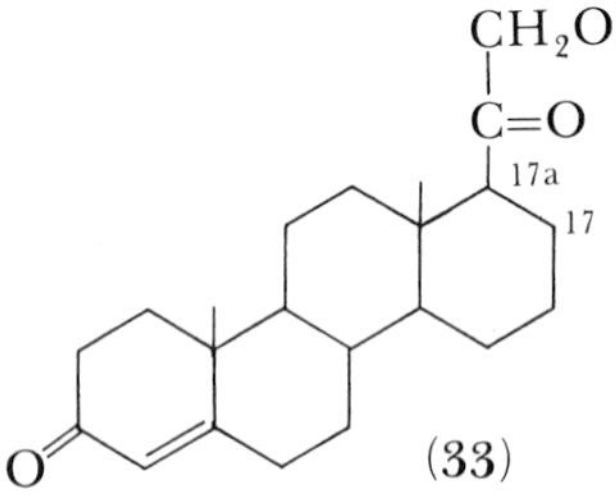

$$CH_2OH$$
$$|$$
$$C=O$$

(33)

Finally, some D-homodesoxycorticosterones (33) were used as starting substrates for the production of further structures with an anti-inflammatory activity (Table 5). Microbiological transformations are known to have many parallels with changes in organic tissue, so that deliberate production of a suitable metabolite is possible by these methods. This advantage was used to produce a 16α,18,21-tri-hydroxy-4-pregnene-3,20-dione which was obtained as a 29% yield by hydroxylation of an 18-hydroxydesoxycorticosterone with Strepto-myces roseochromogenes A.T.C.C. 13400 in addition to another byproduct that was not enzymically produced (13%) (Braham et al., 1975).

F. Hydroxylation of Compound S Structures

The majority of anti-inflammatory steroids on the market are prednisolone structures. Hence the very great importance of hydroxylation of compound S and its derivatives, which has been going on in industry since the 1950s. Highest yields of hydrocortisone are known to be obtained by using Reichstein S-17-acetate for hydroxylation with Curvularia lunata. In this reaction, the ester provides stereochemical protection against undesired side hydroxylations to the α-side of the molecule (de Flines and Waard, 1967). Another fungus is Tieghemella orchidis which gives parallel 11α- and 11β-hydroxylations, and on which a great many further experiments have been described (All-Union Chemical and Pharmaceutical Research Institute, 1973; Mogilnitzky, 1973; Mogilnitzky *et al.*, 1977; Garcia-Rodriguez *et al.*, 1977a, b).

For 11α-hydroxylation of compound S, *Aspergillus ochraceus* still appears to be the most suitable fungus, although further investigations have been carried out with the following strains, which were already known in some quarters: *Beauveria* sp. (Ordzhonikdze, 1976), *Monosporium olivaceum*, *Aspergillus ustus*, *Fusarium graminearum* and *Verticillium glaucum* (Sedlaczek *et al.*, 1973a, b; Table 5). In 11β-

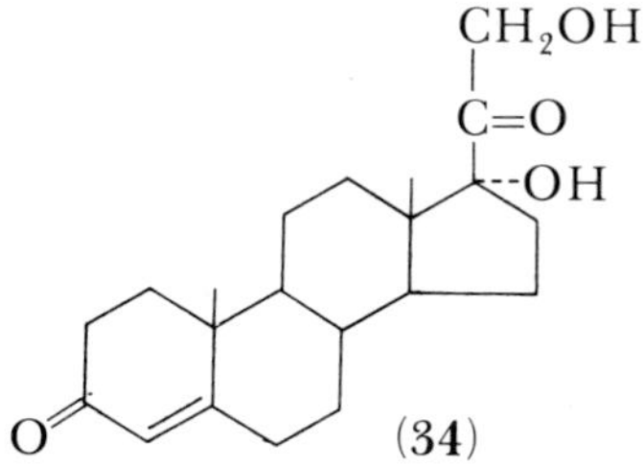

$$CH_2OH$$
$$|$$
$$C=O$$
$$\cdots OH$$

O   (34)

hydroxylation of the *D*-homo compounds, only very small quantities of the usual 14α-hydroxy byproducts appear, as the presence in the 6-membered D ring of a link with the C ring affords some protection to the 14α-position. Moreover, the slight side reaction of 20-oxo reduction is hardly discernible (Kieslich *et al.*, 1976). The 1-dehydrogenation required for synthesis of prednisolone structures is generally carried out after the 11-hydroxylation, since the progress of 11-hydroxylation of RS structures that are already 1-dehydrogenated is much less satisfactory in most cases. So the first two of the following

Table 6

Microbial dehydrogenation of derivatives of compound S

Derivative	Organism	Group introduced and yield	Reference
17α,21-Dihydroxy-4-pregnene-3,20-dione	*Glomerella cingulata* A.T.C.C. 10534	Δ^1 (86%)	Costa Pla (1973)
6,6-Difluoro-16α-methyl-17α,21-dihydroxy-4-pregnene-3,20-dione	*Arthrobacter simplex*	Δ^1	Scribner (1973)
14α,17α-Dihydroxy-4,9(11)-pregnadiene-3,20-dione	*Arthrobacter simplex*	Δ^1 (88%)	Marx *et al.* (1973)

three new examples have little practical significance. The third example shows dehydrogenation of a $\Delta^{9(11)}$-structure which is the next step chemically to the already completed 11-hydroxylation (Table 6).

G. 1-Dehydrogenation of Cortisol Structures

The final transformation of hydrocortisone into commercial products with prednisolone-like structures is also carried out microbiologically as this is superior to the chemical method. The most suitable organisms for introducing the 1,2-double bond are, *Arthrobacter simplex*, *Bacillus sphericus*, *Bacillus lentus*, *Nocardia corallina* and *Septomyxa affinis* which have all been recognized for some time. Recently, investigations were carried out with *Bacillus cereus* (Sallam *et al.*, 1974b; El-Refai *et al.*, 1975; Sallam and El-Refai, 1975) and more particularly with *Mycobacterium globiforme* (Golovtzeva *et al.*, 1977; Borman *et al.*, 1975; Union of Soviet Socialist Republics Academy of Sciences, 1972, 1973a, b; Klimontovich *et al.*, 1975; Koshcheenko *et al.*, 1975; Skryabin *et al.*, 1974a, b, 1976; Sterkin *et al.*, 1973); discussion of these is not possible in this context.

Some fungus strains, including certain *Fusarium* spp. as well as *Gliocladium* and *Helminthosporium* spp., are constitutively adapted to 1-dehydrogenation (El-Refai *et al.*, 1974a), while the unwanted side reaction, particularly noticeably with *Mycobacterium globiforme*, is carried out as an exclusive reaction in various cortisol structures through reduction of the 20-keto group to 20β-alcohol with *Actinomyces roseoviridis* (Balashova *et al.*, 1975a, b) and *Bacillus megaterium* (Krasnova *et al.*, 1975). In a process similar to that already patented with *Arthrobacter simplex* (Shionogi Co. Ltd., 1957), substrate concentrations of 400 g of hydrocortisone/l can be dehydrogenated with *Mycobacterium globiforme* if a microcrystalline substrate is used (particle size <5 μm). Yields of from 80 to 90% of the Δ^1-compound are claimed using cortisone, progesterone and compound S in substrate concentrations of 25 g/l (Borman *et al.*, 1975).

Finally, microbiological 1-dehydrogenation of hydrocortisone was the first conversion in which the immobilized-cell technique was applied to the steroid field. So far, *Mycobacterium globiforme* (Skryabin *et al.*, 1974a, b, 1976) or *Arthrobacter simplex* (Larsson *et al.*, 1976) have been entrapped on polyacrylamide gel for this process. Optimum

Table 7

Microbial dehydrogenation of steroids

Compound	Micro-organism	Group introduced and yield	Reference
(a). HYDROCORTISONE DERIVATIVES			
6-Chloro-11β,17α,21-trihydroxypregna-4,6-diene-3,20-dione	*Arthrobacter simplex* N.R.R.L.-B-8055	Δ^1	Gallegra (1976)
2α,6β,9α-Trifluoro-11β,17α,21-trihydroxy-4-pregnene-3,20-dione	*Nocardia corallina* A.T.C.C. 999	Δ^1 (55%)	Riva and Toscano (1975)
2α,6β,9α-Trifluoro-11β,17α,21-trihydroxy-16α-methyl-4-pregnene-3,20-dione	*Nocardia corallina* A.T.C.C. 999	Δ^1	Riva and Toscano (1976)
2α,6β,9α-Trifluoro-11β,17α,21-trihydroxy-16β-methyl-4-pregnene-3,20-dione	*Nocardia corallina* A.T.C.C. 999	Δ^1	Riva and Toscano (1976)
11β,17α,21-Trihydroxy-15α,16α-methylene-4-pregnene-3,20-dione	*Bacillus lentus* A.T.C.C. 13805	Δ^1	Wiechert *et al.* (1974)
11α,17α,21-Trihydroxy-15α,16α-methylene-4-pregnene-3,20-dione	*Bacillus lentus* A.T.C.C. 13805 *Arthrobacter simplex* A.T.C.C. 6946	Δ^1	Wiechert *et al.* (1974)
(b). D-HOMOHYDROCORTISONE DERIVATIVES			
11β,17aα,21-Trihydroxy-*D*-homo-4-pregnene-3,20-dione (**36**)	*Bacillus lentus* A.T.C.C. 13805	Δ^1	Kieslich *et al.* (1976)
6α-Fluoro-11β,17aα,21-trihydroxy-*D*-4-pregnene-3,20-dione	*Bacillus lentus* A.T.C.C. 13805	Δ^1	Kieslich *et al.* (1976)
6α-Methyl-11β,17aα,21-trihydroxy-*D*-homo-4-pregnene-3,20-dione	*Bacillus lentus* A.T.C.C. 13805	Δ^1	Kieslich *et al.* (1976)
17α-Methyl-11β,17aα,21-trihydroxy-*D*-homo-4-pregnene-3,20-dione	*Bacillus lentus* A.T.C.C. 13805	Δ^1	Kieslich *et al.* (1976)
11β,17aα,21-Trihydroxy-*D*-homo-4,16-pregnadiene-3,20-dione	*Bacillus lentus* A.T.C.C. 13805	Δ^1	Kieslich *et al.* (1976)
17α-Methyl-11β,21-dihydroxy-*D*-homo-4-pregnene-3,20-dione	*Bacillus lentus* A.T.C.C. 13805	Δ^1	Kerb *et al.* (1975a)
6α-Fluoro-17α-methyl-11β,21-dihydroxy-*D*-homo-4-pregnene-3,20-dione	*Bacillus lentus* A.T.C.C. 13805	Δ^1	Kerb *et al.* (1975a)

Table 7 (contd.)

Compound	Micro-organism	Group introduced and yield	Reference
6α-Fluoro-17α-methyl-11α,21-dihydroxy-D-homo-4-pregnene-3,20-dione	*Bacillus lentus* A.T.C.C. 13805	Δ^1	Kerb *et al.* (1975a)
17aα,21-Dihydroxy-D-homo-4-pregnene-3,11,20-trione and 6α,9α or 17α-substituted structures	*Bacillus lentus* A.T.C.C. 13805 *Arthrobacter simplex* A.T.C.C. 6946 and other strains	Δ^1	Alig *et al.* (1976b)
(c). 4,17(20)-PREGNADINE DERIVATIVES			
11β,21-Dihydroxy-4,17(20)-pregnadiene-3-on-hemi-succinate (potassium salt) (or hemicitrate, hemi-phthalate, hemialeate, hemiglutarate)	*Septomyxa affinis* A.T.C.C.	OAc OH Δ^1 (87%)	Kominek (1973)
6α-Methyl-11β,21-dihydroxy-4,17(20)-pregnadien-3-on-21-hemisuccinate (potassium salt)	*Septomyxa affinis* A.T.C.C.	OAc OH Δ^1	Kominek (1973)

enzyme levels can be achieved by special culturing processes (Sterkin *et al.*, 1973; Udvardy, 1974; Ryu *et al.*, 1973). However, the economic advantages of this technique will need to be confirmed in future optimizations. In addition to further developmental work on de-hydrogenation of simple hydrocortisone (35), this reaction has also

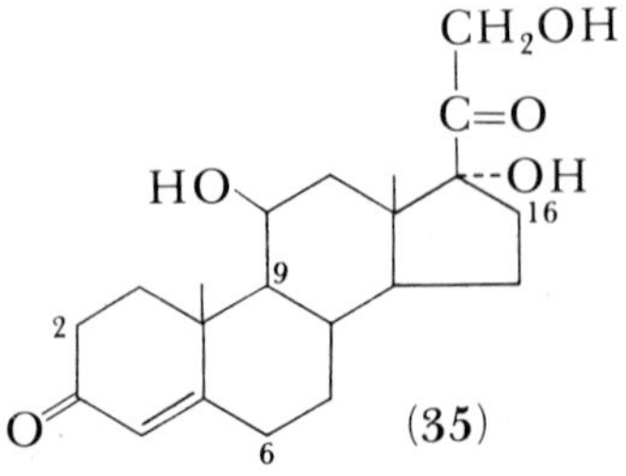

(35)

been described with the following substituted substrates (Table 7). The 21-hemiesters of certain dicarboxylic acids are more suitable as substrates in potassium-salt formulations because of their greater water solubility (Kominek, 1973).

H. Saponification of Various Hydroxypregnane Esters

Since esterases and hydrolysing enzymes are relatively common, ester saponifications may frequently be observed as precursors of the type of reaction actually aimed at, and may be of practical use. Because of the mild conditions involved, the microbiological method has advantages over chemical processes, especially with acyl groups that are hard to saponify like 11β-or 17α-OAc (Table 8).

With precise adjustment of the pH to a value of 6.6, it is possible to achieve selective saponification of the 3β- and 21-acetate groups in a $3\beta,17\alpha,21$-triacetate (Koninklijke Nederlandsche Gist-en Spiritusfabriek, 1967) and this has been used industrially for several years to produce compound S-17-acetate. This compound has special significance as the best tailor-made substrate for production of hydrocortisone (de Flines and Waard, 1967). By contrast, *Bacillus megaterium* saponifies a 21-acetoxy-$3\beta,17\alpha$-dihydroxy-16α-methyl-5-pregnen-20-one without oxidizing the 3-hydroxy group (Ryzhkova *et al.*, 1974).

With a suitable esterase from, for example, *Penicillium lilacinum* N.R.R.L. 895 (Carlström, 1974a, b; Carlström and Krook, 1973), it might be an advantage to work with immobilized enzymes, especially as these hydrolases do not need cofactors. Some preliminary work with a steroid esterase from *Corynebacterium* sp. N.R.R.L.-B-3791 bound to porous glass beads are already available (Grove *et al.*, 1971).

I. Reduction and Oxidation of the Oxygen Function in Pregnane Structures

Investigations in this field, on the one hand, served to stimulate metabolic processes or fundamental enzyme studies and, on the other, led to attempts to use this type of reaction for analytical and diagnostic purposes. Since cell-free or purified enzymes are generally used, with consequently little relevance to preparative work, only a few recent examples are listed (Table 9). Only a reduction of a 21-acetoxy-5α-pregnane-3,11-20-trione with *Saccharomyces cerevisiae* to the anaesthetizing 3α-hydroxy structure (Alfaxalon) can be considered of practical value, and this structure can only be achieved chemically by a synthetic process (Glaxo Lab. Ltd., 1973).

Table 8

Microbial hydrolysis of steroid esters and oxidation of steroid hydroxy groups

Substrate	Micro-organism	Conversion	Reference
6-Azido-9α-fluoro-16α-methyl-11β,21-diacetoxy-17α-hydroxy-4,6-pregnadiene-3,20-dione	Flavobacterium dehydrogenans A.T.C.C. 13930	11β,21-Diac. → 11β,21-Diol.	Rausser and Shapiro (1972), Teutsch et al. (1970)
6-Azido-9α-fluoro-16β-methyl-11β,21-diacetoxy-17α-hydroxy-4,6-pregnadiene-3,20-dione	Flavobacterium dehydrogenans A.T.C.C. 13930	11β,21-Diac. → 11β,21-Diol.	Rausser and Shapiro (1972), Teutsch et al. (1970)
6-Azido-9α-halogeno-11β-acetoxy-4,6-pregnadiene-3,20-dione	Flavobacterium dehydrogenans A.T.C.C. 13930	11β-OAcyl → 11β-OH	Rausser and Shapiro (1972), Teutsch et al. (1970)
6-Azido-9α-fluoro-11β-acetoxy-17α-hydroxy-4,6-pregnadiene-3,20-dione	Flavobacterium dehydrogenans A.T.C.C. 13930	11β-OAcyl. → 11β-OH	Rausser and Shapiro (1972), Teutsch et al. (1970)
17α-Oenanthoxy-21-acetoxy-9α-fluoro-16-methylene-prednisolone	Flavobacterium dehydrogenans A.T.C.C. 13930	17α,21-OAcyl. → 17α,21-OH	E.-Merck AG (1972)
11β,21-Dioenanthoxy-9α-fluoro-16-methylene-prednisolone	Flavobacterium dehydrogenans A.T.C.C. 13930	17α,21-OAcyl. → 17α,21-OH	E.-Merck AG (1972)
17α-Oenanthoxy-21-acetoxy-9α-fluoro-16-methylene-hydrocortisone	Bacillus sphaericus	21-OAcyl. → 21-OH	E.-Merck AG (1972)
3β,21-Diacetoxy-17α-methyl-D-homo-5-pregnene-20-one	Flavobacterium dehydrogenans A.T.C.C. 13930	3β,21-OAcyl. → 21-OH,3C=OΔ⁴	Kerb et al. (1975a)
3β,21-Diacetoxy-17α-hydroxy-15α,16α-methylene-5-pregnen-20-one	Flavobacterium dehydrogenans A.T.C.C. 13930	3β,21-OAcyl. → 21-OH,3C=OΔ⁴	Wiechert et al. (1974)

Saponification of a 3β,21-diacetate is used with advantage in connection with transformation of the 3βOH-Δ^5-system to the 3-keto-Δ^4-structure.

Table 9

Microbial oxidations and reductions of steroids

Conversion	Micro-organism	Reference
OXIDATIONS		
3-OHΔ⁵ → 3-ketoΔ⁴	Streptomyces hydrogenans	Palmowski and Träger (1974); Kohler and Träger (1975)
3C=OΔ⁵ → 3-ketoΔ⁴	Flavobacterium luceoloratum	Udvardy et al. (1976)
3β-OH → 3-keto	Pseudomonas testosteroni	Weintraub et al. (1973)
17β-OH → 17-keto	Brevibacterium sterolicum	Terada et al. (1973)
	Brevibacterium sterolicum A.T.C.C. 21387	
3α-OH → ⎫	Streptomyces hydrogenans	Markert and Träger (1975)
3β-OH → ⎪	Candida albicans	Ghrai et al. (1975)
17β-OH → ⎬ C=O	Candida albicans	
20α-OH → ⎪	Candida albicans	
20β-OH → ⎭	Candida albicans	
	Candida albicans	
20β-OH → 20 C=O	Streptomyces hydrogenans	Chin and Warren (1972)
3α-OH → 3 C=O	Pseudomonas testosteroni A.T.C.C. 11996	Skalhegg (1973, 1974)
REDUCTIONS		
Progesterone → Pregnenolone	Tetrahymena piriformis	Lamontagne et al. (1977)
Desoxycorticosterone → 3α,21-dihydroxy-5β-pregnan-20-one 20ξ,21-dihydroxy-5ξ-pregnan-3-one	Human fecal flora	Bokkenheuser et al. (1976)
Progesterone → 3α-hydroxy-5β-pregnan-20-one	Clostridium paraputrificum	Bokkenheuser et al. (1976)
16α,17α-Epoxy-4-pregnene-3,20-dione → 16α,17α-epoxy-20α-hydroxy-4-pregnen-3-one (further side-chain degradation)	Fusarium solani	Plourde et al. (1972)

J. Pregnane Degradation

The well-known degradation of the progesterone side chain by enzymic Bayer–Villiger reaction to 17β-hydroxy-, or 17-oxo or testolactone structures is generally an undesirable oxidative consequence of preceding oxidations, so investigations with free preparations of *Penicillium lilacinum* N.R.R.L. 895 (Carlström, 1973, 1974a, b) with spores of *Cylindrocarpon radicicola* (Zedan and El-Tayeb, 1977; El-Tayeb *et al.*, 1976) and *Fusarium solani* (Zedan and El-Tayeb, 1977), as well as *Fusarium lini* IFO 7156 (Abdel-Fattah and Badawi, 1975b), can be regarded as fundamental research. Only exceptionally does degradation of artificial progesterone structures imply a simplified way of producing special 17-ketones as, for example, a 9α-fluoro-11β-hydroxy-Δ^3-A-nor-androstene-2,17-dione (David *et al.*, 1973).

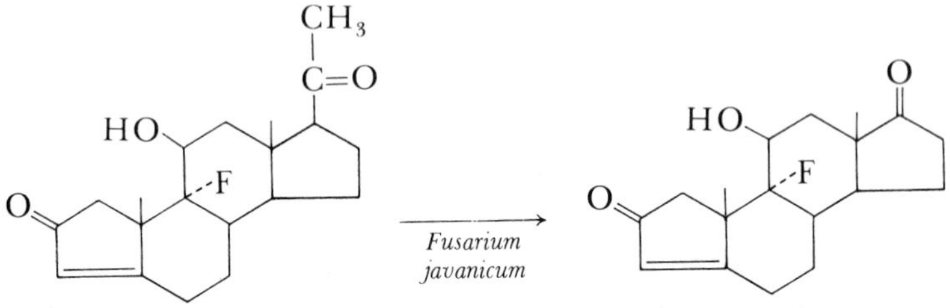

(37)   (38)

The second pathway of pregnane degradation begins with oxidative opening of the B ring, and was more closely investigated during transformation of pregnene-$3\beta,20\beta$-diol (39) to 3[5-oxo-$7\alpha\beta$-methyl $\beta('\beta$-hydroxy)-ethyl-$3\alpha\alpha$-perhydroindane-4α]-propionic acid (pregnenediol-seco acid (40)) with *Nocardia* sp. (Tan *et al.*, 1972).

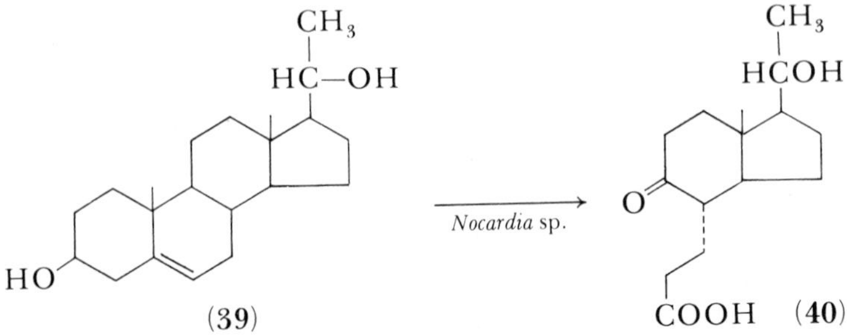

(39)   COOH (40)

5-Pregnene-$3\beta,20\alpha$-diol and progesterone can be oxidized in a similar way. Yields of 55–95% of the seco acids are obtained depending on the culture medium (Probst *et al.*, 1973). *Corynebacterium* sp. A.T.C.C. 14887 yields the analogue of these seco acids from 20β-acetoxy-5-pregnen-3β-ol, but this has found no practical application so far (Probst, 1974).

III. ANDROSTANE AND ESTRANE STRUCTURES

In addition to the pregnane structures important for synthesizing corticoids, gestagens and steroid anaesthetics, androstane and estrane are highly important in commercial production of sexual hormones or mineralocorticoid compounds.

A. Testosterone

Production of the simplest male sexual hormone, testosterone (**41**), by microbiological reduction of 4-androstene-3,17-dione with yeast strains has advantages over chemical reduction because of the specific course of the reaction with no formation of byproducts.

The opposite reaction, oxidations of testosterone to the 17-ketone which has been reported with numerous micro-organisms, only shows inactivation of the mineral. More recent work with *Streptomyces hydrogenans* (Markert *et al.*, 1975), *Pseudomonas* sp. (A.T.C.C. 27330; Kundsin *et al.*, 1972), *Ps. testosteroni* (Antonini, 1974), *Cylindrocarpon radicicola* (A.T.C.C. 11011; Al Rawi *et al.*, 1974) and *Candida albicans* (Cremonesi *et al.*, 1975) is only of limited interest because of the use of special techniques, for example, cell-free enzymes in an organic/aqueous two-phase system (Antonini, 1974; Cremonesi *et al.*, 1975), or because of sequential (Al Rawi *et al.*, 1974) or opposite reactions (Ghraf *et al.*, 1975).

Hydrogenations of the 4-double bond to 5α-H structures, which are also well known, could be useful in special steroid structures, since chemical hydrogenations lead to 5α-H- and 5β-H- mixtures. For the specific 5α-hydrogenation (**42**), cell-free extracts from *Mycobacterium smegmatis* (Hörhold *et al.*, 1975, 1977; Lestrovaja *et al.*, 1977) were used with testosterone and 4-androstene-3,17-dione (Hörhold *et al.*, 1975; Lestrovaja *et al.*, 1977) and with progesterone, 17α-methyltestosterone and 19-nortestosterones (Hörhold *et al.*, 1977). With *Penicillium*

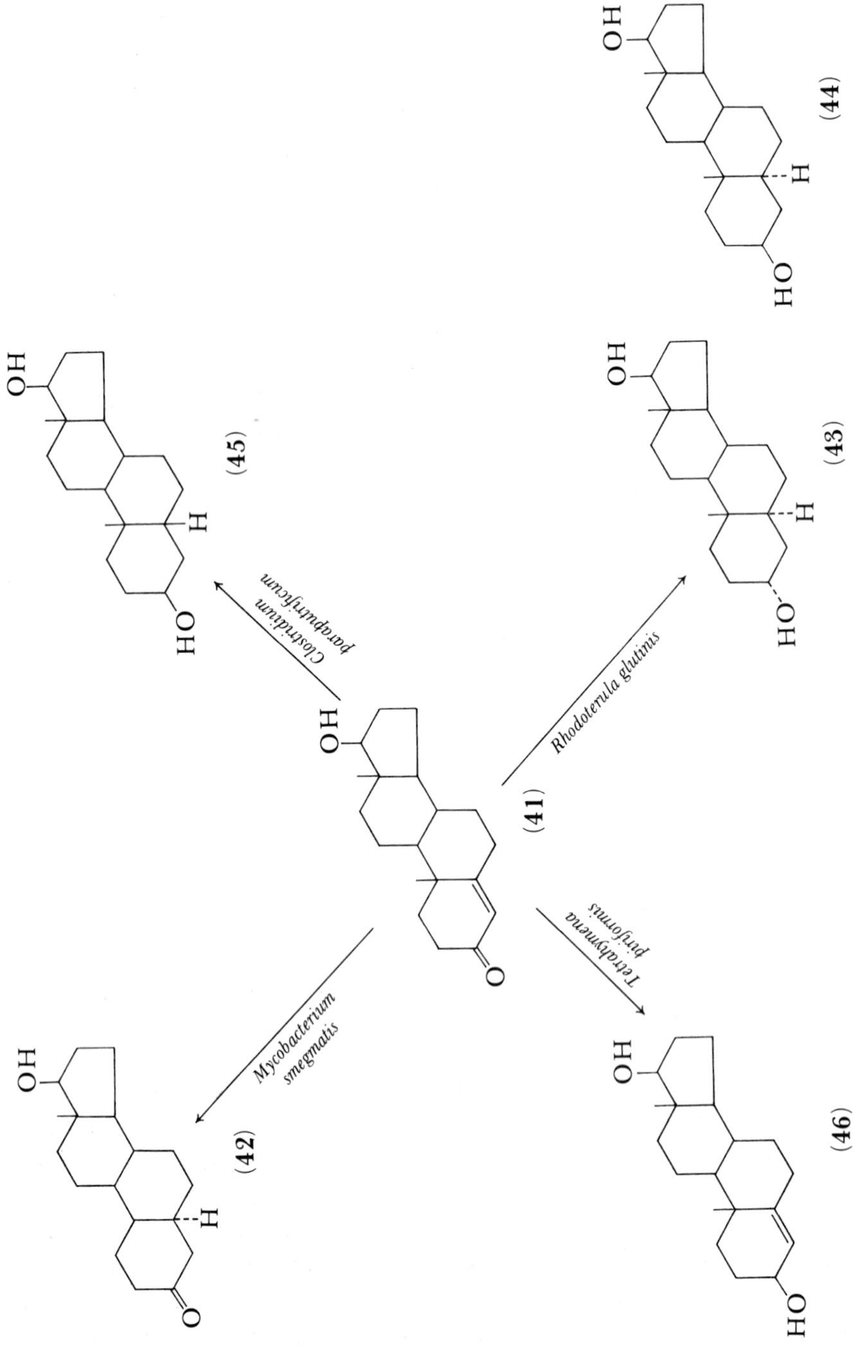
(41)
Clostridium paraputrificum
Rhodoterula glutinis
Mycobacterium smegmatis
Tetrahymena piriformis
(42)
(43)
(44)
(45)
(46)

decumbeus (Abdel-Fattah and Badawi, 1975a), 4α-*D*-5α-androstane-3,17-dione evolved from 4*D*-labelled androstene dione and covered the attack from the β-side (Nambarra *et al.*, 1977). *Rhodotorula glutinis* yields 40% 3α- (**43**) and 30% 3β-hydroxy-5-androstan-17-one (**44**) during a simultaneous reduction of the keto group (Schubert *et al.*, 1972).

By contrast, *Clostridium paraputrificum* hydrogenates uniformly the 5β-H structure with 73% yield of 3α-hydroxy-5β-androstan-17-one (**45**) (Schubert *et al.*, 1972). Only the protozoan *Tetrahymena piriformis* reduces the 3-keto group without attacking the double bond to the 3β-alcohol (**46**) (Lamontagne *et al.*, 1977).

Under anaerobic conditions, *Mycobacterium phlei* transforms a 4-hydroxy-4-androstene-3,17-dione to the 3β- and 3α-hydroxy-4-ones and to a 3β,4α-diol, an interesting example of a coupling of isomerization and reduction (Törmörkény *et al.*, 1975). Finally, hydroxylations of testosterone with *Aspergillus niger* (N.R.R.L. 599; A.T.C.C. 9142) at 16β-position (Yamashita *et al.*, 1976a, b) and with *Aspergillus flavus* at 6β- and 7β- position have been described but these were accompanied by isomerizations or sequential oxidations (Schubert *et al.*, 1972).

B. Substituted Testosterones and Derivatives

When incubating 4-chlorotestosterone (**47**) and 4-chloro-17α-methyl-testosterone (**48**) and its 1-dehydro structure with *Nocardia opaca* (strain iMET 7030, A.T.C.C. 11996), the products of ring cleavage were isolated with 4-chloro-3-hydroxy-10-methyl-9(10)-seco-1,3,5(10)-estratrien-9-one (**50**). The 9α-hydroxy compounds of the 3-keto-Δ⁴-substrates are also held to be known intermediates (**49**) (Hörhold *et al.*, 1976). The anti-aldosterone spironolactones 17-hydroxy-7α-methyl-carbonyl-3-oxo-17α-4-pregnene-21-carboxylic acid lactone (**51**) (Karim and Marsheck, 1976) and 3-(3-oxo-7α-methylthio-4-androsten-17α-yl)-propionic acid-γ-lactone (**52**; Marsheck and Karim, 1973) are hydroxylated at the 6β- position by *Chaetomium cochliodes*. The 7α-methylthio compound yields 17α-methylsulphinyl structures (Marsheck and Karim, 1973) in sequential or parallel reactions.

All of the transformations quoted were the far-reaching results of investigations into the metabolic pathways of synthesis of the cor-

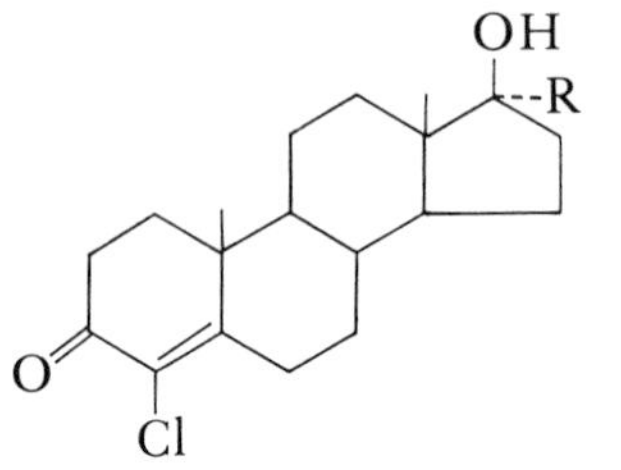

R
(47) H
(48) CH₃

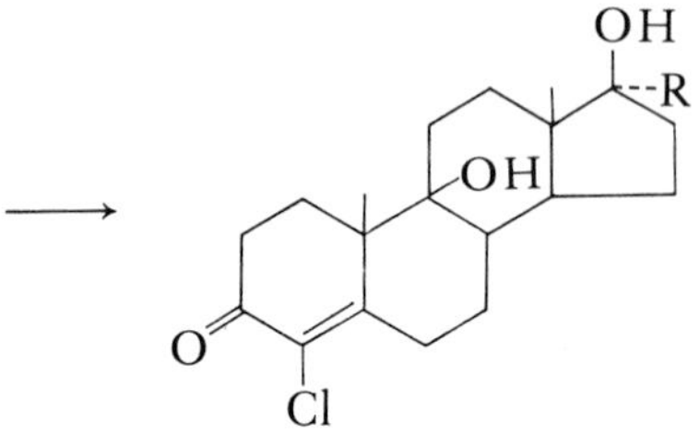

(49)

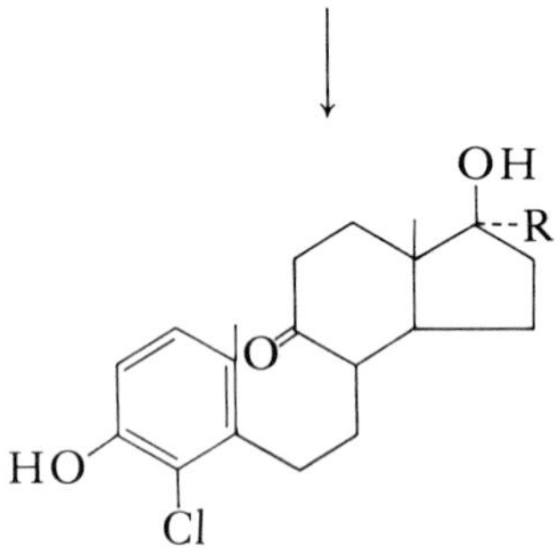

(50)

responding pharmaceuticals, which were carried out with micro-organisms as inadequate substitutes for expensive animal models (Schubert *et al.*, 1972). Critical evaluation of this method indicates the main sites of enzymic attack, opening the door to further chemical variations of the drug to give increased effectiveness.

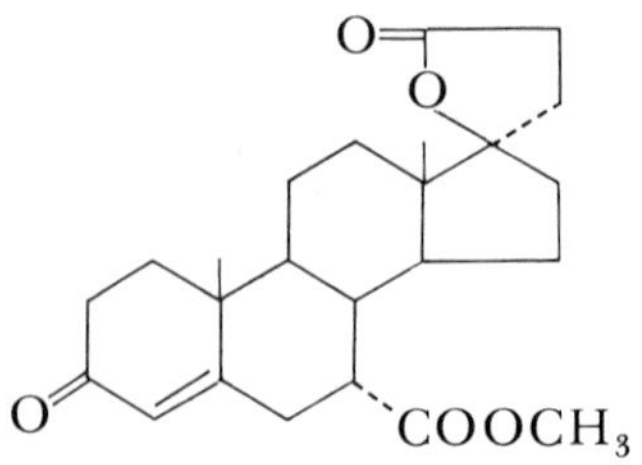

(51) 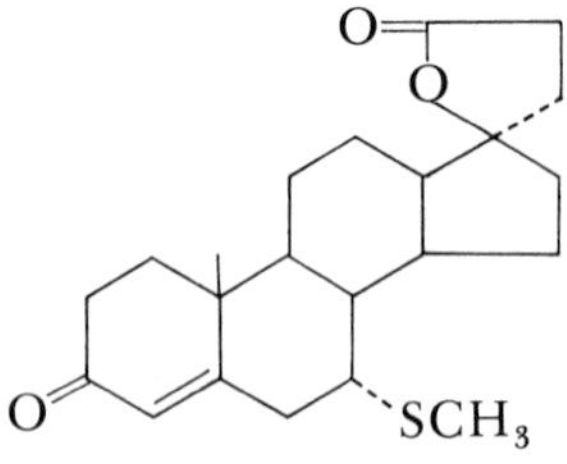   (52)

C. Dehydrogenation of 5α- and 5β-Androstan-3-ones

The reaction mechanism of Δ^4- and Δ^1-dehydrogenations was investigated as a follow-up to earlier experimental work with 4α- and

Table 10

Microbial hydroxylation of 17a-methyl-substituted testosterones. From Schubert
et al. (1972)

Substrate	Micro-organism	Groups introduced	Yield (%)
17α-Methyl-17β-hydroxy-4-androsten-3-one	Aspergillus flavus	15α-OH	15
		15β-OH	15
		6β-OH	5
17α-Methyl-17β-hydroxy-1,4-androstadien-3-one	Aspergillus flavus	15α-OH	10
		15β-OH	60
4-Chloro-17α-methyl-17β-hydroxy-4-androsten-3-one	Aspergillus flavus	15α-OH	80
4-Chloro-17α-methyl-17β-hydroxy-1,4-androstadien-3-one	Aspergillus flavus	15α-OH	40
		15β-OH	40

The reactions described for the basic substances follow a similar course in the
following substituted testosterones (Schubert et al., 1972).

17α-Methyl-17β-hydroxy-4-androsten-3-one	Clostridium paraputrificum	3α-OH 5β-H	70
	Rhodotorula glutinis	3α-OH 5α-H	45
		3β-OH 5α-H	30
17α-Methyl-17β-hydroxy-1,4-androstadien-3-one	Clostridium paraputrificum	3-on 5β-H	70
	Rhodotorula glutinis	no hydrogenation	
4-Chloro-17α-methyl-17β-hydroxy-4-androsten-3-one	Clostridium paraputrificum	3α-OH 5β-H	65
	Rhodotorula glutinis	3β-OH-Δ⁴	30
		3β-OH 4αCl 5α-H	45
4-Chloro-17α-methyl-17β-hydroxy-1,4-androstadien-3-one	Clostridium paraputrificum	3-on 4βa 5β-H	35
		3α-OH 4βCl 5β-H	25
	Rhodotorula glutinis	3-on 4αCl 5α-H	5
		3β-OH 4αCl 5α-H	65

4β-deuterated 5α- (Nambara et al., 1975b) and 5β- (Nambara et al.,
1973), and with 2α- and 2β-deuterated 5β- (Nambara et al., 1973)
androstane-3,17-diones using cell-free extracts of Nocardia restrictus
(A.T.C.C. 14887). In the 5α- system, the 4β- and in the 5β- system the
4α-deuterium was eliminated during dehydrogenation. As during
1-dehydrogenation, with the exception of the 1α- and 2β- hydrogen
atoms (Nicholas, 1975; Ikegawa and Nambara, 1973), this always
indicates a transdiaxial loss of hydrogen atoms. Further conditions for
1-dehydrogenation with Arthrobacter simplex were worked out with 20

different steroids (Penasse and Nomine, 1974) and consist of: (1) the presence of a 3-oxo group, a Δ^4 double bond and axial hydrogens at 1α- and 2β- positions; (2) the absence of bulky substituents on the α-side of the molecule centre and at the 4-position; and (3) limitation of the size of the alkyl substituent at 10β- and 13β- to two carbon atoms.

D. Hydroxylation of Various Androstane-Mono- and Diketones and of Some D-Homo and Estrane Analogues

During research into the direction of hydroxylation in simple unsubstituted androstane structures with one or two keto groups at various positions, systematic studies of the site of enzymic attack were carred out in relation to the substrate structure, this for the first time in steroid research. An account of over 300 individual results obtained with the following fungal strains would be beyond the scope of this paper: *Penicillium urticae* (Blunt *et al.*, 1971), *Calonectria decora* (Bell *et al.*, 1972a, b; Ashton *et al.*, 1974; Evans *et al.*, 1975; Chambers *et al.*, 1975b; Jones *et al.*, 1975b), *Aspergillus ochraceus* (Blunt *et al.*, 1971; Bell *et al.*, 1972b, c; Clegg *et al.*, 1973; Bell *et al.*, 1974; Evans *et al.*, 1975; Jones *et al.*, 1975b), *Rhizopus nigricans* (Browne *et al.*, 1973; Chambers *et al.*, 1973; Ashton *et al.*, 1974; Jones *et al.*, 1975b), *Rhizopus arrhizus* (Bell *et al.*, 1973), *Rhizopus circinans* (Bell *et al.*, 1973), *Wojnowicia graminis* (Chambers *et al.*, 1975a), *Ophiobolus herpotrichus* (Chambers *et al.*, 1975a), *Daedalea rufescens* (Bell *et al.*, 1975a), *Diaporthe celastrina* (Bell *et al.*, 1975b), *Acromyrmex fungus* (Jones *et al.*, 1975a), *Absidia orchidis* (Bell *et al.*, 1975c), *Syncephalastrum racemosum* (Bell *et al.*, 1975c), *Leptoporus fissilis* (Denny *et al.*, 1976).

Because of exceptionally good yields of the isolated reaction product, only the transformations described in Table 11 have been selected for possible use in preparing syntheses of far-reaching new types of steroids.

All research results show that the principal directions of attack by *Calonectria decora* (12β, 15α), *Aspergillus ochraceus* (6β, 11α), *Rhizopus nigricans*, *R. arrhizus* and *R. circinans* (11α) can be completely diverted to new hydroxylation directions by changing the substrate structure. This variability in the hydroxylations can be traced back to different bindings of the substrate to the enzyme, as shown in a model with

Table 11

Hydroxylation of androstane mono- and diketones and some analogues

Ring system	Keto groups	Hydroxy groups	Substituents	Main hydroxylation	Yield (%)	Micro-organism	Reference
5α-Androstane	6,17	—	—	1β	64	*Calonectria decora*	Bell *et al.* (1972b)
5α-Androstane	7,17	—	—	1β	55	*Calonectria decora*	Bell *et al.* (1972b)
5α-Androstane	17	3α-OH	3 OCH₃	1β,6β	60	*Calonectria decora*	Bell *et al.* (1972b)
5α-Androstane	16	—	16 ⎡–O– ⎤ –O–	6α,12β	71	*Calonectria decora*	Evans *et al.* (1975)
5α-Androstane	17	—	17 ⎡–O– ⎤ –O–	7β,12β,15α	47	*Calonectria decora*	Evans *et al.* (1975)
5α-Androstane	17	1β,6α	—	19	56	*Calonectria decora*	Chambers *et al.* (1975b)
5α-Androst-2-ene	17	—	—	1C=O,6α	52	*Calonectria decora*	Chambers *et al.* (1975b)
5α-Androst-2-ene	1,17	—	—	6α	54	*Calonectria decora*	Chambers *et al.* (1975b)
3α,5-Cyclo-androstane	17	6α	—	11α	44	*Calonectria decora*	Chambers *et al.* (1975b)
3α,5-Cyclo-androstane	17	6β	—	11α	50	*Calonectria decora*	Chambers *et al.* (1975b)
5α-Androstane	3	—	—	6β,11α	51	*Aspergillus ochraceus*	Bell *et al.* (1972c)
5α-Androstane	2,16	—	—	11α	61	*Aspergillus ochraceus*	Bell *et al.* (1972c)

Table 11—*continued*

Ring system	Keto groups	Hydroxy groups	Substituents	Main hydroxylation	Yield (%)	Micro-organism	Reference
5α-Androstane	2,17	—	—	11α	57	*Aspergillus ochraceus*	Bell *et al*. (1972c)
5α-Androstane	3,6	—	—	11α	60	*Aspergillus ochraceus*	Bell *et al*. (1972c)
5α-Androstane	3,7	—	—	11α	74	*Aspergillus ochraceus*	Bell *et al*. (1972c)
5α-Androstane	3,16	—	—	11α	55	*Aspergillus ochraceus*	Bell *et al*. (1972c)
5α-Androstane	16	3β	—	11α	71	*Aspergillus ochraceus*	Bell *et al*. (1972c)
5α-Androstane	3,17	—	—	11α	75	*Aspergillus ochraceus*	Bell *et al*. (1972c, 1974)
5α-Androstane	3	17β	—	11α	79	*Aspergillus ochraceus*	Bell *et al*. (1972c, 1974)
5α-Androstane	17	3α	—	11α	81	*Aspergillus ochraceus*	Bell *et al*. (1972c)
5α-Androstane	17	3β	—	11α	53	*Aspergillus ochraceus*	Bell *et al*. (1972c)
5α-Androstane	17	3β	3 OAc	11α	61	*Aspergillus ochraceus*	Bell *et al*. (1972c)
5α-Androstane	17	3β,7β	—	11α	78	*Aspergillus ochraceus*	Bell *et al*. (1974)
Androst-4-ene	3	—	—	6β,11 *et al*.	75	*Aspergillus ochraceus*	Bell *et al*. (1972c)
Androst-4-ene	3	17β	—	11α	73	*Aspergillus ochraceus*	Bell *et al*. (1972c)

Table 11—*continued*

Ring system	Keto groups	Hydroxy groups	Substituents	Main hydroxylation	Yield (%)	Micro-organism	Reference
Androst-5-ene	17	3β	—	11α	87	*Aspergillus ochraceus*	Bell *et al.* (1972c)
Estr-4-ene	3	—	—	$6\beta,11\alpha$	53	*Aspergillus ochraceus*	Bell *et al.* (1972c)
5α-Estrane	3	—	—	$6\beta,11\alpha$	50	*Aspergillus ochraceus*	Bell *et al.* (1972c)
5α-Pregnane	20	3β	—	$1\beta,11\alpha$	60	*Aspergillus ochraceus*	Bell *et al.* (1972b), Clegg *et al.* (1973)
5α-Pregnane	3,20	—	—	11α	68	*Aspergillus ochraceus*	Clegg *et al.* (1973)
5α-Estr-4-ene	3	—	—	$10\beta,16\beta$	58	*Rhizopus nigricans*	Browne *et al.* (1973)
5α-Androstane	16	—	—	$3\beta,7\alpha$	46	*Rhizopus nigricans*	Ikegawa and Nambara (1973)
5α-Androstane	2,16	—	—	$2\alpha,7\alpha$	55	*Rhizopus nigricans*	Chambers *et al.* (1973)
5α-Androstane	3,16	—	—	7α and 3 OH	50	*Rhizopus nigricans*	Chambers *et al.* (1973)
5α-Androstane	3	16β	—	11α	71	*Rhizopus nigricans*	Chambers *et al.* (1973)
5α-Androstane	16	3β	—	7α	90	*Rhizopus nigricans*	Chambers *et al.* (1973)
5α-Androstane	17	3β	—	7β	48	*Rhizopus nigricans*	Chambers *et al.* (1973)
5α-Androstane	3	7β	—	16β	52	*Rhizopus nigricans*	Chambers *et al.* (1973)

Table 11—*continued*

Ring system	Keto groups	Hydroxy groups	Substituents	Main hydroxylation	Yield (%)	Micro-organism	Reference
5α-Androstane	7	3β	—	16β	55	*Rhizopus nigricans*	Chambers et al. (1973)
5α-Androstane	3,11	—	—	16β	55	*Rhizopus nigricans*	Chambers et al. (1973)
5α-Androstane	3	11α	—	16β	45	*Rhizopus nigricans*	Chambers et al. (1973)
5α-Androstane	6	17β	—	3β and 3α	44	*Rhizopus nigricans*	Chambers et al. (1973)
5α-Androstane	7,17	—	—	3α and 3β	44	*Rhizopus nigricans*	Chambers et al. (1973)
5α-Androstane	17	6α	—	11α	58	*Rhizopus nigricans*	Chambers et al. (1973)
5α-Androstane	17	7α	—	3β	77	*Rhizopus nigricans*	Chambers et al. (1973)
5α-Androstane	17	7β	—	3β	57	*Rhizopus nigricans*	Chambers et al. (1973)
5α-Androstane	11,16	—	—	3α and 3β	73	*Rhizopus nigricans*	Chambers et al. (1973)
5α-Androstane	11	17β	—	3α	47	*Rhizopus nigricans*	Chambers et al. (1973)
5α-Androst-4-ene	3	6β	—	16β	61	*Rhizopus nigricans*	Chambers et al. (1973)
5α-Androstane	17	3β	—	7β	57	*Rhizopus arrhizus*	Bell et al. (1973)
5α-Androstane	17	6α	—	11α	72	*Rhizopus arrhizus*	Bell et al. (1973)
5α-Androstane	17	6α	—	11α	56	*Rhizopus circinans*	Bell et al. (1973)
5α-Androstane	11,17	—	—	4α	54	*Rhizopus circinans*	Bell et al. (1973)

Table 11—continued

Ring system	Keto groups	Hydroxy groups	Substituents	Main hydroxylation	Yield (%)	Micro-organism	Reference
5α-Androstane	7	3α	—	17β	55	*Wojnowicia graminis*	Chambers *et al.* (1975a)
5α-Androstane	7	3β	—	17β	63	*Wojnowicia graminis*	Chambers *et al.* (1975a)
5α-Androstane	11	3β	—	16β	48	*Wojnowicia graminis*	Chambers *et al.* (1975a)
5α-Androstane	7,17	—	—	2α	64	*Wojnowicia graminis*	Chambers *et al.* (1975a)
5α-Androstane	7	17β	—	2α	58	*Wojnowicia graminis*	Chambers *et al.* (1975a)
5α-Androstane	11	17β	—	3α + 3β	46	*Wojnowicia graminis*	Chambers *et al.* (1975a)
5α-Androstane	3,7	—	—	17β	62	*Ophiobolus herpotrichus*	Chambers *et al.* (1975a)
5α-Androstane	3,11	—	—	16 + 16C=O	49	*Ophiobolus herpotrichus*	Chambers *et al.* (1975a)
5α-Androstane	7,17	—	—	3β	75	*Ophiobolus herpotrichus*	Chambers *et al.* (1975a)
5α-Androstane	11,17	—	—	3α	91	*Ophiobolus herpotrichus*	Chambers *et al.* (1975a)
5α-Androstane	11	17β	—	3α	95	*Ophiobolus herpotrichus*	Chambers *et al.* (1975a)
5α-Androstane	7	—	3⟨—O—, —O—⟩	3β,16β	57	*Daedalea rufescens*	Bell *et al.* (1975a)
5α-Androstane	3,7	—	—	16β	66	*Daedalea rufescens*	Bell *et al.* (1975a)
5α-Androstane	3	7α	—	16β	53	*Daedalea rufescens*	Bell *et al.* (1975a)

Table 11—continued

Ring system	Keto groups	Hydroxy groups	Substituents	Main hydroxylation	Yield (%)	Micro-organism	References
5α-Androstane	7,11	—	—	3β,16β	42	*Daedalea rufescens*	Bell *et al.* (1975a)
5α-Androstane	3,7	—	—	16α	48	*Diaporthe celastrina*	Bell *et al.* (1975b)
5α-Androstane	3	9α	—	16α	47	*Diaporthe celastrina*	Bell *et al.* (1975b)
5α-Androstane	3,11	—	—	16α	55	*Diaporthe celastrina*	Bell *et al.* (1975b)
5α-Androstane	3	12α	—	16α	48	*Diaporthe celastrina*	Bell *et al.* (1975b)
5α-Androstane	7,17	—	—	3α	52	*Diaporthe celastrina*	Bell *et al.* (1975b)
5α-Androstane	6	3β	—	17β	40	*Acromyrex fungus*	Jones *et al.* (1975a)
5α-Androstane	7	3β	—	12α	51	*Syncephalastrum racemosum*	Bell *et al.* (1975c)

non-steroid substrates (Fonken and Johnson, 1972). Since polar groups in the substrate are responsible for the binding, the site and steric position of such substituents are critical for the speed and direction of the reaction. Whereas only one hydroxy group is introduced into substrates already polar in character, mono-oxo-androstanes are predominantly dihydroxylated and this is reinforced by the di-methylsulphoxide and a longer fermentation period.

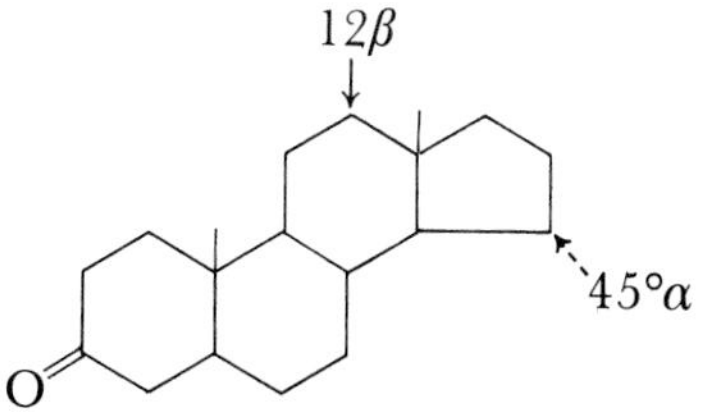

12β
45°α
O
(53)

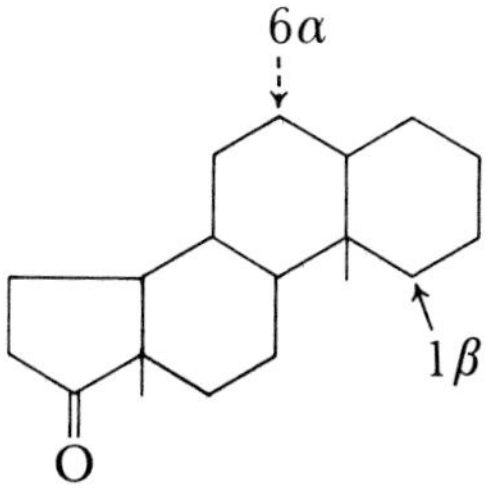

6α
1β
O
(54)

Calonectria decora always introduces two hydroxyl groups simultaneously at equatorial positions. 3- and 4-oxo-5α-androstanes yield 12β,15α-(OH)$_2$ products (53) in contrast to 16- and 17-oxo-5α-androstanes which yield the 1β,6α-(OH)$_2$ products (54; Bell *et al.*,

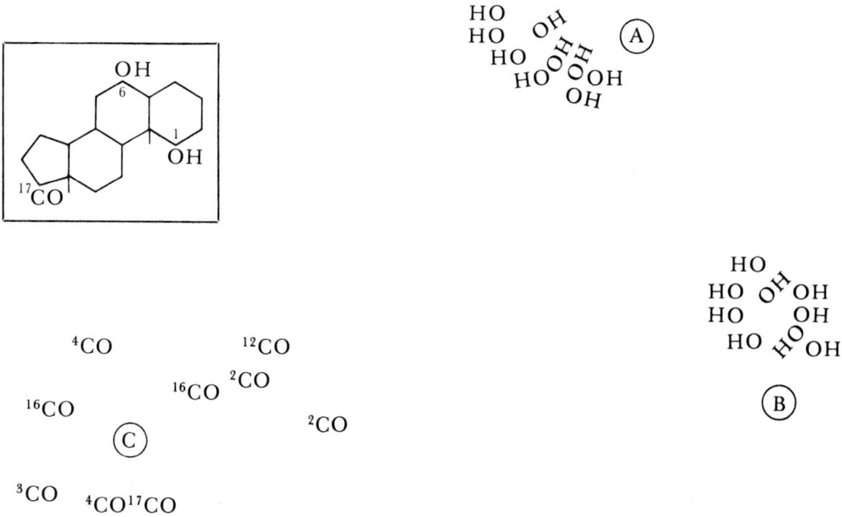

OH
6
1
OH
^{17}CO

HO
HO O
HO
HO OOH
OH
(A)

^{4}CO ^{12}CO
^{16}CO ^{2}CO
^{16}CO
^{2}CO
(C)

^{3}CO ^{4}CO^{17}CO

HO
HO O OH
HO   OH
HO   OH
(B)

Fig. 1. Scheme for dihydroxylation of androstane-ones. From Jones (1973). A is a point of enzymic attack, B is a second point of enzymic attack and C is the binding of the keto function at the enzyme.

1972a). If each pair of substrates is rotated through 180° about a vertical axis between 0–8 and 0–9, we can see that the two sites of attack, $1\beta,6\alpha$ and $12\beta,15\alpha$, are very similar. The distance between the carbon atoms under paired attack ($12\beta\leftrightarrow15\alpha$ and $6\alpha\leftrightarrow1\beta$) amounts at any given time to 0.4 nm and that of the hydroxyl groups from the keto functions to about 0.6 nm.

This regularity is repeated in many analogous dihydroxylations of mono-oxo-5α-androstanes and, in the case of the results from *Calonectria decora* with the 2,3,4,12,16 and 17-monoketones, led to the configuration shown in Fig. 1 (Jones, 1973). The nine possibilities illustrated in the figure for binding of the six substrates all show the points of attack in two close circles located at A and B, whereas the binding of the keto function at the enzyme (C) shows a less clearly defined location.

Although unlimited validity should not be expected from this figure, it nevertheless gives a useful insight into the conditions of steric substrates, and helps the practitioner to comprehend wider applications.

E. 3β-Hydroxy-5-Androsten-17-one and 4-Androstene-3,17-Dione

The significance of a 16α-hydroxyl group in estriol gave rise to production and investigation of other 16α-hydroxylated 17-keto steroids. Microbiological 6α-hydroxylation using as substrate compound S or 9α-fluorohydrocortisone to synthesize triamcinolone is a technical advance that has long been in use with strains of *Streptomyces roseochromogenes*. One strain (N.R.R.L.-B-1233) can also hydroxylate 3β-hydroxy-5-androsten-17-one (dehydro-epiandrosterone; **55**) at the 16α- position with 81% yield (**56**; Iida *et al.*, 1975a; PERMACHEM Asia Ltd., 1976). With longer fermentation times, 3β,16α,17β-triol (**57**) is formed as a byproduct (Iida *et al.*, 1975b). When dehydro-epiandrosterone-3-succinate undergoes 16α-hydroxylation with strain J.F.O.-1308, it suffers partial loss of the ester (16% 3β,16α(OH)$_2$ and 15% 3-succinyloxy-16αOH; Yamashita and Kurosawa, 1975). While 4-androstene-3,17-dione (**58**) with strain N.R.R.L.-B-1223 also gives 41% hydroxy product (**59**; Iida *et al.*, 1975a), a 16β-hydroxylation with *Aspergillus niger* N.R.R.L. 599 leads through isomerization or reduction to the byproducts 17β-hydroxy-

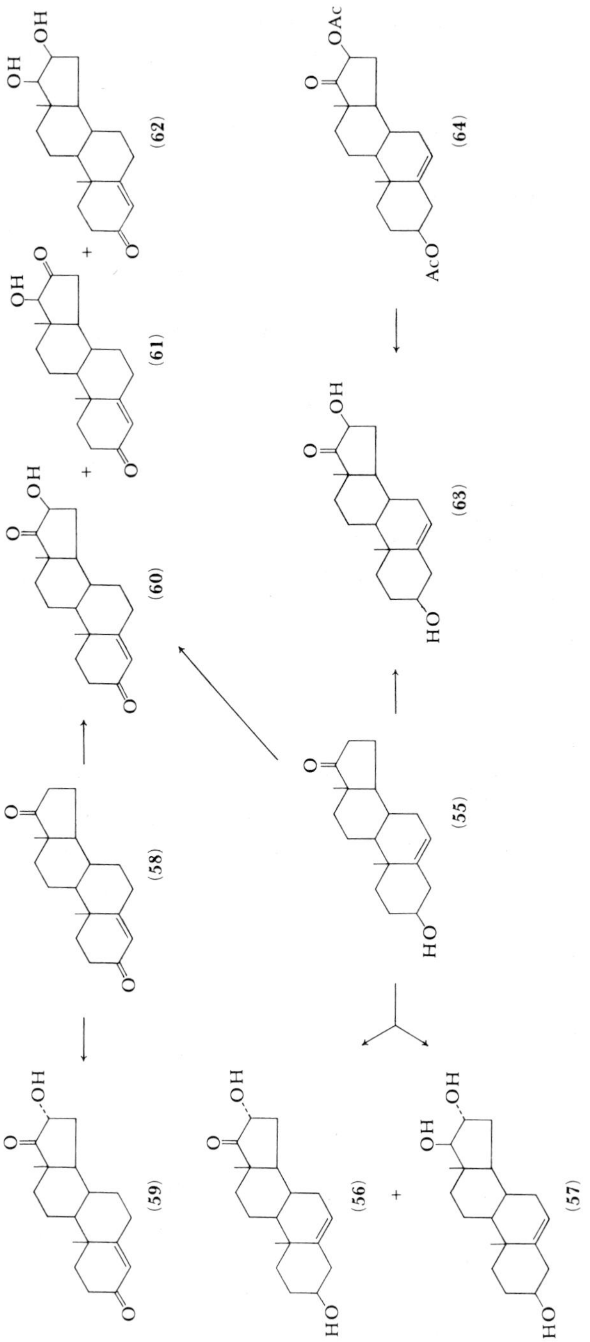
OH
OH
O
OH
O
OH
OAc
AcO
HO
HO
HO
HO
(55)
(56)
(57)
(58)
(59)
(60)
(61)
(62)
(63)
(64)

16-ketone (**61**) and 16β,17β-diol (**62**; Yamashita *et al.*, 1976a) in addition to 25% of the desired structure (**60**). Fermentation of dehydro-epiandrosterone (**55**) under the same conditions is always accompanied by simultaneous oxidation to the 3-oxo-4-ene system (Kurosawa *et al.*, 1975) giving rise to the same products as 4-androstene-3,17-dione (Yamashita *et al.*, 1976b). 3β,16β-Dihydroxy-5-androstene-3,17-dione (**63**) could hitherto only be obtained by 16β-hydroxylation with an unidentified strain of a *Mucor* sp., giving a yield of 1% in addition to the main product 7α-OH and several byproducts (K. Kieslich, unpublished observations).

This mineral corticoid compound, however, can be successfully obtained in high yield by 'saponification of a 3β,16β-diacetoxy-5-androsten-17-one (**64**) with a liver esterase (Boehringer Mannheim) or with *Flavobacterium esteraromaticum* A.T.C.C. 8091. Both oxidation of the 3β-hydroxy group, which succeeds quantitatively in saponification of 3β,16β-diacetoxy-5α-androstan-17-one with *Flavobacterium* var. *hydrolyticum* (Fishman, 1960), and also isomerization of the 16β-hydroxy-17-oxo system are thereby excluded (Kieslich, 1978). With pancreatic amylase and human erythrocyte acetylcholinesterase, yields of 60% are obtained (Synne and Rennick, 1972).

Hydroxylations with *Absidia orchidis* using 4-androstene-3,17-dione deliver the 7α-hydroxy compound as well as some 6β-hydroxy product (Engelfried *et al.*, 1976); using 3β-hydroxy-5-androsten-17-one and a 17α-methyl-5-androstene-3β,17β-diol, mixtures of 7α- and 7β-hydroxy structures are obtained. Sequential oxidation with *Flavobacterium peregrinum* to give the 3-oxo-4-ene structures made it easier to separate the two components (Schwartz and Protiva, 1974).

F. Substituted or Modified DA and AD Structures

A 6β-Hydroxy-3α,5-cyclo-5α-androstan-17-one (**65**) gives 26% 11α- in addition to 1.5% 2α- and 1% 7β-hydroxy compounds with *Rhizopus nigricans* (Procházka *et al.*, 1974). This change in the structure of the dehydro-epiandrosterone seemed to be necessary as the 3β-hydroxy-5-ene structure (**55**) is hydroxylated predominantly at 7α-position by this fungus. Analogous 40% hydroxylation with *Rhizopus arrhizus* A.T.C.C. 11145 using 4-androstene-3,17-dione (**58**) takes place at 6β-position, whereas the 6α- and 6β-methyl-substituted structures give

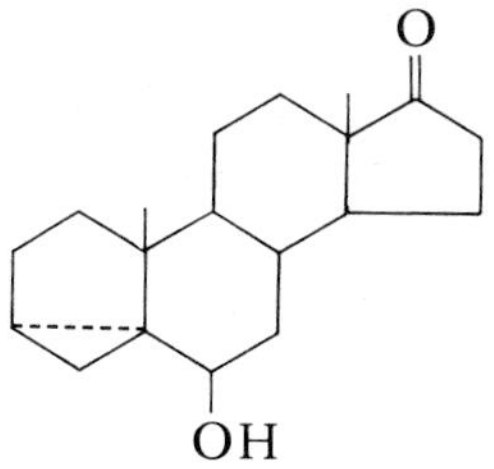

(65)

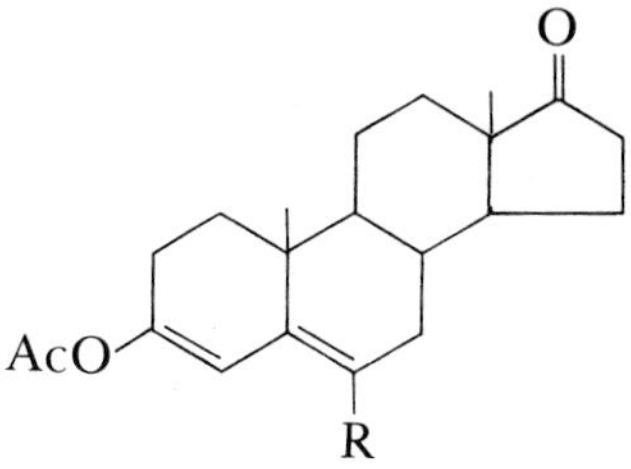

(66) H
(67) CH₃

$$\begin{array}{cc} & R \\ (66) & H \\ (67) & CH_3 \end{array}$$

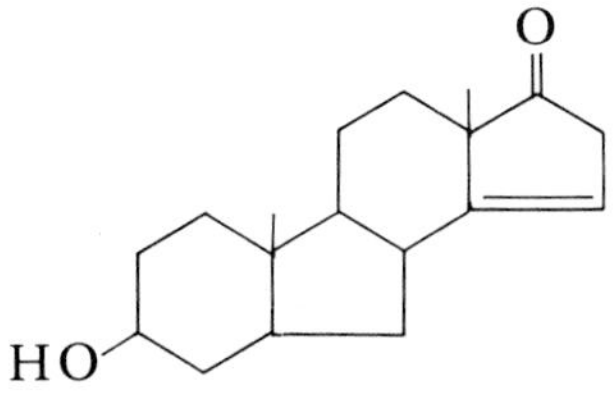

(68)

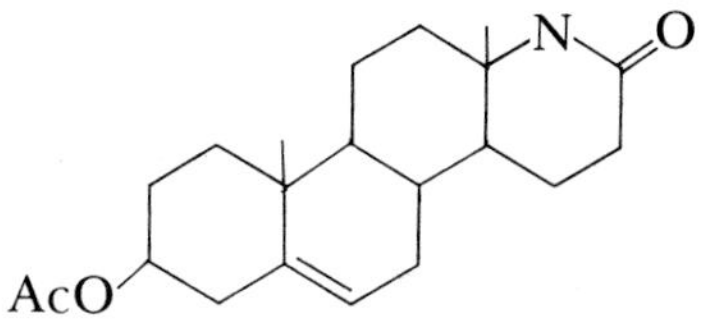

(69)

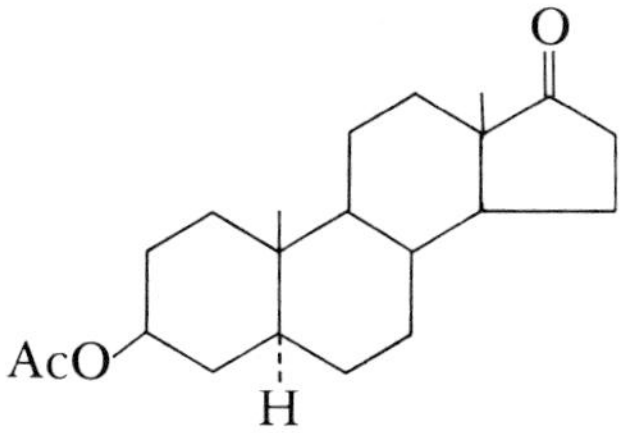

(70)

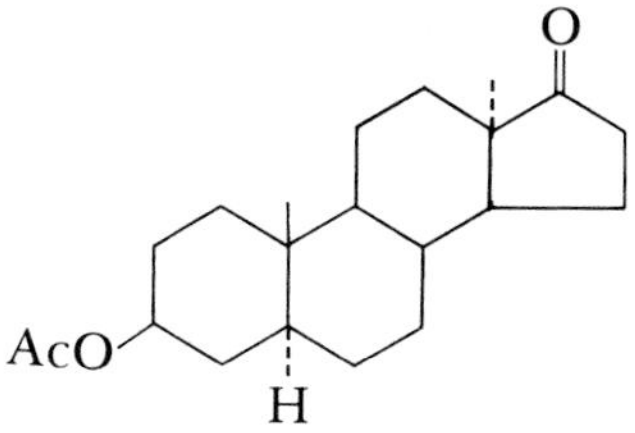

(71)

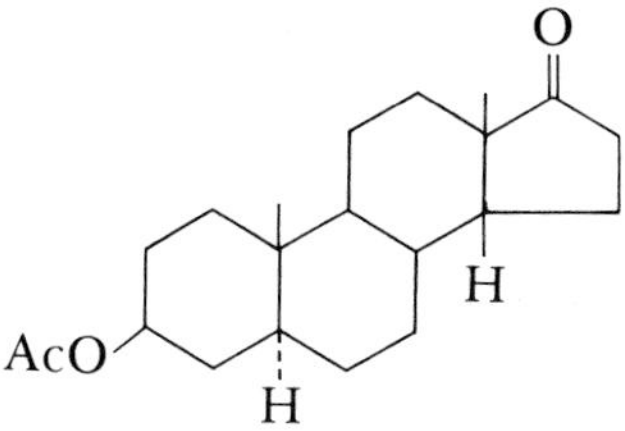

(72)

only 11α-hydroxy products (Holland and Auret, 1975a). Since the corresponding $\Delta^{3,5}$-3-enol acetates (66) (67) give mixtures of 6β- and 11α-hydroxylated structures, the resulting free $\Delta^{3,5}$-3-enol form is confirmed as the condition for a 6β-hydroxylation. Easy development of this intermediate determines whether 6β- or 11α- attack has precedence (Holland and Auret, 1975b). Dehydro-epiandrosterone

Table 12

Percentage conversions of dehydroepiandrosterone acetates and of
3β-acetoxy-17α-aza-D-homo-5-androstene-17-ones

Dehydroepiandrosterone acetates		3β-Acetoxy-17α-aza-D-homo-5-androstene-17-ones	
Derivative	Per cent conversion	Derivative	Per cent conversion
7α-Hydroxy	26.0	7α-Hydroxy	6.7
7β-Hydroxy	9.0	7β-Hydroxy	6.7
7-Oxo	5.4	7-Oxo	1.7
14α-Hydroxy	1.1		
5β,1β-Epoxy-17α-hydroxy	4.0	5β,6β-Epoxy	4.3

and 4-androstene-3,17-dione, however, can be hydroxylated with
good yields at the 11α- position with *Aspergillus ochraceus* (K. Kieslich,
unpublished observations).

11α-Hydroxylation of a 3β-hydroxy-*B*-nor-5-androsten-17-one
(**68**) with *Rhizopus nigricans* is accompanied by formation of a 5α,6α-
epoxide with partial hydrolysis to 5β,6α-diol (Joská *et al.*, 1973). The
customary 7α-hydroxylation in the 3β-hydroxy-5-ene system becomes
an α-epoxidation here, in accordance with the Bloom and Shull rule
(Bloom and Shull, 1955). *B*-Nor-4-androstene-3,17-dione gives only
6α- and 11α- hydroxylation with *Absidia orchidis* (Joská *et al.*, 1973).

Modification of the D ring in dehydro-epiandrosterone-3-acetate to
3β-acetoxy-17α-aza-D-homo-5-androsten-17-one (**69**) has only a
slight influence on the direction of the attack by *Cunninghamella elegans*
(CBS 16753) (Crabb *et al.*, 1975; Table 12).

In a saturated 3β-acetoxy-5α-androstan-17-one system (**70**),
complete inhibition of the 7- and 14α- hydroxylation by *Mucor
griseocyanus* was observed when 13- (**71**) and 14- (**72**) isostructures were
brought in (Valcavi *et al.*, 1975).

G. 19-Hydroxyandrostene Structures (73)

Oxidative transformation of 19-hydroxy compounds to ring-A
aromatic structures has long been known (Sih and Wang, 1965) and,

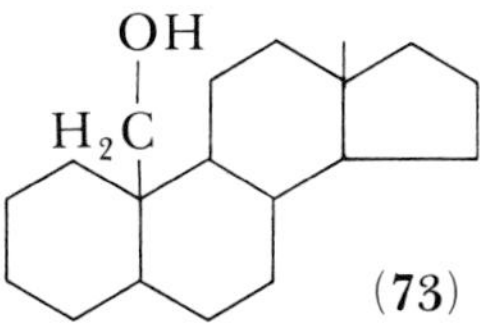

OH
|
H₂C

(73)

because of the high yields described in recent examples (70–80%), it offers a preparative production (Table 13).

With various *Clostridia* spp. (Goddard *et al.*, 1975; Goddard and Hill, 1972) or with *Escherichia coli* EB 417 (Goddard and Hill, 1972), 4-androstene-3,17-dione can be transformed into estradiol or 12-methoxyestradiol. These results, however, correspond to known human metabolic pathways and, because of very low yields, are unfortunately still far from being of preparative use.

Table 13

Transformations of 19-hydroxyandrostenes to 1,3,5(10)-estratrienes

Compound	Organism	Per cent conversion	Reference
17α-Ethinyl-3β,17β,19-trihydroxy-5-androstene	*Proactinomyces globerula*	71.0	Protiva and Schwartz (1974)
19-Hydroxy-4-androstene-3,17-dione	*Corynebacterium equi* *Arthrobacter ureafaciens*	up to 80.0	Shirasaka *et al.* (1972)
3β,19-Dihydroxy-5-androstene-17-one	*Corynebacterium equi* *Arthrobacter ureafaciens*		Shirasaka *et al.* (1972)
17α-Ethinyl-17β,19-dihydroxy-4-androstene-3-one	*Corynebacterium equi* *Arthrobacter ureafaciens*		Shirasaka *et al.* (1972)

H. 19-Nortestosterone Structures

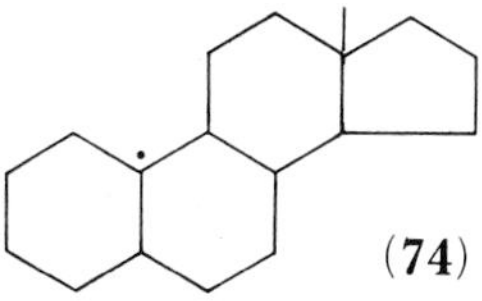

(74)

Table 14

Aromatization of 19-norandrostenes

Compound	Organism	Per cent conversion	Reference
7α-Methyl-19-nortestosterone	*Corynebacterium simplex* A.T.C.C. 6946 *Septomyxa affinis* A.T.C.C. 6737		Babcock and Campbell (1972)
3β-Hydroxy-19-nor-1(10),5-androstadien-17-one	*Proactinomyces globerula*	65	Protiva and Schwartz (1974)
17α-Ethinyl-3β,17β-dihydroxy-19-nor-1(10),5-androstadiene	*Proactinomyces globerula*	78	Protiva and Schwartz (1974)

The known aromatizations of 19-nor structures have been supplemented by a few examples (Table 14). This reaction is of interest only for metabolic research as 19-nor structures are produced industrially from ring-A aromatics by chemical methods.

19-Nor structures are frequently hydroxylated at the 10- position and, for this, a further example is described of transformation of 19-nortestosterone with *Rhizopus arrhizus* Fischer (byproduct 6β-hydroxy-19-nortestosterone; Favero *et al.*, 1977). 10-Hydroxy-19-nortestosterone can be hydroxylated only with difficulty at the 11β-position, as can be seen from the 2.7% yield with *Curvularia lunata* N.R.R.L. 2380 (Kulig and Smith, 1974). A 10-hydroperoxy-19-nortestosterone is reduced without hydroxylation only to 10-hydroxy-19-nortestosterone (Kulig and Smith, 1974).

A 15α-hydroxylation of 18-methyl-19-nor-4-androstene-3,17-dione with *Penicillium* spp. or *Fusarium* spp. (Petzoldt and Wiechert, 1974) acquired preparative significance for introducing a Δ^{15}-double bond. Introduction of this function, the progress of which is unsatisfactory by chemical means, leads to a significant increase in the contraceptive effect of corresponding 17α-ethinylnortestosterone structures (Hofmeister *et al.*, 1977).

Oestrogen, anti-oestrogen or contraceptive properties are ascribed to 1α- and 1β-hydroxylated 17α-ethinyl-19-nortestosterones (Greenspan *et al.*, 1974; Greenspan and Rees, 1975; Ambrus *et al.*, 1975). These compounds are difficult to synthesize chemically but are easily produced by microbiological hydroxylation. Norethisterone (**75**) is hydroxylated by *Botryodiplodia malorum* C.B.S. 1340 at the 1β-position (Greenspan *et al.*, 1974; Greenspan and Rees, 1975) and by *Acremonium strictum* or *A. kiliense* (Ambrus *et al.*, 1975) or *Calonectria decora* A.T.C.C. 14767 (Petzoldt and Elger, 1976) or, better, by *Mucor spinosus* C.B.S. 29563 (Petzoldt and Elger, 1977) at the 1α- position. While 1β-hydroxylation of norethisterone gives larger quantities of 11β-hydroxy compound as a byproduct, this side reaction is excluded using D-norgestrel (**76**) because of the increased 13-alkyl substitution (Greenspan *et al.*, 1974).

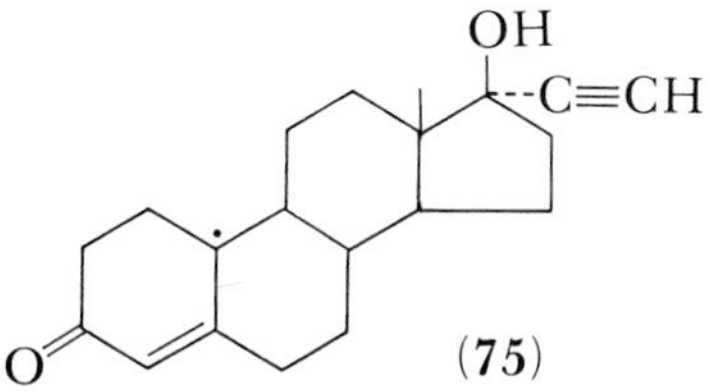

D-Norgestrel hydroxylated in the 1α position can also be formed by *Calonectria decora* (Petzoldt and Elger, 1976), but much better yields are achieved with *Aspergillus clavatus* A.T.C.C. 9598 (Petzoldt and Elger, 1976). *Mucor griseocyanus* hydroxylates D-norgestrel (**76**) at the 6β-position (Engelfried *et al.*, 1976).

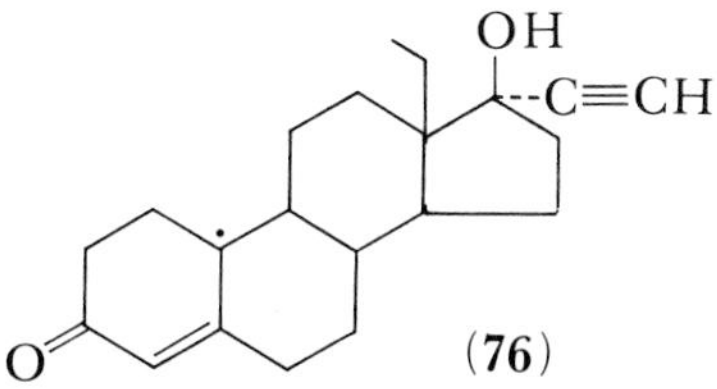

I. Estranes and Ethinylestradiols

Ring-A aromatic structures are often less suitable as substrates. Thus, 16α-hydroxylation of estrone (**77**) with *Streptomyces roseochromogenes* N.R.R.L.-B-1233 gives a yield of only 18% (Iida and Iizuka, 1976). Hydroxylation of *D*-13β-ethylgona-1,3,5(10)-triene-3,17β-diol (**78**) with *Bacillus megaterium* N.R.R.L.-B-938 yields 40% of the 16α-hydroxy derivative at a product (Buzby and Greenspan, 1973). Hydroxylation of the aromatic ring with *Aspergillus flavus* on 17α-methylestradiol at the 2- and 4- positions was detected, and the main product was the 6α-hydroxy derivative (Schubert *et al.*, 1971).

Using *Clostridium paraputrificum*, a 6-dehydro-estrone (**80**) can be dehydrogenated to equilenin (Babcock and Campbell, 1972). Equilenin can be hydroxylated by *Neurospora crassa* during reduction of the 17-keto group to the 12β-hydroxy compound (Maugras *et al.*,

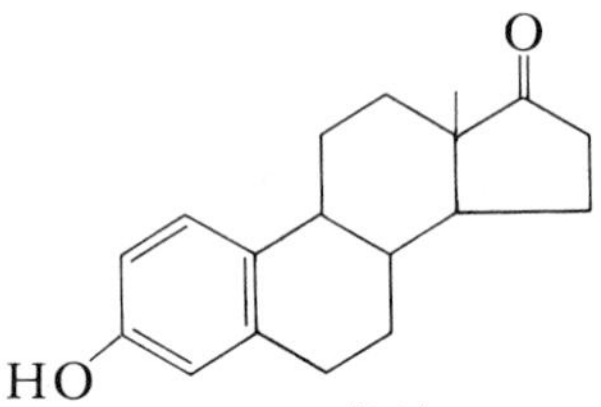

(77)

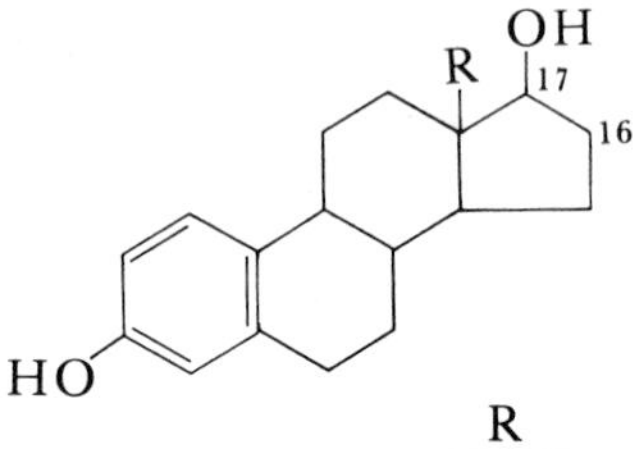

(78) $\overline{CH_3}$

(79) C_2H_5

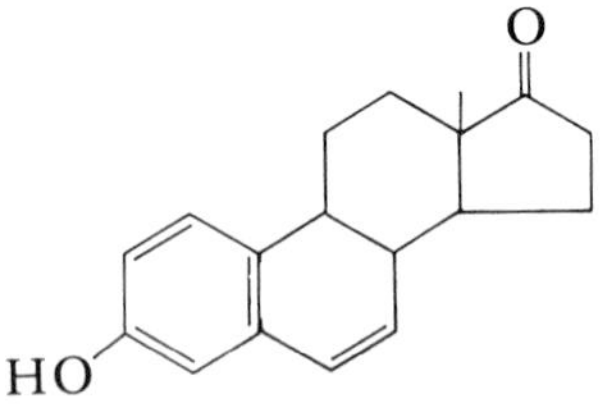

(80)

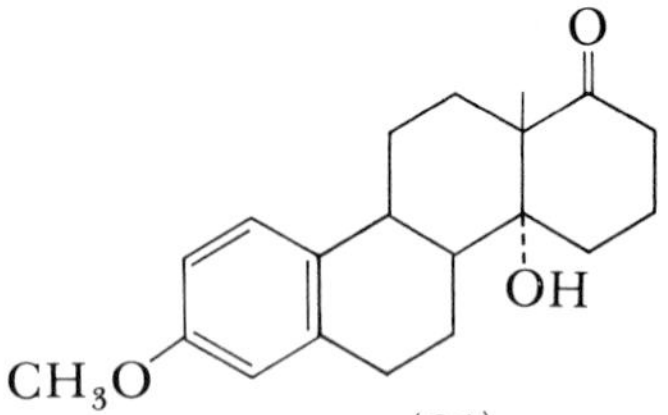

(81)

1973). A racemic mixture of 3-methoxyl-14α-hydroxy-D-homo-1,3,5(10)-estratrien-17-one (81) was separated into two diastereomeric 17β-alcohols with *Saccharomyces cerevisiae* or *Rhizopus arrhizus* (Eignerová and Procházka, 1974). However, conversion of estradiol- or estrone-3-methyl ether and of some of their derivatives by *Corynebacterium* sp., *Bacterium mycoides* or *Nocardia erythropolis* to give the free compounds (Schubert *et al.*, 1975a) could be interesting because of the mild conditions used, provided that yields of 30–50% could be increased further and that sensitive structures could also be formed in the same way. Formation of a β-glucoside of the 3-hydroxy-substituted compound by *Rhizopus* spp. is of equal interest and, to date, estradiol (78), estrone (77), 7-dehydroestradiol and equilin have been suggested as suitable substrates (Petzoldt *et al.*, 1974). A further preparative formation of glucoside is known on the 16α-hydroxy group of the estriol, the 17-epi-estriol, 16α-hydroxytestosterone and 16α-hydroxy-19-nortestosterone with *Aspergillus niger* N.R.R.L. 337 (Pan and Lerner, 1972). A vicinal 16α,17-diol structure is a requirement for glucoside formation.

Hydroxylations on 17α-ethinylestradiol (82) played a part in bringing about greater effectiveness. *Diplodia natalensis* A.T.C.C. 9055, and a few other fungi, hydroxylate at the 7α-position (Petzoldt *et al.*, 1976). The 6α- and 6β- hydroxy product can be obtained by insertion of the 3-methyl ether with *Penicillium chrysogenum* (Schubert *et al.*, 1975b). A 17α-ethinyl-8α-estra-1,3,5(10)-triene-3β,17-diol is hydroxylated by *Botryodiplodia malorum* (C.B.S. 13450) and a range of other fungi at the 6αβ- position (Petzoldt *et al.*, 1977a) and by *Glomerella cingulata* (A.T.C.C. 12097) at the 12α-position (Petzoldt *et al.*, 1977b).

The first microbiological hydroxylation of an aromatic steroid was achieved by transformation of 17α-ethinylestradiol (82) to 17α-ethinyl-10β,17β-dihydroxy-4-estren-3-one (83) with *Aspergillus flavus* (Schubert *et al.*, 1973, 1974). The reaction mechanism was made clear when the hypothetical intermediate, norethisterone (75), was transformed under the same conditions as the 10-hydroxy compound, and was demonstrated under anaerobic conditions, where 10-hydroxylation is absent.

J. 8,14-Seco-1,3,5(10),9(11)-estratetraenes

Further suitable yeast strains and reaction conditions are being described or patented for selective stereospecific reductions of 3-methoxy-8,14-seco-$\Delta^{1,3,5(10),9(11)}$-estratetraene-14,17-dione (84) on an industrial scale, the purpose of which is to introduce the first asymmetrical centre in total synthesis of 19-norsteroids. *Saccharomyces*

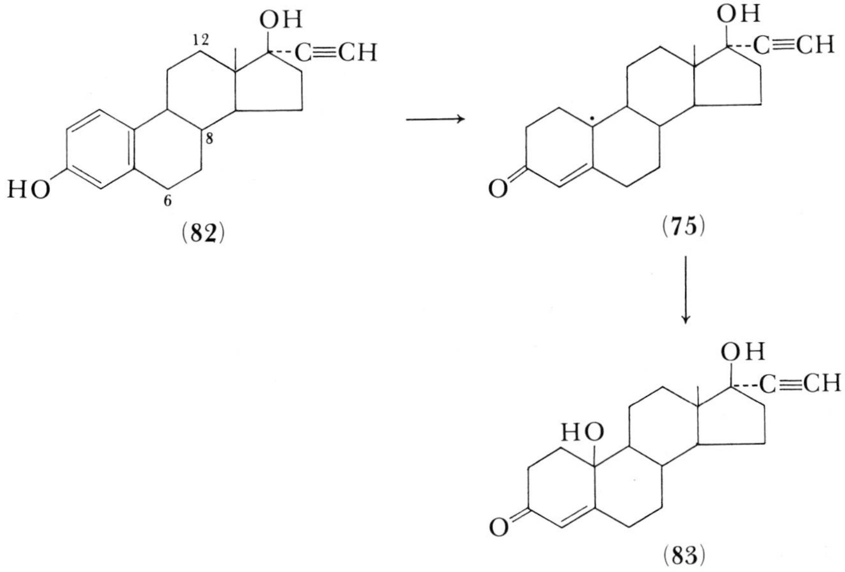

(82)

(75)

(83)

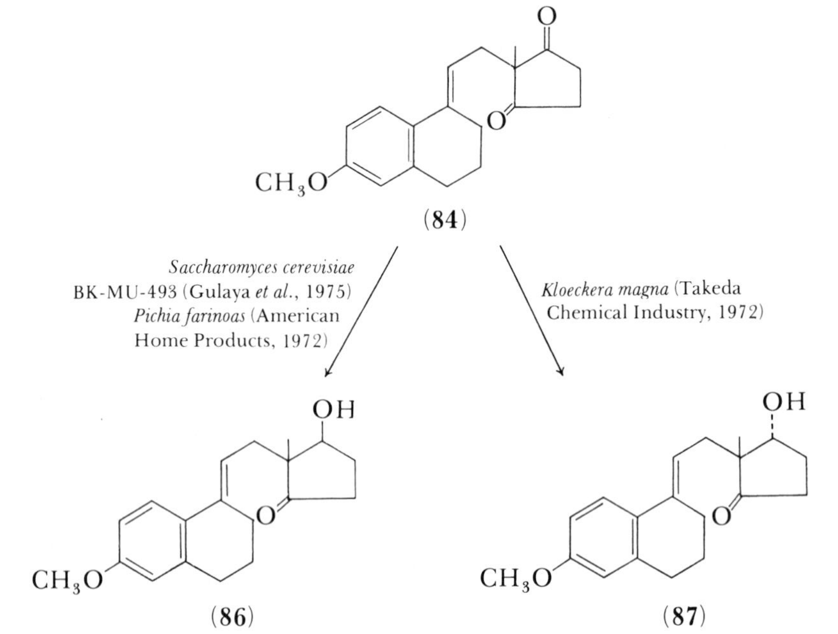
O
CH_3O
O
(84)
Saccharomyces cerevisiae
BK-MU-493 (Gulaya et al., 1975)
Pichia farinoas (American
Home Products, 1972)
Kloeckera magna (Takeda
Chemical Industry, 1972)
OH
O
CH_3O
(86)
OH
O
CH_3O
(87)

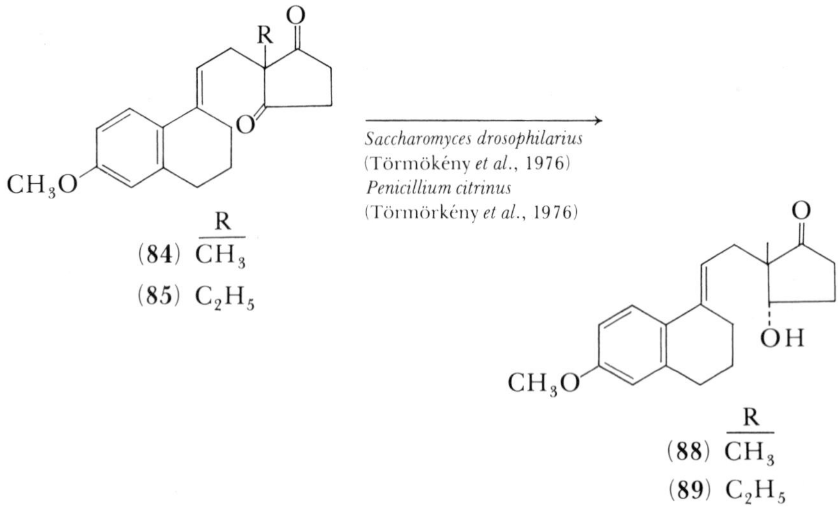
R
O
O
CH_3O
R
(84) CH_3
(85) C_2H_5
Saccharomyces drosophilarius
(Törmökény et al., 1976)
Penicillium citrinus
(Törmörkény et al., 1976)
O
OH
CH_3O
R
(88) CH_3
(89) C_2H_5

elongosporus BK My-426 effects an 11α-hydroxylation in addition to reducing the 14-keto group to a 14α-hydroxy structure. Known chemical techniques can then be used to effect ring formation and give a 9α,14α-oxide structure (Gulaya *et al.*, 1976).

IV. BILE ACIDS

Microbiological transformations of bile acids were reviewed in detail by Hayakawa (1973). In the last few years, these data have been broadened and deepened, and a new trend has developed with experimental work on degradation to 17-keto structures.

A. Oxidation of Hydroxyl Groups

Oxidation of cholic acid (McDonald *et al.*, 1975a) and glycocholic acid (McDonald *et al.*, 1973) or taurocholic acid (McDonald *et al.*, 1974) to 7-keto structures with cell-free extracts of *E. coli* was developed as a method for analytical determination of these compounds. The reaction is specific for the 7α-hydroxy group, since 3α, 12α and 7β-hydroxy groups are not attacked (Haslewood and Haslewood, 1976). However, additional 3α-hydroxysteroid hydrogenases are present in *Clostridium perfringens* (McDonald *et al.*, 1975b). The 7α-hydroxy group is not oxidized if sulphate-ester groups are present in the molecule (Haslewood and Haslewood, 1976) or if the side chain is shortened to fewer than four carbon atoms (Haslewood and Haslewood, 1976). *Eubacterium lentum* A.T.C.C. 25559 forms only 3α- and 12α-hydroxy-steroid dehydrogenases (McDonald *et al.*, 1977). Oxidation by cell suspensions of *Eubacterium aerofaciens* and *Bacteroides fragilis* can be increased a hundred-fold by addition of lysozyme (Hylemon and Stellwag, 1976). Various strains of *Bacteroides fragilis* and *Bacteroides thetaiotaomicron* oxidize chenodeoxycholic acids to the 7-keto compound even under anaerobic conditions (Edenharder *et al.*, 1976).

B. Dehydroxylations

Reductive elimination of the 7α-hydroxy group (Hayakawa, 1973) with anaerobic intestinal bacteria (Lewis and Gorbach, 1972; Lefebvre *et al.*, 1977) has been known for many years. With *Clostridium perfringens* (A.T.C.C. 19574), the 12α group of cholic acids can also be removed by reduction (Saltsman, 1976). Lithocholic acid, on the other hand, can only be converted into the ethyl ester during partial isomerization of the 3α-hydroxy group with a rat intestinal flora (Kelsey and Sexton, 1976).

C. Etio Acids and Analogues

Unlike the transformations described so far, which are only of interest as metabolic conversions, the following degradation reactions might possibly be useful in production of starting material for synthesizing unnatural steroid structures. 4-Androsten-3-one-17-carboxylic acid (90) can, like progesterone, be degraded with *Arthrobacter simplex* to 3(1β-carboxy-3aα-hexahydro-7aβ-methyl-5-oxoindan-4α-yl)-propionic acid (91) with 80% yield (Shionogi Co. Ltd., 1972; Hayakawa *et al.*, 1977). Similarly, 17α-hydroxy-4-androsten-3-one-17β-carboxylic acid (92) gives the corresponding degradation structure (93) (Hayakawa

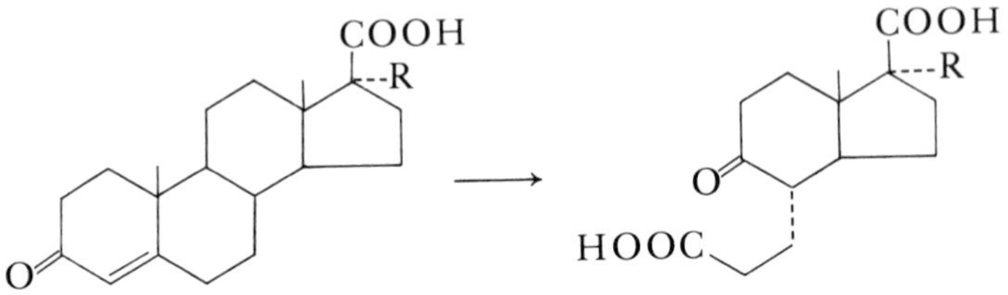

	R		R
(90)	H	**(91)**	H
(92)	OH	**(93)**	OH

et al., 1977). The corresponding *D*-homostructure, 17aα-hydroxy-3-oxo-*D*-homo-androst-4-ene-17aβ-carboxylic acid (94), however, can be hydroxylated with *Curvularia lunata* N.R.R.L. 2380 at 11β-position and then converted with *Arthrobacter simplex* into the anti-inflammatory 1,2-dehydrostructure (Alig *et al.*, 1976b). 23,24-Bisnorchol-4-en-3-on-22-ol (95) is dihydroxylated by *Rhizopus arrhizus* at 6β,11α-position (Smith *et al.*, 1973).

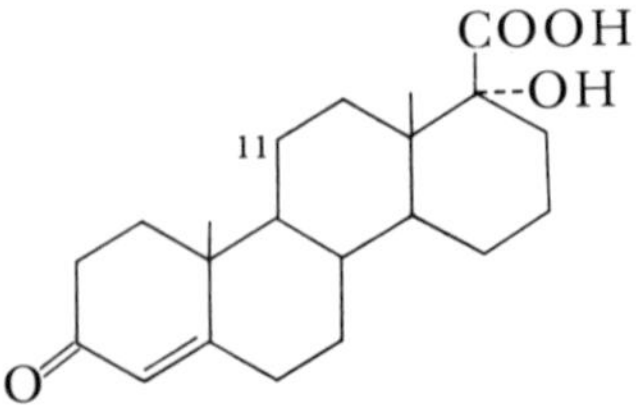

(94) 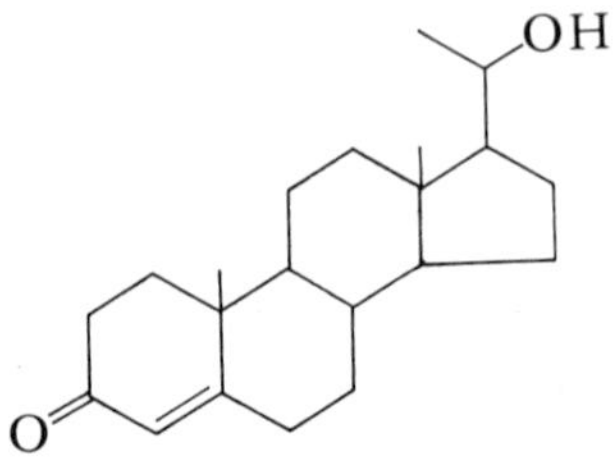   **(95)**

D. Degradation of Cholic Acid Structures

By analogy with the partial degradation of etio acids, cholic acids, too, can be used to produce 4α(2-carboxyethyl-5-oxo-7aβ-methyl-3aα-hexahydroindan)-1β-alkyl-carboxylic acids (97) by oxidative opening of the B ring with *Corynebacterium simplex*. Substrates with modified side-chain structures shown in Table 15 were used for this (Hayakawa *et al.*, 1972).

In 3-oxo-5β, 3-oxo-4-ene or 3β-hydroxy-5-ene structures, the oxidation follows the same course. 3-Oxo-24-nor-5β-cholan-23-oic

Table 15

Oxidative degradation of cholic acids

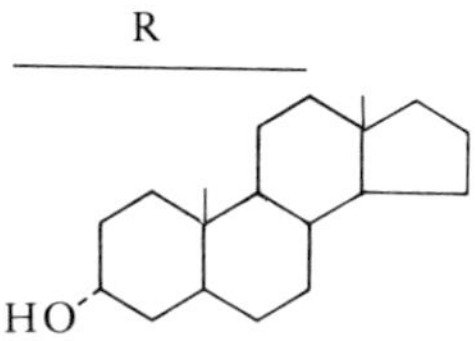

Substrates	Products
Lithocholic acid	
24-Norcholan acid	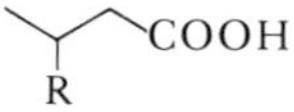
Bisnorcholan acid	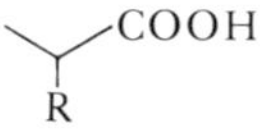
Pregnan-21-oate	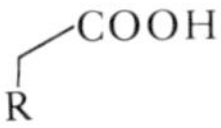
17α-Hydroxypregnane-20-carboxylate	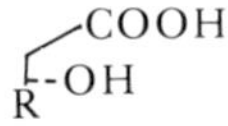
17α-Hydroxy-17β-carboxylate	

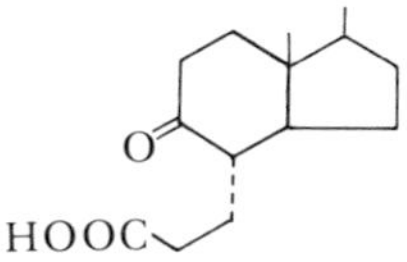

acid yields 70% and (20S)-3β-hydroxy-5-pregnene-20-carboxylic acid 50% of the corresponding acids (Hayakawa *et al.*, 1976a; Alig *et al.*, 1976b).

The metabolite (4R)-4-(4α-(2-carboxyethyl)-3aα-hexahydro-7aβ-methyl-5-oxoindan-1β-yl)valeric acid (98), which is not capable of further degradation by *Arthrobacter* (*Corynebacterium*) *simplex*, is transformed with *Corynebacterium equi* to conjugates of glycine, L-alanine, L-glutamic acid and glutamine (Hayakawa *et al.*, 1977). The substrate, however, is broken down further by this bacterium (Hayakawa and Fujiwara, 1977). Depending on the conditions of incubation and development, degradation products (99) and (100) or (101) and (102) as well as a monohydroxy- product were isolated from compound (98). The metabolic pathway following initial cleavage of ring B was the subject of parallel investigations on the analogous 17-keto structure, 3-(3aα-hexahydro-7aβ-methyl-1,5-dioxoindan-4α-yl)propionic acid (103) with *Streptomyces rubescens*. In addition to analogous (104) or similar structures (102), new degradation products (105–109) were isolated (Hashimoto and Hayakawa, 1977). In the end, metabolites of even simpler structure were demonstrated in the form of laevulinic acid (111) and bernstein acid (112) after degradation of double-labelled 7α-methyl,5,6,7,7α-tetrahydroindane-[5-¹⁴C]-1,5-dione-4-(3-propionic acid)-[1-¹⁴C] (10) (Schubert *et al.*, 1975c).

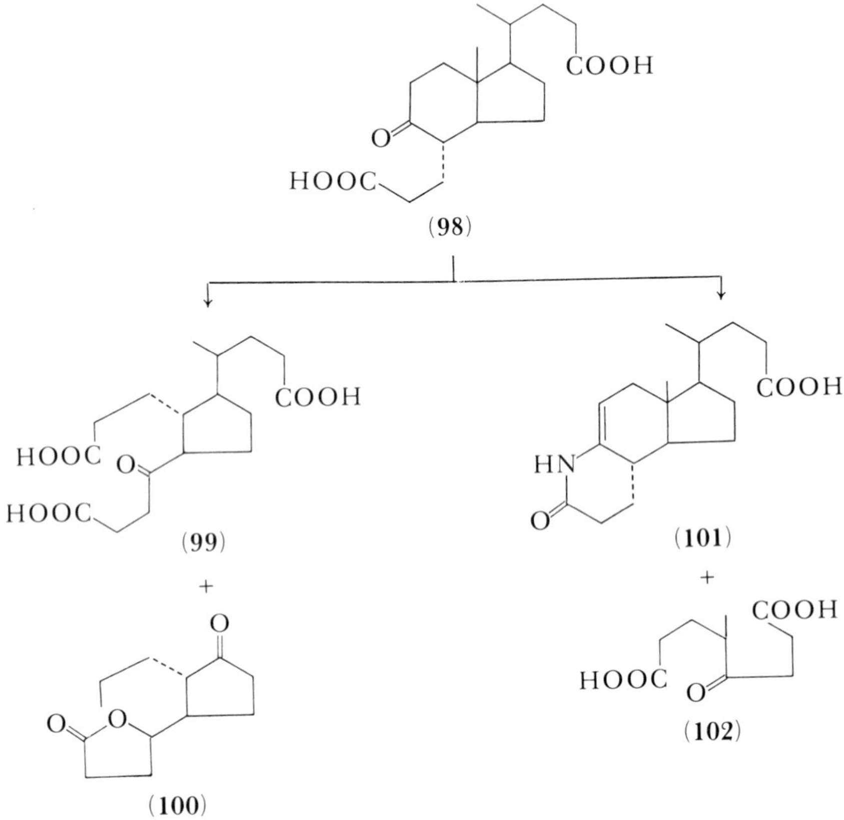

A second pathway for degradation of cholic acids is oxidative degradation of the side chain, leading to 1,4-androstadiene-3,17-dione

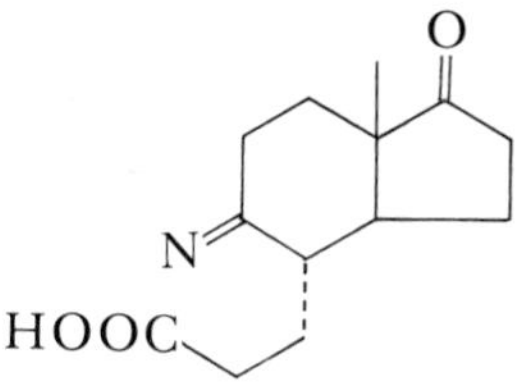

(103)

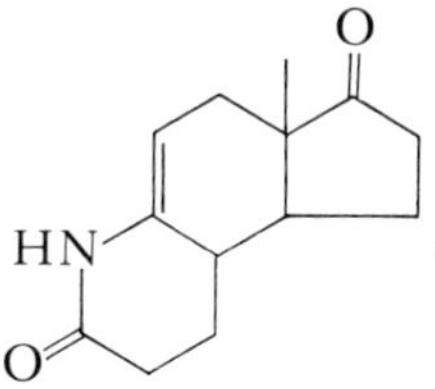

(104) 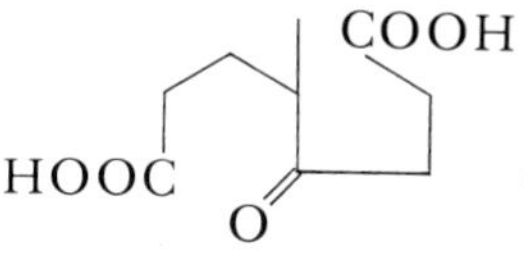 + (102) 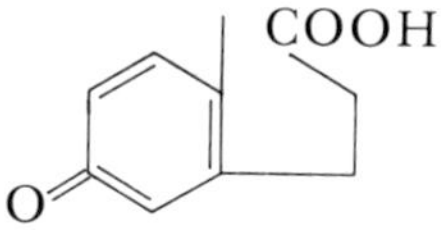 + (105)

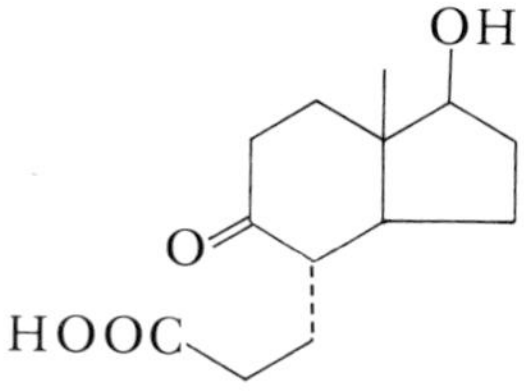

(106) 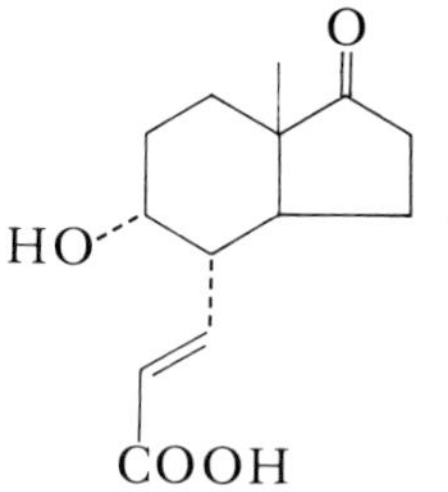 + (107) + (108) H
(109) OH

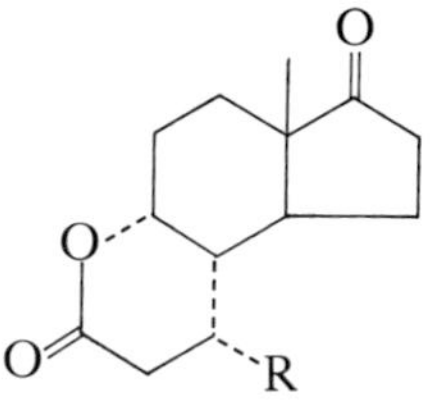

R

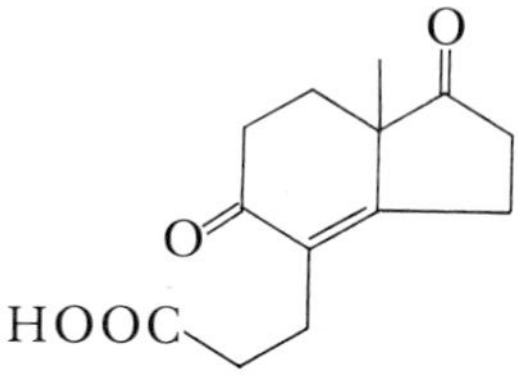

(110) $\xrightarrow{\textit{Nocardia opaca}}$ (111) 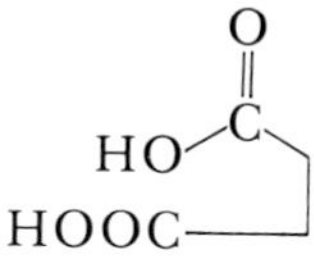 (112)

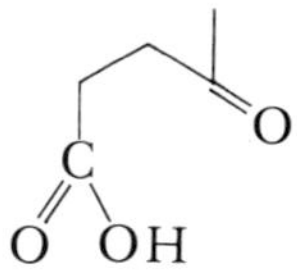

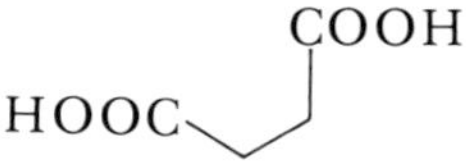

(112)

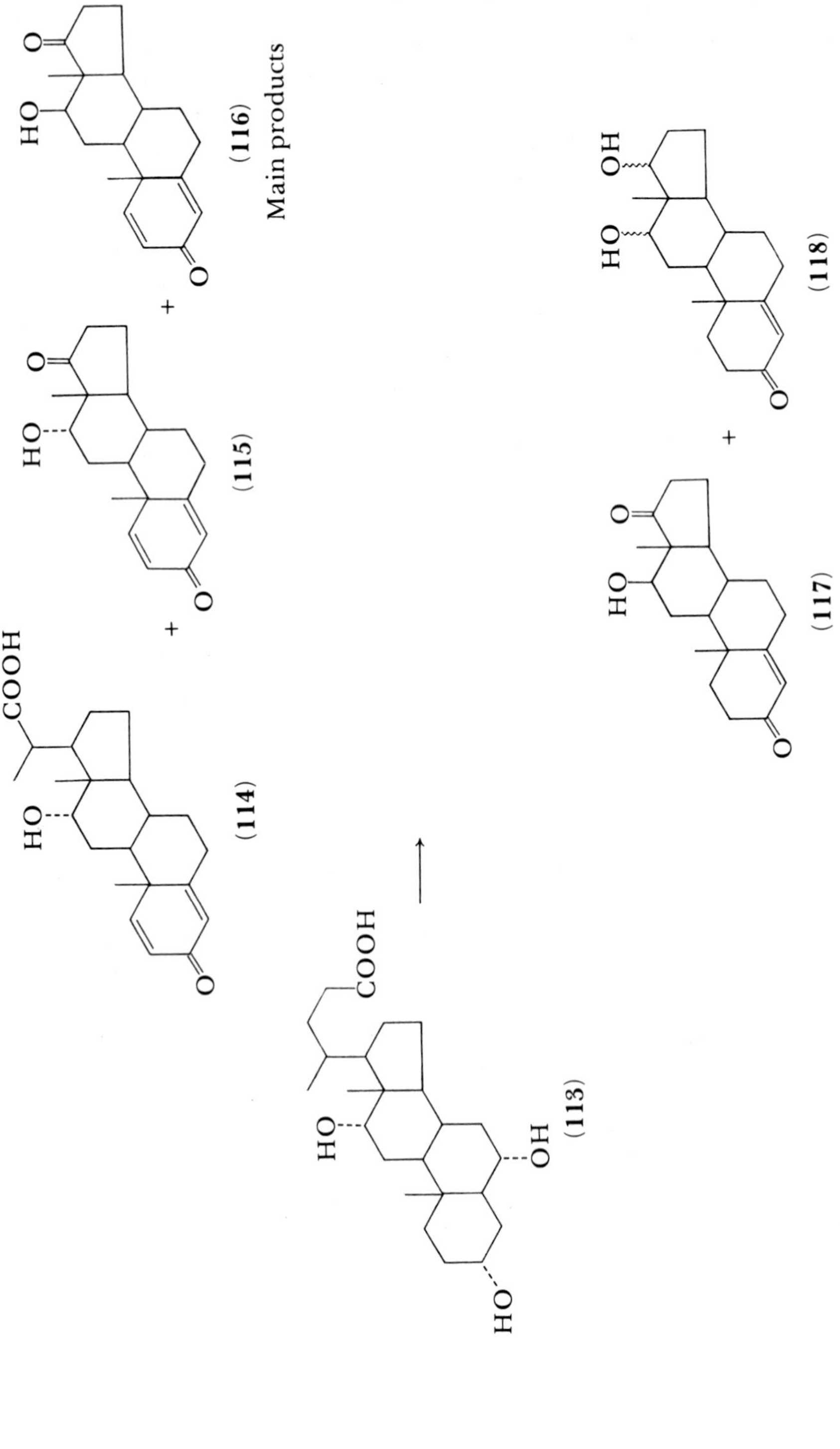

(116)
Main products

(115)

(114)

(113)

(118)

(117)

structures (Barnes *et al.*, 1975). A *Pseudomonas* strain, isolated from animal faeces, degrades lithocholic acid to the 1,4-androstadiene-3,17-dione, and desoxycholic acid (113) to the corresponding 12ξ-hydroxy compound (Barnes *et al.*, 1974). In the preparative transformation of desoxycholic acid (113), 12β-hydroxy-1,4-androstadiene-3,17-dione was obtained, surprisingly, as the main product and this must have arisen through epimerization of the 12α-hydroxy compound (115). The following byproducts were isolated: 12α-hydroxy-1,4-androstadiene-3,17-dione (115), 12β-hydroxy-4-androstenedione (117), 12,17-dihydroxy-1,4-androstadien-3-one (118) and the intermediary stage 12α-hydroxy-1,4-pregnadien-3-one-20-carboxylic acid (114) (Barnes *et al.*, 1976).

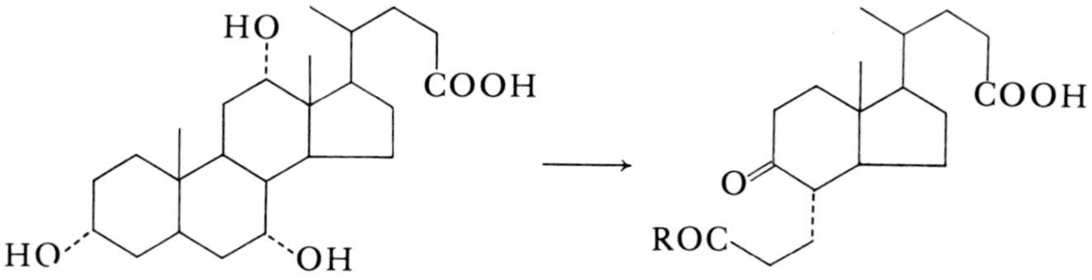

(119)

(120) OH

(121) NH₂

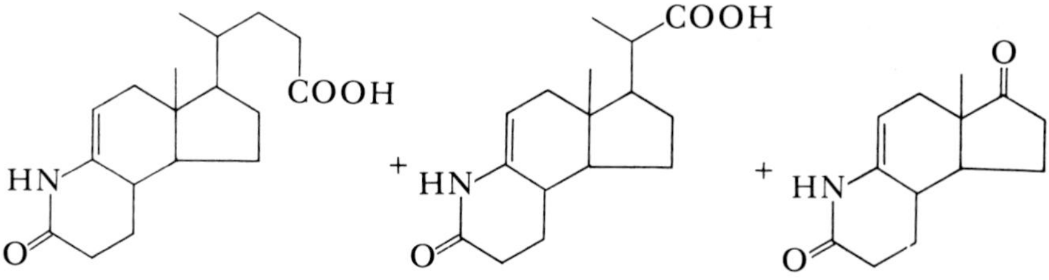

(122)   (123)   (124)

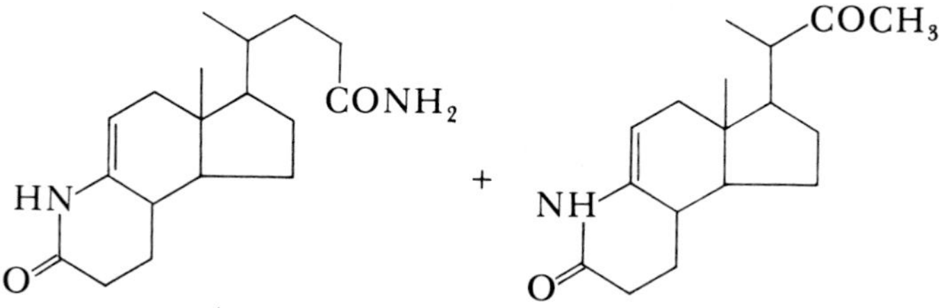

(125)   (126)

Simultaneous degradation of the side chain and steroid skeleton was observed and investigated during long-term incubations of cholic acid (119) with *Streptomyces rubescens* (Hayakawa *et al.*, 1976). The following compounds were isolated as metabolites: (4*R*)-4-[4α-(2-carboxyethyl)-3aα-hexahydro-7aβ-methyl-5-oxoindan-1β-yl]-valeric acid (120) and its mono-amide (121), and 2,3,4,6,6aβ,7,8,9bβ-decahydro-6aβ-methyl-1,4-cyclopenta[f]-quinoline 3,7-dione (124) and its homologues with the β-oriented side chains valeric acid (122), valeramide (125), butanone (126) and propionic acid (123). Unfortunately, however, all of these results are only of value in elucidating pathways for degradation of cholic acid structures, and they have found no practical application.

V. STEROLS

A. Saponification or Formation of Sterol Esters

Saponification of cholesterol esters is only of interest as the prescribed reaction for analytical determination of cholesterol by cholesterol-3-hydroxydehydrogenase. Processes for production of cholesterol ester hydrolase from *Pseudomonas fluorescens* (Sugiura *et al.*, 1976) A.T.C.C. 31156 (Terada and Uwajima, 1976), *Nocardia cholesterolicum* N.R.R.L. 5767 or 5768 (Masureka and Goodhue, 1976) and *Candida rugosa* A.T.C.C. 14380 (Boehringer Mannheim, 1974; Eastman Kodak, 1977b) have been patented.

Conversely, formation of 3-hydroxy esters of various sterols was demonstrated with *Staphylococcus epidermis*, propionibacteria (Puhvel, 1975), *Mycobacterium smegmatis* (Schubert *et al.*, 1969), *Pseudomonas* nov. sp. Nr. 109 (Nagase Sangyo, 1977) and with a cell-free extract from *Phycomyces blakesleeanus* (Bartlett *et al.*, 1974). Saponification and ester formation would be unlikely to be of any preparative value as opposed to chemical processes.

B. Hydrogenation of Cholesterol and Analogues

Eubacterium sp. A.T.C.C. 21048 hydrogenates cholesterol to coprostanol (Eyssen and Parmentier, 1974). Campestrol, β-sitosterol, stimasterol and 7-dehydrocholesterol all gave equally good yields of 5β-H-structures, whereas cholesterol esters, cholesterol halides, 5-cholestene and 3α-hydroxy-sterols were not attacked; 5α- and 5β-cholestan-3-one were reduced to the 3β-alcohols. Since a 4-double bond can also be hydrogenated, 4-cholesten-3β-ol and 4-cholesten-3-one (**127**) also produce coprostanol (**128**) (Eyssen *et al.*, 1973, 1974). From experiments with double-labelled substrates (Parmentier and Eyssen, 1974), the sequence 3β-OH Δ^5 3-ketoΔ^4 3-oxo-5β-H 3β-OH 5β-H is postulated. Because of the stereoselectivity of this hydrogenation, with good yields, the reaction may possibly have other applications.

The corresponding 5α-H structures can also be arrived at by means of microbiological hydrogenation, and *Nocardia corallina* A.T.C.C. 13259 has been described as the most suitable strain (Lefebvre *et al.*, 1974a). While this micro-organism dehydrogenates 4-androstene-3,17-dione at the 1-position in a medium rich in nitrogenous nutrients but with a low concentration of glucose, after induction with progesterone, the same substrate is hydrogenated with 32% yield to

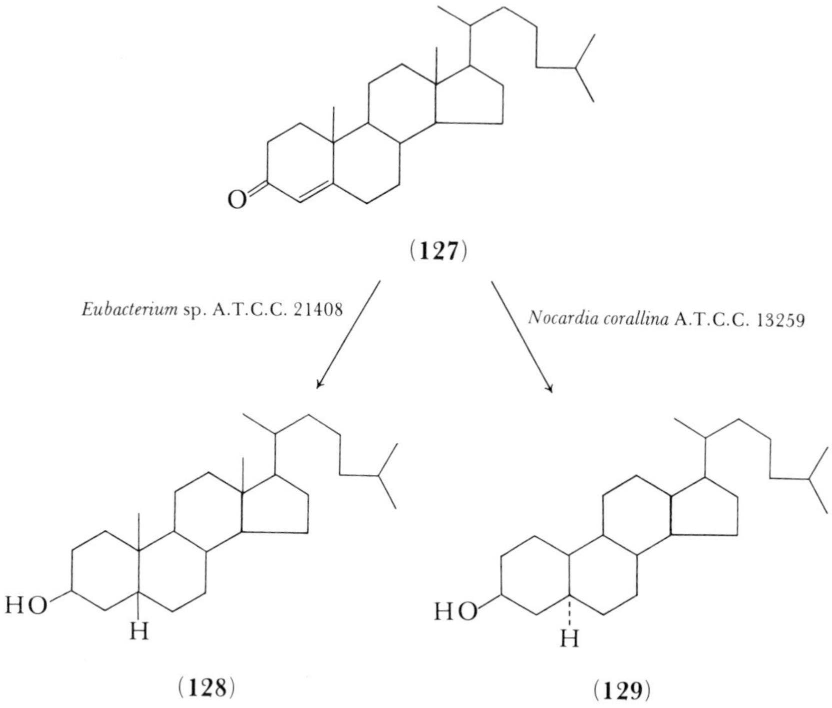

(**127**)

Eubacterium sp. A.T.C.C. 21408

Nocardia corallina A.T.C.C. 13259

(**128**)

(**129**)

5α-androstane-3,17-dione without steroid induction in a glucose-rich medium containing a low concentration of nitrogenous nutrients. Under the same conditions, 5α-cholestan-3β-ol (**129**) is obtained from 4-cholesten-3-one (**127**), whereas 3,5-cholestadien-7-one and 4-cholestene-3,6-dione are not attacked. *Nocardia opaca* A.T.C.C. 4176 gives similar results (Lefebvre *et al.*, 1974b).

C. Oxidation of Cholesterol to Cholestenone and Accompanying Reactions

Oxidation of the 3β-hydroxy group is used analytically to detect cholesterol in body fluids. For this reaction, corresponding cholesterol oxidase preparations were produced. Starting sources are *Nocardia erythropolis* A.T.C.C. 17895 and A.T.C.C. 4277 (Bergmeyer *et al.*, 1974b), *Nocardia formica* A.T.C.C. 14811 and *Proactinomyces* N.C.I.B. 9158 (Boehringer Mannheim 1975; Gruber *et al.*, 1975) and A.T.C.C. 17895 (Bergmeyer *et al.*, 1974a) and other strains of these species (Whitehead and Flegg, 1974), *Nocardia rhodochorus* N.C.I.B. 10544 (Buckland *et al.*, 1975, 1976) and N.C.I.B. 10555 (Richmond, 1973), *Nocardia cholesteriolicum* N.R.R.L. 5762 and N.R.R.L. 5768 (Masureka and Goodhue, 1976; Goodhue and Hugh, 1975; Eastman Kodak, 1977a), *Corynebacterium cholesterolicum* Nr. 3134 (Kikkoman Shoyu, 1977), *Arthrobacter simplex* I.F.O. 12069 (Nagase Sangyo, 1976), *Schizophyllum commune* I.F.O. 4928 (Sugiura *et al.*, 1975), *Streptomyces griseocarnus* (Kerényi *et al.*, 1975), *Brevibacterium sterolicum* A.T.C.C. 21387 (Uwajima *et al.*, 1973, 1974; Kyowa Hakko Kogyo, 1975; Terada *et al.*, 1974) or *Streptomyces violascens* (Fukuda *et al.*, 1973). In addition, 3-oxo-cholestan-Δ⁵-Δ⁴-isomerase was separated from *Nocardia erythropolis* (Roeschlau *et al.*, 1975).

Transformation of cholesterol to cholestenone with *Nocardia rhodochrous* N.C.I.B. 10555 can only be carried out, curiously enough, in the presence of large quantities of a water-immiscible solvent. If cell pastes are used, after being stored at −20°C and then thawed out, 7 g of cholestenone/hour can be obtained with 100 g of cells in 200 ml of an 8% solution of cholesterol in carbon tetrachloride (Buckland *et al.*, 1976). As this micro-organism also oxidizes pregnenolone, 17α-hydroxypregnenolone and dehydroepiandrosterone to the 3-oxo-Δ⁴ structures, and testosterone and estradiol to the 17-ketones, it offers a relatively simple process for preparative oxidation of 3β- and 17β-alcohols (Lilly et al., 1976b). For an analogous oxidation of 19-hydroxy-β-sitosterol to the corresponding sitostenone, *Corynebacterium hydrocarboclastus* (FERM-P-2487) has been recommended (Iizuka, 1974).

When incubating cholesterol with *Coriolus hirsutus* (I.F.O. 497), the intially formed 7α- and 7β-hydroxy compounds are oxidized to the 7-ketone (Wada and Ishida, 1974). On the other hand, an alcohol dehydrogenase oxidizes 5β-cholestane-$3\alpha,7\alpha,12\alpha,26$-tetrol only to $3\alpha,7\alpha,12\alpha$-trihydroxy-5β-cholestanoic acid (Björkhem et al., 1973).

D. Degradation of Cholesterol and Native Sterols

It has been known for 65 years that mycobacteria can utilize cholesterol and phytosterol as carbon and energy source (Söhngen, 1913). Subsequently, this property was shown to be shared with a number of species as norcadiae, corynebacteria, pseudomonads and streptomycetes (Martin, 1977). During the last five years, the list has been extended to further strains of mycobacteria and related species (Imshenetskii and Mavrina, 1972a, b; Imshenetskii et al., 1975) as well as of *Aeromonas* sp. (Voets and Lamot, 1974), agarbacteria and pseudomonads (Kozlova and Fonina, 1972). A cell-free extract with high degradative activity has already been obtained from *Streptomyces lavendulae* (Imshenetskii et al., 1976).

Total degradation of sterols is of no biotechnological interest. However, oxidative degradation of the side chain to C_{19} steroids is of outstanding importance as, by this process, low-priced sterols may be utilized as new starting compounds for production of suitable primary products in steroid synthesis. However, side-chain degradation is in competition with cleavage of ring 3 and steroid skeleton degradation. As degradation of the steroid skeleton basically is initiated by 9-hydroxylation, it is necessary to check that this hydroxylation does not affect desired oxidizing enzymes. The following methods have been known for some time. (a) Chemical alteration of the substrate to a structure that cannot be affected for steric reasons (Eder et al., 1977) or because of lack (Koninklijke Nederlandsche Gist-en Spiritusfabriek, 1966, 1972a, b) or blocking of the 3-hydroxy group for the initial 3-keto formation in the 9-position. (b) Inhibition of 9-hydroxylation by displacement of the central Fe^{2+} atom by Ni^{2+} or CO^{2+} (van der

Waard *et al.*, 1968; Koninklijke Nederlandsche Gist-en Spiritusfabriek, 1967, 1968). (c) Inactivation of 9-hydroxylation by removal of the central atom with a chelating agent, such as α,α-dipyridyl or 8-hydroxyquinoline (Wix *et al.*, 1968; Nagasawa *et al.*, 1969). (d) Mutation of the micro-organism (Marsheck and Krachy, 1972).

With addition of αα-dipyridyl as the preferred chelating agent, newer processes were patented. With *Arthrobacter simplex* I.A.M. 1660 or *Nocardia corallina* I.F.O. 3338, a yield of 31% was achieved using a substrate concentration of 2 g of cholesterol/l, and adding antibiotics (Kanebo, 1974a). A technique using mixed cultures of both strains is described as favourable (Kanebo, 1974b). Special processes for building up populations of these mixed fermentations, by adding calcium chloride (Kanebo, 1974c) to refine the culture broth, or acidifying to pH 2.0 in order to precipitate 1,4-androstadiene-3,17-dione (**134**) (Kanebo, 1974d), resulted in much smaller yields of the isolated pure product. In mixed fermentations of *Arthrobacter simplex* I.F.O. 12069 and *Pseudomonas aeruginosa* I.F.O. 3447, 55% of the substrate can be recovered as 4-cholesten-3-one (**127**) in addition to a yield of over 30% 1,4-androstadiene-3,17-dione (**134**) and 3% androstene-dione (Kanebo, 1974e). In all of the processes quoted, substrate suspensions in sorbitan and sesquioleate or polyoxyethylene sorbitan were used.

Cholesterol sulphate (Kanebo, 1975a) or cholesterol acetate (Kanebo, 1975b) are also degraded to 1,4-androstadiene-3,17-dione with *Arthrobacter simplex* I.A.M. 1660 alone (Arima *et al.*, 1968) or in mixed culture with *Nocardia corallina* but with poorer yields. By contrast, a 3-hydroxy group protected by ether formation proved stable, and the corresponding dehydro-epi-androsteron-3-alkyl ether was obtained (Eder *et al.*, 1977). All of these processes, as well as the one originally patented, failed to achieve satisfactory yields and returns (Wix *et al.*, 1968).

It was a variant, patented barely five years ago (Nishikawa *et al.*, 1975), of the fermentation with *Arthrobacter simplex* (I.A.M. 1660), *Brevibacterium lipolyticum* (I.A.M. 1398) or *Mycobacterium phlei* (I.F.O. 3158) that increased yields substantially to over 50% 1,4-androstadiene-3,17-dione using substrate concentrations of 3 to 4 g/l and contact times of 100 to 110 hours and with addition of approximately 0.01–0.02% 2,2-dipyridyl, 8-hydroxyquinoline or 1,10-phenanthroline. The variation consists in addition of glycerides, fats, oilseeds or oil fruits at concentrations of 0.5 to 2%. Of the oils tested, linseed or soya oil were among the most effective, although experiments with rape oil, soybean oil, groundnut oil and others, or with ground seeds of various plants combined, also worked well. When fatty-acyl salts (C_{12} to C_{18}) (Mitsubishi Chemicals Industries, 1977a) or molasses ash (Mitsubishi Chemical Industries, 1977c) are added, 17β-hydroxy-androsta-1,4-dien-3-one as well as 1,4-androstadiene-3,17-dione can be obtained from sterol degradation.

Chelating agents occasionally display adverse effects on further development and stability of the culture, even when they are added to fully grown cultures after about 30 hours. The toxic effect was diminished by adsorption of the inhibitor, which was partially bound to added Amberlite XAD-2 or XAD-4 (Martin, 1977; Martin and Wagner, 1976a, b). By adding the ion-exchanger 1,4-androstadiene-3,17-dione, yields are raised to 50–60% because the concentration of the intermediate is lowered and this increases inhibition of end-product formation. 3-Oxo-23,24-dinor-1,4-choladien acid and 3-oxo-23,24-dinor-1,4-choladien acid methyl ester were successfully isolated as intermediary products, and occasionally reached concentrations of 30% (Martin, 1977). When the preferred styrene divinyl benzene resins were used, only 1,4-andro (**133**) and 1,4-androstadiene-3,17-dione (**134**) were selectively bound, whereas the starting material and the byproducts were not adsorbed (Mitsubishi Chemical Industries, 1977b).

It is much more advantageous to use stable mutants rather than the processes just described, and a methodology has been developed for their production and selection (Cargile and McChesney, 1974; Wovcha and Biggs, 1977). The first promising examples were discovered with mutants of *Mycobacterium* sp. N.R.R.L. 3683 (Krachy *et al.*, 1972) or N.R.R.L. 3805 (Marsheck and Krachy, 1979). The last-named has only slight dehydrogenase activity so that only 4-androstene-3,17-dione is formed. Mixtures of 1,4-androstadiene-3,17-dione (**134**) 1,4-androstadiene (**133**) are obtained with the mutants *Mycobacterium fortuitum* N.R.R.L. 8153 and N.R.R.L. 8154 (Wovcha and Biggs, 1978). Transformation of the added sterols unfortunately remained incomplete at first, despite a long fermentation period lasting up to 240 hours during which 20α-hydroxymethyl-1,4-pregnadien-3-one (**135**) appeared us an intermediate product (Marscheck and Krachy, 1973; Wovcha and Biggs, 1978). Conversion of 1,4-androstadiene-3,17-

dione (1 g/l) to 85% in 110 hours has been described with strain N.R.R.L. 3683 during fermentation in a medium consisting of mineral salts and 20 g of yeast extract/l (Lilly *et al.*, 1976a).

Initial investigations with different substrates (Nagasawa *et al.*, 1970) led to the conclusion that high yields of 17-keto steroids can only be obtained from cholesterol. Indeed stigmasterol (Marsheck and Krachy, 1972, 1973; Krachy *et al.*, 1972) or various stigmasterones (Marscheck and Krachy, 1972; Krachy *et al.*, 1972), as well as the notably inexpensive sitosterol (Kanebo, 1974c; Martin and Wagner, 1976a, b; Krachy *et al.*, 1972; Marscheck and Krachy, 1972, 1973) or sitosterol (132)/campesterol (131) mixtures, were successfully degraded to 17-keto steroids. Even a mixture of campestanol (1%), campesterol (8%), stigmasterol (4%) and sitosterol (72%), which is present in tall oil, can be used as an economical starting substrate (Conner *et al.*, 1976).

The reaction mechanism of sterol degradation has been understood for some time (Sih and Whitlock, 1968). Supplementary work on the epimers 2- and 4-deuterocholesterolene (Nambara *et al.*, 1975a) with *Arthrobacter simplex* and on three ^{3}H and ^{14}C-double-labelled cholesterols (Philips and Ross, 1974) with *Mycobacterium phlei* has shown that, at the stage of primary oxidation to cholestenone (Nambara *et al.*, 1975a), the 4β-atom is eliminated while, at the later 1-dehydrogenation stage (Nambara *et al.*, 1975a; Philips and Ross, 1974), the 1α- and 2β-H atoms are eliminated.

With new mutants, e.g. N.R.R.L.-B-8119 from *Mycobacterium fortuitum*, it is even possible to obtain accumulation of 9α-hydroxy-4-androstene-3,17-dione (137) by incubating sitosterol (132), cholesterol (130), stigmasterol or campesterol (131) (Wovcha, 1977).

A further product, isolated in small quantities, is 9α-hydroxy-3-oxo-bisnorchol-4-en-22-oic acid (136) (Antosz *et al.*, 1977). This confirms the assumption that degradation of the steroid skeleton begins with a formation of a 9α-hydroxy-3-oxo-1,4-diene structure. Consequently, the undesirable cleavage of the B-ring can be suppressed by inhibiting the 1-dehydrogenation. Finally, a 3-oxo-pregna-4,17(20)-diene-21-carboxylic acid can be assumed to be a further intermediate in degradation of the side chain with *Mycobacterium* sp. N.R.R.L.-B-8054. This compound presents partly as a methyl ester (139) (Jiu and Marsheck, 1976, 1977).

Next to 1,4-androstadiene (133) and 1,4-androstadiene-3,17-dione (134), 9α-hydroxy-1,4-androstadiene (137) is the most interesting

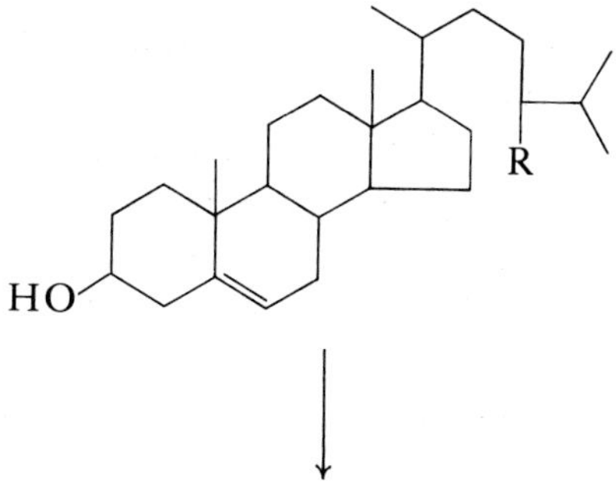

	R
(130)	H
(131)	CH_3
(132)	C_2H_5

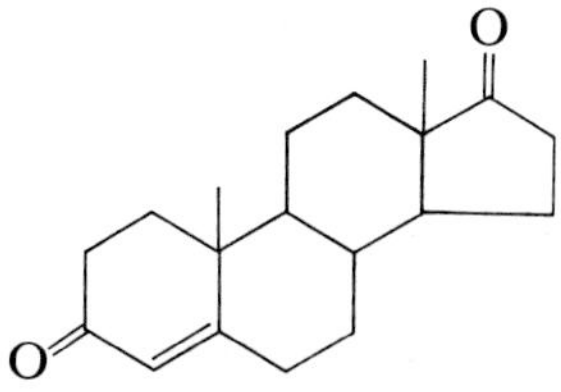

(133)

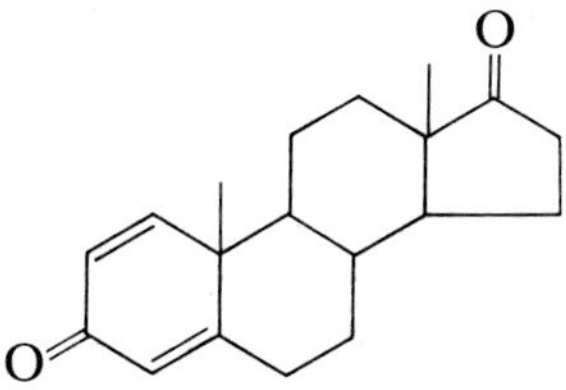

(134)

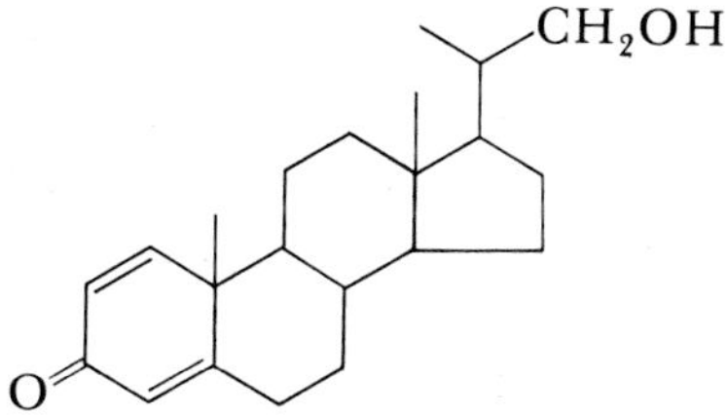

CH$_2$OH

(135)

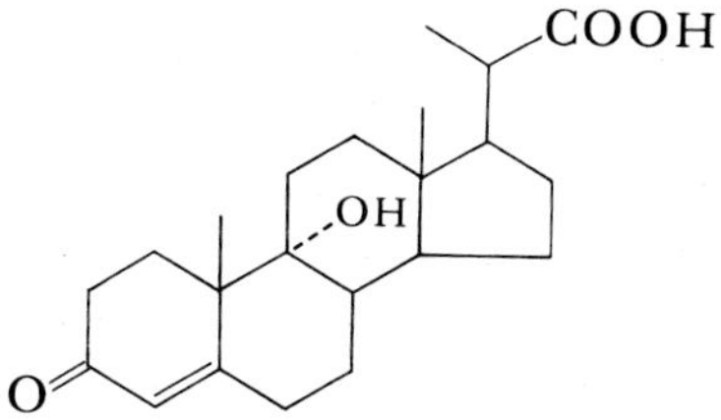

COOH

OH

(136)

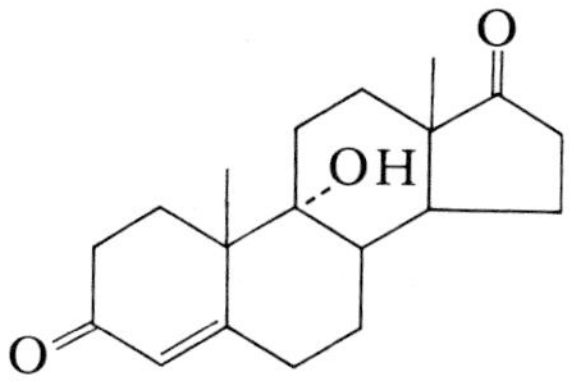

OH

(137)

COOR

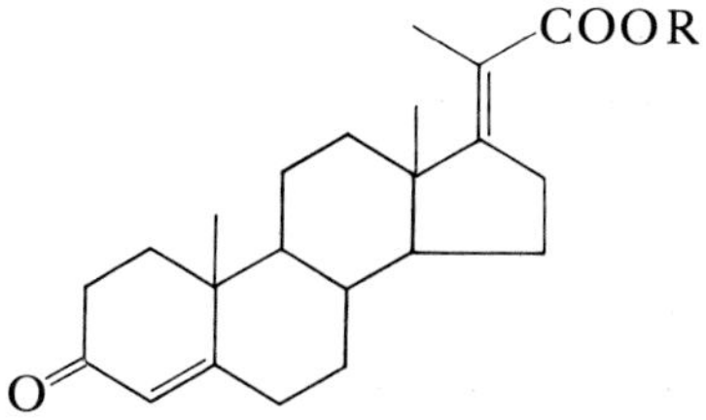

	R
(138)	H
(139)	CH_3

breakdown product from a technical point of view. Since the
9α-hydroxy group can be dehydrogenated to give a $\Delta^{9,11}$-double bond,
and since several possibilities are known for chemical construction of

the 17α-hydroxyprogesterone side chain, the compound represents a cheap starting material for synthesizing 9α-halogen-substituted corticoids without a microbiological 11-hydroxylation. While 1,4-androstadiene (133) serves as starting product for production of male and oestrogen sex hormones, 1,4-androstadiene-3,17-dione (134) can be aromatized to estrone by known pyrolysis procedures.

A mutant (N.R.R.L.-B-8128) was used to produce further mutations of the mother strain of *M. fortuitum* A.T.C.C. 6842. This mutant breaks down sterol substrates such as 1,4-androstadiene and its 3,17-dione, dehydroepiandrosterone and testosterone to the 7aβ-methyl-hexahydro-1(5)-indan-one structures (140 to 143) (Knight and Wovcha, 1977a, b).

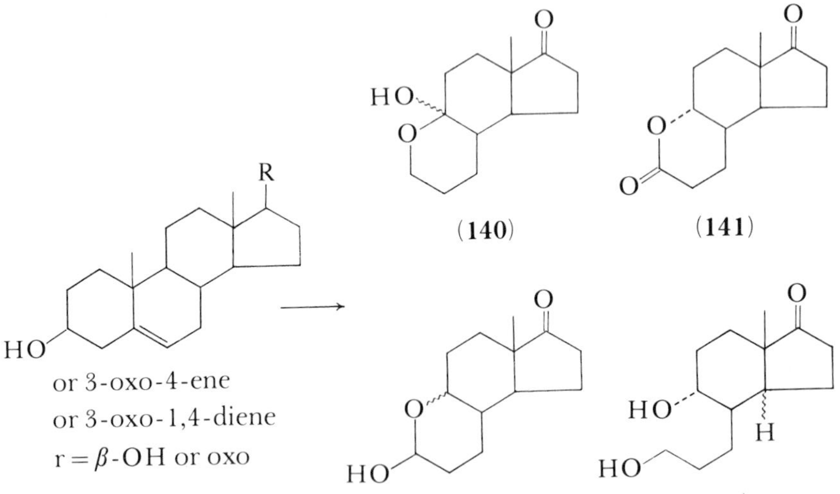

(140)   (141)

R

HO

or 3-oxo-4-ene
or 3-oxo-1,4-diene
r = β-OH or oxo

(142)   (143)

Transformation of cholesterol to pregnenolone by way of the intermediate 20S-hydroxy- and 20R,22R-dihydroxy- structures (van Liev and Rousseau, 1976; Morisaka *et al.*, 1976), known in human metabolism, constitutes a fundamentally different breakdown mechanism, which has not so far been detected in micro-organisms.

E. Degradation of Various Sterol Structures

Various substituted sterols can be degraded in a similar way to 17-keto structures. Suitable substrates for *Mycobacterium* sp. N.R.R.L.-B-3805 were the 6α-fluoro-, 6α-methyl- and 7α-methyl-4-cholesten-3-one (Weber et al., 1977a), 1α-methyl- and 1α,2α-methylene-5α-cholestan-3-one (Weber et al., 1977b) and 1α-methyl-4-cholesten-3-one derivatives as well as 1α,2α-methylene-4,6-cholestadien-3-one (Weber et al., 1977c). Since the substituents can generally be introduced more successfully in the 17-keto structures, these processes are of only limited interest. However, it is desirable to obtain the 3β-hydroxy-5-ene system during degradation, and this is achieved by etherization and ketalization to 3-methoxy-, 3-ethoxy-, 3,3-ethylenedioxy- or 2,2-dimethylenepropylene-dioxy- cholesterols or stigmasterols (Eder et al., 1977) or by acetal formation as 3β-methoxymethoxy groups (Weber et al., 1978). Incubation of 4-hydroxy-4-cholestene-3-one with Mycobacterium phlei surprisingly yields 3β- and 3α-hydroxy-androstane-4,17-dione as well as the further reduced compound 3β,4α-di-hydroxy-5α-androstan-17-one (Törmörkény et al., 1975). Equally unexpected, but without practical significance, was the isolation of a 9,11-dehydroesterone as byproduct of the known transformation to estrone of 19-hydroxy cholesterine with Corynebacterium sp. N.R.R.L.-B-3931 (Mallett, 1973).

In addition to the 17-keto structures, corresponding structures with a progesterone side chain were obtained from some insect moulting hormones, such as crustecdyson (144), makisterone (145) and muristerone (148) with Rhizopus arrhizus (A.T.C.C. 11145), R. nigricans (A.T.C.C. 6261) or Curvularia lunata (A.T.C.C. 12017) (Canonica et al., 1974). Hitherto, degradation products of the pregnane type could only be obtained from sterol side chains with a 20-hydroxy group.

Crustecdyson (144) gives a 72% yield of the pregnane-20-one structure poststerone (146), as well as 3% of the 17-keto compound rubrosterone (147) (Canonica et al., 1974). Rubrosterone itself can be obtained in 15% yield by oxidation of ponasterone A (149) with Fusarium lini A.T.C.C. 9593, whereas crustecdyson (144) only yields 5% of the 17-keto compound with this strain. α-Ecdyson, with the hydroxyl group missing, is not degraded (Tom et al., 1975).

Fucosterol and isofucosterol (152) are converted by Nocardia restrictus A.T.C.C. 14887 only into fucostenone and isofucostenone. 1,4-Androstadiene-3,17-dione can be detected only in traces during fermentation with additional chelating agents (Soliman, 1975).

Transformations of ergosterol and various similar compounds with Sacch. cerevisiae (Freyberg et al., 1973; Neals and Parks, 1977; Anding et

	R
(**144**)	H
(**145**)	CH$_3$

(**146**)   +   (**147**)

(**148**)   (**149**)

(**150**)

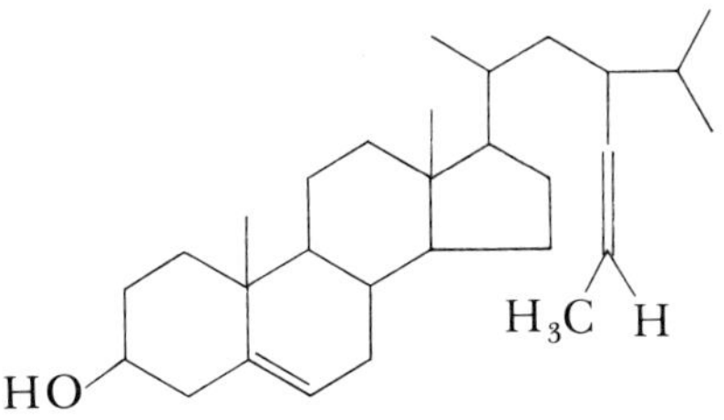

H_3C H

(151)

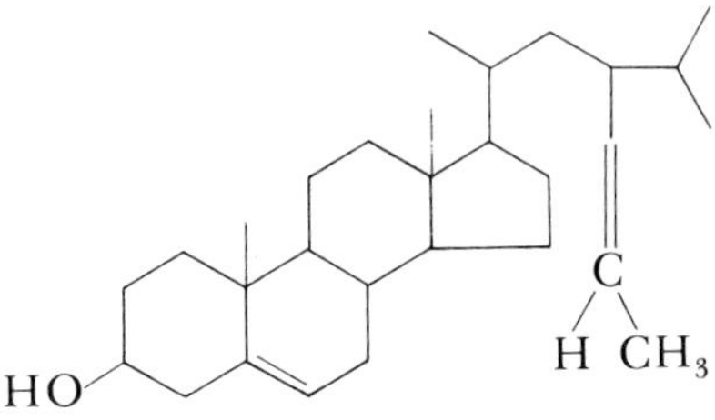

(152)

al., 1974), *Mucor* spp. (Atherton *et al.*, 1972), *Penicillium rubrum*, *Gibberella fujikuroi* (Bates *et al.*, 1976) and with the algae *Ochromonas malhamensis* (Knapp *et al.*, 1977) and *Trebouxia* spp. (Largeau *et al.*, 1977) lie in the field of biosynthesis studies and have no practical significance.

VI. SAPOGENINS AND CARDENOLIDS

A. Diosgenin and Analogues

As with sterols, so with sapogenins, an oxidative attack on the steroid skeleton at the 9- position with degradation of the A and B rings is possible, as is degradation of the E and F rings. Oxidation of diosgenin (153) with an unidentified bacterium leads to (25R)-9-oxo-A-B-spiro-stane-8α-propionic acid (154), while *Nocardia globula*, in addition, forms small quantities of 17α-hydroxy derivative (Howe *et al.*, 1973).

Hecogenin is converted, surprisingly, into the same keto acid with loss of the 12-keto function, and the corresponding 12α-hydroxy structure is in evidence as a byproduct (Howe *et al.*, 1973).

With *Fusarium solani*, on the other hand, oxidative cleavage of the F and E rings is known to be accompanied by formation of a 1,4-androstadiene-3,16-dione (Kondo and Mitsugi, 1972). A host of analogous sapogenins can be transformed in a similar way (Kondo and Mitsugi, 1972). Yields of up to 65% are obtained from diosgenin (Kondo and Mitsugi, 1973a), while 16α- and 16β-hydroxy-1,4-androstadien-3-one can be isolated as byproducts (Kondo and Mitsugi, 1973a).

In addition, a partial degradation of sapogenins to the 16-oxo-20-hydroxypregnane structure (157) by strains of *Verticillium* spp. or

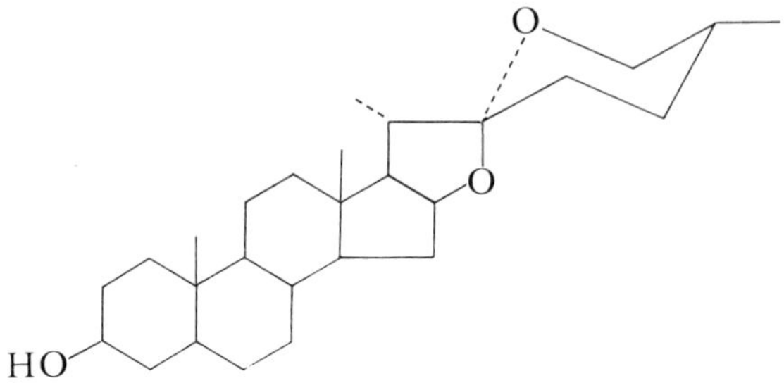

Diosgenin
(153)

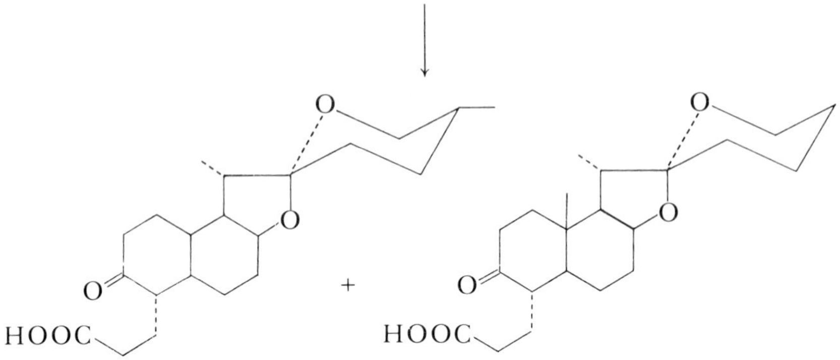

(154)   +   (155)

Stachylidum bicolor is possible (Shionogi Co. Ltd., 1972; Kondo and
Mitsugi, 1973b).

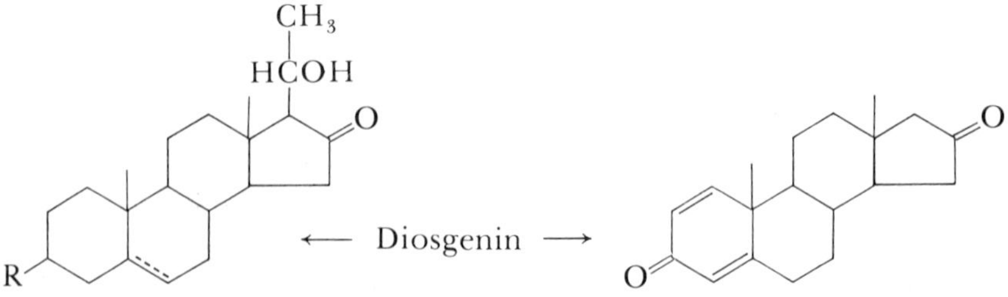

(157)   ← Diosgenin →   (156)
                      (153)

B. Scillarenins, Scilloglaucosidins and Whithaferins

Transformations of cardenolids and bufanolids are valuable as
attempts to bring about a positive change in the effects of these

compounds on the heart. Scillarenin (158) and scillarenone are transformed into 1-dehydroscillarenone with yields of almost 70% by *Arthrobacter simplex* (A.T.C.C. 6946), *Nocardia restrictus* (A.T.C.C. 14887) or *Nocardia asteroides* (A.T.C.C. 3308) (Görlich, 1973a). Δ^6-Dehydroscillarenone and Δ^6-dehydrocanaringenone (159) are 55% dehydrogenated with *Arthrobacter simplex* at 1,2-position (Görlich, 1973b).

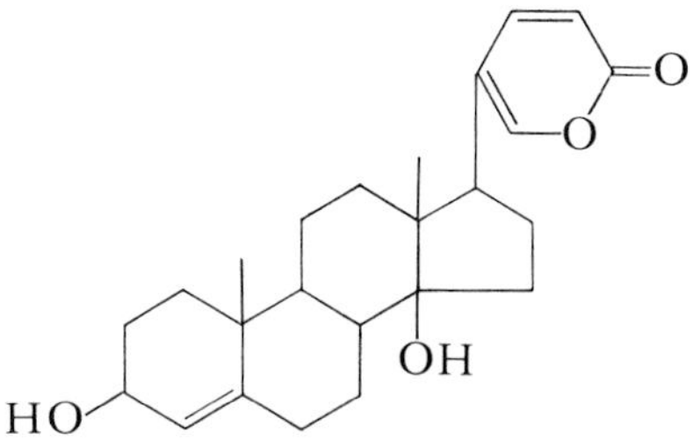

(158) 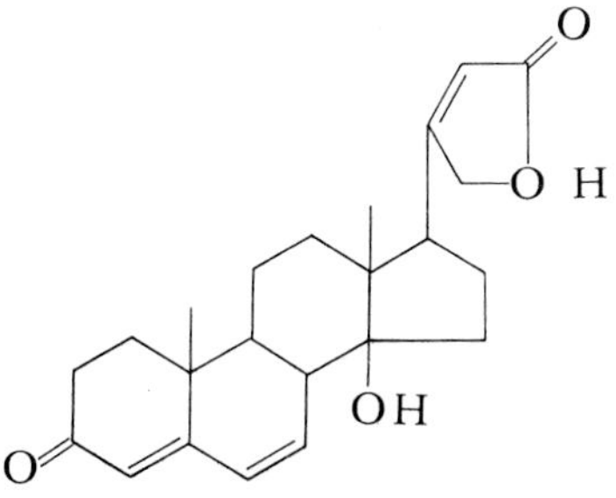　　　　　　　　　(159)

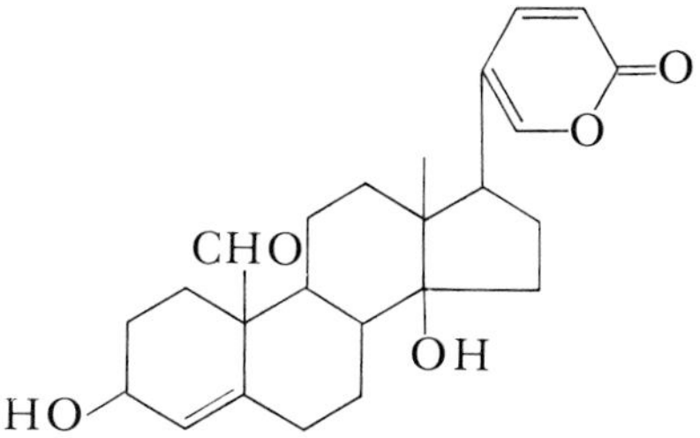

(160) 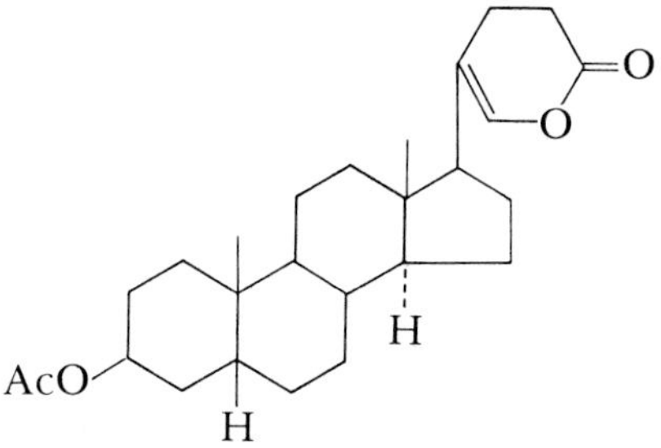　　　　　　　　　(161)

Nocardia corallina gives a 48% yield of 9α-hydroxy-scillarenone from scillarenin, and a 66% yield from scillarenone (Görlich, 1973c). Scilloglaucosidone (160) with its 19-aldehyde group, on the other hand, yields 6.5% of the corresponding A ring aromatics with *Arthrobacter simplex* in addition to the main products, 19-hydroxy-scillarenone (22%) and scilloglaucosidone (30%) (Görlich, 1973d). *Nocardia restrictus* (A.T.C.C. 14887), *Nocardia asteroides* (A.T.C.C. 3308) and *Nocardia corallina* (A.T.C.C. 4237) yield the aromatic predominantly. A review (Görlich, 1973d) links these and older results in cardenolids and bufadienolids with stages in chemical synthesis. For formation of the 14-oxy structures, chemical synthesis can be supported by microbiological 15α- (Vidic *et al.*, 1974) or 14α-hydroxylation (Dias and Pettit, 1972). The 14α-hydroxylation reaction with *Helminthosporium buchloes* C.B.S. on 3β-acetoxy-5β-,14α-bufa-20,22-dienolid-3-acetate (161) can be used to introduce the 14β,15β-epoxide function in order to produce resibufogenin (Dias and Pettit, 1972).

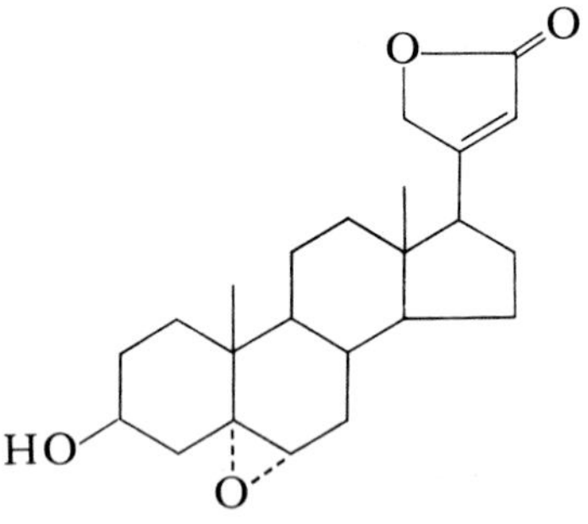

(162)

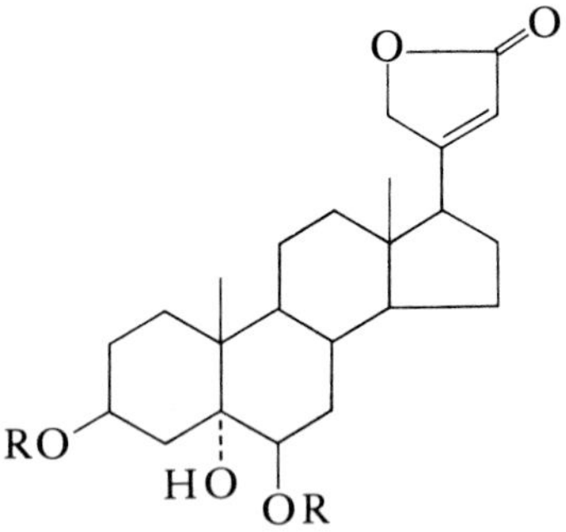

	R
(163)	H
(164)	Ac

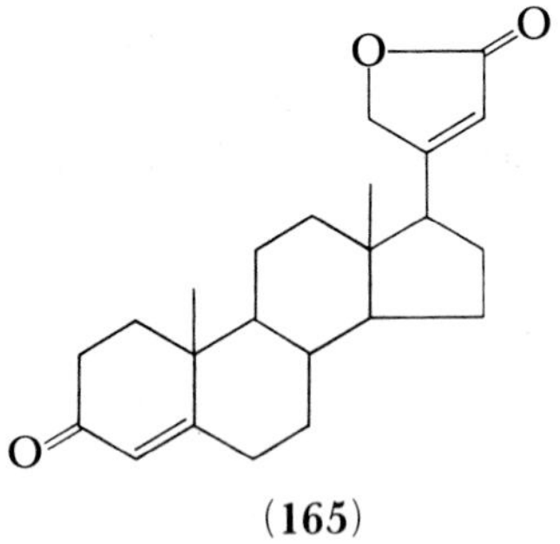

(165)

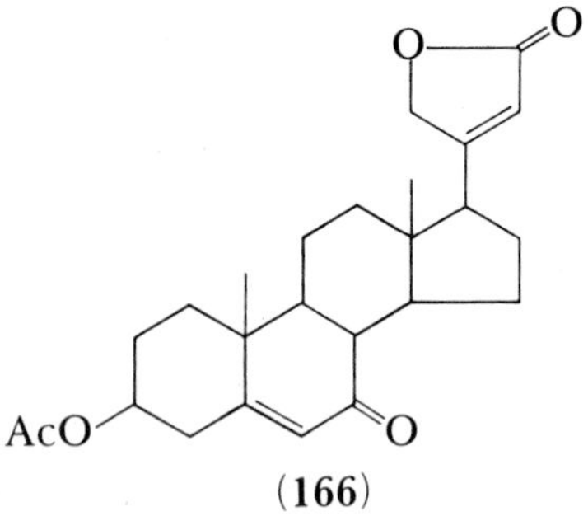

(166)

C. Cardenolid-Type Structures

In attempts to introduce an 11-hydroxy group into synthetic 14α-*H*-steroid butenolids by microbiological hydroxylation (Wagner *et al.*, 1976), insertion of a 5α,6α-epoxide structure (162) with *Curvularia lunata* resulted only in a 17α-hydroxy compound (26%) together with 7% of a 5α,6β-diol (163); this was in contrast to previous experiences with these structures which are sterically appropriate for 11β-hydroxylations, and arose through hydrolytic opening of the epoxide.

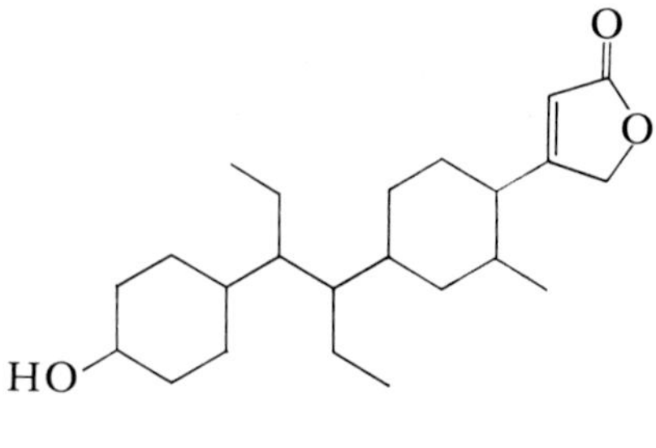

(167) 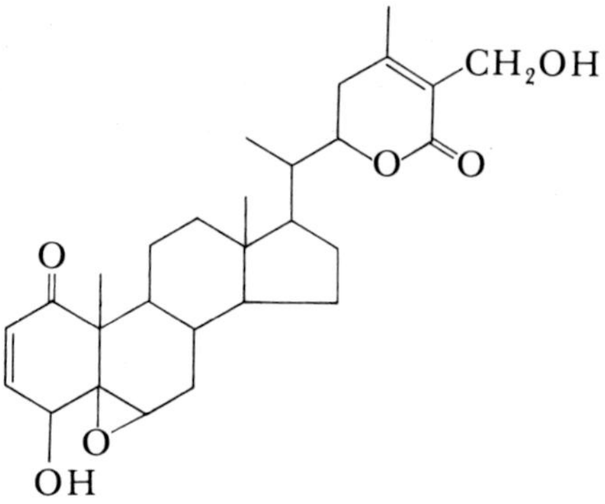 (168)

Aspergillus ochraceus, on the other hand, gave a 55% yield of the 11α-hydroxy compound in addition to 13% of the 5α,6β-diol. The free diol (163) proved inert to both fungus strains, while the 3β,6β-diacetate (164) yielded 26% of the 11α-hydroxy compound, in addition to partial saponification of the 3β-acetyl group with the *Aspergillus* sp. and, surprisingly, with *Curvularia lunata* (23% yield) as well. *Curvularia lunata* produced an additional 20% of the 17α-hydroxy compound. As expected, the 3-oxo-4-ene structure (165) formed 11α-mono- and 6β,11α-dihydroxy compounds with *Aspergillus ochraceus,* and the 3β-acetoxy-7-oxo-5-ene substrate (166) formed 11α-hydroxy compound (56%) exclusively. *Curvularia lunata,* on the other hand, hydroxylated the last-named substrate at the 17α- (39%) and 14α- (8%) positions.

In addition to these results, fermentation of a synthetic butenolid (167) with *Curvularia lunata* was described, showing only tertiary hydroxylations (Lang, 1975; Kreiser and Lang, 1976). First attempts to transform the natural product whithaferin with *Cunninghamella elegans* N.R.R.L. 1393 have also been described (Rosazza *et al.*, 1976). The structures of the dehydroxy compound (68%) obtained and of a hydroxylation product have not yet been fully elucidated.

D. Tomatidine

Tomatidine (169) is oxidized by *Nocardia restrictus* (C.B.S. 15745) to the 1,4-diene-3-one compound (21%); the saturated 3-ketone, the Δ¹-3-one and the similar Δ⁴-3-one are formed as byproducts (Belič and Sočič, 1972).

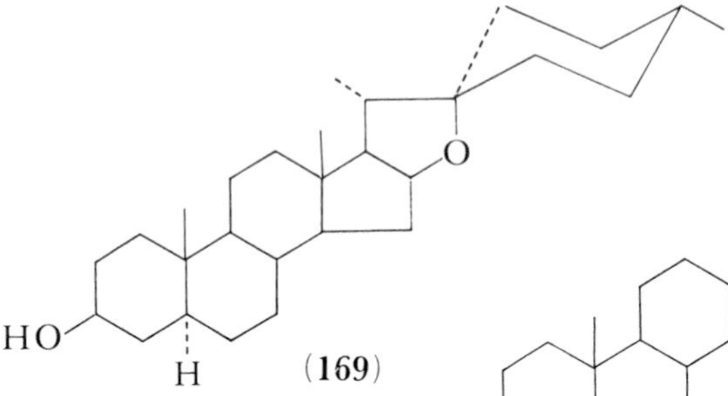

In the same way, it was also possible to oxidize and dehydrogenate two dihydrotomatidines (**170**, **171**) with an open E ring to the 3-keto $\Delta^{1,4}$ compounds. While the dihydro structure, with an open F ring (**172**), and the acetylated amino group also formed the $\Delta^{1,4}$-3-one structure, the corresponding substrate with a free amino group was only *N*-acetylated with no further reaction (Belič *et al.*, 1973). *N*-Acetyl-tomatidine (*N*-acetyl-5α-tomatanin-3β-ol) (**173**) is oxidized by *Nocardia restrictus* only to the 3-ketone, whereas *N*-methyltomatidine (*N*-methyl-5α-tomatanin-3β-ol) (**174**), 5α-conanin-3β-ol (**175**) and demissidine (**177**) are normally transformed to 3-oxo-1,4-diene structures (Belič *et al.*, 1977a).

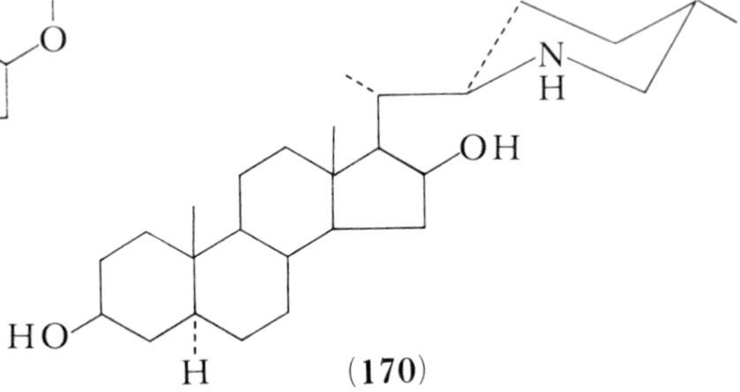

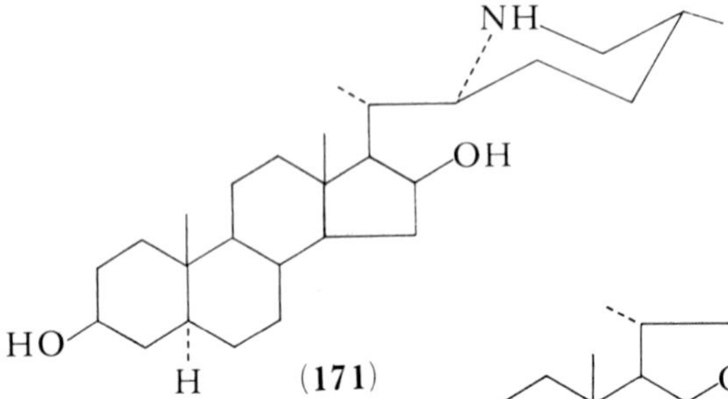

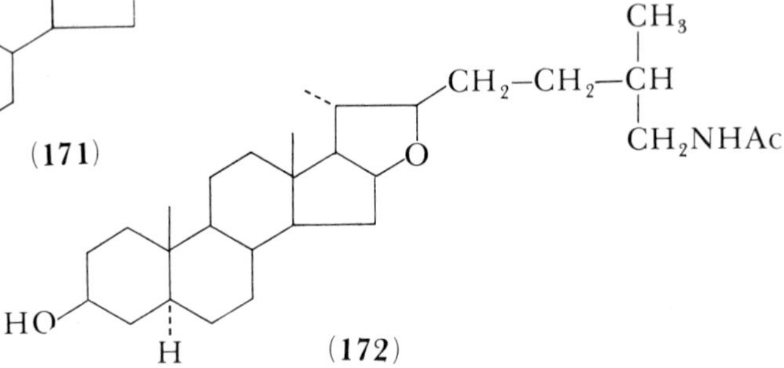

In a fermentation of tomatidine (**169**) with *Nocardia restrictus*, with a low starting pH value of 4 and a short incubation period of four hours, an unexpected epimerization (Belič *et al.*, 1977b) of the 3β-hydroxy group to the epimer 3α-tomatidine was observed (up to 70%). It was possible to transfer this epimerization to solasodanol (**175**) (20–25%), *N*-acetyltomatidine (**173**) (15%) and dihydrotomatidine B (**171**) (5%). Epimerization was not observed with demissidine (**177**).

Since tomatidine can also be used as a starting material for steroid syntheses, a microbiological degradation of the E and F rings was tested, and yielded 1,4-androstadiene-3,17-dione with *Arthrobacter simplex*. However, despite addition of α,α-dipyridyl, so far the yield has not exceeded 2.5–4% (Belič *et al.*, 1975), so with these structures chemical processes are superior.

E. Hydrolysis of Saponins

Because of the manifold reaction products and a lack of identification of the soil bacteria used, experiments for hydrolysis of jegosaponin (Yosika *et al.*, 1972a), a few ginsenoids (Yosika *et al.*, 1972b), saponin mixtures from *Metanarthecium luteo-viride* (Yosika *et al.*, 1973; Kitagawa *et al.*, 1977) and from *Madhuca longifolia* (Yosika *et al.*, 1974) can be mentioned only as preliminary attempts to make various sapogenols. For conversion of α-tomatine into tomatidines and into the tetra-saccharide lyotetrose, an inducable extracellular enzyme from *Fusarium oxysporum* f. sp. *lycopersici* is mentioned (Ford *et al.*, 1977). Normally, dioscin is hydrolysed in a fermentation using soil bacteria attached to *Dioscorea* tubers. This release of diosgenin can be carried out as a directed fermentation with *Aspergillus terreus*, giving a 90-95% yield (Nogai *et al.*, 1970).

REFERENCES

Abdel-Fattah, A. F. and Badawi, M. A. (1974). *Journal of General and Applied Microbiology* **20**, 363.

Abdel-Fattah, A. F. and Badawi, M. A. (1975a). *Journal of General and Applied Microbiology* **21**, 217.

Abdel-Fattah, A. F. and Badawi, M. A. (1975b). *Journal of General and Applied Microbiology* **21**, 225.

Akhrem, A. A. and Titow, Yu A. (1965). 'Microbial Transformations of Steroids'. Nauka Press, Moscow.

Akhrem, A. A. and Titow, Yu A., (1970). 'Steroids and Micro-organisms'. Nauka Press, Moscow.

Alig, L., Fürst, A. and Müller, M. (1976). United States Patent 3,939,193.

Alig, L., Müller, M., Wiechert, R., Nicolson, R., Fürst, A., Kerb, U. and Kieslich, K. (1976b). German Patent 2614,079.

Alig, L., Müller, M. and Fürst, A. (1977). German Patent 2,646,994.

Alig, L., Müller, M., Fürst, A., Wiechert, R., Kerb, U. and Kieslich, K. (1978). German Patent 27,3863.

All-Union Chemical and Pharmaceutical Research Institute (1973). Soviet Patent 370,229.

Al Rawi, J. M. A., Elvidge, J. A., Thomas, R. and Wright, B. J. (1974). *Journal of the Chemical Society Chemical Communications* 24, 2031.

Ambrus, G., Szarka, E. and Barta, I. (1975). *Steroids* 25, 99.

American Home Products (1972). United States Patent 3,697,379.

Anding, C., Parks, L. W. and Qurisson, G. (1974). *European Journal of Biochemistry* 43, 459.

Anner, G. and Wieland, P. (1973). United States Patent 3,763,145

Antonini, E. (1974). British Patent 810,170

Antosz, F. J., Haak, W. J. and Wovcha, M. G. (1977). United States Patent 4,029,549.

Arima, K., Tamura, G., Nagasawa, M. and Bae, M. (1968). United States Patent 3,388,042.

Ashton, M. J., Bailey, A. S. and Jones, E. R. H. (1974). *Journal of the Chemical Society Perkin Transactions I*, 1658.

Atherton, L., Duncan, M. M. and Safe, S. (1972). *Journal of the Chemical Society,* 882.

Babcock, J. C. and Campbell, J. A. (1972). United States Patent 3,642,992.

Balashova, K. N., Gabinskaya, Alekseeva, L. A., Shner, V. F., Messinova, O. V. and Surorov, N. P. (1975a). *Khimiya Prirodnykh Soedinenii* 3, 360.

Balashova, E. G., Gabinskaya, K. N., Alekseeva, L. M., Shner, V. F., Messinova, O. V. and Suvorov, N. P. (1975b). *Chemistry of Natural Compounds* 11, 373.

Barnes, P. J., Baty, J. D., Bilton, R. F. and Mason, A. N. (1974). *Journal of the Chemical Society Chemical Communications* 3, 115.

Barnes, P. J., Bilton, R. F. and Mason, A. F. (1975). *Biochemical Society Transactions* 3, 299.

Barnes, P. J., Baty, J. D., Bilton, R. F. and Mason, A. N. (1976). *Tetrahedon* 32, 89.

Barlett, K., Keat, M. J. and Mercer, E. I. (1974). *Phytochemistry* 13, 1107.

Bates, M. L., Reid, W. W. and White, J. D. (1976). *Journal of the Chemical Society* 44.

Belič, I. and Sočič, H. (1972). *Journal of Steroid Biochemistry* 3, 843.

Belič, I., Ghero, E. and Pertot, E. (1972). *Steroids and Lipids Research* 3, 201.

Belič, I., Kramer, V. and Sočič, H. (1973). *Journal of Steroid Biochemistry* 4, 363.

Belič, I., Socic, H. and Vranek, B. (1975). *Journal of Steroid Biochemistry* 6, 1211.

Belič, I., Mervic, M. and Kastelic-Suhadolc, T. (1977a). *Journal of Steroid Biochemistry*, 8, 311.

Belič, I., Komel, R. and Sočič, H. (1977b). *Steroids* 29, 271.

Bell, A. M., Cherry, P. J., Clark, T. M., Denny, W. A., Jones, E. R. H. and Meakins, G. D. (1972a). *Journal of the Chemical Society Perkin Transactions*, 2081.

Bell, A. M., Denny, W. A., Jones, E. R. H., Meakins, G. D. and Müller, W. E. (1972b). *Journal of the Chemical Society Perkin Transactions* I, 2759.

Bell, A. M., Browne, J. W., Denny, W. A., Jones, E. R. H., Kasal, A. and Meakins, G. D. (1972c). *Journal of the Chemical Society Perkin Transactions I*, 2930.

Bell, A. M., Clark, I. M., Denny, W. A., Jones, E. R. H., Meakins, G. D., Müller, W. E. and Richards, E. E. (1973). *Journal of the Chemical Society Perkin Transactions I*, 2131.

Bell, A. M., Chambers, V. E. M., Jones, E. R. H., Meakins, G. D., Müller, W. E. and Pragnell, J. (1974). *Journal of the Chemical Society Perkin Transactions I* 312.

Bell, A. M., Jones, E. R. H., Meakins, G. D., Miners, J. O. and Pendlebury, A. (1975a). *Journal of the Chemical Society Perkin Transactions I*, 357.

Bell, A. M., Boul, A. D., Jones, E. R. H. and Meakins, G. D. (1975b). *Journal of the Chemical Society Perkin Transactions I*, 1364.

Bell, A. M., Jones, E. R. H., Meakins, G. D., Miners, J. O. and Wilkins, A. L. (1975c). *Journal of the Chemical Society Perkin Transactions I*, 2040.

Bergmeyer, H. U., Nellböck-Hochstetter, M., Beaukamp, K., Holz, G., Gruber, W., Gramsall, J., Gavertshausen, L. and Lang, G. (1974a). German Patent 2,224,133.

Bergmeyer, H. U., Nellböck-Hochstetter, Beaukamp, K., Holz, G., Gruber, W., Gramsall, J. and Lang, G. (1974b). German Patent 2,224,131.

Beukers, R., Marx, A. F. and Zuidweg, M. H. J. (1972). In 'Drug Design' (E. J. Ariens, ed.), vol. III, pp. 1–131. Academic Press, New York.

Björkhem, J., Jörnvall, H. and Zeppezauer, E., (1973). *Biochemical and Biophysical Research Communications* 52, 413.

Bloom, B. M. and Shull, G. M. (1955). *Journal of the American Chemical Society* 77, 5767.

Blunt, J. W., Clark, I. M., Evans, J. M. and Jones, E. R. H. (1971). *Journal of the Chemical Society (C)*, 1136.

Boehringer Mannheim (1974). Belgian Patent 812,858

Boehringer Mannheim (1975). British Patent 1,412,244.

Bokkenheuser, V. D., Winter, J., Dehazya, P., Leon de, O. and Kelly, W. G. (1976). *Journal of Steroid Biochemistry* 7, 837.

Borman, E. A., Koshcheenko, K. A., Sokolova, L. V. and Kovylkina, N. F. (1975). *Ivestiya Akademii Nauk S.S.R. Seriya Biologicheskaya* 1, 143.

Boswell, G. A. and Ripka, W. C. (1972). United States Patent 3,641,005.

Boswell, G. A. and Ripka, W. C. (1973). Unites States Patent 3,718,673.

Braham, R. L., Dale, S. L. and Melby, J. C. (1975). *Steroids* 26, 697.

Browne, J. W., Denny, W. A., Jones, E. R., Meakins, G. D., Morisawa, Y., Pendlebury, A. and Pragnell, J. (1973). *Journal of the Chemical Society Perkin Transactions I*, 1493.

Buckland, B. C., Hill, M. D. and Dunill, P. (1975). *Biotechnology and Bioengineering* 17, 815.

Buckland, B. C., Hill, M. D. and Dunill, P. (1976). *Biotechnology and Bioengineering* 18, 601.

Buzby, G. C. and Greenspan, S. (1973). United States Patent 3,733,254.

Canonica, L., Danieli, B., Palmisano, G., Rainoldi, G. and Rauzi, B. M. (1974). *Journal of the Chemical Society Chemical Communications*, 656.

Čapek, A. and Fassitová, O. (1974). *Folia Microbiologica* 19, 378.

Čapek, A., Hanc, O. and Tadra, M. (1966). 'Microbial Transformations of Steroids'. Akademia Press, Prague.

Čapek, A., Fassitová, O. and Hanc, O. (1975a). *Folia Microbiologica* 20, 166.

Čapek, O., Fassitová, O. and Hanc, O. (1975b). *Folia Microbiologica* 20, 517.

Čapek, A., Fassitová, O. and Hanc, O. (1976). *Folia Microbiologica* 21 70.

Cargile, N. L. and McChesney, J. D. (1974). *Applied Microbiology* 27, 991.

Carlström, K. (1973) *Acta Chemica Scandinavia* 27, 1622.

Carlström, K. (1974a). *Acta Chemica Scandinavia B* 28, 23.

Carlström, K. (1974b). *Acta Chemica Scandinavia B* 28, 832.

Carlström, K. and Krook, L. (1973). *Acta Chemica Scandinavia B* 27, 1240.

Chambers, V. E. M., Denny, W. A., Evans, J. M., Jones, E. R. H., Kasal, A., Meakins, G. D. and Pragnell, J. (1973). *Journal of the Chemical Society Perkin Transactions I*, 1500.

Chambers, V. E. M., Jones, E. R. H., Meakins, G. D., Miners, J. O. and Wilkins, A. L. (1975a). *Journal of the Chemical Society Perkin Transactions I*, 55.

Chambers, V. E. M., Denny, W. A., Jones, E. R. H., Meakins, G. D., Miners, J. O., Piney, J. T. and Wilkins, A. L. (1975b). *Journal of the Chemical Society Perkin Transactions I*, 1359.

Charney, W. and Herzog, H. L. (1967). 'Microbial Transformations of Steroids'. Academic Press, New York.

Chin, Ch. Ch. and Warren, J. C. (1972). *Biochemistry, New York* 11, 2720.

Ciba-Geigy (1970). Belgian Patent 771,900.

Clegg, A. S., Denny, W. A., Jones, E. R. H., Meakins, G. D. and Piney, J. T. (1973). *Journal of the Chemical Society Perkin Transactions I*, 2137.

Collingsworth, D. R., Brunner, M. P. and Haines, W. J. (1952). *Journal of the American Chemical Society* 74, 2381.

Conner, A. H., Nagaoka, M., Rowe, J. W. and Perlman, D. (1976). *Applied Environmental Microbiology* 32, 310.

Costa Pla, L. (1973). *Biochemical Journal* 136, 501.

Crabb, T. A., Dawson, P. J. and Williams, R. O. (1975). *Tetrahedron Letters* 42, 3623.

Cremonesi, P., Carrea, G., Ferrara, L. and Antonini, E. (1975). *Biotechnology and Bioengineering* 17, 1101.

Cross, A. D. (1972). United States Patent 3,644,421.

David, S., Diassi, P. A. and Principe, P. A. (1973). British Patent 1,315,342.

Denny, W. A., Fredericks, P. M., Ghilezan, I., Jones, E. R. H., Meakins, G. D. and Miners, J. O. (1976). *Journal of the Chemical Society Chemical Communications*, 900.

Dias, J. R. and Pettit, G. R. (1972). United States Patent 3,661,941.

Dulaney, E. L., McAller, W. J., Koslowski, M., Stapley, E. O. and Jaglom, J. (1955). *Applied Microbiology* 3, 336.

Eastman Kodak (1977a). Belgian Patent 848,070.

Eastman Kodak (1977b). Unites States Patent 4,042,461.

Edenharder, R., Stubenrauch, S. and Slemrova, J. (1976). *Zentralblatt für Bakteriologie, Parasitenkunde, Infektionskrankheiten und Hygiene, Abteilung I, Originate* 162, 506.

Eder, U., Sauer, G., Haffer, G., Neef, G., Wiechert, R., Weber, A., Popper, A., Kennecke, M. and Müller, R. (1977). German Patent 2,534,911.

Edwards, J. A., Fried, J. H. and Mills, J. S. (1972). United States Patent 3,705,182.

Eignerová, L. and Procházka, Z. (1974). *Collection of Czechoslovak Chemical Communications* 39, 2828.

El-Kady, I. A. and Allam, A. M. (1973). *Journal of General Microbiology* 77, 465.

El-Kady, I. A., Sallam, L. A. R. and El-Refai, A. M. (1972). *Pakistan Journal of Biochemistry* 5, 39.

El-Monem, A., El-Refai, A. M., Sallam, L. A. R. and Geith, H. (1972). *Acta Microbiologica Polonica Series B* 4, 31.

El-Rafai, A. M., Sallam, L. A. R. and Naim, N. (1974a). *Journal of General and Applied Microbiology, Tokyo* 20, 111.

El-Rafai, A. M., Sallam, L. A. R. and El-Kady, I. A. R. (1974b). *Journal of General and Applied Microbiology, Tokyo* 20, 129.

El-Rafai, A. M., Sallam, L. A. R. and Naim, N. (1975). *Zeitschrift für Allgemeine Microbiologie, Morphologie, Physiologie. Ökologie und Microorganismen* 15, 59.

El-Tayeb, O. M., Murad, F. E., Zedan, H. H. and Abdel-Aziz, A. (1977). *Planta Medica* 31, 40.

Engelfried, O., Nieuweboer, O., Petzoldt, K., Kerb, U. and Lübke, K. (1976). German Patent 2,330,159.

Evans, J. M., Jones, E. R. H., Meakins, G. D., Miners, J. O., Pendlebury, A. and Wilkins, A. O. (1975). *Journal of the Chemical Society Perkin Transactions I*, 1356.

Eyssen, H. J. and Parmentier, G. (1974). *American Journal of Clinical Nutrition* 27, 1329.

Eyssen, H. J., Parmentier, G., Compernolle, F. C., de Pauw, G. and Piessens-Denef, M. (1973). *European Journal of Biochemistry* 36, 411.

Eyssen, H. J., de Pauw, G. and Parmentier, G. (1974). *Journal of Nutrition* 104, 605.

Favero, J., Marchand, J. and Winternitz, F. (1977). *Bulletin de la Société Chimique France,* 310.

Fishman, J. (1960). *Journal of the American Chemical Society* 82, 6143.

Flines, J. de and Waard, F. v. d. (1967). Netherlands patent 6,605,514.

Fonken, G. S. and Johnson, R. A. (1972). 'Chemical Oxidations with Micro-organisms'. Marcel Dekker, New York.

Ford, J. E., McCance, D. J. and Drysdale, R. B. (1977). *Phytochemistry* 16, 545.

Freyberg, M., Oehlschläger, A. C. and Unrau, A. M. (1973). *Biochemical and Biophysical Research Communications* 51, 219.

Fukuda, H., Kawakami, Y. and Nakamura, S. (1973). *Chemical and Pharmaceutical Bulletin* 21, 2057.

Gabinskaya, K. N., Shpingis, A. A., Kovalenko, T. A., Polievktov, M. K. and Messinova, O. V. (1974). *Mikrobiolgiya* 43, 917.

Gallegra, P. G., (1976). United States Patent 3,956,349.

Garcia-Rodriquez, L. K., Gabrinskaya, K. N., Shner, V. F. and Messinova, O. M. (1977a). *Khimiko-Farmatsevticheskii Zhurnal* 11, 62.

Garcia-Rodriquez, L. K., Gabrinskaya, K. N., Krasnova, L. A., Shner, V. F. and Messinova, O. M. (1977b). *Khimiko-Farmatsevticheskii Zhurnal* 11, 66.

Ghraf, A., Lax, E. R., Oza, S., Schriefers, H., Wildfeuer, A. and Haferkamp, O. (1975). *Journal of Steroid Biochemistry* 6, 1534.

Ghrai, R., Lax, E. R., Oza, S. and Schriefers, H. (1975). *Journal of Steroid Biochemistry* 6, 1531.

Glaxo Laboratories Ltd. (1973). British Patent 1,317,185.

Goddard, P. and Hill, M. J. (1972). *Biochimica et Biophysica Acta* 250, 336.

Goddard, P., Fernandez, F., West, B., Hill, M. J. and Barnes, P. (1975). *Journal of Medical Microbiology* 8, 429.

Görlich, B. (1973a). German Patent 2,226,846.

Görlich, B. (1973b). German Patent 2,226,930.

Görlich, B. (1973c). German Patent 2,226,997.

Görlich, B. (1973d). *Planta Medica* 23, 39.

Golovtzeva, V. A., Korovkina, A. S. and Skovotseva, L. F. (1977). *Chimia Farmaceutica Journal*, 150.

Goodhue, G. T. and Hugh H. A. (1975). German Patent 2,512,606.

Gulaya, V. E., Kryutchenko, E. G., Rattel, S. N., Messinova, O. V., Anachenko, S. N. and Torgov, I. V. (1975). *Prikladnaya Biokhimia Mikrobiologiya* 11, 657.

Gulaya, V. E., Kryutchenko, E. G., Anachenko, S. N. and Torgov, I. V. (1976). *Khimiko Farmatsevticheskii Zhurnal* 10, 41.

Greenspan, G. and Rees, R. W. (1975). United States Patent 3, 880,895.

Greenspan, G., Rees, R. W., Link, G. D., Boyd, C. P., Jones, R. C. and Alburn, H. E. (1974). *Experientia* 30, 328.

Grove, M. J., Strandberg, G. W. and Smiley, K. L. (1971). *Biotechnology and Bioengineering* 13, 709.

Gruber, W., Bergmeyer, H. U., Nellböck-Hochstetter, M., Beaukamp, K., Holz, G., Gramsall, J. and Lang, G. (1975). German Patent 2,456,686.

Hashimoto, S. and Hayakawa, S. (1977). *Biochemical Journal* 164, 715.

Haslewood, E. S. and Haslewood, G. A. D. (1976). *Biochemical Journal* **157**, 207.

Hayakawa, S. (1973). *Advances in Lipid Research* **11**, 143.

Hayakawa, S. and Fujiwara, T. (1977). *Biochemical Journal* **162**, 387.

Hayakawa, S., Kanematsu, Y. and Fujiwara, T. (1972). United States Patent 3,674,842.

Hayakawa, S., Kanematsu, Y., Fujiwara, T. and Kako, H. (1976a). *Biochemical Journal* **154**, 577.

Hayakawa, S., Hashimoto, S. and Onaka, T. (1976b). *Biochemical Journal* **160**, 745.

Hayakawa, S., Fujiwara, T. and Kako, H. (1976c). *Biochemical Journal* **160**, 757.

Hayakawa, S., Takata, T., Fujiwara, T. and Hashimoto, S. (1977). *Biochemical Journal* **164**, 709.

Hörhold, C., Groh, H., Dänhardt, S., Lestrovaja, N. N. and Schubert, K. (1975). *Zeitschrift für Allgemeine Mikrobiologie, Morphologie, Physiologie, Ökologie und Mikroorganismen* **15**, 563.

Hörhold, C., Rose, G., Müller, E. and Schubert, K. (1976). *Journal of Steroid Biochemistry* **7**, 199.

Hörhold, C., Groh, H., Dänhardt, S., Lestrovaja, N. N. and Schubert, K. (1977). *Journal of Steroid Biochemistry* **8**, 701.

Hoffmann-La Roche A. G. (1976). British Patent 1,445,945.

Hofmeister, H., Wiechert, R., Annen, K., Laurent, H. and Steinbeck, H. (1977). German Patent 2,546,062.

Holland, H. L. and Auret, B. J. (1975a). *Canadian Journal of Chemistry* **53**, 845.

Holland, H. L. and Auret, B. J. (1975b). *Tetrahedron Letters* **44**, 3787.

Howe, R., Moore, R. H., Rao, B. S. and Gibson, D. T. (1973). *Journal of the Chemical Society Perkin Transactions I*, 1940.

Hylemon, P. B. and Stellwag, E. J. (1976). *Biochemical and Biophysical Research Communications* **69**, 1088.

Iida, M. and Iizuka, H. V. (1976). International Fermentation Symposium, Berlin, p. 324.

Iida, M., Matsuhashi, K. and Nakayama, T. (1975a). *Zeitschrift für Allgemeine Mikrobiologie* **15**, 181.

Iida, M., Matsuhashi, K. and Nakayama, T. (1975b). *Zeitschrift für Allgemeine Mikrobiologie* **15**, 189.

Iizuka, H. (1974). Japanese Patent 4,044,346.

Iizuka, H. and Naito, A. (1967). 'Microbial Transformations of Steroids and Alkaloids'. University of Tokyo Press.

Ikegawa, S. and Nambra, T. (1973). *Chemical Industry* **5**, 230.

Imshenetskii, A. A. and Mavrina, L. A. (1972a). *Mikrobiologiya* **41**, 399.

Imshenetskii, A. A. and Mavrina, L. A. (1972b). *Mikrobiologiya* **41**, 598.

Imshenetskii, A. A., Nikitin, L. E. and Nazarova, T. S. (1975). *Mikrobiologiya* **44**, 210.

Imshenetskii, A. A., Nazarova, T. S. and Nikitin, L. E. (1976). *Mikrobiologiya* **45**, 298.

Irmscher, K., Werder, v., F., Bork, K. H. and Kraft, H. G. (1973). United States Patent 3,718,542.

Jiu, J. and Marsheck, W. J. (1976). United States Patent 3,994,933.

Jiu, J. and Marsheck, W. R. (1977). United States Patent 4,032,408.

Jones, E. R. H. (1973). *Pure and Applied Chemistry* **33**, 39.

Jones, E. R. H., Meakins, G. D. and Clegg, A. S. (1972). United States Patent 3,692,629.

Jones, E. R. H., Meakins, G. D., Miners, J. O., Pragnell, J. H. and Wilkins, A. L. (1975a). *Journal of the Chemical Society Perkin Transactions I*, 1552.

Jones, E. R. H., Meakins, G. D., Miners, J. O. and Wilkins, A. L. (1975b). *Journal of the Chemical Society Perkin Transactions*, 2308.

Jones, E. R. H., Meakins, G. D., Miners, J. O., Mirrington, R. N. and Wilkins, A. L. (1976). *Journal of the Chemical Society Perkin Transactions*, 1842.

Joská, J., Procházka, Z., Fajkos, J. and Sorm, F. (1973). *Collection of Czechoslovak Chemical Communications* **38**, 1398.

Kanebo, K. K. (1974a). Japanese Patent 4 900-490.

Kanebo, K. K. (1974b). Japanese Patent 4 9050-192.

Kanebo, K. K. (1974c). Japanese Patent 4 9094-894.

Kanebo, K. K. (1974d). Japanese Patent 4 9094-895.

Kanebo, K. K. (1974e). Japanese Patent 4 910-591.

Kanebo, K. K. (1975a). Japanese Patent 5 005-593.

Kanebo, K. K. (1975b). Japanese Patent 5 0100-290.

Karim, A. and Marsheck, W. J. (1976). United States Patent 3,972,871.

Kelsey, M. I. and Sexton, S. A. (1976). *Journal of Steroid Biochemistry* **7**, 641.

Kerb, U., Wiechert, R., Petzoldt, K., Kieslich, K. and Kolb, K. H. (1974). United States Patent 3,787,454.

Kerb, U., Wiechert, R., Kieslich, K. and Petzoldt, K. (1975a). German Patent 2,349,023.

Kerb, U., Wiechert, R., Kieslich, K., Petzoldt, K., Hoyer, G. A. and Thiel, M. (1975b). *Chemische Berichte* **108**, 54.

Kerb, U., Wiechert, R., Kieslich, K., Petzoldt, K., Wachtel, H., Palenschat, D., Horowski, R., Paschelke, G. and Kehr, W. (1976). German Patent 2,330,159.

Kerényi, G., Szentirmai, A., Natonek, M. (1975). *Acta Microbiologica Academiae Scientiarum Hungaricae* **22**, 487.

Kieslich, K. (1978). German Patent 2,705,917.

Kieslich, K. and Raspé, G. (1967). German Patent 1,226,575.

Kieslich, K. Wieglepp, H., Petzoldt, K. and Hill, F. (1971). *Tetrahedron* **27**, 445.

Kieslich, K., Kerb, U., Mengel, K. and Domenico, A. (1972). German Patent 2,059,310.

Kieslich, K., Kerb, U., Nickolson, R., Wiechert, R., Alig, L., Fuerst, A. and Müller, M. (1976). *Vth. International Fermentation Symposium, Berlin 2976*, 322.

Kieslich, K., Wieglepp, H. and Hoyer, G. A. (1978). *Chemische Berichte* **110**; German Patent 2,656,575.

Kikkoman Shoyu, K. K. (1977). Japanese Patent 5 2007-484.

Kitagawa, I., Nakanishi, T., Morii, Y. and Yosika, I. (1977). *Chemical Pharmaceutical Bulletin* **25**, 2343.

Klimontovich, N. N., Kovalenko, T. A., Acheva, A. F. U., Polievktov, M. K. and Messinova, O. V. (1975). *Khimiko-Farmatsevticheskii Zhurnal* **9**, 33.

Knapp, F. F., Goad, L. J. and Goodwin, T. W. (1977). *Phytochemistry* **16**, 1683.

Knight, J. C. and Wovcha, M. G. (1977a). German Patent 2,652,376.

Knight, J. C. and Wovcha, M. G. (1977b). German Patent 2,652,377.

Koester, D. H. (1937). United States Patent 2,236,574.

Kohler, B. and Träger, L. (1975). *Naturwissenschaften* **62**, 299.

Kominek, L. A. (1973). United States Patent 3,770,586.

Kondo, E. and Mitsugi, T. (1972). United States Patent 3,665,022.

Kondo, E. and Mitsugi, T. (1973a). United States Patent 3,740,317.

Kondo, E. and Mitsugi, T. (1973b). *Tetrahedron* **29**, 823.

Koninklijke Nederlandsche Gist-en Spiritusfabriek N. V. (1966). Netherlands Patent 6,605,738.

Koninklijke Nederlandsche Gist-en Spiritusfabriek N. V. (1967). Netherlands Patent 6,513,718.

Koninklijke Nederlandsche Gist-en Spiritusfabriek N.V. (1968). Netherlands Patent 6,705,450.

Koninklijke Nederlandsche Gist-en Spiritusfabriek N. V. (1972a). German Patent 1,618,582.

Koninklijke Nederlandsche Gist-en Spiritusfabriek N. V. (1972b). German Patent 1,618,598.

Koninklijke Nederlandsche Gist-en Spiritusfabriek N. V. (1973). British Patent 1,314,191.

Koshcheenko, K. A., Borman, E. A., Sokolova, L. W., Suvovov, N. N. and Skryabin, G. K. (1975). *Izvestiya Akademii Nauk S.S.R.*, Seriya Biologicheskaya 1, 25.

Kozlova, V. Kh. and Fonina, N. A. (1972). *Mikrobiologiya* 14, 602.

Krachy, S., Marsheck, W. J. and Muir, R. D. (1972). United States Patent 3,684,657.

Krasnova, L. A., Sokolova, L. V., Korzinkina, N. A. and Suvorov, N. P. (1975). *Khimiya Prirodnykh Soedinenii* 6, 742.

Kreiser, W. and Lang, S. (1976). *Chemische Berichte* 109, 3318.

Kulig, M. J. and Smith, L. L. (1974). *Journal of Steroid Biochemistry* 5, 485.

Kundsin, R. B., Underwood, R. H. and Rose, L. I. (1972). *Applied Microbiology* 24, 665.

Kurosawa, Y., Yamashita, Y., Shibata, K., Yamakoshi, N. and Mori, H. (1975). *Transactions in Mycology, Japan* 16, 245.

Kyowa Hakko Kogyo (1975). Japanese Patent 5,006,773.

Lamontagne, N. S., Johnson, D. F. and Holmlund, Ch. E. (1977). *Journal of Steroid Biochemistry* 8, 329.

Lang, S. (1975). Dissertation: Technical University of Braunschweig, Germany.

Largeau, C., Goad, L. J. and Goodwin, T. W. (1977). *Phytochemistry* 16, 1931.

Larsson, P. O., Ohlson, S. and Mosbach, K. (1976). *Nature, London* 263, 796.

Lee, B. K., Thoma, R. W., Ryu, D. Y. and Brown, W. E. (1969). *Bacteriological Proceedings* A, 35.

Lefebvre, G., Germain, P. and Gray, P. (1974a). *Tetrahedron Letters*, 127.

Lefebvre, G., Germain, P. and Gray, P. (1974b). *Comptes Rendus des Seances de la Société de Biologie et de ses Filiales* 274, 449.

Lefebvre, G., Ratelle, S., Chartrand, L. and Rov, C. C. (1977). *Experientia* 33, 616.

Lestrovaja, N. N., Groh, H., Hörhold, C., Dänhardt, S. and Schubert, K. (1977). *Journal of Steroid Biochemistry* 8, 313.

Lewis, R. and Gorbach, S. (1972). *Archives of Internal Medicine* 130, 545.

Liev, J. E. and Rousseau, J. (1976). *Federation of European Biochemical Societies Letters* 70, 23.

Lilly, M. D., Smith, S. W. and Dunill, P. (1976a). *Vth. International Fermentation Symposium, Berlin* 301.

Lilly, M. D., Cheetham, P. S. J., Lewis, D. J., Yates, J. and Dunill, P. (1976b). *Vth. International Fermentation Symposium, Berlin* 327.

McDonald, I. A., Williams, C. N. and Mahony, D. E. (1973). *Biochimica et Biophysica Acta* 309, 243.

McDonald, I. A., Williams, C. N. and Mahony, D. E. (1974). *Analytical Biochemistry* 57, 127.

McDonald, I. A., Williams, C. N. and Mahony, D. E. (1975a). *Journal of Lipid Research* 16, 244.

McDonald, I. A. Bishop, J. M., Mahony, D. E. and Williams, C. N. (1975b). *Applied Microbiology* 30, 530.

McDonald, I. A., Mahony, D. E., Jellet, J. F. and Meyer, C. E. (1977). *Biochimica et Biophysica Acta* 489, 466.

Mallett, G. E. (1973). United States Patent 3,741,870.

Mamoli, L. (1937). United States Patent 2,186,906.

Markert, C. and Träger, L. (1975). *Hoppe-Seyler's Zeitschrift für Physiologische Chemie* **356**, 1843.

Markert, C., Betz, B. and Träger, L. (1975). *Zeitschrift für Naturforschung* **30**, 266.

Marsheck, W. J. (1971). *In* 'Progress in Industrial Microbiology' (D. J. D. Hockenhull, ed.), vol. 10, p. 49. Livingstone, London.

Marsheck, W. J. and Karim, A. (1973). *Applied Microbiology* **25**, 647.

Marsheck, W. J. and Krachy, S. (1972). *Applied Microbiology* **23**, 72.

Marsheck, W. J. and Krachy, S. (1973). Unites States Patent 3,759,791.

Martin, C. K. (1977). *Advances in Applied Microbiology* **22**, 29.

Martin, C. K. and Wagner, F. (1976a). *European Journal of Applied Microbiology* **2**, 243.

Martin, C. K. and Wagner, F. (1976b). *Proceedings of the Vth. International Fermentation Symposium, Berlin*, 5328.

Marx, A. F., Kooreman, H. J., Jaitley, K. D. and Sijdke Vakder, D. (1973). *Journal of Medical Chemistry* **16**, 1302.

Masureka, P. G. and Goodhue, Ch. T. (1976). German Patent 2,656,063.

Maugras, M., Hartemann, D., Granger, P. and Lematre, J. (1973). *Comptes Rendus des Séances de al Société de Biologie et de Ses Filiales* **277**, 681.

E.-Merck AG. (1972). Belgian Patent 788,355.

Micková, R., Protiva, J. and Schwarz, V. (1976). *Folia Microbiologica* **21**, 144.

Mitsubishi Chemical Industries, K. K. (1977a). Japanese Patent 5 2148-688.

Mitsubishi Chemical Industries, K. K. (1977b). Japanese Patent 5 2070-082.

Mitsubishi Chemical Industries, K. K. (1977c). Japanese Patent 5 2070-161.

Mogilnitzky, G. M. (1973). *Prikladnaya Biochimia i Microbiologia* **9**, 240.

Mogilnitzky, G. M., Kostrichenko, K. A., Ponomareva, E. N., Bukhar, M. I. and Skryabin, G. K. (1977). *Khimiko-Farmatsevticheskii Zhurnal* **11**, 94.

Morisaka, M., Sato, S., Ikekawa, N. and Shikita, M. (1976). *Federation of European Biochemical Societies Letters* **72**, 337.

Morozova, L. S., Gabinskaya, K. N. and Grinenko, G. S. (1973). *Chemistry of Natural Compounds* **9**, 39 (Engl. translation).

Murray, H. C. and Peterson, D. H. (1950). United States Patent 2,602,769.

Nagasawa, M., Bae, M., Tamura, G. and Arima, K. (1969). *Agricultural and Biological Chemistry* **33**, 1644.

Nagasawa, M., Hashiba, H., Watanabe, N., Bae, M., Tamura, G. and Arima, K. (1970). *Agricultural and Biological Chemistry* **34**, 801–804.

Nagase Sangyo, K. K. (1976). Japanese Patent 1,067,786.

Nagase Sangyo, K. K. (1977). Japanese Patent 5 2007-483.

Nambara, T., Ikegawa, S. and Hosada, H. (1973). *Chemical and Pharmaceutical Bulletin* **21**, 2794.

Nambara, T., Ikegawa, S. and Kato, M. (1975a). *Chemical and Pharmaceutical Bulletin* **23**, 2164.

Nambara, T., Ikegawa, S. and Takahashi, C. (1976b). *Chemical and Pharmaceutical Bulletin* **23**, 2358.

Nambara, T., Ikegawa, S., Ishida, H. and Hosoda, H. (1977). *Chemical and Pharmaceutical Bulletin* **25**, 3415.

Neals, W. D. and Parks, L. W. (1977). *Journal of Bacteriology* **129**, 1375.

Nicholas, A. W. (1975). *Dissertation Abstracts* **35**, 3839.

Nishikawa, D., Imada, Y., Kinoshita, M., Takahashi, K., Machida, M. and Nagasawa, M. (1975). German Patent 2,625,333.

Nobile, A. (1958). United States Patent 2,837,464.

Nogai, Y., Sawai, M. and Jurosawa, Y. (1970). *Journal of the Agricultural Chemical Society of Japan* **44**, 15.

Ordzhonikdze, S. (1976). Soviet Union Patent 511,946.

Oxlea, P. and Rosindale, J. (1972). United States Patent 3,639,434.

Palmowski, B. and Träger, L. (1974). *Hoppe-Seyler's Zeitschrift für Physikalische Chemie* 355, 1070.

Pan, S. C. and Lerner, L. J. (1972). British Patent 1,267,534.

Parmentier, G. and Eyssen, H. (1974). *Biochimica et Biophysica Acta* 348, 279.

Penasse, L. and Nomine, G. (1974). *European Journal of Biochemistry* 47, 555.

PERMACHEM Asia Ltd. (1976). Japanese Patent 5 1057-886.

Petzoldt, K. and Elger, W. (1976). German Patent 2,450,106.

Petzoldt, K. and Elger, W. (1977). German Patent 2,539,261.

Petzoldt, K. and Wiechert, R. (1974). German Patent 2,456,068.

Petzoldt, K., Laurent, H. and Steinbeck, H. (1972). German Patent 2,041,071.

Petzoldt, K., Kieslich, K. and Steinbeck, H. (1974). German Patent 2,326,084.

Petzoldt, K., Vidic, H. J., Nishino, K., Wiechert, R. and Laurent, H. (1976). German Patent 2,449,327.

Petzoldt, K., Prezewoski, K., Steinbeck, H. and Wiechert, R. (1977a). Belgian Patent 852,527.

Petzoldt, K., Prezewowski, K., Steinbeck, H. and Wiechert, R. (1977b). German Patent 2,621,646.

Philips, G. T. and Ross, F. P. (1974). *European Journal of Biochemistry* 44, 603.

Plourde, R., El-Tayeb, O. M. and Hafez-Zedan, H. (1972). *Applied Microbiology* 23, 601.

Probst, M. (1974). Dissertation: Technical University of Braunschweig.

Probst, M., Strijewski, A., Tan, T. L. and Wagner, F. (1973). *Biotechnology and Bioengineering* 4, 217.

Procházka, Z., Budesinský, M. and Prekajski, P. (1974). *Collection of Czechoslovak Chemical Communications* 39, 982.

Protiva, J. and Schwartz, V. (1974). *Folia Microbiologica* 19, 151.

Puhvel, S. M. (1975). *Journal of Investigative Dermatology* 64, 397.

Raspé, G., Kieslich, K. and Kerb, U. (1964). *Arzneimittelforschung* 14, 450.

Rausser, R. C. and Shapiro, E. L. (1972). United States Patent 3,707,484.

Richmond, W. (1973). German Patent 2,246,695.

Ripka, W. C. (1972). Belgian Patent 774,754.

Riva, M. and Toscano, L. (1976). German Patent 2,533,377.

Roeschlau, P., Lang, G., Beaukamp, K. and Bernt, E. (1975). German Patent 2,418,978.

Rosazza, J. P., Nicholas, A. W., Fyfe, L. and Loebig, D. (1976). *Proceedings of the Vth. International Fermentation Symposium, Berlin.* p. 33.

Ryu, D. Y., Lee, B. K. and Thoma, R. W. (1973). United States Patent 3,734,830.

Ryzhkova, V. M., Sokolova, L. V., Shpingis, A. A. and Korylkina, N. V. (1974). *Chemistry of Natural Compounds* 10, 117.

Sallam, L. A. R. and El-Refai, A. M. (1975). *Acta Biologica et Medica Germanica* 34, 21.

Sallam, L. A. R., El-Refai, A. M. H., Nada, S. and Abdel-Fattah, A. F. (1973). *Journal of General and Applied Microbiology* 19, 155.

Sallam, L. A. R., El-Kady, I. A. and El-Refai, A. M. (1974a). *Microbiologia Espanola* 27, 299.

Sallam, L. A. R., El-Refai, A. M. and Naim, N. (1974b). *Annals of Microbiology and Enzymology* 23, 173.

Samanta, T. B., Chattophadyay, S. and Nandi, P. (1976). *Proceedings of the Vth. International Fermentation Symposium, Berlin,* 323.

Sanda, V., Fajkos, J. and Protiva, J. (1977). *Collection of Czechoslovak Chemical Communications* 42, 3646.

Saltsman, W. H. (1976). United States Patent 3,954,562.

Schneider, J. J. (1974). *Journal of Steroid Biochemistry* 5, 9.

Schoeller, W. (1937). United States Patent 2,184,167.

Schubert, K., Kaufmann, G. and Hörhold, C. (1969). *Biochimica et Biophysica Acta* 176, 163.

Schubert, K., Groh, H. and Hörhold, C. (1971). *Journal of Steroid Biochemistry* 2, 281.

Schubert K., Schlegel, J., Groh, H., Rose, G. and Hörhold, C. (1972). *Endokrinologie* 59, 99.

Schubert, K., Rose, G. and Hörhold, C. (1973). *Journal of Steroid Biochemistry* 4, 283.

Schubert, K., Rose, G. and Hörhold, C. (1974). *Pharmazie* 29, 74.

Schubert, K., Phái, L. D., Kaufmann, G. and Knöll, R. (1975a). *Acta Biologica et Medica Germanica* 34, 167.

Schubert, K., Phái, L. D., Kaufmann, G. and Hörhold, C. (1975b). *Acta Biologica et Medica Germanica* 34, 173.

Schubert, K., Ritter, F., Sorkina, T., Höhme, K. H. and Hörhold, C. (1975c). *Journal of Steroid Biochemistry* 6, 1501.

Schwartz, V. and Protiva, J. (1972). *Collection of Czechoslovak Chemical Communications* 37, 1577.

Schwartz, V. and Protiva, J. (1974). *Folia Microbiolgica* 19, 156.

Scribner, R. M. (1973). United States Patent 3,767,684.

Sedlaczek, L., Pajkowska, H., Jaworski, A. and Czerniawski, E. (1973a). *Acta Microbiologica Polonica Series B* 5, 103.

Sedlaczek, L., Jaworski, A., Pajkowska, H., Zych, L. and Czerniawski, E. (1973b). *Acta Microbiologica Polonica Series B* 5, 163.

Shionogi, Co. Ltd. (1957). British Patent 833,595.

Shionogi Co. Ltd. (1972). Japanese Patent 72,20,958.

Shionogi Co. Ltd. (1975). Japanese Patent 5,016,438.

Shirasaka, M., Tanabe, K., Naito, A. and Icki, M. (1972). United States Patent 3,639,212.

Sih, C. J. and Rosazza, J. P. (1976). *In* 'Microbial Transformations in Organic Synthesis in Applications of Biochemical Systems in Organic Chemistry' (B. Jones, C. J. Sih and D. Perlman, eds.), p. 69. John Wiley, New York.

Sih, C. J. and Wang, C. K. (1965). *Journal of the American Chemical Society* 87, 1387.

Sih, C. J. and Whitlock, H. W. (1968). *Annual Review of Biochemistry* 37, 661.

Skalhegg, B. A. (1973). German Patent 2,230,320.

Skalhegg, B. A. (1974). *European Journal of Biochemistry* 46, 117.

Skryabin, G. K. and Golovteva, L. A. (1976). 'Microorganisms in Organic Chemistry'. Nauka Press, Moscow (in Russian).

Skryabin, G. K., Koshcheenko, K. A. and Surovtsev, V. T. (1974a). *Doklady Akademii Nauk S.S.S.R.* 215, 737.

Skryabin, G. K., Koshcheenko, K. A., Mogilnitski, G. M., Surovtsev, V. T., Tyurin, V. S. and Fikhte, B. A. (1974b). *Izvestiya Akademii Nauk S.S.S.R., Seriya Biologicheskaya* 857 (6).

Skryabin, G. K., Koshcheenko, K. A., Sukhololskaya, G. V. and Arinbasarova, A. Yu. (1976). *V. International Fermentation Symposium, Berlin 1976*, 326.

Smith, L. L. (1974). *In* 'Terpenoids and Steroids' (K. H. Overton, ed.), vol. 4. pp. 394. Academic Press, New York.

Smith, P. S., Poole, N. J. and Jawott, W. F. A. (1973). *Phytochemistry* 12, 561.

Smithies, J. R. and Eskins, K. F. (1973). United States Patent 3,732,261.

Söhngen, N. L. (1913). *Zentralblatt für Bakteriologische Parasitenkunde Abteilung II* **37** , 595.

Soliman, F. M. (1975). *Dissertation Abstracts* **35**, 921.

Sterkin, V. E., Morozowa, G. R., Zyakun, A. M., Chigalcichik, A. G. and Skryabin, G. K. (1973). *Izvestiya Akademii Nauk S.S.R., Seriya Biologicheskaya* 233.

Sugiura, M., Shimizu, H., Sugiyama, M., Kuratsu, T. and Hiraoto, F. (1975). German Patent 2,557,499.

Sugiura, M., Isobe, M., Oikawa, T. and Oono, H. (1976). *Chemical and Pharmaceutical Bulletin* **24**, 1202.

Synne, K. N. and Rennick, A. G. C. (1972). *Steroids* **19**, 293.

Szentirmai, A. (1977). *Proceedings of the 4th Enzyme Engineering Conference, Bad Neuenahr,* Poster T 26.

Takeda Chemical Industry (1972). United States PAtent 3,689,564.

Tan, L. (1972). Canadian Patent 890,901.

Tan, T. L., Strijewski, A. and Wagner, F. (1972). *Archives of Microbiology* **87**, 249.

Terada, O. and Uwajima, T. (1976). German Patent 2,527,068.

Terada, O., Yai, H. and Nakamura, S. (1973). United States Patent 3,776,816.

Terada, O., Nakamura, S., Yagi, H. and Machido, T. (1974). German Patent 2,021,465.

Teutsch, J. G., Rausser, R. C., Shapiro, E. L. and Herzog, H. C. (1970). United States Patent 3,665,017.

Tom, W. H., Abul-Jojj, Y. J. and Koveeda, M. (1975). *Journal of the Chemical Society Chemical Communications* 24.

Törmörkény, E., Toth, G., Horvarth, Gy. and Buki, K. G. (1975). *Acta Chimica Academiae Hungaricae* **87**, 409.

Törmörkény, E., Makk, N., Toth, G., Szabo, A. and Szentirmai, A. (1976). *Acta Chimica Academiae Hungaricae* **91**, 81.

Udvardy, E. N. (1974). *Acta Micriobiologica Academiea Scientiarum Hungaricae* **21**, 237.

Udvardy, E. N., Hantos, G. and Trinn, M. (1976). *Proceedings of the Vth. International Fermentation Symposium, Berlin.* p. 325.

Union of Soviet Socialist Republics Academy of Sciences (1972). Soviet Patent 323,975.

Union of Soviet Socialist Republics Academy of Sciences (1973a). Soviet Patent 376,440.

Union of Soviet Socialist Republics Academy of Sciences (1973b). Soviet Patent 366,712.

Uwajima, T., Yagi, H., Nakamuva, S. and Terada, O. (1973). *Agricultural and Biological Chemistry* **37**, 2345.

Uwajima, T., Yagi, H. and Terada, O. (1974). *Agricultural and Biological Chemistry* **38**, 1149.

Valcavi, U. (1971). *Farmaco-Edizione Scientifica* **25**, 105.

Valcavi, U. and Innocenti, S. (1973). *Farmaco-Edizione Scientifica* **29**, 194.

Valcavi, U., Corsi, B. and Tedeschi, S. (1973a). *Farmaco-Edizione Scientifica* **28**, 873.

Valcavi, U., Corsi, B., Marotta, V. and Tedeschi, S. (1973b). *Farmaco-Edizione Scientifica* **28**, 1024.

Valcavi, U., Japelj, J. and Tedeschi, S. (1973c). *Croatica Chemica Acta* **45**, 611.

Valcavi, U., Corsi, B., Innocenti, S. and Martelli, P. (1974). *Farmaco-Edizione Scientifica* **30**, 597.

Valcavi, U., Martelli, P., Sironi, U. C. and Tedeschi, S. (1975). *Farmaco-Edizione Scientifica* **30**, 464.

Vezina, C. and Rakhit, S. (1974). *In* 'Handbook of Microbiology' (A. I. Laskin and H. Lechevalier, eds.), vol. 4 p. 117. C.R.C. Press, Ohio.

Vezina, C., Seghal, S. N., Singh, K. and Kluepfel, K. (1971). *In* 'Progress in Industrial Microbiology', (D. J. D. Hockenhull, ed.), vol. 10, p. 1, Churchill Livingstone, London.

Vidic, H. J., Kieslich, K. and Lehmann, H. G. (1974). German Patent 2,306,529.

Voets, J. P. and Lamot, E. (1974). *Zeitschrift für Allgemeine Mikrobiologie* **14**, 77.

Waard, Van der, W. F., Doodeward, J., Flines de, J. and Weele van der, S. (1968). *Abhandlungen der Deutschen. Akademie der Wissenschaften, Berlin.* 101.

Wada, K. and Ishida, T. (1974). *Phytochemistry* **13**, 2755.

Wagner, F., Lang, S. and Kreiser, W. (1976). *Chemische Berichte* **109**, 3304.

Wallen, L. L., Stodola, F. H. and Jackson, R. W. (1959). 'Type Reactions in Fermentation Chemistry'. A.R.S.-71-13. Agricultural Research Service US Department of Agriculture, Peoria, Illinois.

Weaver, E. A., Kenney, H. E. and Wall, M. E. (1960). *Applied Microbiology* **8**, 345.

Weber, A., Müller, R., Kennecke, M., Eder, U. and Wiechert, R. (1977a). German Patent 2,558,088.

Weber, A., Müller, R., Kennecke, M., Eder, U. and Wiechert, R. (1977b). German Patent 2,558,089.

Weber, A., Müller, R., Kennecke, M., Eder, U. and Wiechert, R. (1977c). German Patent 2,558,090.

Weber, A., Kennecke, M. and Dahl, H. (1978). German Patent 2,632,677.

Weintraub, H., Vincent, F. and Baulieu, E. E. (1973). *Federation of European Biochemical Societies Letters* **37**, 82.

Whitehead, T. P. and Flegg, H. M. (1974). German Patent 2,305,232.

Wiechert, R., Kieslich, K. and Koch, H. (1974). Belgian Patent 835, 427.

Wix, G., Büki, K. G., Törmörkeny, E. and Ambrus, G. (1968). *Steroids* **11**, 401.

Wovcha, M. G. (1977). German Patent 2,647,895.

Wovcha, M. G. and Biggs, C. G. (1977). German Patent 2,703,645.

Wovcha, M. G. and Biggs, C. G. (1978). German Patent 2,746,383.

Yamashita, H. and Kurosawa, Y. (1975). *Agricultural and Biological Chemistry* **39**, 2243.

Yamashita, H., Shibata, K., Yamakoshi, N., Jurosawa, Y. and Mori, H. (1976a). *Agricultural and Biological Chemistry* **40**, 505.

Yamashita, H., Kurata, S. and Kursosawa, Y. (1976b). *Journal of the Agricultural Chemical Society of Japan* **50**, 61.

Yosika, I., Saijoh, S. and Kitagawa, I. (1972a). *Chemical and Pharmaceutical Bulletin* **20**, 564.

Yosika, I., Sugawara, S., Imai, K. and Kitagawa, I. (1972b). *Chemical and Pharmaceutical Bulletin* **20**, 2418.

Yosika, I., Marli, Y. and Kitagawa, I. (1973). *Chemical and Pharmaceutical Bulletin* **21**, 2092.

Yosika, I., Inada, A. and Kitagawa, I. (1974). *Tetrahedron* **30**, 707.

Zedan, H. H. and El-Tayeb, O. M. (1977). *Planta Medica* **31**, 163.

Zedan, H. H., El-Tayeb, O. M. and Absel-Aziz, M. (1976). *Planta Medica* **30**, 251.

Zelinski Organic Chemistry Institute (1975). Soviet Patent 457,722.

9. Penicillin Acylases and *Beta*-Lactamases

ERICK J. VANDAMME

Laboratory of General and Industrial Microbiology, University of Ghent, Coupure, 533, Ghent, Belgium.

I. PENICILLIN ACYLASE

A. Introduction

A wide range of microbial enzymes that are able to modify antibiotics have been described. Indeed, microbial bioconversions of antibiotics can vary from extensive degradations to well-defined reactions. Most bioconversion reactions result in inactivation or degradation of the antibiotic compound, thereby excluding a practical application for the products obtained (Sebek and Perlman, 1971; Sebek, 1975; Benveniste and Davies, 1973). In a few cases, however, useful compounds are formed, which can subsequently be used for industrial production of semisynthetic antibiotics or new antibiotic analogues (Jarvis and Berridge, 1969; Vandamme and Voets, 1974b; Sebek, 1974, 1975). Bioconversion of penicillins into 6-aminopenicillanic acid (6-APA)—the penicillin nucleus—and the side-chain acid is one of these important reactions (Fig. 1), as 6-APA is the starting compound for industrial production of the semisynthetic penicillins with superior clinical effectiveness. The microbial enzymes that hydrolyse penicillins into 6-APA have been given different names, but the name penicillin acylase (penicillin amidohydrolase EC 3.5.1.11) is now generally accepted (Vandamme and Voets, 1974b).

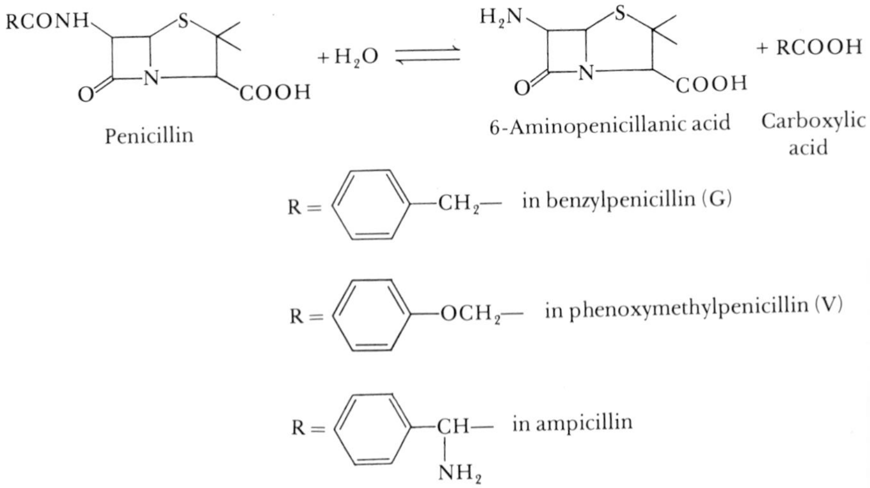

Fig. 1. Reactions involved in hydrolysis of penicillins to 6-aminopenicillanic acid, and re-acylation of the acid by penicillin acylases to give penicillins.

The commercial importance of 6-APA for production of semi-synthetic penicillins has forced the development of penicillin acylase research and application. Originally, the penicillin nucleus–6-APA—has been produced directly by fermentation, culturing *Penicillium chrysogenum* in a medium to which no precursor had been added (Kato, 1953; Ballio *et al.*, 1959; Batchelor *et al.*, 1959). Low yields and a complex 6-APA isolation procedure soon made this process obsolete (Carrington, 1971; Lee, 1974). Instead, 6-APA is now commercially produced by chemical or enzymic hydrolysis of penicillin G or V, readily obtained by fermentation. Already in 1950, it was claimed in Japan that 6-APA could be produced more easily by hydrolysing penicillin with an enzyme found in *Penicillium chrysogenum* Q176 and in *Aspergillus oryzae* but, at that time, these findings did not elicit full attention (Sakaguchi and Murao, 1950, 1955). However, as a result of the preparation of 6-APA by direct fermentation, the enzymic deacylation procedure of Sakaguchi and Murao revived interest. Soon, it was reported that penicillin acylase activity occurred in a wide range of bacteria, actinomycetes, yeasts and moulds, and that this simple bioconversion process allowed high yields of 6-APA to be obtained (Cole, 1967; Vandamme and Voets, 1974b).

At the moment, highly productive penicillin acylase strains are used to produce 6-APA on an industrial scale, starting from the bio-synthetic penicillins G and V. In addition, some penicillin acylases catalyse the reverse reaction, and can acylate 6-APA to form penicillin compounds (Fig. 1). Furthermore, certain cephalosporins—structural analogues of the penicillins—can also be deacylated or acylated into useful compounds.

B. Assay Methods for Penicillin Acylase Activity

At present, the most reliable method of assaying penicillin acylase consists in detection and measurement of the end-product 6-APA, formed from penicillins when incubated in the proper reaction conditions in the presence of the enzyme. Relevant properties of penicillins and of 6-APA have been summarized by Vandamme and Voets (1974b).

Usually, after paper- or thin-layer chromatographic resolution of media or reaction mixtures containing penicillins and 6-APA, the 6-APA is treated with phenylacetylchloride and thus converted again into penicillin G, which is then demonstrated by bioautographic methods (Cole et al., 1975). Despite the unstable nature of the products involved, this time-consuming bioautographic phenylacetylation technique is, until now, the only one widely applied that allows quantitative determination of 6-APA (Batchelor et al., 1961a; Uri and Sztaricskai, 1961; Lemke and Nash, 1972). Bondareva et al. (1969a, b), Vandamme et al. (1971a, b), Nara et al. (1971a) and Okachi et al. (1973a, b) detected 6-APA qualitatively in the presence of intact penicillins and other degradation products by means of thin-layer chromatography. Serova et al. (1973) used this technique for quality control of 6-APA samples. A semiquantitative method for determination of 6-APA, based on thin-layer chromatography, was also reported by Korchagin et al. (1971). Solvent systems, allowing a fast thin-layer chromatographic separation of substrate penicillins and 6-APA, have been described by Vandamme and Voets (1972a, b).

The colorimetric hydroxylamine method of Ford (1947) and Boxer and Everett (1949) for penicillin determination is also applicable to 6-APA determination if residual penicillins are previously extracted from the mixture at pH 2 with n-butylacetate. This quantitative, although time-consuming, 6-APA determination method is still widely used (Nara et al., 1971a, b; Cole et al., 1975). Ivashkiv (1964), Bomstein and Evans (1965) and Nys et al. (1972, 1973) described a colorimetric 6-APA dosage method suitable for fermentation media and reaction mixtures in the presence of residual penicillin; the estimation is based on the reaction of the 6-APA amino group with p-dimethylamino-benzaldehyde in acid medium (Balasingham et al., 1972). A similar method was described whereby 6-APA can be determined spectro-photometrically, D-glucosamine being used as reagent, in the presence of other penicillin-like compounds without prior separation (Shaikh et al., 1973; Gang and Shaikh, 1976).

A new colorimetric method for 6-APA estimation was recently reported by Kornfield (1978). The procedure is based on formation of a 2,4-pentanedione derivative of 6-APA, followed by reaction with p-dimethylaminobenzaldehyde. 6-Aminopenicillanic acid yields a red product and its absorbance is linear from 0 to 350 µg quantities at 538 nm. Penicillins do not interfere with this sensitive and quick assay, but 6-aminopenicilloic acid does.

Cole (1964, 1966, 1969a, b, c, d), Sjöberg et al. (1967), Vanderhaeghe et al. (1968) and Kutzbach and Rauenbusch (1974) estimated 6-APA quantitatively by titration of the liberated side-chain acid. Findlater and Orsi (1973) described an indicator method for penicillin acylase determination; in this procedure, proton release is measured during the hydrolysis process. These fast methods are useful for following the course of large-scale bioconversion reactions, unless penicilloic acid is present in the reaction mixtures.

Oostendorp (1972) described a microbiological plate assay for quantitative determination of 6-APA, using *Serratia marcescens* A.T.C.C. 27117 as the test organism. Intact biosynthetic penicillins do not interfere with the assay, so extraction procedures are no longer needed. Ampicillin interferes with the assay, however.

The penicillin acylase assay methods so far mentioned are most widely used and have proved their usefulness. Several other assays have been reported, and are summarized below, but they have not found a general use up to the moment. Niedermayer (1964) and Chiang and Bennett (1967) determined the liberated side chain of the penicillin molecules by means of gas chromatography as a measure of penicillin acylase activity. An iodometric titration method has been proposed, but it was useful only for quantitative determination of pure 6-APA preparations (Alicino, 1961). Chapman *et al.* (1964) used infrared spectroscopy to determine 6-APA. Nathorst-Westfelt *et al.* (1963) and Pruess and Johnson (1965, 1967), working with ^{35}S-labelled penicillin, could detect ^{35}S-labelled 6-APA by means of radioautography. Huang *et al.* (1960, 1963) extracted the nonconverted penicillin from the reaction mixture and treated the 6-APA with a β-lactamase. Subsequently, the resulting 'penicic acid' (6-aminopenicilloic acid) was colorimetrically determined by the ninhydrin reaction. Marrelli's (1968) chromophore method was found to be suitable for quantitative colorimetric determinations of 6-APA and has been used for screening penicillin acylase-producing strains (Shimizu *et al.*, 1975a, b). Gas–liquid chromatographic determination of 6-APA was described by Silingardi *et al.* (1977).

A simplified determination of the enzyme penicillin G acylase is based on the ability of the acylase to hydrolyse phenylacetyl-L-aspara-gine to L-asparagine and phenylacetic acid. As quickly as these products are formed, L-asparagine is hydrolysed to aspartic acid and ammonia by an excess of added L-asparaginase. The released ammonia is easily determined with Nessler's reagent (Bauer *et al.*, 1971). Kutzbach and Rauenbusch (1974) described also a new colorimetric assay, using

6-nitro-3-phenylacetamidobenzoic acid as a substrate. This test is similar to that applied by Walton (1964a, b) using phenylacetyl-4-nitroaniline as a colour indicator in broths screened for penicillin acylase-producing bacteria. It is not clear whether these indirect determinations of penicillin acylase are applicable to all penicillin acylases. Probably, this procedure is useful for acylases from *Escherichia coli* A.T.T.C. 9637 and A.T.C.C. 11105 and for acylases whose specificity is not closely connected with the nuclear structure of the penicillin molecule.

However, in view of the unstable nature of the products involved and because a convenient and rapid quantitative assay for 6-APA is not yet available, indirect quantitative assay methods should deserve full attention (Vandamme and Voets, 1974b). Units of penicillin acylase activity are generally expressed as μmoles of 6-APA produced per minute at an optimal pH value and temperature.

C. Screening Procedures for Penicillin Acylase-Producing Micro-Organisms

Originally, penicillin acylase was thought to occur preferentially in penicillin-producing fungi, but this enzymic activity was soon reported also in fungi that do not produce penicillins and in yeasts, actinomycetes and bacteria (Vandamme, 1977). After a period of random screening, several authors tried to select, on well-defined principles, micro-organisms that produce penicillin acylase. Holt and Stewart (1964a, b) argued that penicillin acylase activity is correlated with penicillin resistance or degradation. However, penicillin acylase activity is not so widespread as β-lactamase activity, as is shown by the results of the screening programmes elaborated by Ayliffe (1963, 1965), Demain et al. (1963), Cole and Sutherland (1966), Hamilton-Miller (1966), Rozansky et al. (1969), Nara et al. (1971a, b), Vandamme (1972), Vandamme and Voets (1971) and Shimizu et al. (1975a, b). Furthermore, co-existence of acylase and β-lactamase in the same microbial strain is a well-known phenomenon (Vandamme and Voets, 1974b). A selective isolation procedure proposed by Kameda et al. (1961) is based on growth on a mineral medium containing benzyl-penicillin or phenylacetic acid as the sole source of carbon; they correlated hydrolysis of acyl-DL-amino acids with the presence of penicillin acylase. Batchelor et al. (1961b) isolated 215 species of fungi, yeasts and actinomycetes and selected 38 penicillin acylase-producing strains after growth on corn-steep liquor media and related penicillin fermentation broths. Huang *et al.* (1963) detected 60 species of bacteria and actinomycetes displaying penicillin acylase activity, among 329 isolates grown on media containing corn-steep liquor, molasses or yeast extract. *Streptomyces* species displaying amide synthetase activity were isolated as possible penicillin acylase-producing strains by Haupt and Thrum (1967). On the assumption that compounds with a structural similarity, or an identical stereo-configuration to the penicillin molecule or to its side chain β-lactam ring part, might behave as selective agents in screening for penicillin acylase-producing micro-organisms, Vandamme and Voets (1973) were able to isolate acylase-producing micro-organisms from soil. They used a mineral medium to which *N*-acetylglycine, *N*-glycine-glycylglycine or phenylacetamide was added as the sole source of carbon and nitrogen. Mutants of *E. coli* with damaged glucose transport were described by Golub *et al.* (1973) as organisms that overproduce penicillin acylase.

The results of these test programmes clearly showed that penicillin acylase is a rather scarce enzymic activity of micro-organisms, and that, as yet, selective detection of a penicillin acylase-producing strain is difficult to predict. However, as none of the above-mentioned methods appears to be promising, the most productive screening programmes are still effected by the trial-and-error method. The most satisfactory procedures for selecting micro-organisms that produce penicillin acylase are those involving direct detection of 6-APA, by one of the methods described above, when the organism, after isolation and culturing, is incubated in the presence of different penicillins. However, the indirect determination of penicillin acylase activity from *E. coli*, as described by Bauer *et al.* (1971) and Kutzbach and Rauenbusch (1974), needs full attention and might be extended and applied to detect new acylase-producing micro-organisms.

D. Penicillin Acylase-Producing Micro-Organisms

Large-scale surveys indicate the rather small incidence of penicillin acylase-producing micro-organisms; nevertheless, this enzymic activity is not restricted to certain genera of micro-organisms.

Starting from 'natural' penicillin acylase-producing strains, the

Table 1

Micro-organisms that produce benzylpenicillin acylase

Micro-organism	Reference
BACTERIA	
Rhodopseudomonas spheroides K.Y. 4112	Nara *et al.* (1971a)
Pseudomonas aeruginosa K.Y. 3591, K.Y. 8501	
Pseudomonas cruciviae K.Y. 3960,	
Pseudomonas desmolytica K.Y. 3981	Okachi *et al.* (1973a, b)
Pseudomonas sp.	Huang *et al.* (1960)
Xanthomonas sp.	Huang *et al.* (1963)
Alcaligenes faecalis A-9424	Claridge *et al.* (1963)
Alcaligenes faecalis B.R.L. 1237, 1238	Cole and Sutherland (1966)
Bacterium faecalis alcaligenes 415	Gotovtseva *et al.* (1965)
Bordetella sp.	Huang *et al.* (1960)
Escherichia sp.	Rolinson *et al.* (1960), Gang and Shaikh (1976)
Escherichia coli A.T.C.C. 9637 (N.C.I.B. 8666)	Kaufmann and Bauer (1960), Vojtisek and Slezak (1975a, b, c), Sato *et al.* (1976)
Escherichia coli N$_y$I/3-67	Szentirmai (1964), Nyiri (1967)
Escherichia coli N.C.I.B. 9465	Holt and Stewart (1964a)
Escherichia coli B.M.N., K.Y. 8219, K.Y. 8268, K.Y. 8275, K.Y. 8289	Okachi *et al.* (1973a, b)
Escherichia coli I 187	Cole and Sutherland (1966)
Escherichia coli B.R.L. 351, B.R.L. 1360	Sjöberg *et al.* (1967)
Escherichia coli N.C.I.B. 8743 (B.R.L. 1040) 8743A	Cole (1969a), Plaskie *et al.* (1978)
Escherichia coli A.T.C.C. 11105 (N.C.I.B. 8878)	Bauer *et al.* (1971), Kutzbach and Rauenbusch (1974)
Escherichia coli XG3A 9455	Claridge *et al.* (1963)
Escherichia coli OIII. B$_4$	Dulong de Rosnay *et al.* (1970)
Escherichia coli N.C.I.B. 8134, 8878, 8879, 8949	Cole (1967)
Escherichia coli N.C.I.B. 8741, 8742, 8744	Cole and Sutherland (1966)
Kluyvera citrophila K.Y. 3641, PL-10, PL-21,	Nara *et al.* (1971a), Okachi *et al.* (1973a, b)
Kluyvera noncitrophila K.Y. 3642, K.Y. 8991	
Aerobacter cloacae	Claridge *et al.* (1960)
Erwinia sp.,	Huang *et al.* (1963)
Serratia sp.	
Proteus morganii K.Y. 4035, K.Y. 4051	Okachi *et al.* (1973a, b)
Proteus rettgeri F.D. 13424, A.T.C.C. 9919, 9250	Huang *et al.* (1963), Cole (1967)
Flavobacterium sp., K.Y. 4082	Huang *et al.* (1963), Shimizu *et al.* (1975a, b)
Micrococcus lysodeikticus	Claridge *et al.* (1960), Huang *et al.* (1960)

Table 1—*continued*

Micro-organism	Reference
Micrococcus roseus A.T.C.C. 515	Pruess and Johnson (1965)
Micrococcus luteus K.Y. 3781	Nara *et al.* (1971a), Shimizu *et al.* (1975a, b)
Sarcina sp.	Huang *et al.* (1963)
Bacillus subtilis var. *niger*	Claridge *et al.* (1960)
Bacillus megaterium A.T.C.C. 14945	Chiang and Bennett (1967)
Bacillus circulans	Abbott (1976)
Corynebacterium sp.,	Huang *et al.* (1963)
Cellulomonas sp.	
Arthrobacter sp.	Cole (1967)
Mycobacterium phlei	Claridge *et al.* (1960)
Nocardia F.D. 46973, A.T.C.C. 13635	Huang *et al.* (1960)
Streptomyces ambofaciens S.P.S.L.-15	Nara *et al.* (1971a)
FUNGI	
Neurospora crassa F.S.C. 987, D.G.C. 757, R.F. 424, R.W.B. 622, F.G.S.C. 262, 3a6A	Rossi *et al.* (1973)

penicillin-producing industry surely has obtained high-yielding constitutive mutants; once a suitable micro-organism has been isolated, mutation and genetic techniques have been employed to select strains with improved abilities.

The division of penicillin acylases into two classes, based on the type of micro-organism producing them, as proposed by Claridge *et al.* (1963), is no longer valid. Instead, penicillin acylases are now classified according to the type of penicillin that is preferentially hydrolysed irrespective of the type of micro-organism involved (Vandamme and Voets, 1974a, b, 1975b).

So far, three types of penicillin acylase can be clearly recognized; these are phenoxymethylpenicillin (penicillin V) acylase, benzylpenicillin (penicillin G) acylase, and D-α-aminobenzylpenicillin (ampicillin) acylase. However, other unusual acylases have been described but, as none of them has been purified to elucidate their substrate spectrum, it is not clear whether they interfere with the above-mentioned classification or whether they should be added as novel types (Vanderhaeghe *et al.*, 1968; Cole, 1966, 1969a; Pruess and Johnson, 1965; Vandamme *et al.*, 1971b; Vandamme and Voets, 1974b). A survey of strains reported to produce penicillin acylase is given in Tables 1, 2 and 3.

Table 2

Micro-organisms that produce phenoxymethylpenicillin acylase

Micro-organism	Reference
FUNGI	
Penicillium chrysogenum Q176	Sakaguchi and Murao (1950)
Penicillium chrysogenum A-9342	Claridge *et al.* (1963)
Penicillium chrysogenum W5120, W501247	Batchelor *et al.* (1959, 1961a, b)
W48701, W39133	
Penicillium chrysogenum Wis. 49-408,	Erickson and Bennett (1965)
Penicillium chrysogenum SC 3576	Erickson and Dean (1966)
Penicillium chrysogenum 51-20F3	Spencer and Maung (1970)
Penicillium B.R.L. 807, B.R.L. 733, 736, 737	Cole (1966)
Penicillium chrysogenum 50935, 1951,	Gatenbeck and Brunsberg (1968),
47-638, 49-2166	Pruess and Johnson (1967)
Penicillium chrysogenum SC-6041	Fawcett *et al.* (1975)
Emericellopsis minima (Stolk) I.M.I. 69015,	Cole and Rolinson (1961)
Cephalosporium salmosynnematum M.D.H.	
3590A	
Cephalosporium C.M.I. 49137	Claridge *et al.* (1963)
Cephalosporium acremonium A.T.C.C. 11550	Dennen *et al.* (1971)
Aspergillus niger	Vandamme *et al.* (1971a)
Aspergillus ochraceus B.R.L. 731	Cole (1966)
Epidermophyton interdigitale, *Epidermophyton*	Uri *et al.* (1963, 1964), Cole (1966)
floccosum B.R.L. 623, B.R.L. 722	
Trichophyton gypseum, *Trichophyton*	Uri *et al.*(1963, 1964), Cole (1966)
mentagrophytes B.R.L. 569, B.R.L. 579,	
Trichophyton interdigitale	
Alternaria, *Epicoccum*, *Mucor*, *Phoma*,	Batchelor *et al.* (1961b)
Trichoderma spp.	
Malbranchea pulchella	Rode *et al.* (1947), Kitano *et al.* (1974,
	1975)
Thermoascus, *Gymnoascus* and *Polypaecilum*	Kitano *et al.* (1975)
spp.	
Botrytis cinerea	Batchelor *et al.* (1961b)
Fusarium sp. 75-5	Thadhani *et al.* (1972)
Fusarium avenaceum	Vanderhaeghe *et al.* (1968)
Fusarium semitectum	Waldschmidt-Leitz and Bretzel (1964)
Fusarium semitectum B.C. 805	Baumann *et al.* (1971)
Giberella fujikuroi	Vasilescu *et al.* (1969)
Fusarium conglutinans A.Y.F. 254 conidia	Singh *et al.* (1969)
Fusarium moniliforme A.Y.F. 255, C.B.S.	Vandamme *et al.* (1971a)
24064, C.B.S. 44064, C.B.S. 26654 conidia	
Pleurotus ostreatus	Brandl (1965)
Bovista plumbea	Schneider and Rohr (1976)
Phialomyces macrosporus,	Kondo and Mitsugi (1975)
Leptosphaerulina australis, *Robillarda* sp.	

Table 2—*continued*

Micro-organism	Reference
YEASTS	
Torulopsis, Zygosaccharomyces, Debaryomyces and *Torula* spp., B.R.L. 809	Cole (1966, 1967)
Cryptococcus, Saccharomyces and *Trichosporon* spp.	Batchelor *et al.* (1961b)
Rhodotorula glutinis var. *glutinis*	Vandamme and Voets (1973)
BACTERIA	
Erwinia aroideae N.R.R.L. B-138	Voets and Vandamme (1972)
Achromobacter sp. B.R.L. 1755 (N.C.I.B. 9424)	Cole (1964)
Pseudomonas acidovorans	
Micrococcus ureae K.Y. 3767	Nara *et al.* (1971a)
Nocardia globerula K.Y. 3901	Nara *et al.* (1971a)
Streptomyces lavendulae B.R.L. 198	Batchelor *et al.* (1961b)
Streptomyces netropsis 2814, *Streptomyces erythreus* J.A. 4143	Haupt and Thrum (1967)
Streptomyces ambofaciens S.P.S.L.-15	Nara *et al.* (1971a)
Actinoplanes utahensis	Dennen *et al.* (1971)

Table 3

Micro-organisms that produce D- α-aminobenzylpenicillin acylase

Micro-organisms	Reference
Pseudomonas melanogenum K.Y. 3987, K.Y. 4030, K.Y. 4031	Nara *et al.* (1972), Okachi *et al.* (1973b)
Pseudomonas ovalis K.Y. 3962	Okachi *et al.* (1973b), Okachi and Nara (1973), Shimizu *et al.* (1975a, b)

E. Fermentation Processes for Production of Penicillin Acylase

For industrial use, high-yielding penicillin acylase-producing strains have been derived from natural isolates by mutation and selection techniques. For bacteria, media based on nutrient broth or yeast extract, corn-steep liquor, soyabean meal or malt extract and hydrolysed proteins or amino acids, with or without the addition of various mineral salts, have been described (Huang *et al.*, 1963; Cole and Sutherland, 1966; Savidge and Cole, 1975; Plaskie *et al.*, 1978). It is well known that the presence of carbohydrates in the media suppresses enzyme formation. With *E. coli* strains A.T.C.C. 9637, N.C.I.B. 8743 and A.T.C.C. 11105, the nitrogen (and carbon) content of the medium may be provided by complex organic sources such as corn-steep liquor, yeast extract, casamino acids or by simple chemically defined sources such as ammonium sulphate or sodium glutamate. Phenylacetic acid is also an important constituent of the fermentation medium as the most effective inducer. Enzyme production is greatest at temperatures between 24°C and 30°C at rather low rates of aeration. Enzyme synthesis is fully repressed by an increased concentration of dissolved oxygen (Kleiner and Lopatnev, 1972; Vojtisek and Slezak, 1975a, b, c). High levels of enzyme have been reported when phenylacetate (0.1%) is added gradually during the fermentation process which is terminated when the pH value rises to 8.0 after about 20–28 hours of incubation. Cells are then collected as a paste or slurry by centrifugation. It can be desirable to kill the cells before harvesting by adding *n*-butylacetate (1%) to the fermentation broth. Chemical and mechanical methods (ultrasonic disruption, Manton-Gaulin homogenization) can be applied for releasing the enzyme from cells. Debris can be removed by vacuum-filtration techniques or centrifugation. The clarified solution can then further be used for enzyme isolation and characterization.

Kaufmann and Bauer (1960) were the first to demonstrate that the penicillin G-acylase activity of *E. coli* A.T.C.C. 9637 could be stimulated by culturing the bacteria in the presence of phenylacetic acid or ammonium phenylacetate (Sikyta and Slezak, 1964). Also, Levitov *et al.* (1967) examined the stimulating effect of phenylacetic acid and its derivatives on biosynthesis of penicillin acylase by *E. coli* A.T.C.C. 9637 and found that a concentration of 1 mg of phenylacetic acid/ml resulted in an eight-fold stimulation of the acylase activity. Derivatives of phenylacetic acid and analogous compounds were inferior in this respect. This was confirmed by Vojtisek and Slezak (1975a, b, c) who found that the highest rate of enzyme synthesis was reached when phenylacetic acid was the only source of carbon and energy. Synthesis of the enzyme is subjected to complete catabolite repression by glucose and partial repression by acetate. Both types of repression are substantially influenced by cAMP. Growth diauxie in a medium containing glucose and phenylacetic acid serving as carbon and energy sources can be overcome by cAMP. Lactate can serve as a catabolically neutral source of carbon for maximal enzyme production, a fact demonstrated by Devcic and Divjak (1968). Similarly, Gang and Shaikh (1976) found that cAMP stimulated penicillin acylase activity, overcame the repression of glucose and restored enzyme synthesis to the non-repressed levels in *E. coli*. Penicillin acylase synthesis by *Kluyvera citrophila* K.Y. 3641 and K.Y. 7844 is also controlled by catabolite repression, which is diminished by addition of cAMP (Takasawa *et al.*, 1972; Shimizu *et al.*, 1975a, b). Golub *et al.* (1973) isolated mutants of *E. coli* with an impaired transport of glucose, in which synthesis of penicillin acylase was partially insensitive to repression by glucose.

Szentirmai (1964) isolated *E. coli* NY 1/3-67 with a low acylase activity. By addition of aromatic carbonic acid and phenylacetic acid and its derivatives, and phenoxyacetic acid, which themselves strongly inhibit the function of this enzyme, biosynthesis of penicillin G acylase was increased 10-fold. Production of this enzyme was effectively repressed by metabolic carbohydrates and polyalcohols (glucose, fructose, lactose, maltose and glycerol). The specific activity of the bacterial cells also depended on temperature, low temperature being favourable for acylase production.

The penicillin G acylase activity of *Bacterium faecalis alcaligenes* No. 415 cells also depends on the composition of the growth medium, abundant growth being invariably accompanied by low enzymic activity of the culture. Acetic and pyruvic acids supplied to the nutrient medium greatly decreased production of penicillin acylase, whereas peptone and casein hydrolysate had a stimulatory effect. More particularly, tryptophan, other indole derivatives, and anthranilic acid stimulated acylase biosynthesis (Gotovtseva and Levitov, 1965; Gotovtseva *et al.*, 1965, 1968).

Penicillin acylases of *Ps. melanogenum* and *Kl. citrophila* were produced in a medium containing peptone (1%), yeast extract (1%), meat extract (0.5%) and sodium chloride (0.25%) at pH 7.2. Sodium glutamate (0.5%) and 15 mM phenylacetic acid were added as inducers (Okachi and Nara, 1973). Maximal enzyme levels were obtained after 24 hours fermentation at 30°C, with an aeration of 150 litres per minute at 650 rev./min in 300-litre fermentors. With *Bacillus megaterium* A.T.C.C. 14945, a medium consisting of enzyme-hydrolysed casein (3%) and glucose (0.5%) is used. After incubation at 30°C for eight hours, 0.15% phenylacetate is added and incubation is continued for about 70 hours. The whole broth is then treated with a flocculant and toluene. After adjustment of the pH value to 7.0–7.5, the cells are removed by centrifugation and the supernatant, which contains the enzyme, further purified.

The enzyme system of *Bacillus megaterium* A.T.C.C. 14945 was used as a model system to study the effect of various growth-limiting nutrients on enzyme production in continuous culture. Preferential excretion of penicillin acylase over other proteins during sulphur-limited growth was demonstrated. The enzyme is produced as a growth-associated metabolite (Acevedo and Cooney, 1973). With fungi, actinomycetes and yeasts, media based on corn-steep liquor and lactose (Cole, 1966), peptone and malt extract and Lab Lemco peptone (Batchelor *et al.*, 1961a, b) have been shown to support good production of enzyme.

A spore suspension is normally used to inoculate precultures, based on corn-steep liquor, sucrose and calcium carbonate at pH 5.5. Incubation occurs at 26°C for three days. Production can also occur in the Jarvis and Johnson (1950) medium supplemented with 0.2% phenoxyacetic acid. The mycelium is collected after three days at 26–28°C by filtration and can further be dried and extracted.

The penicillin V acylase of strains of *Fusarium* sp., *Trichophyton* sp., *Epidermophyton* sp. and *Aspergillus orchraceus* could be stimulated by addition of phenoxyacetic acid or its derivatives to the growth medium (Uri *et al.*, 1963, 1964; Cole, 1966; Vandamme and Voets, 1974a, b; Vanderhaeghe, 1975). More particular, the enzyme from *Fusarium* sp. was formed adaptively when the organism was grown in the presence of phenoxyacetate or its derivatives. The amount of enzyme induced depended on the nature of the substitution of phenoxyacetate and on the amino-acid moiety of the phenoxyacetyl-L-alanine, and decreased with higher homologues of aliphatic amino acid. The presence of phenyl, heterocyclic or hydroxyl groups on the side chain or the absence of the α-carboxyl group in the amino-acid moiety did not decrease the inducer effect of the compound. Phenoxyacetyl-D-alanine failed to induce enzyme formation, indicating that enzyme formation exhibited marked stereospecificity toward the amino-acid moiety of phenoxyacetyl-amino acids. Although the phenoxyacetyl-amino acids served as good inducers of the enzyme, none of them served as substrate for the enzyme (Thadhani *et al.*, 1972).

F. Characterization and Properties of Penicillin Acylase

1. Phenoxymethylpenicillin acylase

Many investigations based on the use of intact cells or crude enzyme preparations as catalysts have led to wrong conclusions about the substrate specificity of these enzymes. Although rather few phenoxymethylpenicillin (penicillin V) acylases have been fully characterized, enough data have accumulated by now to confirm that these acylases are highly specific enzymes and are not aspecific deacylating enzymes as is the case for the other types of penicillin acylase. Relevant properties of phenoxymethylpenicillin acylases are summarized in Table 4. This type of enzyme is produced as well by fungi, including basidiomycetes and yeasts, as by bacteria and actinomycetes. Penicillin V acylases are mainly intracellular enzymes, with optimal hydrolytic acitivity at pH 7 to 8, but exceptions to these general characteristics are known (Table 4). In particular, *Streptomyces* sp. produces the enzyme also extracellularly.

Waldschmidt-Leitz and Bretzel (1964) detected in *Fusarium semitectum* an intracellular penicillin acylase with a pH optimum of 7.5 with penicillin V as substrate. They liberated the enzyme by extracting acetone-dried mycelium with 0.2 N sodium acetate solution at pH 6. After treatment with DEAE-cellulose at pH 8, separation on Sephadex G-25 and lyophilization, the extract was redissolved in 0.005 N sodium citrate. After precipitation with acetone and chromatography on Amberlite IRC-50 and Sephadex G-25, a 300-fold purified enzyme extract was obtained. Ultracentrifugation experiments indicated a molecular weight of about 65,000. The nitrogen content was 16.7%, and two atoms of zinc per molecule were found, indicating the role of this metal as a cofactor. The enzyme was strongly inhibited by metal-chelating agents such as 8-hydroxyquinoline. Its activity was fully restored by addition of zinc sulphate and partially restored with sulphates of manganese, magnesium, cobalt and ferrous ion. Brandl (1965, 1972) extended these studies and confirmed the preferential and specific conversion of penicillin V into 6-APA. The activation energy for cleavage of phenoxymethylpenicillin was 38.5 kJ/mole. After testing 52 different biosynthetic and semisynthetic penicillins as possible substrates for the purified enzyme, he confirmed that bulky substituents and the presence of double bonds in the side chain of the

Table 4

Relevant properties of phenoxymethylpenicillin acylases

| | Optimum pH value | | Optimum temperature (°C) | K_m Value (mM) | Molecular weight | References |
	Hydrolysis	Synthesis				
MOULDS						
Penicillium chrysogenum Wis. 49408,	8.5	6.8	30	—	—	Erickson and Bennett (1965)
Penicillium chrysogenum P5009						Spencer and Maung (1970)
Penicillium chrysogenum 51-20F3 (enzyme)	8	7.4	28	16.7	—	
Penicillium chrysogenum sp.	8	—	20	—	—	Cole and Rolinson (1961), Cole (1966)
Cephalosporium sp.	8	—	—	—	—	
Emericellopsis minima (Stolk) I.M.I. 69015	8	—	—	—	—	Uri et al. (1963)
Trichophyton mentagrophytes	8	—	—	—	—	
Epidermophyton interdigitale						
Fusarium avenaceum	7.5	—	37	—	—	Vanderhaeghe et al. (1968)
Fusarium conglutinans A.Y.F. 254	8	—	28	—	—	Singh et al. (1969)
Fusarium moniliforme A.Y.F. 255	8	—	28	5.75	—	Vandamme et al. (1971a)
Fusarium semitectum (intact cells)	7.5	—	—	2.50–2.75	—	Brandl (1965, 1972)
(enzyme)	8	—	50		65,000	Waldschmidt-Leitz and Bretzel (1964)
Fusarium semitectum B.C. 805 (enzyme)	7.5	—	37	4.75	67,000	Baumann et al. (1971)
Pleurotus ostreatus	8	—	50	—	—	Brandl (1965)
Bovista plumbea (enzyme)	7.5	—	52	1.67	80,000	Schneider and Röhr (1976)

						Reference
YEASTS						
Rhodotorula glutinis	6.5	—	28	5.1	—	Vandamme and Voets (1973)
BACTERIA						
Achromobacter B.R.L. 1755	—	—	—	—	—	Cole (1964)
Erwinia aroideae (enzyme)	5.5	—	28	35	62,000	Vandamme (1972), Voets and Vandamme (1972), Vandamme and Voets (1975a)
Micrococcus ureae K.Y. 3767	7.4	—	35	—	—	Nara *et al.* (1971a)
Streptomyces lavendulae B.R.L. 198	9	—	28	10.3	—	Batchelor *et al.* (1961b)
Streptomyces erythreus J.A. 4143,						
Streptomyces netropsis 2814	7.5	—	28	—	—	Haupt and Thrum (1967)
Streptomyces ambofaciens S.P.S.L.-15	7.4	—	35	—	—	
Nocardia globerula K.Y. 3901	7.4	—	35	—	—	Nara *et al.* (1971a)

penicillin molecule cause steric hindrance, which results in a lowered binding capacity to the active site of the enzyme.

Baumann *et al.* (1971) reported on the substrate specificity of the penicillin acylase from *Fusarium semitectum* B.C. 805. When cultivated in media that contain phenoxyacetic acid as an inducer, the intact mycelium is able both to hydrolyse penicillin V into 6-APA and to deacylate a number of N-phenoxyacetylamino acids. They showed that this pattern of substrate specificity is due to the action of two different enzymes. By extraction of the mycelium with distilled water and 0.2 M sodium chloride, and by precipitation with acetone, ultracentrifugation and chromatography on calcium phosphate gel, the two enzymes were separated. The penicillin V acylase was purified 20.5-fold with a recovery of 43%. It displayed most pronounced specificity for hydrolysis of penicillin V, but lacked the ability to split either N-phenoxyacetylamino acids of benzylpenicillin. The K_m value for hydrolysis of phenoxymethylpenicillin was 4.75 mM. The other enzyme, called phenoxyacetylase, was purified 15-fold and preferentially deacylated N-phenoxyacetyl-L-alanine and other N-acetylamino acids at pH 7–8 at 40°C but displayed no hydrolytic action on penicillins. These results were confirmed by Thadhani *et al.* (1972) who proved that phenoxyacetylamino acids serve as good inducers but are certainly not substrates for the enzyme.

The penicillin V acylase formed constitutively by the basidiomycete *Bovista plumbea* has been purified 220-fold with a recovery of 40% by a combination of two filtration runs on Sephadex G-25, ion-exchange chromatography on DEAE-cellulose, ultrafiltration and final chromatography on hydroxyapatite (Schneider and Röhr, 1976). The molecular weight was estimated as 80,000; kinetic properties are given in Table 4. The activation energy was calculated at 19.45 kJ/mole. Neither 8-hydroxyquinoline, EDTA, iodoacetate nor the products of enzymic cleavage of 6-APA or phenoxyacetic acid, displayed any characteristic inhibitory effect. Metal ions or thiol groups are here not necessary for enzyme action, which is in contrast with the enzyme from *F. semitectum* (Waldschmidt-Leitz and Bretzel, 1964; Brandl, 1972). Phenoxymethylpenicillin was found to be the best substrate followed by p-hydroxyphenoxymethylpenicillin, o-kresoxymethylpenicillin, benzylpenicillin and thymoxymethylpenicillin and N-acylamino acids; dipeptides and tripeptides were not hydrolyzed. Affinity occurred only towards penicillins lacking a nitrogen atom in the side-chain acid. Penicillins with aryloxy residues possessing hydrophilic groups are favoured over aryl residues and short side chains over bulky ones. The penicillin acylase preparation, purified from *Penicillium chrysogenum* 51-20F3, also displayed a high specificity towards the penicillin molecule (Spencer and Maung, 1970).

Penicillin V acylases produced by yeasts (Batchelor *et al.*, 1961; Cole, 1966, 1967; Vandamme and Voets, 1973) and by actinomycetes (Batchelor *et al.*, 1961b; Dennen *et al.*, 1971; Haupt and Thrum, 1967; Nara *et al.*, 1971a) have not yet been purified, but they all display preferential hydrolysis of penicillin compounds.

Few bacterial penicillin V acylases have been described (Cole, 1964; Nara *et al.*, 1971a; Vandamme *et al.*, 1971b). Vandamme *et al.* (1971b) detected a penicillin V acylase in cultures of *Erwinia aroideae*. Cell suspensions of this bacterium rapidly hydrolyse penicillin V and pheneticillin into 6-APA at pH 5.6, while penicillin G is hardly affected. At pH 8, intact cells rapidly hydrolyse different amides, while the acylase activity is very low at this pH value. The enzyme was extracted and purified 98-fold by repeated gel filtration. The purified enzyme, with a molecular weight of 62,000, was highly specific for penicillin V and was inhibited by phenoxyacetic acid. The purified acylase displays neither synthetic, amidohydrolase nor transacylase activity (Vandamme, 1972; Voets and Vandamme, 1972; Vandamme and Voets, 1975a).

Crude preparations of penicillin V acylase from *Penicillium chrysogenum* Wis. 49-308, S.C. 3576, B.R.L. 807, 51-20F3, a yeast strain B.R.L. 809 and *Erwinia aroideae* have been claimed to be able to carry out—in addition to hydrolysis of penicillin V into 6-APA—the reverse reaction, namely synthesis of penicillin V and G from 6-APA and phenoxy- or phenylacetic acid and their derivatives (Fig. 1, p. 468) (Erickson and Dean, 1966; Cole, 1966; Vandamme and Voets, 1974a, b, 1975a, b). However, none of the above-mentioned purified penicillin V acylases displayed this activity. Furthermore, Brunner *et al.* (1968a) and Spencer (1968) obtained evidence that resynthesis of 6-APA into penicillin by *Penicillium* sp. is effected by enzymes quite different from penicillin acylase. It follows that the reversible character of penicillin V acylases is still doubtful (Vandamme, 1977) although Spencer and Maung (1970) have claimed the identity of these hydrolytic and synthetic enzymes in *Penicillium chrysogenum* 51-20F3.

Table 5

Relevant properties of benzylpenicillin acylases

| | Optimum pH value | | Optimum temperature (°C) | K_m Value (mM) | Molecular weight | Reference |
	Hydrolysis	Synthesis				
BACTERIA						
Alcaligenes sp.	8	—	40	—	—	Claridge *et al.* (1960)
Alcaligenes faecalis	7.5	—	37	—	—	Claridge *et al.* (1963)
Escherichia coli A.T.C.C. 9637						
(intact cells)	7.5	4.5–5.5	30	—	—	Kaufmann and Bauer (1960)
(enzyme)	7.8–8	—	50–52	—	—	Bondareva *et al.* (1969a, b)
(immobilized enzyme)	7.5	5	37	7.7	—	Self *et al.* (1969)
Escherichia coli N.C.I.B. 8734A						
(intact cells)	8.2	5	50	30	—	Cole (1969a, b, c, d)
(enzyme)	8.2	—	37	0.67	—	Balasingham *et al.* (1972)
(immobilized enzyme)	7.65	—	—	0.63	—	Warburton *et al.* (1972, 1973), Lilly *et al.* (1972)
Escherichia coli	7	—	30–35	1.35–1.59	—	Brandl (1965)
Escherichia coli	7.5	—	—	17.5	—	Badr-Eldin and Attia (1973)
Escherichia coli C15 (N.C.I.B. 9465)	5.5	—	—	4.00	—	Holt and Stewart (1964a)
Escherichia coli A.T.C.C. 11105 (enzyme)	8.1	—	—	0.02	70,000	Kutzbach and Rauenbusch (1974)
Kluyvera citrophila K.Y. 3641, K.Y. 7844	7.5	6.5	35	—	63,000	Nara *et al.* (1971a), Okachi *et al.* (1973b), Shimizu *et al.* (1975a, b)

Proteus rettgeri F.D. 13424	8	—	—	—	Huang et al. (1963)
Micrococcus roseus A.T.C.C. 516	9	35	—	—	Pruess and Johnson (1965)
Bacterium faecalis alcaligenes 415	7.8	42	—	—	Gotovtseva et al. (1965)
Bacillus megaterium A.T.C.C. 14945 (enzyme)	8.5	—	4.5	120,000	Chiang and Bennett (1967), Acevedo and Cooney (1973)
(immobilized enzyme)	8	—	6.0	—	Ryu et al. (1972a, b)
Nocardia sp. F.D. 46973	8	28	—	—	Huang et al. (1960)
Streptomyces ambofaciens S.P.S.L.-15	7.4	28	—	—	Nara et al. (1971a)
FUNGI					
Neurospora crassa	7.0	37	—	—	Rossi et al. (1973)

2. Benzylpenicillin acylase

Benzylpenicillin (penicillin G) acylases are produced as well by bacteria, including actinomycetes (Claridge *et al.*, 1960; Huang *et al.*, 1960; Nara *et al.*, 1971a) and fungi (Rossi *et al.*, 1973). In sharp contrast to the penicillin V acylases, the substrate spectrum of penicillin G acylases is rather broad and these enzymes can be considered as aspecific deacylating enzymes. Again, relatively few penicillin G acylases have been purified to homogeneity, but sufficient data are now available to confirm the above statement.

Relevant properties of benzylpenicillin acylases are summarized in Table 5. Penicillin G acylases again are mainly intracellular enzymes, with optimal hydrolytic action at pH 7 to 9 (Table 5). So far, only *Bacillus megaterium* is known to produce the enzyme extracellularly.

The penicillin acylase activity of *E. coli* A.T.C.C. 9637 is so far the most intensively studied. From numerous experiments it is clear that this bacterial strain contains an enzyme that is specific for phenyl-acetylated compounds and is not a specific penicillin G acylase (Kaufmann and Bauer, 1960, 1964a, b; Sikyta and Slezak, 1964; Lucente *et al.*, 1965; Pruess and Johnson, 1965; Levitov *et al.*, 1967; Devic and Divjak, 1968; Kleiner and Lopatnev, 1972; Self *et al.*, 1969; Marconi *et al.*, 1973; Rossi *et al.*, 1977).

Bondareva *et al.* (1969a, b) isolated the penicillin G acylase of *E. coli* A.T.C.C. 9637 by ultrasonic disruption and obtained a 100-fold purified extract. The preparation was electrophoretically homogeneous and catalysed not only hydrolysis of penicillin G and other phenylacetylated substrates, but also their synthesis. Maximal hydrolytic activity was found in the pH range 7.8–8 and the optimal temperature was 50–52°C. It was found that, both in reactions leading to hydrolysis and resynthesis, the purified enzyme is specific with regard to phenyl- and phenylacetylated compounds. All alterations in the structure of the acyl part of the substrate molecule (benzylpenicillin) resulted in a decreased acylase activity. Besides penicillin G and ampicillin, the purified enzyme also hydrolysed a number of phenyl- and phenoxyacetylated amino acids, and therefore seems not to be an enzyme of narrow specificity.

Cole (1969a, b, c, d) reported in detail on the penicillin acylase activity of *E. coli* N.C.I.B. 8743A, a mutant derived from *E. coli* A.T.C.C. 9637. This strain produced, after induction with phenylacetic

Table 6.

Hydrolysis of semisynthetic cephalosporins into 7-aminocephalosporanic acid (7-ACA) by penicillin G acylases

Micro-organism	Substrate	References
Escherichia coli B.R.L. 351	Benzylcephalosporin	Sjöberg *et al.* (1967)
Escherichia coli N.C.I.B. 8743A	Cephalothin	Sjöberg *et al.* (1967)
Escherichia coli B.R.L. 351	Cephaglycin	Cole (1969a)
Nocardia sp. F.D. 4697, *Proteus rettgeri* F.D. 13424	Phenoxyacetyl-7ACA, Phenylmercapto-7ACA	Huang *et al.* (1963)
Kluyvera citrophila K.Y. 3641, K.Y. 7844, *Kluyvera noncitrophila* K.Y. 3642, *Micrococcus luteus* K.Y. 3781, *Achromobacter* sp. K.Y. 3048, *Flavobacterium* sp. K.Y. 4082	Cephalothin	Nara *et al.* (1971b), Shimizu *et al.* (1975a, b)

acid, an intracellular acylase with a pH optimum of 8.2 at 50°C. A suspension of *E. coli* treated with 1% *n*-butyl acetate was used as enzyme preparation. The specificity was directed to the acyl group of the penicillin molecule, rather than to the nucleus; *p*-hydroxybenzyl-penicillin was the best substrate, followed by benzylpenicillin. Also, ampicillin was hydrolysed into 6-APA, and some semisynthetic cephalosporins were transformed into 7-aminocephalosporanic acid (Table 6). In addition to penicillin hydrolysis, a wide range of amides, *N*-acylglycines and acylated L-α-amino acids were hydrolysed, indicating that the enzyme involved is an amidohydrolase. A K_m value of 30 mM for penicillin G hydrolysis was found. A 100-fold purification process resulted in an enzyme preparation revealing that the penicillin acylase activity is still associated with amidohydrolase activity. Using a purified preparation of penicillin acylase from *E. coli* N.C.I.B. 8743, Plaskie *et al.* (1978) essentially confirmed these findings. In these aspects, the enzyme from *E. coli* N.C.I.B. 8743A closely resembles that from *E. coli* A.T.C.C. 9637. Penicillin acylase was extracted on a large scale from *E. coli* N.C.I.B. 8743A by mechanical disruption in a Manton-Gaulin homogenizer and purified as described by Self *et al.* (1969). The enzyme was shown to be inhibited by benzylpenicillin and by both of the products of hydrolysis (Balasingham *et al.*, 1972). Inhibition by phenylacetic acid was competitive, whereas 6-APA inhibits noncompetitively, an inhibition pattern similar to that found for the enzyme from *Bacillus megaterium*, described by Chiang and Bennett (1967) and Ryu *et al.* (1972a, b).

The penicillin acylase from *E. coli* A.T.C.C. 11105 has been purified to maximum specific activity and crystallized (Kutzbach and Rauenbusch, 1974): the molecular weight of the enzyme is 70,000. In 1% sodium dodecyl sulphate, the enzyme is partially and reversibly dissociated, yielding units with a molecular weight of 20,500. The purified enzyme contains several active forms distinguished by their electrophoretic behaviour; two main types, with isoelectric points of 6.7–6.8 and 6.3–6.4, have been identified. The enzyme hydrolyses amides of phenylacetic acid and some closely related aryl aliphatic acids with little preference for the structure of the amine component, which can be ammonia, an aromatic amine, an amino acid or 6-aminopenicillanic acid. If the α-carbon atom of the amine is asymmetric, there is a strong specificity for the natural L-configuration. Small changes in the phenylacetyl moiety of the substrates cause much lower rates of hydrolysis. The enzyme is weakly inhibited by the reaction products, phenylacetic acid (competitive inhibition; K_i, 0.2 mM) and 6-APA (noncompetitive inhibition; K_i, 1.5 mM). The enzyme is irreversibly inhibited by phenylmethylsulphonylfluoride, but not by di-isopropylfluorophosphonate, indicating that serine is not located in the active centre. The properties of this penicillin acylase are very similar to those of a high-yielding mutant strain of *E. coli* N.C.I.B. 8743A, which is derived from *E. coli* A.T.C.C. 9637 (Kutzbach and Rauenbusch, 1974); it thus seems that the mutations carried out to increase the yield of penicillin acylase did not affect the structural gene involved.

Nevertheless, this enzyme from *E. coli* is clearly different from the extracellular penicillin G acylase purified from *Bacillus megaterium* A.T.C.C. 14945 with respect to molecular weight, K_m value and stability at different pH values. Indeed, Chiang and Bennett (1967) purified the extracellular penicillin amidase of *B. megaterium* A.T.C.C. 14945 approximately 96-fold by means of two cycles of adsorption and elution from Celite, followed by further fractionation on CM-cellulose. From ultracentrifugation data, they concluded that they had obtained a homogeneous preparation with an apparent molecular weight of approximately 120,000. The enzyme is rather specific for benzylpenicillin and has a pH optimum between 8 and 9. Other penicillins, and non-penicillin derivatives of phenylacetic acid such as phenylacetamide and derivatives, were also hydrolysed although at a slower rate. The enzyme displayed no activity on other amides, dipeptides, polypeptides and proteins. At high concentrations of substrate, hydrolysis of benzylpenicillin was inhibited by the reaction products; 6-APA acts as a noncompetitive inhibitor, whereas phenylacetic acid is a competitive inhibitor. A K_m value of 4.5 mM was found.

Shimizu *et al.* (1975a) reported on purification and properties of the penicillin acylase from *Kluyvera citrophila* K.Y. 7844. The enzyme was purified about 120-fold by DEAE-cellulose and hydroxyapatite chromatography, followed by isoelectrofocusing fractionation. The molecular weight was calculated as 63,000. 6-Aminopenicillanic acid was formed from penicillin G and ampicillin, and to a lesser extent from penicillin V and carboxybenzylpenicillin. 7-Aminocephalosporanic acid (7-ACA) was formed from cephalothin, and cephalexin was hydrolysed into 7-aminodeacetoxycephalosporanic acid (7-ADCA). The K_m value was estimated to be 1.4 mM with cephalexin. This acylase seemed to be as specific for cephalosporins as for penicillins, but data were not given for hydrolysis of other phenylacetic acid derivatives.

Penicillin G acylases from actinomycetes and fungi have not yet been fully characterized but, in addition to penicillin G, they all hydrolysed a range of N-phenylacetyl-L-α-amino acids (Nara *et al.*, 1971a; Rossi *et al.*, 1973, 1977).

Many penicillin acylase-producing bacterial strains not only catalyse hydrolysis of penicillin G, but also the reverse reaction starting from 6-APA and phenylacetic acid or its derivatives. This synthetic reaction occurs at acid pH values (4.5–5.5), especially when the 6-APA concentration is raised. This phenomenon has been demonstrated with cell suspensions or crude extracts of *E. coli* A.T.C.C. 9637 (Bauer *et al.*, 1960; Kaufmann and Bauer, 1960; Bondareva *et al.*, 1969a, b; Self *et al.*, 1969), *E. coli* (Rolinson *et al.*, 1960), *Alcaligenes faecalis* (Claridge *et al.*, 1960), *E. coli* N.C.I.B. 8743A (Cole, 1969c, d) and *Kluyvera citrophila* (Nara *et al.*, 1971b). As the substrate specificity of the hydrolytic and synthetic reaction was identical, many authors concluded that only one enzyme is involved in the processes of hydrolysis and synthesis. The synthetic activity of the acylase of *E. coli* N.C.I.B. 8743A was studied in more detail by Cole (1969c, d). The optimum pH value for resynthesis was 5.0. The rate of penicillin G synthesis improved in the presence of energy-rich compounds when phenylacetylglycine was added to 6-APA instead of phenylacetic acid. According to Brunner *et al.* (1968a, b), resynthesis can occur also in a non-enzymic way, if at least the appropriate energy-rich side-chain group is available. Penicillin G, ampicillin and hydroxypenicillin were synthesized in this way. This acylase also catalysed acylation of hydroxylamine by acids or amides to yield hydroxamic acids. Whether this transacylase activity and the resynthetic activity are due to the action of different enzymes, or to the multiple activities of non-specific amidase, will not be known until extensive enzyme purification and characterization have been carried out. However, Bondareva *et al.* (1969b) were able to demonstrate both hydrolytic and synthetic activity with a purified, electrophoretically homogeneous acylase preparation from *E. coli* A.T.C.C 9637 cells. The rates of synthetic reaction—acylation of 6-APA with carboxylic acids—increased significantly when derivatives of these acids were used with amino acids and thioglycollic acid.

Nara *et al.* (1971a, b, 1972) reported also that *Kl. citrophila* K.Y. 3641 and its β-lactamase-deficient mutant K.Y.-7844 were able to produce good yields of D-(−)-α-aminobenzylpenicillin from 6-APA and phenylglycine derivatives. The optimum pH value for synthesis was 6.5.

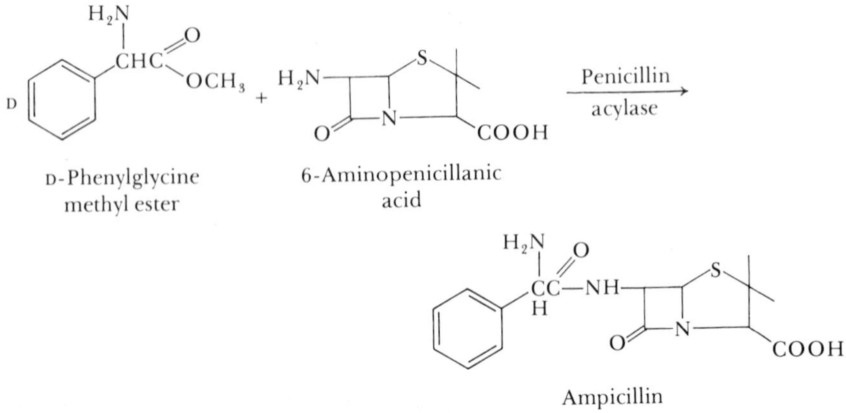

Fig. 2. Reactions involved in enzymic synthesis of ampicillin from 6-aminopenicillanic acid and D-phenylglycine methyl ester.

Among various phenylglycine derivatives examined as substrates, D-phenylglycine methyl ester was the best, yielding high levels of ampicillin (Okachi *et al.*, 1973a, b; Takasawa *et al.*, 1972; Shimizu *et al.*, 1975b; Fig. 2). The purified acylase from *Kl. citrophila* K.Y. 7844 displayed a potent synthetic ability; 6-APA, 7-ACA and 7-ADCA (Figs. 3, 4, 5) could be acylated with 1-(1*N*)-tetrazolylacetate methyl esters to produce the corresponding β-lactam antibiotics (Shimizu *et al.*, 1975a, b).

Profiting from this synthetic activity of penicillin G acylases, enzymic synthesis of many semisynthetic penicillins and cephalosporins has been achieved (Kaufmann and Bauer, 1960; Bondareva *et al.*, 1969b; Cole, 1969c, d; Nara *et al.*, 1972; Okachi *et al.*, 1973a, b; Shimizu *et al.*, 1975a, b). However, the rather low yield and the reversibility of these processes have precluded so far industrial application of this enzymic synthesis of semisynthetic β-lactam antibiotics, although the use of immobilized acylases or cells for this synthesis looks promising.

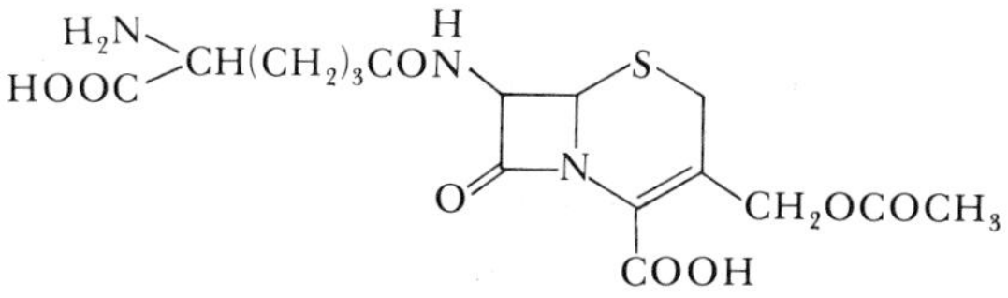

Cephalosporin C

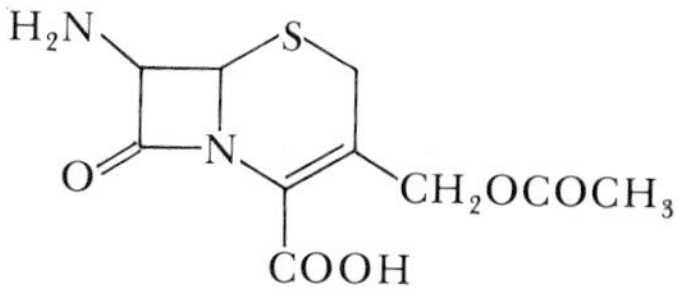

7-Aminocephalosporanic acid

Fig. 3. Structures of cephalosporin C and 7-aminocephalosporanic acid.

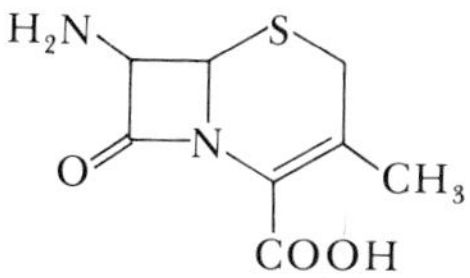

Fig. 4. Structure of 7-aminodeacetoxycephalosporanic acid.

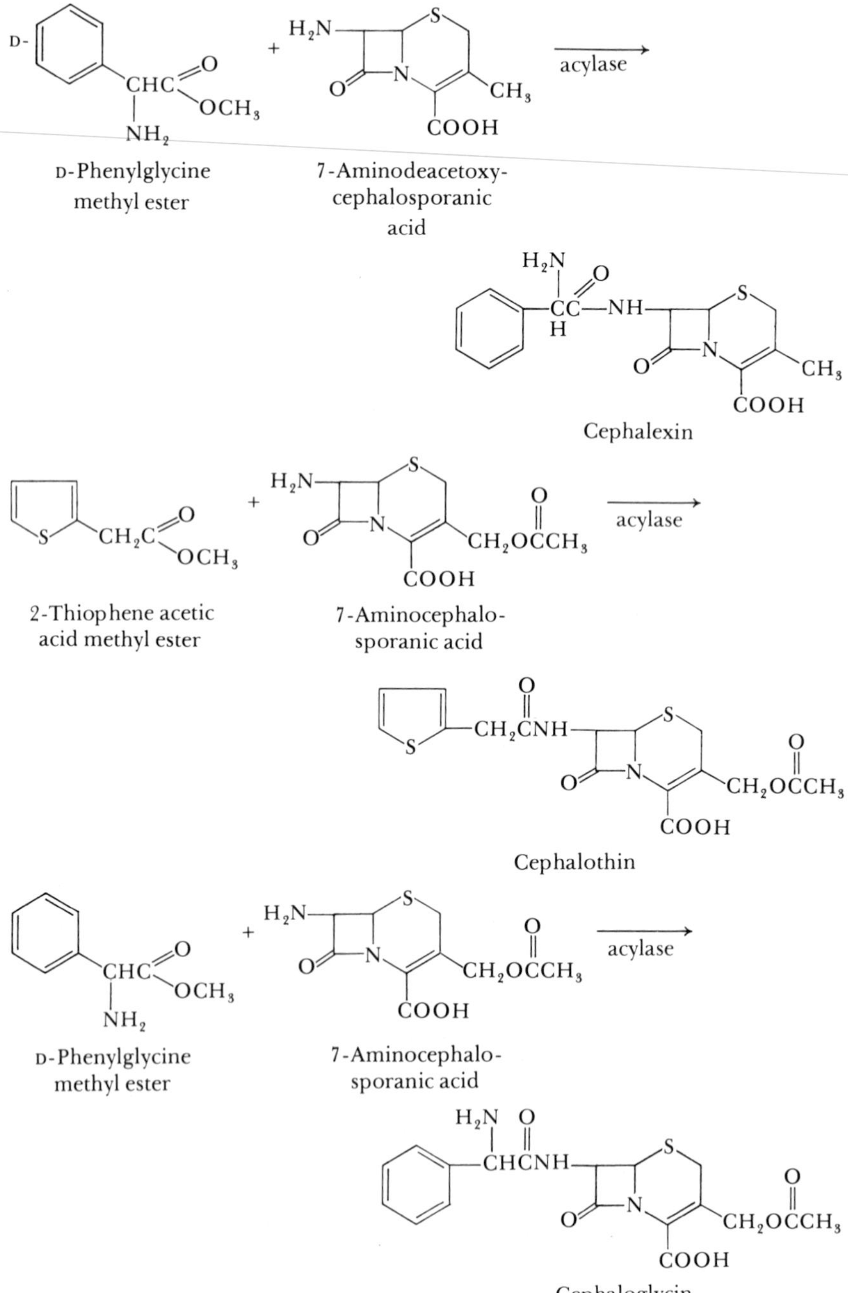

D-Phenylglycine
methyl ester

7-Aminodeacetoxy-
cephalosporanic
acid

Cephalexin

2-Thiophene acetic
acid methyl ester

7-Aminocephalo-
sporanic acid

Cephalothin

D-Phenylglycine
methyl ester

7-Aminocephalo-
sporanic acid

Cephaloglycin

Fig. 5. Reactions involved in enzymic acylation of 7-aminocephalosporanic acid or 7-amino-deacetoxycephalosporanic acid to give semisynthetic cephalosporins.

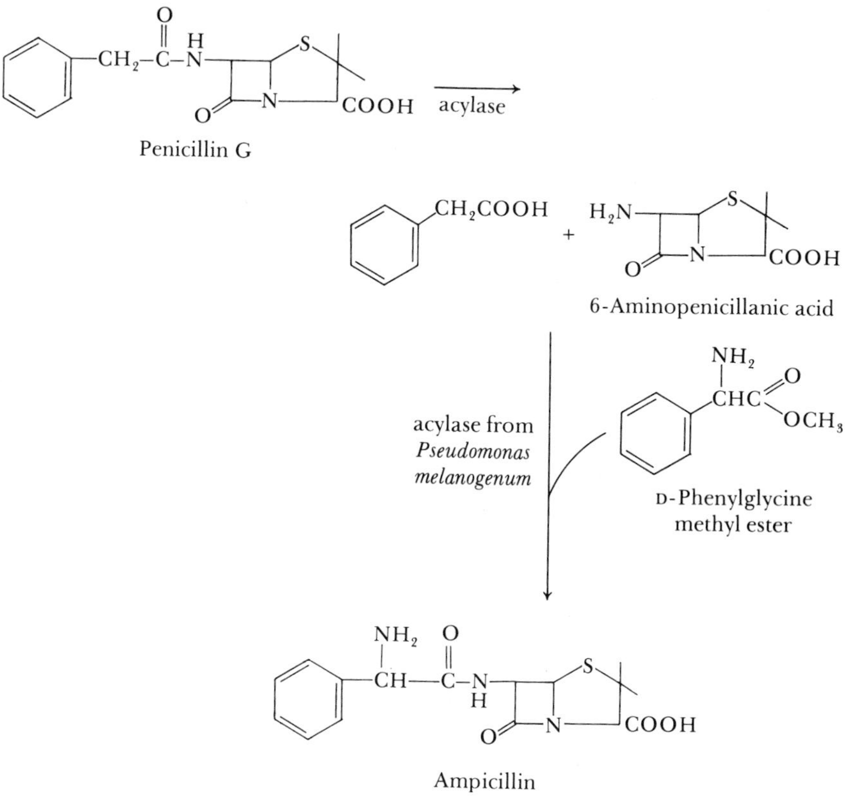

Penicillin G

6-Aminopenicillanic acid

acylase from
Pseudomonas
melanogenum

D-Phenylglycine
methyl ester

Ampicillin

Fig. 6. Reactions involved in a two-step enzymic synthesis of ampicillin from penicillin G.

3. D-*Alpha-Aminobenzylpenicillin Acylases*

In 1972, micro-organisms were selected producing penicillin acylases that display a novel substrate spectrum. Nara *et al.* (1972) and Okachi *et al.* (1973a, b) reported on a D-α-aminobenzylpenicillin (ampicillin) acylase so far encountered particularly among *Pseudomonas* strains. These enzymes form 6-APA only from ampicillin, but not from penicillin G, V or other semisynthetic penicillins. Cephalexin was also hydrolysed, however. Hence, this enzyme can be considered as a novel type of penicillin acylase. At pH 7.5 and 34°C, specific hydrolysis of ampicillin into 6-APA was observed with four strains, namely *Ps. melanogenum* K.Y. 3987, K.Y. 4030 and K.Y. 4031 and *Ps. ovalis* K.Y. 3962.

Pseudomonas melanogenum K.Y. 3987 in particular received detailed attention (Nara *et al.*, 1972; Okachi *et al.*, 1973a, b). The enzyme was purified by the use of Sephadex chromatography and electrofocusing methods (Okachi and Nara, 1973). In addition to a β-lactamase, this strain possessed an ampicillin acylase, producing rather high levels of ampicillin at pH 5.5–6 at 35°C from 6-APA and DL-phenylglycine methyl ester. Okachi *et al.* (1973a, b) found that β-lactamase-deficient mutants derived from the parent strain produced better yields of ampicillin. A penicillin G hydrolysate, containing 6-APA and phenyl-acetic acid, obtained by the acylase from *Kl. citrophila* could be used as a substrate for enzymic ampicillin production, after addition of α-aminophenylacetic acid or its derivatives, together with the acylase from *Ps. melanogenum* (Okachi *et al.*, 1973a, b) (Fig. 6).

G. Application of Penicillin Acylases

1. *Bioconversion of Penicillins and Analogues*

a. *Free cells or enzyme preparations as catalyst.* 6-Aminopenicillanic acid production has been carried out mostly with suspensions of intact cells as an enzyme source. In addition, washed intact cells or mycelium (Rolinson *et al.*, 1960; Kaufmann and Bauer, 1960; Cole and Rolinson, 1961; Batchelor *et al.*, 1961a, b; Claridge *et al.*, 1963; Uri *et al.*, 1963; Erickson and Bennett, 1965; Cole, 1966, 1969a, d; Nara *et al.*, 1971a), acetone-dried cells (Pruess and Johnson, 1965; Haupt and Thrum, 1967), lyophilized cells suspended in toluene-saturated buffer (Huang *et al.*, 1963) and fungal spore suspensions (Singh *et al.*, 1969; Vandamme *et al.*, 1971a) have been used as catalysts. Crude cell extracts (Claridge *et al.*, 1963; Vanderhaeghe *et al.*, 1968) and extracted purified enzyme (Waldschmidt-Leitz and Bretzel, 1964; Bondareva *et al.*, 1969a, b; Baumann *et al.*, 1971; Kutzbach and Rauenbusch, 1974; Vandamme and Voets, 1974a, b, 1975a, b; Shimizu *et al.*, 1975a, b; Schneider and Röhr, 1976) have been used for 6-APA production only in a limited number of investigations. Culture filtrates (Batchelor *et al.*, 1961b; Uri *et al.*, 1963; Holt and Stewart, 1964a; Haupt and Thrum, 1967; Nara *et al.*, 1971a) and purified extracellular acylase (Chiang and Bennett, 1967) could also serve as enzyme source in the case of acylases from *Streptomyces* sp. and *B. megaterium* acylases. Optimum conditions for maximal production of 6-APA depend on the properties of the strain or enzyme preparation used (Tables 4, p. 482 and 5, p. 486).

b. *Immobilized enzymes as catalyst.* Several attempts have now been made to isolate and purify penicillin acylases, and methods to stabilize and immobilize these enzyme preparations have been described (Self *et al.*, 1969; Dinelli, 1972; Ryu *et al.*, 1972a, b; Warburton *et al.*, 1972; Marconi *et al.*, 1973; Abbott, 1976; Brodelius, 1978).

So far, few penicillin V acylases have been immobilized. A crude acylase preparation from *Rhodotorula glutinis* var. *glutinis* was entrapped in polyacrylamide gel and tested for its capacity to act as a continuous immobilized enzyme reactor (Vandamme and Voets, 1973). By contrast, several penicillin G acylases have been purified and immobilized. Self *et al.* (1969) extracted the penicillin acylase from *E. coli* A.T.C.C. 9637 grown in a medium containing phenylacetic acid and glutamate, and they purified the enzyme by fractionation with streptomycin sulphate, ammonium sulphate and poly(ethylene glycol), followed by chromatography on DEAE-cellulose. The purification factor was about 100–200, and the overall yield was about 35%. The enzyme was chemically linked to derivatives of cellulose, and the kinetics of these immobilized penicillin acylase preparations were investigated in a penicillin acylase reactor. An optimum pH value of 7.5 was found at a temperature of 37°C. A DE52-cellulose reactor and a cellulose-sheet reactor were active during 11 weeks of operation, and no loss of activity was observed. The free enzyme, under the same conditions, lost 42% of its activity in one day. The ability to carry out the reverse reaction (resynthesis) was checked by adjusting effluent from the preceding experiment to pH 5, and passing it through the reactor again. The presence of benzylpenicillin in the effluent confirmed that the reverse reaction was indeed occurring.

The enzyme from *E. coli* N.C.I.B. 8743A has been immobilized by covalent binding to DEAE-celluose, using 2-amino-4,6-dichloro-*s*-triazine. The preparation retained 45–81% of the activity observed before attachment. There was no evidence of diffusional limitation of the reaction rate. Using the immobilized enzyme, a K_m value of 0.63 mM was determined at 37°C at pH 8. The optimum pH value for penicillin G hydrolysis was 7.65, whereas the free enzyme displayed maximal activity at pH 8.2 (Lilly *et al.*, 1972; Warburton *et al.*, 1972). The use and stability of this immobilized enzyme preparation have been described in batch and continuous-flow stirred-tank reactors (Warburton *et al.*, 1973). Several other immobilization methods involving covalent-bond formation have been described for penicillin acylase from *E. coli*. Ekström *et al.* (1974) tested several coupling procedures involving polysaccharide (cyanogen bromide-activated Sephadex G-100) or polyacrylamide supports for this enzyme. Other methods include coupling of the acylase to a resin with glutaraldehyde (Savidge *et al.*, 1974) or to copolymers containing acrylamide and maleic acid anhydric (Huper, 1973). A penicillin G acylase from *B. circulans* was succinoylated and then adsorbed to DEAE-Sephadex (Otsuka Seiyaku Company, 1972). The enzyme also catalysed synthesis of ampicillin from 6-APA and D-phenylglycine methyl ester (Fig. 2, p. 492). Ryu *et al.* (1972a, b) compared the kinetic properties of the soluble and the bentonite or diatomaceus earth-immobilized forms of the extracellular enzyme of *B. megaterium* A.T.C.C. 14945, and found that they exhibited significantly different inhibition constants. A continuous enzyme reactor was designed, consisting of a continuous-flow stirred tank and an ultrafiltration unit; it enabled recirculation of the enzyme and continuous removal of the end-products. A kinetic model was derived based on the inhibition effects of the end-products on the enzymic action. This model was then used, with the aid of a computer, to simulate the performance of a continuous enzyme-reactor system and to optimize the productivity of the reactor in terms of process variations (Ryu *et al.*, 1972b).

A purified preparation of the penicillin acylase from *E. coli* A.T.C.C. 9637 was entrapped in cellulose triacetate fibres, which were packed in a column in order to obtain an immobilized enzyme reactor (Marconi *et al.*, 1973). The acylase fibres were found to be quite stable. No loss of activity by casual contamination was found, because the entrapped enzyme was protected against microbial proteolytic attack. A remarkable lowering of the hydrolysis time was observed, and a conversion yield higher than 90% was reached. It was found that acylase fibres were a better catalyst at low temperature than the free enzyme. This reactor was also able to synthesize penicillins (ampicillin, amoxycillin) and cephalosporins (cephalexin) from 6-APA and 7-ADCA (Figs. 4 and 5) respectively (Marconi *et al.*, 1975).

An immobilization process involving a combination of physical entrapment and covalent-bond formation has been described for *E. coli* that was coupled to a water-soluble sucrose epichlorohydrin-co-polymer (Beecham Group Ltd., 1974).

c. *Whole-cell immobilization.* Immobilization of intact microbial cells has been the subject of increased interest (Abbott, 1976; Jack and Zajic, 1977; Vandamme, 1976; Chibata and Tosa, 1977; Brodelius, 1978). Such systems obviate the need for enzyme isolation before immobilization, and can even catalyse a series of sequential reactions without the need for a supply of cofactors. However, intact cell walls or membranes may act as permeability and diffusion barriers, and unwanted side reactions can occur. Several attempts are now being made to apply this cell-immobilization technology for (de)acylation of penicillins (and cephalosporins).

In view of the development of chemical procedures for industrial preparation of 6-APA (Weissenburger and Vanderhoeven, 1970; Fosker *et al.*, 1971), the introduction of methods for immobilization of enzymes or cells, which give stability and prolonged high activity that can result in a continuous bioconversion process, seems bound to improve the economics of microbial processes that may approach the borderline of success by competition with chemical procedures (Faith *et al.*, 1971; Smiley and Strandberg, 1972; Wingard, 1972).

The first attempt to immobilize penicillin acylase-producing cells was made by Dinelli (1972), who entrapped *E. coli* A.T.C.C. 9637 cells in wet-spun cellulose triacetate fibres. The fibres, containing 15 mg of wet cells per gram of polymer, exhibited 80% of the activity of a similar amount of cells in free suspension. This novel system displayed a high and lasting activity over a period of four months. Sato *et al.* (1976) immobilized *E. coli* A.T.C.C. 9637 cells by entrapment in a poly-acrylamide-gel lattice, and used an immobilized reactor column for continuous production of 6-APA. The half-life of the enzyme was 17 days at 40°C and 42 days at 30°C. From the effluent of the column, 6-APA was obtained with a yield of 78%. They concluded that this procedure is more advantageous than continuous methods using immobilized penicillin acylase. Glutaraldehyde-treated cells of *Proteus rettgeri*, bound to glycidyl–methacrylate polymer, could also transform penicillin G into 6-APA (Nelson, 1976). Relatively few data have been published on enzymic acylation of 6-APA, using immobilized cells as catalysts. Commercial realization of this process has until now been postponed by the availability of chemical acylation methods, and by the reversibility of the reaction which is, furthermore, subject to product or substrate inhibition (Vandamme and Voets, 1974b). Intact cells of *Achromobacter* sp. or *B. megaterium*, immobilized by adsorption on DEAE-cellulose, were used to synthesize ampicillin from 6-APA and D-phenylglycine methyl ester (Fig. 2, p. 492, Fujii *et al.*, 1973). These cells could also acylate the cephalosporin nucleus 7-ACA and 7-ADCA (Figs. 4, 5 and 6).

2. Acylases and Cephalosporin Compounds

a. *Free cells or enzyme as catalyst.* Microbial acylases capable of hydrolysing cephalosporin C (Fig. 3) specifically into 7-amino-cephalosporanic acid and D-α-aminoadipic acid have not been detected so far. Although cephalosporin C acylases, splitting off α-aminoadipic acid, have been found in species of *Brevibacterium*, *Achromobacter* and *Flavobacterium*, during their hydrolytic action the acetyl group of cephalosporin is also lost, so that deacetyl-7-aminocephalosporanic acid is produced rather than 7-ACA (Walton, 1964a).

Certain semisynthetic cephalosporins can be split into their side-chain acid and 7-ACA by a few bacterial penicillin G acylases, but generally the rate of hydrolysis is lower than that of the corresponding penicillins (Huang *et al.*, 1963; Sjöberg *et al.*, 1967; Cole, 1969a; Nara *et al.*, 1971b; Shimizu *et al.*, 1975a, b; see Table 6, p. 489).

In the preparation of cephalosporin compounds, the penicillin V acylase of *Erwinia aroideae* (Vandamme and Voets, 1971b) has been used for deacylation of the 7-β-phenoxyacetyl derivative of 7-ADCA (Fleming *et al.*, 1974). Similarly, the extracellular penicillin G acylase from *B. megaterium* was found to hydrolyse the phenylacetyl derivative of 7-ADCA into the side chain and 7-ADCA (Abbott, 1976; Figs. 4 and 5).

The penicillin G acylases of *Kl. citrophila* K.Y. 7844 and K.Y. 3641, *Kl. noncitrophila* K.Y. 3642, *M. luteus* K.Y. 3781, *Achromobacter* K.Y. 3048 and *Flavobacterium* K.Y. 4082 also hydrolysed cephalexin (Fig. 7) into 7-ADCA, a fact also observed with the ampicillin acylase of *Ps. melanogenum* K.Y. 3987.

Purified penicillin G acylase of *Kl. citrophila* K.Y. 7844 hydrolysed cephalexin at pH 7.3 into 7-ADCA, while pH 6.5 was the optimum for synthesis at 55°C. Cephaloridine was also hydrolysed. As well as 6-APA, both 7-ACA and 7-ADCA could be acylated to produce β-lactam antibiotics with this acylase (Figs. 1, p. 468; and 7, p. 502; Shimizu *et al.*, 1975b). This penicillin G acylase also catalyses tetrazoyl acetylation of 7-ACA and 6-APA. This process might be used in the production of cefazolin and the corresponding penicillin derivative (Shimizu *et al.*, 1975a, b). The enzyme displays no activity towards cephalosporin C, however. In fact, most of these penicillin acylase-producing bacteria, as well as *E. coli* A.T.C.C. 9637, displayed synthetic ability for cephalexin and other cephalosporins from 7-aminocephem compounds and organic acid esters (Fig. 5, p. 494).

Micrococcus ureae K.Y. 3767, *Rhosopseudomonas spheroides* K.Y. 4112 and *Nocardia globerula* K.Y. 3901 preferentially hydrolysed cephalexin into 7-ADCA, above other cephalosporins and penicillins, but displayed no detectable synthetic activity (Figs. 5, p. 494, and 7, p. 502; Shimizu *et al.*, 1975a, b). Takahashi *et al.* (1972) showed that bacterial α-amino acid esterases could synthesize cephalosporins starting from 7-ACA or 7-ADCA and an appropriate side-chain ester. As side-chain structures, α-D-phenylglycine methyl ester and its analogues (glycine ethyl ester, D-alanine ethyl ester, D-leucine ethyl ester, and the methyl esters of D-α-cyclohexenylglycine, D-α-*p*-hydroxyphenylglycine, and D-α-cyclohexylglycine) were coupled to 7-ACA or 7-ADCA at pH 6 by enzymic action of micro-organisms including *Xanthomonas oryzae* I.F.O.3995, *Xanthomonas citri* I.F.O.3835, *Acetobacter pasteurianus* A.T.C.C. 6033, *Acetobacter turbidans* A.T.C.C. 9325, *Gluconobacter suboxydans* A.T.C.C. 621, *Pseudomonas melanogenum* I.F.O.2020, *Mycoplana dimorpha* I.F.O.13212, and *Protaminobacter alvoflavus* I.F.O.13221 (Takahashi *et al.*, 1972, 1973).

The enzyme from *Acetobacter turbidans* A.T.C.C. 9325 displayed a strong ester hydrolase activity with α-amino-acid esters only. No activity was found with α-amino-acid amides or *N*-(α-aminoacyl)-glycine, although cephalosporins and penicillins having an α-aminoacyl side chain were slightly hydrolysed. The enzyme displayed α-aminoacyltransferase activity too in the presence of suitable acyl receptors, such as 7-ADCA, to produce cephalexin. The enzyme from *X. citri* I.F.O.3835 was purified 390-fold after CM-cellulose adsorption and Sephadex G-100 gel filtration. The molecular weight was estimated as 270,000. The enzyme hydrolyses α-amino-acid esters (such as D-phenylglycine methyl ester) and cephalexin, but could also synthesize cephalexin from 7-ADCA and D-phenylglycine methyl ester in high yield (Figs. 5 and 7). The optimum pH value for these reactions was 5.5–6.5 at an optimal temperature of 35–40°C. The K_m value for cephalexin deacylation was 3.5 mM (Takahashi *et al.*, 1972, 1973). The enzymic activity was clearly different from that of penicillin

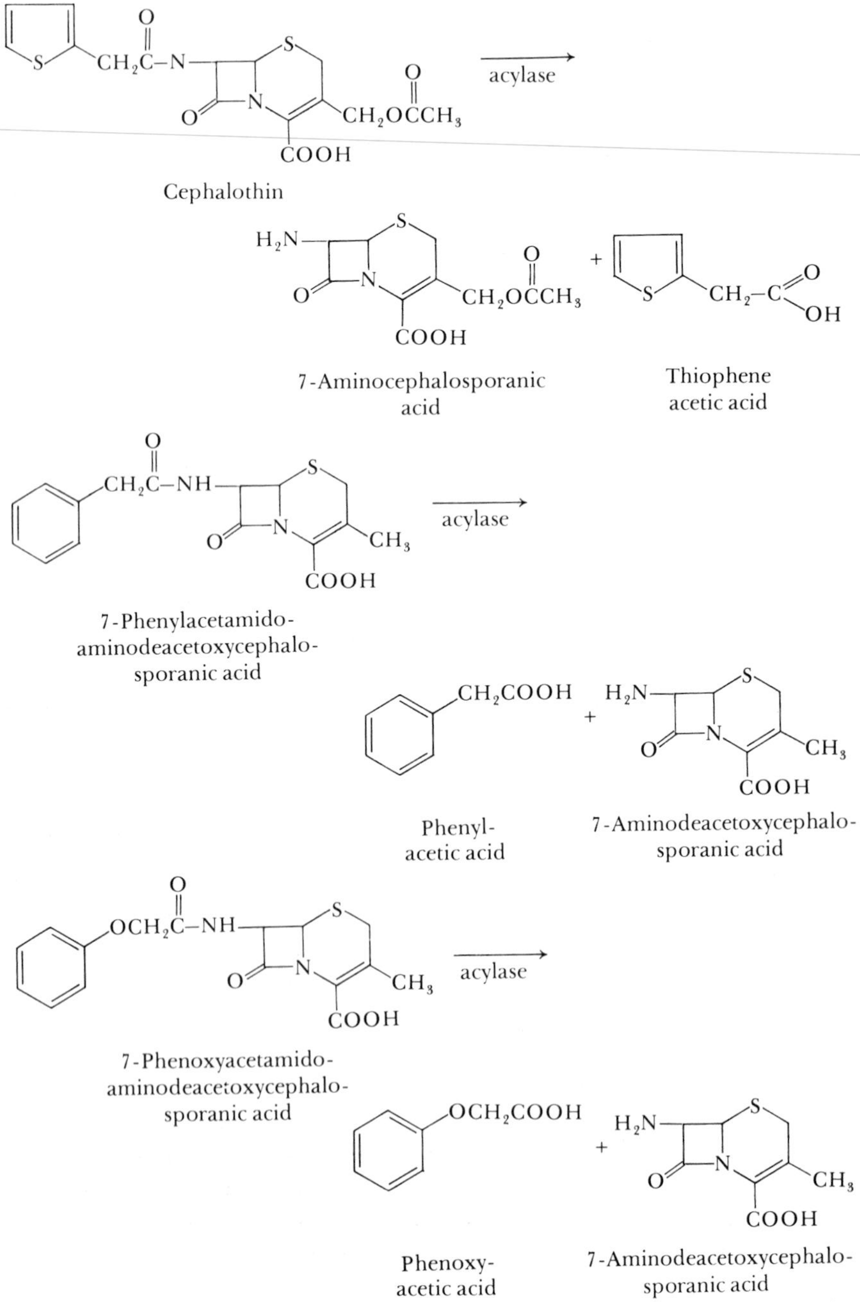

Cephalothin

7-Aminocephalosporanic   Thiophene
acid   acetic acid

7-Phenylacetamido-
aminodeacetoxycephalo-
sporanic acid

Phenyl-   7-Aminodeacetoxycephalo-
acetic acid   sporanic acid

7-Phenoxyacetamido-
aminodeacetoxycephalo-
sporanic acid

Phenoxy-   7-Aminodeacetoxycephalo-
acetic acid   sporanic acid

Fig. 7. Reactions involved in enzymic hydrolysis of semisynthetic cephalosporins to give 7-aminocephalosporanic acid or 7-aminodeacetoxycephalosporanic acid.

acylase, however, although several penicillin acylases are also able to acylate 7-ADCA to form cephalexin.

b. *Immobilized enzymes*. Immobilized acylases have also been used for deacylation or acylation of cephalosporins. Extracellular penicillin G acylases from *B. megaterium* and *P. rettgeri* were adsorbed on Celite, activated carbon, carboxymethylcellulose or an Amberlite ion-exchange resin (Toyo Jozo Company, 1973a, b, 1975) and hydrolysed 7-phenylacetamide-ADCA to 7-ADCA. A partially purified penicillin V acylase from *Erwinia aroideae*, entrapped in cellulose triacetate fibres, catalysed hydrolysis of 7-phenoxyacetamido-ADCA to 7-ADCA (Fig. 7).

The immobilized *E. coli* acylase, used by Marconi *et al.* (1975) to acylate 6-APA, could be applied for cephalexin synthesis from 7-ADCA and D-phenylglycine methyl ester with a yield of 75%. An extracellular acylase from *B. megaterium* adsorbed on Celite, and a cell-free acylase from *Achromobacter* sp. adsorbed on hydroxyapatite, performed the same reaction with a 80–85% yield (Abe *et al.*, 1973; Toyo Jozo Company, 1974). Immobilized forms of acylase from *B. megaterium* have also been used to acylate 7-ACA, the cephalosporin C nucleus, with D-phenylglycine methyl ester into cephaloglycin. Cephalothin was obtained by acylating 7-ACA with 2-thiophene acetic acid methyl ester (Fig. 5).

c. *Immobilized cells*. Most studies on enzymic deacylation of cephalo-sporins are directed towards producing the 7-ADCA nucleus (Fig. 4). The substrates for this reaction, 7-phenyl- and 7-phenoxyacetamido-deacetoxycephalosporanic acid (Fig. 4), are readily obtained by ring expansion of penicillin G and V. Although several immobilized-enzyme processes have been reported to hydrolyse the above compound into 7-ADCA (Toyo Jozo Company, 1973a, b; Fig. 7), only one report on hydrolysis by *Erwinia aroideae*, entrapped in cellulose triacetate fibres, has been published (Fleming *et al.*, 1974).

A novel immobilized-cell system was described by Fukushima *et al.* (1976) for preparing the cephalosporin nucleus, 7-aminocephalo-sporanic acid. Cells of *Pseudomonas* sp. or *Comamonas* sp., entrapped in cellulose-acetate capsules, hydrolyse 3-acetoxymethyl-7-(4-carboxy-butaneamide)-3-cephem-4-carboxylic acid into 7-ACA. This substrate compound can be obtained from culture broths of strains of

Cephalosporium chrysogenum (Kitano *et al.*, 1976) or it can be obtained from cephalosporin C by oxidative deamination of the cephalosporin C side chain with D-amino-acid oxidase in the presence of perborate (Mazzeo and Romeo, 1972).

Several immobilized-cell systems for synthesizing the clinically useful antibiotic cephalexin have been developed (Abe *et al.*, 1973; Toyo Jozo Company, 1974; Fig. 7). Acetone-dried cells of *Achromobacter* sp., *Beneckea hiperoptica*, *Alcaligenes faecalis* or *Flavobacterium aquatile*, adsorbed to DEAE-, TEAE- or CM-cellulose, could synthesize cephalexin from 7-ADCA and D-phenylglycine methyl ester with a 50 to 80% yield. It thus seems that cell-immobilization technology is rapidly finding applications in synthesis of semisynthetic cephalosporins.

II. *BETA*-LACTAMASE

A. Introduction

Probably the first antibiotic bioconversion (and inactivation) was, ironically enough, observed with penicillin, almost parallel with the demonstration of the chemotherapeutic value of this antibiotic. In 1940, Abraham and Chain (1940) found, in extracts of an *E. coli* strain which was relatively insensitive to penicillin, an enzyme activity which they called penicillinase. At that time, they could not determine the nature of the enzymic reaction, but it is now almost certain that the

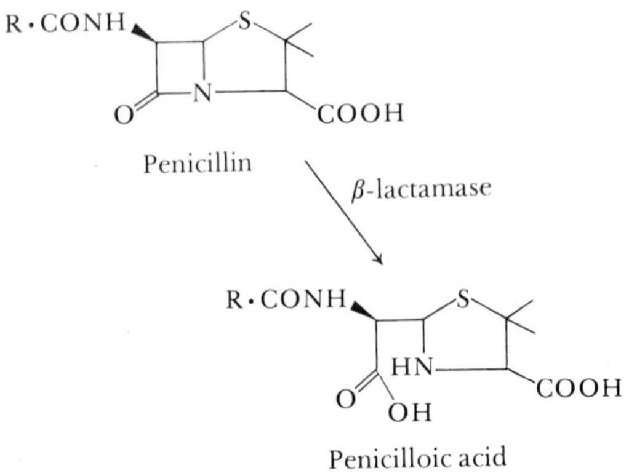

Penicillin

β-lactamase

Penicilloic acid

Fig. 8. Reaction involved in hydrolysis of penicillins to give penicilloic acids catalysed by β-lactamase.

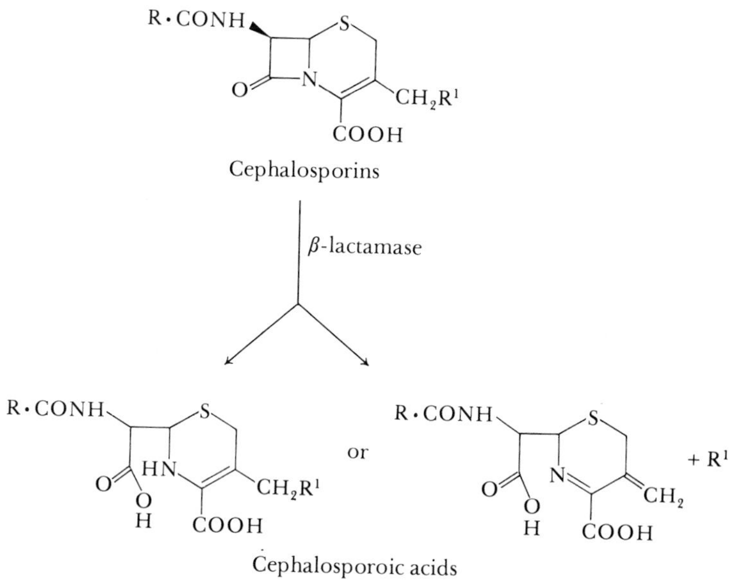

R·CONH

O

N

S

CH$_2$R^1

COOH

Cephalosporins

β-lactamase

R·CONH

O

HN

O

H

S

CH$_2$R^1

COOH

or

R·CONH

O

N

O

H

S

CH$_2$

COOH

+ R^1

Cephalosporoic acids

Fig. 9. Reactions involved in hydrolysis of cephalosporins to give cephalosporanic acids catalysed by β-lactamase.

enzyme had opened the β-lactam ring of the penicillin molecule and was thus a β-lactamase (Abraham, 1974). It is now well known that β-lactamase (penicillin β-lactam amidohydrolase; EC 3.5.2.6.) action transforms penicillins into their corresponding penicilloic acids, which are completely devoid of antibacterial activity (Fig. 8; Abraham *et al.*, 1949; Citri and Pollock, 1966; Richmond and Sykes, 1973). Cephalosporins, too, are susceptible to inactivation by β-lactamases, the nature of the degradation products formed being dependent upon the cephalosporin side chain structure at C-3 (Fig. 9) (Hamilton-Miller *et al.*, 1970; Richmond and Sykes, 1973). The term β-lactamase now denotes an enzyme that catalyses hydrolysis of the amide bond in the β-lactam ring of 6-APA or 7-ACA and of their *N*-acyl derivatives.

Many micro-organisms produce β-lactamases, which often play a crucial role in the resistance of pathogenic bacteria towards β-lactam antibiotics. Resistance to penicillin and cephalosporin antibiotics is indeed commonly encountered in clinical isolates of Gram-negative and Gram-positive bacteria. One of the major mechanisms of this bacterial resistance is production of β-lactamases. A number of these resistances, especially in Gram-negative species, have been specified by R-plasmids (Mitsuhashi, 1971).

Production of new semisynthetic penicillins and cephalosporins, which are insensitive to β-lactamase inactivation, has become possible since the availability of the penicillin and cephalosporin nucleus. These processes are hence of immense importance in the continuing battle against penicillin- and cephalosporin-resistant micro-organisms.

The majority of reports on β-lactamases are concerned with the clinical significance of the enzyme, and relatively little is published on pilot- or larger-scale production of this enzyme. Nevertheless, an attempt is made here to select data useful for β-lactamase production and its medical, industrial or laboratory application.

B. Assay Methods for *Beta*-Lactamase Activity

Different assay techniques have been widely used for measuring β-lactamase activity. Apart from the manometric technique and the hydroxamate assay (Pollock, 1952; Hamilton-Miller *et al.*, 1963), which are now obsolete, the most widely applied assay is the iodometric method, originally devised by Perret (1954) but since refined and modified (Novick, 1962; Sawai *et al.*, 1978). It relies on the reaction of eight equivalents of iodine with the penicilloic acid produced by the action of β-lactamase on penicillins. Iodometric assay of β-lactamase action on cephalosporins is less reliable than with penicillins, because of the complex character of the break-down mechanism.

Destruction of the β-lactam bond in cephalosporins causes a change in absorption of 260 nm. This spectrophotometric assay is very sensitive compared with the iodometric technique (O'Callaghan *et al.*, 1968). O'Callaghan *et al.* (1972) devised a simple chromogenic assay, specific for cephalosporin β-lactamase. A spectrophotometric assay of β-lactamase action on penicillins was reported by Waley (1974). Titrimetric assays were described by Sabath *et al.* (1965) and Hou and Poole (1972), and an analytical isolectric-focussing assay for detection and identification of β-lactamase by Matthew *et al.* (1975). More details about these assays and their limitations are given by Citri and Pollock (1966), Citri (1972), Richmond and Sykes (1973) and by Ross and O'Callaghan (1975). One β-lactamase unit is equivalent to one μmole of substrate hydrolysed per minute at the optimal pH value and temperature.

C. Screening Procedures for *Beta*-Lactamase-Producing Micro-Organisms

In screening for activity, cell suspensions or culture liquids are incubated in the presence of penicillins or cephalosporins after which one of the above-mentioned assay methods is applied. Penicillin- or cephalosporin-resistant strains can be isolated first, before the normal trial-and-error procedure, to enhance the yield of the screening programme. Rapid detection of β-lactamase-producing strains, directly from colonies growing on the primary culture plates, is also possible (Thornsberry and Kirven, 1974; Jorgensen et al., 1977). A simple method for detecting constitutive mutants of β-lactamase producers consists in growing bacteria that survive mutagenic treatment on nutrient agar containing 1% soluble starch, and then staining colonies with a solution of benzylpenicillin (20 mg/ml) in iodine (0.1 N I_2 in 0.4 M KI). The reagent is poured onto the plates for about ten seconds, and the excess is removed. The agar plates are coloured deep purple from the reaction between iodine and starch. Constitutive β-lactamase synthesis is detected by a rapidly spreading white zone around a colony. The principle of the method is that penicilloic acid, formed from penicillin by β-lactamase action, reacts with iodine, thus decolourizing the agar plates (Novick, 1963). The great majority of microbial cultures examined contained β-lactamase.

Ayliffe (1965) screened 148 bacterial strains and found 55 of them able to synthesize β-lactamase, although no acylase activity could be detected. Among 201 micro-organisms tested, Demain et al. (1963) detected 52 producers of β-lactamase but no producers of acylase. Cole and Sutherland (1966) found 70 β-lactamase-positive strains among 148 bacterial strains. Nara et al. (1971a, b) screened 251 strains of bacteria, 229 of actinomycetes, two of yeasts and 37 of fungi, and concluded that most of the strains decomposed penicillin, the hydrolysis product mostly being penicilloic acid. The author tested 184 different micro-organisms, of which 37 displayed β-lactamase activity (Vandamme and Voets, 1971; Vandamme, 1972). Ogawara (1975) screened 100 strains of *Streptomyces* spp. isolated from soil for production of β-lactamase and found 72 of them to be positive. Newsom et al. (1974) tested 1,050 strains of enterobacteria, isolated during clinical practice, and almost 50% of them produced β-

lactamase. Of 292 bacterial strains tested by Shimizu *et al.* (1975a, b) for penicillin acylase and β-lactamase synthesis, 81 of them were found to produce a β-lactamase and only 16 produced a penicillin acylase. From these and many other screening programmes, it can be concluded that β-lactamase is really widespread among micro-organisms compared with the rather scarce occurrence of penicillin acylase.

D. *Beta*-Lactamase-Producing Micro-Organisms

Production of β-lactamase seems to be confined to bacteria, including actinomycetes, although recently it has been demonstrated in yeasts (Mehta and Nash, 1978). The enzyme has been detected in all micro-organisms examined for it so far, as long as the assay procedure was sensitive enough (Matthew *et al.*, 1975). Typical producing strains are summarized in Table 7.

E. Fermentation Processes for *Beta*-Lactamase Production

Few published data are available on pilot-plant scale or even larger-scale production of β-lactamase (Kogut *et al.*, 1956; Miller *et al.*, 1965; Pollock, 1965; Kuwabara, 1970; Melling and Scott, 1972). Details are known for production of the extracellular β-lactamase of *B. cereus* 569/H (Miller *et al.*, 1965). Seed cultures for shake-flask or fermentor production were prepared from a spore stock. Production of β-lactamase occurred in the following medium (pH 7.0): Casamino acids (Difco), 10 g (or N-Z-Amine type B, Sheffield); KH_2PO_4, 2.72 g; sodium citrate. 5.5 H_2O, 5.88 g; $MgSO_4 \cdot 7H_2O$, 0.41 g; $FeSO_4 \cdot 7H_2O$, 0.014 g, to which antifoam was added. Agitation in a 65-litre fermentor was by a fully baffled system at 210 rev./min and 2 litres of air per minute were supplied. The process was conducted at 35°C. Growth continued for 12 to 14 hours at which time β-lactamase activity had reached its maximal level. Dynamics of enzyme synthesis revealed that growth and enzyme formation proceed in parallel, with the pH value rising gradually to 7.8 at the end of the fermentation process. An initial dose of 50 p.p.m. antifoam depressed foaming during the whole process. After centrifugation of the culture, the supernatant, containing the enzyme, was stored at 4°C after the pH value was adjusted to 4.5. Greater than 11,000 units of β-lactamase per ml could be obtained.

Table 7.

Some *Beta*-Lactamase producing micro-organisms

Micro-organisms	References
BACTERIA	
Bacillus subtilis	Woodruff and Foster (1945)
Bacillus cereus 569/H, 5B	Kuwabara (1970)
Bacillus anthracis	Barnes (1947)
Bacillus licheniformis 749/C, 6346/C	Pollock (1965)
Staphylococcus albus	Bondi and Dietz (1945)
Staphylococcus aureus A, B, C,	Abraham and Jennings (1941),
	Richmond (1965)
Micrococcus lysodeikticus	Abraham and Chain (1940)
Streptomyces sp.	Welsch (1949), Johnson *et al.* (1975),
	Ogawara (1975), Ogawara *et al.*
	(1978)
Nocardia sp.	Nyiri (1963)
Mycobacterium tuberculosis,	Elwan (1964)
Mycobacterium phlei	
Mycobacterium smegmatis,	
Mycobacterium fortuitum	Muftic (1962)
Escherichia coli	Abraham and Chain (1940), Melling
	and Scott (1972)
Escherichia coli W3110, K12, D.B. 103,	Neu (1968), Yamagishi *et al.* (1969),
W3630	Lindstrom *et al.* (1970)
Enterobacter freundii,	Hennessey and Richmond (1968),
	Ross (1975)
Enterobacter aerogenes,	Holt and Stewart (1964a, b), Percival
Aerobacter cloacae, Aerobacter aerogenes	*et al.* (1963), Jack and Richmond (1970)
Klebsiella aerogenes	
Serratia sp.	Fleming *et al.* (1963)
Erwinia amylovora	
Shigella sp.	
Salmonella typhimurium	Chang and Weinstein (1964)
Proteus morgani, Proteus vulgaris,	Fleming *et al.* (1963), Sabath and
Proteus mirabilis, Proteus inconstans, Proteus	Abraham (1964)
rettgeri	
Alcaligenes faecalis	
Pseudomonas aeruginosa	
Aeromonas liquefaciens,	Elwan (1964), Percival *et al.* (1963)
Aeromonas proteolytica	Jack and Richmond (1970)
Bacteroides fragilis,	
Bacteroides fundiformis,	Salyers *et al.* (1977), Britz and
Bacteroides oralis, Bacteroides melaninogenicus	Wilkinson (1978)
Haemophilus influenzae	
Neisseria gonorrhoeae	Saunders and Sykes (1977)
Yersinia enterocolitica	Elwell *et al.* (1977)
YEASTS	Cornelis and Abraham (1975)
Candida albicans	
MOULDS	Mehta and Nash (1978)
	Citri and Pollock (1966)

A simplified 6–8-fold purification and concentration procedure consists in adsorption on Celite for 45 minutes at 300 rev./min with a 95% efficiency, followed by elution with 1 M NaCl and 0.1 M sodium citrate, dissolved in 0.1 M phosphate, borate or Tris buffer at pH 8.0 (Thatcher, 1975a). Freeze-dried enzyme preparation displayed good stability on storage.

Kuwabara (1970) gives also some details about large-scale production of this β-lactamase in 142-litre fermentors, using the media described by Miller *et al.* (1965) but supplied with 1 mM zinc sulphate. The aeration rate was 140 litres per minute at 520 rev./min. During growth, the pH value was kept lower than 7.5 by addition of a citric acid solution. Kuwabara and Abraham (1967) demonstrated that two different extracellular β-lactamases were produced by this strain, namely, β-lactamase I (penicillinase) and β-lactamase II (cephalosporinase). Maximum titres of β-lactamase I were obtained after 10 hours of culturing when the growth rate began to decline. Formation of β-lactamase II reached a peak after six hours of growth in the middle of the exponential-growth phase.

The substrate profile of β-lactamase I was similar to that of many other β-lactamases from Gram-positive bacteria. Phenoxymethylpenicillin and ampicillin are hydrolysed at rates similar to that of benzylpenicillin, whereas degradation of methicillin, oxacillin, cloxacillin, 6-APA and cephalosporins occurs at low rates. Cephalosporin was hydrolysed by β-lactamase II at rates approaching that of the hydrolysis of benzylpenicillin.

Bacillus licheniformis β-lactamase has been produced on a 200-litre scale by Pollock (1965) in a medium based on 1% Casamino acids (Difco), 0.02 M potassium phosphate (pH 7.2) and salts. The enzyme could be isolated from cells, the whole culture, or the culture supernatant of magnoconstitutive mutants of the strains 749 and 6346. Purification of this β-lactamase type has been described by Thatcher (1975b).

The β-lactamases from *Bacillus* species are now commercially available from a wide number of firms (Newsom and Walsingham, 1973), including Whatman Biochemicals (β-lactamase I–II, as described by Davies *et al.*, 1974), Ricker Laboratories (Neutrapen), Baltimore Biological Laboratories (Penicillinase concentrate), Difco Laboratories (Penase concentrate), A.V.M. Pharmaceutical Ltd (Penicillinase), Wellcome Reagents Ltd. (Penicillinase), Koch–Light

Laboratories (Penicillinase) and others. These freeze-dried or liquid preparations are manufactured for administration to man, for addition to milk and dairy products, or for laboratory use.

Melling and Scott (1972) gave details about the production and purification of the intracellular β-lactamase from *E. coli* K12-W3110. A culture medium with the following composition was used: sodium β-glycerophosphate, 0.12 M; $MgSO_4 \cdot 5H_2O$, 1.0 mM; yeast extract (Oxoid), 1%; Casamino acids (Oxoid), 1%; glucose, 0.4%; trace-metal solution: $CuSO_4 \cdot 5H_2O$, 0.5%; $ZnSO_4 \cdot 7H_2O$, 0.5%; $FeSO_4 \cdot 7H_2O$, 0.5%; $MnCl_2 \cdot 4H_2O$, 0.2%; conc. HCl, 10%. Growth occurred at 37°C in a 400-litre fermentor fitted with four baffles as described by Elsworth and Stockwell (1968). The pH value was automatically maintained at 6.8 with phosphoric acid. Aeration was at 300 litres of air per minute at 250 rev./min. After 15 hours of incubation, β-lactamase titres reached a maximum. The cells were collected by centrifugation and washed in 0.1 M sodium phosphate buffer (pH 7.0). Cell breakage was performed with a Manton-Gaulin homogenizer, operated at 50 litres per hour. Further purification was obtained by centrifugation at 13,000 *g*, ammonium-sulphate fractionation and dialysis, absorption on DEII- and DE52-cellulose and separation through Sephadex G-75.

Staphylococcus aureus β-lactamase has also been produced on a 200-litre level, using the above-described medium composition (Richmond, 1975b). Addition of 0.5 μg of methicillin/ml is the common induction procedure. Alternatively, 6-APA (100 μg/ml) or cloxacillin (0.2 μg/ml) can be used as inducers. Constitutive mutants have been obtained by mutagenesis with ethylmethylsulphonate (EMS) or with *N*-methyl-*N*-nitro-*N*-nitrosoguanidine (MSG).

The β-lactamases produced by *Enterobacter* species display little activity towards penicillins, but hydrolyse many of the cephalosporins (Ross, 1975). Production of these 'cephalosporinases' has not yet been carried out on a large scale, but details about the fermentation process were given by Marshall *et al.* (1972). Pilot-scale production of *Streptomyces* β-lactamase has been described by Johnson *et al.* (1975).

F. Characterization and Properties of *Beta*-Lactamase

The β-lactamases are known to differ from each other in their substrate specificities, physical and biochemical properties and reactions with antisera. Unlike the enzymes from Gram-positive bacteria, which are inducible, most of the β-lactamases from Gram-negative species are constitutive. There is also a clear difference in location between β-lactamases from Gram-negative and Gram-positive bacteria. The β-lactamases from Gram-positive bacteria are liberated into the medium and are extracellular enzymes, whereas those from Gram-negative bacteria are cell-bound. Their physiological function seems to be to protect the micro-organisms that produce them against the deleterious effect of penicillins and cephalosporins. In this role, they often provide a major obstacle to successful use of β-lactam antibiotics for thereapy, which can be overcome in many cases by using semisynthetic analogues.

Apart from its clinical aspects, β-lactamase has been studied very extensively from the point of view of enzyme induction (Citri and Pollock, 1966; Richmond, 1968), enzyme secretion (Lampen, 1967; Sawai *et al.*, 1973) and transfer of genetic elements (Novick, 1969).

Several β-lactamases have been purified and shown to be homogeneous. Their molecular properties have been determined (Citri, 1972; Abraham, 1974). The amino-acid composition of several β-lactamases is known, and complete absence of cysteine is characteristic in most cases (Richmond and Sykes, 1973). The amino-acid sequence of the β-lactamases from *Staph. aureus* and *B. licheniformis* has been determined (Ambler and Meadway, 1969).

Substrate profiles of various β-lactamases differ, depending on whether the 6-APA or 7-ACA nucleus is the substrate and also on the nature of the side chains of the molecules (Citri, 1972). An almost continuous spectrum extending from extreme cephalosporinase to enzymes predominantly active against penicillins has now been described.

Typical pH-activity curves obtained with Gram-positive β-lactamases show maxima in the range of pH 6.0–7.0 with a sharp decline in the alkaline region (Citri and Pollock, 1966). The β-lactamases from Gram-negative bacteria show optimal values scattered over the range of pH 5.0 to 8.5. Optimal temperatures reported range from 30°C (Goldner and Wilson, 1961) through 35–40°C for β-lactamases from *B. cereus* (Manson *et al.*, 1954) to 45–55°C.

Normally, β-lactamase activity does not depend on specific activators or cofactors, although there are exceptions to this general rule, such as the β-lactamase II of *B. cereus* 569/H, which requires Zn^{2+} for stability and activity (Kuwabara and Abraham, 1967; Sabath and

Table 8.

Relevant properties of *beta*-lactamases

Producing strains	Molecular weight	pH Optimum	Optimum temperature (°C)	Isoelectric point	Cofactor requirements	References
Bacillus cereus						
569/4 I	28,000	6.0–6.6	36	9.2–9.7	—	Kogut *et al.* (1956), Abraham (1974), Kuwabara (1970)
II	22,500	—	—	8.45	Zn^{2+}	
III	18,000	—	—	8.4–9.0	Zn^{2+}	Abraham (1974)
Bacillus licheniformis						
6346/C	28,000	6.0–7.0	40	—	—	Pollock (1965), Citri and Pollock (1966)
Staphylococcus aureus DC I	29,600	5.9	55	8.9	—	Richmond (1963, 1965)
Escherichia coli W3110 (R-TEM)	25,000	7.0	30	5.7	—	Datta and Richmond (1968), Melling and Scott (1972)
Escherichia coli R1818	44,600	—	—	8.3	—	Dale (1971)
Enterobacter cloacae P99	39,000	8.2	—	7.9	—	Hennessey and Richmond (1968), Ross (1975)
Yersinia enterocolitica						
W222 A	22,000	6.5	30	8.1	—	Cornelis and Abraham (1975)
B	34,000	7.5	—	5.4	—	
Pseudomonas aeruginosa 8203	42,000	8.0–8.5	—	7.5	—	Abraham (1974)

Finland, 1968). Relevant properties of β-lactamases are summarized in Table 8.

Many substrate analogues, such as semisynthetic penicillins and cephalosporins, act as powerful competitive inhibitors (Cole *et al.*, 1972). Generally, thiol reagents have no effect on the activity, which follows from a lack of cysteine residues in the amino-acid composition of β-lactamases (Citri, 1972), but here also exceptions are found (Cornelis and Abraham, 1975; Abraham, 1974).

Recently, a specific β-lactamase inhibitor, named clavulanic acid and produced by *Streptomyces clavuligerus*, has been described by Reading and Cole (1977). Other β-lactamase inhibitors are produced by other *Streptomyces* species, and have been described by Hata *et al.* (1972), Umezawa *et al.* (1973), Tally *et al.* (1978) and English *et al.* (1978). The immunology of β-lactamases has also been studied extensively and reviewed (Pollock, 1963; Zyk and Citri, 1968; Richmond, 1975a). The properties of clinically important β-lactamases from Gram-positive and Gram-negative bacteria have been reviewed in detail by Citri and Pollock (1966) and Richmond and Sykes (1973). Their physiological role, and genetic regulation and evolutionary origin, have been discussed extensively by Pollock (1971) and Richmond and Sykes (1973).

G. Application of *Beta*-Lactamases

Apart from their interest from the medical viewpoint, β-lactamases are used for routine identification of penicillin in food (milk) supplies and biological specimens such as blood serum, for inactivation of penicillin in milk so as to prevent allergic reactions in consumers, or inhibition of starter cultures. They are also used in sterility testing of penicillin preparations, in automatic assays of penicillin and other cases where inactivation of penicillins or cephalosporins is desired, such as inactivation of penicillin in sensitized patients and recovery of micro-organisms from clinical specimens containing β-lactam antibiotics (Goodall and Roseda, 1961; Sizer, 1964; Jarvis and Berridge, 1969; Peromet *et al.*, 1974; Enfors and Molin, 1978; Pache, 1978).

III. ACKNOWLEDGMENTS

The author gratefully acknowledges the advice of E. P. Abraham, Sir William Dunn School of Pathology, University of Oxford, Oxford,

England and of A. L. Demain, Massachusetts Institute of Technology, Cambridge, U.S.A. on various aspects of the manuscript. He is indebted to T. Nara, Kyowa Hakko Kogyo Cy, Ltd, Tokyo, Japan for providing unpublished data and translation of Japanese literature; and is thankful to M. Cole, Beecham Research Labs., Betchworth, Surrey, England and to H. Vanderhaeghe, Rega Institute, University of Louvain, Belgium, for their continuous interest.

REFERENCES

Abbott, B. J. (1976). *Advances in Applied Microbiology* 20, 203.

Abe, J., Watanabe, T., Yamaguchi, T. and Matsumoto, K. (1973). United States Patent 3,761,354.

Abraham, E. P. (1974). 'Biosynthesis and Enzymic Hydrolysis of Penicillins and Cephalosporins'. University of Tokyo Press, Tokyo, Japan.

Abraham, E. P. and Chain, E. (1940). *Nature, London* 146, 837.

Abraham, E. P. and Jennings, M. A. (1941). *The Lancet* 2, 177.

Abraham, E. P., Baker, W., Boon, W. R., Calam, C. T., Carrington, H. C., Chain, E., Florey, H. W., Freeman, G. G., Robinson, R. and Saunders, H. G. (1949). *In* 'The Chemistry of Penicillin' (H. T. Clarke, J. R. Johnson and R. Robinson, eds.), p. 10. Princeton University Press, Princeton, New Jersey, U.S.A.

Acevedo, F. and Cooney, C. L. (1973). *Biotechnology and Bioengineering* 15, 493.

Alicino, J. F. (1961). *Analytical Chemistry* 33, 648.

Ambler, R. P. and Meadway, R. J. (1969). *Nature, London* 222, 24.

Ayliffe, G. A. J. (1963). *Journal of General Microbiology* 30, 339.

Ayliffe, G. A. J. (1965). *Journal of General Microbiology* 40, 119.

Badr-Eldin, S. M. and Attia, M. M. (1973). *Acta Microbiologica Polonica Series B* 5, 43.

Balasingham, K., Warburton, D., Dunnill, P. and Lilly, M. D. (1972). *Biochimica et Biophysica Acta* 276, 250.

Ballio, A., Chain, E. B., Dentice Di Accadia, F., Rolinson, G. N. and Batchelor, F. R. (1959). *Nature, London* 183, 180.

Barnes, J. M. (1947). *Journal of Pathological Bacteriology* 59, 113.

Batchelor, F. R., Doyle, F. P., Nayler, J. H. C. and Rolinson, G. N. (1959). *Nature, London* 183, 257.

Batchelor, F. R., Chain, E. B., Hardy, T. L., Mansford, K. R. L. and Rolinson, G. N. (1961a). *Proceedings of the Royal Society Series B* 154, 498.

Batchelor, F. R., Chain, E. B., Richards, M. and Rolinson, G. N. (1961b). *Proceedings of the Royal Society Series B* 154, 522.

Bauer, K., Kaufmann, W. and Offe, H. A. (1960). *Naturwissenschaften* 47, 469.

Bauer, K., Kaufmann, W. and Ludwig, G. A. (1971). *Hoppe Seyler's Zeitschrift für Physiologische Chemie* 352, 1723.

Baumann, F., Brunner, R. and Rohr, M. (1971). *Hoppe Seyler's Zeitschrift für Physiologische Chemie* 352, 853.

Beecham Group Ltd. (1974). German Patent 2,356,630.

Benveniste, R. and Davies, J. (1973). *Annual Review of Biochemistry* 42, 471.

Bomstein, J. and Evans, W. G. (1965). *Analytical Chemistry* 37, 578.

Bondareva, N. W., Levitov, M. M. and Goryachenkova, E. V. (1969a). *Biokhimiya* 34, 96.

Bondareva, N. W., Levitov, M. M. and Rabinovich, M. S. (1969b). *Biokhimiya* **34**, 478.
Bondi, A. and Dietz, C. C. (1945). *Proceedings of the Society for Experimental Biology and Medicine* **60**, 55.
Boxer, G. E. and Everett, P. M. (1949). *Analytical Chemistry* **21**, 670.
Brandl, E. (1965). *Hoppe Seyler's Zeitschrift für Physiologische Chemie* **342**, 86.
Brandl, E. (1972). *Scientia Pharmazeutica* **40**, 89.
Britz, M. L. and Wilkinson, R. G. (1978). *Antimicrobial Agents and Chemotherapy* **13**, 373.
Brodelius, P. (1978). *In* 'Advances in Biochemical Engineering', (T. K. Ghose, A. Fiechter and N. Blakeborough, eds.), vol. 10, p. 75. Springer-Verlag, Berlin.
Brunner, R., Rohr, M. and Zinner, M. (1968a). *Hoppe Seyler's Zeitschrift für Physiologische Chemie* **349**, 95.
Brunner, R., Rohr, M., Hampel, W. and Zinner, M. (1968b). *Monatshefte für Chemie* **99**, 2244.
Carrington, T. R. (1971). *Proceedings of the Royal Society Series B* **179**, 321.
Chang, T. W. and Weinstein, L. (1964). *Antimicrobial Agents and Chemotherapy* **1963**, 278.
Chapman, M. J., Holt, R. J., Mattocks, A. R. and Stewart, G. T. (1964). *Journal of General Microbiology* **36**, 215.
Chiang, D. and Bennett, R. E. (1967). *Journal of Bacteriology* **93**, 302.
Chibata, I. and Tosa, T. (1977). *Advances in Applied Microbiology* **22**, 1.
Citri, N. (1972). *In* 'The Enzymes' (P. D. Boyer, ed.), vol. 4, p. 23. Academic Press, New York.
Citri, N. and Pollock, M. R. (1966). *Advances in Enzymology* **28**, 237.
Claridge, C. A., Gourevitch, A. and Lein, J. (1960). *Nature, London* **187**, 337.
Claridge, C. A., Luttinger, J. R. and Lein, J. (1963). *Proceedings of the Society for Experimental Biology and Medicine* **113**, 1008.
Cole, M. (1964). *Nature, London* **203**, 519.
Cole, M. (1966). *Applied Microbiology* **14**, 98.
Cole, M. (1967). *Process Biochemistry* **2**, 35.
Cole, M. (1969a). *Biochemical Journal* **115**, 733.
Cole, M. (1969b). *Biochemical Journal* **115**, 741.
Cole, M. (1969c). *Biochemical Journal* **115**, 747.
Cole, M. (1969d). *Biochemical Journal* **115**, 757.
Cole, M. and Rolinson, G. N. (1961). *Proceedings of the Royal Society Series B* **154**, 490.
Cole, M. and Sutherland, R. S. (1966). *Journal of General Microbiology* **42**, 354.
Cole, M., Elson, S. and Fullbrook, P. D. (1972). *Biochemical Journal* **127**, 295.
Cole, M., Savidge, T. and Vanderhaeghe, H. (1975). *In* 'Methods in Enzymology' (J. H. Hash, ed.), vol. 43, p. 698. Academic Press, New York.
Cornelis, G. and Abraham, E. P. (1975). *Journal of General Microbiology* **87**, 273.
Dale, J. W. (1971). *Biochemical Journal* **123**, 501.
Datta, N. and Richmond, M. H. (1968). *Biochemical Journal* **98**, 204.
Davies, R. B., Abraham, E. P. and Melling, J. (1974). *Biochemical Journal* **143**, 115.
Demain, A. L., Walton, R. B., Newkirk, J. F. and Miller, I. M. (1963). *Nature, London* **199**, 909.
Dennen, D. W., Allen, C. C. and Carver, D. D. (1971). *Applied Microbiology* **21**, 907.
Devcic, B. and Divjak, S. (1968). *Acta Biologica Jugoslavia Mikrobiologiya* **5**, 189.
Dinelli, D. (1972). *Process Biochemistry* **7**, 9.
Dulong De Rosnay, C., Castagnou, F. and Latrille, J. (1970). *Annales de l'Institut Pasteur, Paris* **118**, 277.
Ekström, B., Lagerlof, E., Nathorst-Westfelt, L. and Sjoberg, B. (1974). *Svenska Farmaca Tidskrift* **78**, 531.

Elsworth, R. and Stockwell, F. T. E. (1968). *Process Biochemistry* 3, 229.

Elwan, S. H. (1964). *Archive für Microbiologie* 47, 286.

Elwell, L. P., Roberts, M., Mayer, L. W. and Falkow, S. (1977). *Antimicrobial Agents and Chemotherapy* 11, 528.

Enfors, S. O. and Molin, N. (1978). *Process Biochemistry* 13, 9.

English, A. R., Retsema, J. A., Civard, A. E., Lynch, J. E. and Barth, W. E. (1978). *Antimicrobial Agents and Chemotherapy* 14, 414.

Erickson, R. C. and Bennett, R. E. (1965). *Applied Microbiology* 13, 738.

Erickson, R. C. and Dean, L. D. (1966). *Applied Microbiology* 14, 1047.

Faith, W. F., Neubeck, C. E. and Reese, E. T. (1971). *In* 'Advances in Biochemical Engineering' (T. K. Ghose and A. Fiechter, eds.), vol. 1, p. 78. Springer Verlag, Berlin.

Fawcett, P. A., Usher, J. J. and Abraham, E. P. (1975). *Biochemical Journal* 151, 729.

Findlater, J. D. and Orsi, B. A. (1973). *Federation of European Biochemical Societies Symposium*, Dublin, Ireland, Abstracts, p. 181.

Fleming, P. C., Goldner, M. and Glass, D. G. (1963). *The Lancet* 1, 1399.

Fleming, I. D., Turner, M. K. and Napier, E. J. (1974). German Patent 2,422,374.

Ford, J. H. (1947). *Analytical Chemistry* 19, 1004.

Fosker, G. R., Hardy, K. D., Nayler, J. H. C., Seggery, P. and Stove, E. R. (1971). *Journal of the Chemical Society (C)* 10, 1917.

Fujii, T., Hanamitsu, K., Izumi, R., Yamaguchi, T. and Watanabe, T. (1973). Japanese Patent 7,399,393.

Fukushima, M., Fujii, T., Matsumoto, K. and Morishita, M. (1976). Japanese Patent 7,670,884.

Gang, D. M. and Shaikh, K. (1976). *Biochimica et Biophysica Acta* 425, 110.

Gatenbeck, S. and Brunsberg, U. (1968). *Acta Chemica Scandinavica* 22, 1059.

Goldner, M. and Wilson, R. J. (1961). *Canadian Journal of Microbiology* 7, 45.

Golub, E. I., Garaev, M. M. and Romanova, N. B. (1973). *Antibiotiki, Moscow* 18, 882.

Goodall, R. R. and Roseda, D. (1961). *The Analyst* 86, 326.

Gotovtseva, V. A. and Levitov, M. M. (1965). *Mikrobiologiya* 34, 857.

Gotovtseva, V. A., Yudina, O. D. and Levitov, M. M. (1965). *Microbiologiya* 37, 180.

Gotovtseva, V. A., Gorchenko, G. L. and Suvorov, N. N. (1968). *Microbiologiya* 37, 683.

Hamilton-Miller, J. M. T. (1966). *Bacteriological Reviews* 30, 761.

Hamilton-Miller, J. M. T., Smith, J. T. and Knox, R. (1963). *Journal of Pharmacy and Pharmacology* 15, 81.

Hamilton-Miller, J. M. T., Newton, G. G. F. and Abraham, E. P. (1970). *Biochemical Journal* 116, 371.

Hata, T., Omura, S., Iwai, Y., Ohno, H., Takeshima, H. and Yamaguchi, N. (1972). *Journal of Antibiotics* 25, 273.

Haupt, I. and Thrum, H. (1967). *Zietschrift für Allgemeine Mikrobiologie* 7, 343.

Hennessey, T. D. and Richmond, M. H. (1968). *Biochemical Journal* 109, 469.

Holt, R. J. and Stewart, G. T. (1964a). *Nature, London* 201, 824.

Holt, R. J. and Stewart, G. T. (1964b). *Journal of General Microbiology* 36, 203.

Hou, J. P. and Poole, J. W. (1972). *Journal of Pharmaceutical Sciences* 61, 1594.

Huang, H. T., English, A. R., Seto, T. A., Shull, G. M. and Sobin, B. A. (1960). *Journal of the American Chemical Society* 82, 3790.

Huang, H. T., Seto, T. A. and Shull, G. M. (1963). *Applied Microbiology* 11, 1.

Huper, F. (1973). German Patent 2,157,970.

Ivashkiv, E. (1964). *Analytical Chemistry* 36, 2506.

518   E. J. VANDAMME

Jack, G. W. and Richmond, M. H. (1970). *Journal of General Microbiology* **61**, 43.
Jack, T. R. and Zajic, J. E. (1977). *In* 'Advances in Biochemical Engineering' (A. Fiechter, ed.), vol. 5, p. 125. Springer Verlag, Berlin.
Jarvis, B. and Berridge, N. J. (1969). *Chemistry and Industry* p. 1721.
Jarvis, F. G. and Johnson, M. J. (1950). *Journal of Bacteriology* **59**, 50.
Johnson, K., Duez, C., Frere, J. M. and Ghuysen, J. M. (1975). *Methods in Enzymology* **43**, 687.
Jorgensen, J. H., Lee, J. C. and Alexander, G. A. (1977). *Antimicrobial Agents and Chemotherapy* **11**, 1087.
Kameda, Y., Kimura, Y., Toyoura, E. and Omori, T. (1961). *Nature, London* **191**, 1122.
Kato, K. (1953). *Journal of Antibiotics* **6**, 130.
Kaufmann, W. and Bauer, K. (1960). *Die Naturwissenschaften* **47**, 474.
Kaufmann, W. and Bauer, K. (1964a). *Journal of General Microbiology* **35**, IV.
Kaufmann, W. and Bauer, K. (1964b). *Nature, London* **203**, 520.
Kitano, K., Kintaka, K., Suzuki, S., Katamoto, K., Nara, K. and Nakao, Y. (1974). *Journal of Fermentation Technology* **52**, 785.
Kitano, K., Kintaka, K., Katamoto, K., Nara, K. and Nakao, Y. (1975). *Journal of Fermentation Technology* **53**, 339.
Kitano, K., Fujisawa, Y., Katamoto, K., Nara, K. and Nakao, Y. (1976). *Journal of Fermentation Technology* **54**, 712.
Kleiner, G. I. and Lopatnev, S. V. (1972). *Priklady Biokhimiya Mikrobiologiya* **8**, 554.
Kondo, E. and Mitsugi, T. (1975). United States Patent 3,926,728.
Kogut, M., Pollock, M. R. and Tridgell, E. Y. (1956). *Biochemical Journal* **62**, 391.
Korchagin, V. B., Serova, L. I. and Kotova, N. I. (1971). *Antibiotiki (Moscow)* **16**, 702.
Kornfield, J. M. (1978). *Analytical Biochemistry* **86**, 118.
Kutzbach, C. and Rauenbusch, E. (1974). *Hoppe Seyler's Zietschrift für Physiologische Chemie* **355**, 45.
Kuwabara, S. (1970). *Biochemical Journal* **118**, 457.
Kuwabara, S. and Abraham, E. P. (1967). *Biochemical Journal* **103**, 27c.
Lampen, J. O. (1967). *Journal of General Microbiology* **48**, 249.
Lee, M. A. (1974). *Chemistry and Industry* p. 399.
Lemke, P. A. and Nash, C. H. (1972). *Canadian Journal of Microbiology* **18**, 155.
Levitov, M. M., Klapovskaya, K. I. and Kleiner, G. I. (1967). *Mikrobiologiya* **36**, 768.
Lilly, M. D., Balasingham, K., Warburton, D. and Dunnill, P. (1972). *In* 'Fermentation Technology Today' (G. Terui, ed.), p. 379. Society of Fermentation Technology, Kyoto, Japan.
Lindström, E. B., Boman, H. G. and Steele, B. B. (1970). *Journal of Bacteriology* **101**, 218.
Lucente, G., Romeo, A. and Rossi, D. (1965). *Experientia* **21**, 317.
Manson, E. E. D., Pollock, P. R. and Ridgell, E. J. (1954). *Journal of General Microbiology* **11**, 493.
Marconi, W., Cecere, F., Morisi, F., Della Penna, G. and Rappuoli, B. (1973). *Journal of Antibiotics* **26**, 228.
Marconi, W., Bartoli, F., Cecere, F., Galli, G. and Morisi, F. (1975). *Agricultural and Biological Chemistry* **39**, 277.
Marrelli, L. P. (1968). *Journal of Pharmacological Sciences* **57**, 2172.
Marshall, M. J., Ross, G. W., Chanter, K. V. and Harris, A. M. (1972). *Applied Microbiology* **23**, 765.
Matthew, M., Harris, A. M., Marshall, M. J. and Ross, G. W. (1975). *Journal of General Microbiology* **88**, 169.

Mazzeo, P. and Romeo, A. (1972). *Journal of the Chemical Society, Perkin Transactions I* **20**, 2532.

Mehta, R. J. and Nash, C. H. (1978). *Journal of Antibiotics* **31**, 239.

Melling, J. and Scott, G. K. (1972). *Biochemical Journal* **130**, 55.

Miller, G., Bach, G. and Markus, Z. (1965). *Biotechnology and Bioengineering* **7**, 517.

Mitsuhashi, S. (1971). 'Transferable Drug Resistance Factor R'. University Park Press, Baltimore, London, Tokyo.

Muftic, M. K. (1962). *Experientia* **18**, 17.

Nara, T., Misawa, M., Okachi, R. and Yamamoto, M. (1971a). *Agricultural and Biological Chemistry* **35**, 1676.

Nara, T., Okachi, R. and Misawa, H. (1971b). *Journal of Antibiotics* **24**, 321.

Nara, T., Okachi, R. and Kato, F. (1972). *Abstracts of the Sixth International Fermentation Symposium, Kyoto, Japan*, p. 207.

Nathorst-Wesfelt, L., Schatz, B. and Thelin, H. (1963). *Acta Chemica Scandinavica* **17**, 1164.

Nelson, R. P. (1976). United States Patent 3,957,580.

Neu, H. C. (1968). *Biochemical and Biophysical Research Communications* **32**, 258.

Newsom, S. W. B. and Walsingham, B. M. (1973). *Journal of Medical Microbiology* **6**, 59.

Newsom, S. W. B., Marshall, M. J. and Harris, A. M. (1974). *Journal of Medical Microbiology* **7**, 473.

Niedermayer, A. O. (1964). *Analytical Chemistry* **36**, 938.

Novick, R. P. (1962). *Biochemical Journal* **83**, 236.

Novick, R. P. (1963). *Journal of General Microbiology* **33**, 121.

Novick, R. P. (1969). *Bacteriological Reviews* **33**, 210.

Nyiri, L. (1963). *Acta Microbiologica Academiae Scientiarum Hungaricae* **10**, 261.

Nyiri, L. (1967). *Nature, London* **214**, 1347.

Nys, P. S., Savitskaya, E. M. and Kolygina, T. S. (1972). *Antibiotiki (Moscow)* **17**, 313.

Nys, P. S., Savitskaya, E. M. and Kolygina, T. S. (1973). *Antibiotiki (Moscow)* **18**, 270.

O'Callaghan, C. H., Muggleton, P. and Ross, G. (1968). *In* 'Antimicrobial Agents and Chemotherapy' (G. Hobby, ed.), p. 57. Academic Press, London.

O'Callaghan, C. H., Morris, A., Kirby, S. M. and Shingler, A. H. (1972). *Antimicrobial Agents and Chemotherapy* **1**, 283.

Ogawara, H. (1975). *Antimicrobial Agents and Chemotherapy* **8**, 402.

Ogawara, H., Horikawa, S., Shima-Miyoshi, S. and Yasazawa, K. (1978). *Antimicrobial Agents and Chemotherapy* **13**, 865.

Okachi, R. and Nara, T. (1973). *Agricultural and Biological Chemistry* **37**, 2797.

Okachi, R., Kawamoto, I., Yamamoto, M., Takasawa, S. and Nara, T. (1973a). *Agricultural and Biological Chemistry* **37**, 335.

Okachi, R., Kato, F., Miyamura, Y. and Nara, T. (1973b). *Agricultural and Biological Chemistry* **37**, 1953.

Oostendorp, J. G. (1972). *Antonie Van Leeuwenhoek* **38**, 201.

Otsuka Seiyaku Company. (1972). Japanese Patent 7,228,187.

Pache, W. (1978). *European Journal of Applied Microbiology and Biotechnology* **5**, 171.

Percival, A., Brumfitt, W. and De Louvois, J. (1963). *Journal of General Microbiology* **32**, 77.

Peromet, M., Schoutens, E., Vanderlinden, M. P. and Yourassowsky, E. (1974). *Chemotherapy* **20**, 1.

Perret, C. J. (1954). *Nature, London* **174**, 1012.

Plaskie, A., Roets, E. and Vanderhaeghe, H. (1978). *Journal of Antibiotics* **31**, 783.

Pollock, M. R. (1952). *British Journal of Experimental Pathology* **33**, 587.

Pollock, M. R. (1963). *Annals of the New York Academy of Sciences* **103**, 969.

Pollock, M. R. (1965). *Biochemical Journal* **94**, 666.

Pollock, M. R. (1971). *Proceedings of the Royal Society Series B* **179**, 385.

Pruess, D. L. and Johnson, M. J. (1965). *Journal of Bacteriology* **90**, 380.

Pruess, D. L. and Johnson, M. J. (1967). *Journal of Bacteriology* **94**, 1502.

Reading, C. and Cole, M. (1977). *Antimicrobial Agents and Chemotherapy* **11**, 852.

Richmond, D. R. (1975a). *Methods in Enzymology* **43**, 86.

Richmond, M. H. (1975b). *Methods in Enzymology* **43**, 664.

Richmond, M. H. (1963). *Biochemical Journal* **88**, 452.

Richmond, M. H. (1965). *Biochemical Journal* **94**, 584.

Richmond, M. H. (1968). *Essays in Biochemistry* **4**, 105.

Richmond, M. H. and Sykes, R. B. (1973). *Advances in Microbial Physiology* **9**, 31.

Rode, L. G., Foster, V. W. and Schuhardt, V. T. (1947). *Journal of Bacteriology* **53**, 565.

Rolinson, G. N., Batchelor, F. R., Butterworth, D., Cameron-Wood, J., Cole, M.,
Eustace, G. C., Hart, V., Richards, M. and Chain, E. B. (1960). *Nature, London* **187**,
236.

Ross, G. W. (1975). *Methods in Enzymology* **43**, 678.

Ross, G. W. and O'Callaghan, C. H. (1975). *Methods in Enzymology* **43**, 69.

Rossi, D., Romeo, A., Lucente, G. and Tinti, O. (1973). *Il Farmaco Edizione Scientifica* **28**,
262.

Rossi, D., Lucente, G. and Romeo, A. (1977). *Experientia* **3**, 1557.

Rozansky, R., Biano, S., Clejan, L., Frenkel, N., Bogokovsky, B. and Altmann, G.
(1969). *Israel Journal of Medical Science* **5**, 297.

Ryu, D. Y., Bruno, C. F. and Lee, B. K. (1972a). *Abstracts of the Sixth International
Fermentation Symposium, Kyoto, Japan*, p. 53.

Ryu, D. Y., Bruno, C. F., Lee, B. K. and Venkatasubramanian, K. (1972b). *In*
'Fermentation Technology Today' (G. Terui, ed.), p. 307. Society of Fermentation
Technology, Kyoto, Japan.

Sabath, L. D. and Abraham, E. P. (1964). *Nature, London* **204**, 1066.

Sabath, L. D. and Finland, M. (1968). *Journal of Bacteriology* **95**, 1513.

Sabath, L. D., Jago, M. and Abraham, E. P. (1965). *Biochemical Journal* **96**, 739.

Sakaguchi, K. and Murao, S. (1950). *Journal of the Agricultural and Chemical Scoiety, Japan*
23, 411.

Sakaguchi, K. and Murao, S. (1955). *Journal of the Agricultural and Chemical Society, Japan*
29, 404.

Salyers, A. A., Wong, J. and Wilkins, T. D. (1977). *Antimicrobial Agents and Chemotherapy*
11, 142.

Sato, T., Tosa, T. and Chibata, I. (1976). *European Journal of Applied Microbiology* **2**, 153.

Saunders, J. R. and Sykes, R. B. (1977). *Antimicrobial Agents and Chemotherapy* **11**, 339.

Savidge, T. A. and Cole, M. (1975). *Methods in Enzymology* **43**, 705.

Savidge, T., Powell, L. W. and Warren, K. B. (1974). German Patent 2,336,829.

Sawai, T., Crane, L. J. and Lampen, J. O. (1973). *Biochemical and Biophysical Research
Communications* **53**, 523.

Sawai, T., Takahashi, I. and Yamagishi, S. (1978). *Antimicrobial Agents and Chemotherapy*
13, 910.

Schneider, W. J. and Röhr, M. (1976). *Biochimica et Biophysica Acta* **452**, 177.

Sebek, O. K. (1974). *Lloydia* **37**, 115.

Sebek, O. K. (1975). *Acta Microbiologia Academiae Scientiae Hungaricae* **22**, 381.

Sebek, O. K. and Perlman, D. (1971). *Advances in Applied Microbiology* **14**, 123.

Self, D. A., Kay, G., Lilly, M. D. and Dunnill, P. (1969). *Biotechnology and Bioengineering*
11, 337.

Serova, L. I., Korchagin, V. B. and Kotova, N. I. (1973). *Antibiotiki (Moscow)* 18, 687.

Shaikh, K., Talati, P. G. and Gang, D. M. (1973). *Antimicrobial Agents and Chemotherapy* 3, 194.

Shimizu, M., Masuike, T., Fujita, H., Kimura, K., Okachi, R. and Nara, T. (1975a). *Agricultural and Biological Chemistry* 39, 1225.

Shimizu, M., Okachi, R., Kimura, K. and Nara, T. (1975b). *Agricultural and Biological Chemistry* 39, 1655.

Sikyta, B. and Slezak, J. (1964). *Biotechnology and Bioengineering* 6, 309.

Silingardi, S., Dibitetto, M. and Mangia, A. (1977). *Journal of Pharmaceutical Sciences* 66, 1769.

Sizer, I. W. (1964). *Advances in Applied Microbiology* 6, 207.

Singh, K., Seghal, S. N. and Vezina, D. (1969). *Applied Microbiology* 17, 643.

Sjöberg, G., Nathorst-Westfelt, L. and Ortengren, B. (1967). *Acta Chemica Scandinavica* 21, 547.

Smiley, K. L. and Strandberg, G. W. (1972). *Advances in Applied Microbiology* 15, 13.

Spencer, B. (1968). *Biochemical and Biophysical Research Communications* 31, 170.

Spencer, B. and Maung, G. (1970). *Biochemical Journal* 118, 29.

Szentirmai, A. (1964). *Applied Microbiology* 12, 185.

Takahashi, T., Yamazaki, Y., Kato, K. and Isono, M. (1972). *Journal of the American Chemical Society* 94, 4035.

Takahashi, T., Yamazaki, Y. and Kato, K. (1973). *Abstracts of the Annual Meeting of the Agricultural Chemical Society of Japan, Tokyo*, p. 287.

Takasawa, S., Okachi, R., Kawamoto, I., Yamamoto, M. and Nara, T. (1972). *Agricultural and Biological Chemistry* 36, 1701.

Tally, F. P., Jacobus, N. V. and Gorbach, S. L. (1978). *Antimicrobial Agents and Chemotherapy* 14, 436.

Thadhani, S. B., Borkar, P. S. and Ramachandran, S. (1972). *Biochemical Journal* 128, 49.

Thatcher, D. R. (1975a). *Methods in Enzymology* 43, 640.

Thatcher, D. R. (1975b). *Methods in Enzymology* 43, 653.

Thornsberry, C. and Kirven, L. A. (1974). *Antimicrobial Agents and Chemotherapy* 6, 653.

Toyo Jozo Company (1973a). Belgian Patent 801,044.

Toyo Jozo Company (1973b). Belgian Patent 1,324,159.

Toyo Jozo Company (1974). British Patent 1,347,665.

Toyo Jozo Company (1975). Japanese Patent 7,588,694.

Umezawa, H., Mitsuhashi, S., Hamada, M., Iyobe, S., Takahashi, S., Utahara, R., Osato, Y., Yamazaki, S., Ogawara, H and Maeda, K. (1973). *Journal of Antibiotics* 26, 51.

Uri, J. and Sztaricskai, F. (1961). *Nature, London* 191, 1223.

Uri, J., Valu, G. and Bekesi, I. (1963). *Nature, London* 200, 896.

Uri, J., Valu, G. and Bekesi, I. (1964). *Naturwissenschaften* 51, 298.

Vandamme, E. J. (1972). Doctoral Thesis: University of Ghent.

Vandamme, E. J. (1976). *Chemistry and Industry* 24, 1070.

Vandamme, E. J. (1977). *Advances in Applied Microbiology* 21, 89.

Vandamme, E. J. and Voets, J. P. (1971). *Zeitschrift für Allgemeine Mikrobiologie* 11, 155.

Vandamme, E. J. and Voets, J. P. (1972a). *Mededelingen van de Faculteit Landbouwwetenschappen, University of Ghent* 37, 1185.

Vandamme, E. J. and Voets, J. P. (1972b). *Journal of Chromatography* 71, 141.

Vandamme, E. J. and Voets, J. P. (1973). *Zeitschrift für Allgemeine Mikrobiologie* 13, 701.

Vandamme, E. J. and Voets, J. P. (1974a). *Mededelingen van de Faculteit Landbouwwetenschappen, University of Ghent* 39, 1463.

Vandamme, E. J. and Voets, J. P. (1974b). *Advances in Applied Microbiology* 17, 311.

Vandamme, E. J. and Voets, J. P. (1975a). *Experientia* 31, 140.

Vandamme, E. J. and Voets, J. P. (1975b). *Revue des Fermentations et des Industries Alimentaires* **30**, 101.

Vandamme, E. J., Voets, J. P. and Beyaert, G. (1971a). *Medelingen van de Faculteit Landbouwwetenschappen, University of Ghent* **36**, 577.

Vandamme, E. J., Voets, J. P. and Dhaese, A. (1971b). *Annales de l'Institut Pasteur, Paris* **121**, 435.

Vanderhaeghe, H. (1975). *Methods in Enzymology* **43**, 721.

Vanderhaeghe, H., Claesen, M., Vlietinck, A. and Parmentier, G. (1968). *Applied Microbiology* **16**, 1557.

Vasilescu, I., Vociu, M., Voinescu, R., Birdladeanu, R., Sasarman, E. and Rafirotu, I. (1969). *In* 'Antibiotics' (M. Herold and Z. Gabriel, eds.), p. 518. Butterworths, London.

Voets, J. P. and Vandamme, E. J. (1972). *Abstracts of the Fourth International Fermentation Symposium, Kyoto, Japan*, p. 246.

Vojtisek, V. and Slezak, J. (1975a). *Folia Microbiologica* **20**, 224.

Vojtisek, V. and Slezak, J. (1975b). *Folia Microbiologica* **20**, 289.

Vojtisek, V. and Slezak, J. (1975c). *Folia Microbiologica* **20**, 298.

Waldschmidt-Leitz, E. and Bretzel, G. (1964). *Hoppe Seyler's Zeitschrift für Physiologische Chemie* **337**, 222.

Waley, S. G. (1974). *Biochemical Journal* **139**, 789.

Walton, R. B. (1964a). *Developments in Industrial Microbiology* **5**, 349.

Walton, R. B. (1964b). *Science, New York* **143**, 1138.

Warburton, D., Balasingham, K., Dunnill, P. and Lilly, M.D. (1972). *Biochimica et Biophysica Acta* **284**, 278.

Warburton, D., Dunnill, P. and Lilly, M. D. (1973). *Biotechnology and Bioengineering* **25**, 13.

Weissenburger, H. W. O. and Vanderhoeven, M. G. (1970). *Receuil des Travaux de Chime des Pays-Bas* **89**, 1081.

Wingard, L. E. (1972). *In* 'Advances in Biochemical Engineering' (T. K. Ghose, A. Fiechter and N. Blakebrough, eds.), vol. 2, p. 1. Springer Verlag, Berlin.

Woodruff, H. B. and Foster, J. W. (1945). *Journal of Bacteriology* **49**, 7.

Welsch, M. (1949). *Abstracts of the Fourth International Congress of Microbiology, Copenhagen, Denmark*, p. 21.

Yamagishi, S., O'Hara, K., Sawai, T. and Mitsuhashi, S. (1969). *Journal of Biochemistry, Tokyo* **66**, 11.

Zyk, N. and Citri, N. (1968). *Biochimica et Biophysica Acta* **159**, 327.

Note Added in Proof

Mayer *et al.* (1979) were the first to apply genetic engineering techniques by subcloning the *E. coli* ATCC 11105 penicillin G acylase gene (using 'cosmid packaging' for initial gene isolation) on multicopy plasmids such as pOP203-3 and pBR322. A cosmid-hybrid *E. coli* 5K (pHM12) strain was obtained with ten-fold higher activity: the hybrid strain did not need any more phenylacetate as an inducer, nor was the acylase sensitive to repression by carbohydrate carbon sources.

Reference

Mayer, H., Collins, J. and Wagner, F. (1979). *In* 'Plasmids of Medical, Environmental and Commercial Importance' (K. N. Timmis and A. Dichler, eds.), pp. 459–470. Elsevier, North Holland Biomedical Press.

10. Conversions of Alkaloids and Nitrogenous Xenobiotics

L. C. VINING

Biology Department, Dalhousie University Halifax, Nova Scotia B3H 4JI, Canada.

I. INTRODUCTION

Publications dealing with microbial transformation of alkaloids reached a peak during the 1960s and the centre of interest since appears to have shifted towards achieving a better understanding of the interactions between xenobiotics, man and micro-organisms in the environment. Much of the work on alkaloid transformations was motivated by the desire to produce more effective drugs from naturally available compounds. No doubt it was encouraged by the success of similar efforts in the steroid field. Unfortunately, the rapid progress made towards understanding the action of steroid hormones has not been matched by a similarly expanding knowledge of the mechanism of drug action generally. As a result, many of the attempts to prepare new alkaloids by fermentation have been limited to exploring the range of transformations available. Since they have not been designed to fulfill specific needs, the modified alkaloids so discovered, unlike their steroid counterparts, have not found immediate commercial application.

Although present knowledge of the scope of microbial transformations is considerable, most of the several-thousand accessable nitrogenous compounds of natural origin have not yet been tested as substrates. The comprehensive survey of type reactions in fermentation chemistry by Wallen et al. (1959), which predates the period of most active interest in alkaloids, lists three transformations of these substances. The more recent compilation of Iizuka and Naito (1968) still contains only 54 entries compared with 652 on steroids. However, microbial transformation of a variety of other nitrogenous compounds, many derived from chemical rather than biological synthesis, has received a good deal of attention. Much of this work is covered in a recent account of oxidative microbial reactions (Fonken and Johnson, 1972) but the older survey by Wallen and his colleagues (1959) contains additional transformations of interest. Several reviews dealing specifically with microbial conversions of plant alkaloids are available (Tsuda, 1964b; Vining, 1969; Gröger, 1969).

This chapter is primarily concerned with the types of reactions that occur when naturally occurring nitrogenous substances are attacked by micro-organisms. Clearly, all such compounds are capable of being degraded, and a distinction between reactions that are part of a catabolic sequence and those that are adventitious and do not lead to assimilation is not always apparent. Although the emphasis is on conversions that have been observed to occur with a high yield of product, and therefore could be of practical value if the properties of the substance prove interesting, reactions that are unquestionably catabolic have been included. Many of the latter type are finding application in studies on drug metabolism in man. The products of microbial action frequently co-incide with those formed in the body. Since metabolites can be prepared in good yield by fermentation, they then become accessible for characterization and study.

Some of the more interesting microbial conversions of nitrogenous compounds have been carried out with synthetic drugs. Here the interest may also lie in obtaining adequate quantities of a mammalian metabolite so that its properties may be studied. However, a subsidiary, if not the prime, motive is often the prospect of specifically modifying the drug to improve its activity or decrease undesirable side effects. Transformations of this type have been included, as have others concerned with specific synthetic reactions, or with discovering the fate of xenobiotics in the environment.

II. STEROIDAL ALKALOIDS

Steroidal alkaloids are abundant in some solanaceous plants and represent a potential source of steroids for drug synthesis. As a result, strong interest developed in the range of conversions brought about by micro-organisms. The most commonly reported are hydroxylations at non-activated ring carbons. Solasodine (**I**) and tomatidine (**II**), the aglycones of alkaloids present in numerous *Solanum* species, are hydroxylated in this manner by the fungus *Helicostylum piriforme* (Sato and Hayakawa, 1961). The alkaloids differ in the stereochemistry of the N-heterocyclic ring as well as in the $\Delta^{5,6}$-unsaturation of solasodine. The consequent change in substrate shape markedly affects the pattern of hydroxylation. Solasodine gives mainly 9α-hydroxysolasodine, small amounts of 11α- and 7β-hydroxysolasodine and still less of a product suspected to be 7ξ,11α-dihydroxysolasodine. With tomatidine, the main product is 7α,11α-dihydroxytomatidine. This is

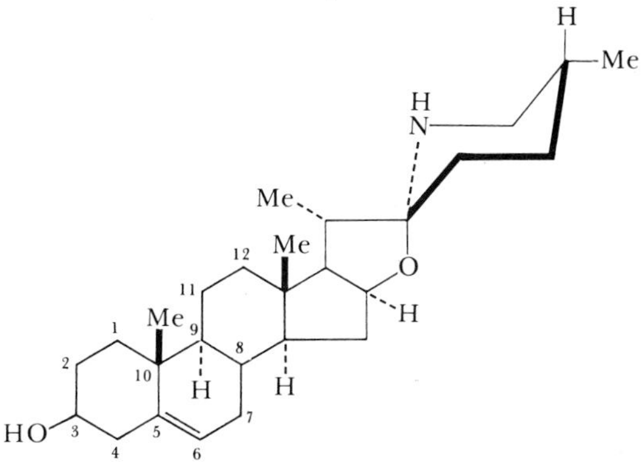

I

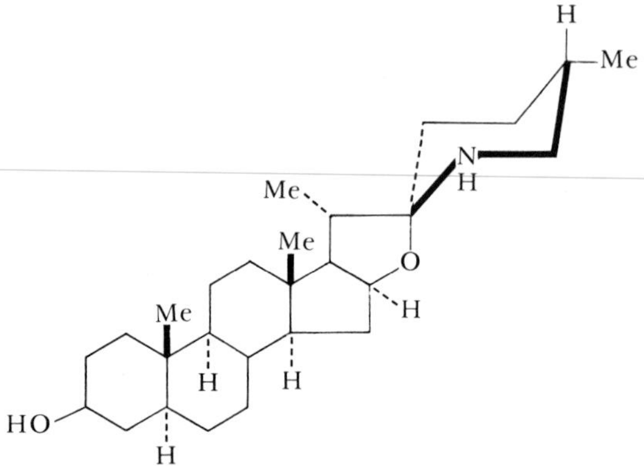

II

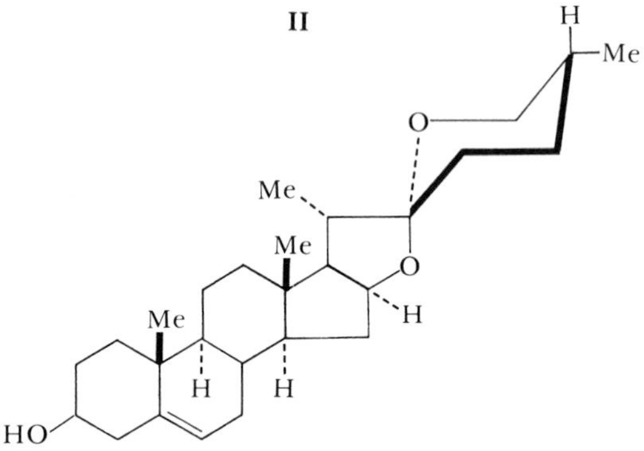

III

accompanied by 7α-hydroxytomatidine but very little of the 9α-hydroxylated compound is formed (Sato and Hayakawa, 1963). 7α-Hydroxylation is an uncommon reaction in steroids of the C-5 allo series, and thus offers a potentially useful route for preparing hormones with modified activity (Sato and Hayakawa, 1964). Interestingly, *H. pyriforme* also hydroxylates the saponin diosgenin (**III**), which bears considerable structural resemblance to solasodine, to give a mixture of 7β,11α-dihydroxy- and 11α-hydroxy-7-oxo-diosgenins. Tigogenin (5,6-dihydrodiosgenin), on the other hand, is not attacked (Hayakawa and Sato, 1963). Since 5,6-dihydrosolasodine is hydroxylated as readily as the parent compound, giving a mixture of 9α- and 7β-hydroxy derivatives (Sato *et al.*, 1964), the result suggests that different enzymes with selectivity towards the *N*-heterocyclic ring orientation are involved in oxygenation of solasodine and tomatidine.

Solanidine (**IV**), an alkaloid related to solasodine but possessing a fused heterocyclic ring system in place of the spiro structure, is

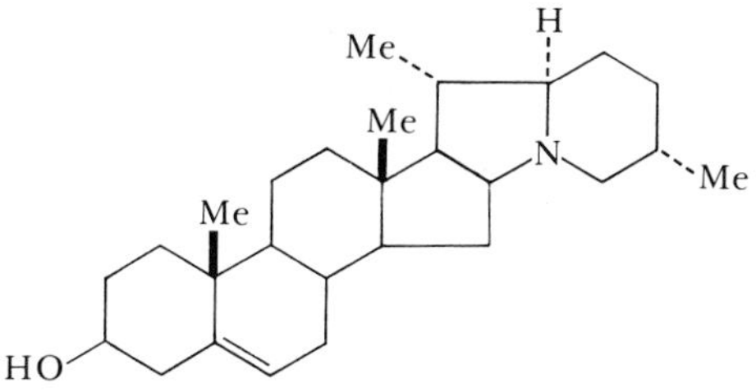

IV

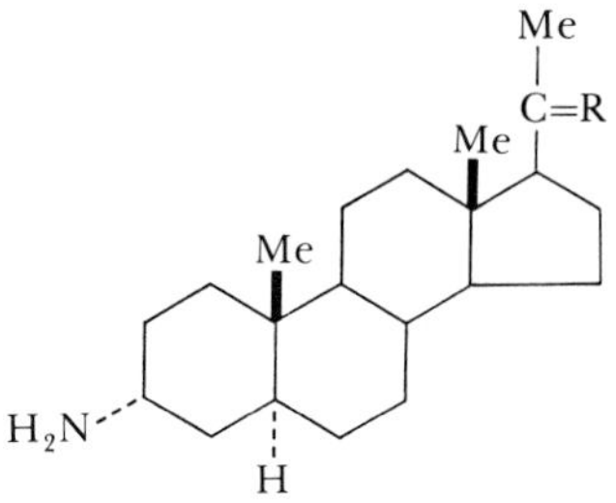

V

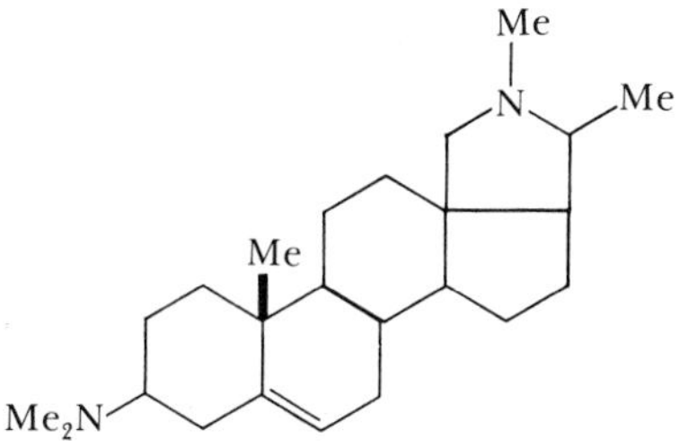

VI

converted into 11α-hydroxysolandine by *H. pyriforme*. The reaction is much slower than with solasodine; optimum yields require 10 days instead of 24 h (Sato *et al.*, 1967). Solanidine is reported to yield an antibacterial product by incubation with *Fusarium lini*. Jervine is converted into similarly active products in cultures of *Piricularia oryzae* but, in neither case, has the substance been identified (Wolters, 1970).

The 3α-amino steroids funtumine (**V**, R = O) and funtumidine (**V**, R = H, OH) from leaves of *Funtumia latifolia* are hydroxylated by *Aspergillus ochraceus* at C-11α or C-12β positions. Differences in the C-17 substituent which distinguish the two alkaloids influence the yield of each product (Greenspan *et al.*, 1965).

The dibasic alkaloid conessine (**VI**), from the bark of *Holarrhena* species, is hydroxylated at C-7α and C-7β by *A. ochraceus* and also by *Cunninghamella echinulata*. In cultures of *A. ochraceus*, the isomeric products are formed independently and not by epimerization (Kupchan *et al.*, 1963; Patterson *et al.*, 1964). The presence of 11α-hydroxyconessine was observed in fermentations with *Gloeosporium*, *Colletotrichum* and *Myrothecium* species by Koninklijke (1966). Marx *et al.* (1966b) noted that *Gloeosporium* species also hydroxylate non-nitrogenous steroids at C-11α. They found 11α-

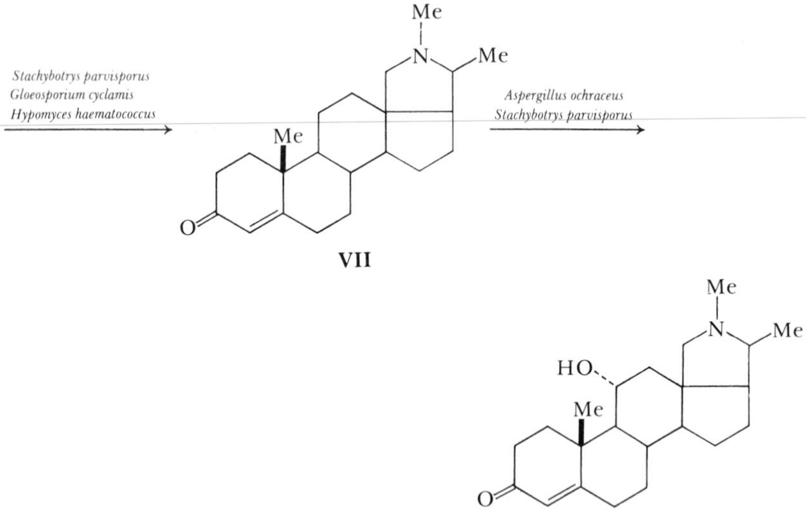

VII

VIII

hydroxylation as well when Δ⁴-conenine-3-one (**VII**) was incubated with *A. ochraceus* or with *Stachybotrys parvisporus*. The latter fungus belongs to the Dematiaceae which include several genera that introduce an 11α-hydroxyl group into steroids. The hydroxyconessines, besides being potentially valuable as synthetic intermediates, show antiphlogistic properties (Koninklijke, 1966).

Hydroxylation of conessine followed a different pattern with *Botryodiplodia theobromae*, giving a mixture of 9α-hydroxyconessine and 12α-hydroxyconessine. Whereas 9α-hydroxylation of steroids is not unusual, formation of a 12α-hydroxy derivative lacking a double bond between C-9 and C-11 had not previously been observed (Marx *et al.*, 1966a).

Compared with hydroxylating reactions, those causing dehydrogenation are infrequent among steroid alkaloids. One of the few that has been reported is conversion of conessine into Δ⁴-conenine-3-one by *Gloeosporium cyclamis* and *Hypomyces haematococcus* (de Flines *et al.*, 1962). As already noted, *S. parvisporus*, which also brings about the transformation, hydroxylates the initial product to form 11α-hydroxy-Δ⁴-conenine-3-one (**VIII**).

In a search for steroid hormone analogues with useful properties, Mazur and Muir (1963) prepared a group of azasteroids with the nitrogen atom inserted into ring C. They were unable chemically to desaturate ring A in these compounds, but microbial transformations were successful. Cultures of a *Nocardia* species converted 12a-aza-3β-hydroxy-C-homo-5α-pregnane-12,20-dione (**IX**, $R_1 = R_2 = H$) into 12a-aza-C-homo-1,4-pregnadiene-3,12,20-trione (**X**, R = H). The corresponding 17α-hydroxy derivative (**X**, R = OH) could be obtained by using 3β-acetoxy-12a-aza-17α-hydroxy-C-homo-5α-pregnane-12,20-dione (**IX**, $R_1 = COCH_3$, $R_2 = OH$) as the starting material. If an *Arthrobacter* species was used, 3β-acetoxy-12a-aza-C-homo-5α-pregnane-12,20-dione (**IX**, $R_1 = COCH_3$, $R_2 = H$) was hydrolysed and dehydrogenated in an unexpected way to give 12a-aza-C-homo-5α-pregn-1-ene-3,12,20-trione (**XI**).

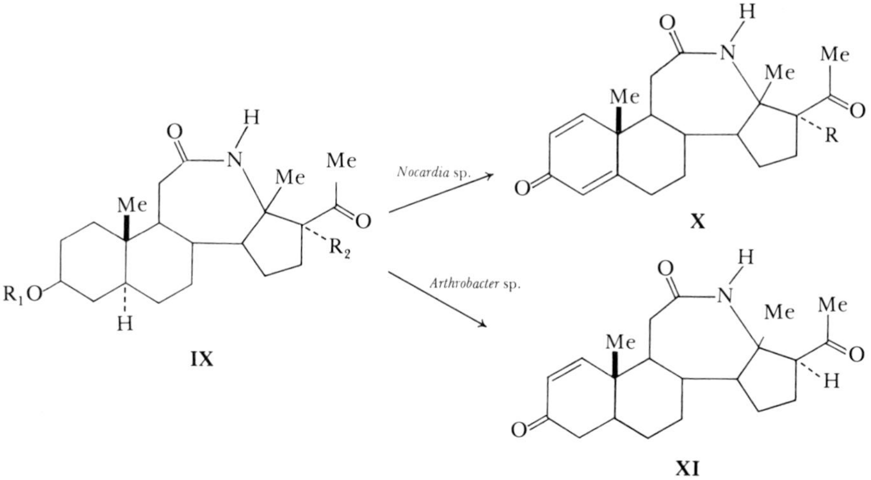

The resistance to dehydrogenation associated with introducing nitrogen atoms into the steroid structure is exemplified by the failure of *Fusarium solani*, which efficiently dehydrogenates many steroids, to metabolize tomatidine (Belič and Sočič, 1972). Two other widely used dehydrogenating organisms, *Nocardia restrictus* and *Mycobacterium phlei*, were able to metabolize the alkaloid and the nature of the products recovered indicated that there were alternative pathways proceeding as far as 1,4-tomatadien-3-one (**XII**). In other steroids, similar reactions represent the early stages of an extensive degradation to low molecular-weight compounds (Hörhold *et al.*, 1969). When side chain-splitting enzymes were induced by adding suitable non-nitrogenous steroids to the fermentation, 1,4-tomatadiene-3-one was degraded further. Belič and Sočič (1975) concluded that nitrogenous steroids such as tomatidine fail to induce the side chain-metabolizing enzymes. Of related interest is the observation of Maturova *et al.* (1967) that, in *Pseudomonas testosteroni*, Δ^5-3-ketosteroid isomerase and several hydroxysteroid dehydrogenases are strongly inhibited by reserpine, and by alkaloids from *Vinca* species; other alkaloids were either less effective or ineffective. Transcription was identified as the site of reserpine action.

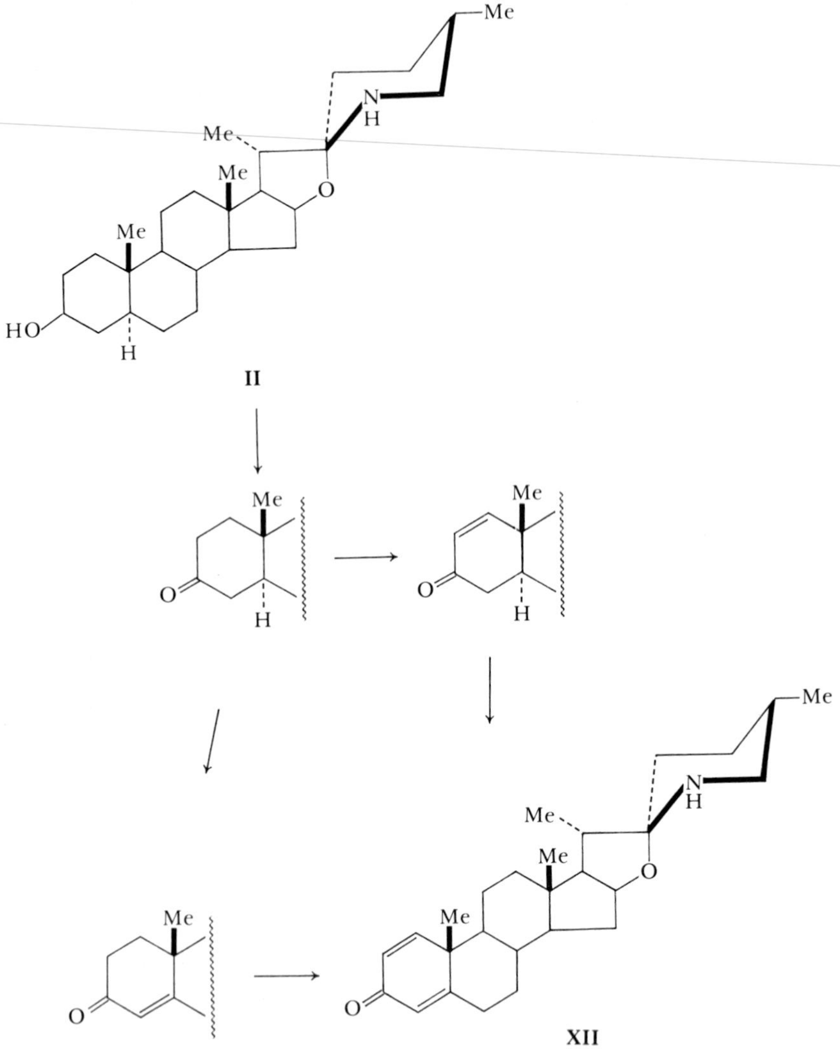

III. ALKALOIDS FROM *RAUWOLFIA* SPECIES

Hydroxylation is a prominent reaction during microbial modification of alkaloids from *Rauwolfia* spp. The aromatic ring is particularly susceptible. *Cunninghamella blakesleena* converts apoyohimbine (**XIII**), 3-*epi*-apoyohimibine and β-yohimbine methyl ether (**XIV, R = Me**) into the 10-hydroxy derivatives (Gotfredsen *et al.*, 1958). Transformation of β-yohimbine (**XIV, R = H**) to 10-hydroxy-β-yohimbine has been observed using *Streptomyces rimosus* (Patterson *et al.*, 1963) and yohimbine (**XV**), 3-*epi*-yohimbine, α-yohimbine (**XVI**) and its C-17 epimer (alloyohimbine) respond similarly with *Streptomyces platensis* (Loo and Reidenberg, 1959). *Streptomyces platensis* displays a measure of selectivity, since reserpine and other indolic substances are not metabolized. It also fails to hydroxylate ajmalicine and gives only a 5% yield with corynanthine (16-*epi*-yohimbine). Both of these alkaloids are 10-hydroxylated by *Gongronella urceolifera*. With corynanthine, the yield is 90% (Bellet and Thuong, 1970). Meyers and Pan (1961) surveyed a large number of organisms for their ability to 10-hydroxylate yohimbine and found the property to be widely but not universally distributed. It was present in actinomycetes, phycomycetes and imperfect fungi, but absent from the cultures of yeasts and bacteria examined. It was also strain-specific, suggesting the possibility that

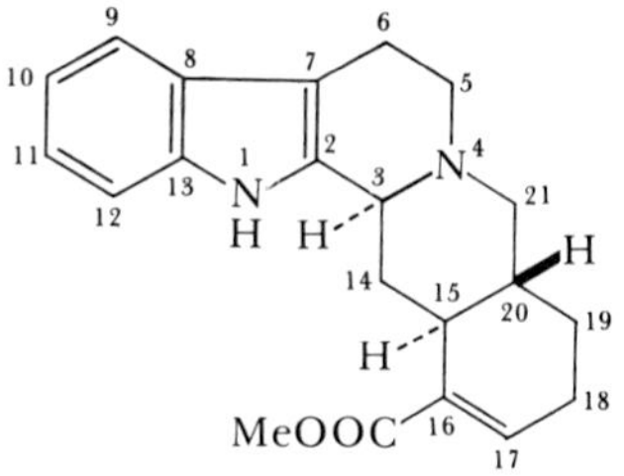

XIII

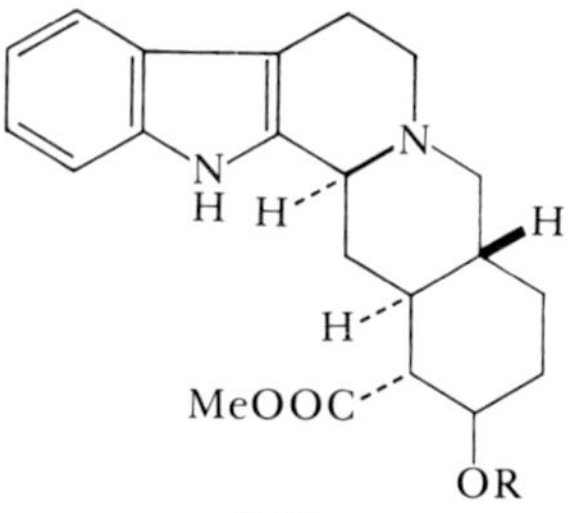

XIV

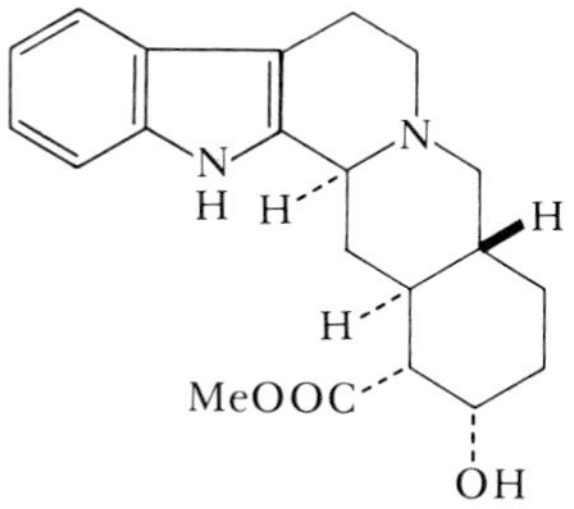

XV

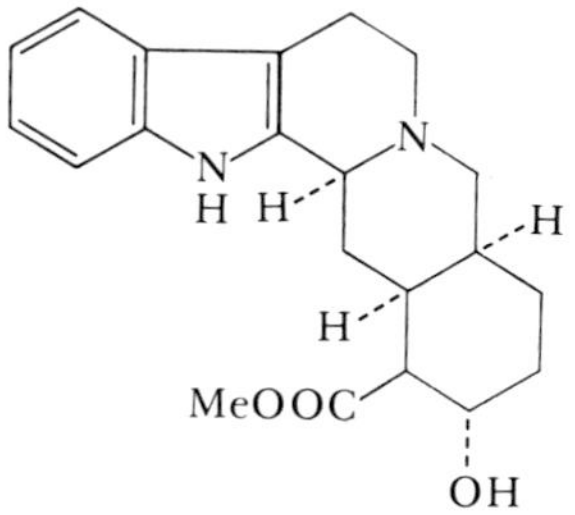

XVI

extrachromosomal genes might be involved. The enzyme has not been isolated but is probably a mixed-function oxygenase.

Hydroxylation is not restricted to C-10. Yohimbine incubated with *Cunninghamella bertholetiae* gave the 11-hydroxy derivative. Similar results were obtained with α-yohimbine (Patterson *et al.*, 1963). Much subsequent study has shown that the locus of attack varies with the species and with the substrate chosen. Thus, *Cunninghamella echinulata*, unlike *C. bertholetiae*, gave the 10-hydroxy derivatives of yohimbine and α-yohimbine. Ajmalicine (**XVII**) was metabolized by both species to give a mixture of 10- and (probably) 11-hydroxyajmalicines (Bellet and Thuong, 1970), but neither species affected β-yohimbine (Patterson *et al.*, 1963), and introduction of an 18α-hydroxyl group into yohimbine prevented hydroxylation in the aromatic ring (Hartman *et al.*, 1964). *Cunninghamella blakesleeana* attacked only yohimbine and α-yohimbine; *Streptomyces rimosus* attacked yohimbine, β-yohimbine and corynanthine, whereas *Calonectria* spp. and species of *Cunninghamella* and *Streptomyces* converted all of these substrates into the 10-hydroxy derivatives. *Streptomyces fulvissimus* hydroxylated at C-11, and attacked only yohimbine and corynanthine but numerous other 11-hydroxylating organisms showed little evidence of substrate specificity.

Besides depending on substrate and species, the site at which hydroxylation occurs can also vary with the strain used. Hartman and his coworkers discovered that some strains of *Cunninghamella bainieri* and *C. echinulata* hydroxylated only at C-10, and others only at C-11. As an added complication, fermentation conditions also influence the results obtained. In general, yohimbine is the preferred substrate in these oxygenation reactions although, as noted by Patterson and his colleagues (1963), fermentation yields are poor and rarely reach 20%. Only monohydroxylated products have been found.

Because susceptibility of the indole ring to attack by chemical oxidation limits opportunities to modify chemically other parts of the

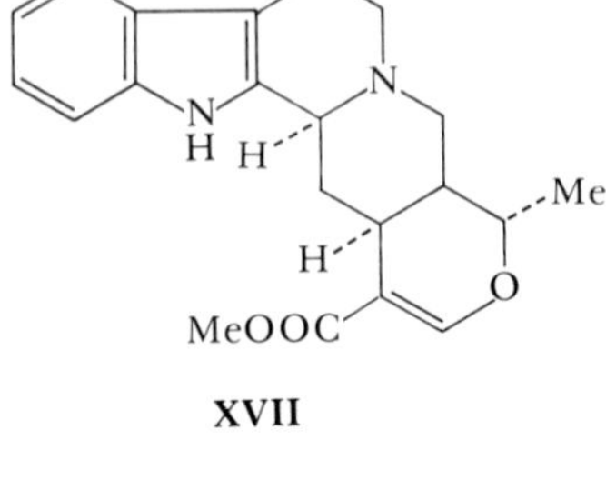

XVII

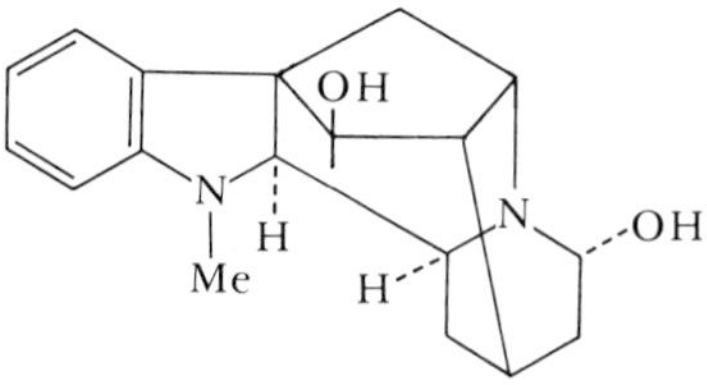

XVIII

indole alkaloids, discovery of microbial hydroxylation in the non-aromatic portion is of particular interest. *Streptomyces aureofaciens* and *S. rimosus*, two organisms known to carry out a variety of specific oxidation, hydroxylation and reduction reactions from their synthesis of tetracycline antibiotics, introduce an α-hydroxyl group into yohimbine and α-yohimbine at C-18 (Pan and Weisenborn, 1958). The yields reported are low and probably depend on fermentation conditions (Hartman *et al.*, 1964). *Calonectria decora* converts yohimbine into 18α-hydroxyyohimbine in a substrate-specific reaction where α- and β-yohimbine are not affected (Patterson *et al.*, 1963). 18-Hydroxyyohimbine may also be one of the products formed when *Cunninghamella blakesleeana* transforms apoyohimbine (Gotfredsen *et al.*, 1958).

In a different type of reaction, the *N*-methyl group is removed from ajmaline (**XVIII**) by several *Streptomyces* species (Bellet and Thuong, 1970). The *N*-demethylajmaline so formed is less toxic than its parent alkaloid. Prolonged incubation with *S. platensis* yields an additional, though minor, product, believed to be 10-hydroxy-*N*-demethylajmaline. Absence of 10-hydroxyajmaline suggests that hydroxylation occurred only after demethylation.

IV. ERGOT ALKALOIDS

Of the numerous transformations of ergot alkaloids now demonstrated, a high proportion is brought about by cultures of *Claviceps* spp. They represent steps in the complex branched biosynthesis process by which the alkaloids are formed. Certain members of the clavine series, notably chanoclavine-I (**XIX**), agroclavine (**XX**), elymoclavine (**XXI**) and probably also paspalic acid (**XXII**), are mainline intermediates *en route* to substituted lysergic acids (**XXIII**). Conversions along this pathway in a forward direction have been shown to occur in alkaloid-

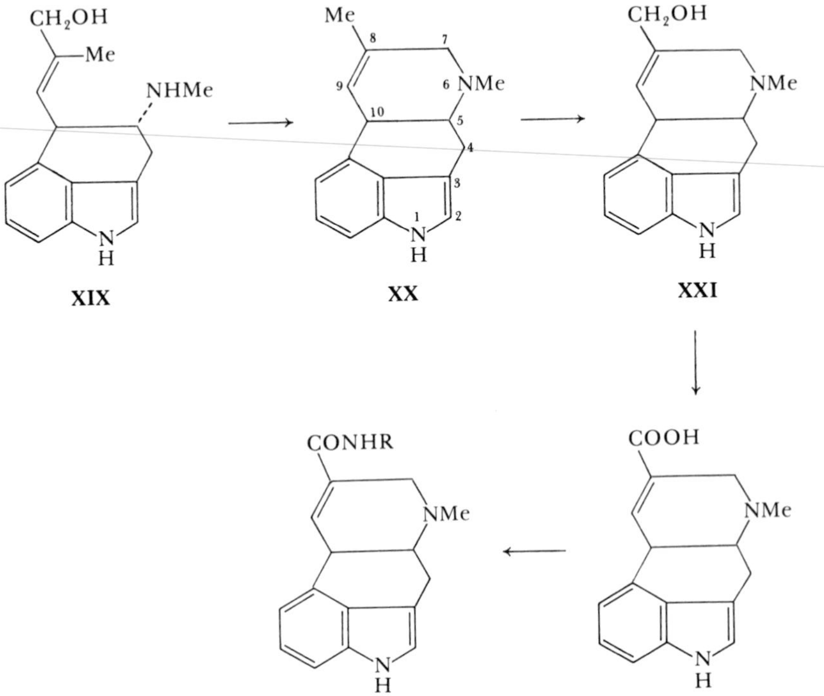

XIX → XX → XXI

XXIII ← XXII

producing cultures (e.g. Tyler *et al.*, 1965) but not with other micro-organisms. The enzymes used in these conversions have narrow specificity and a restricted distribution.

Extending earlier experiments (Voigt and Bornschein, 1964) in which ergot sclerotia injected *in vivo* with the early biosynthetic intermediate, chanoclavine, markedly increased production of the pathway end-products, ergometrine and ergotamine, Voigt and Bornschein (1965) incubated powdered sclerotia in a solution of chanoclavine and detected a 238% increase in formation of the water-insoluble terminal alkaloids. Submerged growing cultures of *Claviceps paspali* can also double production of amide derivatives of lysergic acid if supplemented with elymoclavine, another early inter-mediate (Eich, 1973). Elymoclavine disappears from the culture as the product forms so that it is not merely stimulating alkaloid synthesis. Among various strains of *C. paspali* tested, those that were good alkaloid producers were the best candidates for increasing yield. Since elymoclavine does not boost the yield from low producers to that of high-producing strains, enzyme capacity rather than precursor supply must be the limiting factor in the former.

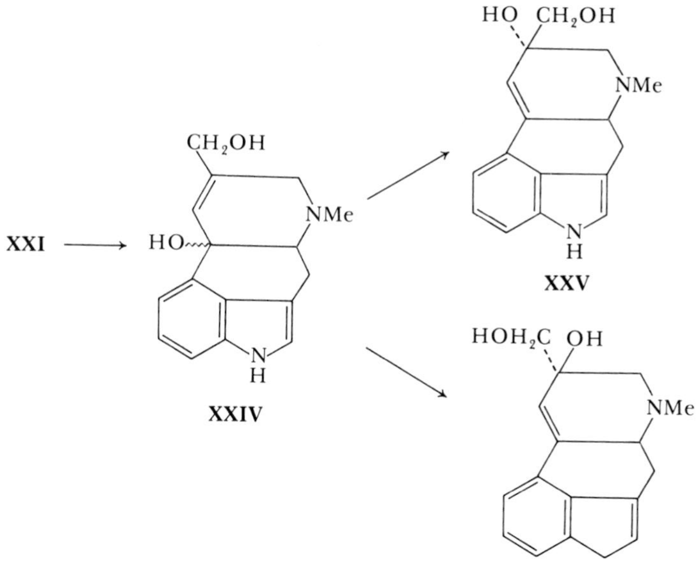

The variety of clavine alkaloids which have been found in *Claviceps* spp. and other ergot alkaloid-producing fungi is, in a large measure, due to the action of enzymes that direct elymoclavine and agroclavine from the main biosynthetic route. Such shunt activity is particularly evident in strains that lack the complete pathway and are unable to synthesize lysergic acid amides. Accumulated $\Delta^{8,9}$ intermediates may be reduced to dihydroalkaloids or converted by allylic hydroxylation into a variety of products (Agurell and Ramstad, 1962). The oxidative conversions are catalysed by peroxidase (Lin and Ramstad, 1967). Elymoclavine first forms a 4:1 mixture of 10α- and 10β-hydroxy-elymoclavines (**XXIV**). Under prevailing acidic culture conditions, these rearrange to give penniclavine (**XXV**) and isopenniclavine (**XXVI**). A similar series of events probably accounts for the conversion of agroclavine into setoclavine and isosetoclavine. Horseradish peroxidase has been shown to catalyse the conversions *in vitro* (Taylor and Shough, 1967); thus it is no surprise to find reports of their occurrence in plant homogenates (Taylor *et al.*, 1966) as well as in other microbial cultures. In the psilocybin-producing fungus *Psilocybe semperviva*, conversion of $\Delta^{8,9}$ into 8-hydroxy-$\Delta^{9,10}$ alkaloids is about

80% and some stereoselectivity is observed (Brack *et al.*, 1962). Although elymoclavine gives a 3 : 2 mixture of isopenniclavine and penniclavine, agroclavine is exclusively converted into setoclavine. The fungus rapidly oxidizes the lysergic acid derivatives ergometrine and ergotamine in the indole ring. No identifiable products have been recovered, but Yamatodani *et al.* (1962) were able to isolate 2-hydroxyagroclavine from cultures of *Corticium sasakii* to which agroclavine was added.

In other fungi and bacteria, several transformations unrelated to the dominant biosynthetic conversions of producing cultures have been observed. Strains of *Aspergillus fumigatus*, *Corticium praticola* and *C. sasakii* are reported to reduce elymoclavine to agroclavine; *A. fumigatus* also forms lysergol (**XXVII**), lysergene (**XXVIII**) and chanoclavine in a series of reversible reactions (Abe *et al.*, 1963) but, since this fungus is known to be capable of producing ergot alkaloids, some of these conversions may not be unrelated to its biosynthetic potential. Lysergol is slowly converted by allylic oxygenation in cultures of *Acetobacter aceti* into penniclavine along with traces of isopenniclavine, elymoclavine and an unidentified product. *Pseudomonas aeruginosa* metabolizes agroclavine to penniclavine, presumably by two allylic oxygenations. Chanoclavine gives a product believed to be the corresponding acid during incubation with *Streptococcus faecalis* (Abou-Chaar and Basmadjian, 1968).

Radiotracer studies with cultures of *Claviceps* spp. have indicated that *N*-methylation and *N*-demethylation reactions are common (Floss, 1976). Removal of methyl groups may, at least in part, be due to the activity of peroxidases. Noragroclavine is formed from agroclavine by several *Streptomyces* species (Yamatodani *et al.*, 1962) and accumulation

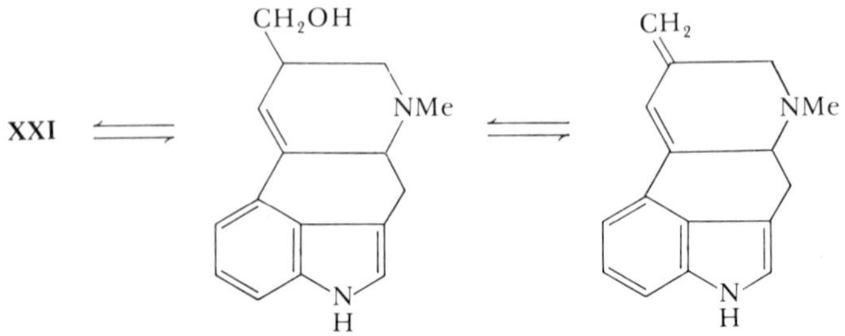

XXI   XXVII   XXVIII

of norsetoclavine (Ramstad *et al.*, 1967) and *N*-demethylchanoclavine-II (Cassady *et al.*, 1973) occurs under conditions where allylic hydroxylation is also prominent.

V. OTHER INDOLE COMPOUNDS

Although the variety of indole alkaloids modified by microbial action is not large, a substantial number of other indole compounds has been examined. At least one such transformation, namely oxidation of 1-acetyl-2,3-cyclo-octano-indoline (**XXIX**) by *Calonectria decora* to the isomeric C-8 and C-9 ketones, introduces functionality into a potentially useful synthetic intermediate (Lemke *et al.*, 1971).

Indole acetic acid (**XXX**) is hydroxylated by *Claviceps fusiformis* to 5-hydroxyindole acetic acid (**XXXI**) (Teuscher and Teuscher, 1965). This capability is not widely distributed in species of *Claviceps*. *Claviceps purpurea* transforms indole acetic acid to indole isopropionic acid (**XXXII**) (Yamano *et al.*, 1962). The fungus *Omphalia flavida* converts it into 3-methyldioxindole (**XXXIII**) (Ray and Thimann, 1956) and, in

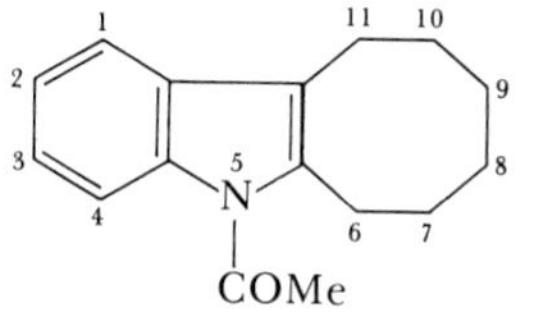

XXIX

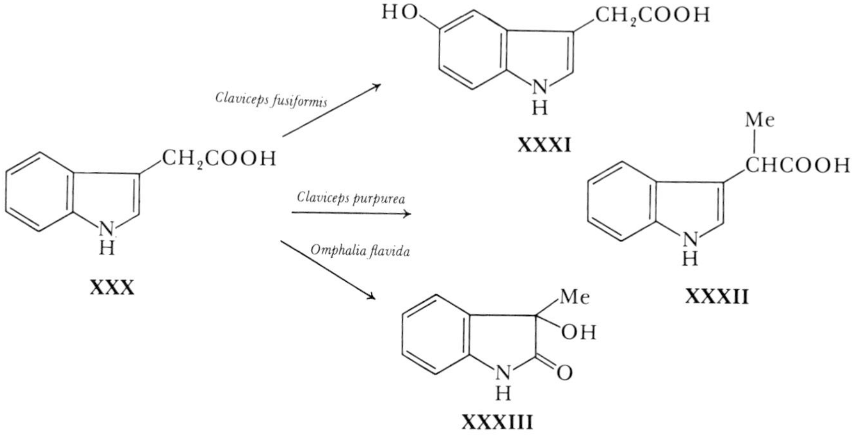

XXXI

XXX

XXXII

XXXIII

Bacillus megatherium it is conjugated with glucose or amino acids (Tabone, 1958).

Tryptophan is oxygenated by *Chromobacterium violaceum* to 5-hydroxytryptophan in a low-yield reaction associated with biosynthesis of the purple pigment violacein (Mitoma *et al.*, 1956). Because of the interest in serotonin-related compounds, attempts have been made to improve the yield by blocking the α-amino group in the expectation that this would decrease competing metabolic reactions (Davis *et al.*, 1975). Unfortunately, the substrate requirements for hydroxylation appear to be very narrow, and 2′,3′-dehydrotryptophan is formed instead. In a similar reaction, *C. violaceum* dehydrogenates indole propionic acid to indole acrylic acid. Indole, indole acetic acid and tryptamine are not metabolized (Mitoma *et al.*, 1956). In a search for alternative 5-hydroxylating organisms, Daum and Kieslich (1973) screened 100 bacterial species and selected a strain of *Bacillus subtilis* that carried out the reaction with a 99% yield of product in medium containing a high concentration of ferrous sulphate. Either L-tryptophan or *N*-acetyl-L-tryptophan is a suitable substrate.

Claviceps spp. and many other micro-organisms metabolize tryptophan to indole acetic acid (Teuscher, 1965). Yeasts use their normal route for degrading amino acids, forming tryptophol and a small amount of indole acetic acid. Yields of tryptophol approaching 70% have been obtained in cultures of *Zygosaccharomyces priorianus* (Rosazza *et al.*, 1973). 5-Hydroxytryptophan and other aromatic amino acids are converted into the corresponding alcohols in similar yields. In *Claviceps paspali* and *Aspergillus niger*, tryptophan is metabolized via 2,3-dihydroxybenzoic acid, and this intermediate may accumulate in the culture (Arcamone *et al.*, 1961; Subba Rao *et al.*, 1967). Other organisms give indole lactic acid as the main product (Schuytema *et al.*, 1966).

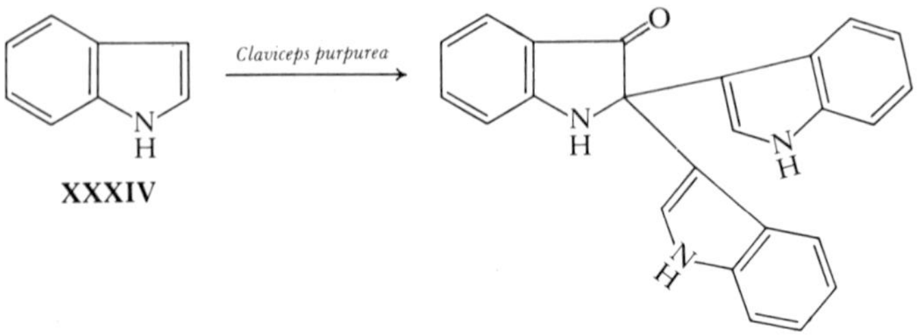

XXXIV   XXXV

Tryptamine is metabolized by numerous bacteria and fungi in the same manner as tryptophan, yielding indole acetic acid (Schuytema *et al.*, 1966) or tryptophol and indole acetic acid (Kaper and Veldstra, 1958; Larsen *et al.*, 1962; Gordon and Gentile, 1962). In *Aspergillus niger* and *A. awamori*, an additional product is 5-hydroxyindole acetic acid (Dvornikova *et al.*, 1968, 1970) and a number of micro-organisms are known to oxidize tryptophol to indole acetic acid or 5-hydroxyindole acetic acid (Schuytema *et al.*, 1966). In *Hygrophorus conicus*, where 3-substituted indoles are oxidized to 2-oxindole acetic acid, the highest yield is obtained with tryptamine (Siehr, 1961). In contrast to these degradative processes, conversion of tryptamine in *Zygosacch. priorianus* is by N-acetylation (Rosazza *et al.*, 1973).

Indole (**XXXIV**) is oxidized by *Claviceps purpurea*, as by plant peroxidase, to give 2,2-*bis*-(3-indolyl)indoxyl (**XXXV**) (Loo and Woolf, 1957). In *C. fusiformis*, it is possible to replace indole in the pathway for tryptophan biosynthesis by 5-hydroxyindole and so obtain a mixture of 5-hydroxytryptophan and its N-acetyl derivative (Mutschler *et al.*, 1968b). Under similar conditions, *Cordyceps militaris* forms N-acetyl-5-hydroxytryptophan and N-acetylserotonin, whereas *A. oryzae* forms only N-acetylserotonin (Mutschler *et al.*, 1968a). A species of *Coryne-bacterium* and certain other bacteria will produce methyl-, hydroxy-, halogeno- and trifluoromethyl-substituted tryptophans by supplementing cultures with the appropriate indole derivative (Watanabe *et al.*, 1973a). 7-Azatryptophan has also been produced in this way (Watanabe *et al.*, 1973b).

VI. TOBACCO ALKALOIDS

Most studies on microbial conversion of nicotine (**XXXVI**) and related compounds have been designed to elucidate the pathway of degradation, and have not been concerned with high-yield transformations to useful products. In the course of this work, however, interesting conversions have come to light. One of the more intriguing has been the repeated appearance in nicotine-metabolizing cultures of a deep-blue pigment. Some cultures of *Arthrobacter* spp. accumulate 6-hydroxynicotine (**XXXVII**) in yields reaching 80% of the substrate added. Most of this compound is formed during rapid growth when the rate of synthesis exceeds the rate of further metabolism; the optimum yield thus depends on the age of the culture (Griffith *et al.*,

1961). Some bacteria accumulate 3-succinylpyridine (**L**) in good yield (Tabuchi, 1954) while *Achromobacter nicotinophagum* and some soil pseudomonads convert about 35% of the nicotine supplied into 3-succinyl-6-hydroxypyridine (**LI**) (Wada, 1957; Hylin, 1959). An explanation can be provided for these and other conversions in terms of a general pathway for microbial degradation of nicotine with superimposed variations in individual organisms.

Examination of nicotine-metabolizing bacteria isolated from the surface of tobacco seeds has suggested a subdivision into three main groups, distinguished by the kinds of metabolites that accumulate in cultures (Frankenburg, 1955; Frankenburg and Vaitekunas, 1955). Group-1 organisms bring about an initial 6-hydroxylation of the pyridine ring. Nicotine oxidase, the soluble metalloflavoprotein enzyme that catalyses the reaction in strains of *Arthrobacter oxydans* metabolizing nicotine (Hochstein and Rittenberg, 1959a, b; Gries *et al.*, 1961), is a dehydrogenase (Fig. 1) with a relatively low degree of substrate specificity (Hochstein and Dalton, 1967). It will hydroxylate nornicotine (**LVI**), anabasine (**LVIII**), D-nicotine and *N*-methyl-myosmine (**XLVIII**) (Decker *et al.*, 1961; Gries *et al.*, 1961). The next plausible intermediate isolated from cultures degrading nicotine is 6-hydroxypseudo-oxynicotine (**XXXIX**). It is evident that this is the result of two reactions and several alternatives have been postulated (Decker *et al.*, 1960; Gries *et al.*, 1961). The most plausible sequence (Hochstein and Rittenberg, 1960) is an initial oxidation of 6-hydroxynicotine to 6-hydroxy-*N*-methylmyosmine (**XXXVIII**) which spontaneously hydrates in an aqueous medium to 6-hydroxypseudo-oxynicotine (Fig. 2). The dehydrogenating enzyme has been isolated and shown to include nicotine as well as 6-hydroxynicotine in its substrate range.

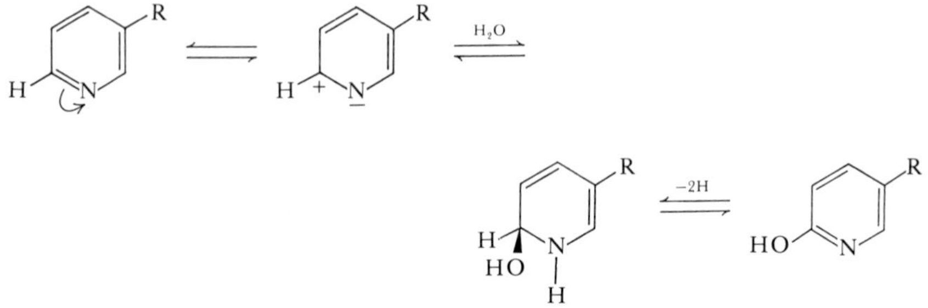

Fig. 1. Pathway for 6-hydroxylation of pyridine derivatives.

6-Hydroxypseudo-oxynicotine is hydroxylated by cell extracts of *A. oxidans* to 2,6-dihydroxypseudo-oxynicotine (**XL**). *In vitro*, this compound rearranges irreversibly to 2,6-dihydroxy-*N*-methyl-myosmine (**XLI**) but, since the rearranged product is not found in cultures metabolizing nicotine, 2,6-dihydroxypseudo-oxynicotine is probably converted into further products without leaving the enzyme surface (Richardson and Rittenberg, 1961). Subsequent reactions include removing the aliphatic substituent as α-methylaminobutyric

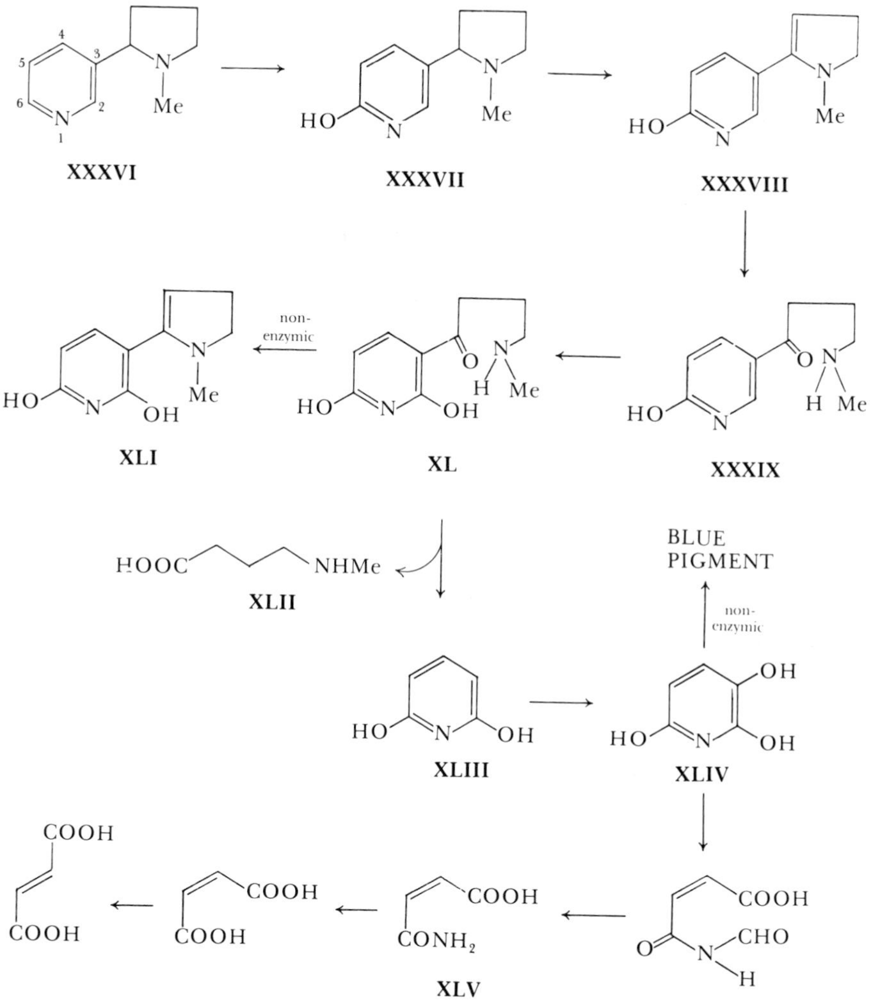

Fig. 2. Pathways for metabolism of nicotine by group-1 bacteria.

acid (**XLII**); the reaction, which is analagous to thiolysis of β-keto-acylcoenzyme-A, is facilitated by the lactam character of 2,6-di-hydroxypyridine (**XLIII**; Gherna *et al.*, 1965). Evidence that 2,3,6-tripyridol (**XLIV**) is the next intermediate is not conclusive, but is plausible since extracts from cells grown on nicotine, but not in its absence, oxidize the postulated further metabolite maleamic acid (**XLV**).

Gherna and his colleagues (1965) considered that the blue pigment formed during nicotine metabolism is a spontaneous oxidation product of the unstable 2,3,6-tripyridol. The pigment is a mixture of compounds, and the structure **XLVI** has been proposed for one component (Niemer *et al.*, 1964). It bears a close resemblance and may be identical to the blue pigment obtained when *Arthrobacter crystallopoietes* is grown on 2-pyridone. This is produced only if substrate is added in excess and only after growth stops. It appears to be a "shunt" product that forms non-enzymically from an accumulated unused metabolite, which Ensign and Rittenberg (1963) suggest is 2,3,6-tripyridol. The product formed from 2-pyridone has been assigned the structure **XLVII** (R = OH), related to that of indigoidine (**XLVII**, R = NH$_2$) and other blue bacterial pigments (Kuhn *et al.*, 1965; Starr *et al.*, 1967).

The group-2 nicotine-metabolizing bacteria of Frankenburg and Vaitekunas (1955) differ from those of group 1 in not forming 6-hydroxynicotine in the initial step. Within group 2, several sub-groups can be distinguished; members of one sub-group accumulate the same metabolites as certain soil pseudomonads capable of growing on nicotine as the sole carbon and nitrogen source (Wada and Yamazaki, 1954). The pathway suggested from formation of pseudo-oxynicotine (**XLIX**), 3-succinylpyridine (**L**) and 3-succinyl-6-hy-

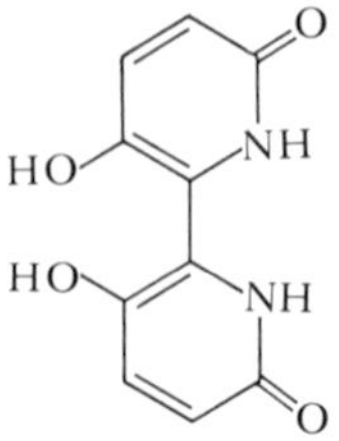

XLVI 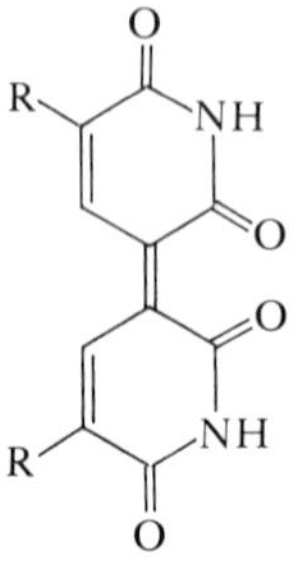   XLVII droxypyridine (**LI**) by these organisms (Fig. 3) is supported by simultaneous-adaptation studies (Wada, 1957). Differences between the pathways in group-1 and group-2 organisms are probably due to differences in the substrate range or inducibility of the 6-hydroxylating enzyme; the group-1 enzyme is able to attack nicotine whereas the group-2 enzyme requires 3-succinylpyridine. However, these differences may sometimes depend on the condition of the culture. Hylin (1959) found that *Achromobacter nicotinophagum* metabolized nicotine via 6-hydroxynicotine when growing fast whereas, in cells that had stopped dividing, enzymes were induced to convert the substrates via pseudo-oxynicotine and 3-succinylpyridine into the dead-end metabolite 3-succinyl-6-hydroxypyridine.

The distinction between subgroups of group-2 organisms also relates to hydroxylating activity. One subgroup (2A; Fig. 3) will metabolize nicotine but is unable to grow on nornicotine or anabasine. A second

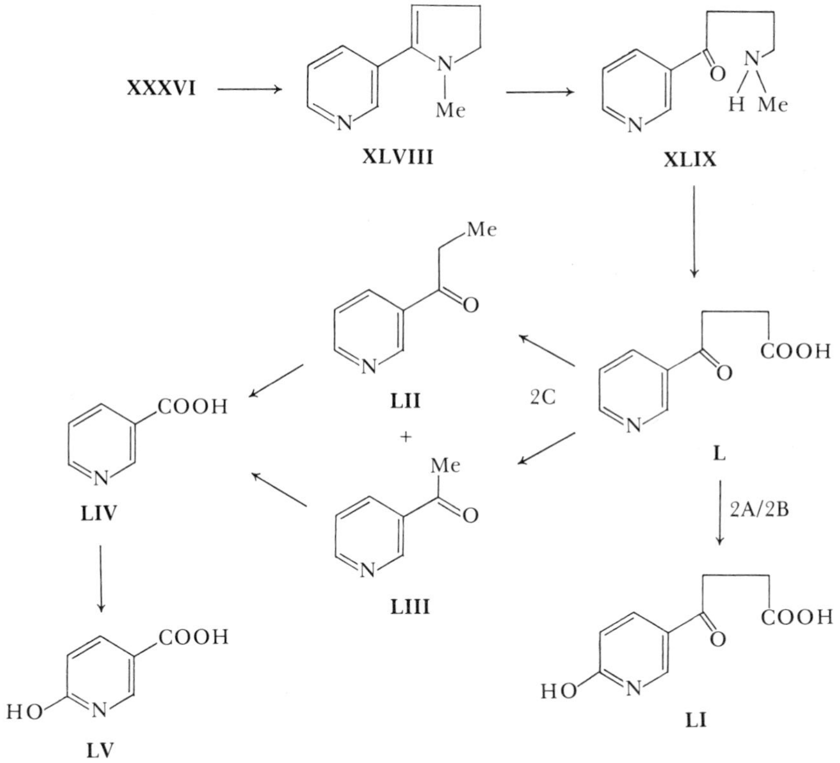

Fig. 3. Pathways for metabolism of nicotine by group-2 bacteria.

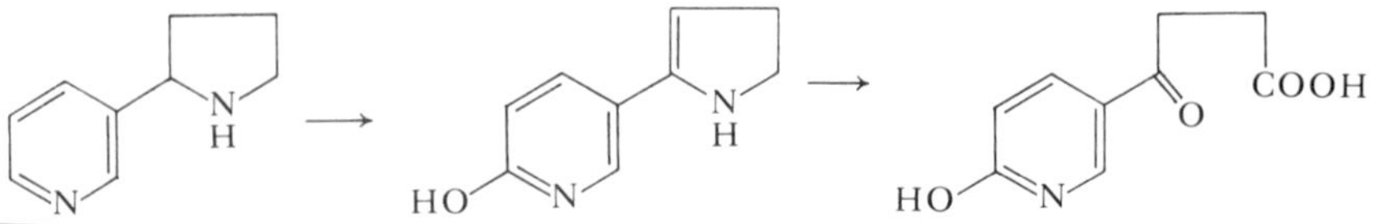

LVI   LVII   LI

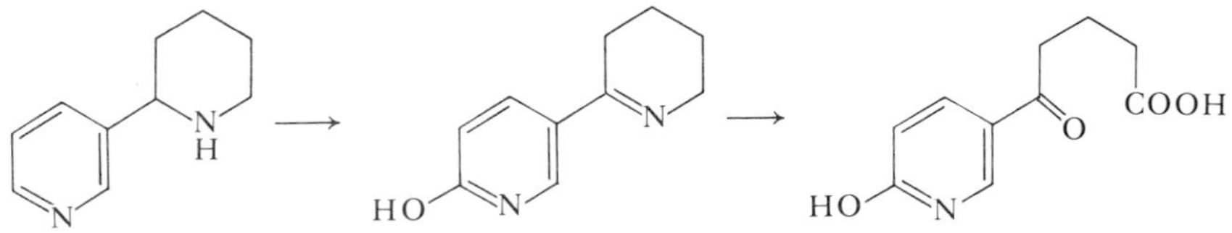

LVIII   LIX   LX

(2B; Fig. 3) will metabolize all three alkaloids. The latter group accumulates 6-hydroxymyosmine (**LVII**) and 3-succinyl-6-hydroxy-pyridine (**LI**) during growth on nornicotine. Adaptation studies show that such cultures metabolize 6-hydroxymyosmine but not myosmine; those grown on 6-hydroxymyosmine are simultaneously adapted to 3-succinyl-6-hydroxypyridine but not *vice versa*. Cultures grown in the presence of anabasine accumulate 1′,6′-dehydro-6-hydroxyanabasine (**LIX**) and 3-glutaryl-6-hydroxypyridine (**LX**). It is apparent that nornicotine and anabasine are hydroxylated at an early stage, suggesting that the success of group-2B organisms lies in the ability of their 6-hydroxylating enzyme to accept the alkaloids as substrates.

Frankenburg and Vaitekunas (1955) distinguished a third subgroup of organisms (2C; Fig. 3) by the absence of 3-succinyl-6-hydroxy-pyridine from nicotine-metabolizing cultures. Instead, 3-propionyl-pyridine (**LII**), 3-acetylpyridine (**LIII**), nicotinic acid (**LIV**) and 6-hydroxynicotinic acid (**LV**) were formed. The plausible pathway (Fig. 3) suggested by this accumulation of metabolites again reflects a difference in the 6-hydroxylating enzyme.

VII. OTHER PYRIDINE COMPOUNDS

The nitrile group of ricinine (**LXI**) can be hydrolysed directly to the carboxylic acid by an enzyme purified from a soil pseudomonad that metabolizes the alkaloid (Robinson and Hook, 1964). A small amount of amide forms during the hydrolysis, but this compound is not a substrate for the enzyme. From the range of synthetic 2-pyridone-

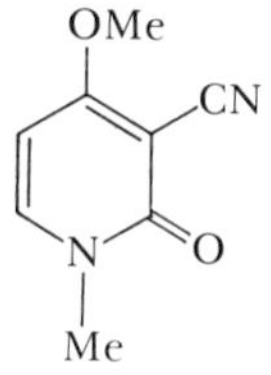

OMe

CN

N   O

|

Me

LXI

COO⁻

⁺N

|

Me

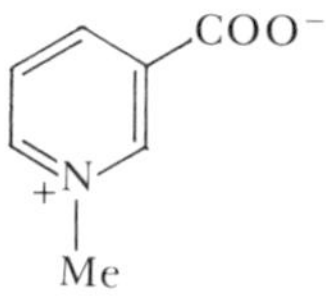

LXII

3-nitriles acceptable as substrates, neither the methoxyl group at C-4 nor the N-alkyl substituent is essential. Hydrolysis of ricinine by the bacterium is similar to direct hydrolysis of indole acetonitrile to indole acetic acid in barley leaves (Thimann and Mahadevan, 1964). In contrast, *Nocardia rhodocrous* metabolizes aliphatic nitriles such as acetonitrile and propionitrile *via* the amides (Di Geronimo and Antoine, 1976). Hook and Robinson (1964) have suggested a general mechanism to account for both routes.

Trigonellin (**LXII**), an alkaloid present in coffee beans and many other plant sources, is oxidatively demethylated by *Torula cremoris* to nicotinic acid which accumulates in the culture (Joshi and Handler, 1962). The reverse process, in which a pyridine is N-methylated, can be demonstrated in cell extracts of *Claviceps purpurea* supplemented with S-adenosylmethionine. Nicotinamide, but not tryptamine, tyramine or serotonin, is a substrate for the enzyme (Audette *et al.*, 1964). Mycobacteria can convert 3-hydroxymethylpyridine or pyridine-3-carboxaldehyde into nicotinic acid. In *M. tuberculosis*, pyridine-4-carboxaldehyde is also converted into the corresponding carboxylic acid (Gross *et al.*, 1968; Youatt, 1962).

Metabolism of nicotinic acid has attracted attention, in part, because of its association with nicotine metabolism. Bacteria growing on nicotinic acid commonly produce a blue pigment similar to that formed during growth on nicotine. However, adaptation of many bacteria to nicotine does not enable them to metabolize nicotinic acid (Wada, 1957). The initial step in nicotinic acid utilization by species of *Pseudomonas*, *Clostridium* and *Bacillus* is 6-hydroxylation (Hughes, 1952; Harary, 1957; Ensign and Rittenberg, 1964; Jones and Hughes, 1972). With *Ps. fluorescens*, 6-hydroxynicotinic acid accumulates if cultures have not been adapted to growth on nicotinic acid, indicating that the hydroxylating enzyme may be constitutive. Adapted cultures oxidize 2- and 5- but not 6-fluoronicotinic acid and also metabolize

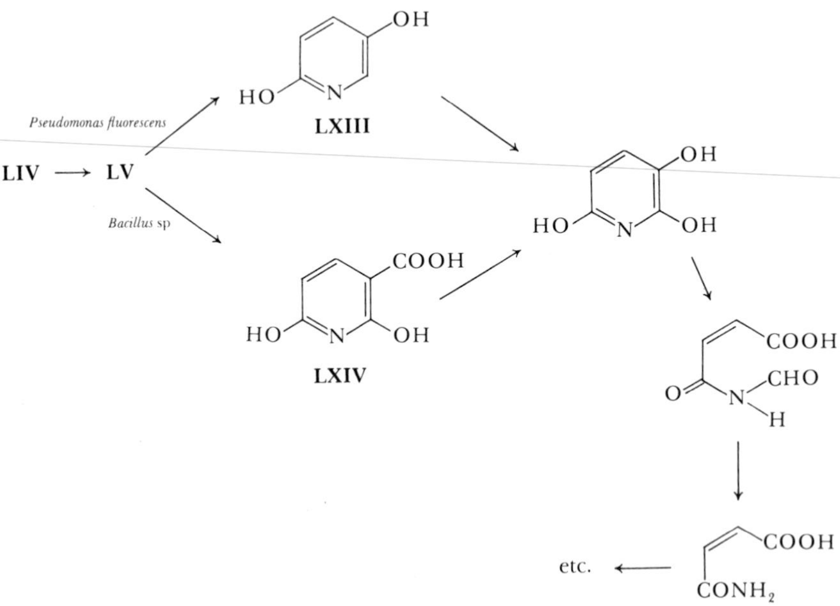

Fig. 4. Pathways for metabolism of nicotinic acid by *Pseudomonas fluorescens* and a *Bacillus* species.

6-hydroxynicotinic acid without a lag (Hughes, 1952). The hydroxylating enzyme is a dehydrogenase similar to the nicotine-hydroxylating enzyme of *Arthrobacter oxidans* in that the hydroxyl group is derived from water (Hurst *et al.*, 1958). However, the enzymes are differently located; the *Ps. fluorescens* dehydrogenase is in the cell envelope and is not soluble (Hunt *et al.*, 1959). 6-Hydroxynicotinic acid is subsequently catabolized in *Ps. fluorescens* by soluble enzymes. The pathway (Fig. 4) includes 2,5-dihydroxypyridine (**LXIII**) and maleamic acid (Behrman and Stanier, 1954).

In the *Bacillus* species examined by Ensign and Rittenberg (1964), the 6-hydroxylase was induced as a soluble enzyme and its product was hydroxylated again to form 2,6-dihydroxynicotinic acid (**LXIV**). The latter accumulated to concentrations of 3% in cultures and was the end-product in cell extracts. However, resting cells rapidly oxidized it and maleamic acid, suggesting an alternative pathway to 2,3,6-tripyridol (Fig. 4).

Metabolism of picolinic acid (**LXV**) is of interest because this molecule is related to commercial herbicides or their photolytic products (Orpin *et al.*, 1972). Like other pyridine derivatives, it is

LXV   LXVI

degraded by initial 6-hydroxylation through the agency of a dehydrogenase. The enzyme in *Arthrobacter picolinophilis* and another
Gram-negative bacterium is particulate (Tate and Ensign, 1974; Orpin
et al, 1972) and resembles, in this respect, the 6-hydroxylating enzyme
of *Ps. fluorescens*, *A. oxydans* and a *Clostridium* species (Holcenberg and
Stadtman, 1969). Species of *Rhodotorula* and *Aerococcus* also metabolize
picolinic acid to 6-hydroxypicolinic acid but, as with other organisms
studied, the pathway of further metabolism is not completely known
(Dagley and Johnson, 1963).

In 2- and 3-hydroxypryridines, a second hydroxyl group is inserted
by a variety of bacteria; the site of hydroxylation depends on the
location of the existing substituent (Ensign and Rittenberg, 1963;
Houghton and Cain, 1972). The more heavily substituted pyridine ring
of vitamin B_6 is opened by specific mixed-function oxygenases in
pseudomonads after modification or removal of substituent groups
(Ikawa *et al.*, 1958; Burg *et al.*, 1960). In 2,6-dipicolinic acid (**LXVI**),
where 6-hydroxylation is prevented, Kobayashi and Arima (1962)
observed trace accumulation of a 3-hydroxylated product.

VIII. MORPHINE ALKALOIDS

Because of its multiple biological effects, morphine has been a
favourite drug for studies on structure-activity relationships; the
value of using micro-organisms to modify drug structures specifically
has not been overlooked. Tsuda and his colleagues screened 1,700
bacteria and fungi for ability to metabolize thebaine and found
120, mostly basidiomycetes. *Trametes sanguinea* was selected to examine
the range of transformations possible with a variety of morphine
alkaloids (Iizuka *et al.*, 1960). (−)-Thebaine (**LXVII**) is converted by
allylic oxygenation and demethylation into (−)-14β-hydroxy-
codeinone (**LXVIII**) which can then be reduced to (−)-14β-hydroxy-
codeine (**LXIX**). Very little reduction occurs under acidic conditions so
that the pH value of the medium determines the type of product
accumulated (Iizuka *et al.*, 1962). (+)-Thebaine is acted on similarly to

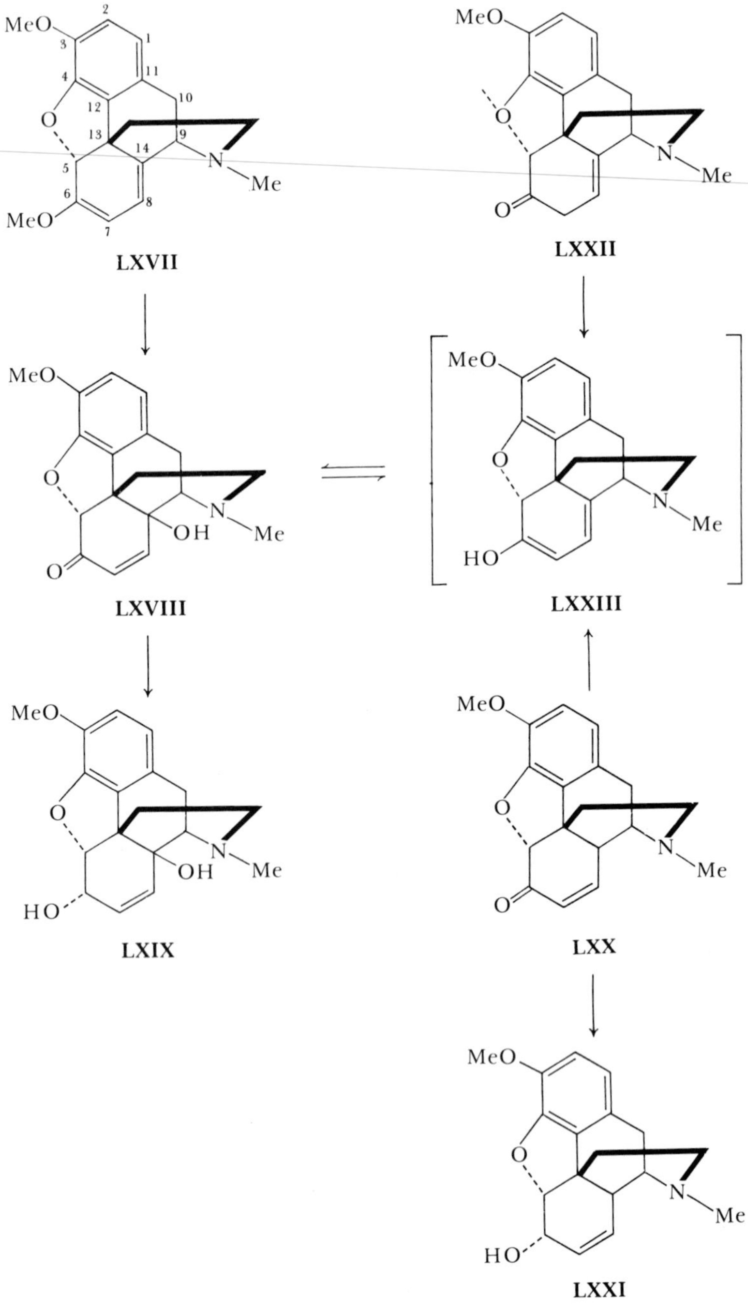

Fig. 5. Pathways for metabolism of (−)-thebaine, (−)-codeinone and neopinone by *Trametes sanguinea.*

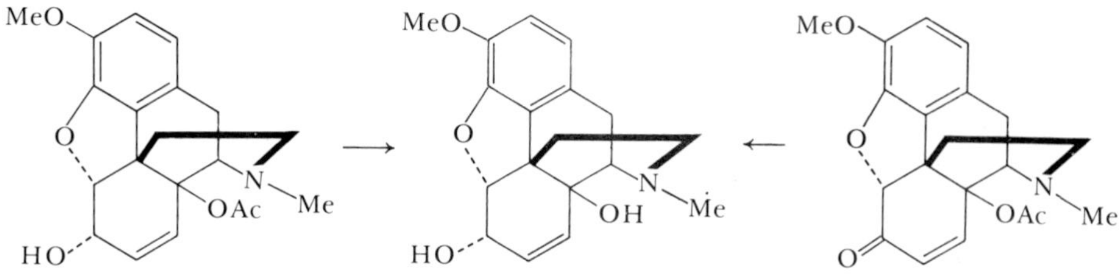

LXXIV   LXIX   LXXV

give the corresponding dextrorotatory products (Tsuda, 1964a). The wood-rotting basidiomycete *Trametes cinnabarina* also converts thebaine into 14β-hydroxycodeinone but only a trace of 14β-hydroxycodeine is formed. Instead, the codeinone derivative is converted in part into its *N*-oxide (Gröger and Schauder, 1969). Experiments with $^{18}O_2$ indicate that oxygen in the 14β-hydroxy group is introduced from molecular oxygen (Aida *et al.*, 1966).

In cultures of *T. sanguinea* to which (−)-codeinone (**LXX**) is added, (−)-codeine (**LXXI**) as well as the 14β-hydroxylated derivatives of (−)-codeine and (−)-codeinone are formed (Yamada *et al.*, 1962). Tsuda (1964a) reported similar results for (+)-codeinone, except that (+)-dihydrocodeine, (+)-dihydroisocodeine and (+)-dihydrocodeinone are additional minor products. Neopinone (**LXXII**) gives only 14β-hydroxycodeine and 14β-hydroxycodeinone; neopine was not detected. Transformations of (−)-codeinone, neopinone and thebaine are summarized in Figure 5. Since codeine is not formed from (−)-thebaine, demethylation is postulated to accompany or follow insertion of the hydroxyl group at C-14 and the enolic intermediate (**LXXIII**) is excluded. Morphine is metabolized by *T. sanguinea*, but the unprotected phenolic ring is probably degraded and products could not be identified.

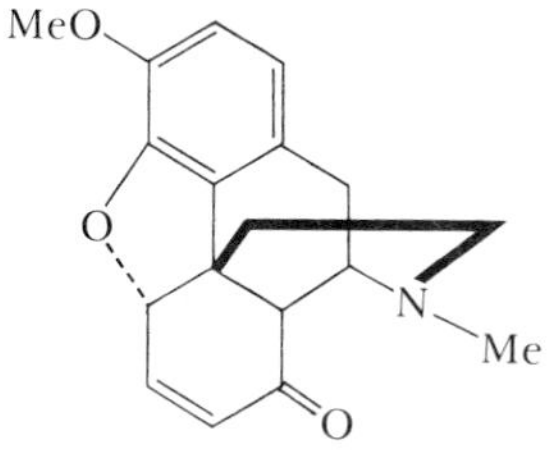

LXXVI

The enzymes that reduce the α,β-unsaturated ketone of codeinone and 14β-hydroxycodeinone are not unspecific dehydrogenases since typical steroid-4-en-3-ones are not attacked. Reduction of the ketones is not prevented by acetylation of the 14β-hydroxyl group, but the acyl group may be removed first since 14β-hydroxycodeine is recovered from incubations with 14β-acetoxycodeine (**LXXIV**) as well as with 14β-acetoxycodeinone (**LXXV**) (Yamada *et al.*, 1963).

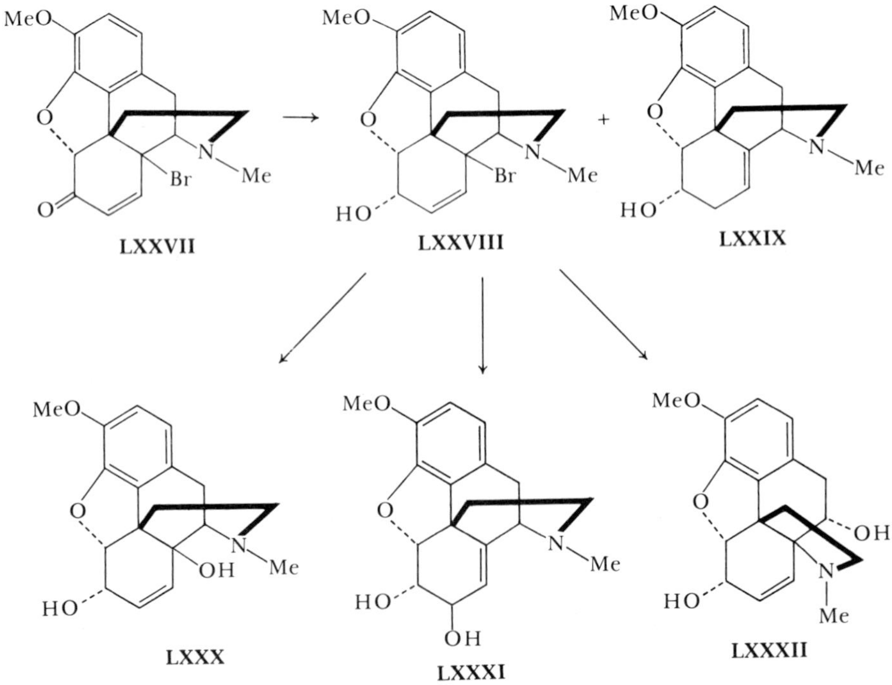

Fig. 6. Pathways for metabolism of (1)-14β-bromocodeinone by *Trametes sanguinea*, and subsequent non-enzymic decomposition of (−)-14β-bromocodeine.

Pseudocodeinone (**LXXVI**) is not attacked, but (−)-14β-bromo-codeinone (**LXXVII**) gives a mixture of products (Yamada *et al.*, 1963). Closer examination (Abe *et al.*, 1970) showed that, initially, (−)-14β-bromocodeine (**LXXVIII**) and a little (−)-neopine (**LXXIX**) are formed. The former slowly and non-enzymically changes (Fig. 6) to a mixture of (−)-14β-hydroxycodeine (**LXXX**), (−)-7β-hydroxyneopine (**LXXXI**) and 9α-hydroxyindolinocodeine (**LXXXII**). *Trametes sanguinea* slowly reduces (−)-dihydrocodeinone (**LXXXIII**), where the carbonyl group is in a saturated ring, to the mixed stereoisomers (−)-di-

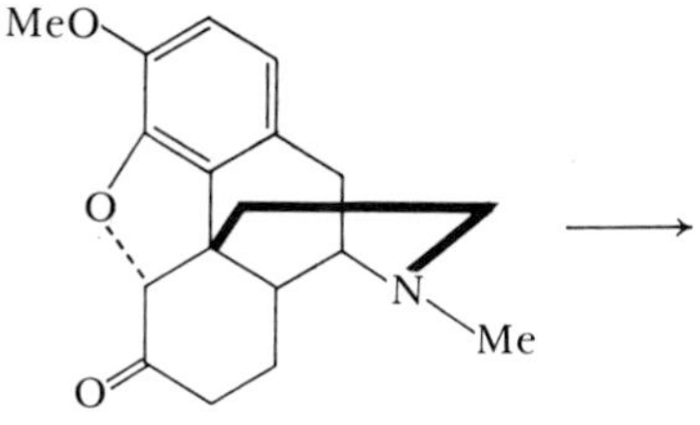

LXXXIII

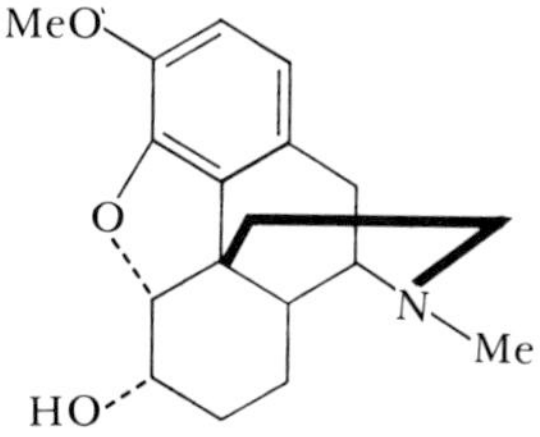

LXXXIV 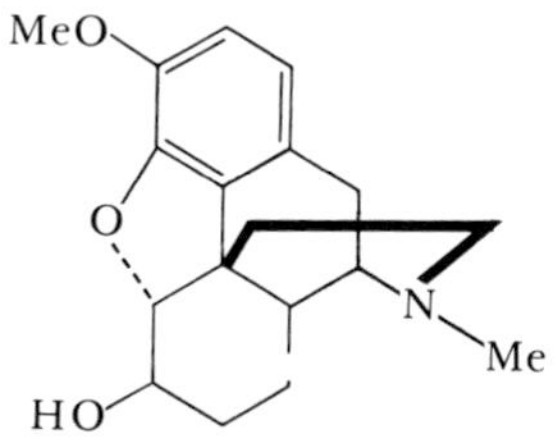   **LXXXV**

hydrocodeine (**LXXXIV**) and (−)-dihydroisocodeine (**LXXXV**) (Yamada
et al., 1963). (+)-Dihydrocodeinone gives, in low yield, only the
(+)-dihydrocodeine isomer, but (+)- and (−)-14-hydroxydihydroco-
deinone give a much better yield of the respective 14-hydroxy-
dihydrocodeines (Tsuda, 1964a). Dihydrothebainone methyl ether
(**LXXXVI**, R = Me) is probably also reduced in low yield, but
4-deoxydihydrothebainone (**LXXXVII**), sinomenin (**LXXXVIII**, R = H)
and sinomenin methyl ether (**LXXXVIII**, R = Me) are not attacked.
These latter results indicate that absence of the cyclic ether bridge
severely limits susceptibility to attack by enzymes of *T. sanguinea*.
Dihydrothebainone (**LXXXVI**, R = H), like morphine, is rapidly
degraded to unidentifiable products and phenolic oxidation is
suspected (Yamada, 1963).

   In a successful attempt to synthesize morphine analogues with high
analgesic activity but lacking undesirable side effects, Bentley and
Hardy (1963) prepared derivatives with a more rigid structure by
Diels–Alder addition of α,β-unsaturated ketones to thebaine. The
ketone group in these 6,14-*endo*-ethenotetrahydrothebaines (e.g.
LXXXIX) can be selectively reduced to an alcohol (**XC**) by species of
Cunninghamella and *Xylaria* (Mitscher et al., 1968). These fungi also
demethylate the compounds. Since nor-alkaloids such as **XCI** have
different analgesic properties from the parent, they may be formed at
receptor sites *in vivo* and are difficult to prepare chemically. Mitscher

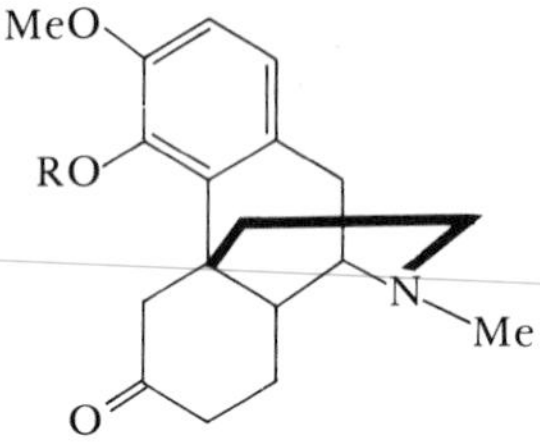

LXXXVI

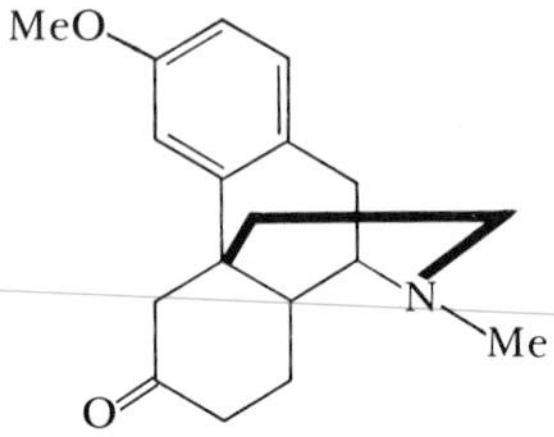

LXXXVII

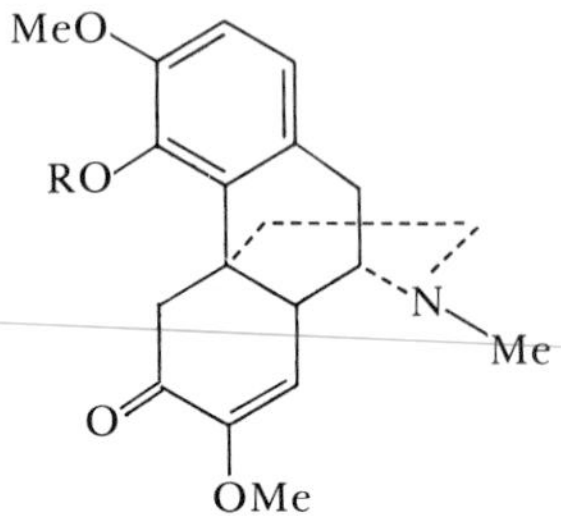

LXXXVIII

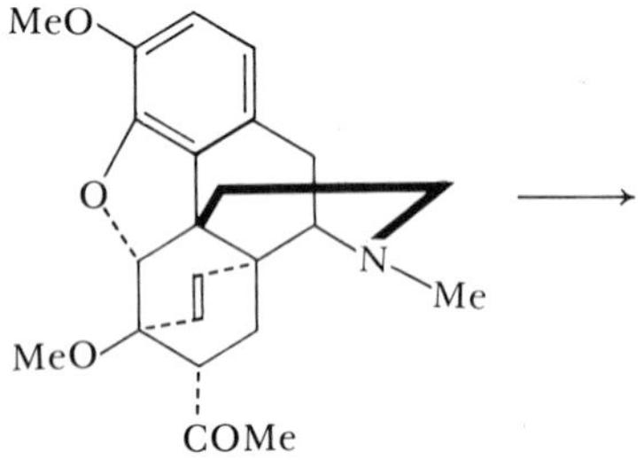

LXXXIX

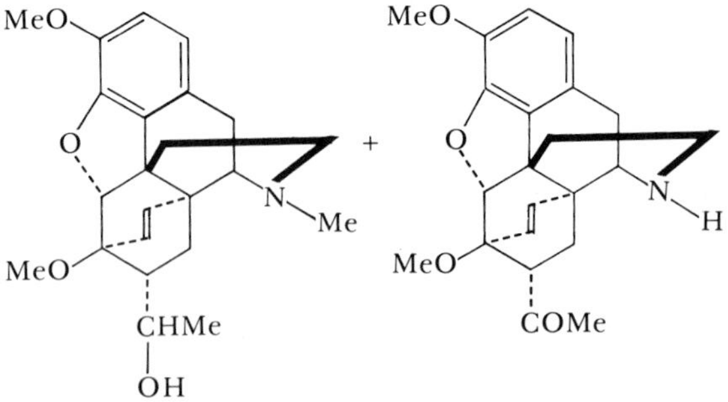

$\longrightarrow$

XC

XCI

and his colleagues explored the reaction further. They demonstrated that reduced compounds are attacked only at the N-alkyl group and that the nature of the alkyl group is not critical. The restricted action of the fungi was attributed to the rigidly prescribed shape of the molecule.

IX. APORPHINE ALKALOIDS

Apomorphine (**XCII**, $R_1 = R_2 = H$), with structural features relating it to dopamine, is of interest in the treatment of Parkinsonism. It is metabolized in the liver by methylation of the phenolic groups and

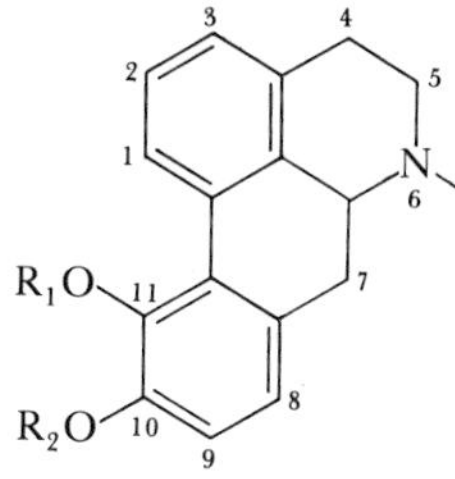

XCII 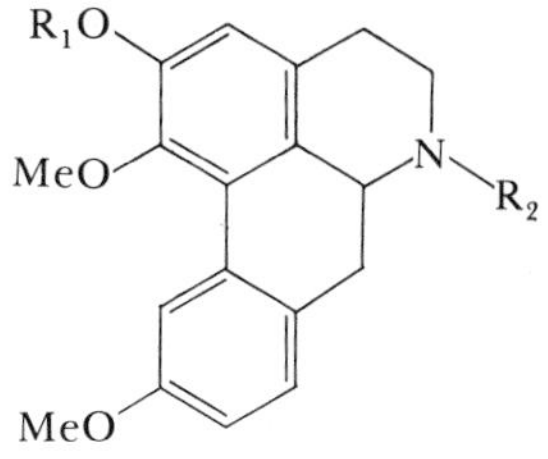   XCIII

also by N-dealkylation. Microbial dealkylation of the fully methylated 10,11-dimethoxyaporphine (**XCII**, $R_1 = R_2 = CH_3$) has been explored as a means of obtaining the individual monomethyl ethers for study. Of 65 organisms screened, three *Cunninghamella* species, *Microsporeum gypseum*, *Mucor mucedo* and *Penicillium duclauxii*, removed the least-hindered methyl group to give isoapocodeine (**XCII**, $R_1 = CH_3$, $R_2 = $ H). With *Cunninghamella blakesleeana*, the conversion was essentially quantitative. *Streptomyces rimosus* yielded apocodeine (**XCII**, $R_1 = $ H, $R_2 = CH_3$), whereas *Helicostylum pyriforme* and another *Streptomyces* sp. gave a mixture of both products (Rosazza *et al.*, 1975).

Incubation of $(+)$-(S)-glaucine (**XCIII**, $R_1 = R_2 = CH_3$) with *Streptomyces griseus* causes dealkylation to a mixture of predicentrine (**XCIII**, $R_1 = $ H, $R_2 = CH_3$) and norglaucine (**XCIII**, $R_1 = CH_3$, $R_2 = $ H) (Davis *et al.*, 1977). Although removal of methyl groups is not completely selective, O-demethylation is restricted to one unhindered position, indicating that steric factors are probably important. In the converse reaction where apomorphine is methylated using a liver catechol-O-methyl transferase, the product is a mixture of monomethyl ethers with the 10-O-methyl ether predominating (Cannon *et al.*, 1972).

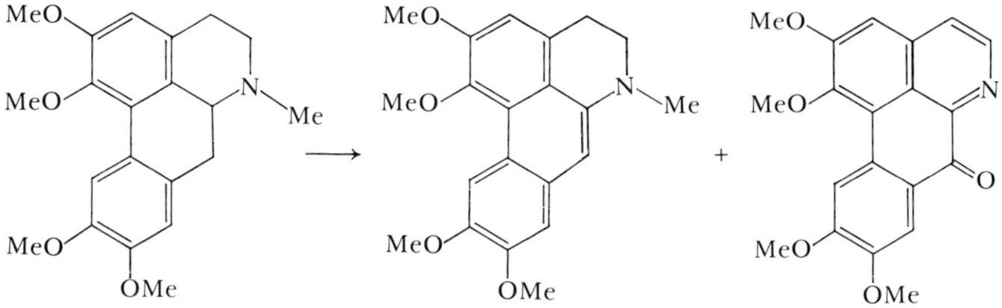

XCIII $R_1 = R_2 = $ Me   XCIV   XCV

Fusarium solani converts (+)-(S)-glaucine mainly into dehydro-glaucine (**XCIV**) with some accumulation of noraporphinone (**XCV**) as a probable artefact. If racemic glaucine is used, the (+)-isomer is preferentially metabolized. However, the (−)-isomer is slowly degraded and, by the time it is optimally enriched, the recovery is poor (Davis *et al.*, 1977).

X. TROPANE ALKALOIDS

Solanaceous alkaloids possessing the tropane ring system resist degradation by most micro-organisms but a soil isolate of *Coryne-bacterium belladonnae* can metabolize atropine, tropine, tropinone, hyoscyamine and scopolamine. It will not attack scopine and nortropine (Niemer *et al.*, 1959). The initial attack on atropine (**XCVI**)

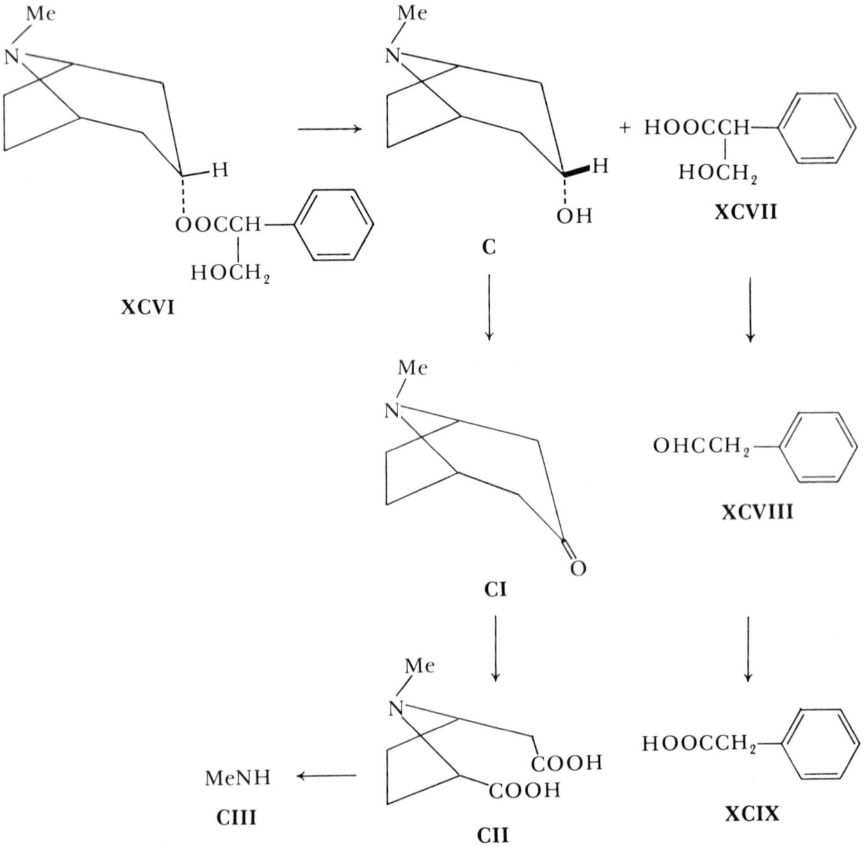

Fig. 7. Pathways for metabolism of atropine by *Corynebacterium belladonnae*.

removes tropic acid (**XCVII**) by hydrolysis (Fig. 7). The tropic-acid fragment is metabolized via phenylacetaldehyde (**XCVIII**) and phenyl-acetic acid (**XCIX**). The latter product accumulates and inhibits growth as the concentration increases. The tropine moiety (**C**) is metabolized by an initial oxidation to tropinone (**CI**); the enzyme catalysing this reaction can be demonstrated in cell extracts. Among the products of further metabolism, tropinic acid (**CII**) and methylamine (**CIII**) have been identified (Niemer and Bucherer, 1961).

In contrast to the oxidative attack which dominates in *C. belladonnae*, reduction of tropinone occurs in *Fusarium lini*. The product is a mixture of tropine (**C**) and ψ-tropine (**CIV**; Tamm, 1960). Atropine, homotropine, *l*-hyoscyamine and scopolamine are hydrolysed by *Arthrobacter atropini* and a *Pseudomonas* species isolated from soil where *Datura stramonium* had been cultivated. Tropine is demethylated by *Arthrobacter atropini* to tropigenin (Kaczkowski and Mazejko-Toczko, 1960).

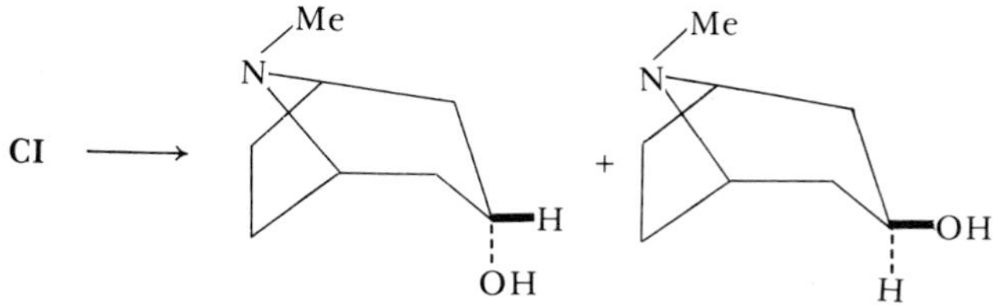

CI ⟶

C   CIV

XI. MISCELLANEOUS ALKALOIDS

Several alkaloids, in addition to those already mentioned, are demethylated by micro-organisms. Colchicine (**CV**) and some derivatives are converted by *Streptomyces griseus* into products with a higher therapeutic index for antimitotic activity than *N*-desacetyl-

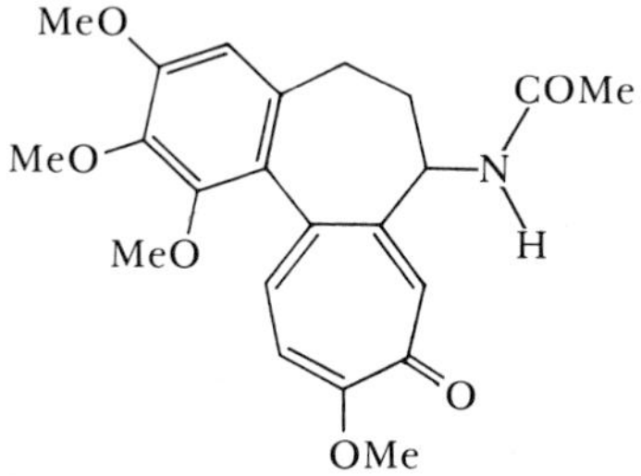

CV 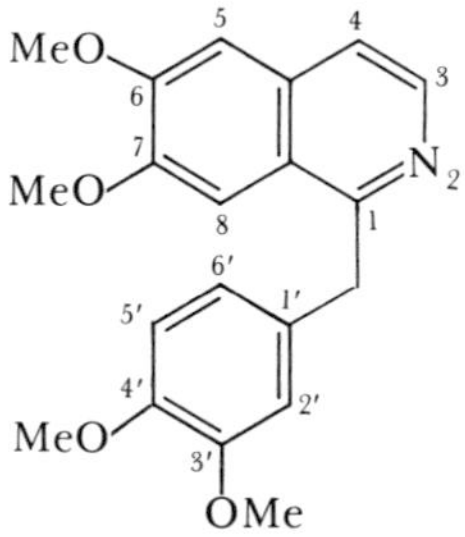   CVI

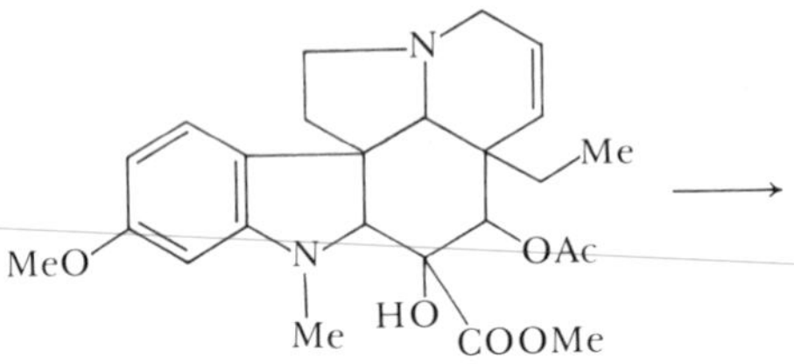

CVII

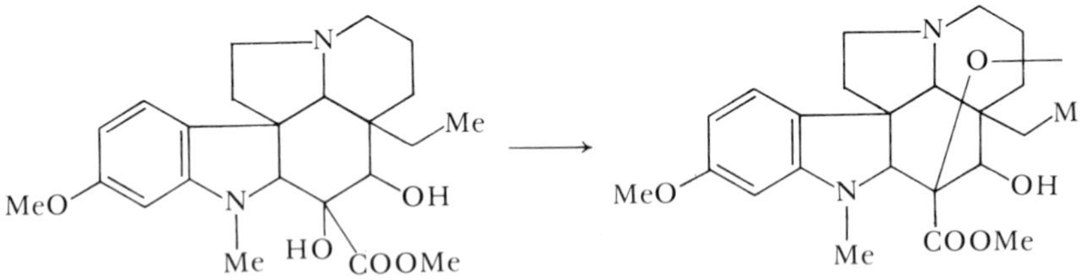

CVIII   CIX

colchicine, the best available compound. The products have not been fully identified, but at least one is a phenolic substance that has lost one methyl group (Roussel-Uclaf, 1963).

The alkaloid papaverine (**CVI**), a coronary and peripheral vasodilator, appears to be extensively metabolized in mammals. Virtually none of the drug is excreted unchanged, and 4′-O-desmethyl-papaverine has been identified among the products. In an attempt to prepare each of the O-demethylated isomers for study, the alkaloid has been incubated with *Cunninghamella echinulata* which converts it into four phenolic products. Two of them have been identified as 6-hydroxy-7-methoxy-(3′,4′-dimethoxybenzyl)-isoquinoline and 7-hydroxy-6-methoxy-(3′,4′-dimethoxybenzyl)-isoquinoline. A different isomer, either 6,7-dimethoxy-(3′-hydroxy-4′-methoxybenzyl)- or 6,7-dimethoxy-(4′-hydroxy-3′-methoxybenzyl)-isoquinoline, is produced from papaverine by *Cunninghamella blakesleeana* (Rosazza *et al.*, 1975).

The related alkaloids vindoline (**CVII**) and vinblastine, isolated from *Vinca* species, are N-demethylated by *Streptomyces albogriseolus* (Brannon and Neuss, 1975). Vindoline is also hydrolysed by numerous streptomycetes to give desacetylvindoline (**CVIII**) and some desacetyl-dihydrovindoline ether (**CIX**). O-Acetylvindoline gives vindoline as well as its transformation products, but desacetyldihydrovindoline is not attacked (Mallett *et al.*, 1964). Demethylation is the dominant reaction when aconitine (**CX**) is incubated with *Streptomyces paucisporo-genes* although the products have not been identified (Bellet and

Penasse, 1960). Incubation of hordenine (**CXI**) with *Geotrichum candidum* or *Hansenula anomala* replaces the entire N-dimethyl substituent with a hydroxyl group to form tyrosol (**CXII**) (Ehrlich, 1916).

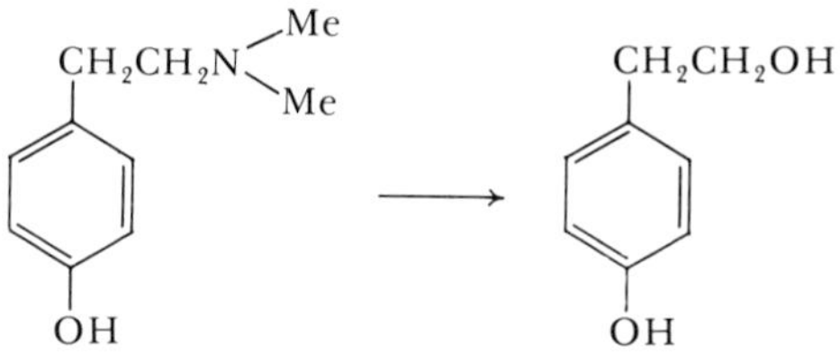

Acronycine (**CXIII**), an antitumour alkaloid from *Acronychia baueri*, is modified by hydroxylation in cultures of *Cunninghamella* species (Betts *et al.*, 1974). *Cunninghamella echinulata* gives 9-hydroxyacronycine as the sole product, but additional metabolites, tentatively identified as 11-hydroxy-, 9,11-dihydroxy- and 3-hydroxymethyl-11-hydroxyacronycine, are formed as minor components in cultures of other *Cunninghamella* species. 9-Hydroxylation also occurs when the alkaloid is administered to mammals. Lupanine (**CXIV**), one of the numerous lupin alkaloids, is metabolized by *Pseudomonas lupanini* to 17-hy-

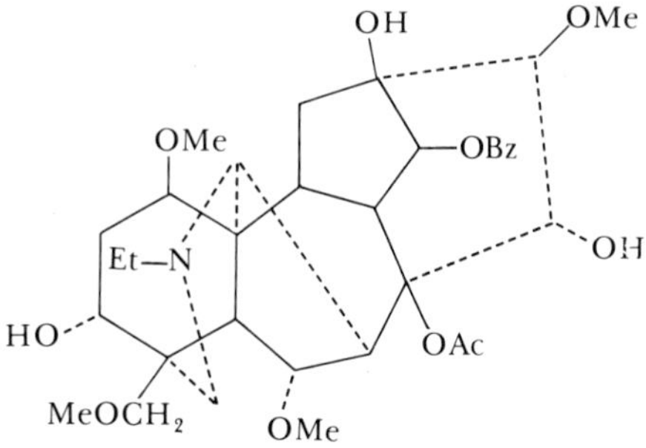

CX

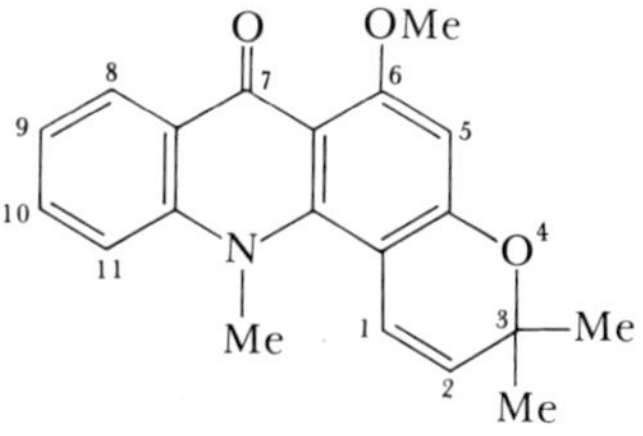

CXIII

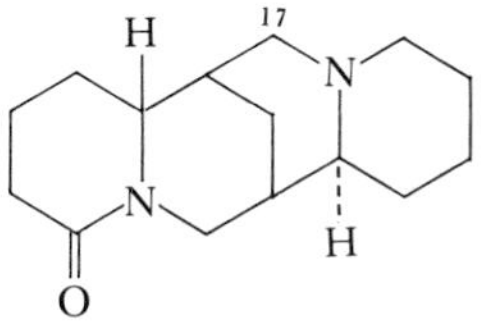

CXIV

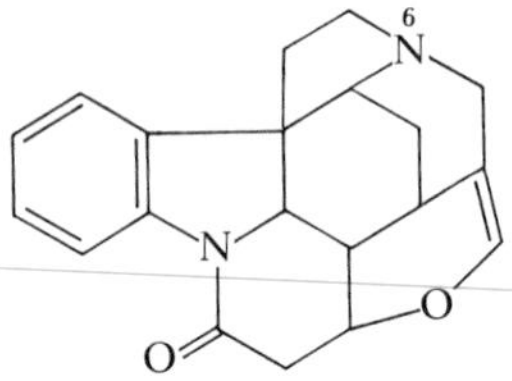

CXV 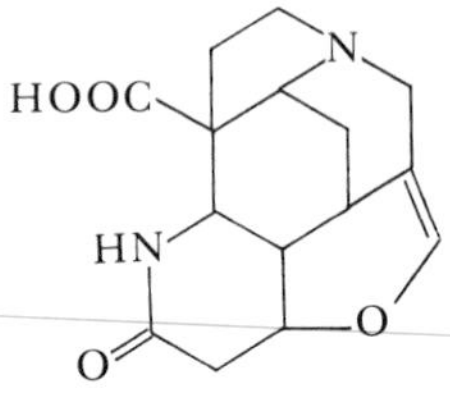   CXVI

droxylupinane. The conversion is probably mediated by a dehy-drogenase with the hydroxyl group being donated by water (Toczko, 1966). *Bacillus thuringiensis* converts strychnine (**CXV**) into its *N*-6 oxide. A similar reaction occurs with brucine (Bellet and Gerard, 1962). A more drastic change, in which the aromatic ring is opened to produce C_{16}-hanssenic acid (**CXVI**), occurs with strychnine but not brucine in cultures of *Achromobacter strychnovorum*, (Niemer and Bucherer, 1962).

Unsaturated members of the pyrrolizidine alkaloids, such as heliotrine (**CXVII**), which are present in the plant *Heliotropium europaeum*, are responsible for severe liver damage in grazing animals. They are probably converted by microsomal enzymes into unstable *N*-conjugated esters which act as alkylating agents (Mattocks and White, 1971). However, the alkaloids can be detoxified by microbial transformation in the rumen of sheep. Two bacteria isolated from rumen contents, one a small Gram-negative coccus (Russell and Smith, 1968) and the other a Gram-positive coccus characterized as *Peptococcus heliotrinreducans* (Lanigan, 1976), bring about reductive fission of the cytotoxic alkaloids under anaerobic growth conditions. Heliotrine is converted into 7α-hydroxy-1-methylene-8α-pyrrolizidine (**CXVIII**) and heliotric acid (**CXIX**). Growth of the bacterium is proportional to the amount of heliotrine detoxified. In the reaction, a mole of hydrogen or formate is used per mole of alkaloid reduced. The alkaloid serves as the terminal acceptor in an electron-transport system

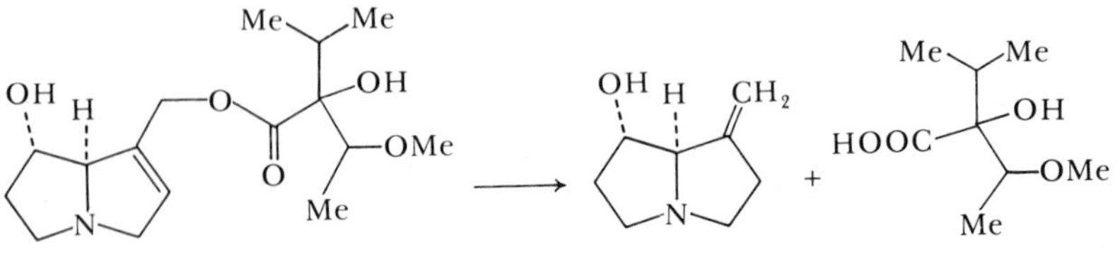

CXVII   CXVIII   CXIX

supplying energy to the organism. The amount of alkaloid metabolized by the mixed microbial population of the sheep's rumen can be increased by the inhibiting competition for metabolic hydrogen from methanogenic bacteria. Lanigan (1972) found that adding vitamin B_{12} enzyme inhibitors, such as chloral hydrate, to the diet markedly decreased the toxicity due to pyrrolizidine alkaloids, and suggested practical application in protecting animals at risk from grazing *Heliotropium europaeum*.

XII. NITROGENOUS XENOBIOTICS

Much recent research on microbial metabolism of nitrogenous compounds that are not of natural origin has been aimed at finding appropriate microbial models for pathways normally encountered in animals. Progress in this area has been reviewed by Smith and Rosazza (1974, 1975a, b). The compounds studied are normally those used as drugs or likely to be ingested unintentionally by man. The microbial system is not only more accessible to experimentation but can be scaled up to produce adequate amounts of interesting products for further study.

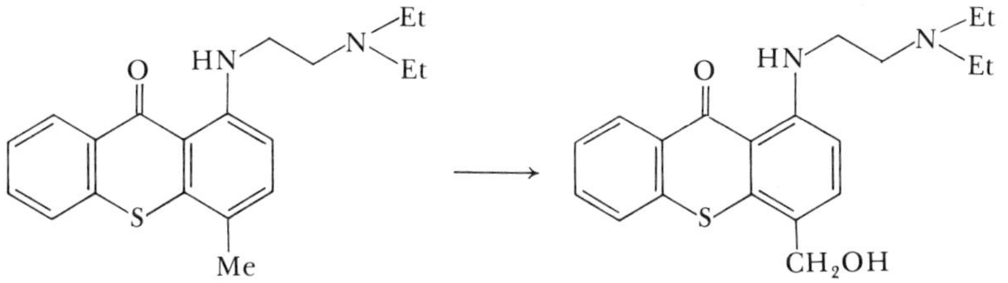

CXX   **CXXI**

Metabolic studies in animals showed that the schistosomicidal agent lucanthone (**CXX**) is converted in animals into the more active hycanthone (**CXXI**). The same hydroxylation can be obtained with high yields in cultures of *Aspergillus sclerotiorum*. The corresponding aldehyde and acid are formed as minor products (Rosi *et al.*, 1967a). Similar transformations with this organism enhance the activity of other synthetic schistosomicidal agents bearing a substituent methyl group (Rosi *et al.*, 1967b). The synthetic antibacterial agent, nalidixic acid (**CXXII**, R = Et), is also known to undergo C-7 methyl group hydroxylation to an active product in humans and the reaction can be mimicked by *Penicillium adametzi* and other fungi (Hamilton *et al.*,

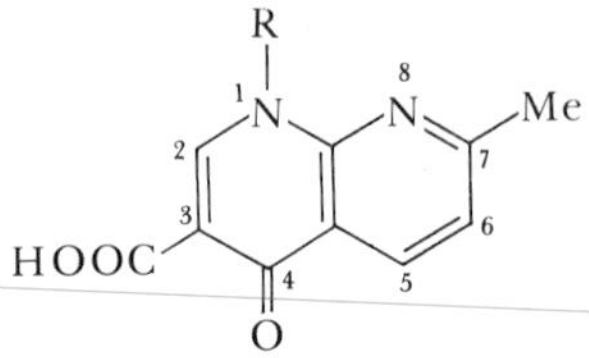

CXXII

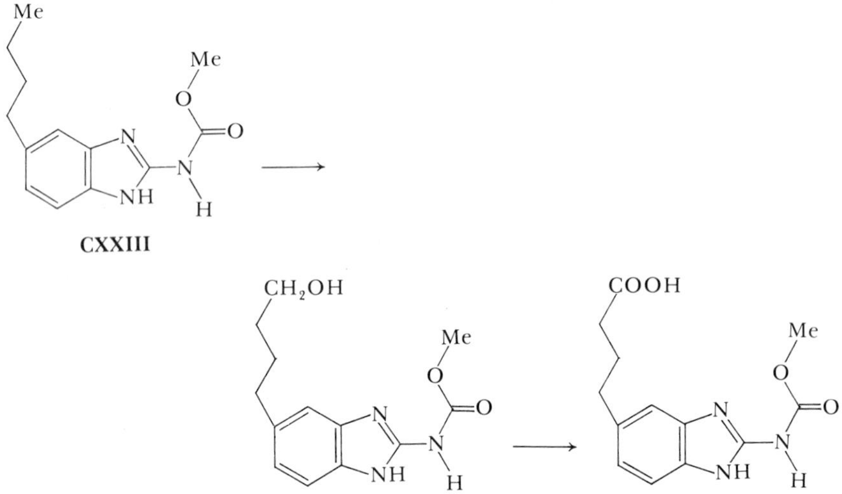

CXXIII

CXXIV   **CXXV**

1969). The yield is affected by altering the size of the alkyl substituent attached to N-1 (Nielson *et al.*, 1967). Hydroxylation at the tertiary carbon of an isobutyl substituent by *Streptomyces roseochromogenes* has also been used to prepare sulphonamide drugs (Siewert and Kieslich, 1973).

The anthelmintic agent parbendazole (**CXXIII**) is metabolized in farm animals, and several products have been identified in urine (Dunn *et al.*, 1973). Many of the 100 bacteria, yeasts and fungi screened for their ability to metabolize the drug also decomposed parbendazole, and a strain of *Cunninghamella baineri* accumulated a good yield of two of the animal metabolites. These were isolated and tested as substrates for *C. baineri*. Methyl-5(6)-(4-hydroxybutyl)-2-benzimidazole carbamate (**CXXIV**) was found to be converted irreversibly into methyl-5(6)-(3-carboxypropyl)-2-benzimidazolecarbamate (**CXXV**), verifying the predicted transformation sequence (Valenta *et al.*, 1974).

The *N*-benzoyl derivatives of mecylamine (**CXXVI**), an effective ganglionic blocking agent which relieves hypertension, is hydroxylated by *Sporotrichum sulfurescens* to give a substantial yield of the *exo*-6-hydroxy derivative. The corresponding 7-hydroxy derivative, also present in fermentation extracts, can be recovered as the keto compound by oxidizing the mixture. In this example, knowledge of microbial metabolism has preceded animal studies and it is not known whether mecylamine is metabolized to the same compounds in mammals (Herr *et al.*, 1971). However, the oxygenation products obtained by incubating several sulphonylureas with *S. sulfurescens* have been identified among metabolites found when the drugs are administered to animals. They are also reported to have promising hypoglycaemic activity (Fonken *et al.*, 1967). *Pellicularia filamentosa* converts the tranquilizer diazepam (**CXXVII**) into an N-4 oxide, a C-3 hydroxy derivative and an *N*-demethyl derivative; only the last of these compounds is detected during diazepam metabolism in mammals (Greenspan *et al.*, 1969).

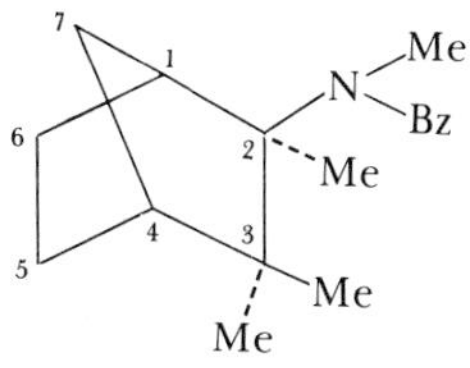

CXXVI

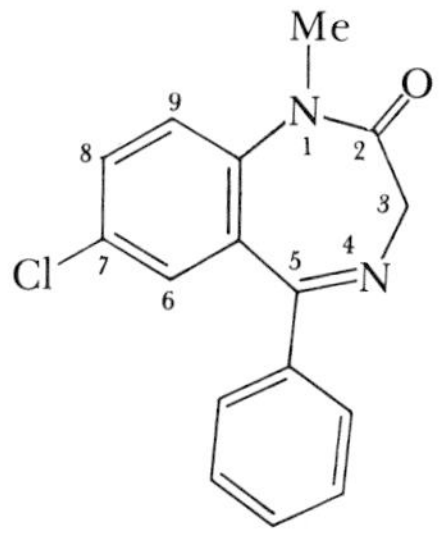

CXXVII

Metabolism of fenclosic acid (**CXXVIII**), a compound that has potent anti-inflammatory activity but causes unacceptable side effects in humans, is entirely different in animals and micro-organisms. The predominant reaction in animals is to oxygenate the aromatic ring whereas, in a wide range of micro-organisms, only the non-aromatic portion of the molecule is attacked. Even cultures known for vigorous hydroxylation of the aromatic ring, after exposure to inducing substrates, fail to modify the benzene ring (Howe *et al.*, 1972). Most micro-organisms convert the drug into several products and many produce similar mixtures in different relative amounts. By monitoring rates of formation and testing isolated products for further conversion, the probable sequence of transformations (Fig. 8) was

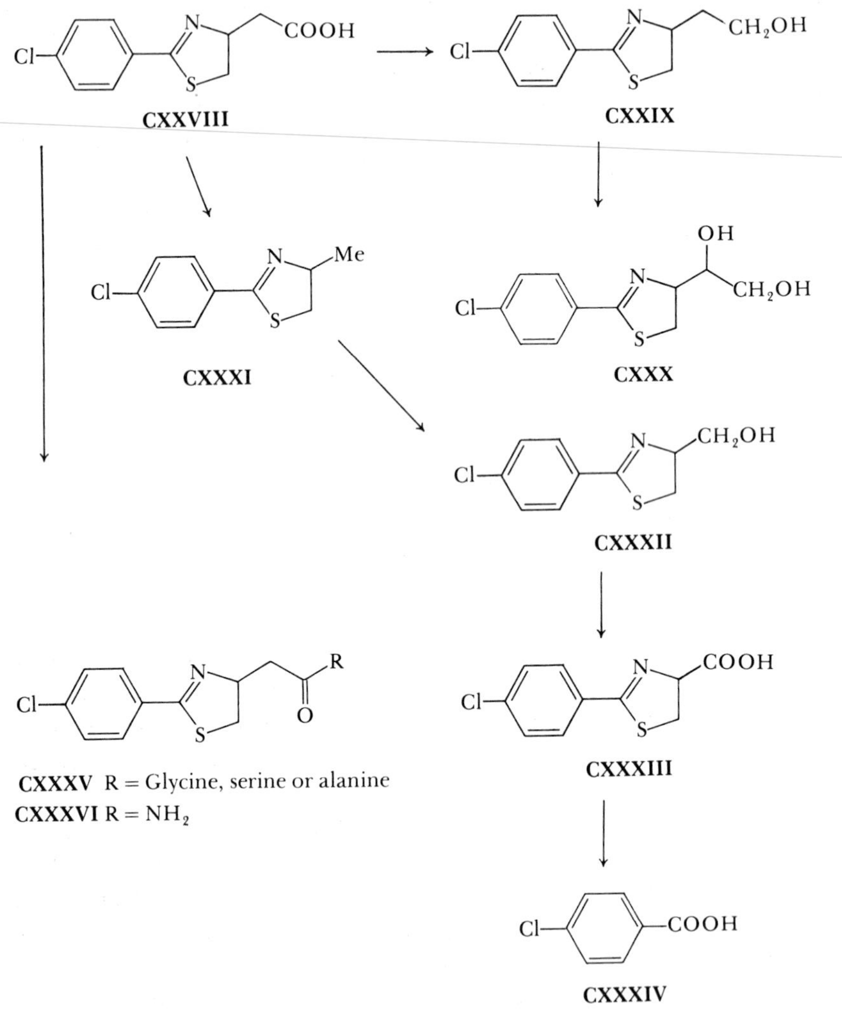

Fig. 8. Pathways for microbial metabolism of fenclosic acid.

established. Reduction to the alcohol (**CXXIX**) is a common route and is reversible. The unusual hydroxylation of the alcohol to the diol (**CXXX**) is stereospecific. Since fenclosic acid decarboxylates slowly in aqueous solution, even at neutral pH values, it was usually added as the methyl ester. However, the ester is readily hydrolysed and formation of the intermediate (**CXXXI**) may be entirely non-enzymic. Its subsequent transformation via **CXXXII** to **CXXXIII** is plausible and, again, conversions of the alcohol into the acid are reversible. *Mucor ramannianus* was the only organism that attacked the thiazole ring and thus gave *p*-chlorobenzoic acid (**CXXXIV**). It was also alone in forming the glycine, L-serine and L-alanine conjugates (**CXXXV**), although several other organisms formed the unsubstituted amide (**CXXXVI**). Similar conjugates of anthranilic acid and indole acetic acid with amino acids have been reported from cultures of *Bacillus megatherium* supplied with the free acid (Robert and Tabone, 1952; Tabone, 1958). Among the various metabolites obtained from fenclosic acid, several showed a useful degree of anti-inflammatory activity and were worthy of further examination (Howe *et al.*, 1972).

In the oral administration of slowly absorbed drugs to man and animals, or where drugs are excreted from the tissues into the intestine with the bile, microbial modification by the intestinal flora may be a factor (Scheline, 1968; Williams, 1972). Variation in the route of elimination with dose indicates that gut microflora are involved. For example, the drug methamphetamine (**CXXXVII**) administered to guinea pigs appears increasingly in the faeces as a mixture of amphetamine (**CXXXVIII**), norephedrine (**CXXXIX**) and an unidentified metabolite when the dose is increased. Investigation of intestinal micro-organisms showed that the drug was metabolized by *N*-demethylation and probably also by one additional pathway (Caldwell and Hawksworth, 1973). Among individual gut bacteria, lactobacilli, enterococci and some clostridia were able to promote *N*-demethylation. Although methamphetamine is absorbed rapidly, and in man little would reach the microflora of the lower gut directly, it is unsafe to assume that the liver is the sole site of metabolism. It has also been established that gut bacteria such as *Proteus vulgaris* and *Streptococcus faecalis* are responsible for reducing various azo compounds used as food-colouring material. The reduction is non-

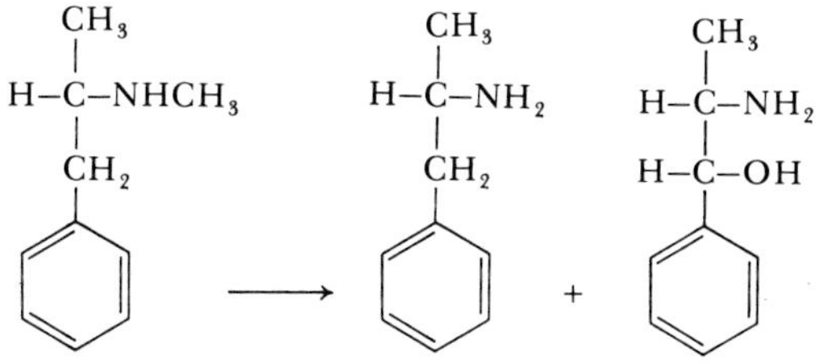

CXXXVII	**CXXXVIII**	**CXXXIX**

enzymic but is mediated by enzymically-generated reduced flavins acting as the principal electron donors (Gingell and Walker, 1971).

The accelerated interest in the fate of man-made chemicals in the environment has inevitably led to better insights into the ways by which micro-organisms can dispose of a wide range of chemical compounds. The scope of oxidative (Large, 1971; Fonken and Johnson, 1972), reductive (Mitchard, 1971) and other (Callely, 1978) transformations which include many nitrogenous substances has been surveyed fairly recently, and an older compilation of type reactions (Wallen *et al.*, 1959) is still useful. There is also a large literature on the metabolism of herbicides and pesticides which may be consulted (Sykes and Skinner, 1971).

XIII. MISCELLANEOUS TRANSFORMATIONS

In the synthesis of nitrogenous drugs as in the synthesis of steroids, micro-organisms have proved useful for modifying sensitive molecules in a highly specific manner. 5-Anilino-1,2,3,4-thiatriazole (**CXL**), a potentially valuable hypertensive agent, was expected to form aniline by metabolism in humans. Since *p*-hydroxyaniline is less toxic than aniline, the corresponding 5-(*p*-hydroxyanilino)-1,2,3,4-thiatriazole was predicted to be a more suitable drug. A chemical synthesis was attempted but proved difficult, and the compound was eventually obtained by microbial hydroxylation of the parent compound (Theriault and Longfield, 1973). Out of 285 organisms screened, 11 carried out the desired transformation and, with *Aspergillus tamari*, conversions as high as 79% were achieved. Cultures of *Aspergillus terreus* contained 5-(*o*-hydroxyanilino)-1,2,3,4-thiatriazole as an additional product.

Extensive studies of hydroxylation of amides by *Torulopsis gropengiesseri* and *Sporotrichum sulfurescens* enable the site of oxygen insertion to be predicted with reasonable confidence (Fonken and Johnson, 1972). The substrates include acylamine derivatives of acyclic,

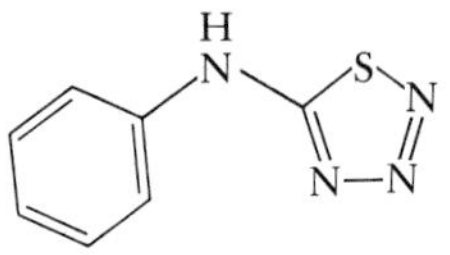

CXL

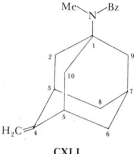

CXLI

monocyclic, bicyclic and polycyclic carbon compounds, as well as a variety of heterocyclic amides. The hydroxylation processes represent a valuable adjunct to chemical synthesis, and have been interpreted in terms of a model defining the site of attack by its distance from electron-rich binding groups. Among the uses to which the reactions have been put are syntheses of the N-benzoyl derivatives of the alkaloids sedridine and isopelletierine.

The foregoing procedures have also been used to prepare adamantanes with antiviral activity. For many of these compounds, microbial oxidation of the N-acylated adamantanamines represents the only feasible route for introducing suitable functionality. Thus, incubation of N-benzoyl-N-methyl-4-methyleneadamantan-1-amine (**CXLI**) with *Sporotrichum sulfurescens* gives the 7-hydroxyepoxide in high yield as the major product (Fonken and Johnson, 1972). The same fungus transforms N-acetyl-1-adamantanamine to a 1:4 mixture of the 3α- and 4α-hydroxy derivatives, whereas N-phenylacetyl-1-adamantanamine gives the 4α,6α-diol (Herr *et al.*, 1968). Although the amino substituent influences the site of oxygenation, the hydroxyl groups are invariably inserted *trans* to the nitrogen. If the preferred *trans* position is already occupied, *trans*-hydroxylation may occur at an adjacent site but no *cis* derivative is formed (Herr *et al.*, 1970). This result conforms with the more general observation (Johnson *et al.*, 1971) that a methyl substituent on a cyclic substrate facilitates oxygenation of the ring carbon to which it is attached unless the methyl group is *trans* to an amide function.

Whereas reduction of the nitro group is a common step in microbial degradation of the antibiotic chloramphenicol (Smith and Worrell, 1950), or of xenobiotics such as trinitrotoluene (McCormick *et al.*, 1976) and the herbicide 3,5-dinitro-*o*-cresol (Tewfic and Evans, 1966),

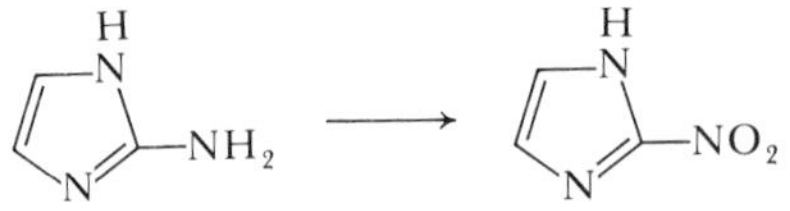

CXLII   CXLIII

the reverse reaction occurs in certain *Streptomyces* species that synthesize aromatic nitro compounds. Both oxygen atoms introduced into the nitro group are derived from molecular oxygen (Kawai *et al.*, 1965; Lancini *et al.*, 1968). The enzyme responsible is moderately flexible since *S. thioluteus* will not only convert 2-amino-imidazole (**CXLII**) into azomycin (**CXLIII**) but will also oxidize the 4-alkyl and 4,5-dimethyl homologues and several *p*-aminophenyl compounds. Neither *p*-aminophenol nor *o*- or *m*-aminophenyl compounds is a suitable substrate, but *p*-dimethylaminobenzoic acid gives the *N*-oxide, indicating a relationship between this enzyme and those that convert alkaloids into their *N*-oxides.

Many naturally occurring aromatic or quasi-aromatic nitrogenous-heterocyclic compounds are known to undergo hydroxylation. In most instances, this is the first step leading to extensive catabolism. Thus, *Pseudomonas aeruginosa* converts urocanic acid into 2,5-dihydroxy-urocanic acid (Ichihara *et al.*, 1957); it also hydroxylates variously substituted purines at C-2, C-6 or C-8, depending on the location of substituents. However, the alkaloid theophylline with a methyl group at N-1 is not attacked (Bergmann *et al.*, 1962). The initial step in metabolism of riboflavin and related compounds by a *Pseudomonas* species is hydroxylation at C-8 of the benzenoid ring (Smyrniotis *et al.*, 1958), whereas pterins, such as xanthopterin, lumazine and 6-oxylumazine are hydroxylated by a soil bacterium at C-7 (Levy and McNutt, 1962). A *Pseudomonas* species able to metabolize hydroxy-pyrazine as the sole carbon and nitrogen source begins assimilation by hydroxylating the molecule, probably forming 2,6-dihydroxypyrazine (Mattey and Harle, 1976).

Microbial oxidations have been experimented with for synthesis of biotin and biotin vitamers. Strains of *Rhodotorula*, *Penicillium* and *Endomycopsis* remove two carbon atoms from the alkanoic acid substituent of biotin by β-oxidation and also from the sulphoxide (Yang *et al.*, 1968). Six out of 300 *Corynebacterium* strains, isolated from soil for their ability to grow on hydrocarbons, were able to co-oxidize

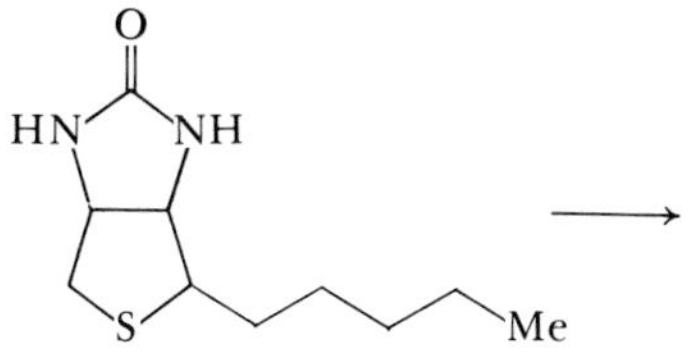

CXLIV

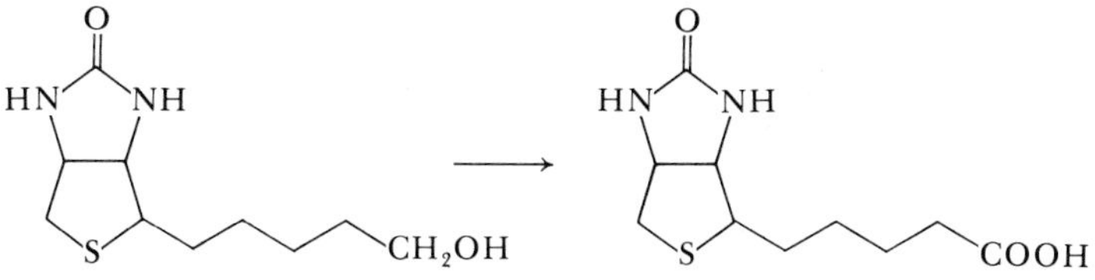

CXLV   **CXLVI**

dl-cis-tetrahydro-2-oxo-4n-pentylthieno[3,4-d] imidazoline (**CXLIV**), a chemically-synthesized precursor, to dl-biotinol (**CXLV**) and dl-biotin (**CXLVI**) (Ogimo et al., 1974; Izumi and Ogata, 1977).

Since L-3,4-dihydroxyphenylalanine (dopa) has become important in treatment of Parkinson's disease there has been a renewed interest in its synthesis by micro-organisms. In melanin-forming organisms, o-hydroxylation of L-tyrosine is the initial step, but a second hydroxylation giving L-2,4,5-trihydroxyphenylalanine is associated with this reaction (Lunt and Evans, 1976) and dopa does not normally accumulate. Other o-hydroxylating enzymes are present in micro-organisms and function in the general dissimilation of aromatic substances. Surmizing that such enzymes might not always be narrowly substrate specific, Sih et al. (1969) screened a large number of organisms for the ability to convert N-substituted tyrosine into the corresponding dopa derivative. With Aspergillus ochraceus and Gliocladium deliquescens, yields of 25–30% were achieved; the latter fungus also formed 3,4-dihydroxyphenylethyl alcohol. For optimal yields, L-ascorbic acid and hydrocinnamic acid were added to the culture, the latter as an inhibitor of peptidase activity to prevent loss of the protecting group (Rosazza et al., 1974).

In an alternative approach to that used by Sih and his coworkers, a Japanese group selected melanin-producing organisms that accumulate dopa when supplemented with tyrosine under controlled culture conditions (Yoshida et al., 1973). Conversions of near 90% by Vibrio tyrosinaticus and Pseudomonas melanogenum depend on moderate and frequent additions of tyrosine and the presence of a reductant such as ascorbate or hydrazine to preserve the product. With *Ps. melanogenum*, addition of excess sulphur-containing amino acids needed for growth depresses the yield (Yoshida *et al.*, 1974). Although success has been achieved in production of dopa by biotrans- formation, biosynthesis *de novo* is, in principle, a more attractive microbiological method. Mutant strains of *Ps. maltophila*, selected for resistance to 3-aminotyrosine, accumulate dopa to a level of 50 mg per litre of culture without supplements of phenylalanine or tyrosine (Nakayama *et al.*, 1974).

Finally, a microbial transformation of particular interest is the detoxication of potent mycotoxin contaminants of grain during the beer-brewing process. Gjertsen *et al.* (1973) have reported that any ochratoxin or citrinin present in barley is degraded; a later study (Chang *et al.*, 1975) in which ochratoxin and aflatoxin were added at a concentration of 1 µg per g to the raw materials showed that detoxication was not complete, and approximately 15% of the amount added could be detected in the beer.

REFERENCES

Abe, M., Yamatodani, S., Yamano, T., Kozu, Y. and Yamada, S. (1963). *Annual Reports of the Takeda Research Laboratories* **22**, 116.
Abe, K., Onda, M., Isaka, H. and Okuda, S. (1970). *Chemical and Pharmaceutical Bulletin* **18**, 2070.
Abou-Chaar, C. I. and Basmadjian, G. P. (1968). *Herba Hungarica* **7**, 105.
Aida, K., Uchida, K., Iizuka, K., Okuda, S., Tsuda, K. and Uemura, T. (1966). *Biochemical and Biophysical Research Communications* **22**, 13.
Aqurell, S. and Ramstad, E. (1962). *Archives of Biochemistry and Biophysics* **98**, 457.
Arcamone, R., Chain, E. B., Ferretti, A and Pennella, P. (1961). *Nature, London* **192**, 552.
Audette, R. C. S., Worthen, L. R. and Youngken, Jr., H. W. (1964). *Journal of Pharmaceutical Sciences* **53**, 117.
Behrman, E. J. and Stanier, R. Y. (1957). *Journal of Biological Chemistry* **228**, 923.
Belič, I. and Sočič, H. (1972). *Journal of Steroid Biochemistry* **3**, 843.
Belič, I. and Sočič, H. (1975). *Acta Microbiologica Academica Scientia Hungarica* **22**, 389.
Bellet, P. and Gerard, D. (1962). *Annales Pharmaceutiques Francaises* **20**, 928.
Bellet, P. and Penasse, L. (1960). *Annales Pharmaceutiques Francaises* **18**, 337.
Bellet, P. and Thuong, T. V. (1970). *Annales Pharmaceutiques Francaises* **28**, 245.
Bentley, K. W. and Hardy, D. G. (1963). *Proceedings of the Chemical Society, London*, 220.
Bergmann, F., Ungar-Warm, H., Kwietny-Govrin, H. and Leon, S. (1962). *Biochimica et Biophysica Acta* **55**, 512.
Betts, R. E., Walters, D. E. and Rosazza, J. P. (1974). *Journal of Medicinal Chemistry* **17**, 599.

Brack, A., Brunner, R. and Kobel, H. (1962). *Helvetica Chimica Acta* **45**, 276.

Brannon, D. R. and Neuss, N. (1975). German Patent 2,440,931.

Burg, R. W., Rodwell, V. W. and Snell, E. E. (1960). *Journal of Biological Chemistry* **235**, 1164.

Caldwell, J. and Hawksworth, G. M. (1973). *Journal of Pharmacy and Pharmacology* **25**, 422.

Callely, A. G. (1978). *Progress in Industrial Microbiology* **14**, 205.

Cannon, J. G., Smith, R. V., Modiri, A., Snood, S. P., Borgman, R. J., Aleem, M. A. and Long, J. P. (1972). *Journal of Medicinal Chemistry* **15**, 273.

Cassady, J. M., Abou-Chaar, C. I. and Floss, H. G. (1973). *Lloydia* **36**, 390.

Chang, F. S., Chang, C. C., Asoor, S. H. and Prentice, N. (1975). *Applied Microbiology* **29**, 313.

Dagley, S. and Johnson, P. A. (1963). *Biochimica et Biophysica Acta* **78**, 577.

Daum, J. and Kieslich, K. (1973). Japanese Patent 73 44,489.

Davis, P. J., Gustafson, M. and Rosazza, J. P. (1975). *Biochimica et Biophysica Acta* **385**, 133.

Davis, P. J., Weise, D. and Rosazza, J. P. (1977). *Journal of the Chemical Society, Perkin Transactions I* p. 1.

Decker, K., Eberwein, H., Gries, F. A. and Brühmüller, M. (1960). *Zeitschrift für Physiologische Chemie* **319**, 279.

Decker, K., Gries, F. A. and Brühmüller, M. (1961). *Zeitschrift für Physiologische Chemie* **323**, 249.

Di Geronimo, M. J. and Antoine, A. D. (1976). *Applied and Environmental Microbiology* **31**, 900.

Dunn, G. L., Gallagher, G., Davis, L. D., Hoover, J. R. E. and Stedman, J. (1973). *Journal of Medicinal Chemistry* **16**, 996.

Dvornikova, P. V., Skryabin, G. K. and Suvorov, N. N. (1968). *Mikrobiologiya* **37**, 228.

Dvornikova, P. V., Skryabin, G. K. and Suvorov, N. N. (1970). *Mikrobiologiya* **39**, 42.

Ehrlich, F. (1916). *Biochemische Zeitschrift* **75**, 417.

Eich, E. (1973). *Planta Medica* **23**, 330.

Ensign, J. C. and Rittenberg, S. C. (1963). *Archiv für Mikrobiologie* **47**, 137.

Ensign, J. C. and Rittenberg, S. C. (1964). *Journal of Biological Chemistry* **239**, 2285.

de Flines, J., Marx, A. F., van der Waard, W. F. and van der Sijde, D. (1962). *Tetrahedron Letters* p. 1257.

Floss, H.-G. (1976). *Tetrahedron* **32**, 873.

Fonken, G. and Johnson, R. S. (1972). 'Chemical Oxidations with Micro-organisms'. Marcel Dekker, Inc., New York.

Fonken, G. S., Herr, M. E. and Murray, H. C. (1967). United States Patent 3,352,884.

Frankenburg, W. G. (1955). *Nature, London* **175**, 945.

Frankenburg, W. G. and Vaitekunas, A. A. (1955). *Archives of Biochemistry and Biophysics* **58**, 509.

Gherna, R. L., Richardson, S. H. and Rittenburg, S. C. (1965). *Journal of Biological Chemistry* **240**, 3669.

Gingell, R. and Walker, R. (1971). *Xenobiotica* **1**, 231.

Gjertsen, P., Myken, F., Krogh, P. and Hald, B. (1973). 'Proceedings of the European Brewery Convention Congress, Salzburg', p. 373. Elsevier, London.

Gordon, B. and Gentile, A. C. (1962). *Plant Physiology* **37**, 439.

Gotfredsen, W. O., Korsby, G., Lorck, H. and Vangedal, S. (1958). *Experientia* **14**, 88.

Greenspan, G., Rees, R., Smith, L. L. and Alburn, H. E. (1965). *Journal of Organic Chemistry* **30**, 4215.

Greenspan, G., Ruelius, H. W. and Alburn, H. E. (1969). United States Patent 3,453,179.

Gries, F. A., Decker, K. and Brühmüller, M. (1961). *Zeitschrift für Physiologische Chemie* **325**, 229.

Griffith, G. D., Byerrum, R. U. and Wood, W. A. (1961). *Proceedings of the Society of Experimental Biology and Medicine* **108**, 162.

Gröger, D. (1969). *In* 'Biosynthese der Alkaloide' (K. Mothes, ed.), pp. 678–698. VEB Deutscher Verlag der Wissenschaften, Berlin.

Gröger, D. and Schauder, H. P. (1969). *Experientia* **25**, 95.

Gross, D., Feige, A. Zureck, A. and Schutte, H. R. (1968). *European Journal of Biochemistry* **4**, 28.

Hamilton, P. B., Rosi, D., Peruzzotti, G. P. and Nielson, E. D. (1969). *Applied Microbiology* **17**, 237.

Harary, I. (1957). *Journal of Biological Chemistry* **277**, 823.

Hartman, R. E., Krause, E. F., Andres, W. W. and Patterson, E. L. (1964). *Applied Microbiology* **12**, 138.

Hayakawa, S. and Sato, Y. (1963). *Journal of Organic Chemistry* **28**, 2742.

Herr, M. E., Johnson, R. A., Murray, H. C., Reineke, L. M. and Fonken,G. S. (1968). *Journal of Organic Chemistry* **33**, 3201.

Herr, M. E., Johnson, R. A., Krueger, W. C., Murray, H. C. and Pschigoda, L. M. (1970). *Journal of Organic Chemistry* **35**, 3607.

Herr, M. E., Murray, H. C. and Fonken, G. S. (1971). *Journal of Medicinal Chemistry* **14**, 842.

Hochstein, L. I. and Dalton, B. P. (1967). *Biochimica et Biophysica Acta* **139**, 56.

Hochstein, L. I. and Rittenberg, S. C. (1959a). *Journal of Biological Chemistry* **234**, 151.

Hochstein, L. I. and Rittenberg, S. C. (1959b). *Journal of Biological Chemistry* **234**, 156.

Hochstein, L. I. and Rittenberg, S. C. (1960). *Journal of Biological Chemistry* **235**, 795.

Holcenberg, J. S. and Stadtman, E. R. (1969). *Journal of Biological Chemistry* **244**, 1194.

Hook, R. H. and Robinson, W. G. (1964). *Journal of Biological Chemistry* **239**, 4263.

Hörhold, C., Böhme, K. H. and Schubert, K. (1969). *Zeitschrift fur Allgemeine Mikrobiologie* **9**, 235.

Houghton, C. and Cain, R. B. (1972). *Biochemical Journal* **130**, 879.

Howe, R., Moore, R. H., Rao, B. S. and Wood, A. H. (1972). *Journal of Medicinal Chemistry* **15**, 1040.

Hughes, D. E. (1952). *Biochimica et Biophysica Acta* **9**, 226.

Hunt, A. L., Roger, A. and Hughes, D. E. (1959). *Biochimica et Biophysica Acta* **34**, 354.

Hurst, A. L., Hughes, D. E. and Lowenstein, J. M. (1958). *Biochemical Journal* **69**, 170.

Hylin, J. W. (1959). *Archives of Biochemistry and Biophysics* **83**, 528.

Ichihara, K., Santani, H., Okada, N., Takago, Y. and Sakamoto, Y. (1957). *Proceedings of the Japanese Academy* **33**, 105.

Iizuka, H. and Naito, A. (1968). 'Microbial Transformation of Steroids and Alkaloids'. University Park Press, Baltimore, Maryland.

Iizuka, K., Okuda, S., Aida, K., Asai, T., Tsuda, K., Yamada, M. and Seki, I. (1960). *Chemical and Pharmaceutical Bulletin* **8**, 1056.

Iizuka, K., Yamada, M., Suzuki, J. Seki, I., Aida, K., Okuda, S., Asai, T. and Tsuda, K. (1962). *Chemical and Pharmaceutical Bulletin* **10**, 67.

Ikawa, M., Rodwell, V. W. and Snell, E. E. (1958). *Journal of Biological Chemistry* **233**, 1555.

Izumi, Y. and Ogata, K. (1977). *Advances in Applied Microbiology* **22**, 145.

Johnson, R. A., Herr, M. E., Murray, H. C. and Reineke, L. M. (1971). *Journal of the American Chemical Society* **93**, 4880.

Jones, M. V. and Hughes, D. E. (1972). *Biochemical Journal* **129**, 755.
Joshi, J. G. and Handler, P. (1962). *Journal of Biological Chemistry* **237**, 3185.
Kaczkowski, J. and Mozejko-Toczko, M. (1960). *Acta Microbiologica Polonica* **9**, 173.
Kaper, J. B. and Veldstra, H. (1958). *Biochimica et Biophysica Acta* **30**, 401.
Kawai, S., Kobayashi, K., Oshima, T. and Egami, F. (1965). *Archives of Biochemistry and Biophysics* **112**, 537.
Kobayashi, Y. and Arima, K. (1962). *Journal of Bacteriology* **84**, 765.
Koninklijke, N. V. (1966). Netherlands Patent 6,405,471.
Kuhn, R., Starr, M. P., Kuhn, D. A., Bauer, H. and Knackmuss, H. J. (1965). *Archiv für Mikrobiologie* **51**, 175.
Kupchan, S. M., Sih, C. J., Kubota, S. and Rahim, A. M. (1963). *Tetrahedron Letters* p. 1767.
Lancini, C. C., Lazzari, E. and Sartori, G. (1968). *Journal of Antibiotics* **21**, 387.
Lanigan, G. W. (1972). *Australian Journal of Agricultural Research* **23**, 1085.
Lanigan, G. W. (1976). *Journal of General Microbiology* **94**, 1.
Large, P. J. (1971). *Xenobiotica* **1**, 457.
Larsen, P., Harbo, A., Klungsöyr, S. and Aasheim, T. (1962). *Physiologia Plantarum* **15**, 552.
Lemke, T. L., Johnson, R. A., Murray, H. C., Duchamp, D. J., Chidester, C. G., Hester, Jr., J. B. and Heinzelman, R. V. (1971). *Journal of Organic Chemistry* **36**, 2823.
Levy, C. C. and McNutt, W. S. (1962). *Biochemistry, New York* **1**, 1161.
Lin, W-N. C. and Ramstad, E. (1967). *Lloydia* **30**, 202.
Loo, Y. H. and Reidenberg, M. (1959). *Archives of Biochemistry and Biophysics* **79**, 257.
Loo, Y. H. and Woolf, D. O. (1957). *Chemistry and Industry* p. 1123.
Lunt, D. O. and Evans, W. C. (1976). *Biochemical Society Transactions* **4**, 491.
Mallett, E. G., Fukuda, D. A. and Gorman, M. (1964). *Lloydia* **27**, 334.
Marx, A. F., Beck, H. C., van der Waard, W. F. and de Flines, J. (1966a). *Steroids* **8**, 391.
Marx, A. F., Beck, H. C., van der Waard, W. F. and de Flines, J. (1966b). *Steroids* **8**, 421.
Mattey, M. and Harle, E. M. (1976). *Biochemical Society Transactions* **4**, 492.
Mattocks, A. R. and White, N. H. (1971). *Chemico-Biological Interactions* **3**, 383.
Maturova, M., Beckmann, H. and Wacker, A. (1967). *Zeitschrift für Naturforschung* **22B**, 621.
Mazur, R. H. and Muir, R. D. (1963). *Journal of Organic Chemistry* **28**, 2442.
McCormick, N. G., Feeherry, F. E. and Levinson, H. S. (1976). *Applied and Environmental Microbiology* **31**, 949.
Meyers, E. and Pan, S. C. (1961). *Journal of Bacteriology* **81**, 504.
Mitchard, M. (1971). *Xenobiotica* **1**, 469.
Mitoma, C., Weissbach, H. and Udenfriend, S. (1956). *Archives of Biochemistry and Biophysics* **63**, 122.
Mitscher, L. A., Andres, W. W., Morton, G. O. and Patterson, E. L. (1968). *Experientia* **24**, 133.
Mutschler, E., Rochelmeyer, H. and Wölffling, D. (1968a). *Archiv der Pharmazie* **301**, 287
Mutschler, E., Rochelmeyer, H. and Wölffling, D. (1968b). *Archiv der Pharmazie* **301**, 291.
Nakayama, K., Tanaka, Y. and Yoshida, H. (1974). Japanese Patent 74,100,290.
Nielson, E. D., Hamilton, P. B., Rosi, D. and Peruzzotti, G. P. (1967). United States Patent 3,317,401.
Niemer, H. and Bucherer, H. (1961). *Zeitschrift für Physiologische Chemie* **326**, 9.
Niemer, H. and Bucherer, H. (1962). *Zeitschrift für Physiologische Chemie* **328**, 108.

Niemer, H., Bucherer, H. and Kohler, A. (1959). *Zeitschrift für Physiologische Chemie* **317**, 238.

Niemer, H., Bucherer, H., Zeitler, H.-J. and Stadtler, E. (1964). *Zeitschrift für Physiologische Chemie* **337**, 282.

Ogino, S., Fujimoto, S. and Aoki, Y. (1974). *Agricultural and Biological Chemistry* **38**, 275.

Orpin, C. G., Knight, M. and Evans, W. C. (1972). *Biochemical Journal* **127**, 819.

Pan, S. C. and Weisenborn, F. C. (1958). *Journal of the American Chemical Society* **80**, 4749.

Patterson, E. L., Andres, W. W., Krause, E. F., Hartman, R. E. and Mitscher, L. A. (1963). *Archives of Biochemistry and Biophysics* **103**, 117.

Patterson, E. L., Andres, W. W. and Hartman, R. E. (1964). *Experientia* **20**, 256.

Ramstad, E., Chan-Lin, W.-N., Shough, H. R., Goldner, K. J., Parikh, R. P. and Taylor, E. H. (1967). *Lloydia* **30**, 441.

Ray, P. M. and Thimann, K. V. (1956). *Archives of Biochemistry and Biophysics* **64**, 175.

Richardson, S. H. and Rittenberg, S. C. (1961). *Journal of Biological Chemistry* **236**, 964.

Robert, D. and Tabone, J. (1952). *Bulletin de la Société de Chemie Biologique* **34**, 1009.

Robinson, W. G. and Hook, R. H. (1964). *Journal of Biological Chemistry* **239**, 4257.

Rosazza, J. P., Juhl, R. and Davis, P. (1973). *Applied Microbiology* **26**, 98.

Rosazza, J. P., Foss, P., Lemberger, M. and Sih, C. J. (1974). *Journal of Pharmaceutical Sciences* **63**, 544.

Rosazza, J. P., Stocklinski, A. W., Gustafson, M. A. and Adrian, J. (1975). *Journal of Medicinal Chemistry* **18**, 791.

Rosi, D., Peruzzotti, G. P., Dennis, E. W., Berberian, D. A., Freele, H., Tullar, B. F. and Archer, S. (1967a). *Journal of Medicinal Chemistry* **10**, 867.

Rosi, D., Lewis, T. R., Lorenz, R., Freele, H., Berberian, D. A. and Archer, S. (1967b). *Journal of Medicinal Chemistry* **10**, 877.

Roussel-Uclaf (1963). British Patent 923,421.

Russell, G. R. and Smith, R. M. (1968). *Australian Journal of Biological Sciences* **21**, 1277.

Sato, Y. and Hayakawa, S. (1961). *Journal of the American Chemical Society* **26**, 4181.

Sato, Y. and Hayakawa, S. (1963). *Journal of Organic Chemistry* **28**, 2739.

Sato, Y. and Hayakawa, S. (1964). *Journal of Organic Chemistry* **29**, 198.

Sato, Y., Waters, J. A. and Kaneko, H. (1964). *Journal of Organic Chemistry* **29**, 3732.

Sato, Y., Sato, Y. and Tanabe, K. (1967). *Steroids* **9**, 553.

Scheline, R. R. (1968). *Journal of Pharmaceutical Sciences* **57**, 2021.

Schuytema, E. C., Hargie, M. P., Merits, I., Schenk, J. R., Siehr, D. J., Smith, M. S. and Varner, E. L. (1966). *Biotechnology and Bioengineering* **8**, 275.

Siehr, D. J. (1961). *Journal of the American Chemical Society* **83**, 2401.

Siewert, G. and Kieslich, K. (1973). German Patent 2,202,410.

Sih, C. J., Foss, P., Rosazza, J. P. and Lemberger, M. (1969). *Journal of the American Chemical Society* **91**, 6204.

Smith, R. V. and Rosazza, J. P. (1974). *Archives of Biochemistry and Biophysics* **161**, 551.

Smith, R. V. and Rosazza, J. P. (1975a). *Journal of Pharmaceutical Science* **64**, 785.

Smith, R. V. and Rosazza, J. P. (1975b). *Biotechnology and Bioengineering* **17**, 1737.

Smith, G. N. and Worrell, C. S. (1950). *Archives of Biochemistry and Biophysics* **28**, 232.

Smyrniotis, P. Z., Miles, H. T. and Stadtman, E. R. (1958). *Journal of the American Chemical Society* **80**, 2541.

Starr, M. P., Knackmuss, H. J. and Cosens, G. (1967). *Archiv für Mikrobiologie* **59**, 287.

Subba Rao, P. V., Moore, K. and Towers, G. H. N. (1967). *Biochemical and Biophysical Research Communications* **28**, 1008.

Sykes, G. and Skinner, F. A., eds. (1971). Microbial Aspects of Pollution, 289 pp. Academic Press, London.

Tabuchi, T. (1954). *Journal of the Agricultural Chemical Society of Japan* **29**, 222.
Tabone, D. (1958). *Bulletin de la Société de Chimie Biologique* **40**, 965.
Tamm, Ch. (1960). *Planta Medica* **8**, 331.
Tate, R. L. and Ensign, J. C. (1974). *Canadian Journal of Microbiology* **20**, 695.
Taylor, E. H. and Shough, H. R. (1967). *Lloydia* **30**, 197.
Taylor, E. H., Goldner, K. J., Pong, S. F. and Shough, H. R. (1966). *Lloydia* **29**, 239.
Teuscher, E. (1965). *Phytochemistry* **4**, 341.
Teuscher, G. and Teuscher, E. (1965). *Phytochemistry* **4**, 511.
Theriault, R. J. and Longfield, T. H. (1973). *Applied Microbiology* **25**, 606.
Tewfic, M. S. and Evans, W. C. (1966). *Biochemical Journal* **99**, 31P.
Thimann, K. V. and Mahadevan, S. (1964). *Archives of Biochemistry and Biophysics* **105**, 133.
Toczko, M. (1966). *Biochimica et Biophysica Acta* **128**, 570.
Tsuda, K. (1964a). *In* 'Chemistry of Microbial Products'. Sixth Symposium of the Institute of Applied Microbiology, University of Tokyo, pp. 167–178.
Tsuda, K. (1964b). *Kogyo Kagaku Zasshi* **67**, 657.
Tyler, V. E., Erge, D. and Gröger, D. (1965). *Planta Medica* **13**, 315.
Valenta, J. R., Di Cuollo, C. J., Fare, L. R., Miller, J. A. and Pagano, J. F. (1974). *Applied Microbiology* **28**, 995.
Vining, L. C. (1969). *In* 'Fermentation Advances' (D. Perlman, ed.), pp. 715–743. Academic Press, New York.
Voigt, R. and Bornschein, M. (1964). *Die Pharmazie* **19**, 342.
Voigt, R. and Bornschein, M. (1965). *Die Pharmazie* **20**, 521.
Wada, E. (1957). *Archives of Biochemistry and Biophysics* **72**, 145.
Wada, E. and Yamazaki, K. (1954). *Journal of the American Chemical Society* **76**, 155.
Wallen, L. L., Stodola, F. H. and Jackson, R. W. (1959). 'Type Reactions in Fermentation Chemistry', 496 pp. Agricultural Research Service, United States Department of Agriculture.
Watanabe, S., Kitajima, N. and Takeda, I. (1973a). Japanese Patent 73 48,684.
Watanabe, S., Kitajima, N. and Takeda, I. (1973b). Japanese Patent 73 48,686.
Williams, R. T. (1972). *Toxicology and Applied Pharmacology* **23**, 769.
Wolters, B. (1970). *Planta Medica* **19**, 189.
Yamada, M. (1963). *Chemical and Pharmaceutical Bulletin* **11**, 356.
Yamada, M., Iizuka, K., Okuda, S., Asai, T. and Tsuda, K. (1962). *Chemical and Pharmaceutical Bulletin* **10**, 981.
Yamada, M., Iizuka, K., Okuda, S., Asai, T. and Tsuda, K. (1963). *Chemical and Pharmaceutical Bulletin* **11**, 206.
Yamano, T., Kishino, K., Yamatodani, S. and Abe, M. (1962). *Annual Reports of the Takeda Research Laboratories* **21**, 83.
Yamatodani, S., Kozu, Y., Yamada, S. and Abe, M. (1962). *Annual Reports of the Takeda Research Laboratories* **21**, 88.
Yang, H.-C., Kusumoto, M., Iwahara, S., Tochikura, T. and Ogata, K. (1968). *Agricultural and Biological Chemistry* **32**, 399.
Yoshida, H., Tanaka, Y. and Nakayama, K. (1973). *Agricultural and Biological Chemistry* **37**, 2121.
Yoshida, H., Tanaka, Y. and Nakayama, K. (1974). *Agricultural and Biological Chemistry* **38**, 455.
Youatt, J. (1962). *Australian Journal of Experimental Biology and Medical Science* **40**, 191.

11. Microbial Transformations of Antibiotics

O. K. SEBEK

Infectious Diseases Research, The Upjohn Company, Kalamazoo, Michigan 49001, U.S.A.

'He that will not apply new Remedies,
must expect new Evils; for Time is
the greatest Innovator'.

SIR FRANCIS BACON
(1561–1626)

I. INTRODUCTION

Antibiotics are microbial products of considerable impact on our society. As described in Volume 3 of this series, they have been used with an outstanding success as antimicrobial and antineoplastic agents in human medicine and animal health care, as supplements in animal and poultry feed, as growth stimulants and as effective substances for control of plant disease. They have challenged biologists and chemists alike because of the uncommon pathways of their biosynthesis, unusual chemical structures and modes of action.

To date, more than 5,500 new antibiotics have been isolated from microbial sources (Berdy, 1979) but only less than 100 of them are manufactured by fermentation. In 1978, their production volume as bulk chemicals in the United States alone was 25.7 million pounds and valued at US $851 million (United States International Trade Commission, 1979).

Practically all the antibiotics known to date have been found by systematic screening of micro-organisms isolated from soil and other sources. In time, this approach was supplemented by chemical modification of the existing ones with the overall objectives (a) to broaden their antimicrobial spectra, (b) to increase their potency, (c) to lower their toxicity and undesirable side-reactions, (d) to synthesize analogues that would resist inactivation by the target bacteria, (d) to improve the route of administration, and (f) to extend their biological half-lives. These efforts have been outstandingly successful in the chemical preparation of new semisynthetic penicillins, cephalosporins, aminoglycosides, tetracyclines, rifamycins, macrolides and lincosaminides.

The early literature also reported that penicillin (Abraham and Chain, 1940), streptomycin (Perlman and Langlykke, 1948; Pramer and Starkey, 1951), chloramphenicol (Smith and Worrel, 1949, 1950, 1953), actinomycin (Katz and Pienta, 1957), or tetracyclines and others (Brian, 1958; Pramer, 1958) were subject to microbial transformation

Table 1

Microbial transformations of antibiotics (For abbreviations of antibiotics, see the text.)

Reaction	Antibiotic
1. Hydrolysis	
(a) β-Lactam	Cephalosporins and penicillins (cf. Huber *et al.*, 1972; Sykes, 1979)
(b) Amide	Bleomycin B₂ (Umezawa *et al.*, 1973), cephalosporins and penicillins (Fujii *et al.*, 1976; Shimizu *et al.*, 1975; Takahashi *et al.*, 1972; Vandamme and Voets, 1974), chloramphenicol (Lingens *et al.*, 1966), colistin (Ito *et al.*, 1966), gramicidin S (Yukioka *et al.*, 1966), nocardicin C (Komori *et al.*, 1978), novobiocin (Sebek and Hoeksema, 1972).
(c) Lactone	Actinomycin (Hou and Perlman, 1970), antimycin (Lennon and Vézina, 1973), cordycepin (Rottman and Guarino, 1964), echinomycin, etamycin, staphylomycin S, stendomycin, vernamycin B (Hou *et al.*, 1970).
(d) Ester	Cephalosporin C (Abbott and Fukuda, 1975), leucomycin (Inouye *et al.*, 1971), maridomycin (Nakahama *et al.*, 1974, 1975).
(e) Glycoside	Mannosidostreptomycin (Perlman and Langlykke, 1948; Demain and Inamine, 1970), validamycin A (Kameda *et al.*, 1975).
2. Hydroxylation	Daunomycin (Arcamone *et al.*, 1969), 5α,11α-dehydrochlortetracycline (Martin *et al.*, 1967), 12α-deoxytetracycline (Holmlund *et al.*, 1959), griseofulvin (Andres *et al.*, 1969; Bod *et al.*, 1973), josamycin and maridomycin (Nakahama *et al.*, 1974), narbomycin (Maezawa *et al.*, 1973), novobiocin (Sebek and Dolak, 1977).
3. Epoxidation	*cis*-Propenylphosphonic acid (White *et al.*, 1971), carbomycin B, leucomycin A₃ (Suzuki *et al.*, 1977).
4. Oxidation	Formycin B (Sawa *et al.*, 1968), fusidic acid (Dvonch *et al.*, 1966; von Daehne *et al.*, 1968), lankamycin (Goldstein *et al.*, 1978), rifamycin B (Lancini and White, 1973)
5. Sulphoxidation	Lincomycin, clindamycin (Argoudelis *et al.*, 1969).
6. Acylation	Aminoglycosides (see Table 2), 6-aminopenicillanic acid, 7-aminocephalosporanic acid (Carrington, 1971; Fujii *et al.*, 1976; Takahashi *et al.*, 1972; Shimizu *et al.*, 1975), chloramphenicol (Argoudelis and Coats, 1971; El-Kersh and Plourde, 1976; Nakano *et al.*, 1977; Sands and Shaw, 1973), daunorubicin and daunorubicinol (Hamilton *et al.*, 1977), leucomycin C (Nakahama *et al.*, 1975), leucomycin A₁, (Ōmura *et al.*, 1976), spiramycin I (Kitao *et al.*, 1979; Ninet and Verrier, 1962).

Table 1 (cont.)

Reaction	Antibiotic
7. Deacylation	Amphomycin (Weber and Perlman, 1978), chromomycin and olivomycin (Schmitz and Claridge, 1977), josamycin, leucomycins, maridomycin, middamycin (Ōmura and Nakagawa, 1975; Singh and Rakhit, 1979), lankamycin (Goldstein et al., 1978), T-2636 (lankacidins; Higashide et al., 1971).
8. Phosphorylation	Aminoglycosides (see Table 2), clindamycin and lincomycin (Argoudelis and Coats, 1969; Argoudelis et al., 1977; Coats and Argoudelis, 1971).
9. Adenylylation (Nucleotidylation)	Aminoglycosides and spectinomycin (see Table 2), clindamycin (Argoudelis et al., 1977).
10. Amination and Deamination	Blasticidin S (Yamaguchi et al., 1975), formycins (Ochi et al., 1975; Sawa et al., 1968), ketomycin (Jackson and Umbarger, 1973; Keller-Schierlein et al., 1969).
11. Glycosylation and Transglycosylation	Erythronolide A oxime (LeMahieu et al., 1976), narbomycin, picromycin (Maezawa et al., 1973, 1976), validamycins (Kameda et al., 1975, 1978).
12. Hydration	Toyocamycin (Uematsu and Suhadolnik, 1974).
13. Isomerization	Showdomycin (Ozaki et al., 1972).
14. Methylation	
N-and-O-Carboxymethylation	Ribostamycin (Kojima et al., 1975), rifamycin SV (Lancini et al., 1969)
Demethylation	Clindamycin and lincomycin (Argoudelis et al., 1969), griseofulvin (Boothroyd et al., 1961), lankamycin (Goldstein et al., 1978).
Hydroxymethylation	N-Demethylclindamycin (Argoudelis et al., 1972a), griseofulvin (Bod et al., 1973).
Transmethylation	Erythromycins C and D (Majer et al., 1977).
15. Reduction	Aclacinomycin A (Yoshimoto et al., 1979), albocycline (Slechta et al., 1978), ascochitine (Oku and Nakanishi, 1964), carbomycins A and B (Suzuki et al., 1977), chloramphenicol (Egami et al., 1951), 7-chloro-5α,11α-dehydrotetracycline (McCormick et al., 1958), daunomycinone (Karnetová et al., 1976; Marshall et al. 1978), dehydrogriseofulvin (Andres et al., 1969), maridomycin (Nakahama and Igarasi, 1974), midecamycin A₃ (Matsuhashi et al., 1979), nisin and subtilin (Jarvis and Farr, 1971), tylosin (Feldman et al., 1973).

and degradation. These and similar findings led to the notion that desirable antibiotic modifications might be realized also by application of microbial enzymes. The support for this approach came also from the highly successful use of micro-organisms in synthesis of steroid hormones, but the work was slow in starting. Nevertheless many micro-organisms have been described to date to affect a host of different antibiotics (see Table 1) and microbial transformations have been a topic of several reviews (Jarvis and Berridge, 1969; Perlman, 1971; Perlman and Sebek, 1971; Sebek and Perlman, 1971; Sebek, 1974, 1975, 1977; Shibata and Uyeda, 1978).

Most of the products obtained by this microbial methodology are only of theoretical interest but a few became economically important, including 6-aminopenicillanic acid, 6-demethyltetracyclines and rifamycins. In general, they have been generated by (a) directed fermentation, (b) blocked mutants, (c) mutasynthesis, and (d) biotransformation.

II. FORMATION OF NEW ANTIBIOTIC ANALOGUES

A. Directed Fermentation

New bioactive compounds have been formed by antibiotic-producing organisms in the presence of suitable precursors or metabolic inhibitors. The early work on penicillin demonstrated that penicillin-producing *Penicillium chrysogenum* can incorporate not only phenyl-acetic acid (into benzylpenicillin) but also other precursors (into new penicillin analogues) in preference to synthesis *de novo* of the natural side-chain precursors. Other antibiotics were also shown to be modified by this method but have not yielded marketable products. For example, 5-fluorouracil was incorporated into two 5-fluoro-polyoxins by the polyoxin-producing *Streptomyces cacaoi* (Isono *et al.*, 1973). Salicylic acid, a component of celesticetin, was replaced by 4-aminosalicylic acid when the latter had been added to the celesticetin-producing *S. caelestis*, and yielded desalicetin-2′-(4-amino-salicylate) (Argoudelis *et al.*, 1974) (Fig. 1).

The addition of several amines to cultures of *Streptomyces verticillus* resulted in their incorporation into new bleomycins (Fujii *et al.*, 1974). 4-Methylprolines were metabolized by *Streptomyces parvulus* in pre-ference to L-proline and yielded new actinomycins (Katz *et al.*, 1977). Echinomycin-producing streptomycetes incorporated 4-quinazolone-3-acetate into two new echinomycin analogues (Dhar *et al.*, 1971). Pyrrolnitrin-producing *Pseudomonas aureofaciens* converted tryptophan analogues into correspondingly substituted pyrrolnitrins (Hamill *et al.*,

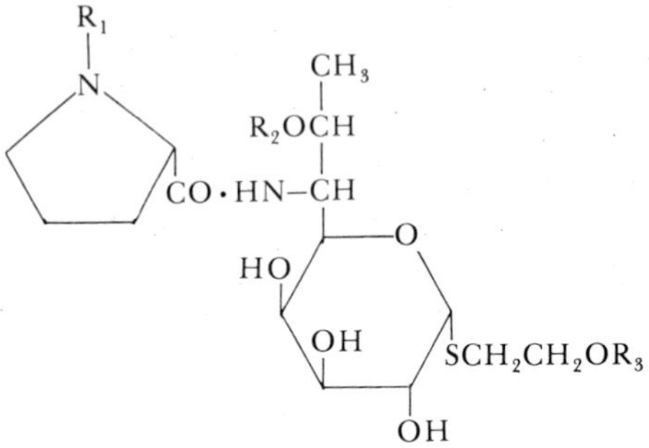

Antibiotic	R_1	R_2	R_3
Celesticetin	$-CH_3$	$-CH_3$	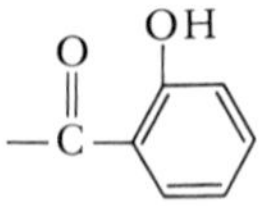 (o-hydroxybenzoyl)
Desalicetin-2'-(4-amino salicylate)	$-CH_3$	$-CH_3$	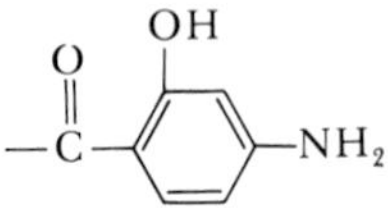 (o-hydroxy-amino-benzoyl, NH_2)
7-O-Demethylcelesticetin	$-CH_3$	$-H$	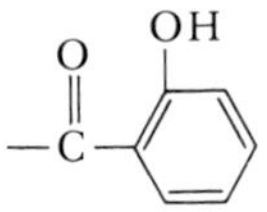 (o-hydroxybenzoyl)
N-Demethylcelesticetin	$-H$	$-CH_3$	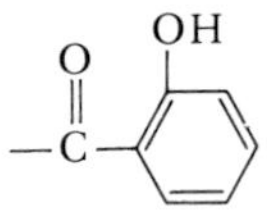 (o-hydroxybenzoyl)
N-Demethyl-7-O-demethyl-celesticetin	$-H$	$-H$	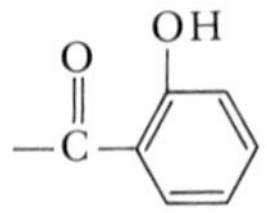 (o-hydroxybenzoyl)

Fig. 1. Structure of celesticetin and its analogues.

1970), and novobiocin-producing *Streptomyces spheroides* incorporated substituted benzoic acid derivatives into new novobiocins (Walton *et al.*, 1962).

Antibiotics were modified also by metabolic inhibitors. Thus 98 selected compounds inhibited chlorination in *Streptomyces aureofaciens*.

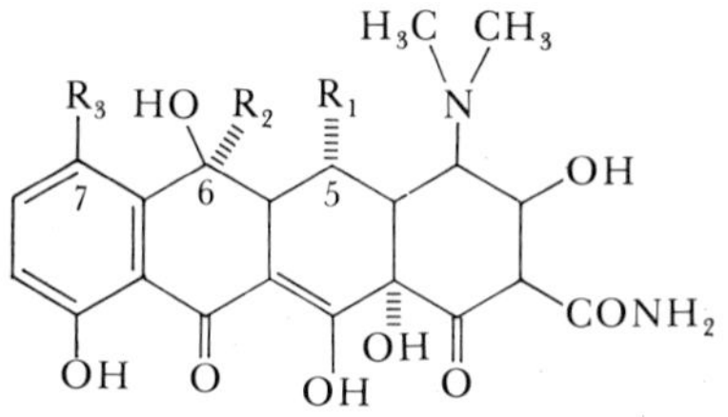

Antibiotic	R_1	R_2	R_3
7-Chlortetracycline	...H	...CH_3	$-Cl$
6-Demethyl-7-chlor-tetracycline	...H	...H	$-Cl$
5-Oxytetracycline	...OH	...CH_3	$-H$
Tetracycline	...H	...CH_3	$-H$

Fig. 2. Structure of tetracyclines.

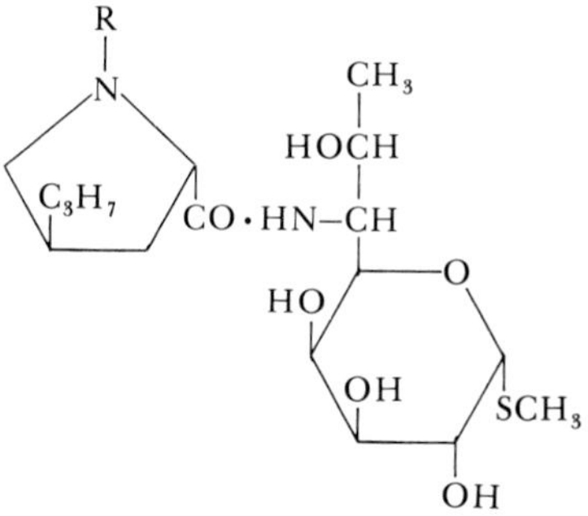

Antibiotic	R
Lincomycin	$-CH_3$
N-Demethyl-lincomycin	$-H$

Fig. 3. Structure of lincomycins.

Their addition to the fermentation resulted in formation of tetracycline instead of chlortetracycline in amounts that in many instances
exceeded 90% of the total tetracylines produced (Goodman *et al.*, 1959;
Evans, 1968). Since the 6-methyl and the two *N*-methyl groups of
tetracyclines are derived from L-methionine, the inhibition of
methylation in chlortetracycline fermentations either by L-methionine
analogues (D-methionine, ethionine) or folic acid antagonists
(aminopterin, sulphonamides) led to the formation of 6-demethyl-
7-chlortetracycline (Goodman and Matrishin, 1961; Hendlin *et al.*,
1962; Neidleman *et al.*, 1963; Perlman *et al.*, 1961) (Fig. 2).

In similar vein, *N*-demethylstreptomycin was formed when the
streptomycin fermentation was supplemented with DL-ethionine
(Hedig, 1968; see Fig. 13). Addition of sulphonamides and sulph-
anilamide to the lincomycin-producing *Streptomyces lincolnensis* yielded
N-demethyllincomycin (Argoudelis *et al.*, 1973a) (Fig. 3).

B. Blocked Mutants

Mutants of antibiotic producers have been useful in elucidating the
biosynthetic pathways of antibiotics and in related genetic investigations (Hopwood and Merrick, 1977; Queener *et al.*, 1978). Some of
the intermediates that accumulate were biologically active, or in turn
served as substrates for biosynthesis of new analogues. Thus mutants
derived from the celesticetin-producing *Streptomyces caelestis* were found
to accumulate three bioactive *O*- and *N*-demethylated celesticetins
(Argoudelis *et al.*, 1972b, 1973b). A mutant of the butirosin-producing
B. circulans, which lacks the ability to acylate the 2-deoxystreptamine
(2-DOS) moiety of the antibiotic, produced ribostamycin (Fujiwara *et
al.*, 1978), an antibiotic normally synthesized by *S. ribosidificus*
(Shomura *et al.*, 1970). Ribostamycin was also formed by a mutant of
the neomycin-producing *S. fradiae* which was blocked in glycosylation
of the ribose moiety of neomycin (Baud *et al.*, 1977) (Fig. 4).

Mutants of the rifamycin-producing *Nocardia mediterranei*
accumulated rifamycin B precursors which served as substrates for the
chemical preparation of many hundreds of semisynthetic rifamycins
(see Fig. 19 and Section III.E, p. 599).

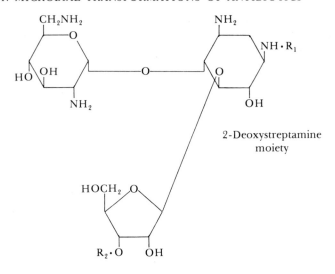

2-Deoxystreptamine
moiety

Antibiotic (Producer)	R_1	R_2
Ribostamycin (*Streptomyces ribosidificus*)	–H	–H
Butirosin (*Bacillus circulans*)	$-CO \cdot CH \cdot CH_2 \cdot CH_2NH_2$ with OH	–H
Neomycin B (*Streptomyces fradiae*)	–H	(sugar moiety: CH_2NH_2, HO, OH, O^-, NH_2)

Fig. 4. Structure of neomycin and related aminoglycosides.

C. Mutasynthesis

Some blocked mutants cannot complete antibiotic synthesis unless they have been supplemented with precursors that are normally produced by enzymes beyond the metabolic block. Since the enzymes involved in secondary metabolism generally possess low substrate specificity,

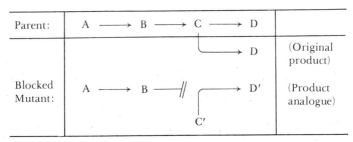

Fig. 5. Biosynthetic sequence in the parent and blocked mutant.

analogues of antibiotic precursors have been converted by such mutants into analogues of the original antibiotics (Fig. 5).

This technique was originally envisioned by Birch (1963), and its feasibility first demonstrated by Shier *et al.* (1969) who prepared new biologically active analogues of neomycin by means of mutants of the neomycin-producing *S. fradiae*. Since these mutants were blocked in synthesis of the 2-DOS moiety of the neomycin molecule, they produced neomycin only when this moiety had been supplied preformed to the mutant (Fig. 4). When in turn analogues of 2-DOS had been supplied, they were incorporated into the corresponding analogues of neomycin (Shier *et al.*, 1969, 1973; Cleophax *et al.*, 1976). Subsequently, many analogues of other aminoglycosides have been prepared by this technique which is known as 'mutational biosynthesis' (Nagaoka and Demain, 1975) or 'mutasynthesis' (Rinehart, 1977). As with *S. fradiae*, all mutants used in this work have been blocked in synthesis of 2-DOS and hence yielded new analogues of the following aminoglycosides modified in their 2-DOS moiety: paromomycin (*S. rimosus*; Shier *et al.*, 1973), ribostamycin (*S. ribosidificus*) and kanamycin (*S. kanamyceticus*; Kojima and Satoh, 1973), sisomicin (*M. inyoensis*; Testa and Tilley, 1975), gentamicin (*M. purpurea*; Daum *et al.*, 1977) and butirosin (*B. circulans*; Claridge *et al.*, 1974; Takeda *et al.*, 1978a, b, c, d). Similarly, streptomycin was modified in the streptidine moiety by a mutant of the streptomycin-producing *S. griseus* which had been blocked in streptidine biosynthesis (Fig. 13; Nagaoka and Demain, 1975). Antibiotics of other classes were also modified by this method. Thus, a new penicillin was formed when a mutant of *Cephalosporium acremonium* was fed with L-S-carboxymethylcysteine (Troonen *et al.*, 1976) and a new chlorine-containing novobiocin was generated by a *S. niveus* mutant blocked in synthesis of the ring B moiety when

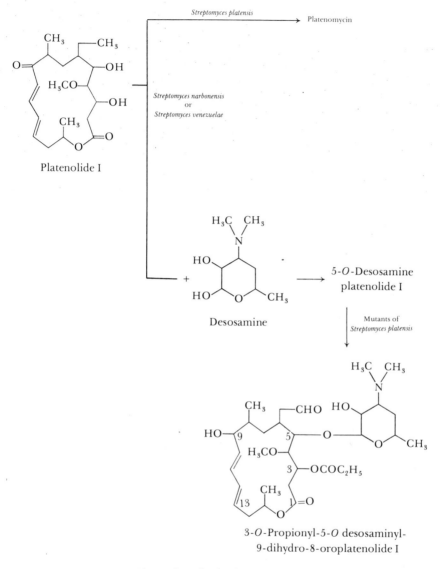

Fig.6. Glycosylation of platenolide I.

3-amino-4-hydroxy-8-chlorcoumarin had been supplied (see Fig. 24; Lemaux and Sebek, 1973).

By glycosylating narbonolide (the aglycone of the 14-membered ring of macrolide narbomycin) with mycaminose (a sugar normally present in the 16-membered ring macrolide antibiotics), two new hybrid anti-biotics were formed and identified as 5-O-mycaminosylnarbonolide and 9-dihydro-5-O-mycaminosylnarbonolide (Maezawa *et al.*, 1976). Conversely, platenolide I, the aglycone of platenomycin which is a 16-membered ring macrolide antibiotic, was glycosylated with desos-amine, a sugar found in the 14-membered ring macrolide antibiotics narbomycin and picromycin, by the narbomycin- or picromycin-producers and yielded 5-O-desosaminylplatenolide I. On subsequent incubation with a mutant of S. *platensis* that is blocked at an early step in platenomycin synthesis, 5-O-desosaminylplatenolide I was converted into 3-O-propionyl-5-O-desosaminyl-9-dihydro-18-oxoplatenolide I (Maezawa *et al.*, 1978; Sebek, 1977) (Fig. 6).

D. Biotransformation

In addition to antiobiotic producers and their mutants, many other micro-organisms have been shown to modify antibiotics. As summarized in Table 1 (pp. 577–578) and illustrated by selected ex-amples below, a great deal of work has beedone to discover such modifications in order to assess their potential and to direct these activities to some useful purpose. To date, however, only two areas have been truly successful. One is enzymic hydrolysis of penicillin to 6-aminopenicillanic acid which in turn is a substrate for the manufacture of several semisynthetic and clinically valuable penicillin analogues. The other area is the elucidation of the mechanisms by which the R factor (plasmid)-carrying bacteria inactivate amino-glycoside antibiotics. Such findings in turn have led to the rational design and chemical synthesis of new analogues that resist such inactivations.

The modifications of most of the other antibiotics examined gave completely inactive or only partially active products. Such studies may not remain mere scientific curiosities, however, but may eventually find practical application since they may become substrates for synthesis of new antibiotic analogues. They may also identify those groups and portions of the molecules that are essential for antibiotic activity as already demonstrated with the aminoglycosides.

III. SELECTED BIOTRANSFORMATIONS

A. *Beta*-Lactams

The first inactivation of an antibiotic, namely that of penicillin, was reported by Abraham and Chain (1940) essentially at the same time as the historic disclosure of the therapeutic property of penicillin (Chain *et al.*, 1940). The authors noted that extracts of *Escherichia coli* and of other bacteria produced a substance that destroyed the growth-inhibiting activity of penicillin; they called it 'penicillinase'. This inactivating property soon became a serious clinical problem and, by 1948, as much as 50% of the hospital strains were penicillin-resistant (Barber and Rozwadowska-Dowzenko, 1948). This incidence increased to 80% by the mid 1950s (Ridley *et al.*, 1970). Subsequently, other penicillinases were reported to be produced by many Gram-positive and Gram-negative bacteria, actinomycetes, yeasts and blue–green algae. Since they inactivate not only penicillins but also the structurally related cephalosporins by the same mechanism, namely by hydrolysing the β-lactam ring of the substrate molecule, they are identified as β-lactamases (Figs. 7 and 11).

Elaboration of these enzymes is generally accepted as a major factor in the resistance of infectious bacteria to penicillins and cephalosporins. The enzymes exist in a variety of forms and their different specificities depend on the bacteria that produce them. They are chromosomal and plasmid-determined, constitutive and inducible, extracellular and intracellular (Pollock, 1971; Richmond and Sykes, 1973; Sykes and Matthew, 1976; Sykes, 1979). To counteract this inactivation, vast efforts have been made, resulting in chemical synthesis of a number of potent β-lactamase-resistant analogues

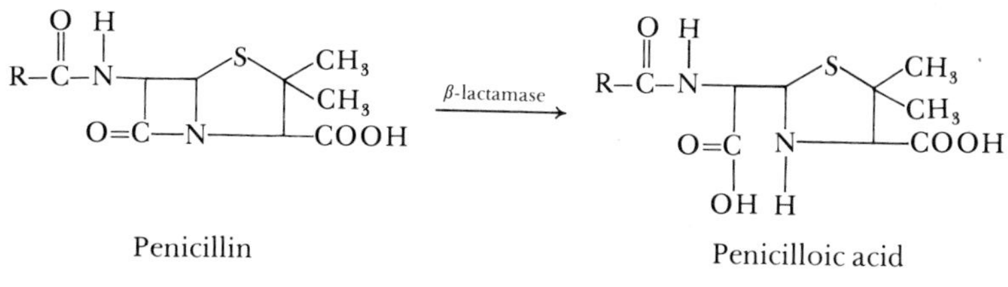

Penicillin

Penicilloic acid

Benzylpenicillin: $R = C_6H_5CH_2-$

Fig. 7. Hydrolysis of penicillin to penicilloic acid.

(ampicillin, carbenicillin, methicillin, cloxacillin) and also in the discovery of naturally occurring β-lactamase inhibitors (clavulanic and olivanic acids).

Penicillin is known to undergo yet another enzymic reaction which has rapidly become one of prime economic significance and involves hydrolysis of the amide bond of the molecule. It yields a product which was termed 6-aminopenicillanic acid in accordance with the nomenclature proposed by Sheehan *et al.* (1953) (Fig. 8).

Bacterial hydrolysis of benzylpenicillin to 6-aminopenicillanic acid has become an essential step in the manufacture of semisynthetic penicillins (Hamilton-Miller, 1966; O'Callaghan and Muggleton, 1972; Vandamme and Voets, 1974). In addition, production of 6-aminopenicillanic acid by immobilized penicillin acylase was developed and is used industrially by several companies in the U.S.A., Japan and Europe (Chibata *et al.*, 1978). An alternative chemical route for the large-scale preparation of 6-aminopenicillanic acid also exists. Since the β-lactam ring is by far the more reactive of the two amide bonds of the penicillin molecule, highly specific reagents, precise reaction times and meticulously anhydrous conditions have to be adhered to in order to achieve a selective and efficient chemical side-chain cleavage (Carrington, 1971). It is estimated that, in 1979, 30,000 kg of penicillin were converted into 6-aminopenicillanic acid and that to date more than 40,000 semisynthetic penicillins have been prepared from this acid. A number of these compounds became important anti-infective agents and are clinically superior to benzylpenicillin.

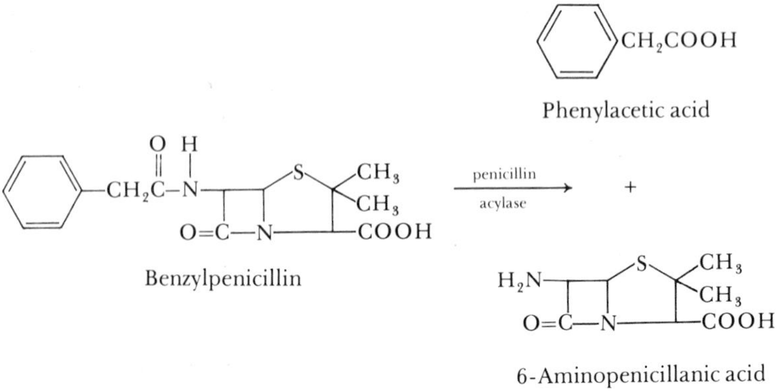

Fig. 8. Hydrolysis of penicillin to 6-aminopenicillanic acid.

The existence of 6-aminopenicillanic acid was revealed in one of the earlier studies on *Penicillium chrysogenum* Q176. When this fungus was grown in the absence of any side-chain precursor, it produced a 'penicillin nucleus' which accumulated in the medium. This nucleus was hydrolysed by penicillinase but, unlike penicillin, was practically devoid of antibacterial activity and was not solvent-extractable at low pH values. When phenylacetic acid was added to the fermentation, it was converted into benzylpenicillin almost quantitatively (Kato, 1953). Pure 6-aminopenicillanic acid was isolated six years later from penicillin fermentations carried out under the same conditions, namely in the absence of the side-chain precursor, and was properly identified (Batchelor *et al.*, 1959). The generation of this compound by the enzymic removal of the penicillin side chain was reported in 1950. Although the physical data of this compound were at variance with those of the authentic 6-aminopenicillanic acid, this substance ('penicin'), behaved like 6-aminopenicillanic acid since it was prepared by incubating benzylpenicillin with washed cells and with a purified enzyme ('penicillinamidase') of the same (Q176) mutant of *P. chrysogenum* and *Aspergillus oryzae* (Sakaguchi and Murao, 1950; Murao, 1955). Attempts to repeat this reaction failed at first, but were successful several years later when two types of penicillin acylase were discovered. One is exocellular and widely distributed among actinomycetes, fungi and certain yeasts, and hydrolyses preferentially phenoxymethyl-, pentyl- and heptyl-penicillins (V, dihydro F and K). The other type is endocellular and is produced only by bacteria (especially species of *Escherichia*, *Alcaligenes*, *Proteus* and *Nocardia*). It readily hydrolyses benzylpenicillin, whereas phenoxymethylpenicillin is affected at a much slower rate. Hydrolysis proceeded at pH 7.5–8.5, but a reverse reaction, acylation of 6-aminopenicillanic acid with a suitable side chain, took place when the reaction mixture was adjusted to pH 5.5 (Rolinson *et al.*, 1960; Claridge *et al.*, 1960; Huang *et al.*, 1960, 1963; Kaufmann and Bauer, 1960; Hamilton-Miller, 1966). The hydrolytic reaction which yields 6-aminopenicillanic acid has been developed into an industrial process by systematic selection of mutant strains, by improvement of media that promote enzyme induction, and by careful regulation of enzymic conditions (substrate and enzyme concentrations, pH value and temperature) to force the equilibrium toward complete deacylation and to cause a minimum of undesirable degradation of the substrate and product. At the completion of the reaction, phenylacetic

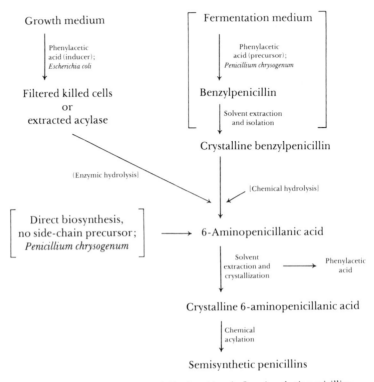

Fig. 9. Production of 6-aminopenicillanic acid and of semisynthetic penicillins.

acid is extracted with an organic solvent from the filtered beer, 6-aminopenicillanic acid is crystallized from the remaining aqueous phase and used for chemical acylation to the desired semisynthetic penicillins (Carrington, 1971) (Fig. 9).

A third type of penicillin acylase also exists. It is specific for ampicillin and neither hydrolyses nor synthesizes benzylpenicillin. This narrow substrate specificity was applied to the development of an interesting two-step microbial transformation of benzylpenicillin into ampicillin without the need of isolating the intermediary 6-amino-penicillanic acid. Benzylpenicillin is first hydrolysed to 6-amino-penicillanic acid at pH 7.5 by a β-lactamase-negative mutant of *Kluyvera citrophila* (KY 3641). A mutant of *Pseudomonas melanogenum* (KY 3987) which elaborates this acylase, and D,L-phenylglycine methyl ester are then added, and the reaction is adjusted to pH 5.5–6.0 to favour synthesis. Since *Ps. melanogenum* does not react with phenyl-acetic acid released in the first (hydrolytic) step and since the K.

citrophila acylase does not function at pH 5.5–6.0 in the second (acylating) one, this operation results in the exclusive formation of D(−)-α-aminobenzylpenicillin (ampicillin) (Okachi *et al.*, 1973a, b) (Fig. 10).

Microbial transformation of cephalosporins presents a different picture. The major product of cephalosporin fermentation is cephalosporin C which consists of 7-aminocephalosporanic acid (7-ACA) and α-aminoadipic acid (Fig. 11).

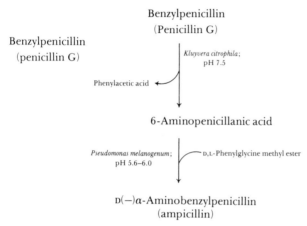

Benzylpenicillin
(Penicillin G)

Benzylpenicillin
(penicillin G)

Kluyvera citrophila;
pH 7.5

Phenylacetic acid

6-Aminopenicillanic acid

Pseudomonas melanogenum;
pH 5.6–6.0

D,L-Phenylglycine methyl ester

D(−)α-Aminobenzylpenicillin
(ampicillin)

Fig. 10. Bacterial synthesis of ampicillin from benzylpenicillin.

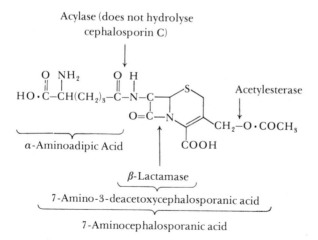

Acylase (does not hydrolyse
cephalosporin C)

Acetylesterase

α-Aminoadipic Acid

β-Lactamase

7-Amino-3-deacetoxycephalosporanic acid

7-Aminocephalosporanic acid

Fig. 11. Structure of cephalosporin C.

Although a two-step microbial conversion of cephalosporin C into 7-ACA was reported (Fujii *et al.*, 1979):

Cephalosporin C $\xrightarrow{\textit{Trigonopsis variabilis} \text{ CBS 4095}}$ Glutaryl-7-ACA $\xrightarrow{\textit{Comamonas} \text{ sp.}}$ 7-ACA

no direct enzymic hydrolysis of this antibiotic is known. Hence the 7-ACA used for the chemical preparation of new semisynthetic cephalosporins is obtained from cephalosporin C or from penicillins by chemical means (Cooper and Spry, 1972; Huber *et al.*, 1972). However, some synthetic *N*-acyl derivatives of 7-ACA have been hydrolysed by several bacteria (Vandamme and Voets, 1974). Furthermore, 7-amino-3-deacetoxycephalosporanic acid (7-ADCA), which is obtained by chemical ring expansion from benzylpenicillin or enzymically from phenylacetyl-7-ADCA (an intermediate in the ring expansion), has been acylated with D-α-phenylglycine methyl ester to cephalexin by a number of bacteria. When 7-ADCA was replaced by 7-ACA, cephaloglycine was produced. Acylases of other bacteria showed variable substrate specificity. Thus, for example, when incubated with 7-aminocephem compounds (7-ACA, 7-ADCA and others) and selected organic acid esters, they produced not only cephalexin but also other cephalosporins such as cephaloridine and

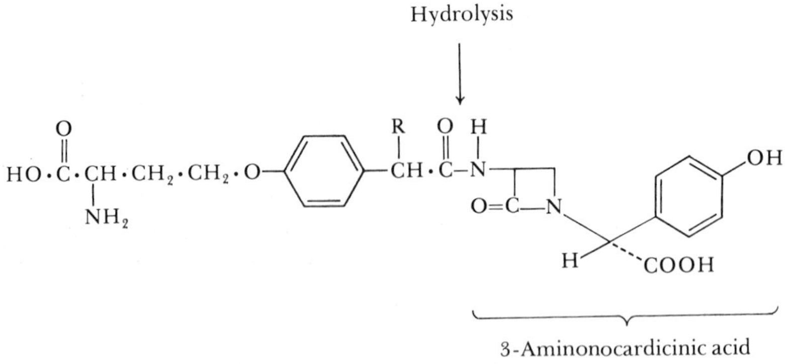

Hydrolysis

$$HO \cdot \underset{\underset{NH_2}{|}}{\overset{\overset{O}{\|}}{C}} \cdot CH \cdot CH_2 \cdot CH_2 \cdot O - \langle\rangle - \underset{\underset{}{}}{CH} \cdot \overset{R}{\underset{}{}} \overset{O}{\underset{}{C}} - \overset{H}{N}$$

3-Aminonocardicinic acid

Antibiotic	R
Nocardicin A	=NOH
Nocardicin C	−NH$_2$

Fig. 12. Structure of nocardicins.

cephalotin (Takahashi *et al.*, 1972; Fujii *et al.*, 1976; Shimizu *et al.*, 1975). In addition, cephalosporins undergo deacetylation by cephalosporin acetylesterase which is widely distributed in Nature (bacteria, rind of citrus fruits, liver, kidney; Fig. 11; Abbott and Fukuda, 1975).

Many bacteria hydrolyse nocardicin C, a minor component of the nocardicin complex (produced by *Nocardia uniformis*), which contains a monocyclic β-lactam nucleus, to 3-aminonocardicinic acid and the corresponding side chain (Fig. 12). On the other hand, nocardicin A, which is the main component of the nocardicin complex, was not affected by any of the 1,220 strains of bacteria, yeasts and fungi tested (Komori *et al.*, 1978).

B. Aminoglycosides and Spectinomycin

Antibiotics belonging to this group are of considerable clinical importance. They contain amino sugars in their molecules and have been divided according to their structures into those that contain residues of streptidine (such as streptomycin) or 2-deoxystreptamine (such as neomycin and butirosin which have their substituents attached at two adjacent hydroxyl groups, or kanamycin and gentamicin in which such substituents are attached at non-adjacent hydroxyl groups). Spectinomycin, another member of this group, does not contain an amino-sugar residue (Figs 13–16; Rinehart, 1969).

Enzymes have been isolated from different Gram-positive and Gram-negative bacteria that inactivate these antibiotics. They were generally found in clinical isolates of resistant bacteria and were identified as determinants of antibiotic resistance. They have been found also in some of the aminoglycoside-producing cultures and may represent a detoxifying mechanism by which the antibiotic producer protects itself against the deleterious effect of its own product (Benveniste and Davies, 1973; Davies *et al.*, 1979). They are classified according to the modification that they bring about (*N*-acetylation of the amino group, and *O*-phosphorylation, *O*-adenylylation or *O*-nucleotidylation of the hydroxyl groups) and according to the site at which the substrate antibiotic is modified. For example, each of two hydroxyl groups of streptomycin is phosphorylated and adenylylated by four different enzymes. The C-6 hydroxyl is phosphorylated by a 6-*O*-phosphotransferase [APH (6)] and adenylated by a 6-*O*-adenylyl-transferase [AAD (6)], the C-3″ hydroxyl by 3″-*O*-phosphotransferase

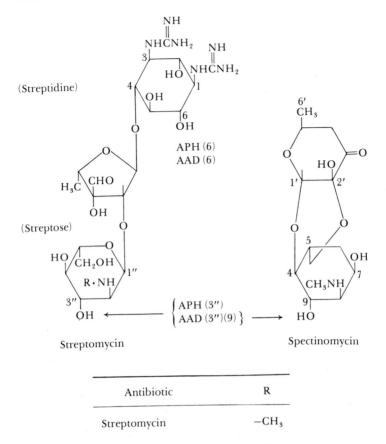

Fig. 13. Structures and inactivation of streptomycin and spectinomycin. For abbreviations of antibiotics, see the text.

Antibiotic	R
Streptomycin	$-CH_3$
N-Demethylstreptomycin	$-H$

[APH (3″)] and adenylylated by 3″(9)-O-adenylyltransferase [AAD (3″) (9)] (Fig. 13). The inactivation of selected antibiotics by N-acetyltransferases [AACs], O-phosphotransferases [APHs] and O-adenylyltransferases [AADs] is shown in Figures 14 and 15. Table 2 lists the transferases known at present, with some representative antibiotics that are subject to such modifications. The emergence of plasmid-determined aminoglycoside resistance has been of considerable clinical concern and has been a subject of intensive biochemical investigations. Elucidation of these inactivating mechanisms led to the successful

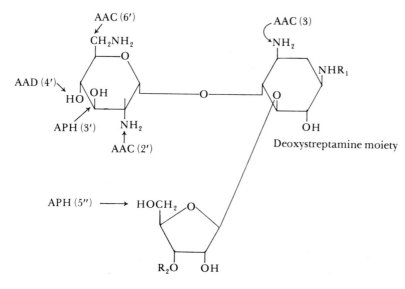

Fig. 14. Structure and inactivation of neomycin, ribostamycin and butirosin. See caption to Fig. 4 for R_1 and R_2. For explanation of other symbols, see the text.

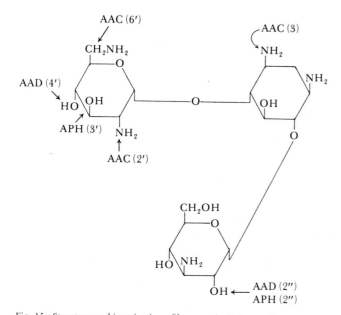

Fig. 15. Structure and inactivation of kanamycin B. For explanation of symbols, see the text.

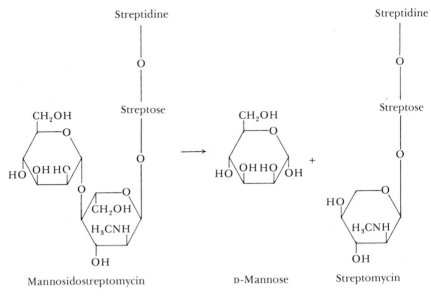

Fig. 16. Hydrolysis of mannosidostreptomycin.

Table 2

Aminoglycoside-modifying transferases

Modification	Modifying enzyme	Transferase substrate
Acetylation	AAC(3)	Kanamycin, gentamicin, tobramycin, sisomicin, ribostamycin.
	AAC(2')	Kanamycin, gentamicin, tobramycin, sisomicin, butirosin.
	AAC(6')	Kanamycin, 3',4'-dideoxykanamycin B, 4'-deoxykanamycin, amikacin, gentamicin, tobramycin, sisomicin, neomycin, ribostamycin, butirosins.
Phosphorylation	APH(6)	Streptomycin
	APH(3')	Kanamycin, neomycin, paromomycin, ribostamycin, butirosin, lividomycin.
	APH(2'')	Gentamicin, sisomicin.
	APH(3'')	Streptomycin
	APH(5'')	Kanamycin, neomycin, paromomycin, ribostamycin, butirosin, lividomycin.
Adenylylation	AAD(6)	Streptomycin.
	AAD(4')	Kanamycin, amakacin, tobramycin.
	AAD(2'')	Kanamycin, 3',4'-dideoxykanamycin, tobramycin, gentamicin, sisomicin.
	AAD(3'') (9)	Streptomycin, spectinomycin.

synthesis of new clinically important analogues which are resistant to the action of these inactivating enzymes.

A different kind of microbial modification of an aminoglycoside antibiotic is the hydrolysis of mannosidostreptomycin to streptomycin (Fig. 16). Both antibiotics are produced as a mixture in streptomycin fermentations (Fried and Titus, 1947) but mannosidostreptomycin is 75–80% less active than streptomycin and may account for as much as 40% of the total streptomycin produced in such fermentations (Hockenhull, 1960). Streptomycin-producing strains that secrete high concentrations of the hydrolysing enzyme (α-D-mannosidase or mannosidostreptomycinase), and hence hydrolyse mannosidostreptomycin to streptomycin, are economically valuable (Perlman and Langlykke, 1948; Demain and Inamine, 1970). Since mannosidostreptomycin does not appear to be an obligatory intermediate in streptomycin synthesis, its formation can be eliminated by selection of appropriate mutants.

C. Anthracyclines

Daunomycin is an anthracycline antibiotic produced by *Streptomyces peucetius*. Although it is weakly active against some micro-organisms, it

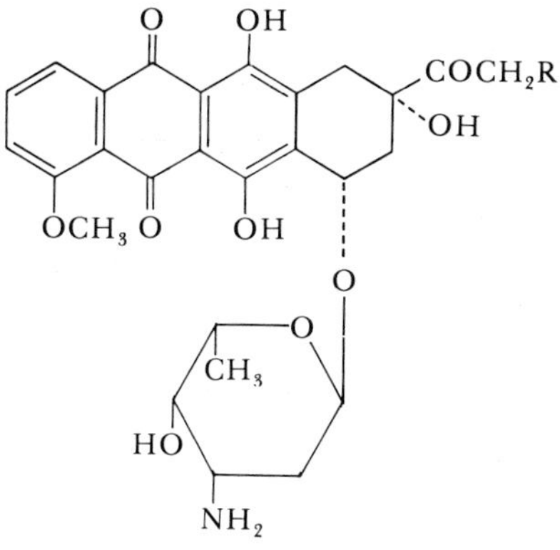

Antibiotic	R
Daunomycin	$-H$
Adriamycin	$-OH$

Fig. 17. Structure of daunomycin and adriamycin.

inhibits multiplication of bacterial and animal viruses. It exhibits a high cytostatic activity against normal and neoplastic cells, and is a powerful inducer of remission in acute lymphocytic leukemia in children. When the streptomycete culture was treated with N-nitroso-N-methyl urethane, a mutant was isolated which produced 14-hydroxydaunomycin (adriamycin or doxorubicin; Fig. 17). The mutant may thus represent a revertant of the parent culture in which the functional hydroxylase has been restored by mutation.

Adriamycin is valuable in the treatment of solid tumours. It is also effective in several childhood tumours and is the most active single drug against breast cancer, seminoma and other soft-tissue sarcomas, carcinoma of the bladder and lung cancer but it also causes depression of leukocytes and degeneration of heart muscles. Both antibiotics are produced on an industrial scale.

D. Chloramphenicol

Some bacteria and the chloramphenicol-producing *Streptomyces venezuelae* metabolize this antibiotic by hydrolysing the amide bond of the side chain to 4-nitrophenylserinol (Lingens *et al.*, 1966; Malik and Vining, 1971). Others carry out not only hydrolysis of the amide bond but also reduction of the nitro group, degradation of the benzene ring and a stepwise oxidation of the side chain (Fig. 18; Smith and Worrel, 1950; Lingens *et al.*, 1966). None of these reactions has produced compounds of biological significance. On the other hand, acetylation of chloramphenicol is a reaction by which many clinically important bacteria inactivate the antibiotic, and is the most common mechanism of resistance to it. It is catalysed by chloramphenicol acetyltransferase, an enzyme that has been well characterized and the structural gene for which is plasmid-coded.

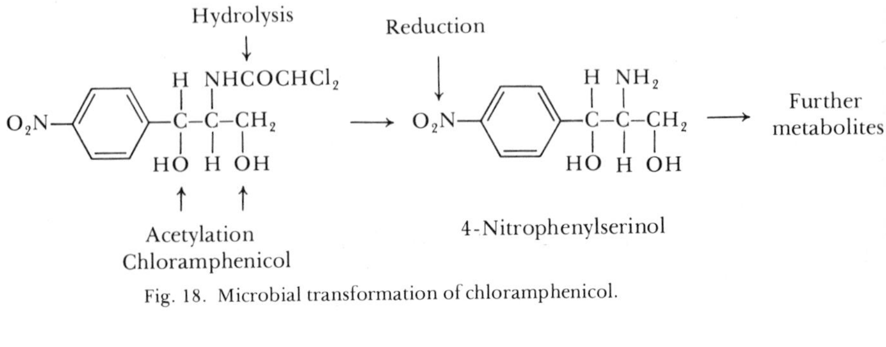

Fig. 18. Microbial transformation of chloramphenicol.

E. Rifamycins

Nocardia mediterranei synthesizes about 20 rifamycins, a family of closely related antibiotics characterized by the presence of an aliphatic ansa chain bridging a naphthalenic chromophore (Lancini and Parenti, 1978). By screening a random selection of chemicals, it was found that barbital (sodium diethyl barbiturate) changed the course of the fermentation drastically and yielded rifamycin B as the main fermentation product with small amounts of rifamycin Y, whereas synthesis of the others was completely suppressed. A mutant that is blocked in the acylation reaction and that accumulates rifamycin SV, a rifamycin B

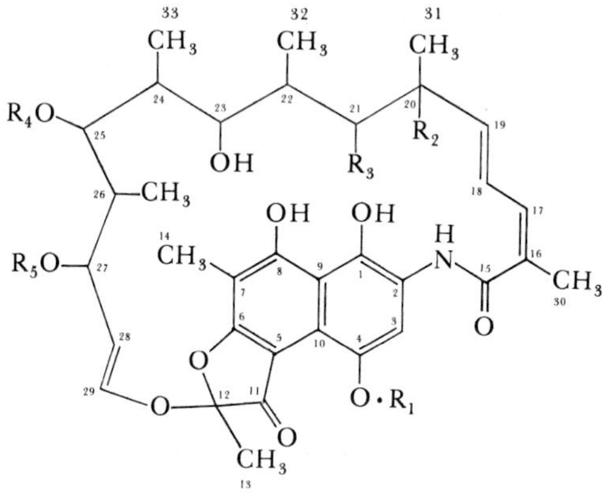

Rifamycins	R_1	R_2	R_3	R_4	R_5
B	$-CH_2COOH$	$-H$	$-OH$	CH_3CO-	$-CH_3$
Y	$-CH_2COOH$	$-OH$	$=O$	CH_3CO-	$-CH_3$
SV	$-H$	$-H$	$-OH$	CH_3CO-	$-CH_3$
27-Demethyl SV	$-H$	$-H$	$-OH$	CH_3CO-	$-H$
27-Demethyl B	$-CH_2COOH$	$-H$	$-OH$	CH_3CO-	$-H$
25-Deacetyl-27-demethyl B	$-CH_2COOH$	$-H$	$-OH$	$-H$	$-H$

Rifampin is 3-N-methyl piperazine derivative of SV

Fig. 19. Structure of selected rifamycins.

precursor, is of considerable economic importance (Lancini and Hengeller, 1969). It has served as an attractive substrate for chemical synthesis of many hundreds of semisynthetic rifamycins. One of them, rifampin (rifampicin), is outstanding as an antituberculosis drug which also has demonstrable antileprosy activity *in vivo*, is effective against trachoma and inhibits certain viruses and experimental tumours (Lester, 1972). Another mutant, blocked in rifamycin methylation, accumulated 27-demethylrifamycin SV (and also 25-deacetyl-27-demethyl-, and 27-demethylrifamycin B). Since 27-demethylrifamycin SV is another valuable substrate for chemical preparation of new semisynthetic rifamycin analogues, and since its chemical preparation (from rifamycin SV) by conventional chemistry is virtually impossible, fermentation is the method of choice for its preparation (Lancini and White, 1973) (Fig. 19). As other mutants are found to accumulate new rifamycins (Ghisalba *et al.*, 1978; Martinelli *et al.*, 1978), some of them may similarly become of practical importance.

F. Tetracyclines

As described in Section II.A. (p. 580), tetracyclines have been modified by inhibitors of chlorination and methylation, and by means of a methylase-deficient mutant (McCormick *et al.*, 1957) (see Fig. 2). The resulting 6-demethyl analogues are therapeutically valuable because of their high resistance to degradation by acid and alkali, and are important chemotherapeutic agents.

IV. METHODOLOGY

A great deal of information exists on preservation and cultivation of antibiotic-producing micro-organisms, biological assays, biochemical and physicochemical detection, and isolation and identification of the antibiotics produced. The literature also describes the selection of suitable organisms and the preparation, properties and application of enzymes that are involved in antibiotic transformations (Hash, 1975; Perlman, 1976).

Such cultures can be obtained by random isolation from different sources (soil, decomposing organic material, natural fermentations) or selected from stock culture collections. They are incubated with the antibiotic of interest and examined for any changes that might have taken place. Thin-layer chromatographic separation and location of the bioactive materials by bioautography is the most reliable and informative way to discover new transformations. Once they have been found, subsequent work involves improvement of transformation efficiency, and isolation and identification of the product(s).

A systematic examination of more than 500 micro-organisms yielded 21 isolates that transformed mycophenolic acid, a fungal metabolite with antitumour properties, at the sites indicated by the arrows in Figure 20. These transformations included oxygenation of the double bond to a hydroxylactone, of methyl groups to an alcohol, aldehyde and a lactol, and lactonization to a Δ-lactone. Reduction of the double bond and subsequent oxidation to a β-hydroxy acid, a methylketone and carboxylic acid, and further hydroxylation to a lactol, a benzyl alcohol and a dihydrobenzofuran were also demonstrated (Jones et al., 1970).

Fosfomycin, an orally active broad-spectrum antibiotic, is (−)-cis-1,2-epoxipropylphosphonic acid. From 645 randomly tested micro-organisms, a number of fungi were selected which carried out stereospecific epoxidation of cis-propenylphosphonic acid to fosfomycin in almost 90% yields (Fig. 21). In contrast, the chemical epoxidation yielded a racemic mixture of the product which had to

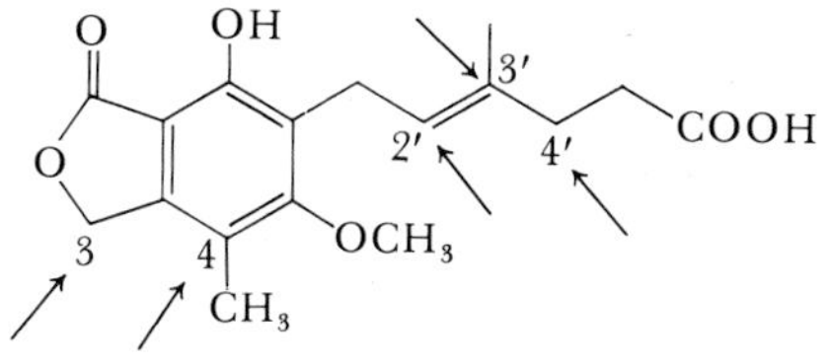

Fig. 20. Structure of mycophenolic acid. The arrows indicate the sites of transformation by micro-organisms.

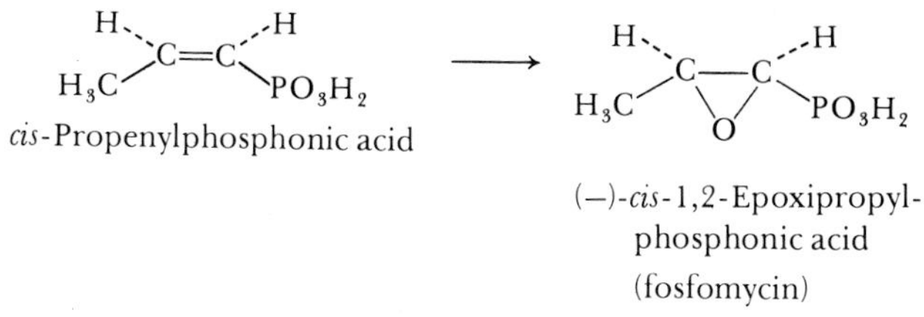

Fig. 21. Formation of fosfomycin.

undergo optical resolution with quinine but the desired bioactive
(−)-*cis*-isomer was formed only in less than 50% yields (White *et al.*,
1971).

Bleomycinic acid is a moiety common to all bleomycins, a group of
related glycopolypeptide antitumour antibiotics which differ from each
other in their terminal amines. It was first obtained by hydrolysis of
bleomycin B_2 by *Fusarium anguioides* found by random screening
(Umezawa *et al.*, 1973). Chemical hydrolysis was subsequently
developed and the acid has served as a substrate for preparation of
many semisynthetic bleomycins. In studies on the biosynthesis and
biotransformation of griseofulvin, 171 micro-organisms were tested.
Of this number, only two fungi were found to convert griseophenone
A into a racemic mixture of dehydrogriseofulvin, and two other carried
out this reaction (lactonization) stereospecifically to the (+)-isomer in
30% yields (Okuda *et al.*, 1967) (Fig. 22).

Micro-organisms have been selected to perform certain reactions on
given antibiotics also because they had been known to catalyse such
reactions of structurally related antibiotics or of structurally unrelated

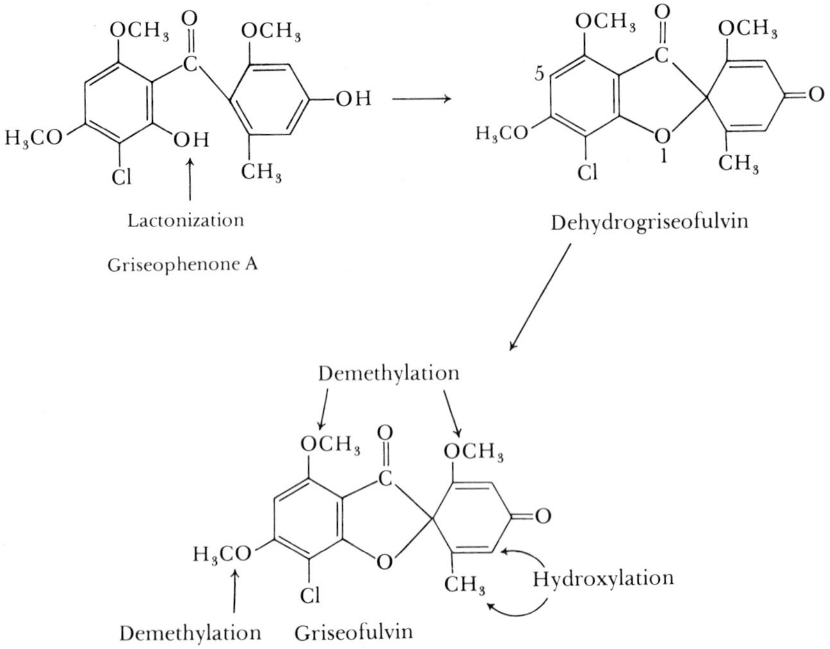

Fig. 22. Biosynthesis and biotransformation of griseofulvin.

compounds. Thus, (+)-dehydrogriseofulvin in the above example (Fig. 22) was reduced to (+)-griseofulvin which was further demethylated and hydroxylated either equatorially or axially by other organisms (Andres *et al.*, 1969; Bod *et al.*, 1973). *Streptomyces cinereocrocatus* was selected for the reduction step not as a result of screening but because it was known to convert a structurally related dehydro-1-thiogriseo-fulvin into (+)-thiogriseofulvin and (+)-5'-hydroxyl-1-thiogriseo-fulvin (Newman *et al.*, 1970). As an illustration of the second example, hydroxylation of 12α-deoxytetracycline to tetracycline was achieved by three fungi which are known to hydroxylate structurally unrelated steroids (Holmlund *et al.*, 1959) (Fig. 23).

As described in Section II (p. 580), certain intermediates of some antibiotics have antibiotic activity. Since these compounds were of practical interest either as new antibiotics or as attractive starting materals for the preparation of new semisynthetic analogues, their accumulation in fermentation beers was accomplished by adding suitable metabolic inhibitors or by isolating appropriate mutants. Thus 6-demethylated tetracyclines, *N*-demethylstreptomycin or *N*-demethyl-lincomycin accumulated on addition of L-methionine analogues or of folic acid antagonists to the cultures that produce the parent antibiotic. Antibiotic-blocked mutants have been used for selective accumulation of certain rifamycins (B, SV, 27-demethyl SV), which in

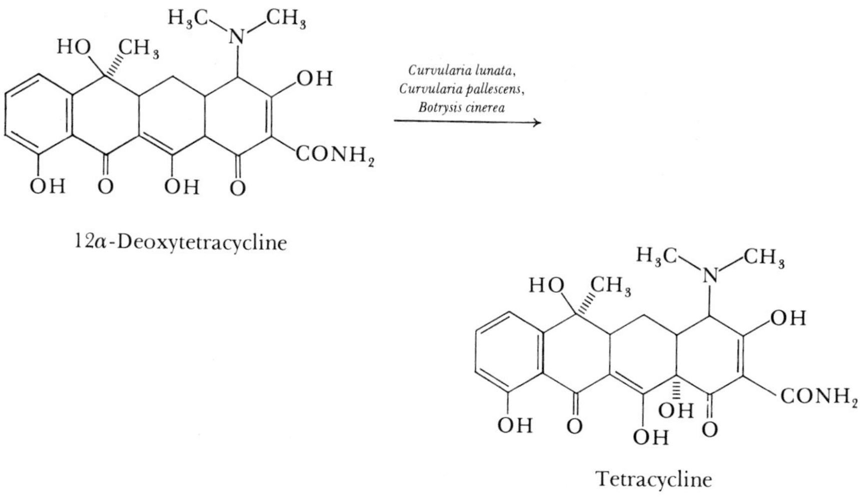

Curvularia lunata,
Curvularia pallescens,
Botrysis cinerea

12α-Deoxytetracycline

Tetracycline

Fig. 23. Hydroxylation of 12α-deoxytetracycline.

turn have served as substrates for chemical synthesis of novel rifamycin analogues. Similar mutants have also been used to generate new antibiotic analogues by mutasynthesis.

The precursor-feeding technique, successful in the early days of penicillin modification, was used to replace the 3-isopentenyl-4-hydroxybenzoic acid moiety of novobiocin by closely related acids, which yielded 19 new novobiocin analogues (Walton et al., 1962). Since preparation of a wider range of such analogues was sought, and since the chemical hydrolysis of the amide bond in question was not possible without adversely affecting other parts of the molecule (hydrolysis of the glycosidic bond, irreversible oxazole formation of the amide), microbial hydrolysis is the only known way in which the amide bond can be broken selectively. This reaction is carried out by a bacterium which was found by the classical enrichment method, and which utilized novobiocin as the sole source of carbon, nitrogen and energy. When the antibiotic was incubated with this bacterium under strictly

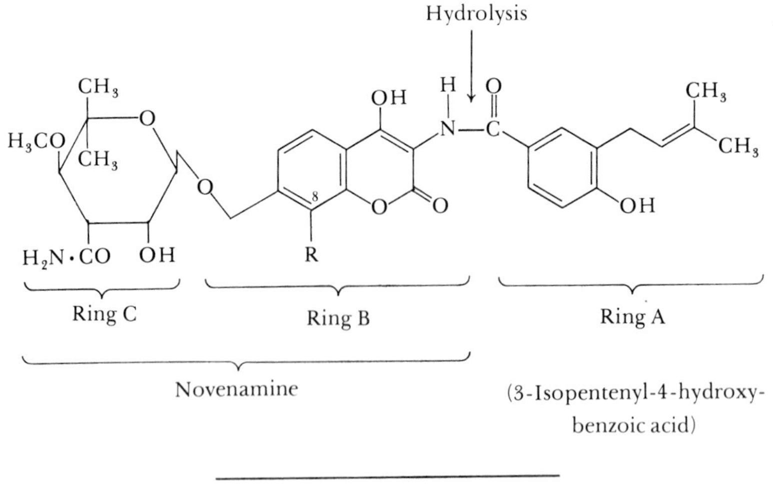

Fig. 24. Structure of novobiocin and its constituents.

Antibiotic	R
Novobiocin	$-CH_3$
8-Demethyl-8-chlor-novobiocin	$-Cl$
Novenamine	$-CH_3$

anaerobic conditions to eliminate all the oxidative reactions, only hydrolytic cleavage of the amide took place and this resulted in formation of the desired novenamine in 38% yields (Fig. 24). Novenamine in turn was chemically acylated to some 180 semi-synthetic novobiocins (Sebek and Hoeksema, 1972; Dolak *et al.*, 1972).

Biotransformations have also been carried out by feeding selected precursors to the antibiotic producers and their mutants. Thus *S. ambofaciens* which produces spiramycins, a group of three 16-membered ring macrolides, was directed to produce either spiramycin II (acetate) or III (propionate) when the fermentations had been supplemented with appropriate aliphatic acid (acetic or propionic) or their esters, amides or alcohols (Ninet and Verrier, 1962). Glycosylation of the aglycone in the 14-and 16-membered ring macrolide antibiotics was successfully carried out both by the antibiotic-producing parents and by their mutants. Thus washed cells of the antibiotic producers formed narbomycin and picromycin by glycosylation of their respective aglycones (Maezawa *et al.*, 1973) (Fig. 25). Antibiotic-blocked mutants were selected by cosynthetic methods

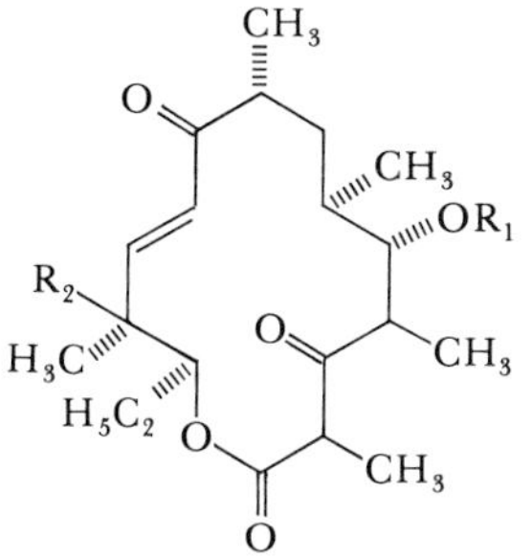

Compound	R_1	R_2
Narbonolide	$-H$	H
Picronolide	$-H$	$-OH$
Narbomycin	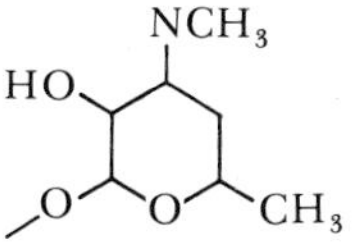	$-H$
Picromycin		$-OH$

Fig. 25. Structure of narbomycin, picromycin and their aglycones.

(Furumai and Suzuki, 1975; Delić et al., 1969; McCormick et al., 1960), which carry out glycosylation of an aglycone of the 16-membered ring macrolide antibiotic platenomycin with a sugar of the 14-membered ring macrolide antibiotics and vice versa as illustrated in Figure 6 (Maezawa et al., 1976, 1978).

Biotransformations of antibiotics and other organic compounds have been carried out by the traditional method of incubating the substrate with whole cells grown in complex fermentation media. In some cases, washed vegetative cells, spores, crude enzyme preparations or purified enzymes have been also used to simplify reaction conditions, to improve reaction efficiency and to obtain cleaner products. In addition, immobilized cells and immobilized enzymes (as in the commercial production of 6-aminopenicillanic acid) have been used for this purpose (Abbott, 1976, 1978; Durand and Navarro, 1978; Chibata et al., 1978; Wingard et al., 1979).

V. CONCLUDING REMARKS

In addition to their considerable value as producers of industrial solvents, amino and organic acids, alkaloids, antibiotics or vitamins, micro-organisms are important agents in catalysing key reactions in the manufacture of L-ascorbic acid (stereospecific dehydrogenation of D-sobitol to L-sorbose), of (1R,2S)-ephedrine (asymmetric condensations of benzaldehyde and acetaldehyde), steroids (regiospecific hydroxylations and 1-dehydrogenation) or sterols (β-oxidation of the C-17 side chain). For the same reason, they have also been used to perform specific reactions in the chemistry of many other natural products (Jones et al., 1976; Kieslich, 1976). Examples of their interaction in the antibiotic field as illustrated in this chapter, should encourage the use of these 'living reagents' (Rose, 1961) in the synthesis of new antibiotic analogues both in the laboratory and in large-scale manufacture.

REFERENCES

Abbott, B. J. (1976). In 'Advances in Applied Microbiology' (Perlman, D., ed.), vol. 20, p. 203. Academic Press, New York.

Abbott, B. J. (1978). In 'Annual Reports on Fermentation Processes' (Perlman, D., ed.), vol. 2, p. 91. Academic Press, New York.

Abbott, B. and Fukuda, D. S. (1975). *In* 'Methods in Enzymology' (Colovick, S. P. and Kaplan, N. O., eds-in-chief; Hash, J. H., ed.), vol. 43, p. 731. Academic Press, New York, San Francisco, London.

Abraham, E. P. and Chain, E. (1940). *Nature (London)* **146**, 837.

Andres, W. W., McGahren, J. J. and Kunstmann, M. P. (1969). *Tetrahedron Letters* 3777.

Arcamone, V., Cassinelli, G., Fantini, G., Grein, A., Orezzi, P., Pol, C. and Spalla, C. (1969). *Biotechnology and Bioengineering* **11**, 1101.

Argoudelis, A. D. and Coats, J. H. (1969). *Journal of Antibiotics* **22**, 341.

Argoudelis, A. D. and Coats, J. H. (1971). *Journal of Antibiotics* **24**, 206.

Argoudelis, A. D., Coats, J. H., Mason, D. J. and Sebek, O. K. (1969). *Journal of Antibiotics* **22**, 309.

Argoudelis, A. D., Coats, J. H. and Johnson, L. E. (1974). *Journal of Antibiotics* **27**, 738.

Argoudelis, A. D., Coats, J. H. and Magerlein, B. J. (1972a). *Journal of Antibiotics* **25**, 191.

Argoudelis, A. D., Coats, J. H., Lemaux, P. G. and Sebek, O. K. (1972b). *Journal of Antibiotics* **25**, 445.

Argoudelis, A. D., Johnson, L. E. and Pyke, T. R. (1973a). *Journal of Antibiotics* **26**, 429.

Argoudelis, A. D., Coats, J. H., Lemaux, P. G. and Sebek, O. K. (1973b). *Journal of Antibiotics* **26**, 7.

Argoudelis, A. D., Coats, J. H. and Mizsak, S. A. (1977). *Journal of Antibiotics* **30**, 474.

Barber, M. and Rozwadowski-Dowzenko, M. (1948). *Lancet* **225/2**, 641.

Batchelor, F. R., Doyle, F. P., Mayler, J. H. C. and Rolinson, G. N. (1959). *Nature, London*, **183**, 257.

Baud, H., Betencourt, A., Peyre, M. and Penasse, L. (1977). *Journal of Antibiotics* **30**, 720.

Benveniste, R. and Davies, J. (1973). *Annual Review of Biochemistry* **42**, 471.

Berdy's Natural Products Data Bank (1979), cf. Bostian, M., McNitt, K., Aszalos, A. and Berdy, J. (1977). *Journal of Antibiotics* **30**, 633.

Birch, A. J. (1963). *Pure and Applied Chemistry* **7**, 527.

Bod, P., Szarka, E., Gyimesi, J., Horváth, G., Vajna-Mehesfalvi, Z. and Horváth, I. (1973). *Journal of Antibiotics* **26**, 101.

Boothroyd, B., Napier, E. J. and Sommerfield, G. A. (1961). *Biochemical Journal* **80**, 34.

Brian, P. W. (1958). *In* 'Microbial Ecology' (Williams, R. E. O. and Spencer, C. C., eds.), p. 168. Cambridge University Press, London and New York.

Carrington, T. R. (1971) *Proceedings of the Royal Society, London, Series B* **179**, 321.

Chain, E., Florey, H. W., Gardner, A. D., Heatley, M. A., Jennings, M. A., Orr-Ewing, J. and Sanders, A. G. (1940). *Lancet* **239/2**, 226.

Chibata, I., Tosa, T., Sato, T. and Mori, T. (1978). *In* 'Immobilized Enzymes', p. 182. Kodansha, Ltd., Tokyo and J. Wiley and Sons, New York.

Claridge, C. A., Bush, J. A., Defuria, M. D. and Price, K. E. (1974). *Developments in Industrial Microbiology* **15**, 101.

Claridge, C. A., Gourevitch, A. and Lein, J. (1960). *Nature, London,* **187**, 237.

Cleophax, J., Gero, S. D., Leboul, J., Aktar, M., Barnett, J. E. G. and Pearce, C. J. (1976). *Journal of American Chemical Society* **98**, 7110.

Coats, J. H. and Argoudelis, A. D. (1971). *Journal of Bacteriology* **108**, 459.

Cooper, R. D. G. and Spry, D. O. (1972). *In* 'Cephalosporins and Penicillins' (Flynn, E. H., ed.), p. 183. Academic Press. New York and London.

Daum, S. J., Rosi, D. and Goss, W. A. (1977). *Journal of Antibiotics* **30**, 98.

Davies, J., Hauk, C., Yagisawa, M. and White, T. J. (1979). *In* 'Genetics of Industrial Microorganisms' (Sebek, O. K. and Laskin, A. I., eds.), p. 166. American Society of Microbiology, Washington, D. C.

Delić, V., Pigac, J. and Sermonti, G. (1969). *Journal of General Microbiology* **55**, 103.

Demain, A. L. and Inamine, E. (1970). *Bacteriological Reviews* **34**, 1.

Dhar, M. M., Singh, C., Khan, A. W., Arif, A. J., Gupta, C. A. and Bhaduri, A. P. (1971). *Pure and Applied Chemistry* **28**, 469.

Dolak, L., Sebek, O. K., Lewis, C. and Hoeksema, H. (1972). *Proceedings of the 12th Interscience Conference on Antimicrobial Agents and Chemotherapy*, 53.

Durand, D. and Navarro, J. M. (1978). *Process Biochemistry* **13**, 14.

Dvonch, W., Greenspan, G. and Alburn, H. E. (1966). *Experientia* **22**, 517.

Egami, F., Ebata, M. and Sato, R. (1951). *Nature, London,* **167**, 118.

El-Kersh, T. A. and Plourde, J. R. (1976). *Journal of Antibiotics* **29**, 292, 1189.

Evans, R. C. (1968). The Technology of the Tetracyclines. Quadrangle Press, New York, 617 p.

Feldman, L. I., Dill, I. K., Holmlund, C. E., Whaley, H. A., Patterson, E. L. and Bohonos, N. (1963). *Antimicrobial Agents and Chemotherapy*, p. 54.

Fried, J. and Titus, E. (1967). *Journal of Biological Chemistry* **168**, 391.

Fujii, A., Takita, T., Shimada, N. and Umezawa, H. (1974). *Journal of Antibiotics* **27**, 73.

Fujii, T., Matsumoto, K. and Watanabe, T. (1976). *Process Biochemistry* **11**, 21.

Fujii, T., Shibuya, T. and Matsumoto. K. (1979). *Proceedings of Annual Meeting of the Agricultural Chemical Society of Japan*, April 1–4.

Fujiwara, T., Tanimoto, T., Matsumoto, K. and Kondo, E. (1978). *Journal of Antibiotics* **31**, 966.

Furumai, T. and Suzuki, M. (1975). *Journal of Antibiotics* **28**, 770.

Ghisalba, O., Traxler, P. and Nüesch, J. (1978). *Journal of Antibiotics* **32**, 1124.

Goldstein, A. W., Egan, R. S., Mueller, S. L. and Martin, J. R. (1978). *Journal of Antibiotics* **31**, 63.

Goodman, J. J. and Matrishin, M. (1961). *Journal of Bacteriology* **82**, 615.

Goodman, J. J., Matrishin, M., Young, R. W. and McCormick, J. R. D. (1959). *Journal of Bacteriology* **78**, 492.

Hamill, R. L., Elander, R. P., Mabe, J. A. and Gorman, M. (1970). *Applied Microbiology* **19**, 721.

Hamilton-Miller, J. M. T. (1966). *Bacteriological Reviews* **30**, 761.

Hamilton, B. K., Sutphin, M. S., Thomas, M. C., Wareheim, D. E. and Aszalos, A. D. (1977). *Journal of Antibiotics* **30**, 425.

Hash, J. H. (ed.), (1975). *Methods in Enzymology*, XLIII. *Antibiotics*, 837 pp. Academic Press, New York, San Francisco and London.

Hedig, H. (1968). *Acta Chemica Scandinavica* **22**, 1649.

Hendlin, D., Dulaney, E. L., Dresher, D., Cook, T. and Chaiet, L. (1962). *Biochimica et Biophysica Acta* **58**, 635.

Higashide, E., Fugono, T., Hatano, K. and Shibata, M. (1971). *Journal of Antibiotics* **24**, 1.

Hockenhull, D. J. D. (1960). *In* 'Progress in Industrial Microbiology' (Hockenhull, D. J. D., ed.), vol. 2, p. 133. Haywood and Co. Ltd., London.

Holmlund, C. E., Andres, W. W. and Shay, A. J. (1959). *Journal of the American Chemical Society* **81**, 4750.

Hopwood, D. A. and Merrick, M. J. (1977). *Bacteriological Reviews* **41**, 595.

Hou, C. T. and Perlman, D. (1970). *Journal of Biological Chemistry* **245**, 1289.

Hou, C. T., Perlman, D. and Schallock, M. R. (1970). *Journal of Antibiotics* **23**, 35.

Huang, H. T., English, A. R., Seto, T. A., Shull, G. M. and Sobin, B. A. (1960). *Journal of the American Chemical Society* **82**, 3790.

Huang, H. T., Seto, T. A. and Shull, G. M. (1963) *Applied Microbiology* **11**, 1.

Huber, F. M., Chauvette, R. R. and Jackson, B. G. (1972). *In* 'Cephalosporins and Penicillins' (Flynn, E. H., ed.), p. 27. Academic Press, New York and London.

Inouye, S., Tsuruoka, T., Omoto, S. and Niida, T. (1971). *Journal of Antibiotics* **24**, 460.

Isono, K., Crain, P. F., Odiorne, T. J., McCloskey, J. A. and Suhadolnik, R. J. (1973). *Journal of the American Chemical Society* **95**, 5788.

Ito, M., Aida, T. and Koyama, Y. (1966). *Agricultural and Biological Chemistry* **30**, 1112.

Jackson, J. H. and Umbarger, H. E. (1973). *Antimicrobial Agents and Chemotherapy* **3**, 510.

Jarvis, B. and Berridge, N. J. (1969). *Chemistry and Industry*, 1721.

Jones, D. F., More, R. H. and Crawley, G. C. (1970). *Journal of the Chemical Society (C)*, 1725.

Jones, J. B., Sih, C. J. and Perlman, D. eds. (1976). *In* 'Techniques of Chemistry', vol. 10. John Wiley, New York, Sidney and Toronto.

Kameda, Y., Horri, S., and Yamano, T. (1975). *Journal of Antibiotics* **28**, 299.

Kameda, Y., Asano, N. and Hashimoto, T. (1978). *Journal of Antibiotics* **31**, 936.

Karnetová, J., Matĕjů, J., Sedmera, P., Vokoun, J. and Vanĕk, Z. (1976). *Journal of Antibiotics* **29**, 1199.

Kato, K. (1953). *Journal of Antibiotics, Ser A.* **6**, 130, 184.

Katz, E. and Pienta, P. (1957). *Science, New York*, **126**, 402.

Katz, E., Williams, W. K., Mason, T. K. and Mauger, A. B. (1977). *Antimicrobial Agents and Chemotherapy* **11**, 1056.

Kaufmann, W. and Bauer, K. (1960). *Naturwissenschaften* **47**, 474.

Keller-Schierlein, W., Poralla, K. and Zähner, H. (1969). *Archiv für Mikrobiologie* **67**, 339.

Kieslich, K. (1976). 'Microbial Transformations of Non-Steroid Cyclic Compounds'. Georg Thieme Publishers, Stuttgart. 162 p.

Kitao, C., Ikeda, H., Hamada, H. and Omura, S. (1979). *Journal of Antibiotics* **32**, 593.

Kojima, M. and Satoh, A. (1973). *Journal of Antibiotics* **26**, 784.

Kojima, M., Inouye, S. and Niida, T. (1975). *Journal of Antibiotics* **28**, 48.

Komori, T., Kunugita, K., Nakahara, K., Aoki, H. and Imanaka, H. (1978). *Agricultural and Biological Chemistry* **42**, 1439.

Lancini, G. and Hengeller, C. (1969). *Journal of Antibiotics* **22**, 637.

Lancini, G. C. and Parenti, F. (1978). *In* 'Antibiotics and Other Secondary Metabolites' (Hütter, R., Leisinger, T., Nüesch, J. and Wehrli, W., eds.), p. 129. Academic Press, London, New York and San Francisco.

Lancini, G. and White, R. J. (1973). *Process Biochemistry* **8**, 14.

Lancini, G. C., Gallo, G. G., Sartori, G. and Sensi, P. (1969). *Journal of Antibiotics* **22**, 369.

Lemaux, P. G. and Sebek, O. K. (1973). *Proceedings of the 13th Interscience Conference on Antimicrobial Agents and Chemotherapy*, Washington, D. C., Abstract 49.

LeMahieu, R. A., Ax, H. A., Blount, J. F., Carson, M., Despreau, C. W., Pruess, D. L., Scannell, J. P., Weiss, F. and Kierstead, R. W. (1976). *Journal of Antibiotics* **29**, 729.

Lennon, R. E. and Vézina, C. (1973). *In* 'Advances in Applied Microbiology' (Perlman, D., ed.), vol. 16, p. 55. Academic Press, New York and London.

Lester, W. (1972). *Annual Review of Microbiology* **26**, 85.

Lingens, F., Eberhardt, H. and Oltmanns, O. (1966) *Biochimica et Biophysica Acta* **130**, 345.

Maezawa, I., Hori, T., Kinumaki, A. and Suzuki, M. (1973). *Journal of Antibiotics* **26**, 771.

Maezawa, I., Kinumaki, A. and Suzuki, M. (1976). *Journal of Antibiotics* **29**, 1203.

Maezawa, I., Kinumaki, A. and Suzuki, M. (1978). *Journal of Antibiotics* **31**, 309.

Majer, J., Martin, J. R., Egan, R. S. and Corcoran, J. W. (1977). *Journal of the American Chemical Society* **99**, 1620.

Malik, V. S. and Vining, L. C. (1971). *Canadian Journal of Microbiology* **17**, 1287.

Marshall, V. P., McGovern, P., Richard, F. A. Richard, R. E. and Wiley, P. F. (1978). *Journal of Antibiotics* **1**, 336.

Martin, J. H., Mitscher, L. A., Miller, P. A., Shu, P. and Bohonos, N. (1967). *Antimicrobial Agents and Chemotherapy*, 563.

Martinelli, E., Antonini, P., Cricchio, R., Lancini, G. and White, R. J. (1978). *Journal of Antibiotics* **31**, 949.

Matsuhashi, Y., Ogawa, H. and Nagaoka, K. (1979). *Journal of Antibiotics* **32**, 777.

McCormick, J. R. D., Sjolander, N. O., Hirsch, U., Jensen, E. R. and Doerschuk, A. P. (1957). *Journal of the American Chemical Society* **79**, 4561.

McCormick, J. R. D., Sjolander, N. O., Miller, D. A., Hirsch, U., Arnold, N. H. and Doerschuk, A. P. (1958). *Journal of the American Chemical Society* **80**, 6460.

McCormick, J. R. D., Hirsch, U., Sjolander, N. O. and Doerschuk, A. D. (1960). *Journal of the American Chemical Society* **82**, 5006.

Murao, S. (1955). *Nippon Nogei-Kagaku Kaishi* **29**, 400, 404.

Nagaoka, K. and Demain, A. L. (1975). *Journal of Antibiotics* **28**, 627.

Nakahama, K. and Igarasi, S. (1974). *Journal of Antibiotics* **27**, 605.

Nakahama, K., Kishi, T. and Igarasi, S. (1974). *Journal of Antibiotics* **27**, 433, 487.

Nakahama, K., Harada, S. and Igarasi, S. (1975). *Journal of Antibiotics* **28**, 390.

Nakano, H., Matsuhashi, Y., Takeuchi, T. and Umezawa, H. (1977). *Journal of Antibiotics* **30**, 76.

Newman, H., Shu, P., and Andres, W. W. (1970). United States Patent 3,532,714.

Neidleman, S. L., Bienstock, E. and Bennett, R. E. (1963). *Biochimica et Biophysica Acta* **71**, 199.

Ninet, L. and Verrier, J. (1962). *Produits Pharmaceutiques* **17**, 155.

O'Callaghan, C. H. and Muggleton, P. W. (1972). *In* 'Cephalosporins and Penicillins' (Flynn, E. H., ed.), p. 438. Academic Press, New York and London.

Ochi, K., Yashima, S. and Eguchi, Y. (1975). *Journal of Antibiotics* **28**, 965.

Okachi, R., Kato, F., Miyamura, Y. and Nara, T. (1973a) *Agricultural and Biological Chemistry* **37**, 1953.

Okachi, R., Kawamoto, I., Yamamoto, M., Takasawa, S. and Nara, T. (1973b). *Agricultural and Biological Chemistry* **37**, 335.

Oku, H. and Nakanishi, T. (1964). *Naturwissenschaften* **51**, 538.

Okuda, S., Isaka, H., Tida, M., Minemura, Y., Iizuka, H. and Tsuda, K. (1967). *Journal of the Pharmaceutical Society of Japan* **87**, 1003.

Ōmura, K., Miyazawa, J., Takeshima, H., Kitao, C., Atsumi, K. and Aizawa, M. (1976). *Journal of Antibiotics* **29**, 1131.

Ōmura, S. and Nakagawa, A. (1975). *Journal of Antibiotics* **28**, 401.

Ozaki, M., Kariya, T., Kato, H. and Kimura, T. (1972). *Agricultural and Biological Chemistry* **36**, 451.

Perlman, D. (1971). *Process Biochemistry* **6** (7), 13.

Perlman, D. (1976). *In* 'Techniques of Chemistry'. Application of Biochemical Systems in Organic Chemistry (Jones, J. B., Sih, C. J. and Perlman, D., eds.), p. 47. John Wiley and Sons, New York, London, Sydney and Toronto.

Perlman, D. and Langlykke, A. F. (1948). *Journal of the American Chemical Society* **70**, 3968.

Perlman, D. and Sebek, O. K. (1971). *Pure and Applied Chemistry* **28**, 637.

Perlman, D., Heuser, L. J., Semar, J. B., Frazier, W. R. and Boska, J. A. (1961). *Journal of American Chemical Society* **83**, 4481.
Pollock, M. R. (1971). *Proceedings of the Royal Society London, Series B* **179**, 385.
Pramer, D. (1958). *Applied Microbiology* **6**, 221.
Pramer, D. and Starkey, R. L. (1951). *Science, New York*, **113**, 127.
Queener, S. W., Sebek, O. K. and Vézina, C. (1978). *Annual Review of Microbiology* **32**, 593.
Richmond, M. H. and Sykes, R. B. (1973). *In* 'Advances in Microbial Physiology' (Rose, A. H. and Tempest, D. W., eds.), vol. 9, pp. 31–88. Academic Press, London and New York.
Ridley, M., Barrie, D., Lynn, R. and Stead, K. C. (1970). *Lancet* **269/1**, 230.
Rinehart, K. L., Jr. (1969). *Journal of Infectious Diseases* **119**, 345.
Rinehart, K. L., Jr. (1977). *Pure and Applied Chemistry* **49**, 1361.
Rolinson, G. N., Batchelor, F. R., Butterworth, D., Cameron-Wood, J., Cole, M., Eustace, G. C., Hart, M. V., Richards, M. and Chain, E. B. (1960). *Nature, London*, **187**, 236.
Rose, A. H. (1961). Industrial Microbiology, p. 255. Butterworths, London.
Rottman, I. and Guarino, A. J. (1964). *Biochimica et Biophysica Acta* **80**, 632.
Sakaguchi, K. and Murao, S. (1950). *Journal of the Agricultural and Chemical Society, Japan*, **23**, 411.
Sands, L. C. and Shaw, W. V. (1973). *Antimicrobial Agents and Chemotherapy* **3**, 299.
Sawa, T., Fukagawa, Y., Homma, Y., Wakashiro, T., Takeuchi, T. and Hori, M. (1968). *Journal of Antibiotics* **21**, 334.
Schmitz, H. and Claridge, C. A. (1977). *Journal of Antibiotics* **30**, 635.
Sebek, O. K. (1974). *Lloydia* **37**, 115.
Sebek, O. K. (1975). *Acta Microbiologica Academiae Scientiarum Hungaricae* **22**, 381.
Sebek, O. K. and Dolak, L. A. (1977). *In* 'Biotechnological Applications of Problems and Enzymes' (Bohak, Z. and Sharon, N., eds.), p. 203. Academic Press, New York, San Francisco and London.
Sebek, O. K. and Dolak, L. A. (1977). *Bacteriological Proceedings*, abstract 26.
Sebek, O. K. and Hoeksema, A. (1972). *Journal of Antibiotics* **25**, 434.
Sebek, O. K. and Perlman, D. (1971). *In* 'Advances in Applied Microbiology' (Perlman, D., ed.), p. 123. Academic Press, New York and London.
Sheehan, J. C., Henery-Logan, K. R. and Johnson, D. A. (1953). *Journal of the American Chemical Society* **75**, 3292.
Shibata, M. and Uyeda, M. (1978). *In* 'Annual Reports on Fermentation Processes' (Perlman, D., ed.), vol. 2, p. 267. Academic Press, New York.
Shier, W. T., Rinehart, K. L., Jr. and Gottlieb, D. (1969). *Proceedings of the National Academy of Sciences of the United States of America* **63**, 198.
Shier, W. T., Ogawa, S., Hichens, M. and Rinehart, K. L., Jr. (1973). *Journal of Antibiotics* **26**, 551.
Shimizu, M., Masuike, T., Fujita, H., Kimura, K., Okachi, R. and Nara, T. (1975). *Agricultural and Biological Chemistry* **39**, 1225.
Shomura, T., Ezaki, N., Tsuruoka, T., Niwa, T., Akita, E. and Niida, T. (1970). *Journal of Antibiotics* **23**, 155.
Singh, K. and Rakhit, S. (1979). *Journal of Antibiotics* **32**, 78.
Slechta, L., Cialdella, J. and Hoeksema, H. (1978). *Journal of Antibiotics* **31**, 319.
Smith, G. N. and Worrel, C. S. (1949). *Archives of Biochemistry* **24**, 216.
Smith, G. N. and Worrel, C. S. (1950). *Archives of Biochemistry* **28**, 232.

Smith, G. N. and Worrel, C. S. (1953). *Journal of Bacteriology* **65**, 313.

Suzuki, M., Takamami, T., Miyagawa, K., Ono, H., Higashide, E. and Uchida, M. (1977). *Agricultural and Biological Chemistry* **41**, 419.

Sykes, R. B. (1979). *In* 'Genetics of Industrial Microorganisms' (Sebek, O. K. and Laskin, A. I., eds.), p. 170. American Society for Microbiology, Washington, D. C.

Sykes, R. B. and Matthew, M. (1976). *Journal of Antimicrobial Chemotherapy* **2**, 115.

Takahashi, T., Yamazaki, Y., Kato, K. and Isono, M. (1972). *Journal of the American Chemical Society* **94**, 4035.

Takeda, K., Okuno, S., Ohashi, Y. and Furumai, T. (1978a). *Journal of Antibiotics* **31**, 1023.

Takeda, K., Kinumaki, A., Hayasaka, H., Yamaguchi, T. and Ito, Y. (1978b). *Journal of Antibiotics* **31**, 1031.

Takeda, K., Kinumaki, A., Okuno, S., Matsushita, T. and Ito, Y. (1978c). *Journal of Antibiotics* **31**, 1039.

Takeda, K., Kinumaki, A., Furumai, T., Yamaguchi, T., Ohshima, S. and Ito, Y. (1978d). *Journal of Antibiotics* **31**, 247.

Testa, R. T. and Tilley, B. C. (1975). *Journal of Antibiotics* **28**, 573.

Troonen, H., Roelants, P. and Boon, B. (1976). *Journal of Antibiotics* **29**, 1258.

Uematsu, T. and Suhadolnik, R. J. (1974). *Archives of Biochemistry and Biophysics* **162**, 614.

Umezawa, H., Takahashi, Y., Fujii, A., Saino, T., Shira, T. and Takita, T. (1973). *Journal of Antibiotics* **26**, 117.

United States International Trade Commission (1979). Synthetic Organic Chemicals, United States Production and Sales, 1978. 153. USITC Publication 1001, U.S. Government Printing Office, Washington, D.C.

Vandamme, E. J. and Voets, J. P. (1974). *In* 'Advances in Applied Microbiology' (Perlman, D., ed.), vol. 17, p. 311. Academic Press, New York and London.

von Daehne, W., Lorch, H. and Gotfredsen, W. C. (1968). *Tetrahedron Letters* No. 47, 4843.

Walton, R. B., McDaniel, L. B. and Woodruff, H. B. (1962). *Developments in Industrial Microbiology* **3**, 370.

Weber, J. M. and Perlman, D. (1978). *Journal of Antibiotics* **31**, 373.

White, R. F., Birnbaum, J., Meyer, R. T., ten Broeke, J., Chemerda, J. M. and Demain, A. L. (1971). *Applied Microbiology* **22**, 55.

Wingard, L. B., Jr., Katchalski-Katzir, E. and Goldstein, L., eds. (1979). *In* 'Applied Biochemistry and Bioengineering'. 2. Enzyme Technology. Academic Press, New York, San Francisco, London.

Yamaguchi, I., Shibata, H., Seta, H., Seto, H. and Misato, T. (1975). *Journal of Antibiotics* **28**, 6.

Yoshimoto, A., Ogasawara, T., Kitamura, I., Oki, T., Inui, T., Takeuchi, T. and Umezawa, H. (1979). *Journal of Antibiotics* **32**, 472.

Yukioka, M., Saito, Y. and Otani, S. (1966). *Journal of Biochemistry, Tokyo,* **60**, 295.

AUTHOR INDEX

Y

SUBJECT INDEX

A

Absidia orchidis
  hydroxylation, of
    androstane by, 410
    of 4-androstene-3,17-dione by, 420
    of steroids by, 422
Acetabularia mediterranea, protein synthesis, 15
Acetobacter sp.
  enzyme attachment to, 362
  glucose dehydrogenase from, 182, 183
Acetobacter aceti, ergot alkaloid conversions by, 536
Acetobacter pasteurianus, cephalosporin conversions by, 501
Acetobacter suboxydans (*see Gluconobacter suboxydans*)
Acetobacter turbidans, cephalosporin conversions by, 501
Acetone, dihydroxy-
  production by biotransformation, 10
  synthesis, biotransformations in, 41
Acetylcholinesterase, 5α-androstan-17-one conversions by, 420
Achromobacter spp.
  acylase conversions, 500
  cephalexin synthesis by, 504
  cephalosporin hydrolysis by penicillin G acylases from, 489
  in soils, 265
  olive spoilage by, 264
  penicillin acylases, immobilized, cephalosporin conversion, 503
  penicillin G acylases, cephalexin hydrolysis by, 500
  phenoxymethylpenicillin acylase from, 477

properties, 483
whole-cell immobilization, 499
Achromobacter liquidun, urocanic acid synthesis and 42
Achromobacter nicotinophagum, tobacco alkaloid conversions by, 540, 543
Achromobacter strychnovorum, strychnine conversions by, 558
Acid proteases, 72–77
  from *Aspergillus oryzae*, properties, 74
  in Takadiastase, 74
Acinetobacter spp., in soils, 265
Aconitine, microbial conversions, 556
Acremonium kiliense, hydroxylation of norethisterone by, 425
Acremonium strictum, hydroxylation of norethisterone by, 425
Acrocylindrium sp., pectic-enzyme synthesis in, regulation, 258
Acrolindrium sp., exopolygalacturonases, 240
Acromyrmex fungus, androstane hydroxylation by, 410
Acronychia baueri, acronycine conversions by, 557
Acronycine, microbial conversions, 557
Acrylamide, β-amylase immobilization on, 334
—. N-(4-carboxyphenyl)-
  polymer, β-galactosidase immobilization on, 351
Actinomucor elegans, use in oriental fermented food, 50
Actinomyces spp., progesterone hydroxylation by, 384
Actinomyces roseoviridis, 1-dehydrogenation of cortisol by, 398

12a-Aza-C-homo-5α-pregnane-12,20-dione, 3β-hydroxy-, microbial conversions, 529
—. 3β-acetoxy-17α-hydroxy-, microbial conversions, 529
Azasteroids, microbial preparation, 528
7-Azatryptophan, microbial conversions, 539
Azotobacter spp.
  glucose dehydrogenase from, 182
  in soils, 265

B

Beauveria sp., 11α-hydroxylation of Compound S by, 397
Beer, brewing, microbial enzymes in, 4
Beneckea hiperoptica, cephalexin synthesis by, 504
Benzaldehyde, acetylphenylcarbinol preparation from, *Saccharomyces cerevisiae* in, 8
Benzene, divinyl-, -styrene copolymer, glucose isomerase immobilization on, 339
Benzylpenicillin acylase 475
  micro-organisms producing, 474–475
  properties, 486–495
Bile acids, microbial conversions, 429–436
Bio 38, 40, 52
Biochemistry, enzyme production by micro-organisms, 15–38
Bioprase, 55
Bioreactors containing whole cells, continuous-flow, 360–361
Biotex, 53
Biotin, microbial conversion, 566
Biotransformations
  antibiotic preparation by, 586
  commercial exploitation, prospects, 40–42
  industrially important, history, 6–15
Biozyme A, 55
Bleomycinic acid, microbial production, 602
Bleomycins, microbial preparation, 579
Blocked mutants in antibiotic preparation, 582

Blood
  glucose determination in, glucose dehydrogenase for, 188
  glucose oxidase for, 181
Blood clots, proteases from *Aspergillus oryzae* for dissolution, 74
Blood stains, removal by enzymic detergents, 94
Bordetella sp., benzylpenicillin acylase production by, 474
Botryodiplodia malorum
  ethinlestradiol conversion by, 427
  hydroxylation of norethisterone by, 425
Botryodiplodia theobromae, hydroxylation of steroidal alkaloids by, 528
Botrytis cinerea
  pectic enzyme synthesis in, regulation, 258
  phenoxymethylpenicillin acylase from, 476
  spoilage of fruit, 262
Bovista plumbia
  penicillin acylase from, properties, 484
  phenoxymethylpenicillin acylase from, 476
  properties, 482
Breadmaking, α-amylases in, 157–158
Brevibacterium spp., acylase conversions, 500
Brevibacterium lipolyticum, degradation of cholesterol by, 439
Brevibacterium pentoaminoacidum, glucose isomerase from, 195
Brevibacterium sterolicum, oxidation of cholesterol by, 438
Brewing
  α-amylases in, 158
  beer, microbial enzymes in, 4
  microbial proteases in, 105–106
Bromelain in brewing, 105
Brucine, microbial conversions, 558
Butirosin
  inactivation, 595
  microbial conversions, 593
  microbial preparation, 584
  structure, 583
Butylene glycol in protease from *Bacillus licheniformis*, 59
Butyrivibrio fibrisolvens in rumen, 266